T0205251

Lecture Notes in Computer Science 12738

More information about this subseries at http://www.springer.com/series/7412

Daniel B. Ennis · Luigi E. Perotti ·
Vicky Y. Wang (Eds.)

Functional Imaging and Modeling of the Heart

11th International Conference, FIMH 2021
Stanford, CA, USA, June 21–25, 2021
Proceedings

 Springer

Editors
Daniel B. Ennis (ID)
Stanford University
Stanford, CA, USA

Luigi E. Perotti (ID)
University of Central Florida
Orlando, FL, USA

Vicky Y. Wang (ID)
Stanford University
Stanford, CA, USA

ISSN 0302-9743   ISSN 1611-3349 (electronic)
Lecture Notes in Computer Science
ISBN 978-3-030-78709-7   ISBN 978-3-030-78710-3 (eBook)
https://doi.org/10.1007/978-3-030-78710-3

LNCS Sublibrary: SL6 – Image Processing, Computer Vision, Pattern Recognition, and Graphics

This Springer imprint is published by the registered company Springer Nature Switzerland AG
The registered company address is: Gewerbestrasse 11, 6330 Cham, Switzerland

Preface

FIMH 2021 was the 11th International Conference on Functional Imaging and Modeling of the Heart. FIMH 2021 was hosted by Stanford University, but conducted entirely online due to the COVID-19 pandemic. The conference took place from June 21 to June 25, 2021, and included both pre-conference and post-conference workshops. This year's publication of the FIMH proceedings followed the success of the past ten conferences held in Helsinki (2001), Lyon (2003), Barcelona (2005), Salt Lake City (2007), Nice (2009), New York (2011), London (2013), Maastricht (2015), Toronto (2017), and Bordeaux (2019).

FIMH 2021 celebrated 20 years of bringing together friends, colleagues, and collaborators to share and discuss the latest advances in cardiac and cardiovascular imaging, electrophysiology, computational modeling, and translational applications. An important goal of this meeting was to openly discuss new ideas, build new collaborations, and foster an inclusive community to promote and discuss the latest developments in the areas of functional cardiac imaging as well as computational modeling of the heart. The topics of the FIMH 2021 conference included: 1) Advanced Cardiac and Cardiovascular Image Processing; 2) Cardiac Microstructure – Measures and Models; 3) Novel Approaches to Measuring Heart Deformation; 4) Cardiac Mechanics – Measures and Models; 5) Translational Cardiac Mechanics; 6) Modeling Electrophysiology, ECG, and Arrhythmia; 7) Cardiovascular Flow – Measures and Models; and 8) Atrial Microstructure, Modeling, and Thrombosis Prediction.

FIMH 2021 attracted submissions from authors in more than ten countries. From the original submissions, and after single-blind review by international experts, 65 papers were accepted for presentation at the meeting and publication in this Lecture Notes in Computer Science Proceedings volume. Before submitting the final version of their manuscripts, authors addressed specific concerns and issues raised by the reviewers and improved the scientific content and the overall quality of the manuscripts.

Three world experts in fields related to cardiac computational modeling, advanced imaging, and translational applications, representing the three major pillars of the FIMH society, were invited to give keynote presentations at the conference. We are very grateful to Dr. Blanca Rodriguez (University of Oxford, UK), Dr. Steffen Petersen (Queen Mary University of London, UK), and Dr. Joseph Woo (Stanford University, USA) for their exceptional keynote presentations.

The FIMH 2021 organizing committee was also committed to promoting the careers of trainees and young scientists and established a trainee committee consisting of enthusiastic post-doctoral researchers that helped coordinating special trainee sessions to foster professional success.

We expect that the FIMH 2021 papers, the keynote speakers' contributions, and the engaging discussion during the conference helped to accelerate progress in several important areas of functional imaging and modeling of the heart.

June 2021

Daniel B. Ennis
Luigi E. Perotti
Vicky Y. Wang

Organization

We would like to thank all organizers, additional reviewers, contributing authors, and sponsor for their time, effort, and financial support in making FIMH 2021 a successful virtual event.

Conference Chairs

Daniel B. Ennis	Stanford University, USA
Luigi E. Perotti	University of Central Florida, USA
Vicky Y. Wang	Stanford University, USA

Trainee Committee

Seraina A. Dual	Stanford University, USA
Alexander Kaiser	Stanford University, USA
Mathias Peirlinck	Stanford University, USA
Alexander Wilson	Stanford University, USA

Program Committee

Daniel Balzani	Ruhr-Universität Bochum, Germany
Peter Bovendeerd	Eindhoven University of Technology, the Netherlands
Oscar Camara	Universitat Pompeu Fabra, Spain
Teodora Chitiboi	Siemens Healthineers, USA
Henry Chubb	Stanford University, USA
Richard Clayton	University of Sheffield, UK
Yves Coudière	Université de Bordeaux, France
Tammo Delhaas	Maastricht University, the Netherlands
Nicolas Duchateau	Université de Lyon, CREATIS, France
Frederick Epstein	University of Virginia, USA
Alberto Figueroa	University of Michigan, USA
Arun Holden	University of Leeds, UK
Daniel Hurtado	Pontificia Universidad Católica de Chile, Chile
Pablo Lamata	King's College London, UK
Cristian Linte	Rochester Institute of Technology, USA
Herve Lombaert	ETS Montreal, Canada
Rob MacLeod	University of Utah, USA
Tommaso Mansi	Siemens Healthineers, USA
Alison Marsden	Stanford University, USA
Steven Niederer	King's College London, UK
Vijay Rajagopal	University of Melbourne, Australia

Frank Sachse	University of Utah, USA
Michael Sacks	University of Texas at Austin, USA
Francisco Sahli Costabal	Pontificia Universidad Católica de Chile, Chile
Andrew Scott	Royal Brompton Hospital, UK
Maxime Sermesant	Inria, France
Shawn Shadden	University of California, Berkeley, USA
Kaleem Siddiqi	McGill University, Canada
Lawrence Staib	Yale University, USA
Regis Vaillant	General Electric, France
Samuel Wall	University of Oslo, Norway
Linwei Wang	Rochester Institute of Technology, USA
Graham Wright	University of Toronto, Canada
Guang Yang	Imperial College London, UK
Nejib Zemzemi	Inria, France
Xiahai Zhuang	Fudan University, China

FIMH Board Members

Leon Axel	New York University, USA
Patrick Clarysse	Université de Lyon, CREATIS, France
Martyn Nash	University of Auckland, New Zealand
Mihaela Pop	Sunnybrook Research Institute, Toronto, Canada
Alistair Young	King's College London, UK

Additional Reviewers

Gaetan Desrues	Inria, France
Rubén Doste	Universitat Pompeu Fabra, Spain
Kathleen Gilbert	University of Auckland, New Zealand
Andrea Guala	Vall d'Hebron Research Institute, Spain
Josquin Harrison	Inria, France
Michael Loecher	Stanford University, USA
Stefano Longobardi	King's College London, UK
Xinzhe Luo	Fudan University, China
Buntheng Ly	Inria, France
Gonzalo Maso Talou	University of Auckland, New Zealand
Viorel Mihalef	Siemens Healthineers, USA
Kevin Moulin	Université de Lyon, CREATIS, France
Abdel Hakim Moustafa	Hospital de la Santa Creu i Sant Pau, Spain
Matthew Ng	Sunnybrook Research Institute, Toronto, Canada
Marta Nuñez Garcia	IHU Liryc, France
Sebastien Piat	Siemens Healthineers, USA
Orod Razeghi	King's College London, UK
Gregory Sands	University of Auckland, New Zealand
Marina Strocchi	King's College London, UK
Felipe Viana	University of Central Florida, USA
Fuping Wu	Fudan University, China
Yuncheng Zhou	Fudan University, China

Administrative Organizers

Katie Pontius	Stanford University, USA
Amy Thomas	Stanford University, USA
Jaqueline Velazquez	Stanford University, USA
Ashley Williams	Stanford University, USA

Sponsor

We are very grateful for the support from Thornton Tomasetti's Life Sciences Team.

Thornton Tomasetti

Contents

Novel Approaches to Measuring Heart Deformation

Cardiac Mechanics: Measures and Models

Cardiovascular Flow: Measures and Models

Atrial Microstructure, Modeling, and Thrombosis Prediction

Advanced Cardiac and Cardiovascular Image Processing

Population-Based Personalization of Geometric Models of Myocardial Infarction

Kannara Mom, Patrick Clarysse, and Nicolas Duchateau[✉]

Univ Lyon, Université Claude Bernard Lyon 1, INSA-Lyon, CNRS, Inserm,
CREATIS UMR 5220, U1294, 69621 Lyon, France
nicolas.duchateau@creatis.insa-lyon.fr

Abstract. We propose a strategy to perform population-based personalization of a model, to overcome the limits of case-based personalization
for generating virtual populations from models that include randomness.
We formulate the problem as matching the synthetic and real populations by minimizing the Kullback-Leibler divergence between their distributions. As an analytical formulation of the models is complex or even
impossible, the personalization is addressed by a gradient-free method:
the CMA-ES algorithm, whose relevance was demonstrated for the case-
based personalization of complex biomechanical cardiac models. The
algorithm iteratively adapts the covariance matrix which in our problem encodes the distribution of the synthetic data.

We demonstrate the feasibility of this approach on two simple geometrical models of myocardial infarction, in 2D, to better focus on the
relevance of the personalization process. Our strategy is able to reproduce the distribution of 2D myocardial infarcts from the segmented late
Gadolinium images of 123 subjects with acute myocardial infarction.

Keywords: Model personalization · Myocardial infarction · Late
Gadolinium enhancement · Cardiac magnetic resonance

1 Introduction

The interest of developing realistic simulations of myocardial infarction is unde-
niable for understanding and validation purposes [1]. As for many biophysical
models, personalization is a necessary milestone for the realism of such simulations. The standard personalization approach consists in adjusting the model
to the data of a given individual. In order to generate large virtual populations
that can for example feed machine learning algorithms, the personalization process should be extended beyond the fit to individuals' data. In this sense, two
approaches may be adopted:

- *Case-based* personalization, which consists in finding the optimal model
  parameters for each individual, from which a range of relevant values can

© Springer Nature Switzerland AG 2021
D. B. Ennis et al. (Eds.): FIMH 2021, LNCS 12738, pp. 3–11, 2021.
https://doi.org/10.1007/978-3-030-78710-3_1

be determined a-posteriori. A synthetic population can therefore be generated by randomly sampling within this range and running the corresponding simulations. This approach is suited for fully deterministic models, but cannot cover models with randomness inside (e.g. generating a lesion at a random location or of random shape).

– *Population-based* personalization (the approach we propose), which can overcome the latter issue by finding the optimal model parameters that best match a virtual population to the real population under study. This strategy is somehow comparable to the matching of distributions pursued in variational auto-encoders [2], except that in our case the virtual population is generated from the personalized model parameters, not from a related latent space that encodes the data.

In this context, we propose a strategy to perform population-based personalization of a model, which we illustrate on simple geometrical models of myocardial infarction.

Tissue-level geometrical models of the lesions have been proposed based on a regional prior about a given coronary territory [3–5] or even up to mimicking the wavefront phenomenon [6] for the lesion propagation around an existing coronary segmentation [7]. Here we rely on two very simple models whose output is rather straightforward to visualize and assess, in 2D, so that we can better focus on the relevance of the personalization process: one that iteratively models an infarct as the union of spheres of random size [4], and one that uses an ellipsoid centered on the endocardium to represent the infarct [8].

Our primary objective is to demonstrate the feasibility of the population-based personalization. In addition, our secondary objective is to state on the relevance of these geometrical models of myocardial infarction to mimic a real population of patients with acute infarcts.

2   Methods

2.1   Data and Pre-processing

The population under study consisted of 123 subjects for which late Gadolinium enhancement images were available and segmented, distributed into 45, 17, and 61 cases for which the LAD, LCX, or RCA coronary arteries were responsible of the infarct. These data came from the Minimalist Immediate Mechanical Intervention (MIMI) study [9], which consists of acute ST-Elevation Myocardial Infarction (STEMI) patients who underwent either immediate or delayed stenting. Cardiac magnetic resonance was performed 5 days (interquartile range 4–6) after inclusion with Avanto 1.5T systems (Siemens, Erlangen, Germany). The infarct patterns were derived from the late Gadolinium enhancement images, performed 10 min after bolus injection, with an inversion time around 240–280 ms. The myocardial contours were manually segmented offline by consensus reading of three experienced observers using commercial software (CVI42 v.5.1.0 Circle Cardiovascular Imaging, Calgary, CA). The LV ranged over 17 ± 2 slices.

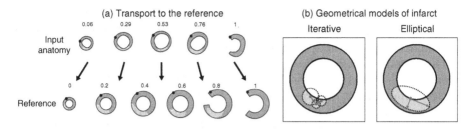

Fig. 1. (a) Transport of the data from an individual to the reference anatomy, illustrated on a subset of slices. The black dot points out the LV-RV junction. (b) Schematic view of the two simple geometrical models of infarct used in this paper.

The infarct zone was determined semi-automatically using the full-width half-maximum (FWHM) method. All contours were controlled and corrected manually if needed.

Radial, circumferential, and long-axis coordinates ranging from 0 to 1 were automatically defined on each stack of slices, after manually identifying the LV-RV junction on each slice and the myocardial borders around the LV outflow tract, when relevant. For this automatic parameterization of the LV geometry, we used images upsampled by a factor of 4 to prevent artifacts that may occur with few pixels covering the myocardium, in particular along the radial direction.

The lesion segmentations were transported to a reference anatomy (defined as a semi-ellipsoid with maximal endocardial and epicardial radii of respectively 30 and 50 pixels, represented as a stack of 21 slices of 80×80 pixels each), using linear interpolation tailored for data defined on a scattered grid (Fig. 1a).

Finally, we rotated all infarct patterns along the circumference such that the centers of mass of the infarcts for each coronary territory are aligned to the LAD one. This allows better focusing on the infarct shape and extent, and lowering the effects of the infarct localization without being restricted to a given coronary territory. Although this is arguable, as wall characteristics and infarct patterns may differ across coronary territories, it allowed us to demonstrate the feasibility of the personalization process on large enough populations.

2.2   Geometric Models of Myocardial Infarction

We focused on two geometrical models that generate synthetic lesions of varying shape, extent, and location. Each model is parameterized by a starting point on the endocardial surface, which conditions the infarct location, and other parameters that encode the shape and extent of the lesion (Fig. 1b):

– The first model approximates a random infarct shape as the union of several spheres of random sizes (intersected with the myocardium) [4]. The first sphere is centered on a given endocardial point, and the model generates

at each iteration a new sphere centered on a random point of the previous sphere. This model requires setting the total number of iterations and the maximal radius of the spheres (2 parameters).
– The second model generates an ellipsoid centered at a given endocardial point, and sets the infarcted region as the intersection between the ellipsoid and the myocardium [8]. It requires setting the short- and long-axes of the ellipsoid, which respectively lie on the radial and circumferential/longitudinal directions of the endocardium. Randomness is introduced in this model by sampling the short- and long-axis values using a uniform distribution (4 parameters: for each axis, the distribution center and its extension).

To better initialize the models and limit the number of parameters to optimize, the starting point was sampled within a small zone around the center of mass of the infarcts obtained from the real subjects, for both models.

In 3D, the synthetic lesions are generated on a template mesh that corresponds to the reference anatomy used to align the real image data (Sect. 2.1). Synthetic slices are then estimated by resampling the mesh data on an image grid. In this paper, to better focus on the personalization process within a reasonable computing time, we focused on a 2D version of the geometrical models at the mid-level of the left ventricle.

2.3   Personalization of the Models

The personalization relied on the CMA-ES algorithm (Covariance Matrix Adaptation Evolution Strategy) [10], a method that consists in adapting across iterations (referred to as generations) the synthetic data distribution parameterized by its covariance matrix. It uses a stochastic search that retains at each iteration a subset of the best cases (in the sense of the energy to minimize) sampled from a multi-normal distribution, and updates the covariance matrix of the distribution with such samples.

CMA-ES is a gradient-free method whose relevance was demonstrated for the personalization of complex models as in cardiac electromechanical simulations [11]. Besides, this generic personalization strategy allows to remain non-specific to a given model.

Let's denote $\{\mathbf{x}_1, \ldots, \mathbf{x}_p\}$ and $\{\mathbf{y}_1, \ldots, \mathbf{y}_q\}$ the sets of synthetic and real infarct images, represented by the distributions P and Q, respectively.

The *case-based personalization* means finding for each $i \in [1, q]$ the optimal parameters $\theta_i \in \mathbb{R}^{m_d}$ such that $\mathbf{x}_i = f_d(\theta_i) \approx \mathbf{y}_i$, where m_d is the number of parameters of a given (deterministic) model f_d (e.g. the number of iterations and the maximal sphere radius for the iterative model).

In contrast, the *population-based personalization* consists in finding the optimal parameters $\theta \in \mathbb{R}^{m_r}$ that lead to comparable synthetic and real distributions P and Q, where $\mathbf{x}_i = f_r(\theta)$ and f_r is the (random) model. From the point of view of the energy to minimize during the personalization process, this can be formulated as minimizing the Kullback-Leibler divergence between the distributions P and Q, approximated from the known samples $\{\mathbf{x}_i\}_{i=1}^p$ and $\{\mathbf{y}_i\}_{i=1}^q$

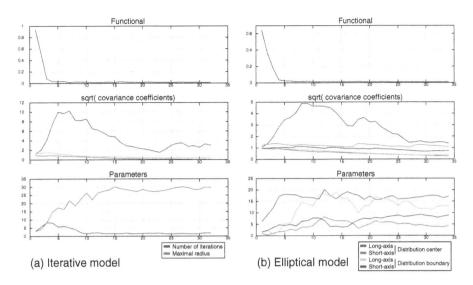

Fig. 2. Minimized energy (top), covariance matrix coefficients (middle), and model parameters (bottom) across iterations of the CMA-ES algorithm, for the iterative (a) and elliptical (b) models.

using a Gaussian kernel of bandwidth σ. The personalization is stopped when one of the three following conditions are met: the energy to minimize is below a given threshold ϵ, or N_{max} iterations have been performed, or the average and covariance of the synthetic population no longer substantially evolve.

3   Experiments and Results

3.1   CMA-ES Parameters and Convergence

For both models, the CMA-ES hyperparameters were empirically set to: 100 samples generated at each new generation of the algorithm, 15 samples retained to estimate the covariance matrix, maximum number of iterations $N_{max} = 500$, and the functional threshold $\epsilon = 0.004$ such that the covariance coefficients and the model parameters have enough iterations to stabilize. The kernel bandwidth σ used for the Kullback-Leibler divergence was set to the average distance between each sample and its 10 nearest neighbors.

The initial values were the identity for the covariance matrix, and 2 for each parameter of the model. However, as recommended by the CMA-ES authors, the long-axis of the elliptical model was rescaled by a factor 3 so that the sensitivity to the long- and short-axis parameters are (a-priori) comparable. This means a wider range of values for the infarct spread along the circumferential direction,

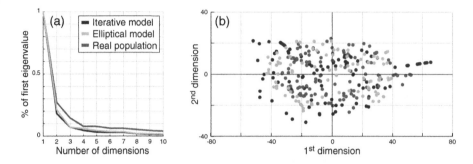

Fig. 3. Relative decrease of the eigenvalues (a) and distribution in the latent space (b) using non-linear dimensionality reduction with Isomap on the real (123 subjects) and synthetic populations generated using the iterative and elliptical models (100 samples for each model).

although this does not explicitly constrain infarcts to spread more along this direction (to better correspond to what is observed on the real data).

Figure 2 reports the evolution of the minimized energy, the covariance matrix coefficients, and the model parameters across the generations of the CMA-ES algorithm. It shows that although the energy is rapidly minimized, the evolution of the covariance matrix coefficients needs to be monitored to ensure a stable solution: the highest coefficient exhibits a first increase necessary to get the energy minimum within the reach of the distribution P, and then decreases and stabilizes while the model parameters also converge, indicating that the personalization has been achieved and is stable. Of note, due to the randomness in the models, we do not expect a perfect stabilization of the model parameters and covariance coefficients fully converging to 0, unless an infinite set of synthetic samples is generated.

3.2   Distribution of the Synthetic and Real Populations

We applied non-linear dimensionality reduction (Isomap) on the infarct patterns from the synthetic and real populations to better examine their distributions and the main variations they encode.

Figure 3 illustrates the relative decrease of the eigenvalues and the distribution of samples in the latent space. It shows that after personalization, the synthetic populations generated from the two models and the real population have similar distribution, although the synthetic populations are slightly more compact (faster decrease of eigenvalues), which is expected given that they originate from rather simple models.

Figure 4 complements these observations by representing the main variations encoded in these latent spaces, obtained by reconstructing the infarct patterns associated to $\{-2, -1, 0, 1, 2\}$ standard deviations along the first two dimensions.

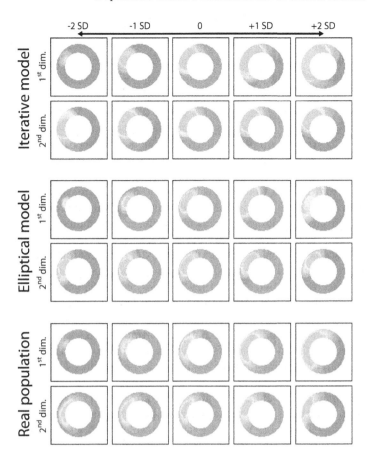

Fig. 4. First modes of variation (obtained through non-linear dimensionality reduction with Isomap on the infarct patterns) for the real and synthetic populations generated using the iterative and elliptical models. Intermediate values are observed between the healthy myocardium (blue) and the infarct (yellow) due to the regression used to reconstruct samples from the latent space. (Color figure online)

Reconstruction was achieved by multi-scale kernel regression [12], with equally weighted regularization and data terms. The regression used for the reconstruction of samples from the latent space may provide intermediate values between the healthy myocardium (blue) and the infarct (yellow): we kept them on purpose to assess the soundness of the reconstructed patterns, and only thresholded the values above/below these limits.

The synthetic and real populations exhibit comparable infarct shape, extent, and location along these modes of variation. Similar results are visible along the next dimensions (not displayed here to fit the page limit), although variations become more subtle as these distributions are rather compact. Slightly unplausible patterns may be observed at extreme values (e.g. +2 standard deviations for the elliptical model), due to the limits of the reconstruction method and the lower density of samples in this region of the latent space. The intrinsic limits of the models are also slightly visible: the iterative model tends to provide fully transmural lesions, while the elliptical model provides simpler infarct shapes.

4 Discussion

We proposed a method for the population-based personalization of simulations, illustrated in 2D on two simple geometrical models of myocardial infarction with respect to a real population of acute infarcts. We demonstrated the feasibility of this personalization strategy, which is therefore promising to generate virtual populations from models that intrinsically contain randomness.

The iterative model provides limited control on the infarct shape and its propagation from the endocardium. This intrinsic randomness, combined with the stochastic search performed by the CMA-ES algorithm, are rather challenging to reach stable personalizations, although the examination of the optimization values (energy, covariance coefficients, and optimized parameters) confirms the convergence of the personalization. Besides, comparable optimal values of the model parameters were obtained by using a large covariance matrix as initialization. Adding relevant constraints on this model may improve realism and the robustness and efficiency of the personalization process.

Examining 2D data allowed focusing on the personalization process in a reasonable computing time (around 1 min for the personalization of each model). The Kullback-Leibler divergence seems viable to compare the synthetic and real populations during personalization, although comparisons are directly performed on the infarct patterns and not on a relevant latent space as in variational auto-encoders [2]. We obtained comparable results with the Maximum Mean Discrepancy [13], although it tended to provide less stable results.

In 2D, the elliptical model seemed to better encode the infarct transmurality but the generated shapes are rather simple. However, generalizing this work to 3D (planned for future work) will help to better state on the realism of such models to generate relevant synthetic populations. In any case, our primary focus here was to assess the feasibility of population-based personalization, and not on developing more complex infarct models. Nonetheless, further work should consider reducing the gap between synthetic and real data, using heterogeneous (instead of binary) infarct patterns, or at least a third region that could represent the border zone, and infarcts that may originate from the occlusion of several arteries. Provided relevant real data are available, our work is also generalizable beyond the acute infarct setting, for example to chronic infarcts, or to consider remodeling that may occur over time.

Acknowledgements. The authors acknowledge the support from the French ANR (LABEX PRIMES of Univ. Lyon [ANR-11-LABX-0063] within the program "Investissements d'Avenir" [ANR-11-IDEX-0007], the JCJC project "MIC-MAC" [ANR-19-CE45-0005]), and the Fédération Francaise de Cardiologie ("MI-MIX" project, Allocation René Foudon). They are also grateful to P Croisille and M Viallon (CREATIS, CHU Saint Etienne) for providing the imaging data for the MIMI population, and M Di Folco (CREATIS) for the preliminary exploration of the data alignment and simulation tools.

References

1. Connolly, A.J., Bishop, M.J.: Computational representations of myocardial infarct scars and implications for arrhythmogenesis. Clin. Med. Insights Cardiol. **10**, 27–40 (2016)
2. Kingma, D.P., Welling, M.: Auto-encoding variational Bayes. arXiv (2014)
3. Kerckhoffs, R.C., McCulloch, A.D., Omens, J.H., et al.: Effects of biventricular pacing and scar size in a computational model of the failing heart with left bundle branch block. Med. Image Anal. **13**, 362–369 (2009). https://doi.org/10.1016/j.media.2008.06.013
4. Duchateau, N., De Craene, M., Allain, P., et al.: Infarct localization from myocardial deformation: prediction and uncertainty quantification by regression from a low-dimensional space. IEEE Trans. Med. Imaging **35**, 2340–2352 (2016). https://doi.org/10.1109/TMI.2016.2562181
5. Leong, C.N., Lim, E., Andriyana, A., et al.: The role of infarct transmural extent in infarct extension: a computational study. Int. J. Numer. Method Biomed. Eng. **33**, e02794 (2017). https://doi.org/10.1002/cnm.2794
6. Reimer, K.A., Lowe, J.E., Rasmussen, M.M., et al.: The wavefront phenomenon of ischemic cell death. 1. Myocardial infarct size vs duration of coronary occlusion in dogs. Circulation **56**, 786–794 (1977). https://doi.org/10.1161/01.cir.56.5.786
7. Pashaei, A., Hoogendoorn, C., Sebastián, R., Romero, D., Cámara, O., Frangi, A.F.: Effect of scar development on fast electrophysiological models of the human heart: in-silico study on atlas-based virtual populations. In: Metaxas, D.N., Axel, L. (eds.) FIMH 2011. LNCS, vol. 6666, pp. 427–436. Springer, Heidelberg (2011). https://doi.org/10.1007/978-3-642-21028-0_54
8. Rumindo, G.K., Duchateau, N., Croisille, P., Ohayon, J., Clarysse, P.: Strain-based parameters for infarct localization: evaluation via a learning algorithm on a synthetic database of pathological hearts. In: Pop, M., Wright, G.A. (eds.) FIMH 2017. LNCS, vol. 10263, pp. 106–114. Springer, Cham (2017). https://doi.org/10.1007/978-3-319-59448-4_11
9. Belle, L., Motreff, P., Mangin, L., et al.: Comparison of immediate with delayed stenting using the minimalist immediate mechanical intervention approach in acute ST-segment-elevation myocardial infarction: the MIMI study. Circ. Cardiovasc. Interv. **9**, e003388 (2016). https://doi.org/10.1161/CIRCINTERVENTIONS.115.003388
10. Hansen, N., Ostermeier, A.: Adapting arbitrary normal mutation distributions in evolution strategies: the covariance matrix adaptation. In: Proceedings ICEC, pp. 312–317 (1996). https://doi.org/10.1109/ICEC.1996.542381
11. Molléro, R., Pennec, X., Delingette, H., Garny, A., Ayache, N., Sermesant, M.: Multifidelity-CMA: a multifidelity approach for efficient personalisation of 3D cardiac electromechanical models. Biomech. Model. Mechanobiol **17**(1), 285–300 (2017). https://doi.org/10.1007/s10237-017-0960-0
12. Duchateau, N., De Craene, M., Sitges, M., Caselles, V.: Adaptation of multiscale function extension to inexact matching: application to the mapping of individuals to a learnt manifold. In: Nielsen, F., Barbaresco, F. (eds.) GSI 2013. LNCS, vol. 8085, pp. 578–586. Springer, Heidelberg (2013). https://doi.org/10.1007/978-3-642-40020-9_64
13. Gretton, A., Borgwardt, K.M., Rasch, M.J., et al.: A kernel two-sample test. J. Mach. Learn. Res. **13**, 723–73 (2012)

Impact of Image Resolution and Resampling on Motion Tracking of the Left Chambers from Cardiac Scans

Orod Razeghi[1]([✉]), Marina Strocchi[1], Cesare Corrado[1], Henry Chubb[2], Ronak Rajani[1], Daniel B. Ennis[3], and Steven A. Niederer[1]

[1] Biomedical Engineering and Imaging Sciences, King's College London, London, UK
orod.razeghi@kcl.ac.uk
[2] Department of Cardiothoracic Surgery, Stanford University, Stanford, CA, USA
[3] Department of Radiology, Stanford University, Stanford, CA, USA

Abstract. Cardiac magnetic resonance (CMR) is an important diagnostic imaging modality in cardiovascular medicine. Estimation of myocardium motion derived from CMR scans is routinely used to measure the cardiac mechanics. However, tracking a rapidly moving organ can be compromised by artefacts or impaired image quality. To assess how in-plane image resolution and slice sampling in short and long axis scans impact errors in motion tracking, we utilised retrospective gated cardiac computed tomography (CCT) imaging as a surrogate groundtruth for motion estimation across 10 clinical datasets, since these scans have a higher isotropic resolution than CMR and the ability to capture full 3D motion. In our work, the left atrial (LA) and ventricular (LV) cavities were first delineated, and then reconstructed short and long axis images were used in a non-rigid registration method with optimised hyperparameters to track endocardial motion. Finally, global and regional functions in the form of area, circumferential, and longitudinal strains were computed. Our findings showed that tracking LA was more sensitive than LV to changes in the in-plane resolution and magnitude of strain was robust to resolution changes in short axis images, when correlated with the groundtruth (r: 0.87–0.99, R^2: 0.75–0.98). We also found that 9 short axis slices could capture the motion of LV almost as accurately as 36 slices captured in long axis (r: 0.89 vs. 0.90, R^2: 0.80 vs. 0.80), illustrating that the cardiac mechanics measured by short axis scans are more likely to be robust to image artefacts and reconstruction parameters.

Keywords: Image resampling · Motion tracking · Strain calculation · Cardiac magnetic resonance

1 Introduction

Cardiac magnetic resonance (CMR) is an important diagnostic imaging modality in cardiovascular medicine. Global cardiac mechanics, routinely measured by

© Springer Nature Switzerland AG 2021
D. B. Ennis et al. (Eds.): FIMH 2021, LNCS 12738, pp. 12–21, 2021.
https://doi.org/10.1007/978-3-030-78710-3_2

ejection fraction, is used in the diagnosis and stratification of patients, but cardiac pathologies do not always manifest as a change in global function. Recent improvements in CMR image reconstruction techniques and progressive enhancements in spatial and temporal resolution have augmented the established global measurements by enabling regional motion estimation of myocardium throughout the cardiac cycle. [12]

Historically, measuring regional cardiac mechanics has focused on characterising motion using two and three dimensional CMR scans [2,10,12], where multiple images of the heart throughout the cardiac cycle are generated and features (tags or anatomy) are tracked between temporal frames. However, insufficient image quality may compromise the diagnostic accuracy. CMR captures the motion of a rapidly moving organ. Thus, several factors may contribute to its artefacts or impaired image quality. Some artefacts are patient dependent, whilst others are due to technical aspects during acquisition, or physical limits of the imaging sequence itself [4].

The standard CMR protocol for the assessment of ventricular function includes 10 to 14 contiguous short axis slices and 3 long axis slices (2-chamber view, 3-chamber view and 4-chamber view). [6] In this study, we aim to assess how image resolution and sampling in short and long axis scans impact errors in motion tracking. Retrospective gated cardiac computed tomography (CCT) imaging has higher isotropic resolution than CMR and the ability to capture full 3D motion throughout the cardiac cycle, thanks to improvements in the temporal resolution of dual-source scanners. [5] We therefore plan to utilise CCT datasets as a surrogate groundtruth for motion estimations.

The high resolution of these datasets allows us to accurately delineate the left atrial (LA) and left ventricular (LV) cavities. In this work, we use a non-rigid registration method with optimised hyperparameters for tracking cardiac motion, named the temporal sparse free-form deformations (TSFFD) method [11], which is based on the widely adopted approach of Rueckert et al. [9] We then measure global and regional function in the form of area, circumferential, and longitudinal strains. We derive area changes using the method described in Pourmorteza et al. [8] and the strains using large strain theory across clinical datasets of 10 patients.

2   Methods

Characterising cardiac wall motion requires a sequence of image processing steps, which we describe in this section. We first introduce the clinical datasets used in our experiments, then review our designed workflow to reconstruct images under different sampling settings, and finally conclude with the description of the image registration technique, which was use to estimate cardiac motion and calculate regional and global strain indices.

2.1 Clinical Data

CCT scans used in this retrospectice study were obtained from 10 patients undergoing cardiac resynchronization therapy (CRT). Scans were performed using a Philips Brilliance iCT 256-slice MDCT scanner (Philips Healthcare, Amsterdam, Netherlands). Intravenous metoprolol was used to achieve a mean heart rate of 64 ± 7 beats/min. The mean radiation dose-area product was 1194 ± 419 mGy·cm^2. A total of 100 ml of intravenous contrast agent (Omnipaque, GE Healthcare, Princeton, NJ, USA) was injected via a power injector into the antecubital vein. Helical scanning was performed with a single breath-hold technique after a 10–12 s delay.

The scanning parameters included: a heart rate dependent pitch of 0.2–0.45, a gantry rotation time of 270 ms, a tube voltage of 100 or 120 kVp depending on the patient's body mass index, and a tube current of 125–300 mA depending upon the thoracic circumference. Retrospectively ECG-gated image reconstruction was used to generate 10 equally spaced images per cardiac cycle.

Dimension of images varied between CCT datasets but the majority were 256×256 voxels with a 0.5–0.9 mm isotropic in-plane resolution. The 3D stacks had 102–191 slices with a through plane thickness of 1 mm. The Hounsfield unit spanned from –1024–3071 and the temporal resolution from 72–120 ms, depending on the patient's heart rate.

Data were collected in accordance with relevant guidelines and regulations as part of two clinical trials, which were approved by the West Midlands Coventry & Warwick REC (14/WM/1069) and the London-Harrow (18/LO/0752) ethics committees. All the patients gave written informed consent and the scans were analysed anonymously. The data was fully anonymised and used with consent from the trusts data governance.

2.2 Preprocessing Data

A reference image was selected from the first frame of the CCT, representing the heart at the end-diastolic phase and the peak of the R-wave in the electrocardiogram (ECG). Next, the blood pools of the LA and LV including the papillary muscles were segmented from the reference image. Segmentations were performed using a grey value based region growing tool. The grey values were determined from all point positions plus/minus a margin of 30 Hounsfield units. The 2D region growing segmentation was applied in 5–10 of the long axis slices. These slices were interpolated to label the cavities in 3D with the option for manual correction. After achieving a full segmentation, a marching cubes process generated a smooth endocardial surface from the segmentation.

Finally, all the cases were annotated with anatomical landmarks by a trained clinician. Six landmarks were selected on the reference image: one on the apex, three on the surface of the mitral valve and two on the septum, delineated by the attachment of the right ventricle. These landmarks defined a coordinate system that could be used to label sections of the LV reference surface with one of the AHA segments [1]. These segments enabled local interpretation of the analysis.

2.3   Resampling Data

To be able to study the effect of image resolution and sampling on strain measurements, we used the CCT datasets to reconstruct short axis images with variable spatial resolution and slice thickness.

An apical basal vector $\mathbf{v}_{ab} = \mathbf{X}_b - \mathbf{X}_a$ was defined, where $\mathbf{X}_a$ and $\mathbf{X}_b$ were the apex and mid point of the mitral valve's annulus, respectively. The cross product between the normalised vector $\hat{\mathbf{v}}_{ab}$ and one of the Cartesian coordinates axis provided us with an axis of rotation, which was then used in a rigid transformation to reshape the scans into contiguous short-axis slices, as seen in Fig. 1.

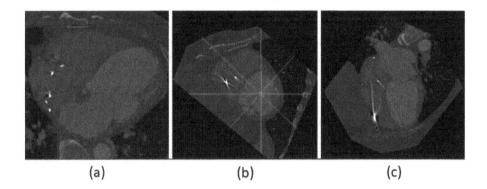

(a)   (b)   (c)

Fig. 1. Reconstruction of short and long axis images: (a) Groundtruth CCT scan used for studying spatial resolution and resampling effects. (b) Reconstructed short axis view. Radially rotational long-axis slices are localised in this view with red lines representing example cutting planes. (C) Reconstructed long axis view using the clinician's defined apical basal direction. (Color figure online)

We resampled the short slices to have a range of 1.5, and 2 mm isotropic in-plane resolutions and a slice thickness of 5, and 10 mm. A cubic interpolator was utilised for reconstruction, when these spatial transformations were applied to the images. As an alternative to short axis images, we also studied the effect of resampling long axis views on the chamber functions by obtaining slices covering both LA and LV from the transformed image with 5°, 10°, 20°, and 45° angular equidistance in a radially rotational manner. A similar cubic interpolator was used for reconstruction of the full 3D stack.

2.4   Endocardium Motion Estimation

Tracking endocardial motion can essentially be defined as the non-rigid registration of cardiac image sequences. We used an optimised registration algorithms to track motion of the endocardium from the series of images sampled.

In free-form deformation (FFD) registration [9], a non-rigid deformation $h = [X \; Y \; Z]^T$ is represented using a B-spline model in which the deformation is parametrised using a set of control points $\boldsymbol{\Phi} = [U \; V \; W]^T$. To be able to deal with large global deformations and to improve the robustness, the classic FFD registration normally uses a multi-level representation. We used the sparse free-form deformation (SFFD) technique [11] to extend the classic FFD approach and recover smoother displacement fields. In our work, we optimised this technique by using a four-level representation and sum of squared differences as the similarity measure. The registration energy function was minimised using a gradient descent approach [7].

2.5 Calculation of Strain on Endocardium

Cardiac deformation is the transformation of the endocardial reference mesh into a configuration representing each phase of the cardiac cycle. To calculate circumferential and longitudinal strains from each of these configurations, we defined an element coordinate system by normalised base vectors:

$$\mathbf{e}_\mathrm{r} = \frac{\mathbf{e}_{21} \times \mathbf{e}_{31}}{\|\mathbf{e}_{21} \times \mathbf{e}_{31}\|}, \tag{1}$$

$$\mathbf{e}_\mathrm{z} = \frac{\mathbf{v}_\mathrm{ab} - (\mathbf{v}_\mathrm{ab} \cdot \mathbf{e}_\mathrm{r})\mathbf{e}_\mathrm{r}}{\|\mathbf{v}_\mathrm{ab} - (\mathbf{v}_\mathrm{ab} \cdot \mathbf{e}_\mathrm{r})\mathbf{e}_\mathrm{r}\|}, \tag{2}$$

$$\mathbf{e}_\theta = \frac{\mathbf{e}_\mathrm{z} \times \mathbf{e}_\mathrm{r}}{\|\mathbf{e}_\mathrm{z} \times \mathbf{e}_\mathrm{r}\|}, \tag{3}$$

where $\mathbf{e}_{21}$ and $\mathbf{e}_{31}$ are the mesh element edge vectors between subscript vertices (note the counter-clockwise triangular element numbering) and $\mathbf{v}_\mathrm{ab} = \mathbf{X}_\mathrm{b} - \mathbf{X}_\mathrm{a}$ represents the left ventricular long axis. The clinically used cylindrical element coordinate system with circumferential and longitudinal components was defined by rotating the Green-Lagrange strain tensor $\mathbf{E}$ from the global Cartesian coordinate system. The cylindrical element coordinate system remained the same for every time frame in the cardiac cycle and the rotation was utilised by the transformation matrix $\mathbf{Q}$ evaluated as:

$$\mathbf{E}_\mathrm{e} = \mathbf{Q}\mathbf{E}\mathbf{Q}^\mathrm{T}, \quad \text{with} \quad \mathbf{Q} = \begin{bmatrix} \mathbf{e}_\mathrm{r} \cdot \mathbf{e}_1 & \mathbf{e}_\mathrm{r} \cdot \mathbf{e}_2 & \mathbf{e}_\mathrm{r} \cdot \mathbf{e}_3 \\ \mathbf{e}_\theta \cdot \mathbf{e}_1 & \mathbf{e}_\theta \cdot \mathbf{e}_2 & \mathbf{e}_\theta \cdot \mathbf{e}_3 \\ \mathbf{e}_\mathrm{z} \cdot \mathbf{e}_1 & \mathbf{e}_\mathrm{z} \cdot \mathbf{e}_2 & \mathbf{e}_\mathrm{z} \cdot \mathbf{e}_3 \end{bmatrix}, \tag{4}$$

where $\mathbf{e}_i$, for $i = 1, 2, 3$, represents the unit direction vector in the Cartesian coordinate system. After the rotation, the radial strain components vanished, i.e. $E_{\mathrm{r}i} = E_{i\mathrm{r}} = 0$, due to the definition of the problem. The other elements in the main diagonal of the strain tensor, i.e. $E_{\theta\theta}$ and E_{zz}, provided us with the circumferential and longitudinal strains for a given element. Regional strains were the mean of the individual elements strains in each AHA segment. We also derived area strains using the method described in Pourmorteza et al.[8] to measure deformation in the LA, as well as the LV. Changes in the area of the endocardial surface was defined as the square root of the area of a triangular patch on the endocardial mesh at time t over the area of the same patch at end diastolic time frame.

2.6 Analysis of Motion Tracking Error

The LV function was analysed regionally using the clinician's defined AHA segments. We selected three measurements from the computed regional strains to run a correlation analysis between the estimated functions after resampling and the calculated functions from the groundtruth. The three measurements were: 1) time to the peak of a time curve for an AHA segment, 2) time from the onset until a curve reaches 50% of its amplitude, and 3) magnitude of a curve defined as the difference between its maximum and minimum peaks. Systolic dyssynchrony index (SDI), defined as the standard deviation of the time from cardiac cycle onset to minimum systolic volume, has been previously used to assess cardiac function in patient's with dissynchronous contraction patterns. [3] We calculated SDI values and used them in the correlation analysis of our CRT patients. As we do not have a standardised way to divide the LA cavity, its function was analysed globally using an average area strain.

3 Results

The patient demographic characteristics are summarised in Table 1. Patients had a high percentage of right ventricular pacing with a mean QRS duration of 170 ± 28 ms.

Table 1. Demographic characteristics

Characteristics	Value	Characteristics	Value
Age (y)	67 ± 17	Atrial fibrillation	30%
Sex: male	70%	Right ventricular pacing	50%
QRS duration (ms)	170 ± 28	Sinus rhythm	20%

3.1 Impact of Spatial Resolution on Short Axis Reconstruction

We used Pearson correlation coefficient r and coefficient of determination from a regression analysis R^2 to compare the estimated LA and LV functions with the groundtruth, derived from isotropic 1mm reference images. The in-plane resolution was set to 1 mm for thickness comparisons, and similarly thickness was set to 1mm for in-plane resolution analysis.

Table 2 represents the correlation between computed global area strain and groundtruth, when short axis slices are reconstructed using a 1.5 and 2 mm

Table 2. Short axis reconstruction

	Resolution						Thickness					
	1.5 mm			2.0 mm			5 mm			10 mm		
LA strains	r	p	R^2	r	p	R^2	r	p	R^2	r	p	R^2
Area T2P	0.32	0.68	0.25	0.20	0.59	0.17	0.65	0.16	0.43	0.14	0.84	0.12
Area TOS	0.56	0.23	0.33	0.47	0.35	0.22	0.89	0.02	0.79	0.12	0.90	0.10
Area MAG	0.80	0.05	0.65	0.54	0.26	0.30	0.60	0.20	0.37	0.32	0.30	0.29
LV strains	r	p	R^2	r	p	R^2	r	p	R^2	r	p	R^2
Area T2P	0.81	0.05	0.66	0.73	0.10	0.53	0.88	0.02	0.78	0.39	0.45	0.15
Area TOS	0.97	0.00	0.94	0.97	0.00	0.95	0.94	0.01	0.88	0.78	0.07	0.60
Area MAG	0.98	0.00	0.97	0.96	0.00	0.92	0.93	0.01	0.86	0.89	0.02	0.80
Circ T2P	0.90	0.02	0.81	0.85	0.03	0.72	0.70	0.12	0.49	0.62	0.19	0.39
Circ TOS	0.77	0.07	0.59	0.70	0.12	0.49	0.49	0.32	0.24	0.10	0.85	0.01
Circ MAG	0.90	0.02	0.80	0.87	0.03	0.75	0.93	0.01	0.86	0.89	0.02	0.79
Long T2P	0.78	0.07	0.60	0.54	0.27	0.29	0.81	0.05	0.66	0.59	0.22	0.35
Long TOS	0.96	0.00	0.91	0.75	0.08	0.57	0.94	0.00	0.89	0.78	0.06	0.62
Long MAG	0.99	0.00	0.97	0.99	0.00	0.98	0.93	0.01	0.86	0.89	0.02	0.78

isotropic in-plane resolutions. Similarly, it summarises the correlation between the two sets, when slice thickness is set to 5, or 10 mm. In the tables in this section, T2P is short for time to the peak, TOS is the time from onset until a curve reaches 50% of its amplitude, and MAG is the magnitude of a curve. Similarly, Circ and Long stand for circumferential and longitudinal strains, respectively.

3.2 Impact of Spatial Resolution on Long Axis Reconstruction

As the alternative to short axis reconstruction, we also studied the effect of resampling long axis views on the chamber functions. The correlation results with the groundtruth is presented in Table 3.

3.3 Discussion

CMR provides a valuable imaging tool for diagnosis of cardiovascular diseases. However, the quality of global and regional cardiac mechanics measured by CMR is sensitive to artefacts and reconstruction parameters. Our results show that high resolution CCT motion tracking, currently in clinical use, can provide low cost datasets for exploring CMR image sequences without the cost of acquisition.

Our findings illustrate that tracking the LA motion was worse than the LV for most metrics. This may be due to its thin wall ($\sim$2 mm), complex topology, or use of tracking hyperparameters specifically optimised for the LV (not LA) motion. This issue was evident in the short axis images, where changes to original

Table 3. Long axis reconstruction

	Slices						Slices					
	36 (5°)			18 (10°)			9 (20°)			4 (45°)		
LA strains	r	p	R^2	r	p	R^2	r	p	R^2	r	p	R^2
Area T2P	0.68	0.14	0.46	0.37	0.47	0.13	0.21	0.69	0.04	0.12	0.82	0.01
Area TOS	0.69	0.13	0.48	0.41	0.41	0.17	0.26	0.62	0.07	0.24	0.65	0.06
Area MAG	0.74	0.09	0.55	0.75	0.09	0.56	0.53	0.28	0.28	0.11	0.84	0.01
LV strains	r	p	R^2	r	p	R^2	r	p	R^2	r	p	R^2
Area T2P	0.89	0.02	0.78	0.74	0.09	0.55	0.68	0.13	0.47	0.57	0.24	0.32
Area TOS	0.29	0.58	0.08	0.19	0.71	0.04	0.16	0.76	0.03	0.07	0.90	0.00
Area MAG	0.90	0.02	0.80	0.87	0.02	0.76	0.82	0.05	0.67	0.55	0.26	0.30
Circ T2P	0.33	0.52	0.11	0.23	0.65	0.05	0.17	0.74	0.03	0.13	0.80	0.02
Circ TOS	0.65	0.16	0.42	0.61	0.20	0.37	0.39	0.44	0.15	0.08	0.89	0.01
Circ MAG	0.93	0.01	0.86	0.93	0.01	0.86	0.81	0.05	0.66	0.35	0.50	0.12
Long T2P	0.86	0.03	0.73	0.68	0.14	0.46	0.65	0.17	0.42	0.42	0.58	0.20
Long TOS	0.94	0.01	0.89	0.92	0.01	0.84	0.89	0.02	0.80	0.34	0.52	0.11
Long MAG	0.99	0.00	0.98	0.92	0.01	0.85	0.92	0.01	0.85	0.64	0.17	0.41

in-plane resolutions were made. The slice thickness had a less severe impact on our ability to track the LA endocardium. Tracking LV was more robust to in-plane resolution changes, and was only affected, when thicker axial slices were resampled. Magnitude of strain was robust to changes in image resolution with a range of r from 0.87–0.99 and R^2 from 0.75–0.98. Temporal based measures did not use interpolation, and they were generally less accurate, although this may have been affected by the lower temporal resolution of CCT images (10 phases throughout the cardiac cycle).

Construction of long axis images with 45° rotation resulted in the poorest correlation with the groundtruth. This was more apparent in the calculated times to peaks of the curves and times to reach 50% of the amplitudes. Calculated magnitudes of curves showed less sensitivity to changes in the resampling angles. However, we found that only 9 short axis slices with 10 mm thickness could capture the motion of LV almost as accurately as 36 slices captured in long axis (r: 0.89 vs. 0.90, R^2: 0.80 vs. 0.80). Longitudinal strains were relatively robust to the degree of resampling in the long axis images.

3.4 Limitations

The outcome of our study is likely to be affected by the dominance of cardiac wall thickening in comparison to longitudinal contractions. It can also be associated with the fact that the short axis motion is easier to resolve for the TSFFD algorithm. Given the empirical nature of this study, it would be desirable to expand on the registration algorithms used in future, as our findings are limited by the choice of methods for motion tracking and strain calculations.

Our calculations were limited to area, circumferential and longitudinal strains. We did not consider radial strain, as segmenting myocardium can be problematic due to image artefacts from existing leads in the CRT datasets. Moreover, segmenting and tracking the feature rich endocardium from CCT datasets is less error prone from adequate number of frames.

4 Conclusion

In this work, we assessed the impact of image resolution and sampling on errors in LA and LV measurements using CCT imaging as a surrogate groundtruth for motion estimations. Our findings illustrated that the cardiac mechanics measured by short axis CMR scans were more likely to be robust than long axis images to artefacts and reconstruction parameters.

Acknowledgements. Dr Niederer acknowledges support from the National Institute of Health (NIH R01-HL152256), and by core funding from the Wellcome/EPSRC Centre for Medical Engineering [WT203148/Z/16/Z].

References

1. Cerqueira, M., et al.: Standardized myocardial segmentation and nomenclature for tomographic imaging of the heart. Circulation **105**(4), 539–542 (2002)
2. Chakraborty, A., Staib, L., Duncan, J.: Deformable boundary finding in medical images by integrating gradient and region information. IEEE Trans. Med. Imaging **15**(6), 859–870 (1996)
3. Kapetanakis, S., Kearney, M., Siva, A., Gall, N., Cooklin, M., Monaghan, M.: Realtime three-dimensional echocardiography. Circulation **112**(7), 992–1000 (2005)
4. Klinke, V., Muzzarelli, S., Lauriers, N., et al.: Quality assessment of cardiovascular magnetic resonance in the setting of the European CMR registry: description and validation of standardized criteria. J. Cardiovasc. Magn. Reson. **15**(1), 55 (2013)
5. Mak, G., Truong, Q.: Cardiac CT: imaging of and through cardiac devices. Curr. Cardiovasc. Imaging Rep. **5**(5), 328–336 (2012)
6. Marchesseau, S., Ho, J., Totman, J.: Influence of the short-axis cine acquisition protocol on the cardiac function evaluation: a reproducibility study. Eur. J. Radiol. Open **3**, 60–66 (2016)
7. Modat, M., et al.: Fast free-form deformation using graphics processing units. Comput. Methods Prog. Biomed. **98**(3), 278–284 (2010)
8. Pourmorteza, A., Schuleri, K., Herzka, D., Lardo, A., McVeigh, E.: A new method for cardiac computed tomography regional function assessment: stretch quantifier for endocardial engraved zones (SQUEEZ). Circ. Cardiovasc. Imaging **5**(2), 243–250 (2012)
9. Rueckert, D., Sonoda, L., Hayes, C., Hill, D., Leach, M., Hawkes, D.: Nonrigid registration using free-form deformations: application to breast MR images. IEEE Trans. Med. Imaging **18**(8), 712–721 (1999)

10. Scatteia, A., Baritussio, A., Bucciarelli-Ducci, C.: Strain imaging using cardiac magnetic resonance. Heart Fail. Rev. **22**(4), 465–476 (2017). https://doi.org/10. 1007/s10741-017-9621-8

11. Shi, W., et al.: Temporal sparse free-form deformations. Med. Image Anal. **17**(7), 779–789 (2013)

12. Tee, M., Noble, J., Bluemke, D.: Imaging techniques for cardiac strain and deformation: comparison of echocardiography, cardiac magnetic resonance and cardiac computed tomography. Exp. Rev. Cardiovasc. Therapy **11**(2), 221–231 (2013)

Shape Constraints in Deep Learning for Robust 2D Echocardiography Analysis

Yingyu Yang and Maxime Sermesant[(✉)]

Inria, Université Côte d'Azur, Sophia Antipolis, Nice, France
{yingyu.yang,maxime.sermesant}@inria.fr

Abstract. 2D Echocardiography is a popular and cost-efficient tool for cardiac dysfunction diagnosis. Automatic solutions that could effectively and efficiently analyse cardiac functions are highly desired in clinical situations. Segmentation and motion tracking are two important techniques to extract useful cardiac indexes, such as left ventricle ejection fraction (LVEF), global longitudinal strain (GLS), etc. However, these tasks are non-trivial since ultrasound images usually suffer from poor signal-to-noise ratio, boundary ambiguity and out of view problem. In this paper, we explore how to introduce shape constraints from global, regional and pixel level into a baseline U-Net model for better segmentation and landmark tracking. Our experiments show that all the three propositions perform similarly as the baseline model in terms of geometrical scores, while our pixel-level model, which uses a multi-class contour loss, reduces segmentation outliers and improves the tracking accuracy of 3 landmarks used for GLS computation. With appropriate augmentation techniques, our models also show a good generalisation performance when testing on a larger unseen cohort.

Keywords: Segmentation · Deep learning · Deformation · Echocardiography

1 Introduction

Echocardiography, a non-invasive and cost-efficient imaging technique, is widely used by cardiologists to evaluate the cardiac function. Segmentation and motion tracking are two essential tasks that can help cardiologists in clinical decision-making. Segmentation offers information of shape and volume while motion tracking provides knowledge of deformation and function.

Deep learning based methods have shown very good performance for medical segmentation and registration. As for segmentation, the U-Net architecture has proved its overwhelming power in large cohort echocardiography segmentation [11]. With appropriate adaptation of U-Net model and data augmentation, the U-Net architecture also demonstrated good generalisation ability in segmenting cardiac magnetic resonance images (CMRI) [5]. However, echocardiography segmentation is still difficult since it usually encounters the problem of out of view, poor signal-to-noise ratio etc., especially for myocardium segmentation.

© Springer Nature Switzerland AG 2021
D. B. Ennis et al. (Eds.): FIMH 2021, LNCS 12738, pp. 22–34, 2021.
https://doi.org/10.1007/978-3-030-78710-3_3

In the field of cardiac motion tracking, unsupervised deep learning are very popular and these schemes reach similar performance or even outperform traditional registration methods. Krebs et al. proposed a conditional variational autoencoder which learned a diffeomorphic transformation from pairwise CMRI in an unsupervised way [10]. Shawn et al. [1] designed a U-Net like network for unsupervised pairwise echocardiography motion tracking. However, the displacement field can be unrealistic without relevant regularisation.

Numerous studies have shown that global longitudinal strain (GLS) is more sensitive than left ventricular ejection fraction (LVEF) as a measure of systolic function and has potential in identifying left ventricle dysfunction in clinical [7,9]. This value can be approximated by measuring the left ventricle length change [14]. Therefore, estimating a dense displacement may not be necessary.

U-Net like deep learning models depend largely on pixel-level classification, which can generate artefacts which are irregular with the organ shape. Researchers are seeking to combine shape constraints with deep learning methods [3]. With the same intention to improve the segmentation consistency with anatomical shapes in 2D echocardiography, in our work, we explore to introduce shape constraints from global, regional and pixel level into a baseline U-Net model. From the segmentation results, useful information such as ejection fraction, landmark based GLS can be extracted. The detailed model architecture will be explained in Sect. 2. We then present the implementation and experiment results for segmentation and landmark detection in Sect. 3.

2   Methods

We use a U-Net model as our baseline model. Its encoder consists of 5 down-sampling (MaxPool + Conv) blocks with ReLU activation after the 3x3 convolution. The corresponding decoder has 5 up-sampling (UpSample + Conv) blocks and is skip-connected with the encoder. Based on this model, we consider to incorporate shape constraints from three levels:

- **Global-level**: estimate a triangle like landmark map in parallel with segmentation (SEG-LM)
- **Regional-level**: add a poly-affine myocardium reconstruction network to constrain the shape of myocardium mask (SEG-AFFINE)
- **Pixel-level**: use a multi-class contour-loss to finely classify the boundary pixel (SEG-CONTOUR)

We will explain the three methods in detail in the following subsections (Fig. 1).

2.1   SEG-LM: Parallel Segmentation and Landmark Detection

We adapt the baseline U-Net model for simultaneous segmentation and landmark prediction by adding a separate branch of decoder for landmark map estimation. The two decoders process the encoder information in parallel. The final layer of

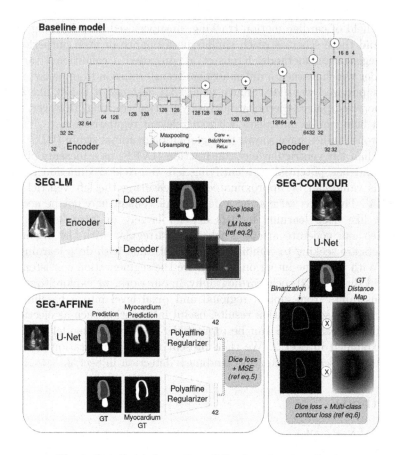

Fig. 1. Detailed information of the 4 explored methods.

segmentation branch and landmark detection have SoftMax activation, Sigmoid activation respectively. In particular, we consider the two end points of mitral valve (basal points) and the apex along the endocardial contour. The basal points are identified as the two end-points of adjacent boundary of both left ventricle and left atrium. The endo-apex point is then calculated as the furthest point to the mid-basal point along the endocardium. The output of landmark detection network is a heatmap of the corresponding target point. From the output heatmap, we extract the landmark position by finding the location of maximum or computing the centroid.

As we have different labels in the ground-truth data (myocardium, blood pool, atrium), a multi-class dice $\mathcal{L}_{dice}$ is used as the segmentation loss. As for landmark detection, we first penalise on the squared error of landmark heat-map (L2 loss: $\mathcal{L}_{l2}$). In order to avoid landmark overlapping on different output layers, we regularise the centre distance loss $\mathcal{L}_{CD}$ of different landmark heat-maps as proposed in [15]:

$$\mathcal{L}_{CD} = \frac{1}{2} \sum_{i=1, j \neq i}^{3} 1/(C(\mathcal{H}_i) - C(\mathcal{H}_j))^2 \tag{1}$$

with C the operation to obtain the centre position and $\mathcal{H}_i$ the landmark heat-map. Finally, a mean squared distance loss between the predicted heatmap centre and the ground truth point $\mathcal{L}_{point}$ is also applied. Thus we have a total loss $\mathcal{L}_{total}$ for optimisation:

$$\mathcal{L}_{total} = \mathcal{L}_{dice} + \alpha \mathcal{L}_{l2} + \beta \mathcal{L}_{CD} + \gamma \mathcal{L}_{point} \tag{2}$$

2.2   SEG-AFFINE: Poly-affine Regulariser for Myocardium

With the intention to constrain the regularity of predicted myocardium mask, we propose to model the myocardium mask as a combination of 6 AHA regions [4]. We first choose a reference myocardium mask R from the training set. All the N training myocardium masks $(M_i)_{i=1}^{N}$ are aligned to the reference mask by an affine transform $(T_i)_{i=1}^{N}$ estimated from the three landmarks (left basal, right basal and endo-apex). Then all aligned masks are averaged to a mean mask $\bar{R}$. The mean mask $\bar{R}$ is threshold-ed ($\bar{R}^f$) and split into 6 AHA regions $(\bar{I}_j)_{j=1}^{6}$. For every myocardium mask M_i, we aim to first find 1 affine matrix A_g that globally transform the reference mask to $\hat{M}_i^g$. The corresponding reference regions become $(\bar{I}_j^g)_{j=1}^{6} = A_g \bar{I}_j$. We then find 6 affine matrices $(A_{ij})_{j=1}^{6}$, that transform the transformed (globally) mean AHA regions $(\bar{I}_j^g)_{j=1}^{6}$ into $\hat{M}_i = \sum_{j=1}^{6} A_{ij} \bar{I}_j^g$ that best reconstructs the target mask M_i, i.e. $\hat{M}_i \approx M_i$. For better fusion of the transformed 6 regions, we use the spatially weighted regions (multi-variate Gaussian) $\tilde{I}_j^w$ instead of $\bar{I}_j$, s.t. $\sum_{j=1}^{6} \tilde{I}_j^w = \bar{R}^f$.

We use a CNN to estimate the affine parameters and reconstruct the given mask. The proposed network has two sub-networks. The first one seeks to estimate global affine parameters for global alignment which consists of two hidden convolutional layers with down-sampling. The second sub-network is to find the 6 regional affine matrix. It begins with an encoder for high level feature extraction. The extracted features are passed through fully connected layers for affine matrix estimation $\hat{A}_{ij}$. By affine transform, we could reconstruct $\hat{M}^0$ from the 6 mean regions $\hat{M}^0 = \sum_{j=1}^{6} \hat{A}_{ij} \tilde{I}_j^{wg}$. Two Conv layers are followed to refine the fusion mask $\hat{M}^0$ and thus we obtain the final output $\hat{M}^f$ (detailed information in Fig. 2).

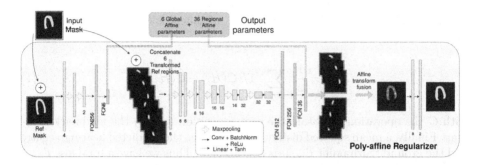

Fig. 2. Architecture of proposed poly-affine regulariser

In order to regularise the value of the affine parameters, we approach the affine parameter estimation problem using *Maximum A Posteriori* (MAP) with prior probabilities on the parameter values $P(A)$. The MAP aims to optimise: $\arg\max[P(A|M)] \propto P(M|A)P(A)$, i.e. $\arg\min[-\log P(M|A) - \log P(A)]$ We use Gaussian distributions for conditional likelihood and priors. For $P(M|A) \propto \exp(-\frac{1}{2}(M - \hat{M}(A))^T \Sigma^{-1}(M - \hat{M}(A)))$, the variance is identity. For $P(A) \sim \mathcal{N}(\hat{\mu}, \hat{\Sigma})$, $\hat{\mu}$ and $\hat{\Sigma}$ are the maximum likelihood estimate (here we only consider diagonal covariance matrices) from the aligned transformation parameters $(T_i)_{i=1}^{N}$. The regularisation for affine parameters is $\mathcal{L}_{affine} = \alpha\mathcal{L}_{l2} + \beta\mathcal{L}_{prior}$, where $\mathcal{L}_{l2}$ is the mean square error between the reconstructed image and the input and

$$\mathcal{L}_{prior} = \sum_{k=1}^{K}\sum_{j=1}^{6} \delta_j \frac{(A_j^k - \hat{\mu}_j)^2}{\hat{\Sigma}_{jj}} \tag{3}$$

where $K = 1$ for global affine parameter and $K = 6$ for regional affine parameters.

Thus the total loss for the poly-affine reconstruction network is:

$$\mathcal{L}_{total} = \underbrace{\mathcal{L}_{dice}(\hat{M}^g) + \beta^g \mathcal{L}_{prior}^g}_{\text{global sub-net}} + \underbrace{\mathcal{L}_{dice}(\hat{M}^0) + \mathcal{L}_{dice}(\hat{M}^f) + \alpha\mathcal{L}_{l2} + \beta^r \mathcal{L}_{prior}^r}_{\text{regional sub-net}} \tag{4}$$

Once the poly-affine regulariser network is trained, the 6+36 affine parameters serve as explicit hidden variables to regularise the shape of myocardium prediction. We train a U-Net model (same as baseline model) which seeks for the best overlapping of mask as well as the minimum distance between the corresponding affine parameters of predicted myocardium and that of ground truth myocardium. The loss function for this method is

$$loss = dice + \alpha MSE(PA(P) - PA(M)), \tag{5}$$

where MSE is the mean squared error, PA is poly-affine regulariser that outputs the 42 affine parameters.

2.3   SEG-CONTOUR: Multi-class Contour-Loss

In order to increase the classification accuracy on the boundary, we choose to use an adapted multi-class contour loss [8]. Firstly, a distance map $D(M)$ is calculated from ground truth mask and it illustrates the shortest euclidean distance of each pixel to the closest border. Then the contour loss is calculated as

$$loss_{contour} = \sum (D(M) \circ contour(B(P))) \tag{6}$$

where $\circ$ performs element-wise multiplication. P is the prediction output after SoftMax activation of U-Net for a certain class. $B(P)$ represents a differentiable thresholded Sigmoid for binarisation

$$B(P) = \frac{1}{1 + \exp^{-\gamma(P-T)}} \tag{7}$$

where $\gamma = 20$ and $T = 0.5$.

The contour of the binarised mask is obtained by applying a 2D Sobel filter

$$contour(P) = |G_x * P| + |G_y * P| \tag{8}$$

where $*$ denotes 2D convolution and G_x, G_y are 2D Sobel kernel in x-,y- dimension:

$$G_x = \begin{bmatrix} 1 & 0 & -1 \\ 2 & 0 & -2 \\ 1 & 0 & -1 \end{bmatrix}, G_y = \begin{bmatrix} 1 & 2 & 1 \\ 0 & 0 & 0 \\ -1 & -2 & -1 \end{bmatrix}.$$

3   Experiments and Results

3.1   Datasets

In this work, we work on two public data sets: CAMUS[1] and ECHONET[2]. CAMUS dataset consist of publicly accessible 2D echocardiographies and the corresponding annotations of 450 patients. For each patient, 2D apical 4-chambers (A4C) and 2-chambers (A2C) view sequences are available. Manual annotation of cardiac structures (left endocardium, left epicardium and left atrium) were acquired by expert cardiologists for each patient in each view, at end-diastole (ED) and end-systole (ES) [11]. Along with the image and annotation data, we also have the following information: image quality (good/medium/poor), left ventricle end-diastole volume (LVedv), left ventricle end-systole volume (LVesv) and left ventricle ejection fraction (LVef).

---

[1] https://www.creatis.insa-lyon.fr/Challenge/camus/databases.html.
[2] https://echonet.github.io/dynamic/.

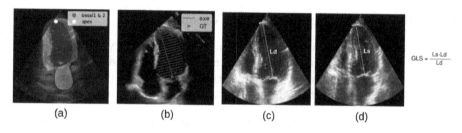

(a)   (b)   (c)   (d)

Fig. 3. (a) An example of segmentation ground truth of CAMUS dataset. The two basal and apex landmarks are extracted following the procedures described in Sect. 2.1. (b) An example of annotation provided by ECHONET dataset. We generate the ground truth mask of LV by linearly connecting the border points. The grand axe (in red) is considered as the line connecting the apex and mid-basal point. The length of grand axe is considered as the LV length. (c–d) GLS calculation illustration (background is one echo image from CAMUS). We calculate the GLS from the LV length change by following the approximation method in [14]. (Color figure online)

ECHONET dataset contains 10 030 apical 4-chambers echocardiography videos as part of routine clinical care at Stanford University Hospital [13]. Segmentation measurements (left endocardium) at end-diastole and end-systole are available for all videos. The corresponding LVedv, LVesv, LVef are also provided for each video.

3.2   Experiments

We trained the baseline UNet, SEG-LM, SEG-AFFINE, SEG-CONTOUR models on the CAMUS dataset (both 2-chamber and 4 chamber ED/ES frames) using 10-fold cross validation. The 450 patients are randomly split-ed into 10 folds, with each fold has similar distribution of image quality and LVef distribution. Every turn we use 8 fold data for training, 1 fold for validation (for model selection) and 1 fold for testing.

From the segmentation result, we first find the largest connected component for each class and applied a closing operation to fill the potential hole that could exist inside the predicted mask. We then extract the basal and apex landmark points following protocol described in Sect. 2.1 for the four models except SEG-LM whose landmark is extracted from the landmark branch. The direction of mid-basal to apex will be regarded as grand axe of left ventricle and is used for left ventricle volume estimation by using modified Simpson's rule [6], thus the calculation of ejection fraction. The distance from mid-basal to apex forms the ventricle length and serves for global longitudinal strain calculation [14]. Because of the dataset limitation, we don't have the all the 3 apical views for left ventricle, so the GLS estimation will not be very accurate. However, we still calculate the GLS from 2 views (CAMUS) and 1 view (ECHONET) as reference to test the landmark detection accuracy.

All the segmentation models are implemented with Pytorch with a batch size of 8 and input image resized to 256×256. In order to avoid over-fitting and improve generalisation performance, we apply random data augmentation at training phase for all the networks. A stack of rotation, random cropping, brightness adjustment, contrast change, sharpening, blurring and speckle noise addition is conducted with each a probability of 0.5 for every input image. CAMUS dataset does not contain the ground truth of desired landmarks. We generated the 'ground truth' landmark position from ground truth segmentation masks. The Gaussian heatmap of ground-truth landmarks is computed with $\sigma = 4$.

The baseline U-Net is trained with a multi-class dice loss. An Adam optimiser is applied with a learning rate of 1^{e-3}. The training is early stopped when the dice loss on validation data shows no increase for more than 5 epochs. For SEG-LM model, we set $\alpha = 0.05, \beta = 0.5, \gamma = 0.5$ in Eq. 2. An Adam optimiser is applied (lr = 10^{-4}). The output heatmap of landmarks is first processed to keep only one point cluster per layer and then the landmark location is extracted as the centroid. The poly-affine regulariser network is first trained with the myocardium ground truth for reconstruction. We set $\alpha = 0.005, \beta^r = \beta^g = 0.01, \sigma_{i=1}^6 = 1$ and learning rate at 1e-4. We then train a baseline U-Net with the polyaffine regulariser using loss function (Eq. 5) with $\alpha = 10$. The parameter for global prior is calculated from the training-specific $(T_i)_{i=1}^N$. The parameter for regional prior is set as $\mu_{1...6} = [1, 0, 0, 0, 1, 0]$ and $[\Sigma_{i=1}^6] = 0.1$. As for SEG-CONTOUR model, contour loss is optimised along with the dice loss. The contour loss is easy to fall into a local minimum of 0 so we set a weight of 100, 1e–4 for dice loss and contour loss respectively.

3.3 Evaluation Metrics

For segmentation results, apart from the most used geometrical metrics: Dice coefficient, Hausdorff distance (HD) and Mean Surface Distance (MSD), we also use two anatomical metrics: Convexity(Cx) and Simplicity(Sp) [12].

$$Convexity(Cx) = \frac{Area(P)}{Area(ConvexHull(P))} \tag{9}$$

$$Simplicity(Sp) = \frac{\sqrt{4\pi * Area(P)}}{Perimeter(P)} \tag{10}$$

Based on these metrics, we calculate the number of outliers for algorithm/model robustness evaluation. The outlier of segmentation prediction for CAMUS dataset is established from the inter-variability tests with the upper limit values for HD and MSD, and lower limit values for the simplicity and convexity [12]. A prediction mask is considered as a geometrical outlier if its $HD > 3.5\,\mathrm{mm}$ or $MSD > 8.2\,\mathrm{mm}$ at ED, if $HD > 4\,\mathrm{mm}$ or $MSD > 8.8\,\mathrm{mm}$ at ES. The corresponding limit for anatomical outlier is if $Cx < 0.529$ or $Sp < 0.741$ for endocardium, if $Cx < 0.694$ or $Sp < 0.960$ for epicardium [12].

Table 1. CAMUS Segmentation Metric. Endo.: endocardium, Epi: epicardium, HD: Hausdorff distance, MSD: mean surface distance, Geo.: Geometrical outlier, Ana.: Anatomical outlier. Values in bold represent the best score.

Method	Endo			Epi			Outlier		
	Dice	HD (mm)	MSD (mm)	Dice	HD (mm)	MSD (mm)	Geo.	Ana.	Both
Baseline	**0.931** ± 0.040	5.04 ± 3.00	1.51 ± 0.83	0.951 ± 0.025	5.75 ± 3.61	1.71 ± 0.91	15%	2.8%	2.5%
SEG-LM	0.928 ± 0.042	5.47 ± 3.04	1.59 ± 0.84	0.950 ± 0.025	6.22 ± 3.74	1.79 ± 0.93	20%	7.6%	5.3%
SEG-AFFINE	0.930 ± 0.042	5.14 ± 3.00	1.53 ± 0.88	0.951 ± 0.027	5.84 ± 3.75	1.72 ± 0.95	14.3%	5.1%	3.7%
SEG-CONTOUR	**0.931** ± 0.041	**4.99** ± 2.95	**1.50** ± 0.73	**0.952** ± 0.026	**5.63** ± 3.32	**1.67** ± 0.87	**13.8%**	**1.5%**	**1.2%**

Table 2. CAMUS Landmark/GLS and EF Prediction. Basal1: the left mitral valve end point, Basal2: the right mitral valve end point, EF: ejection fraction, GLS: global longitudinal strain. Values in bold represent the best score.

Method	Basal1 MAE (mm)	Basal2 MAE (mm)	Apex MAE (mm)	GLS(%) MAE (%)	Corr.	Bias(%) ± std	EF(%) MAE (%)	Corr.	Bias(%) ± std
Baseline	2.36	3.06	4.06	3.97	0.74	−0.73 ± 5.20	**4.76**	**0.86**	0.93 ± 7.07
SEG-LM	2.82	3.45	4.48	4.51	0.70	−1.74 ± 5.27	5.26	0.84	2.51 ± 8.04
SEG-AFFINE	2.35	2.97	4.22	4.03	0.74	−0.9 ± 5.15	4.96	0.85	1.16 ± 7.31
SEG-CONTOUR	**2.25**	**2.80**	**3.97**	**3.96**	**0.75**	−0.36 ± 5.34	5.06	0.84	0.7 ± 7.50

3.4   Results

We show the evaluation results in Tables 1 and 2, computed from 450 patients of CAMUS dataset using 10 fold cross-validation for the four methods detailed in Sect. 2. Compared with the baseline model, the SEG-LM, SEG-AFFINE models demonstrate only a slight decrease in terms of segmentation metric, ejection fraction (EF) prediction accuracy and landmark detection accuracy. The SEG-CONTOUR model shows a similar performance with baseline model in Dice score (Table 1), but reduces greatly the number of geometrical and anatomical outliers (Table 1). It's quite reasonable that with a smaller HD and MSD, the SEG-CONTOUR reduces the classification error along the boundary area thus less outlier predictions. This is consistent with the observations that a good Dice score does not always guarantee a good HD [2] and in our case, not always leads to anatomically-plausible segmentation. The SEG-CONTOUR model is

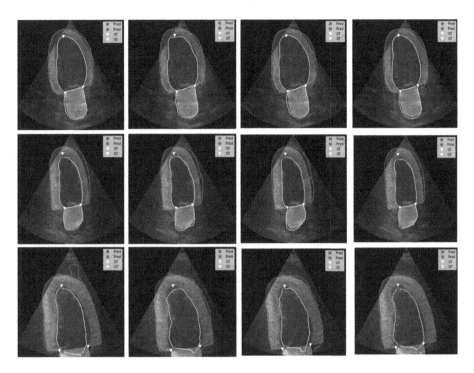

Fig. 4. 3 CAMUS segmentation examples (good/medium/bad in terms of HD). The four columns represent the baseline model, SEG-LM, SEG-AFFINE, SEG-CONTOUR respectively from left to right. The red, green, cyan lines represent the predicted segmentation contours of epicardium, endocardium and left atrium. The transparent green, blue and yellow regions are the ground truth masks of myocardium, left ventricle blood pool and left atrium respectively. (Color figure online)

also capable of tracking more precisely the boundary especially the landmarks (Table 2) thus a smaller bias of GLS prediction. In terms of EF calculation, the baseline U-Net model shows smaller mean absolute error but a larger bias than the SEG-CONTOUR loss. We show three CAMUS test examples (Fig. 4) where each row has a good/medium/bad performance in terms of HD score respectively. Comparing with the the rest 3 models, SEG-CONTOUR has a more fluent border similar as the ground truth annotation.

The evaluation results of applying the trained model on a totally different dataset ECHONET (the same test fold of 1277 patients as in [13]) show the same trend of performance of our 4 methods. The SEG-CONTOUR method demonstrates a good performance on tasks related to boundary information, for example, lower HD and MSD, better GLS estimation. It's less accurate on area based task, i.e. volume estimation thus ejection fraction prediction. It is noticeable that all of the four models demonstrate a nice Dice coefficient on this

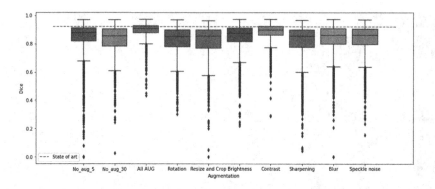

Fig. 5. The baseline U-Net model is trained with only one of the mentioned augmentation methods on CAMUS dataset and then is evaluated on the same test fold of ECHONET segmentation model, whose dice coefficient is 0.92. Contrast adjustment contributes most to the improvement of generalisation result while with all the techniques, we obtain the best Dice score and less variation. No-aug-5: trained model of 5th epoch without augmentation, No-aug-30: trained model of 30th epoch without augmentation. At 30th epoch, the model has already over-fitted the CAMUS data.

Table 3. ECHONET prediction

Method	Endo-cardium (LV)			EF				GLS			
	Dice	HD (pxls)	MSD (pxls)	MAE (%)	R2	Corr.	Bias(%) ± std	MAE (%)	R2	Corr.	Bias(%) ± std
Baseline	0.892 ± 0.069	12.99 ± 7.96	3.7 4± 3.42	**7.97**	**0.11**	0.69	4.46 ± 12.80	4.96	−1.2	0.38	0.36 ± 7.94
SEG-LM	0.886 ± 0.063	13.54 ± 6.86	3.87 ± 2.63	8.79	0.02	**0.72**	6.21 ± 13.39	5.61	−1.49	0.36	−1.9 ± 7.07
SEG-AFFINE	0.887 ± 0.068	13.12 ± 7.25	3.87 ± 2.97	9.35	−0.14	0.67	7.11 ± 14.41	5.88	−2.44	0.29	−0.75 ± 9.29
SEG-CONTOUR	**0.895** ± 0.057	**12.56** ± 6.34	**3.58** ± 2.16	8.39	0.07	0.69	5.03 ± 13.21	**4.39**	−0.43	**0.51**	0.59 ± 6.52

different dataset (the Dice coefficient of models trained on ECHONET data is 0.92% [13]), which proves the importance of appropriate image augmentation techniques (Fig. 5).

4   Conclusion

In this paper, we explored methods to introduce shape constraints into 2D Echocardiography segmentation models from three levels: global-level (SEG-LM), regional-level (SEG-AFFINE), pixel-level (SEG-CONTOUR). From the evaluation results on CAMUS dataset and its generalisation result on a unseen dataset ECHONET, it is more efficient to introduce pixel-level shape constraint than global or regional level constraints for U-Net based models. With a multiclass contour loss, SEG-CONTOUR model achieves better classification on the boundary pixels with a reduced a Hausdorff distance and more accurate land-mark detection result. Our experiments showed the good potential of SEG-CONTOUR for a more robust segmentation and deformation analysis.

Acknowledgments. This work has been supported by the French government through the National Research Agency (ANR) Investments in the Future with 3IA Côte d'Azur (ANR-19-P3IA-0002) and by Inria PhD funding.

References

1. Ahn, S.S., Ta, K., et al.: Unsupervised motion tracking of left ventricle in echocardiography. In: Byram, B.C., Ruiter, N.V. (eds.) Medical Imaging 2020: Ultrasonic Imaging and Tomography, vol. 11319, pp. 196–202 (2020)
2. Bernard, O., Lalande, A., et al.: Deep learning techniques for automatic MRI cardiac multi-structures segmentation and diagnosis: is the problem solved? IEEE Trans. Med. Imaging **37**(11), 2514–2525 (2018)
3. Bohlender, S., Oksuz, I., Mukhopadhyay, A.: A survey on shape-constraint deep learning for medical image segmentation (2021). http://arxiv.org/abs/2101.07721
4. Cerqueira, M.D., Weissman, N.J., et al.: Standardized myocardial segmentation and nomenclature for tomographic imaging of the heart: a statement for health-care professionals from the cardiac imaging committee of the council on clinical cardiology of the American heart association. J. Nucl. Cardiol. **9**(2), 240–245 (2002)
5. Chen, C., Bai, W., et al.: Improving the generalizability of convolutional neural network-based segmentation on CMR images. Front. Cardiovasc. Med. **7**, 105 (2020)
6. Folland, E.D., Parisi, A.F., et al.: Assessment of left ventricular ejection fraction and volumes by real-time, two-dimensional echocardiography: a comparison of cineangiographic and radionuclide techniques. Circulation **60**(4), 760–766 (1979)
7. Hasselberg, N.E., Haugaa, K.H., et al.: Left ventricular global longitudinal strain is associated with exercise capacity in failing hearts with preserved and reduced ejection fraction. Eur. Heart J. - Cardiovasc. Imaging **16**(2), 217–224 (2014)
8. Jia, S., et al.: Automatically segmenting the left atrium from cardiac images using successive 3D U-nets and a contour loss. In: Pop, M., et al. (eds.) STACOM 2018. LNCS, vol. 11395, pp. 221–229. Springer, Cham (2019). https://doi.org/10.1007/978-3-030-12029-0_24
9. Kraigher-Krainer, E., Shah, A.M., et al.: Impaired systolic function by strain imaging in heart failure with preserved ejection fraction. J. Am. Coll. Cardiol. **63**(5), 447–456 (2014)
10. Krebs, J., Delingette, H., et al.: Learning a probabilistic model for diffeomorphic registration. IEEE Trans. Med. Imaging **38**(9), 2165–2176 (2019)
11. Leclerc, S., Smistad, E., et al.: Deep learning for segmentation using an open large-scale dataset in 2D echocardiography. IEEE Trans. Med. Imaging **38**(9), 2198–2210 (2019)
12. Leclerc, S., Smistad, E., et al.: Ru-net: a refining segmentation network for 2D echocardiography. In: 2019 IEEE International Ultrasonics Symposium (IUS), pp. 1160–1163 (2019)
13. Ouyang, D., He, B., et al.: Video-based AI for beat-to-beat assessment of cardiac function. Nature **580**(7802), 252–256 (2020)

14. Støylen, A., Mølmen, H.E., Dalen, H.: Left ventricular global strains by linear measurements in three dimensions: interrelations and relations to age, gender and body size in the HUNT Study. Open Heart **6**(2), 1–9 (2019)
15. Wang, X., Yang, X., et al.: Joint segmentation and landmark localization of fetal femur in ultrasound volumes. In: 2019 IEEE EMBS International Conference on Biomedical Health Informatics (BHI), pp. 1–5 (2019)

Image-Derived Geometric Characteristics Predict Abdominal Aortic Aneurysm Growth in a Machine Learning Model

Jordan B. Stoecker[1]([✉]), Kevin C. Eddinger[1], Alison M. Pouch[2,3],
and Benjamin M. Jackson[1]

[1] Department of Surgery, Hospital of the University of Pennsylvania, Philadelphia,
PA 19146, USA
jordan.stoecker@uphs.upenn.edu
[2] Department of Radiology, Hospital of the University of Pennsylvania, Philadelphia,
PA 19146, USA
[3] Department of Bioengineering, University of Pennsylvania, Philadelphia, PA 19146, USA

Abstract. Abdominal aortic aneurysm (AAA) growth is correlated with rupture risk, but predicting either AAA growth or rupture remains challenging. Global aneurysm geometric properties have been linked with elevated peak AAA wall stress when using finite element analysis (FEA) and may predict AAA growth. We used a machine learning model to evaluate whether image-derived geometric parameters, calculated both globally and locally over the surface of the aneurysm can predict local AAA wall growth, avoiding material property assumptions used in FEA. Sequential CTAs one year apart were collected from 10 patients with AAAs. The luminal and aortic wall were segmented in patient's baseline CTA. In order to calculate local geometric properties, each baseline AAA was divided into 64 regions to define regional geometric aneurysm characteristics from vertices in that region, and into 1,500 sub-regions in order to define sub-regional geometric characteristics. The global and local (regional and sub-regional) aortic geometric properties were all derived from the images and determined from the aortic segmentation and surface mesh. Local AAA growth between CTAs was determined at the sub-regional level using deformable image registration and was the outcome variable for the model. Patient demographics, as well as the global and local geometric aneurysm properties were used to predict local AAA growth using an eXtreme gradient boosted regression tree using a performance metric of root-mean-square error (RMSE) with 80/20 training to testing split. Mean relative error in predicting maximum AAA growth was 10.5% in the testing set. The most impactful predictors were AAA volume, regional maximum diameter, regional maximum Gaussian surface curvature, regional median aneurysm thickness, and patient age. Removal of local geometric properties from the model increased RMSE from 0.5 to 1.1 and decreased model performance by likelihood test (P = 0.01). Utilizing both global and local aneurysm geometric characteristics better predicts local aortic wall growth in AAAs, avoiding assumptions required using FEA.

Keywords: Shape analysis · Machine learning · Aortic aneurysm

© Springer Nature Switzerland AG 2021
D. B. Ennis et al. (Eds.): FIMH 2021, LNCS 12738, pp. 35–45, 2021.
https://doi.org/10.1007/978-3-030-78710-3_4

1 Introduction

The prevalence of asymptomatic abdominal aortic aneurysms (AAAs) has increased as noninvasive screening modalities have become more widely available. Aortic diseases, including aortic aneurysms, are the 12th leading cause of death in the United States [1]. AAA rupture is a major source of morbidity and mortality, with reported mortality rates of 40–50% [2]. Surgical intervention in asymptomatic patients is generally reserved for those with aortic diameters greater than 5.5 cm, or with aneurysm expansion rates greater than 10 mm per year [3]. However, historical autopsy studies have estimated that 10–24% of ruptured AAAs are less than 5.5 cm in diameter [4], and 7% of ruptured AAAs had a diameter of less than 5.0 cm [5]. Conversely, modern studies examining outcomes of AAAs with diameter 5.5–7.0 cm in patients who are unfit for intervention have noted the yearly rupture risk may be below 4%, with the majority of patients expiring from comorbid diseases [6]. Clearly, improved risk stratification of AAAs may prevent aortic rupture and save lives.

Growth of AAA is correlated with rupture risk, but predicting AAA growth or rupture remains challenging [3]. AAA growth and rupture occurs when the peak wall stress (PWS) surpasses wall strength. AAA PWS can be calculated on a patient-specific basis using finite element analysis (FEA). PWS has been shown to be elevated in ruptured aneurysms, and it may be a better indicator than maximal transverse diameter alone for predicting future AAA rupture [7, 8]. However, utilization of PWS for prediction of AAA rupture has never been widely implemented clinically. Determining FEA derived PWS remains cumbersome and has not been implemented in clinical practice. It has been demonstrated that global aneurysm geometric properties, including maximal AAA diameter, total aneurysm volume, maximal surface curvature across the entire AAA, and degree of mural thrombus (which determines effective AAA wall thickness) are highly correlated with PWS [8, 9]. Unfortunately, utilization of these simplified geometric characteristics has not been shown to reliably predict AAA growth or risk of rupture in prior work [10]. Machine-learning prediction models have shown some promise in predicting future AAA growth, but current methods focus on predicting growth in patients who have already undergone several surveillance scans, extrapolating whether patients are at risk for future growth based on their prior growth (requiring sequential imaging scans years apart) [11]. Therefore, while there has been significant research into novel methods of aneurysm surveillance and modeling, maximal transverse diameter and observed AAA growth rate remain the most commonly used criteria for repair in clinical practice.

We hypothesize that image-derived geometric properties at the local level better predicts the future behavior of these aneurysms compared to prior approaches leveraging simplified global geometric characteristics. Additionally, assessing AAA growth and behavior at the local level allows us to maximize the data integrated from each aorta into the prediction algorithm and optimize performance, instead of reducing each AAA to a handful of global geometric data points. By generating a model that predicts local aneurysm growth instead of global growth, we are able to generate an aneurysm heatmap that can display local growth over the entire AAA and assist in identification of specific areas that may be at risk for growth and rupture. Additionally, our technique shows promise in this small sample size for approximating both the magnitude of future

AAA growth and location of growth from a single CTA obtained at a single time point without requiring material properties assumptions as seen in FEA.

2 Methods

2.1 Data Generation and Pre-processing

Ten patients with sequential CTAs acquired 12–24 months apart were selected for inclusion. The aneurysms were all limited to the abdominal aorta and infrarenal in location, but varied considerably in their shape, size, and quantity of mural thrombus. For each patient's baseline CTA, both the aortic wall and aortic lumen were manually segmented using ITK-SNAP [12]. These baseline AAA segmentations were then used to create a surface mesh for each aneurysm. Laplacian smoothing was recursively applied 15 times to each surface mesh to eliminate residual segmentation artifacts prior to calculation of geometric properties. The number of vertices in the ten surface meshes varied between 17,000 and 54,000, with a median of 32,000. Image segmentation, mesh manipulation, and algorithm training were all performed on a Lenovo ThinkPad with 8 gigabytes of RAM and Intel Core i5-8350 CPU.

2.2 Image-Derived Geometric Parameter Calculations

Aneurysm thickness was calculated as the combination of aortic wall thickness and mural thrombus thickness. Locally varying diameter and aneurysm wall thickness were determined using shape diameter functions as shown in Fig. 1, which have been shown previously to be effective in assessing diameter in irregularly shaped surface meshes [13]. Given each vertex on the surface mesh, 128 rays were sent along the inward pointing normal in a 120° cone. The aneurysm diameter at each vertex is determined by the ray lengths required to intersect with the opposite outer aortic wall; specifically, the weighted average of all rays whose lengths fall within one standard deviation of the median of all ray lengths. The inverse of the angle between the rays to the center of the cone determines the ray weight, giving smaller weights to rays that are oriented away from the inward normal. Vertex-based aneurysm wall thickness was defined as the weighted average of ray lengths required to intersect the aortic lumen from the outer aortic wall. Mean and Gaussian curvature were calculated at each vertex on the AAA surface mesh using the standard technique described in [9].

2.3 Image-Derived Predictors and Outcome Calculations

Local aortic wall growth (magnitude measured in mm) at each vertex was calculated using deformable image registration on each patient's baseline and sequential CTA and was determined using the *greedy* toolkit [13]. We have previously shown that this deformable image registration technique achieves sub-millimeter accuracy when measuring local aortic growth in this dataset [14]. Each baseline AAA in the dataset was then spatially divided into 64 regional sections (using a $4 \times 4 \times 4$ grid) and 1,500 sub-regional sections (using a $10 \times 10 \times 15$ grid). Within each baseline AAA, every surface mesh

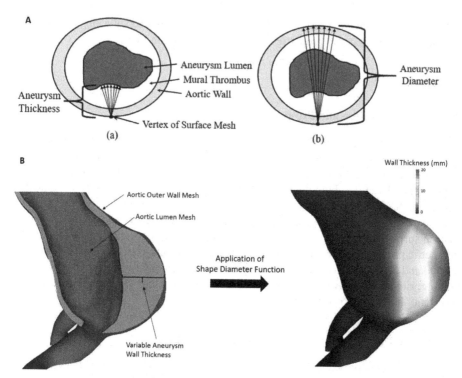

Fig. 1. A. 2-D axial representation displaying method by which aneurysm thickness and aneurysm diameter were calculated using shape diameter functions. **B.** 3D sagittal representation displaying method used to calculate aneurysm wall thickness using aortic outer wall and aortic lumen mesh with application of shape diameter function.

vertex was assigned to a single regional section and a single sub-regional section based on its spatial coordinates. Median and maximal geometric parameters were then calculated for each regional and sub-regional section using the data for the vertices contained within, as displayed in Fig. 2. Maximal properties were calculated as the 97th percentile within each spatial section. The outcome/predicted variable, local aortic growth, was defined at the sub-regional level by taking the median of the local aortic wall growth for each vertex contained within.

2.4 Machine Learning Model Selection

Patient demographics and global, regional, and sub-regional geometric properties were used to predict local aneurysm growth utilizing an eXtreme Gradient Boosted regression tree model (XGB) in RStudio (Fig. 3). Eight patients were used for training (12,000 data points) and two for testing (3,000 points), resulting in a 80/20 training to testing data split. There were 27 predictor variables (11 global, 8 regional, 8 sub-regional) for each of the 15,000 data points in the final dataset that was used to generate the machine learning model. Plots were generated using Tecplot (Bellevue, WA) and Meshlab (Pisa, Italy).

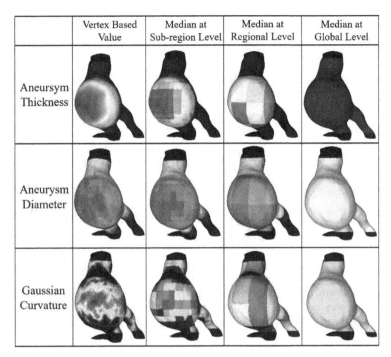

Fig. 2. Display showing initial image-derived geometric characteristics at each surface mesh vertex (left-most column), and subsequent generation of sub-regional, regional, and global geometric parameters from these vertex-based values. Blue sections indicate low values and red sections indicate high levels. (Color figure online)

An XGB machine learning model was selected due to several advantages including: (1) ability to handle different types of predictor variables with no need for prior data transformation, regularization, or elimination of outliers, (2) ability to accurately fit complex nonlinear relationships and automatically handle interaction effects between predictors, (3) fitting multiple trees using XGB overcomes the poor predictive performance that can be seen in single tree models, and (4) ability to easily tune the model parameters to optimize performance using automated methods during training cross-validation.

2.5 Tuning of the Machine Learning Model

We used 5-fold cross-validation and a performance metric of root mean square error (RMSE) to establish model parameters and limit overfitting during model training. Model tuning took 20 h on a Lenovo ThinkPad with 8 gigabytes of RAM and Intel Core i5-8350 CPU with integrated Intel UHD 620 graphics processor while final model training was accomplished in approximately one hour. During model tuning, the XGB learning rate was set to an initial value of 0.1 while other boosting parameters were varied in stepwise fashion with final values selected to minimize cross-validated RMSE. During this process maximum tree depth was varied from 4 to 8, gamma value was varied from 0 to 1.0, minimum child weight was varied from 1 to 3, and column sample and subsample value

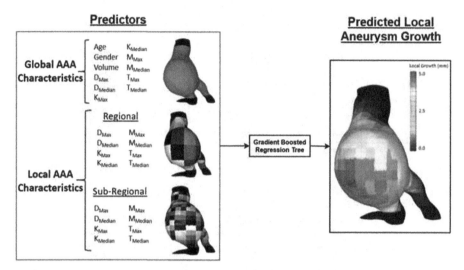

Fig. 3. Patient characteristics and geometric parameters used in the prediction model: Volume (total aneurysm volume), D_{Max} (maximum diameter), D_{Median} (median diameter), K_{Max} (maximum of the Gaussian curvature), K_{Median} (median of the Gaussian curvature), M_{Max} (maximum of the Mean curvature), M_{Median} (median of the Mean curvature), T_{Max} (maximum aortic wall thickness), T_{Median} (median aortic wall thickness).

were varied from 0.5 to 1.0. Once the value these boosting parameters were determined, the learning rate was varied from 0.001 to 0.1 with assessment of cross-validated RMSE in order to establish final parameter values. The tuning and final model were run until the cross-validated error no longer decreased over the subsequent 50 iterations, with maximum iterations capped at 1,000. Figure 4 demonstrates the cross-validated RMSE with variation in the model learning rate. Model performance was assessed using the two hold out AAAs, by visual inspection of the predicted aneurysm growth when mapped to the respective aneurysm mesh compared to the true aneurysm growth and by comparison of predicted maximal growth rate for each testing AAA compared to actual maximal growth rate.

2.6   Assessment of Machine Learning Model Performance

Model performance was assessed by application of the final model to the two hold out AAAs. First by visual inspection of the predicted aneurysm growth when mapped to the respective aneurysm mesh compared to the true aneurysm growth and by comparison of predicted maximal growth rate for each testing AAA compared to actual maximal growth rate.

Predictor importance was determined using permutation feature importance calculations in the training dataset. Permutation importance was chosen given it is intuitive to interpret and independent of model type, allowing eventual comparisons to future models. Permutation feature importance is determined after the final model is fitted to the training dataset and the cross-validated predictive performance has been determined

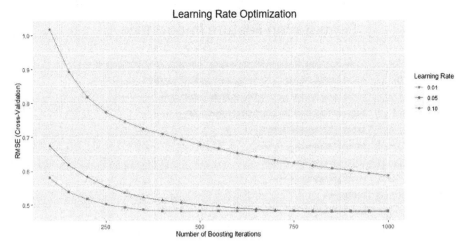

Fig. 4. Plot demonstrating the five-fold cross-validated error (RMSE) of the XGB machine learning model across several learning rates. A learning rate of 0.05 and 800 iterations showed the optimal performance as demonstrated by the RMSE minimum achieved at that point.

(RMSE). Each individual predictor in the training dataset is then randomly altered, and a new predictive performance (RMSE) is assessed in this permuted training data. The feature importance is determined by the difference in RMSE between the original model and the permutated model. Since each predictor is randomly altered, results can vary and the process was repeated 50 times for every predictor. The results were standardized to allow for interpretability and future comparisons.

3 Results

The tuned and final XGB model had a learning rate of 0.05, maximum tree depth of 6, gamma value of 0.1, column sample and subsample value of 1.0, and was run for 800 iterations. The most impactful predictors by permutation feature importance (change in outcome RMSE observed with permutation of a single predictors, while holding all other predictors stable) were: total aneurysm volume, regional maximal diameter, regional median and maximal Gaussian surface curvature, and regional maximal thrombus thickness. The ten most importance predictors with associated error are shown in Fig. 5. The predictions for local aortic growth at the sub-regional level over the entire two AAAs from the testing set are shown as a heat-map Fig. 6.

The prediction model is able to accurately identify regions of AAA growth and regions of AAA stability with appropriate growth magnitude, as demonstrated in Fig. 6. In the first AAA from the testing set, predicted maximal AAA growth was 4.6 mm compared to true maximal growth of 5.3 mm. In the second aneurysm, predicted maximal AAA growth was 5.9 mm compared to true maximal growth of 6.4 mm. Relative error in the testing set was 10.5%. Removal of local geometric properties from the prediction model increased training RMSE from 0.47 to 0.97, with decreased model performance by likelihood test (P = 0.01).

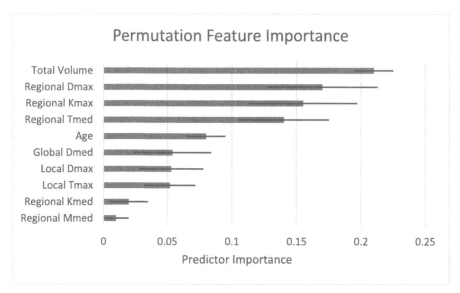

Fig. 5. Importance of each image-derived geometric characteristic in the final machine learning model, as assessed by error induced in the final model when permuting individual predictors. Values standardized to sum to one and error bars indicating two standard deviations. Dmax (maximum diameter), Dmed (median diameter), Kmax (maximum of the Gaussian curvature), Kmed (median of the Gaussian curvature), Mmed (median of the Mean curvature), Tmax (maximum aneurysm thickness), Tmed (median aneurysm thickness).

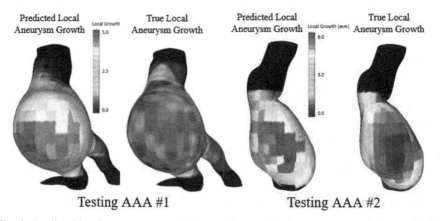

Fig. 6. Predicted local aneurysm growth compared to true local aneurysm growth in mm for the AAAs in the testing set. Note the apparent correlation between predicted and true growth (as determined by deformable image registration).

4   Discussion and Future Direction

Global, regional, and sub-regional image-derived geometric characteristics predict local aortic aneurysm growth in this small sample using a machine learning model. A significant benefit of this process is that it avoids material property assumptions required for FEA-derived peak wall stress calculations, but still allows complex shape analysis [15–17]. We present a novel approach which leverages deformable image registration to train and validate a machine learning prediction model. Based on visual comparison of the predicted and actual local growth heat maps in our testing set, it is apparent that this model is able to accurately predict the regions of the aorta that are likely to grow, in addition to accurately predicting maximal transverse growth rate. Utilization of local geometric parameters (calculated at the regional and sub-regional level) allows for identification of specific AAA wall locations of rapid growth, whereas prior models utilizing global geometric parameters alone are limited to prediction of maximal growth [18, 19]. In our final XGB model, regional and sub-regional geometric parameters were more influential in predicting aneurysm growth compared to global parameters, and removal of these local parameters led to significant decreases in model performance.

Our method does have notable limitations. While the XGB algorithm we used is able to infer and account for interactions between predictors, there is currently no agreed upon method to readily assess and automatically report these. Using permutation feature importance we were able to assess for the impact of predictors on final model accuracy, which accounts for predictor interactions, but we are not able to report explicit interactions. Linear regression analysis could be used to assess and better report discrete predictor interactions, but this method is limited by requiring interactions to be defined prior to analysis. Additionally, we combined both the aortic wall thickness and aneurysm mural thrombus thickness into one predictor, aneurysm thickness. While it may be more accurate to keep these values separate in our model, aortic wall thickness cannot be accurately assessed with CT imaging, with magnetic resonance imaging being the preferred imaging modality used to characterize aortic wall thickness in prior literature [20]. Given that CT is the recommended method for AAA surveillance, we wished to limit our predictors to variables that could be easily and accurately derived from this imaging modality. Further, while there have been studies showing that there are variations in aortic wall thickness between gender and race, these difference are submillimeter in magnitude and below our measurement error rate, with one study demonstrating a mean aortic wall thickness in men of 2.32 mm compared to 2.11 mm in women ($p = 0.023$) [20]. Lastly, while our method does not predict PWS (as FEA does), we are able to directly predict the magnitude of aneurysm growth. It remains unclear whether aneurysm growth rate or PWS is more associated with aneurysm rupture risk. Albeit, aneurysm growth is currently used in clinical practice to guide AAA management and therefore likely more intuitive to many clinicians.

The geometric parameters used as inputs in our machine-learning model are automatically calculated from surface meshes. Although segmentation of the aortic outer wall and generation of the surface mesh still requires user input, our techniques require significantly less effort and domain-specific expertise than FEA approaches (which also require aortic wall segmentation). Future efforts will seek to develop a more robust machine learning model by expanding the dataset, investigating additional predictive geometric parameters, and assessing the interactions present between predictors. Machine learning models such as the one demonstrated, may allow for automated prediction of AAA growth from single time-point CTAs, aiding clinicians in aneurysm rupture risk-stratification and decision making.

References

1. Beckman, J.A., Creager, M.A., Dzau, V.J., Loscalzo, J.: Aortic aneurysms: pathophysiology, epidemiology and prognosis. In: Vascular Medicine. Saunders Elsevier Inc, Philadelphia, PA (2006)
2. Schermerhorn, M.L., et al.: Changes in abdominal aortic aneurysm rupture and short-term mortality, 1995–2008: a retrospective observational study. Ann. Surg. **256**, 651–658 (2012)
3. Chaikof, E.L., et al.: The society for vascular surgery practice guidelines on the care of patients with an abdominal aortic aneurysm. J. Vasc. Surg. **67**, 2-77.e2 (2018)
4. Nicholls, S.C., Gardner, J.B., Meissner, M.H., Johansen, H.K.: Rupture in small abdominal aortic aneurysms. J. Vasc. Surg. **28**, 884–888 (1998)
5. Hong, H., Yang, Y., Liu, B., Cai, W.: Imaging of abdominal aortic aneurysm: the present and the future. Curr. Vasc. Pharmacol. **8**, 808–819 (2010)
6. Parkinson, F., Ferguson, S., Lewis, P., Williams, I.M., Twine, C.P.: Rupture rates of untreated large abdominal aortic aneurysms in patients unfit for elective repair. J. Vasc. Surg. **61**, 1606–1612 (2015)
7. Shang, E.K., et al.: Peak wall stress predicts expansion rate in descending thoracic aortic aneurysms. Ann. Thorac. Surg. **95**, 593–598 (2013)
8. Leemans, E.L., Willems, T.P., van der Laan, M.J., Slump, C.H., Zeebregts, C.J.: Biomechanical indices for rupture risk estimation in abdominal aortic aneurysms. J. Endovasc. Ther. **24**, 254–261 (2017)
9. Urrutia, J., Roy, A., Raut, S.S., Antón, R., Muluk, S.C., Finol, E.A.: Geometric surrogates of abdominal aortic aneurysm wall mechanics. Med. Eng. Phys. **59**, 43–49 (2018)
10. Hua, J., Mower, W.R.: Simple geometric characteristics fail to reliably predict abdominal aortic aneurysm wall stresses. J. Vasc. Surg. **34**, 308–315 (2001)
11. Lee, R., et al.: Applied machine learning for the prediction of growth of abdominal aortic aneurysm in humans. EJVES Short Rep. **39**, 24–28 (2018)
12. Yushkevich, P.A., et al.: User-guided 3D active contour segmentation of anatomical structures: significantly improved efficiency and reliability. Neuroimage **31**, 1116–1128 (2006)
13. Yushkevich, P.A., Pluta, J., Wang, H., Wisse, L.E.M., Das, S., Wolk, D.: IC-P-174: fast automatic segmentation of hippocampal subfields and medial temporal lobe subregions in 3 tesla and 7 tesla T2-weighted MRI. Alzheimer's Dement. **12**, P126–P127 (2016)
14. Eddinger, K.C., Stoecker, J.B., Pouch, A.M., Vrudhula, A., Jackson, B.M.: Local aortic wall expansion measured with automated image analysis. J. Vasc. Surg. **72**, e262 (2020)
15. Soto, B., Vila, L., Dilmé, J.F., Escudero, J.R., Bellmunt, S., Camacho, M.: Increased peak wall stress, but not maximum diameter, is associated with symptomatic abdominal aortic aneurysm. Eur. J. Vasc. Endovasc. Surg. **54**, 706–711 (2017)
16. Shang, E.K., et al.: Impact of wall thickness and saccular geometry on the computational wall stress of descending thoracic aortic aneurysms. Circulation **128**, S157–S162 (2013)
17. Haller, S.J., Azarbal, A.F., Rugonyi, S.: Predictors of abdominal aortic aneurysm risks. Bioengineering (Basel) **7**, 79 (2020)

18. Jalalahmadi, G., Helguera, M., Linte, C.A.: A machine leaning approach for abdominal aortic aneurysm severity assessment using geometric, biomechanical, and patient-specific historical clinical features. In: Proceedings of SPIE International Society for Optical Engineering, vol. 11317 (2020)
19. Hirata, K., et al.: Machine learning to predict the rapid growth of small abdominal aortic aneurysm. J. Comput. Assist. Tomogr. **44**, 37–42 (2020)
20. Li, A.E., et al.: Using MRI to assess aortic wall thickness in the multiethnic study of atherosclerosis: distribution by race, sex, and age. Am. J. Roentgenol. **182**, 593–597 (2004)

Cardiac MRI Left Ventricular Segmentation and Function Quantification Using Pre-trained Neural Networks

Fumin Guo[1,2]($\boxtimes$) (iD), Matthew Ng[1,2] (iD), Idan Roifman[3], and Graham Wright[1,2]

[1] Department of Medical Biophysics, University of Toronto, Toronto, Canada
fumin.guo@utoronto.ca
[2] Sunnybrook Research Institute, University of Toronto, Toronto, Canada
[3] Sunnybrook Health Sciences Center, University of Toronto, Toronto, Canada

Abstract. Deep learning has demonstrated promise for cardiac magnetic resonance image (MRI) segmentation. However, the performance is degraded when a trained model is applied to previously unseen datasets. In this work, we developed a way to employ a pre-trained model to segment the left ventricle (LV) and quantify LV indices in a new dataset. We trained a U-net with Monte-Carlo dropout on 45 cine MR images and applied the model to 10 subjects from the ACDC dataset. The initial segmentation was refined using a continuous kernel-cut algorithm and the refined segmentation was used to fine-tune the pre-trained U-net for 10 min. This process was iterated several times until convergence and the updated model was used to segment the remaining 90 patients in the ACDC dataset. For the test dataset, we achieved Dice-similarity-coefficient of 0.81 ± 0.12 for LV myocardium and 0.90 ± 0.09 for LV cavity. Algorithm LV indices were strongly correlated with manual results ($r = 0.86$–0.99, $p < 0.0001$) with marginal biases of -8.8 g for LV myocardial mass, -0.9 ml for LV end-diastolic volume, -0.2 ml for LV end-systolic volume, -0.7 ml for LV stroke volume, and -0.6% for LV ejection fraction. The proposed approach required 12 min for fine-tuning without requiring manual annotations of the new datasets and 1 s to segment a new image. These results suggest that the developed approach is effective in segmenting a previously unseen cardiac MRI dataset and quantifying LV indices without requiring manual segmentation of the new dataset.

Keywords: Cardiac MRI · Machine learning · Left ventricle segmentation · Cardiac function

1 Introduction

Evaluation of left ventricular (LV) function is central for risk stratification and treatment of cardiovascular disease [3]. Cardiac magnetic resonance imaging (MRI) is the gold standard for quantifying LV function [11], including LV myocardial mass (LVMM), end-systolic volume (LVESV), end-diastolic volume

© Springer Nature Switzerland AG 2021
D. B. Ennis et al. (Eds.): FIMH 2021, LNCS 12738, pp. 46–54, 2021.
https://doi.org/10.1007/978-3-030-78710-3_5

(LVEDV), stroke volume (LVSV), and ejection fraction (LVEF). To derive these LV indices, segmentation of the LV myocardium (LVM) and cavity (LVC) is required. Manual segmentation requires intensive work for experts, introduces user variability, and is not suitable for efficient cardiac imaging workflow [7].

Various computerized cardiac MR image analysis methods have been developed and have demonstrated utility for research and clinical use. These methods mainly relied on machine learning without a deep architecture and generally provided suboptimal performance likely because of the use of a limited number of manually designed image features [14]. Recently, deep learning using convolutional neural networks (CNN) has demonstrated promise for various cardiac imaging applications [10]. This approach leverages deep architecture to automatically learn descriptive features through hierarchical feature abstraction. In particular, recent investigations [1,2,7] showed promising performance using CNNs for cardiac cine MRI segmentation; these studies mainly employed a CNN trained and tested on datasets from the same scanner or centre. Unfortunately, these [1,7] and other investigations also showed that CNNs do not generalize well and direct application of pre-trained CNNs to previously unseen datasets, which represents the vast majority of applications of deep learning in research and clinical settings, often yields degraded performance.

To make this important tool more applicable for broader clinical use, it is urgent to improve CNN generalizability for use in datasets acquired using different imaging protocols on different scanners at different centers in patients with different pathologies. Accordingly, our objective was to develop a way to employ a pre-trained CNN to segment the LV in previously unseen cine MRI datasets and demonstrate its utility in generating clinically relevant LV indices.

2 Materials and Methods

2.1 Cardiac MRI Datasets

In this retrospective study, a dataset of 45 cine MR images at end-diastolic phase provided by the 2009 Left Ventricle Segmentation Challenge (LVSC) [12] was used for U-net pre-training. Another cine MRI dataset of 100 patients provided by the 2017 Automated Cardiac Diagnosis Challenge (ACDC) [2] was randomly divided into clusters of 10 and 90 subjects (end-diastolic and end-systolic volumes for each subject, n = 200 image in total) for the pre-trained U-net fine-tuning and testing, receptively.

2.2 Algorithm Pipeline

We employed a five-level 2D U-net [13] with Monte-Carlo dropout [4] (dropout rate = 0.5) at the bottom three levels to minimize overfitting. The number of feature maps ranged from 16 to 256 from the top to the bottom levels. The network was pre-trained on the LVSC dataset for 200 epochs using cross-entropy loss and an ADAM solver at a learning rate of 10^{-4}. Data augmentation, including random rotation ($-50°$–$50°$), translation (-50–50 pixels), voxel size scaling

(0.7–1.3 times), and intensity scaling (0.7–1.3 times), was performed in parallel. The same settings were used during the subsequent fine-tuning process.

The pre-trained U-net was used to segment the 10 fine-tuning subjects. For each subject, 50 Monte Carlo dropout prediction samples ($S1, S2, \ldots, S50$) were generated and averaged to generate a single "mean" segmentation $\bar{S}$; the pixel-wise segmentation uncertainty was calculated based on the standard deviation of the segmentation prediction for each pixel $\omega(x) \propto \frac{1}{std(\{S1(x),S2(x)...,S50(x)\})}$, $x \in \Omega$. The initial CNN "mean" segmentation was refined using a continuous kernel-cut algorithm [7] implemented in MATLAB 2013a (Mathworks, Natick, Massachusetts, USA). The kernel-cut segmentation algorithm comprises normalized cut for balanced feature partitioning to avoid segmentation shrinking and image grid continuous regularization to promote spatially smooth segmentation. In particular, the kernel cut algorithm enforcedthe similarity between the initial "mean" segmentation $\bar{S}$ and the final mask Y based on the derived pixel-wise segmentation uncertainty, i.e., $\int_{x \in \Omega} \omega(x) \cdot |Y - \bar{S}| dx$. The refined segmentation Y of the 10 subjects was used to fine-tune the pre-trained U-net for another 20 epochs. The updated U-net model was applied to the remaining 90 test subjects for LV segmentation (averaging the 50 prediction samples) and function measurements. To investigate the effectiveness of the proposed approach, Algorithm 1 was compared to Algorithm 2 and Algorithm 3 as follows:

1. Algorithm 2: The pre-trained U-net model was directly applied to the 90 test subjects to generate LV segmentation and functional parameters.
2. Algorithm 3: The pre-trained U-net model was used to segment the 10 fine-tuning subjects and the segmentations were used to update the pre-trained model directly. The updated model was applied to the 90 test subjects.

2.3   Evaluation Methods

LV segmentation accuracy was determined using Dice-similarity-coefficient (DSC) and average symmetric surface distance (ASSD) between algorithm and manual LV myocardium and cavity masks [6,8]. Algorithm segmentation was also used to calculate LVMM, LVESV, LVEDV, LVSV, and LVEF, which were compared with manual measurements. LVMM was calculated using a myocardium density of $1.05\,g/ml$ [5].

2.4   Statistical Analysis

Algorithm 1 DSC and ASSD were compared with Algorithm 2 and Algorithm 3 using paired t-tests. Differences between algorithm and manual LV indices were evaluated using paired t-tests. Relationships between algorithm and manual LV functional parameters were determined using Pearson correlation coefficients (r) and the 95% confidence intervals (95% CI); the agreement was evaluated using Bland-Altman analyses. Comparison of correlation coefficients was performed using Fisher z-transformation [9]. Normality of data distribution was determined using Shapiro-Wilk tests, and when significant, nonparametric t-tests were performed. All statistical analyses were performed using GraphPad Prism V7.00 (GraphPad Software Inc., San Diego, CA, USA). Results were considered significant with the probability of making a two-tailed type I error was less than 5% ($p < 0.05$).

3 Results

Figure 1 shows Algorithm 1 and manual segmentation of three subjects. For the 90 test subjects, Table 1 shows that the proposed approach (Algorithm 1) yielded DSC of {0.81, 0.90} and ASSD of {2.04, 1.82} mm for {LVM, LVC}. In contrast, DSC was {0.74, 0.87} and ASSD of {2.43, 2.40} mm for Algorithm 2, and these were {0.77, 0.89} and {2.35, 2.11} mm for Algorithm 3. In addition, the accuracy yielded by Algorithm 1 was higher than directly applying the kernel cut algorithm to the 90 test subjects without fine-tuning, which yielded DSC = {0.78, 0.89} and ASSD = {2.27, 2.16} mm for {LVM, LVC}.

Table 1. Algorithm 1–3 segmentation accuracy for n = 180 images from 90 subjects.

Methods	Algorithm 1		Algorithm 2		Algorithm 3	
	LVM	LVC	LVM	LVC	LVM	LVC
DSC ([0,1])	0.81 ± 0.09	0.90 ± 0.09	$0.74 \pm 0.12^*$	$0.87 \pm 0.12^*$	$0.77 \pm 0.10^*$	$0.89 \pm 0.09^*$
ASSD (mm)	2.04 ± 1.77	1.82 ± 2.18	$2.43 \pm 2.16^*$	$2.40 \pm 2.58^*$	$2.35 \pm 2.12^*$	$2.11 \pm 2.20^*$

*: $p < 0.05$ for comparison of Algorithm 2–3 and Algorithm 1 LV indices.

Table 2 reveals that LV indices provided by our approach are similar to manual results with not significant differences between the two methods ($p = 0.20$–0.25). In contrast, Algorithm 2 (except for LVSV, $p = 0.07$) and Algorithm 3 (except for LVEF, $p = 0.06$) LV function measurements were significant different from manual results ($p < 0.05$).

Table 2. Algorithm and manual LV indices for n = 180 images from 90 subjects.

LV indices	Manual	Algorithm 1	Algorithm 2	Algorithm 3
LVMM (g)$^¥$	138.1 ± 54.3	$129.3 \pm 49.8_{0.20}$	$110.8 \pm 48.2_{<0.0001}$	$115.2 \pm 47.9_{<0.0001}$
LVEDV (ml)	163.8 ± 75.2	$162.9 \pm 72.0_{0.80}$	$174.6 \pm 74.5_{<0.0001}$	$170.9 \pm 73.7_{<0.0001}$
LVESV (ml)	99.4 ± 80.4	$99.2 \pm 76.7_{0.86}$	$108.2 \pm 80.0_{<0.0001}$	$104.1 \pm 79.1_{<0.0001}$
LVSV (ml)	64.4 ± 24.6	$63.7 \pm 25.8_{0.66}$	$66.5 \pm 31.5_{0.07}$	$66.8 \pm 27.25_{0.03}$
LVEF (%)	46.2 ± 20.4	$45.5 \pm 20.5_{0.25}$	$43.0 \pm 23.6_{0.006}$	$45.1 \pm 20.3_{0.06}$

$^¥$: n = 180 images from 90 patients. Subscripts represent p-value for comparison of Algorithm 1–2 and manual LV indices.

Figure 2 and Table 3 show that there were strong correlations and minimal biases between Algorithm 1 and manual LVMM ($r = 0.86$, $p < 0.0001$; bias =

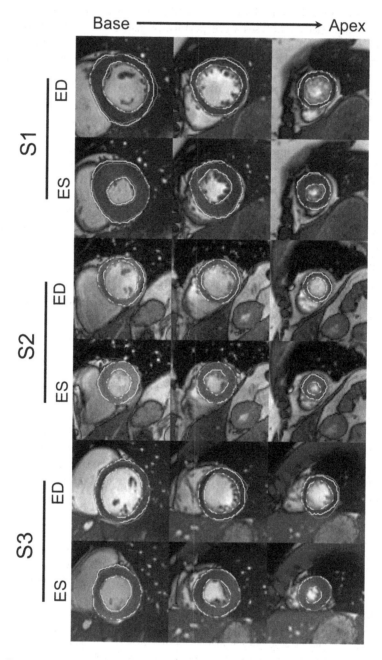

Fig. 1. Representative segmentation of basal to apical slices at end-systole and end-diastole phases for three subjects (S1, S2, and S3). Algorithm 1 and manual segmentation of myocardium are shown in purple and yellow, respectively. ED: end-diastole; ES: end-systole. (Color figure online)

$-8.8 \pm 30.3\,\mathrm{g}$), LVEDV ($r = 0.99$, $p < 0.0001$; bias $= -0.9 \pm 13.1$ ml), LVESV ($r = 0.99$, $p < 0.0001$; bias $= -0.2 \pm 13.8$ ml), LVSV ($r = 0.92$, $p < 0.0001$; bias $= -0.7 \pm 10.0$ ml), and LVEF ($r = 0.86$, $p < 0.0001$; bias $= -0.6 \pm 7.8\%$). While Algorithm 2 LV indices were also correlated with manual results, Fisher z-transformations showed that the correlation coefficients were significantly different from Algorithm 1, i.e., $p = 0.02$ for LVMM, 0.01 for LVEDV, 0.01 for LVESV, 0.01 for LVSV, and < 0.0001 for LVEF. In addition, Algorithm 1 yielded smaller biases and lower standard deviations with narrower 95% limits of agreement compared with Algorithm 2.

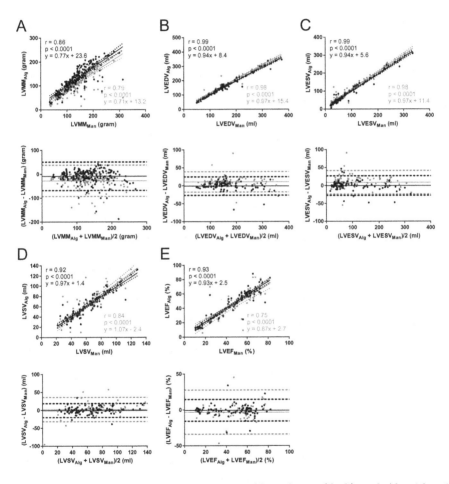

Fig. 2. Relationships and agreement between Algorithm 1 (dark) and Algorithm 2 (gray) vs manual LVMM (A), LVEDV (B), LVESV (C), LVSV (D), and LVEF (E) ($n = 180$ images from 90 subjects). The top plots show the linear regression and the bottom plots indicate the Bland-Altman analyses between Algorithm and manual LV indices.

Table 3. Pearson correlation coefficients (r) and the 95% confidence intervals ([,]) between Algorithm 1–2 and manual LV indices for n = 180 images from 90 subjects. The correlation coefficients were compared using Fisher z-transformation.

Correlation (r)	Algorithm 1 *vs* Manual	Algorithm 2 *vs* Manual	Fisher z-score
LVMM ¥	0.86 ([0.80, 0.90])	0.79 ([0.73, 0.84])	0.02
LVEDV	0.99 ([0.98, 0.99])	0.98 ([0.97, 0.99])	0.01
LVESV	0.99 ([0.98, 0.99])	0.98 ([0.97, 0.99])	0.01
LVSV	0.92 ([0.88, 0.95])	0.84 ([0.76, 0.89])	0.01
LVEF	0.93 ([0.89, 0.95])	0.75 ([0.65, 0.83])	<0.0001

¥: n = 180 images from 90 patients.

Our approach required ~12 min, inducing 10 min for U-net fine-tuning, 10 s to post-process each image using the kernel-cut algorithm, and 1 s to segment a new image using the updated CNN.

4   Discussion

In this work, we developed a way to employ a pre-trained CNN for cine MRI LV segmentation and function evaluation. For a group of patients with diverse cardiac pathologies, we demonstrated: 1) improved segmentation accuracy; 2) strong and significant correlations between our approach and manual LV indices; 3) rapid and fully automated algorithm implementation *without* requiring manual annotations of the unseen datasets.

The poor generalizability of deep learning segmentation models represents a major obstacle that hinders broad use of this technique. Previous studies mainly focused on training and testing a deep learning model on the same dataset, which represents limited applications. For example, Bai et al. [1] trained a U-net using a large cardiac cine MRI dataset of 4,275 subjects from the UK Biobank database. They achieved DSC of 0.65 and 0.74 for LVM and LVC, respectively, when applying the pre-trained model to 20 subjects in the ACDC dataset. Although the sub-optimal accuracies were later improved by fine-tuning the trained model on 80 subjects in the ACDC dataset, this procedure requires manual annotation of the 80 subjects and is not compatible with efficient workflow. In our previous work [7], we showed that direct application of a state-of-the-art deep learning model to the ACDC dataset yielded DSC of 0.78 and 0.86 for LVM and LVC, respectively, similar to Algorithm 2 in this work. Compared with the previous studies, the proposed approach in this work achieved DSC of 0.81 for LVM and 0.90 for LVC *without* requiring manually-annotated fine-tuning subjects, reducing the work for experts. Importantly, the derived LV indices were strongly correlated with expert manual segmentation with no significant differences between the techniques, suggesting that our approach could generate clinically relevant LV indices.

The demonstrated promise of our approach stemmed from a U-net and a recently developed continuous kernel-cut algorithm. The improved segmentation generalizability may be due to the integration of the power of the kernel-cut method, which employed shallower image features and had only a few parameters, potentially alleviating the overfitting issue associated with deep learning methods. Although seems simple, the automated fine-tuning scheme is indeed effective in improving deep learning generalizability for LV segmentation and generating clinically-relevant LV function measurements without increasing the difficulty of algorithm interpretability. The proposed approach may be further improved by: 1) examining the initial segmentation quality of the 10 fine-tuning subjects and applying the kernel cut algorithm to those with relatively low segmentation quality, 2) iterating the fine-tuning procedure several times and adding several new datasets (without manual segmentation) for each iteration; note that the strategies in 1) and 2) could be implemented in parallel. Regardless, these obtained results suggest that our approach may facilitate broader research and clinical use of deep learning for cardiac imaging tasks.

References

1. Bai, W., et al.: Automated cardiovascular magnetic resonance image analysis with fully convolutional networks. J. Cardiovasc. Magn. Reson. **20**(1), 65 (2018)
2. Bernard, O., et al.: Deep learning techniques for automatic MRI cardiac multi-structures segmentation and diagnosis: is the problem solved? IEEE Trans. Med. Imaging **37**(11), 2514–2525 (2018)
3. Flachskampf, F.A., Biering-Sørensen, T., Solomon, S.D., Duvernoy, O., Bjerner, T., Smiseth, O.A.: Cardiac imaging to evaluate left ventricular diastolic function. JACC Cardiovasc. Imaging **8**(9), 1071–1093 (2015)
4. Gal, Y., Ghahramani, Z.: Dropout as a bayesian approximation: representing model uncertainty in deep learning. In: International Conference on Machine Learning, pp. 1050–1059 (2016)
5. Grothues, F., et al.: Comparison of interstudy reproducibility of cardiovascular magnetic resonance with two-dimensional echocardiography in normal subjects and in patients with heart failure or left ventricular hypertrophy. Am. J. Cardiol **90**(1), 29–34 (2002)
6. Guo, F., Krahn, P.R., Escartin, T., Roifman, I., Wright, G.: Cine and late gadolinium enhancement mri registration and automated myocardial infarct heterogeneity quantification. Magn. Reson. Med. **85**, 2842–2855 (2020)
7. Guo, F., et al.: Improving cardiac MRI convolutional neural network segmentation on small training datasets and dataset shift: a continuous kernel cut approach. Med. Image Anal. **61**, 101636 (2020)
8. Guo, F., Ng, M., Wright, G.: Cardiac cine MRI left ventricle segmentation combining deep learning and graphical models. In: Medical Imaging 2020: Image Processing, vol. 11313, p. 113130Z. International Society for Optics and Photonics (2020)
9. Kirby, M., et al.: Hyperpolarized 3he and 129xe MR imaging in healthy volunteers and patients with chronic obstructive pulmonary disease. Radiology **265**(2), 600–610 (2012)

10. Leiner, T., et al.: Machine learning in cardiovascular magnetic resonance: basic concepts and applications. J. Cardiovasc. Magn. Reson. **21**(1), 1–14 (2019)
11. Members, W.C., et al.: Accf/acr/aha/nasci/scmr 2010 expert consensus document on cardiovascular magnetic resonance: a report of the american college of cardiology foundation task force on expert consensus documents. Circulation **121**(22), 2462–2508 (2010)
12. Radau, P., Lu, Y., Connelly, K., Paul, G., Dick, A., Wright, G.: Evaluation framework for algorithms segmenting short axis cardiac MRI. MIDAS J.-Cardiac MR Left Ventricle Segment. Chall. **49** (2009)
13. Ronneberger, O., Fischer, P., Brox, T.: U-Net: convolutional networks for biomedical image segmentation. In: Navab, N., Hornegger, J., Wells, W.M., Frangi, A.F. (eds.) MICCAI 2015. LNCS, vol. 9351, pp. 234–241. Springer, Cham (2015). https://doi.org/10.1007/978-3-319-24574-4_28
14. Shen, D., Wu, G., Suk, H.I.: Deep learning in medical image analysis. Ann. Rev. Biomed. Eng. **19**, 221–248 (2017)

Three-Dimensional Embedded Attentive RNN (3D-EAR) Segmentor for Left Ventricle Delineation from Myocardial Velocity Mapping

Mengmeng Kuang[1]([✉]), Yinzhe Wu[2,3], Diego Alonso-Álvarez[4], David Firmin[2,5], Jennifer Keegan[2,5], Peter Gatehouse[2,5], and Guang Yang[2,5]([✉]) [iD]

[1] Department of Computer Science, The University of Hong Kong, Hong Kong, China
mmkuang@cs.hku.hk
[2] National Heart and Lung Institute, Imperial College London, London, UK
g.yang@imperial.ac.uk
[3] Department of Bioengineering, Imperial College London, London, UK
[4] Research Computing Service, Information and Communication Technologies, Imperial College London, London, UK
[5] Cardiovascular Research Centre, Royal Brompton Hospital, London, UK

Abstract. Myocardial Velocity Mapping Cardiac MR (MVM-CMR) can be used to measure global and regional myocardial velocities with proved reproducibility. Accurate left ventricle delineation is a prerequisite for robust and reproducible myocardial velocity estimation. Conventional manual segmentation on this dataset can be time-consuming and subjective, and an effective fully automated delineation method is highly in demand. By leveraging recently proposed deep learning-based semantic segmentation approaches, in this study, we propose a novel fully automated framework incorporating a 3D-UNet backbone architecture with **E**mbedded multichannel **A**ttention mechanism and LSTM based **R**ecurrent neural networks (RNN) for the MVM-CMR datasets (dubbed 3D-EAR segmentor). The proposed method also utilises the amalgamation of magnitude and phase images as input to realise an information fusion of this multichannel dataset and exploring the correlations of temporal frames via the embedded RNN. By comparing the baseline model of 3D-UNet and ablation studies with and without embedded attentive LSTM modules and various loss functions, we can demonstrate that the proposed model has outperformed the state-of-the-art baseline models with significant improvement.

Keywords: Cardiac MRI · Myocardial Velocity Mapping · Segmentation · Attention mechanism · LSTM

Electronic supplementary material The online version of this chapter (https://doi.org/10.1007/978-3-030-78710-3_6) contains supplementary material, which is available to authorized users.

D. B. Ennis et al. (Eds.): FIMH 2021, LNCS 12738, pp. 55–62, 2021.
https://doi.org/10.1007/978-3-030-78710-3_6

1   Introduction

Healthy functioning of the left ventricle requires complex motion and all parts of the myocardium must perform synergistically in order to ensure the heart to pump efficiently. In certain pathologies, early regional myocardial instability may be compensated for by altered movement in other areas in order to maintain the ventricular function. Global myocardial velocities are widely used in clinical practice; however, the global measurement might only be detectable when the condition has advanced to a point where compensation is no longer possible. By additional measurement of local myocardial dynamics, myocardial stability can be more specifically quantified and potential cardiovascular disease can therefore be detected earlier [1].

Among different cardiac MR (CMR) techniques, there are a few methods that can be used to calculate global and regional myocardial dynamics [2]. Myocardial Velocity Mapping CMR (MVM-CMR) [3] can potentially provide both high spatial and temporal resolution, and has clear advantages compared to the blood velocity scans that are commonly used in clinical.

Accurate segmentation of the left ventricle (LV) is the first step and a prerequisite for robust and reproducible global and regional myocardial velocity estimation. Conventional manual segmentation on the MVM-CMR dataset can be extremely time-consuming considering both the high spatial and temporal resolution of the dataset. Such manual segmentation is also limited by clinician's experience and potential human operator fatigue may also affect the delineation accuracy. Therefore, an efficient and robust automated LV segmentation method for the multichannel (i.e., magnitude and three velocity channels) and multi-frame (i.e., temporal frames of the cardiac "movie" acquired) MVM-CMR data is necessary for the clinical deployment of the global and regional velocities estimation.

Development in deep learning represents a major leap for digital healthcare. Several research studies have demonstrated promising results for the anatomical and pathological segmentation of the heart from CMR images, e.g., whole heart segmentation [4], LV segmentation [5], left atrial segmentation [6] and atrial scar delineation [7]. A more detailed review of the segmentation of cardiac images can be found elsewhere [8].

Despite successful applications of deep learning based techniques for CMR data segmentation, a plain deployment of existing methods for the MVM-CMR data can be challenging, but more informative. This is due to (1) multi-frame temporal MVM-CMR data may require a more complicated network design to ensure both accurate slice-wise delineation and continuity in the temporal dimension and (2) MVM-CMR data have multiple channels, e.g., magnitude and phase channels, that can provide richer information of the LV anatomy, but how to explore such informative multichannel data is still an open question.

Inspired by recent progress on semantic segmentation, e.g., UNet [9], we propose a novel **E**mbedded multichannel **A**ttentive **R**ecurrent neural networks (RNN), abbreviated 3D-EAR segmentor, for LV delineation from MVM-CMR datasets. The proposed 3D-EAR segmentor consists of three major components: (1) a 3D-UNet based backbone network, (2) embedded attention modules to enhance the network skip connections for more accurate localisation of the LV anatomy, and (3) long short-term memory (LSTM) based RNN modules to learn the temporal information of the multi-frame context at the bottom of the U-shaped network. In doing so, the proposed method can leverage the amalgamation of magnitude and phase images as more informative input to realise an effective information fusion of this multichannel dataset. Besides, the temporal dimension continuity can be ensured by exploring the correlations of temporal frames via the embedded LSTM. In addition, by varying different loss functions (i.e., Cross-Entropy loss, Dice loss [10], and Dice-IoU[1] loss [11]), we perform ablation studies to find the optimal architecture of the proposed network.

By validation on MVM-CMR data collected from healthy controls, our proposed 3D-EAR segmentor achieves superior LV segmentation performance compared to state-of-the-art baseline models at the patient-level.

2 Method

2.1 Data Acquisition, Preprocessing and Augmentation

The training and testing datasets contain 26 MVM-CMR datasets with the data size of $50 \times 512 \times 512 \times 4$ which were acquired from 18 healthy subjects (8 of them were acquired twice, giving 26 datasets) at Royal Brompton Hospital. Each of the datasets consists of 3–5 cine slices, giving 121 cine slices in total. There are 50 temporal frames per cardiac cycle and 4 channels reconstructed by a non-Cartesian SENSE reconstruction channel (one magnitude encoding channel and three velocity encoding channels of orthogonal directions), constituting the multi-frame multichannel MVM-CMR data. The MVM-CMR slices have spatial resolutions of $0.85 \, \text{mm} \times 0.85 \, \text{mm}$ that were reconstructed from the acquired $1.7 \, \text{mm} \times 1.7 \, \text{mm}$. The MVM-CMR were acquired in short-axis slices from base to apex of the LV. An experienced cardiac MRI physicist performed manual delineation of the LV myocardium to create the ground truth for this study. In addition, we augmented the data by random rotation (angle = $[90°, 180°, 270°]$) before model training. An example of our multichannel MVM-CMR dataset with the manual segmentation can be found in Fig. 1.

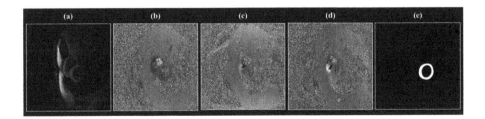

Fig. 1. A sample frame of our multi-channel MVM-CMR dataset with the manual segmentation. From left to right: magnitude channel, three phase channels and the manual LV segmentation.

---

[1] IoU stands for Intersection-Over-Union, which is also known as the Jaccard index.

2.2 Network Architectures

Attention Enhanced Skip Connections. We propose embedded attention modules to realise the enhanced skip connections and synthesise features from the transferred image F of the convolutional layers in the original 3D-UNet structure. The objective is that the relevant features in a single slice can be enhanced during the attention process, while the less relevant ones can be less focused on. To achieve this, an attention map is computed in every frame of the input MVM-CMR images to represent the confidence of the transferred features for each position as expressed in Eq. (1) as follows

$$\text{Layer}_i^{\text{Attention}} = \text{Softmax}(F_i \times \text{Conv}(F_i))) \cdot \text{Conv}(F_i)', \qquad (1)$$

where i donates the i-th frame in the 3D feature map. Conv and Softmax represent the convolutional layer and the Softmax activation operation, respectively (Fig. 2 (a)).

LSTM Based Temporal Feature Extractor. We also develop LSTM layers to capture the cross-frame features at the bottom of the U-shaped structure in the 3D-UNet network. We assume that the LSTM can learn the temporal correlations from the multi-frame MVM-CMR data. For our 3D (2D + t) MVM-CMR data, we need to convert them into sequences and then transfer back into 3D (i.e., 2D + t) images before and after the LSTM layer. The whole operation can be denoted as Eq. (2) that is

$$\text{Layer}_{\text{out}} = \text{Reshape}(\text{LSTM}(\text{Reshape}(\text{Layer}_{\text{in}}))). \qquad (2)$$

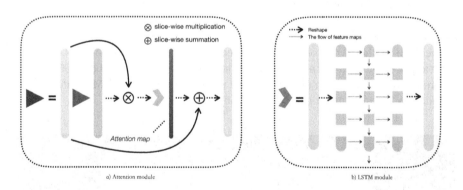

a) Attention module   b) LSTM module

Fig. 2. The structure of the embedded in-slice attention block (a) and cross-frame LSTM module (b).

Figure 2(b) shows the LSTM workflow and Fig. 3 represents the overall structure of the proposed model (i.e., 3D-EAR segmentor). In this figure, we use rounded rectangles to denote the flow of feature maps, triangles of different colours to represent different neural network blocks and arrows to indicate the skip connections and up-sampling.

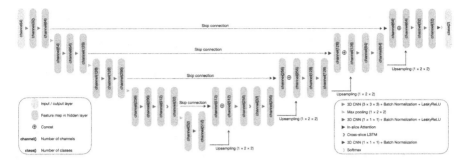

Fig. 3. The overall architecture of our proposed 3D-EAR segmentor.

2.3 Loss Functions

For our 3D-EAR segmentor, we implement various loss functions, e.g., (1) Cross-Entropy loss, (2) Dice loss (with Laplace smoothing) and (3) Dice-IoU loss (with Laplace smoothing) to seek an optimal solution. The standard Cross-Entropy loss can be represented by Eq. (3), that is

$$\mathrm{Loss}_{\mathrm{Cross-Entropy}} = -\frac{1}{n} \sum_{i} \sum_{c=1}^{n} y_{ic} \log (p_{ic}), \tag{3}$$

where y and p represent the true and predicted labels in the nth class, respectively.

The Dice loss (with Laplace smoothing factor f_{smooth}) can be denoted as Eq. (4), that is

$$\mathrm{Loss}_{\mathrm{Dice}} = 1 - \frac{2 \times |\mathrm{GT} \cap \mathrm{Pred}| + f_{\mathrm{smooth}}}{|\mathrm{GT}| + |\mathrm{Pred}| + f_{\mathrm{smooth}}}, \tag{4}$$

where GT stands for the ground truth of the segmentation, Pred donates the prediction of the model and $| \bullet |$ represents the area of $\bullet$.

The Dice-IoU loss, which is a combination of the Dice loss with the IoU calculation, can be represented as Eq. (5)

$$\mathrm{Loss}_{\mathrm{Dice-IoU}} = \frac{1}{n} \times \sum_{i=1}^{n} \mathrm{Loss}_{\mathrm{Dice}} \times \mathrm{IoU}. \tag{5}$$

Equation (6) illustrates the IoU calculation, which is also smoothed by the Laplace factor f_{smooth}, that is

$$\mathrm{IoU} = 1 - \frac{|\mathrm{GT} \cap \mathrm{Pred}| + f_{\mathrm{smooth}}}{|\mathrm{GT} \cup \mathrm{Pred}| + f_{\mathrm{smooth}}}. \tag{6}$$

2.4   Implementation Details

The input of our model (and compared models) was a 4-channel 50-frame 512 × 512 MVM-CMR dataset, and the output prediction was with the same size as the input but had 2 different labels (i.e., LV and non-LV). We divided the MVM-CMR cine slices into two sets for experiments, one consisting of 80% of the subjects for model training and the other one consisting of the remaining 20% as independent testing. During the training process, we also performed the 5-fold cross-validation. The training was carried out on two standard NVIDIA GEFORCE RTX 2080 Ti GPUs. Our implementation was based on Keras and TensorFlow backend. The implementation and pre-trained models will be open source (on Github) for a reproducible study.

3   Results

3.1   Experiments and Evaluation Metrics

We performed the following comparison and ablation studies, including UNet3D (the baseline model 3D-UNet), UNet3D-Attention (3D-UNet with attention) and our proposed 3D-EAR segmentator with various loss functions. We evaluated model performance using (1) Dice scores, (2) Sensitivities and (3) Positive Predictive Values (PPV).

3.2   Quantitative Results

Quantitative results of our comparison study can be found in Table 1.

Table 1. Dice scores of our comparison studies and ablation studies using various loss functions.

Structures	Losses		
	Cross-Entropy	Dice	Dice-IoU
UNet3D	0.84 ± 0.02	0.85 ± 0.03	0.88 ± 0.02
UNet3D-Attention	0.87 ± 0.03	0.87 ± 0.02	0.89 ± 0.02
3D-EAR	0.88 ± 0.03	0.89 ± 0.02	0.91 ± 0.03

Table 1 and Table 2 show outstanding segmentation performance of using our proposed 3D-EAR model. Compared to the baseline model, our proposed 3D-EAR has achieved significantly higher Dice scores, sensitivities and PPV. We can also find that with the LSTM, our 3D-EAR has further improvement on the model with only the attention module.

An automated segmentation example result obtained by our 3D-EAR model is shown in Fig. 4. Followed by a morphological post-processing stage, we were able to generate their LV myocardium global velocity curves from the predicted

Table 2. Sensitivities and PPV of our comparison studies and ablation studies using various loss functions.

Structures	Losses					
	Cross-Entropy		Dice		Dice-IoU	
	Sensitivity	PPV	Sensitivity	PPV	Sensitivity	PPV
UNet3D	0.75 ± 0.03	0.95 ± 0.02	0.81 ± 0.04	0.90 ± 0.05	0.86 ± 0.03	0.91 ± 0.01
UNet3D-Attention	0.84 ± 0.02	0.90 ± 0.02	0.84 ± 0.01	0.91 ± 0.02	0.85 ± 0.02	0.93 ± 0.02
3D-EAR	0.80 ± 0.01	0.98 ± 0.01	0.86 ± 0.01	0.93 ± 0.02	0.87 ± 0.01	0.96 ± 0.02

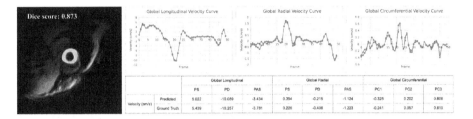

Fig. 4. A typical example of the segmentation results randomly selected from our MVM-CMR datasets with the global longitudinal, radial and circumferential velocity curves and peak velocities per slice of the example frames (More examples of full cardiac cycle segmentations will be provided in the Supplementary Material). For the segmentation: Blue—true positive; Yellow—false positive; Red— false negative; Blue and Yellow regions—automated segmentation results. For the global velocity curves: Blue/Orange curves—derived from automated and manual segmentations. (Color figure online)

results and compare it with the ones derived from the ground truth. We are able to observe close alignments of curves and little differences in the peak velocities generated from these curves.

4   Discussion and Conclusion

In this study, we have developed and validated a novel 3D-EAR segmentor for the delineation of LV from MVM-CMR data. The proposed model incorporated embedded attention enhanced skip connections to filter our irrelevant features from images and LSTM based RNN for accounting correlations among temporal frames of the MVM-CMR data. The experimental results have shown promising quantification and visualisation that can facilitate accurate and reliable estimation of global and local myocardial velocities. More detailed method descriptions, comparison results with and without multichannel input data, ablation studies of network parameters, and velocity comparisons in all the slices will be presented.

Acknowledgement. This study was supported in part by BHF (TG/18/5/34111, PG/16/78/32402), in part by Heart Research UK RG2584, in part by the ERC IMI [101005122], and in part by ERC H2020 [952172].

References

1. Simpson, R., Keegan, J., Gatehouse, P., Hansen, M., Firmin, D.: Spiral tissue phase velocity mapping in a breath-hold with non-cartesian SENSE. Magn. Reson. Med. **72**(3), 659–668 (2014)
2. Simpson, R.M., Keegan, J., Firmin, D.N.: MR assessment of regional myocardial mechanics. J. Magn. Reson. Imaging **37**(3), 576–599 (2013)
3. Simpson, R., Keegan, J., Firmin, D.: Efficient and reproducible high resolution spiral myocardial phase velocity mapping of the entire cardiac cycle. J. Cardiovasc. Magn. Reson. **15**(1), 1–14 (2013)
4. Zhuang, X., et al.: Evaluation of algorithms for multi-modality whole heart segmentation: an open-access grand challenge. Med. Image Anal. **58**, 101537 (2019)
5. Bai, W., et al.: Automated cardiovascular magnetic resonance image analysis with fully convolutional networks. J. Cardiovasc. Magn. Reson. **20**(1), 65 (2018)
6. Yang, G., et al.: Simultaneous left atrium anatomy and scar segmentations via deep learning in multiview information with attention. Fut. Gener. Comput. Syst. **107**, 215–228 (2020)
7. Li, L., et al.: Atrial scar quantification via multi-scale CNN in the graph-cuts framework. Med. Image Anal. **60**, 101595 (2020)
8. Chen, C., et al.: Deep learning for cardiac image segmentation: a review. Front. Cardiovasc. Med. **7**, 25 (2020)
9. Ronneberger, O., Fischer, P., Brox, T.: U-Net: convolutional networks for biomedical image segmentation. In: Navab, N., Hornegger, J., Wells, W.M., Frangi, A.F. (eds.) MICCAI 2015. LNCS, vol. 9351, pp. 234–241. Springer, Cham (2015). https://doi.org/10.1007/978-3-319-24574-4_28
10. Milletari, F., Navab, N., Ahmadi, S.A.: V-net: fully convolutional neural networks for volumetric medical image segmentation. In: 2016 Fourth International Conference on 3D Vision (3DV), pp. 565–571. IEEE (2016)
11. Rahman, M.A., Wang, Y.: Optimizing intersection-over-union in deep neural networks for image segmentation. In: Bebis, G., et al. (eds.) ISVC 2016. LNCS, vol. 10072, pp. 234–244. Springer, Cham (2016). https://doi.org/10.1007/978-3-319-50835-1_22

Whole Heart Anatomical Refinement from CCTA Using Extrapolation and Parcellation

Hao Xu[1]([✉]), Steven A. Niederer[1], Steven E. Williams[1,2], David E. Newby[2], Michelle C. Williams[2], and Alistair A. Young[1]

[1] Department of Bioengineering, King's College London, London, UK
hao2.xu@kcl.ac.uk
[2] University/BHF Centre for Cardiovascular Science,
University of Edinburgh, Edinburgh, Scotland

Abstract. Coronary computed tomography angiography (CCTA) provides detailed anatomical information on all chambers of the heart. Existing segmentation tools can label the gross anatomy, but addition of application-specific labels can require detailed and often manual refinement. We developed a U-Net based framework to i) extrapolate a new label from existing labels, and ii) parcellate one label into multiple labels, both using label-to-label mapping, to create a desired segmentation that could then be learnt directly from the image (image- to-label mapping). This approach only required manual correction in a small subset of cases (80 for extrapolation, 50 for parcellation, compared with 260 for initial labels). An initial 6-label segmentation (left ventricle, left ventricular myocardium, right ventricle, left atrium, right atrium and aorta) was refined to a 10-label segmentation that added a label for the pulmonary artery and divided the left atrium label into body, left and right veins and appendage components. The final method was tested using 30 cases, 10 each from Philips, Siemens and Toshiba scanners. In addition to the new labels, the median Dice scores were improved for all the initial 6 labels to be above 95% in the 10-label segmentation, e.g. from 91% to 97% for the left atrium body and from 92% to 96% for the right ventricle. This method provides a simple framework for flexible refinement of anatomical labels. The code and executables are available at cemrg.com.

Keywords: CCTA · U-Net · Whole heart segmentation

1 Introduction

Coronary computed tomography angiography (CCTA) is a widely used imaging tool for investigation of the coronary artery anatomy in patients with suspected coronary artery disease [1]. However, a lot of anatomical information is also present in these scans [2]. Although CCTA has high signal and spatial resolution relative to MRI or echocardiography, existing segmentation methods are difficult to adapt to different applications. In particular, planning ablation therapy for atrial fibrillation requires the identification of

© Springer Nature Switzerland AG 2021
D. B. Ennis et al. (Eds.): FIMH 2021, LNCS 12738, pp. 63–70, 2021.
https://doi.org/10.1007/978-3-030-78710-3_7

left and right pulmonary veins (LPV, RPV) and their intersection with the left atrium (LA) body. Similarly, anatomical features such as the pulmonary artery valve (PAV) and left atrial appendage (LAA) are important for particular pathologies (such as tetralogy of Fallot and stroke respectively).

Previous Work. Neural network whole heart segmentation methods have previously shown good results with CCTA data [3]. In particular, Baskaran et al. [4] applied a 2D U-Net, using 132 training, 34 validation cases and 17 test cases, to predict 5 labels: left ventricle (LV), right ventricle (RV), LA, right atrium (RA) and LV myocardium (LVMyo). Median Dice scores ranged from 0.915 (RV) to 0.938 (LV). LPV and RPV were excluded from the LA but the LAA was included. In the 2017 Multi-Modality Whole Heart Segmentation (MMWHS) challenge, a variety of methods performed well on CCTA datasets (n = 60) with 7 labels: LV, RV, LA (excluding LPV, RPV and LAA), LVMyo, ascending aorta (AA), and pulmonary artery (PA) [3]. The leading method used a two-step process, with a localization 3D U-Net and heatmap regression and a subsequent 3D U-Net for segmentation [5].

In this paper, we present a method to adapt algorithms to a different label definition, leveraging existing segmentation tools derived from different sources. We apply our method to the problem of PAV localization and LA parcellation into LPV, RPV, LAA and LA body segments.

2 Methods

We describe a multi-stage process (Fig. 1), in which existing segmentations were used to provide ground truth for an initial image-to-label 3D U-Net (*U-Net 1*) using 200/30/30 (train/validation/test) cases, giving 6 regions directly from CCTA images. We then refined the segmentation and manually identified the PAV to separate PA from RV, and the image-to-label network was retrained with the refined 6 labels (*U-Net 1 no PA*,

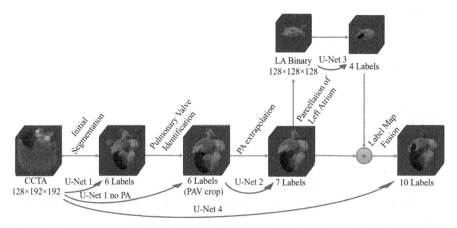

Fig. 1. Refinement approach overview. *U-Net 1*, *U-Net 1 no PA*, and *U-Net 4* are image to label maps; *U-Net 2* is a label-to-label extrapolation map; *U-Net 3* is a label-to-label parcellation map.

200/30/30 cases). A label-to-label 3D U-Net (*U-Net 2*) was trained to extrapolate the PA label (62/9/9 cases), and another label-to label 3D U-Net (*U-Net 3*) was trained to parcellate LA into LA body, LPV, RPV and LAA (38/6/6 cases). The resulting networks were applied to 1770 cases, and results from 260 cases were manually reviewed and used to train and test a final image-to-label 3D U-Net *(U-Net 4*, 200/30/30) to predict the refined labels directly from the CCTA images.

2.1 Data

CCTA exams of 1770 patients (56% male, 58 ± 10 years old) who participated in the Scottish COmputed Tomography of the Heart (SCOT-HEART) trial were included in this study; patient demographics have been reported previously [6]. Briefly, all patients had suspected angina attributable to coronary artery disease and were imaged between 2010 and 2014 at one of three sites using either 64- or 320-detector row scanners (Brilliance 64, Philips Medical Systems, Netherlands; Biograph mCT, Siemens, Germany; Aquilion ONE, Toshiba Medical Systems, Japan). Tube current, voltage, and volume of iodine-based contrast were adjusted based on body mass index. To illustrate how different sources of data can be combined, we also included 20 cases from the MMWHS challenge training dataset and 40 cases from the MMWHS challenge testing dataset [3]. These manually annotated cases were obtained from two 64-slice scanners (both Philips) at two sites in Shanghai, China.

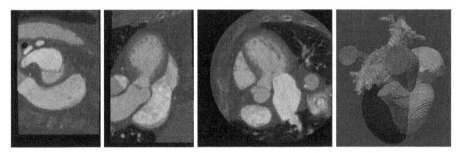

Fig. 2. Coronal, sagittal and axial views of the CCTA image overlapped with initial segmentation labels, and 3D visualization of the anatomies. LV, LVMyo, RV, LA, RA and AA are in purple, dark blue, light blue, green, yellow and orange respectively (Color figure online).

2.2 Image-to-Label Initial Segmentation

For the SCOT-HEART cases an initial segmentation was automatically performed using Siemens AXseg v4.11 prototypical software (Siemens Healthineers, Erlangen, Germany). This method used an atlas combined with marginal space learning and steerable filters [7]. Six regions were labelled: LV, LVMyo, RV, LA, RA, and AA. For the MMWHS cases the manual annotations provided for the corresponding regions were used. The image and label maps were normalized to voxel size of 1 mm^3 and cropped or padded to volumes with size of 128 × 192 × 192 with the heart at the center of the volume. For large hearts, the voxel size was iteratively increased by 10% until the field of view covered all labelled voxels. Typical results of the initial segmentation are shown in Fig. 2.

A 3D U-Net (*U-Net 1*) was designed with 3 max-pooling and deconvolutional stages with a stride of $2 \times 2 \times 2$. The numbers of convolutional kernels were set to be (16, 32), (32, 64), (64,128) for the contraction path and (128, 256) for the bottle neck, with the kernel size of $3 \times 3 \times 3$. The numbers of deconvolutional kernels were set to be (128, 128), (64, 64), (32, N) for the expansion path, where N is the number of layers of the output volume. For the initial segmentation network $N = 7$ (background and 6 labels). The network was trained with cross-entropy loss, using 200 training (180 randomly selected from SCOT-HEART and 20 from MMWHS) cases, with additional 30 cases for validation and 30 cases for testing. The training, validating and testing cases from SCOT-HEART were randomly and equally sampled from three types of scanner. Dice scores were used for evaluation.

2.3 Label Processing

We found that several refinements were necessary to improve the accuracy and consistency of the initial segmentations. Firstly, AXseg tended to visually over-segment the LV cavity in non-Siemens scanners (Fig. 2). We therefore used a morphological operator to dilate the LVMyo mask, and voxels overlapping between dilated LVMyo and LV were transferred to the LVMyo label if i) the mean intensity value of the overlap region was closer to LVmyo than LV, and ii) the voxel intensity was less than the mean plus one standard deviation of the original LVMyo voxel intensities. This process was repeated up to a maximum of three times.

Secondly, a PA label was used for PAV identification, with the PAV defined as the intersection of PA with the RV (all voxels in the RV with a PA neighbor, and vice versa). This method was preferred to the direct segmentation of the PAV, since U-Nets do not work as well on classes with such small number of voxels due to class-imbalance. However, a significant number of SCOT-HEART cases cropped the PA from the field of view, resulting in inconsistent PA segmentations. Furthermore, the AXseg tool did not include a PA label but included a PA section (if present) in the RV label. Therefore, we manually partitioned the initial RV label for the 260 cases used to train and test *U-Net 1* into RV and PA labels using a PAV plane defined using landmarks on the reformatted images. The initial image-to-label network was then retrained without the PA section, using the refined RV label (*U-Net 1 no PA* in Fig. 1).

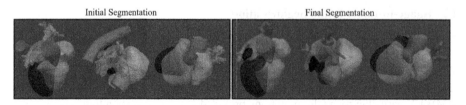

Fig. 3. Examples of image to label results for initial (6 labels) and final (10 labels) maps.

2.4 Label to Label Networks

Extrapolation of PA

In order to enable prediction of a PA label by extrapolating the RV outflow tract, even in cases in which the PA was cropped, we chose 80 cases with PA present (from the original 260 cases used to train *U-Net 1 no PA*, with ground truth partitioned into RV and PA) split into 62 training cases, 9 validation cases and 9 testing cases. We applied same 3D U-Net architecture (Sect. 2.2) to map predictions from *U-Net 1 no PA* to a 7-label segmentation with PA included (*U-Net 2*). The input/output volumes were set to be 128 × 192 × 192.

LA Parcellation

We manually partitioned the LA into body, LPV, RPV and LAA labels (multiple LPV were given the same label, similarly for RPV) using 3DSlicer's cropping boxes (scissors tool). Compared to PA extrapolation, manual annotation was more time-consuming, but the number of cases requiring annotation was smaller as it is simpler for a label-to-label network to learn how to relabel an existing structure (initial LA label) than to predict a new one. Therefore, only 50 cases were required (38 training cases, 6 validating cases and 6 testing cases). The same 3D U-Net architecture (Sect. 2.2) was applied and input/output volumes were set to be 128 × 128 × 128 (*U-Net 3*).

2.5 Image to Label Refinement

The image-to-label 3D U-Net was retrained to directly predict 10-label maps from CCTA input (*U-Net 4*). The ground truth for this network was generated by applying the refined initial segmentation network and label-to-label networks, fusing the predictions, and manually evaluating the result using ITK-snap, until 260 predictions with good quality were identified. The segmentations consisting of one object were cleaned by choosing the largest connected component. This process resulted in 200 training (180 from SCOT-HEART and 20 from MMWHS) cases, with additional 30 validation cases and 30 test cases, evenly split by scanner types. Dice scores were used for evaluation against the output of the label-to-label network results. An example of a pair of initial and refined segmentations are shown in Fig. 3.

3 Results

3.1 Image to Label Initial Segmentation

Dice scores for *U-Net 1* are shown in Fig. 4 for the testing dataset of 30 cases. The network performed well on RV, LA and RA cavities, with median Dice scores above 90%. The network performed less well on LV cavity, LVMyo and AA, with lower median values and higher variations of the Dice scores. For LV and LVMyo, the Dice scores reflect a larger variation around the LV endocardial surface across scanner types, resulting in both a reduction in the overall similarity and a larger variation of the Dice scores. The LVMyo volume was smaller and therefore showed lower Dice than LV cavity. The larger variations of the RV and AA scores reflect the effect of PA and descending aorta variations.

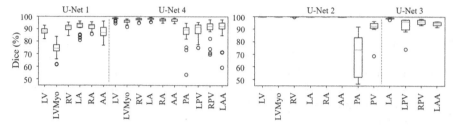

Fig. 4. Dice scores for test cases of image-to-label networks (*U-Net 1* and *U-Net 4*, n = 30) and label-to-label networks (*U-Net 2* n = 9, and *U-Net 3*, n = 6).

3.2   Label to Label Refinement

Dice scores for the label-to-label U-Nets (*U-Net 2* and *U-Net 3*) are shown in Fig. 4. *U-Net 2* extrapolated from 6 labels to 7 labels, and the 6 input labels were almost identically reproduced by the network with Dice scores all >99%. The PA Dice scores were relatively lower reflecting the large variation of PA cropping. However, the goal of the PA label was to identify the PAV, and the resulting PAV Dice was very good considering it has a thickness of just two voxels. The outliers of PAV Dice in *U-Net 2* represent cases with incomplete PAV caused by the limited scanning field of view, which was common within SCOT-HEART. The initial segmentations were constrained by the image size, however, the extrapolation approach mitigated this problem by extrapolating the RV outflow tract as shown in Fig. 5.

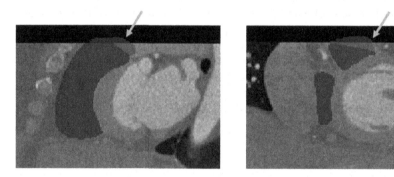

Fig. 5. Coronal and sagittal views of the CCTA image overlapped with reconstructed RV (blue) and PA (red) labels from U-Net 2 predictions for the outlier testing case. The yellow arrows indicate the predicted PA label voxels outside the scanning field of view (Color figure online).

U-Net 3 mapped the initial LA label to LA body, LPV, RPV and LAA labels. The Dice scores showed good similarity between the reference and the predicted labels, especially for LA body. The Dice score was affected by the size of the object, and therefore the Dice scores for LPV, RPV and LAA were relatively lower.

3.3 Image to Label Final Segmentation

Dice scores for *U-Net 4*, calculated against the 10-label predictions for the manually reviewed test cases, are shown in Fig. 4. The median Dice scores are all above 95% for 6 initial labels, with the largest value above 98% for LV cavity. The mean values of Dice scores are also between 95% and 98% with the standard deviations between 1% and 2%. The reduction of outliers suggests that the consistency of the segmentations across scanner types was clearly improved through our refinement pipeline, including LV cavity adjustment, PA extrapolation and LA parcellation. The final segmentation network could also accurately predict the refined labels from CCTA images showing the manual annotations were highly correlated to the image features. As expected, the Dice scores for small regions are not as good as big regions, and PA has the lowest median Dice score (89%) while LPV, RPV and LAA all have median Dice score above 90%. The larger variations and more outliers reflected images with partially cropped field of view superior to the LV. Compared with previous reports [3, 4, 8, 9], our results show high performance even in smaller regions. We also evaluated *U-Net 1* and *U-Net 4* using the testing dataset of the MMWHS challenge, and compared to the top-5 ranked participants of the challenge for CT image segmentation [5]. The Dice scores are shown in Table 1. The Dice score of *U-Net 1* is smaller because it is trained using automatically generated segmentation as the reference, and there is an obvious improvement after our refinement process, giving the performance of *U-Net 4* similar to the best performed challenge participants.

Table 1. Dice on the CT test datasets of the MMWHS challenge for U-Net 1, U-Net 4, and the top-5 ranked participants. The values are the mean of 40 cases in %.

	LV	LVMyo	RV	LA	RA
U-Net 1	88.0	81.5	83.4	81.6	82.2
U-Net 4	90.1	84.7	89.2	91.7	87.7
1	91.8	88.1	90.9	92.9	88.8
2	92.3	85.6	85.7	93.0	87.1
3	90.4	85.1	88.3	91.6	83.6
4	90.1	84.6	85.6	88.4	83.7
5	90.8	87.4	80.6	90.8	85.5

4   Conclusions and Limitations

In this paper we described a flexible method for refining cardiac segmentations using a multi-stage process, by training both image-to-label and label-to label networks to learn task-dependent manual corrections in an iterative fashion. The label-to-label networks were able to learn the anatomical configuration of the refined label maps using only mask information by either extrapolating new labels from background (U-Net 2) or partitioning existing labels (U-Net 3). Taking the advantage of neural networks, which provided a simple way to refine or adapt segmentations to suit the application, we applied this method to the problem of identifying the PAV, as well as the locations of the pulmonary veins and LAA with the LA.

Limitations of this study include i) no cases had metal artefacts which are often present in CCTA, and ii) the PAV could be distorted in some cases due to restricted field of view (Fig. 5) – this could be corrected in future work by manually extrapolating the RV in addition to the LA for subsequent label-to-label refinement.

Acknowledgements. This research was supported by the UKRI London Medical Imaging & Artificial Intelligence Centre for Value Based Healthcare, and core funding from the Wellcome/EPSRC Centre for Medical Engineering [WT203148/Z/16/Z]. We thank Siemens Healthineers for allowing us to use the AXseg prototypical software during this research. SCOT-HEART was funded by The Chief Scientist Office of the Scottish Government Health and Social Care Directorates (CZH/4/588), with supplementary awards from Edinburgh and Lothian's Health Foundation Trust and the Heart Diseases Research Fund. MCW and DEN are supported by the British Heart Foundation FS/ICRF/20/26002 and CH/09/002.

References

1. Knuuti, J., et al.: 2019 ESC Guidelines for the diagnosis and management of chronic coronary syndromes. Eur. Heart J. **41**(3), 407–477 (2020)
2. Hoogendoorn, C., et al.: A high-resolution atlas and statistical model of the human heart from multislice CT. IEEE Trans. Med. Imaging **32**(1), 28–44 (2013)
3. Zhuang, X., et al.: Evaluation of algorithms for multi-modality whole heart segmentation: an open-access grand challenge. Med. Image Anal. **58**, 101537 (2019)
4. Baskaran, L., et al.: Identification and quantification of cardiovascular structures from CCTA: an end-to-end, rapid, pixel-wise, deep-learning method. JACC Cardiovasc. Imaging **13**(5), 1163–1171 (2020)
5. Payer, C., Štern, D., Bischof, H., Urschler, M.: Multi-label Whole Heart Segmentation Using CNNs and Anatomical Label Configurations. In: Pop, M., et al. (eds.) STACOM 2017. LNCS, vol. 10663, pp. 190–198. Springer, Cham (2018). https://doi.org/10.1007/978-3-319-75541-0_20
6. Williams, M.C., et al.: Low-attenuation noncalcified plaque on coronary computed tomography angiography predicts myocardial infarction: results from the multicenter SCOT-HEART trial (Scottish Computed Tomography of the HEART). Circulation **141**(18), 1452–1462 (2020)
7. Zheng, Y., Barbu, A., Georgescu, B., Scheuering, M., Comaniciu, D.: Four-chamber heart modeling and automatic segmentation for 3-D cardiac CT volumes using marginal space learning and steerable features. IEEE Trans. Med. Imaging **27**(11), 1668–1681 (2008)
8. Leventić, H., et al.: Left atrial appendage segmentation from 3D CCTA images for occluder placement procedure. Comput. Biol. Med. **104**, 163–174 (2019)
9. Jin, C., et al.: Left atrial appendage segmentation using fully convolutional neural networks and modified three-dimensional conditional random fields. IEEE J. Biomed. Health Inform. **22**(6), 1906–1916 (2018)

Optimisation of Left Atrial Feature Tracking Using Retrospective Gated Computed Tomography Images

Charles Sillett[1]([⊠]), Orod Razeghi[1], Marina Strocchi[1], Caroline H. Roney[1], Hugh O'Brien[1], Daniel B. Ennis[2], Ulrike Haberland[3], Ronak Rajani[1,4], Christopher A. Rinaldi[1,4], and Steven A. Niederer[1]

[1] School of Biomedical Engineering and Imaging Sciences, King's College London, London, UK
charles.sillett@kcl.ac.uk
[2] Department of Radiology, Stanford University, Stanford, CA, USA
[3] Siemens Healthcare Limited, Camberley, UK
[4] Cardiology Department, Guy's and St Thomas' NHS Foundation Trust, London, UK

Abstract. Retrospective gated cardiac computed tomography (CCT) images can provide high contrast and resolution images of the heart throughout the cardiac cycle. Feature tracking in retrospective CCT images using the temporal sparse free-form deformations (TSFFDs) registration method has previously been optimised for the left ventricle (LV). However, there is limited work on optimising nonrigid registration methods for feature tracking in the left atria (LA). This paper systematically optimises the sparsity weight (SW) and bending energy (BE) as two hyperparameters of the TSFFD method to track the LA endocardium from end-diastole (ED) to end-systole (ES) using 10-frame retrospective gated CCT images. The effect of two different control point (CP) grid resolutions was also investigated. TSFFD optimisation was achieved using the average surface distance (ASD), directed Hausdorff distance (DHD) and Dice score between the registered and ground truth surface meshes and segmentations at ES. For baseline comparison, the configuration optimised for LV feature tracking gave errors across the cohort of 0.826 ± 0.172 mm ASD, 5.882 ± 1.524 mm DHD, and 0.912 ± 0.033 Dice score. Optimising the SW and BE hyperparameters improved the TSFFD performance in tracking LA features, with case specific optimisations giving errors across the cohort of 0.750 ± 0.144 mm ASD, 5.096 ± 1.246 mm DHD, and 0.919 ± 0.029 Dice score. Increasing the CP resolution and optimising the SW and BE further improved tracking performance, with case specific optimisation errors of 0.372 ± 0.051 mm ASD, 2.739 ± 0.843 mm DHD and 0.949 ± 0.018 Dice score across the cohort. We therefore show LA feature tracking using TSFFDs is improved through a chamber-specific optimised configuration.

Keywords: Left atrial feature tracking · Retrospective gated computed tomography · Atrial fibrosis

© Springer Nature Switzerland AG 2021
D. B. Ennis et al. (Eds.): FIMH 2021, LNCS 12738, pp. 71–83, 2021.
https://doi.org/10.1007/978-3-030-78710-3_8

1   Introduction

Background: Atrial fibrillation (AF) is the most common arrhythmia, and leads to an increased risk of stroke, heart failure, and mortality [2]. Histological studies have shown atrial fibrosis to be a hallmark of atrial remodelling that initiates and sustains AF, and its extent can indicate AF presence and persistence [7,12]. Current non-invasive measures of atrial fibrosis rely on late gadolinium enhancement (LGE) cardiac magnetic resonance (CMR) imaging. However, this technique remains controversial and its reproducibility is contested. Atrial strain may provide a measure of fibrotic changes in local biomechanics that can be used for estimating atrial fibrosis. However, there are limited high resolution methods optimised for measuring atrial strain in the complex three dimensional atrial anatomy.

Previous quantification of atrial strain has involved the tracking of features and speckles in CMR imaging and echocardiography techniques, respectively [4]. These methods are limited by suboptimal resolution of the thin wall and complex anatomy of the LA. Furthermore, large variation of strain measurements are reported across scanners. We propose to use high resolution contrast enhanced retrospective gated CCT to capture local atrial strain that could be used to identify LA fibrosis by quantifying akinetic regions. This requires a concomitant improvement in atrial feature tracking methods using CCT images. Improvement in the temporal resolution of dual-source CT scanners has made cardiac motion tracking using CT possible, and offers an alternative for patients contraindicated from LGE CMR imaging due to implanted devices or severe kidney failure.

Previous Work: Feature tracking using CCT images has been used to quantify regional deformation of the left ventricle (LV). Pourmorteza et al. calculated the local area strains, terming it "SQUEEZ", in the LV endocardial surface deformed from end-diastole (ED) to end-systole (ES) using the coherent point drift method [5,8,9]. More recently, temporal sparse free-form deformations (TSFFDs) have been optimised to track the LV using 10 and 20-frame retrospective gated CCT images [10,13]. Circumferential, longitudinal and local area strains were calculated and validated against manual anatomical annotations for each time frame. It was found that the TSFFD hyperparameters used for ventricular registration of CMR images needed to be optimised for CCT images [10].

In this paper, we optimise the TSFFD method with respect to the sparsity weight (SW) and bending energy (BE) hyperparameters to track LA features from ED to ES using retrospective gated CCT images for 10 patients. We do this at two sets of control point (CP) resolutions, a coarser and finer set of resolutions, to investigate if accurate LA feature tracking requires higher resolution registration than accurate LV feature tracking. We use the TSFFD method as it achieves spatially and temporally consistent nonrigid registration of features between a reference image and multiple subsequent image frames within several minutes, in addition to its wide accessibility and use in feature tracking in CMR and CCT images. A point set transformation algorithm is combined with the registration to transform LA surface meshes and segmentations from ED to ES. Semi-automatically generated surface meshes and segmentations at ES serve as a ground truth to evaluate the registered output, and the quality of the feature tracking is evaluated using global surface and volumetric similarity metrics.

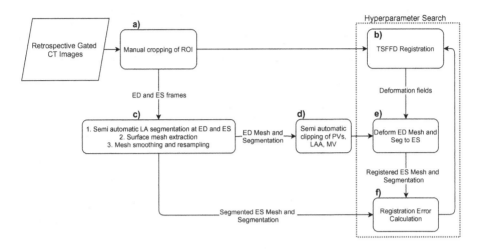

Fig. 1. Workflow diagram. (a) Ten retrospective gated CT images were acquired per patient and cropped to include the LA body and PVs. (b) The 9 subsequent time frames were registered with respect to the initial ED frame using the TSFFD method. (c) The LA bloodpool was segmented at the ED and ES frames using a semi-automatic method provided by CemrgApp. Corresponding surface meshes were then extracted and smoothed, and the edge lengths were resampled. (d) A semi-automatic method was used to clip the ED surface mesh and segmentation. (e) The clipped ED mesh and segmentation were deformed to the ES frame using the calculated deformation field from the registration step. (f) The differences between the ES surface meshes and segmentations registered from ED and the semi-automatic meshes and segmentations at ES was quantified using the ASD, DHD and Dice Score. Steps (b), (e) and (f) were repeated for different combinations of the sparsity weight (SW) and bending energy (BE) parameters that define TSFFD. Abbreviations: PVs - pulmonary veins, LAA - left atrial appendage, MV - mitral valve.

2   Methods

The image processing, LA segmentation and mesh generation steps prior to the registration step are depicted in Fig. 1. Semi-automatic LA segmentations were created at the ED and ES frames using the region growing tool and the medical imaging toolkit (MITK) 3D interpolation method implemented in our open-source platform, CemrgApp [11] (www.cemrgapp.com). Corresponding surface meshes were extracted and smoothed using the medical image registration toolkit (MIRTK), and resampled to have minimum and maximum edge lengths of 0.02 mm and 0.6 mm respectively using meshtool [6]. Prior to deformation to the target ES frame, the PVs, LAA and mitral valve (MV) of the ED segmentations and meshes were clipped using CemrgApp's automatic clipping tool, and manually corrected using Paraview. This ensured the feature tracking of the atrial body is optimised.

The repeated process of registration, deformation and evaluation of the registration from ED to ES for different TSFFD hyperparameters is outlined by the dotted box in Fig. 1. The registration output is used to transform the clipped ED segmentation and mesh to the target ES frame, giving the registered ES segmentation and mesh. These are compared with the mesh and segmentation from the target ES frame to evaluate the feature tracking of the LA endocardium. Two surface dissimilarity metrics were used: average surface distance (ASD) which takes the mean distance along the cell normals of the registered ES mesh to the segmented ES mesh; and the directed Hausdorff distance (DHD) which takes the maximal minimum distance between points in the registered and segmented ES meshes. The DHD, h, is defined:

$$h(R, S) = \max_{r \in R} \left[\min_{s \in S} \left[d(r, s) \right] \right] \tag{1}$$

where r and s are points of the registered ES surface mesh, R, and the segmented ES surface mesh, S, respectively, and $d(r, s)$ represents the Euclidean distance between points r and s. The directive nature of both surface dissimilarity metrics circumvents the need to clip the segmented ES mesh, and avoids possible artefacts due to inconsistent PV, LAA or MV clipping. The Dice score was calculated between the registered segmentations from ED and the semi-automatic segmentation at ES to provide a volumetric similarity measure. Both segmentations were clipped to cover the LA body only.

How does TSFFD Work? The TSFFD method achieves nonrigid registration between images by embedding the source image in a 3 dimensional grid of CPs, and translating CPs according to a deformation field to align the source image with the target image. The CP spacing defines the resolution of the registration. Alignment is achieved by minimising a dissimilarity metric based upon differences in image voxel intensities. The deformation field is parameterised using cubic B-splines functions to ensure smooth deformation in space and time, and the motion is cyclically modelled. Multi-level CP resolutions are used to ensure robust global and local registration, and a sparsity constraint encourages multi-level sparse grouped active CPs. In this study we employ the TSFFD method to utilise all 10 retrospective gated CCT images to construct temporally consistent deformation fields from ED to each of the subsequent time frames.

The TSFFD registration was evaluated for 63 combinations of the bending energy (BE) and sparsity weight (SW) hyperparameters for each case. As the two main hyperparameters that define the TSFFD registration, the BE quantifies the cost of spatial deformation in the registration, and the SW parameterises the sparsity constraint of the number of active control points. These hyperparameters were optimised for LV feature tracking in [10]. Nine values for the BE, including the LV optimised value, were tested spanning four orders of magnitude. Six values for the SW, including the LV optimised value, were tested spanning two orders of magnitude; additionally a SW of zero was tested to investigate the effect of the sparsity constraint. Two sets of CP resolutions were tested: a coarser set of resolutions previously used for optimised LV feature tracking referred to as 'LV-Res', and a finer set of resolutions referred to as 'Hi-Res', as described in Table 1. The TSFFD registration was optimised for a patient cohort of 10 cases, whose demographics and image details are shown in Table 2.

Table 1. Control point (CP) resolutions tested for the TSFFD method. All resolution levels are given in units of [mm × mm × mm].

CP resolution	Level 1	Level 2	Level 3	Level 4
LV-Res	$5 \times 5 \times 5$	$10 \times 10 \times 10$	$20 \times 20 \times 20$	$40 \times 40 \times 40$
Hi-Res	$2 \times 2 \times 2$	$5 \times 5 \times 5$	$10 \times 10 \times 10$	$20 \times 20 \times 20$

Table 2. Patient cohort demographics and image details. Abbreviations: ECG - electrocardiogram, RV - right ventricle, LBBB - left bundle branch block, BiV - biventricular, SR - sinus rhythm.

Case	Age	Sex	NYHA Class	ECG Rhythm	CT resolution [mm × mm × mm]	Slice dimension
Patient 1	83	M	2	RV paced	$0.44 \times 0.44 \times 1.0$	$512 \times 512 \times 127$
Patient 2	66	M	3	LBBB	$0.49 \times 0.49 \times 0.4$	$512 \times 512 \times 359$
Patient 3	79	F	3	SR	$0.49 \times 0.49 \times 0.4$	$512 \times 512 \times 329$
Patient 4	50	M	3	LBBB	$0.38 \times 0.38 \times 0.4$	$512 \times 512 \times 339$
Patient 5	63	M	3	BiV paced	$0.38 \times 0.38 \times 0.5$	$512 \times 512 \times 399$
Patient 6	85	M	3	RV paced & LBBB	$0.49 \times 0.49 \times 1.0$	$512 \times 512 \times 348$
Patient 7	38	M	2	Paced & LBBB	$0.41 \times 0.41 \times 0.4$	$512 \times 512 \times 288$
Patient 8	69	F	2	RV paced & LBBB	$0.38 \times 0.38 \times 0.5$	$512 \times 512 \times 367$
Patient 9	41	M	3	RV paced & LBBB	$0.39 \times 0.39 \times 0.5$	$512 \times 512 \times 272$
Patient 10	76	M	2	RV paced & LBBB	$0.40 \times 0.40 \times 0.5$	$512 \times 512 \times 338$

3   Results

Comparison of LV-Res and Hi-Res CP resolutions: The distributions of the ASD, DHD and Dice score between the registered and segmented ES meshes with respect to the hyperparameter search are shown in Fig. 2. The three error metrics provide a global evaluation of the quality of the registration, and are shown for both the coarser LV-Res and finer Hi-Res CP resolutions. The latter achieved generally superior results for all error metrics across cases, shown by the generally smaller medians for ASD and DHD distributions and larger medians for Dice score distributions per case. Only the median DHD error for Case 2 worsened from LV-Res to Hi-Res, however the magnitude of error increase was small (0.023 mm) and both the ASD and Dice score for Case 2 showed improvement at the higher resolution.

Comparison of Case Specific, Population and LV Optimised Registrations: The case specific optimisations corresponded to the hyperparameter combinations that gave the minimum error per case. As we used three error metrics, we obtained three separate BE SW combinations per case. Depicted by the ASD and DHD minima and Dice score maxima in Fig. 2, the case specific optimisation errors improved upon the LV optimisation errors across all cases for both resolutions, with the largest improvements in ASD of 0.151 mm (Case 9), in DHD of 1.646 mm (Case 5) and in Dice score of 0.013 (Case 5) at the LV-Res resolution. At the Hi-Res resolution, the largest improvements were: 0.077 mm in ASD (Case 4), 0.674 mm in DHD (Case 3) and 0.011 in Dice score (Case 9).

The population optimisation corresponded to the hyperparameter combination that gave the optimal summed error across cases. Three population optimisations were obtained, one for each error metric. The population optimisation errors, displayed by the blue scatter points in Fig. 2, improved upon the LV optimisation errors for most cases, but to a lesser extent than the case specific optimisation, as expected. Due to the variation of anatomy in the patient cohort and the different sensitivities of the error metrics, the population optimisation errors occasionally were worse than the LV optimisation errors for some cases. However, no case's population errors were worse than the LV errors across all three error metrics.

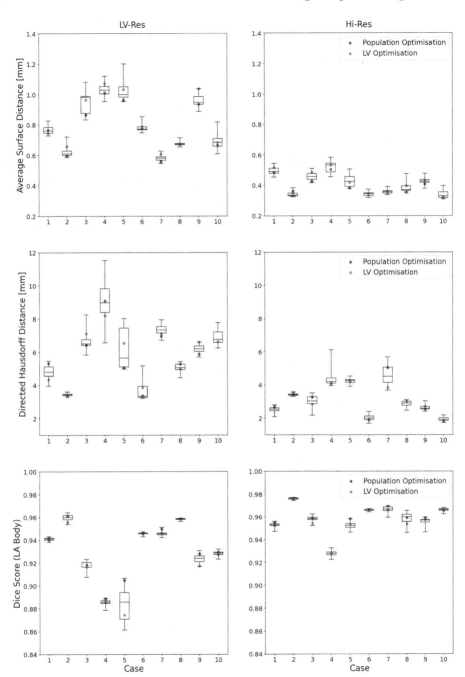

Fig. 2. Error metrics for LA registration quality at ES for two different CP resolutions: LV-Res (left column) and Hi-Res (right column). The distributions of the ASD, DHD and Dice Score with respect to the 63 tested hyperparameter combinations for the TSFFD method are shown for each case. The blue and orange scatter points represent the errors when using the population and LV optimised hyperparameters, respectively. (Color figure online)

The mean and standard deviation of the error distributions for the case specific, population and LV optimisations across cases are shown in Tables 3 and 4 for the LV-Res and Hi-Res resolutions, respectively. There is consistent increased improvement, shown by the mean of all three error metrics across cases, in the population and case specific optimisations upon the LV optimisation. The increased CP resolution has the effect of approximately halving the ASD and DHD errors, and increasing the Dice score by 3%.

Hyperparameter Combinations: Optimal case specific hyperparameter combinations for the TSFFD method varied across cases and between the three error metrics. Hyperparameter values, stated as (BE, SW), that gave population optimisations according to the ASD error metric were $(1 \times 10^{-6}, 3 \times 10^{-3})$ for LV-Res and $(1 \times 10^{-6}, 9 \times 10^{-3})$ for Hi-Res. These compared to the original hyperparameters for optimised LV tracking of $(4 \times 10^{-7}, 9 \times 10^{-3})$.

Table 3. LV-Res error metrics for different hyperparameter optimisations.

Hyperparameter optimisation	ASD [mm]	DHD [mm]	Dice
LV	0.826 ± 0.172	5.882 ± 1.524	0.912 ± 0.033
Population	0.781 ± 0.150	5.625 ± 1.558	0.917 ± 0.028
Case specific	0.750 ± 0.144	5.096 ± 1.246	0.919 ± 0.029

Table 4. Hi-Res error metrics for different hyperparameter optimisations.

Hyperparameter optimisation	ASD [mm]	DHD [mm]	Dice
LV	0.413 ± 0.073	3.070 ± 0.799	0.945 ± 0.020
Population	0.394 ± 0.058	3.008 ± 0.830	0.947 ± 0.020
Case specific	0.372 ± 0.051	2.739 ± 0.843	0.949 ± 0.018

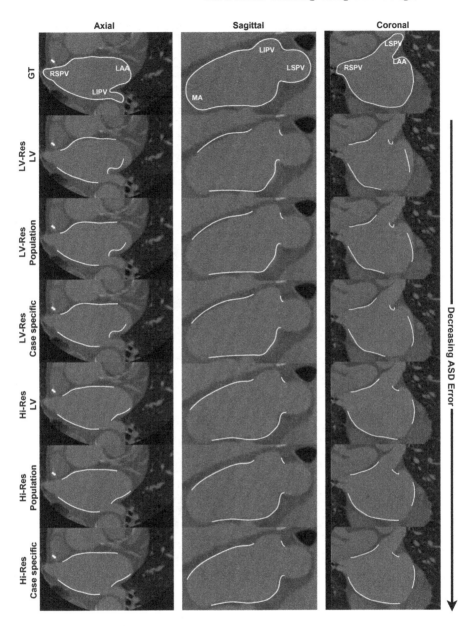

Fig. 3. Qualitative assessment of the ES registered mesh at optimal LV, population and case-specific hyperparameters for LV-Res (2nd-4th rows) and Hi-Res (5th-7th) control point resolutions. The ground truth (GT) is shown in the 1st row. The registered surface mesh outline is overlaid upon axial (1st column), sagittal (2nd column) and coronal (3rd column) slices of the ES frame for Case 3. Optimal case and population shown were those which minimised the ASD metric. For the LV-Res results, the population and case specific hyperparameters track the LA wall below the LAA better than the LV hyperparameters, which is best shown in the coronal slice. The increased resolution for the Hi-Res rows improves the tracking between the LAA and LIPV, and the RSPV and LAA (best seen in axial slice) and provides notably better tracking in the coronal slice.

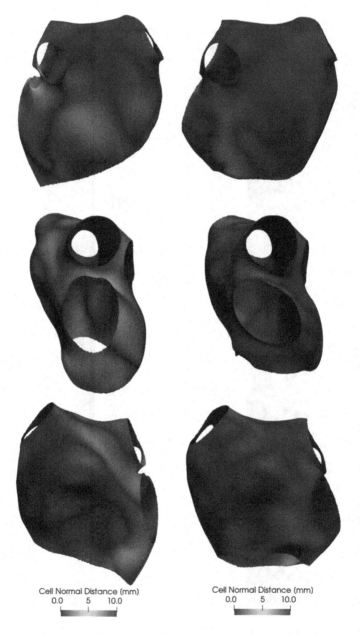

Fig. 4. Two registered meshes at ES for case specific optimisations for Case 3 for LV-Res (left) and Hi-Res (right). The colour-map corresponds to the normal distance of each cell to the segmented ground truth mesh. Red regions are the furthest from the reference mesh and blue regions are closest. The pink region in the top left view showed very poor registration, with cell normal distances greater than 10 mm. The largest distances occur in the regions between the LAA and LSPV and between the LAA and MA. The Hi-Res optimisation shows improved global tracking. (Color figure online)

4   Discussion

Comparison of Dissimilarity Metrics. A study for benchmarking algorithms for LA body segmentation using CCT and CMR images found that the ASD metric best captured the accuracy of LA body delineation [14]. We therefore used the ASD metric to define our case specific and population optimisations in this study. However, we complemented ASD with the DHD and Dice scores to evaluate our registration quality from ED to ES. A universal score that takes into account all three error metrics could be beneficial for identifying case specific and population optimisation hyperparameters.

Comparison with Optimal LV Hyperparameters: Comparing the ASD population optimised hyperparameters to those for optimal LV feature tracking, the BE for both CP resolutions is greater than the LV value within an order of magnitude. The LV-Res SW is smaller than the LV optimised SW by a factor of 3, whereas the Hi-Res SW shows no difference. A weaker sparsity constraint for the LV-Res resolution suggests that more active control points are needed in the registration step to accurately track the LA wall features. This could be satisfied by the Hi-Res resolution, as the finer resolution approximately doubled the number of active CPs.

This may intuitively make sense as the LA is smaller in size compared with the LV, and so its deformations from ED to ES necessitate higher resolution registration. The colour-maps in Fig. 4 show local normal distance errors on the registered ES mesh for case specific optimisations at both resolutions. The largest errors occur at the coarser LV-Res resolution as expected, in regions between the LAA and the MV, and between the LAA and the LSPV. This suggests the features of these regions are hardest to track by the TSFFD method at the LV-Res resolution.

Effect of CP Resolution. This study found that using an increased CP resolution had a greater effect on improving the error metrics than optimising the SW and BE hyperparameters alone. Figure 3 shows very subtle improvements between the different hyperparameter optimisations at the same CP resolution, whereas there is more notable improvement between the two resolutions. An increased CP resolution comes at a cost of increased computational expense and compute time, as the degrees of freedom increase with the number of active CPs. We found that at the Hi-Res resolution, the registration typically took on the order of ~10 min using a laptop CPU with 12 threads and 32 GB memory. The benefit of finer CP resolutions is likely to break down as the CP resolution approaches the image resolution due to high sensitivity to noise.

5   Limitations and Future Work

We optimised the registration hyperparameters in this study by only evaluating similarity metrics between the LA surfaces and volumes. This gives a global validation of feature tracking, but lacks local validation. Augmenting these volumetric and surface error terms with point-to-point error terms of anatomical landmarks would provide more evidence that we were tracking material points, and improve confidence in derived measures of atrial wall motion, such as atrial strain.

In TSFFD, we considered a single, but widely used and accessible registration method. The evaluation of alternative registration methods, such as DEEDS [3] and deep learning methods such as VoxelMorph [1] to determine the optimal method for tracking atrial features may further enhance the accuracy of our atrial motion measurements.

We conducted this investigation using 10 CCT datasets. It would be beneficial to investigate a larger patient cohort with a greater diversity in age, disease and gender and use data from different centres in order to generalise our findings.

6   Conclusion

In this paper, we have optimised the bending energy and sparsity weight hyperparameters used in the TSFFD method at two sets of CP resolutions to track the LA endocardium features from ED to ES in CCT images. Case specific and population hyperparameter optimisations improved upon the LV optimised hyperparameters in tracking features of the LA. An increased CP resolution and optimised hyperparameters further improved tracking performance and showed noticable qualitative improvements in LA wall delineation, which demonstrates the need for TSFFD optimisation according to its application.

References

1. Balakrishnan, G., Zhao, A., Sabuncu, M.R., Guttag, J., Dalca, A.V.: Voxelmorph: a learning framework for deformable medical image registration. IEEE Trans. Med. Imaging **38**(8), 1788–1800 (2019). https://doi.org/10.1109/tmi.2019.2897538
2. Cochet, H., et al.: Age, atrial fibrillation, and structural heart disease are the main determinants of left atrial fibrosis detected by delayed-enhanced magnetic resonance imaging in a general cardiology population. J. Cardiovasc. Electrophysiol **26**(5), 484–492 (2015). https://doi.org/10.1111/jce.12651
3. Heinrich, M.P., Jenkinson, M., Brady, M., Schnabel, J.A.: MRF-based deformable registration and ventilation estimation of lung CT. IEEE Trans. Med. Imaging **32**(7), 1239–1248 (2013). https://doi.org/10.1109/TMI.2013.2246577
4. Lacalzada-Almeida, J., Garciá-Niebla, J.: How to detect atrial fibrosis Journal. J. Geriatric Cardiol. (2017)
5. Myronenko, A., Song, X.: Point set registration: coherent point drifts. IEEE Trans. Pattern Anal. Mach. Intell. **32**(12), 2262–2275 (2010). https://doi.org/10.1109/TPAMI.2010.46
6. Neic, A., Gsell, M.A., Karabelas, E., Prassl, A.J., Plank, G.: Automating image-based mesh generation and manipulation tasks in cardiac modeling workflows using Meshtool. SoftwareX (2020). https://doi.org/10.1016/j.softx.2020.100454
7. Platonov, P.G.: Atrial fibrosis: an obligatory component of arrhythmia mechanisms in atrial fibrillation? (2017). https://doi.org/10.11909/j.issn.1671-5411.2017.04.008

8. Pourmorteza, A., Chen, M.Y., van der Pals, J., Arai, A.E., McVeigh, E.R.: Correlation of CT-based regional cardiac function (SQUEEZ) with myocardial strain calculated from tagged MRI: an experimental study. Int. J. Cardiovasc. Imaging **32**(5), 817–823 (2015). https://doi.org/10.1007/s10554-015-0831-7

9. Pourmorteza, A., Schuleri, K.H., Herzka, D.A., Lardo, A.C., McVeigh, E.R.: A new method for cardiac computed tomography regional function assessment: stretch quantifier for endocardial engraved zones (SQUEEZ). Circ. Cardiovasc. Imaging (2012). https://doi.org/10.1161/CIRCIMAGING.111.970061

10. Razeghi, O., et al.: Hyperparameter optimisation and validation of registration algorithms for measuring regional ventricular deformation using retrospective gated computed tomography images. Nat. Sci. Rep. **11**(1), 5718 (2021)

11. Razeghi, O., et al.: CemrgApp: an interactive medical imaging application with image processing, computer vision, and machine learning toolkits for cardiovascular research. SoftwareX **12** (2020). https://doi.org/10.1016/j.softx.2020.100570

12. Schotten, U., Verheule, S., Kirchhof, P., Goette, A.: Pathophysiological mechanisms of atrial fibrillation: a translational appraisal (2011). https://doi.org/10.1152/physrev.00031.2009

13. Shi, W., et al.: Temporal sparse free-form deformations. Med. Image Anal. **17**(7), 779–789 (2013). https://doi.org/10.1016/j.media.2013.04.010

14. Tobon-Gomez, C., et al.: Benchmark for algorithms segmenting the left atrium from 3D CT and MRI datasets. IEEE Trans. Med. Imaging **34**(7), 1460–1473 (2015). https://doi.org/10.1109/TMI.2015.2398818

Assessment of Geometric Models for the Approximation of Aorta Cross-Sections

Pau Romero[1], Dolors Serra[1], Miguel Lozano[1,2], Rafael Sebastián[1,2],
and Ignacio García-Fernández[1,2(✉)]

[1] Computational Multiscale Simulation Lab (COMMLAB), 46100 Valencia, Spain
[2] Department of Computer Science, Universitat de Valencia, Valencia, Spain
`ignacio.garcia@uv.es`

Abstract. The ellipse can be an appropriate geometry for aorta cross-section fitting on the lumen contour. However, in some regions of the aorta, such as the Sinuses of Valsalva, this approximation can suffer of a relatively high error. Thus, some authors use closed polynomial curves for a better representation of the cross section. This paper presents a detailed comparison between the use of an elliptic cross section model and a spline based model with different number of knots. We use a cohort of 32 thoracic aorta geometries (segmented triangle meshes), obtained using CT scan in the mesosystole phase of the cardiac cycle, for the assessment of both methods. We use the root mean squared error of the fitting of the studied methods to quantify their accuracy. As expected, the spline based model improves the fitting accuracy of the elliptic one and specially in complex aorta cross-sections. However, we have observed that with a high number of knots some cross sections may show high error values due to the adaption of the function to noise.

Keywords: Aorta · Geometric approximation · Segmentation

1 Introduction

Pathologies on the aorta are highly related to the mechanical process that happen therein during the cardiac cycle. Understanding the dynamics of blood flow in the aorta and its relationship to different pathologies such as atherosclerosis or aortic aneurysm can, in turn, help understand and treat these pathologies [5,9,10,14]. The usage of Medical Imaging (MI) such as Computed Tomography (CT) scans or Magnetic Resonance Imaging (MRI) not only provide a unique source of information for diagnosis and clinical decision, but they are also the basis for many *in-silico* studies that range from the creation of population atlases [4,5,8] to patient-specific computer simulation studies [3,6,14]. However, the time required to process the MI to obtain a meaningful representation of the anatomy can suppose a considerable burden to the generalization of computational techniques such as computational fluid dynamics in the clinical practice, and to the execution of large *in-silico* studies. Many methods have been

© Springer Nature Switzerland AG 2021
D. B. Ennis et al. (Eds.): FIMH 2021, LNCS 12738, pp. 84–92, 2021.
https://doi.org/10.1007/978-3-030-78710-3_9

published that aim to completely automate the segmentation of the acquired MI and the generation of a representation of the aorta that is relevant either to the clinical practice or to execute simulation studies [1,2,7,13].

The anatomy of the thoracic aorta can be easily characterized by means of a tubular surface. In this approach, a curve representing the centerline of the aortic lumen is first computed and a sequence of cross sections of the aorta are then obtained [1,8,11,14]. This method is also close to the clinical practice where anatomical analysis of the vessels is often focused on 2D properties such as cross section diameter and area. For the sake of simplicity, many authors use ellipses or even circumferences to fit each cross section to the acquired data. In some cases, ellipses are used to track the aortic lumen and extract its surface [1,7,11,15] or to identify anatomic landmarks [16]. However, other authors use closed polynomial curves both for segmentation and measurement of relevant aorta diameters [13]. In the later case, a decision that has to be made is the number of degrees of freedom that are used for the fitting polynomial. If cubic uniform B-Splines are used, this stands for the number of knots to be used. In this work we present a quantitative analysis of the quality of a surface reconstruction method when approximating the aortic lumen using cubic B-Spline curves for cross-section fitting. We conduct our study using a population of 32 thoracic aortas that suffer of ascending aorta aneurysm.

2   Materials and Methods

We first present the clinical dataset we have used in our study. Then we describe the procedure we have followed to approximate the sections of the aorta by means of both ellipses and closed polynomial curves together with the error function used to assess the quality of each type of approximation.

2.1   Clinical Data

We have used a cohort of 32 thoracic aorta geometries obtained using CT scan in the mesosystole phase of the cardiac cycle and segmented by the physicians at the hospital. The geometries are in the form of triangle meshes representing the aorta wall. From the population of aortas, 6 do not contain the region of the sinuses of Valsalva due to issues during the segmentation process. The cohort was obtained in a retrospective study among patients with ascending aorta aneurysm, with ages ranging from 78 to 89 years old. All the cases have been anonymized.

2.2   Cross-Section Models

When segmenting the thoracic aorta it is common to extract, in the first place, a centerline curve $\alpha : [0, L] \rightarrow \mathbb{R}^3$. This centerline contains the points inside the lumen that maximize the distance to the wall. Then, for each point of the centerline $\alpha(s)$, the cross section of the wall, generated by a plane orthogonal to the tangent to the curve, is identified [11,13,15]. Given a point of the centerline,

$\alpha(s)$, and two vectors $\mathbf{v}_1$ and $\mathbf{v}_2$ forming an orthogonal basis with the tangent at $\alpha(s)$. This cross section can be described as a closed curve in the plane defined by $\mathbf{v}_1$ and $\mathbf{v}_2$.

An ellipse can be a good choice to approximate the shape of the wall after the sino-tubular junction, as it is demonstrated in [11,15]. It has the advantage that it only requires three values: the angle φ of the major semi-axis respect to $\mathbf{v}_1$ and the length of the two semi-axes, r_1 and r_2. Using this definition, the cross section at the centerline point $\alpha(s)$ can be defined as

$$\mathbf{x}(s,\theta) = \alpha(s) + r_1(s)\cos(\theta - \varphi(s))\mathbf{v_1}(s) + r_2(s)\sin(\theta - \varphi(s))\mathbf{v_2}(s). \quad (1)$$

In this formulation, the parameter θ is, indeed, the angle formed by the vector $\mathbf{x}(s,\theta) - \alpha(s)$ with respect to $\mathbf{v}_1$. A more general approximation is to replace the ellipse equation by a differentiable function ρ that depends on θ and represents the distance from $\mathbf{x}(s,\theta)$ to the centerline, leading to a polar representation of the cross section

$$\mathbf{x}(s,\theta) = \alpha(s) + \rho(s,\theta)(\cos(\theta)\mathbf{v_1}(s) + \sin(\theta)\mathbf{v_2}(s)). \quad (2)$$

In this work, we approximate the function $\rho(s,\theta)$ at s with uniform cubic B-Spline depending on θ. We test the fitting with $n = 4, 6, \ldots, 18$ knots. Note that the representation of the closed curve in polar coordinates assumes that the closed curve has a unique value of ρ for each θ. This is equivalent to the cross section of the lumen being a star domain with respect to $\alpha(s)$. This is the segments connecting each point on the wall and $\alpha(s)$ lie inside the domain.

Figure 1 shows some cross sections for one of the aortas, superimposed to the fitted closed curve. Below each cross section the function $\rho(\theta)$ is also presented. The election of *uniform* splines allows a direct comparison between cross sections and between aortas, that would be much harder with non-uniform splines.

2.3   Error Measurement

We consider a sequence of consecutive cross sections and evaluate the error we obtain when fitting each section using the different proposed curves. Given an aorta, we consider a sequence of $N = 1000$ equidistant points along the centerline, $\{\alpha(s_i)\}_{i=0,\ldots,N}$, with $\alpha(s_0)$ located at the aortic valve. Taking the tangent to the centerline curve $\mathbf{t}_i$ at $\alpha(s_i)$, we compute the intersection between the triangle mesh of the aorta geometry and the plane perpendicular to $\mathbf{t}_i$ passing through $\alpha(s_i)$. This task is done using the Visualization Toolkit (VTK) [12]. Using this set of intersecting points we fit the different types of curves for each slice measuring the root mean square error (RMSE).

3   Results

As described in Sect. 2.3, the experiments have been reproduced for uniform B-splines with different number of knots (n). In addition, an elliptic cross section

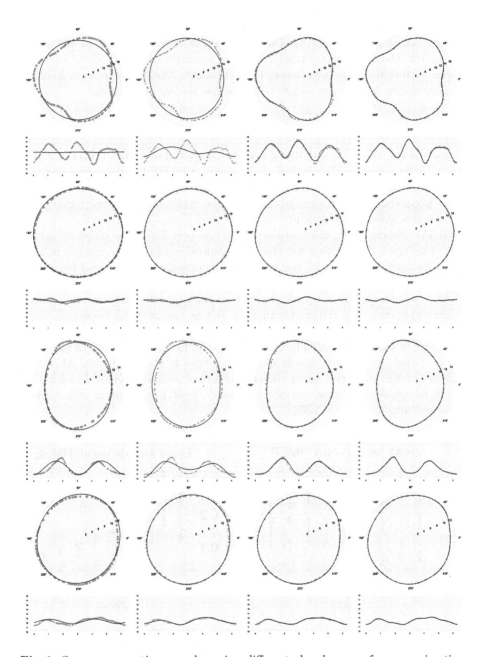

Fig. 1. Some cross section samples using different closed curves for approximation. From left to right, elliptic approximation, and cubic B-Splines with $n = 4$, 8 and 16 knots. From top to bottom, a slice at the Sinuses of Valsalva, the ascending aorta, the aortic arch and at the descending thoracic aorta. Blue dotted line presents points on the wall, black line is the fitting approximation and red lines are the graphical representation of the difference. (Color figure online)

has also been used in order to asses the value added by the generalization to a cubic radius function $\rho(s, \theta)$. First we present the assessment of the error obtained in the slice approximation process for different values of n, followed by the evolution of the error along the centerline. Finally, we present a more detailed insight of the error in the region of the sinuses of Valsalva.

3.1 Slice Fitting Error Distribution

For each aorta, the error distribution in the computed slices has been obtained. In our analysis we consider two values; the highest RMSE for each aorta and the mean RMSE for each aorta, and analyze the distribution of these two magnitudes in our population of 32 aortas. In addition, the evolution of the error along the aorta anatomy is also analyzed.

Figure 2 shows their distribution within the population for different values of n. The error distribution for the slice approximation using ellipses is also shown for comparison. As we can see, both the maximum RMSE and the mean RMSE reduce as the number of the spline knots increases. In the case of the mean RMSE, the interquartile range also narrows as the value of n increases, indicating a clear improvement of the approximation. In the case of the maximum RMSE, on the contrary, we can observe that the interquartile range remains stable and that there is a non-decreasing number of outliers. In some of the aorta samples, this is caused by the existence of spurious vertices in the mesh that, when the number of knots increases, cause the appearance of oscillations in ρ. In addition, a reduced number of cross sections presented slight folds leading to non star domains, causing a high RMSE.

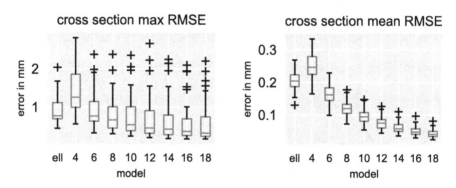

Fig. 2. Distribution of the highest RMSE for a slice in the population of aortas (left) and distribution for the average RMSE for the different fitting models.

To evaluate how different anatomic regions are affected by error, in Fig. 3 the maximum and average RMSE are shown for 20 slices along the centerline. Since each aorta has a different total length, the x axis has been normalized in the range $x \in [0, 1]$. It is clear that the distribution of the RMSE is not independent

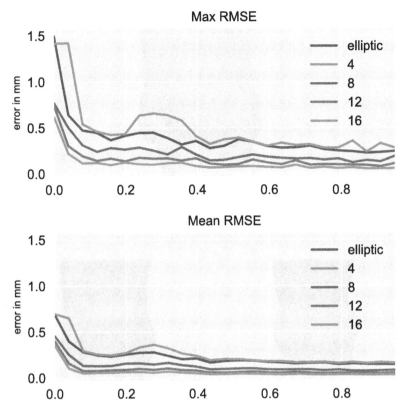

Fig. 3. The maximum (left) and average error for 20 cross sections equidistant along the length of each aorta. The x axis has been normalized.

of the section of the aorta; while in the second half of the aorta the error is low in all cases, regions such as the aortic root, that are less circular, lead to greater RMSE values when approximated with ellipses or splines with fewer number of knots. The region with lower error values between $x = 0$ and $x = 0.2$ corresponds to the region of the aneurysm in the ascending aorta. The cross sections shown in Fig. 1 correspond to a section near $x = 0$, a section near $x = 0.1$, a section around $x = 0.3$ and a section after $x = 0.5$, from top to bottom.

It can be seen that the ascending aorta, the aortic arch and the descending aorta can be approximated as tube-like regions with cross sections close to an ellipse, while the behavior in the Sinuses of Valsalve is substantially different. In that region, ellipses, which can only reproduce two equal local maxima and minima, lead to a poor approximation, while a spline is able to adapt to the irregularities of the contour of the cross section. Figure 4 shows a reconstruction of the aortic root with the elliptic and B-spline with $n = 4$, 8 and 16 knots models. The RMSE is mapped in the color of the meshes with an appropriate scale for

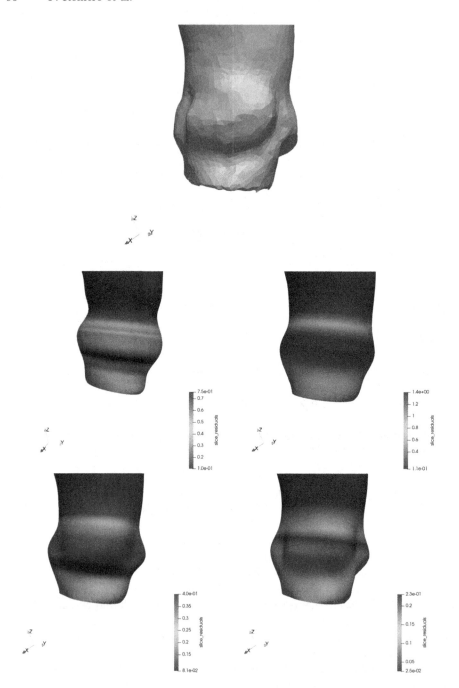

Fig. 4. Comparison of the aortic root reconstructed with different models for one of the aortas. From left to right, and top to bottom; original geometry, elliptic approximation, and cubic B-Splines with $n = 4$, 8 and 16 knots. The reconstructed meshes are colored with the RMSE of each slice. Note the variation of the color scale. (Color figure online)

each one. The elliptic model generates an uniform bulging may underestimate the maximum dilation whilst the splines are capable of reproducing the three sinuses if enough knots are used.

4 Conclusion

In this paper, the accuracy of different models for cross sections of the thoracic aorta have been tested. B-Spline models, with enough number of knots, have proven to be more accurate than an elliptic model. It is noteworthy that ellipses perform even better than B-splines with 4 knots. However, as the number of knots increases, the fitting of anatomical details are better approximated, specially in the aortic root and the Sinuses of Valsalva. Nonetheless, even though the average error clearly reduces as n increases, there are some slices that lead to higher error if too many degrees of freedom are used. A good compromise can be between $n = 8$ and $n = 14$ depending on the properties of the population, leading to a good approximation which is more robust to noise and artifacts inherited from the segmentation process.

References

1. Alvarez, L., et al.: Tracking the aortic lumen geometry by optimizing the 3D orientation of its cross-sections. In: Descoteaux, M., Maier-Hein, L., Franz, A., Jannin, P., Collins, D.L., Duchesne, S. (eds.) MICCAI 2017, Part II. LNCS, vol. 10434, pp. 174–181. Springer, Cham (2017). https://doi.org/10.1007/978-3-319-66185-8_20
2. Antiga, L., Steinman, D.A.: Robust and objective decomposition and mapping of bifurcating vessels. IEEE Trans. Med. Imaging **23**(6), 704–713 (2004). https://doi.org/10.1109/TMI.2004.826946
3. Bianchi, M., et al.: Patient-specific simulation of transcatheter aortic valve replacement: impact of deployment options on paravalvular leakage. Biomech. Model. Mechanobiol. **18**(2), 435–451 (2018). https://doi.org/10.1007/s10237-018-1094-8
4. Cibis, M., Bustamante, M., Eriksson, J., Carlhäll, C.J., Ebbers, T.: Creating hemodynamic atlases of cardiac 4D flow MRI. J. Magn. Reson. Imaging **46**(5), 1389–1399 (2017). https://doi.org/10.1002/jmri.25691. https://onlinelibrary.wiley.com/doi/abs/10.1002/jmri.25691
5. Cosentino, F., et al.: Statistical shape analysis of ascending thoracic aortic aneurysm: correlation between shape and biomechanical descriptors. J. Pers. Med. **10**(2) (2020). https://doi.org/10.3390/jpm10020028. https://www.mdpi.com/2075-4426/10/2/28
6. Ghosh, R.P., Marom, G., Bianchi, M., D'souza, K., Zietak, W., Bluestein, D.: Numerical evaluation of transcatheter aortic valve performance during heart beating and its post-deployment fluid–structure interaction analysis. Biomech. Model. Mechanobiol. **19**(5), 1725–1740 (2020). https://doi.org/10.1007/s10237-020-01304-9
7. Haj-Ali, R., Marom, G., Zekry, S.B., Rosenfeld, M., Raanani, E.: A general three-dimensional parametric geometry of the native aortic valve and root for biomechanical modeling. J. Biomech. **45**(14), 2392–2397 (2012). https://doi.org/10.1016/j.jbiomech.2012.07.017. http://www.sciencedirect.com/science/article/pii/S0021929012004125

8. Medrano-Gracia, P., et al.: A computational atlas of normal coronary artery anatomy. EuroIntervention : J. EuroPCR Collab. Work. Group Intervent. Cardiol. Eur. Soc. Cardiol. **12**(7), 845–854 (2016). https://doi.org/10.4244/eijv12i7a139

9. Moore, J.E., Xu, C., Glagov, S., Zarins, C.K., Ku, D.N.: Fluid wall shear stress measurements in a model of the human abdominal aorta: oscillatory behavior and relationship to atherosclerosis. Atherosclerosis **110**(2), 225–240 (1994)

10. Nerem, R.M.: Vascular fluid mechanics, the arterial wall, and atherosclerosis. J. Biomech. Eng. **114**(3), 274–282 (1992). https://doi.org/10.1115/1.2891384

11. Romero, P., et al.: Reconstruction of the aorta geometry using canal surfaces. In: International Conference on Computational and Mathematical Biomedical Engineering (2019)

12. Schroeder, W., Martin, K., Lorensen, B.: The Visualization Toolkit: An Object-Oriented Approach to 3D Graphics; [visualize data in 3D - medical, engineering or scientific; build your own applications with C++, Tcl, Java or Python; includes source code for VTK (supports Unix, Windows and Mac)], 4th edn. Kitware Inc., Clifton Park (2006). oCLC: 255911428

13. Sedghi Gamechi, Z., et al.: Automated 3D segmentation and diameter measurement of the thoracic aorta on non-contrast enhanced CT. Eur. Radiol. **29**(9), 4613–4623 (2019). https://doi.org/10.1007/s00330-018-5931-z. https://pubmed.ncbi.nlm.nih.gov/30673817. 30673817[pmid]

14. Shahcheraghi, N., Dwyer, H., Cheer, A., Barakat, A., Rutaganira, T.: Unsteady and three-dimensional simulation of blood flow in the human aortic arch. J. Biomech. Eng. **124**(4), 378–387 (2002). https://doi.org/10.1115/1.1487357

15. Tahoces, P.G., et al.: Automatic estimation of the aortic lumen geometry by ellipse tracking. Int. J. Comput. Assist. Radiol. Surg. **14**(2), 345–355 (2018). https://doi.org/10.1007/s11548-018-1861-0

16. Tahoces, P.G., et al.: Automatic detection of anatomical landmarks of the aorta in CTA images. Med. Biol. Eng. Comput. **58**(5), 903–919 (2020). https://doi.org/10.1007/s11517-019-02110-x

Improved High Frame Rate Speckle Tracking for Echocardiography

Marta Orlowska[1]([✉]) [ID], Alessandro Ramalli[2] [ID], and Jan D'hooge[1] [ID]

[1] Laboratory of Cardiovascular Imaging and Dynamics, Department of Cardiovascular
Sciences, KU Leuven, Leuven, Belgium
marta.orlowska@kuleuven.be
[2] Department of Information Engineering, University of Florence, Florence, Italy

Abstract. High frame rate (HFR) speckle tracking echocardiography (STE)
assesses myocardial function by quantifying motion and deformation at high temporal resolution. Our lab recently proposed a two-step HFR STE methodology
based on 1-D cross-correlation [1]. Even if it was proved to be accurate for a global
assessment of the mid-wall myocardial motion, an impaired sensitivity to motion
and lower feasibilities for the apical regions gave higher errors. Thus, the aim of
this study was to improve the speckle tracking algorithm by improving the tracking
quality in the apical region specifically while preserving tracking quality on the
other segments as well as picking up movement transmurally, i.e. the endocardial,
mid-wall and epicardial motion. Hereto, the original algorithm was modified by
improving robustness and accuracy of the lateral motion estimation. Simulation
results showed that the proposed changes resulted in a significantly lower error in
the estimation of the global longitudinal strain (GLS). Moreover, these improvements were mostly visible in the apical region (e.g. strain error $4.64 \pm 2.65\%$ vs
$1.19 \pm 0.71\%$ for the septum) and prioritized the local movements resulting in
lower error ranges between contours ($[0.2–0.46]\%$ vs $[0.33–1.66]\%$). Finally, for
a qualitative comparison, a preliminary in vivo acquisition was performed.

Keywords: High frame rate · Speckle tracking · Echocardiography

1 Introduction

Myocardial deformation imaging is an important echocardiographic method for the
assessment of global and regional myocardial function by measuring the shortening and
lengthening of the cardiac muscle throughout the heart cycle. Among the techniques
proposed so far, speckle tracking echocardiography (STE) is a semi-automatic tool that
gives angle-independent strain measurements [2]. Unfortunately, since STE is based
on standard B-mode imaging, its operational frame rate is usually lower than 80 Hz.
This relatively low frame rate is sufficient only to describe the gross cardiac motion
and deformation but misses the finer detail occurring at higher temporal resolution. To
overcome this issue, high frame rate (HFR) STE has been developed [1, 3–5], which
is based on non-conventional imaging sequences [6–8]. They are typically based on

© Springer Nature Switzerland AG 2021
D. B. Ennis et al. (Eds.): FIMH 2021, LNCS 12738, pp. 93–100, 2021.
https://doi.org/10.1007/978-3-030-78710-3_10

the transmission of defocused waves, i.e. diverging waves (DW), which together with massive parallel beamforming and coherent compounding in reception can obtain high frame rates (typically around 1 kHz). Among others, our lab recently developed a two-step HFR STE methodology based on 1-D cross-correlation [1, 9]. It was optimized and validated on simulated ultrasound data from 3-D electromechanical models. Then, it was tested using in vivo data and compared to commercially available algorithms proving to be accurate for a global assessment of the mid-wall longitudinal strain (GLS). However, an impaired sensitivity to motion and lower feasibilities for the apical regions gave higher errors.

Therefore, the aim of this study was to improve the speckle tracking algorithm by improving tracking quality in the apical region and picking up transmural movements i.e. the endocardial, mid-wall and epicardial motion of the left ventricle (LV). In particular, the improvements concerned the lateral movement estimator, which typically suffers from higher underestimations. The algorithm was upgraded by subtraction of the mean intensity from the estimation kernel and by including a 2-D weighting window. The performance was assessed on simulations and compared with the originally proposed approach. Finally, a preliminary in vivo acquisition was performed by an expert echocardiographer on a 73-year old amyloidosis patient and the motion of the heart was tracked with both methods.

2   Methods

2.1   Simulation Setup

The imaging setup was implemented in Field II [10]. It was based on a 64-element phased array with 60% bandwidth centered at 3.5 MHz and having a 0.22-mm pitch. The excitation pulse was a single-cycle bipolar pulse at 3.5 MHz (duty cycle = 77%) and the pulse repetition frequency (PRF) was set to 5 kHz. A 90°-wide, 144-line sector over a maximum depth of 13 cm was obtained at a frame rate of 833 Hz by implementing a DW-based scan sequence. Specifically, six DWs were transmitted from 21 elements sliding over the aperture. The virtual sources of the different waves were placed behind the subaperture's center at a depth $z = -2.42$ mm and were shifted on the x-direction in a triangular transmission sequence as described in [11]. Thus, the x-coordinates of the 6 virtual sources were -7.82, -1.53, 4.76, 7.82, 1.53, and -4.76 mm, respectively. Finally, the channel data, sampled at 25 MHz, was captured and reconstructed offline using a compounding scheme. In addition, to test the sensitivity of the method to noise, two datasets with different signal-to-noise ratios (SNR) were generated ($+\infty$ and 10 dB) by adding white Gaussian noise to the postbeamformed data.

Five 3-D electromechanical models of the LV (which are available online [12]) were scanned from an apical view. Specifically, one healthy and four ischemic models were used, corresponding to a distal and proximal occlusion of the left anterior descending artery (LADdist and LADprox, respectively) and an occlusion of the right coronary artery (RCA) and the left circumflex (LCX) [12]. Every model consists of a 3-D cloud of point scatters, which moves according to the electromechanical simulations. However, since the provided positions of the scatterers are sampled at 30 Hz, they were interpolated to 5 kHz using cubic splines for this work.

2.2 2-D Motion Estimator

Original Approach. Our previously proposed 2-D HFR STE algorithm [1] is based on a two-step motion estimation using 1-D cross-correlation on 2-D kernels. First, the axial displacement is estimated on radiofrequency (RF) signals using 7°-wide and 4.5 mm high kernels. A pair of kernels is extracted from the same position on consecutive frames and the 1-D cross-correlation, along the axial direction, is calculated between them. Then, the kernels are moved across the whole frame with overlap of 80% in the axial direction and a lateral shift of 1 line. As the second step, the lateral displacement is estimated on the envelope-detected data. Since the resolution in the lateral direction is lower and the block matching better picks up displacement that is larger than one pixel, the kernels (22°-wide and 7 mm high) are extracted from frames with a time lag of 30 ms. In this case, the axial shift between regions of interest, occurring during the 30 ms lag, is compensated according to the estimation in the previous step. To achieve subpixel precision during both steps of the algorithm, a 10:1 spline peak fitting is applied.

Novel Improved Approach. To further improve the estimation of the lateral displacement, i.e. the most error-prone one, two changes were implemented. In particular, on the envelop data inside each kernel:

1. A tapering Hamming window was applied along both directions – width and height, to limit possible border effects on the cross-correlation;
2. the average value of the pixels was subtracted to obtain positive and negative values and, hence, a narrower response on the cross-correlation function.

Then, the optimal parameters of the algorithm were determined by following the same optimization process used in [1], i.e. by minimizing the displacement estimation error on the simulated data of the healthy heart. The optimal parameters for the improved algorithm, hereinafter called upgraded approach, were: 16°-wide and 10 mm high kernel and time lag of 30 ms.

2.3 Accuracy Estimation

Contour and Spatial Filtering. To investigate the accuracy of the method for local motion estimation, an expert echocardiographer manually placed reference points along the myocardial mid-wall on the frame at end-diastole. The mid-wall contour line was obtained by fitting a 106-point spline curve connecting all reference points, corresponding to an approximate distance of 2 mm between points. Then, two additional contours, i.e. the endocardial and epicardial ones, were automatically generated by offsetting the mid-wall contour by 4 mm towards the inside and the outside of the chamber, respectively. The 4 mm offset was chosen empirically for this study, but it can be adjusted to match the specific morphology of the cardiac wall. Moreover, to improve the robustness of the tracking, eight supporting contours were automatically selected by expanding and contracting each contour on a band of 4 mm (Fig. 1), i.e. each contour was offset by 0.5 mm from the original one. For each location (endocardial, mid and epicardial), all nine contours were tracked separately between all frames and the motion of the main contour was defined as an average position of the ROI. Since the points along the contours were tracked independently, unnatural motions could occur. Hence, a Savitzky-Golay filter (second-order polynomial and an 11-point wide window) was used to smooth the curve that connected all the points that belonged to the contour.

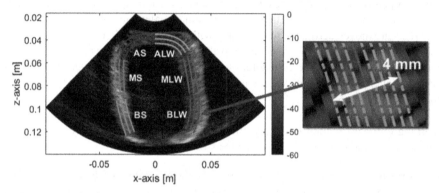

Fig. 1. Endocardial (inner), mid-wall, and epicardial (outer) contour. Each line is obtained as the average position of 8 supporting contour lines (right) to improve the robustness of the algorithm. Each contour is split in six segments: BS – basal septum, MS – mid septum, AS – apical septum, ALW – apical lateral wall, MLW – mid lateral wall and BLW – basal lateral wall.

Global and Segmental Longitudinal Strain. The 3 contours were used for further calculations. Each contour was divided into six segments (BS – basal septum, MS – mid septum, AS – apical septum, ALW – apical lateral wall, MLW – mid lateral wall and BLW – basal lateral wall) according to the recommended guidelines of the European Association of Cardiovascular Imaging (EACVI)/American Society of Echocardiography (ASE) [13]. For each segment, the strain, which is defined as the deformation of a contour, was calculated as follows:

$$S(t) = \frac{L(t) - L_0}{L_0} \qquad (1)$$

where $L(t)$ is the length of the contour at time instance t and L_0 is its initial length, i.e., at end-diastole.

Strain estimates were benchmarked against the reference strain, which was calculated by tracking the contours using the displacement extracted from the position of the scatterers on the heart models. The accuracy of the 2-D motion estimator was computed as the root-mean-square error (RMSE) as follows:

$$\varepsilon = \sqrt{\frac{1}{N} \sum_{i=1}^{N} (R(i) - E(i))^2} \qquad (2)$$

where N is the total number of frames, R is the reference strain and E is estimated strain at frame i.

3 Results

Figure 2 shows the GLS patterns of the healthy heart model obtained with the original and the upgraded methods. This qualitative comparison shows that in each case, the plot estimated by the upgraded method is closer to the reference values and the difference between both methods is bigger in the endocardial tracking than in epicardial. Moreover, in Fig. 2 the endocardial segmental strains suggest that the biggest influence of better tracking lays in the apical region. Figure 3 shows the RMSE values obtained for all heart models and both SNRs as no significant difference was noticed between different cases. In panel (a), the average RMSE for the GLS quantitatively confirms the improvement obtained with the upgraded method. For all three contours, the obtained error is significantly lower for the upgraded method: while the improvement is relatively small for the epicardial contour (0.33% vs 0.2%), it is higher for the mid-wall (0.75% vs 0.33%), and the highest for endocardial contour (1.66% vs 0.46%).

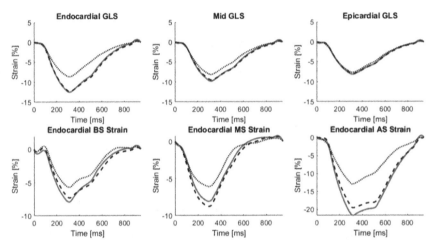

Fig. 2. Example of the reference strain patterns (red, solid) and the one obtained with original (black, dotted) and upgraded (black, dashed) approach in the healthy heart model. The top row shows the endocardial, mid-wall and epicardial GLS, while the bottom row shows the endocardial strain of the basal (BS), mid (MS) and apical (AS) septal segments. (Colour figure online)

Figure 3(b)–(d) shows the distribution of the RMSE values obtained for the three different contours and for the six different segments. For the endocardial contour, a significantly lower error can be observed for both apical and mid segments. The biggest improvement can be observed in the apical region, e.g. the apical septum strain is estimated with an average error of 4.64 ± 2.65% with the original method and 1.19 ± 0.71% with the upgraded one. However, no difference can be observed in the basal segments, e.g. for BLW the error is 1.43 ± 0.35% vs 1.38 ± 0.71% for the original and the upgraded method, respectively. In the mid-wall contour, the same trend can be observed, but with a smaller improvement, 1.86 ± 1.00% vs 0.90 ± 0.66% for AS and 1.27 ± 0.22% vs 1.34 ± 0.49% for BLW. In the epicardial contour only the apical lateral wall and mid

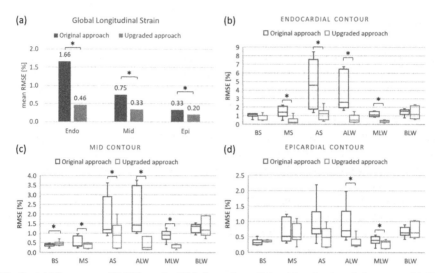

Fig. 3. (a) The average RMSE on the GLS for endocardial, mid and epicardial contour. (b) The distribution of the RMSE for the segmental strain on the endocardial contour. (c) The distribution of the RMSE for the segmental strain on the mid contour. (d) The distribution of the RMSE for the segmental strain on the epicardial contour. * p < 0.05

lateral wall are estimated with a significantly lower RMSE using the upgraded method $(0.93 \pm 0.54\%$ vs $0.33 \pm 0.18\%$ for ALW and $0.67 \pm 0.15\%$ vs $0.73 \pm 0.25\%$ for MLW).

4 Discussion and Conclusions

The aim of this study was to improve our recently proposed speckle tracking algorithm by improving tracking quality in the apical region and picking up local movements, i.e. the endocardial, mid-wall and epicardial motion of the left ventricle. The improvements were implemented to limit possible border effects on the cross-correlation and to obtain a narrower response on the cross-correlation function. The improvements were aimed at improving primarily the lateral motion estimation, i.e. the most error-prone one. The performance was assessed on simulations and compared with the originally proposed approach. The subtraction of mean intensity of pixels around ROI made the peak of the cross-correlation function narrower, thus improved accuracy of the method. Using kernels with a weighting factor allowed prioritizing the movement in the center of the kernel and reducing border effect on the cross-correlation function. These changes resulted in a significantly lower error (~0.91% vs 0.33% for original and upgraded approach respectively). The improvements were most visible in the apical region where the errors were in average four times lower when using the upgraded method. The difference decreases for the mid segments and there is no significant improvement observed for basal segments. This is most probably due to more complex motion (i.e. bigger difference between endocardial and epicardial movement) in the apical part. Nevertheless, GLS on all the contours was estimated with improved accuracy. The difference between GLS error values on the epicardial contour, which is characterized by the global motion of the heart, was small (0.33% for the original method and 0.2% for the upgraded one). However, closer to the endocardial border, the motion and the related strain become bigger and localized. Hence, the GLS of the endocardial contour was estimated with a 3.6 times lower error with the upgraded method (0.46% vs 1.66%). Moreover, this shows that the implemented changes made the method accuracy less dependent on the location of the contour.

Finally, for a qualitative comparison of both the methods, a preliminary in vivo acquisition was performed by an expert echocardiographer on a 73-year old amyloidosis patient. The same imaging setup described in Sect. 2.1 was implemented on the High channel Density Programmable Ultrasound System based on consumer Electronics (HD-PULSE) [14] connected to P2-5AC (Samsung Medison, Seoul, South Korea) probe. Figure 4 shows the position of the tracked contour at the frame during end-systole with the respective GLS curve. It can be seen that even though the difference in the position of the contours seems neglectable, it has a relevant impact on the strain values. Unfortunately, since there is no ground truth present for in vivo recordings, it cannot be demonstrated if the changes are in fact improving the strain computation in this case.

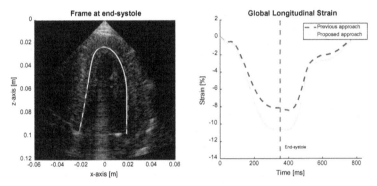

Fig. 4. Example of the position of the contour at the frame during end-systole and the corresponding GLS curve obtained with the upgraded (solid, yellow line) and original method (dashed, red line). (Colour figure online)

In conclusion, the upgraded version of the developed 2-D motion estimator follows closely the motion of the endocardium, improving the tracking accuracy, especially in the apical region, and prioritizing the local movement of the region of interest.

Acknowledgements. This study was supported by The Research Foundation - Flanders (FWO) under Grant G002617N and Grant G092318N.

References

1. Orlowska, M., et al.: A novel 2D speckle tracking method for high frame rate echocardiography. IEEE Trans. Ultrason. Ferroelectr. Freq. Control, 1 (2020). https://doi.org/10.1109/tuffc.2020.2985451

2. D'hooge, J.: Principles and different techniques for speckle tracking. In: Myocardial Imaging: Tissue Doppler and Speckle Tracking, pp. 17–25. Wiley (2008)
3. Grondin, J., Sayseng, V., Konofagou, E.E.: Cardiac strain imaging with coherent compounding of diverging waves. IEEE Trans. Ultrason. Ferroelectr. Freq. Control **64**(8), 1212–1222 (2017). https://doi.org/10.1109/TUFFC.2017.2717792
4. Joos, P., et al.: High-frame-rate speckle-tracking echocardiography. IEEE Trans. Ultrason. Ferroelectr. Freq. Control **65**(5), 720–728 (2018). https://doi.org/10.1109/TUFFC.2018.280 9553
5. Porée, J., Baudet, M., Tournoux, F., Cloutier, G., Garcia, D.: A dual tissue-doppler optical-flow method for speckle tracking echocardiography at high frame rate. IEEE Trans. Med. Imaging **37**(9), 2022–2032 (2018). https://doi.org/10.1109/TMI.2018.2811483
6. Tong, L., Ramalli, A., Jasaityte, R., Tortoli, P., D'hooge, J.: Multi-transmit beam forming for fast cardiac imaging—experimental validation and in vivo application. IEEE Trans. Med. Imaging **33**(6), 1205–1219 (2014). https://doi.org/10.1109/tmi.2014.2302312
7. Denarie, B., et al.: Coherent plane wave compounding for very high frame rate ultrasonography of rapidly moving targets. IEEE Trans. Med. Imaging **32**(7), 1265–1276 (2013). https://doi.org/10.1109/TMI.2013.2255310
8. Hasegawa, H., Kanai, H.: High-frame-rate echocardiography using diverging transmit beams and parallel receive beamforming. J. Med. Ultrason. **38**(3), 129–140 (2011). https://doi.org/10.1007/s10396-011-0304-0
9. Orlowska, M., Ramalli, A., Bézy, S., Meacci, V., Voigt, J.-U., D'hooge, J.: In-vivo comparison of multiline transmission and diverging wave imaging for high frame rate speckle tracking echocardiography. IEEE Trans. Ultrason. Ferroelectr. Freq. Control, 1 (2020). https://doi.org/10.1109/tuffc.2020.3037043
10. Jensen, J.A.: FIELD: a program for simulating ultrasound systems. In: 10th Nordicbaltic Conference on Biomedical Imaging, vol. 4, Supplement 1, Part 1, pp. 351–353 (1996)
11. Poree, J., Posada, D., Hodzic, A., Tournoux, F., Cloutier, G., Garcia, D.: High-frame-rate echocardiography using coherent compounding with doppler-based motion-compensation. IEEE Trans. Med. Imaging **35**(7), 1647–1657 (2016). https://doi.org/10.1109/TMI.2016.252 3346
12. Alessandrini, M., et al.: A pipeline for the generation of realistic 3D synthetic echocardiographic sequences: methodology and open-access database. IEEE Trans. Med. Imaging **34**(7), 1436–1451 (2015). https://doi.org/10.1109/TMI.2015.2396632
13. Voigt, J.-U., et al.: Definitions for a common standard for 2D speckle tracking echocardiography: consensus document of the EACVI/ASE/Industry Task Force to standardize deformation imaging. Eur. Heart J. Cardiovasc. Imaging **16**(1), 1–11 (2015). https://doi.org/10.1093/ehjci/jeu184
14. Ortega, A., et al.: HD-PULSE: high channel density programmable ultrasound system based on consumer electronics. In: 2015 IEEE International Ultrasonics Symposium (IUS), October 2015, pp. 1–3. https://doi.org/10.1109/ultsym.2015.0516

Efficient Model Monitoring for Quality Control in Cardiac Image Segmentation

Francesco Galati and Maria A. Zuluaga[✉][ID]

Data Science Department, EURECOM, Sophia Antipolis, France
{galati,zuluaga}@eurecom.fr

Abstract. Deep learning methods have reached state-of-the-art performance in cardiac image segmentation. Currently, the main bottleneck towards their effective translation into clinics requires assuring continuous high model performance and segmentation results. In this work, we present a novel learning framework to monitor the performance of heart segmentation models in the absence of ground truth. Formulated as an anomaly detection problem, the monitoring framework allows deriving surrogate quality measures for a segmentation and allows flagging suspicious results. We propose two different types of quality measures, a global score and a pixel-wise map. We demonstrate their use by reproducing the final rankings of a cardiac segmentation challenge in the absence of ground truth. Results show that our framework is accurate, fast, and scalable, confirming it is a viable option for quality control monitoring in clinical practice and large population studies.

Keywords: Quality control · Cardiac segmentation · Machine learning

1 Introduction

With the advent of learning-based techniques over the last decade, cardiac image segmentation has reached state-of-the-art performance [5]. This achievement has opened the possibility to develop image segmentation frameworks that can assist and automate (partially or fully) the image analysis pipelines of large-scale population studies or routine clinical procedures.

The current bottleneck towards the large-scale use of learning-based pipelines in the clinic comes from the monitoring and maintenance of the deployed machine learning systems [15]. As shown in [5], despite the very high performances achieved, these methods may generate anatomically impossible results. In clinical practice and population studies, it is of utmost importance to constantly monitor a model's performance to determine when it degrades or fails, leading to poor quality results, as they may represent important risks. A system's continuous performance assessment and the detection of its degradation are challenging after deployment, due to the lack of a reference or ground truth. Therefore,

---

This work was supported through funding from the Monaco Government.

D. B. Ennis et al. (Eds.): FIMH 2021, LNCS 12738, pp. 101–111, 2021.
https://doi.org/10.1007/978-3-030-78710-3_11

translation of models into clinical practice requires the development of monitoring mechanisms to measure a model's segmentation quality, in the absence of ground truth, which guarantee their safe use in clinical routine and studies.

In a first attempt to assess performances of cardiac segmentation models in the absence of ground truth, Robinson et al. [12] proposed a supervised DL-based approach to predict the Dice Score Coefficient (DSC). More recently, Puyol-Antón et al. [11] used a Bayesian neural network to measure a model's performance by classifying its resulting segmentation as correct or incorrect, whereas Ruijsink et al. [14] also use qualitative labels to train a support vector machine that predicts both the quality of the segmentation and of derived functional parameters. The main drawback of these methods is that they require annotations reflecting a large set of quantitative (e.g. DSC) or qualitative (e.g. correct/incorrect) segmentation quality levels, which can be difficult to obtain. The Reverse Classification Accuracy (RCA) [13,18] addresses this problem by using atlas label propagation. This registration-based method relies on the spatial overlap between the predicted segmentation and a reference atlas. It works under the hypothesis that if the predicted segmentation is of good quality, then it will produce a good segmentation on at least one atlas image. However, the atlas registration step may fail. This is often the case for certain cardiac pathologies that introduce significant morphological deformations that the registration step is not able to recover [21]. In such cases, it is necessary to verify the results and manually fine-tune the registration step, limiting the method's scalability.

We present a novel learning framework to monitor the performance of cardiac image segmentation models in the absence of ground truth. The proposed framework is formulated as an anomaly detection problem. The intuition behind this work lies in the possibility of estimating a model of the variability of cardiac segmentation masks from a reference training dataset provided with reliable ground truth. This model is represented by a convolutional autoencoder, which can be subsequently used to identify anomalies in segmented unseen images. Differently from previous learning-based approaches [11,12], we avoid the need of any type of annotations about the quality of a segmentation for training. Our approach also avoids the required spatial alignment between ground truth images and segmentations of RCA [13,18], thus circumventing image registration.

2   Method

Let us denote $X \in \mathcal{C}^{H \times W}$ a segmentation mask of width W and height H, with $\mathcal{C}$ the set of possible label values. A Convolutional Autoencoder (CA) is trained to learn a function $f : \mathcal{C}^{H \times W} \to \mathcal{C}^{H \times W}$, with $X' = f(X) \approx X$, by minimizing a global dissimilarity measure between an input mask X and its reconstruction X'. In an anomaly detection setup, the CA is trained using normal samples, $i.e.$ samples without defects. In our framework, the normal samples are the ground truth (GT) masks associated with the images used to train a segmentation model (Fig. 1a). The CA learns to reconstruct defect-free samples, $i.e.$ the GT, through a bottleneck, the latent space Z.

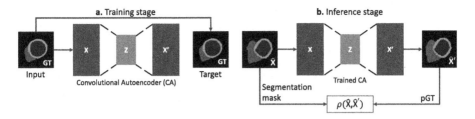

Fig. 1. a. A Convolutional Autoencoder (CA) is trained with ground truth (GT) masks from a cardiac imaging dataset. **b.** At inference, the CA reconstructs an input mask $\widehat{X}$, previously segmented by a model. The reconstructed mask $\widehat{X}'$ acts as pseudo ground truth (pGT) to estimate a function $\rho(\widehat{X}, \widehat{X}')$, a surrogate measure of the segmentation quality and the model's performance.

At inference (Fig. 1b), the CA is used to obtain $\widehat{X}' = f(\widehat{X})$, where $\widehat{X}$ is a segmentation mask, generated by a cardiac segmentation model/method on unseen data, and $\widehat{X}'$ its reconstruction. Since the CA is trained with ground truth data, the quality of the reconstruction will be generally higher for segmentation masks with similar characteristics than those in the ground truth. Poor segmentations, which the CA has not encountered at training, will instead lead to bad reconstructions ($\widehat{X}' \not\approx \widehat{X}$). Autoencoder-based anomaly detection methods exploit the reconstruction error, i.e. $\|\widehat{X}' - \widehat{X}\|_2$, to quantify how anomalous is a sample [1]. We use this principle to establish a surrogate measure of the segmentation quality by quantifying a segmented mask and its reconstruction.

Let us so formalize the function $\rho(\widehat{X}, \widehat{X}')$, a surrogate measure of the segmentation quality of the mask in the absence of GT. In this context, we denote $\widehat{X}'$ a pseudo GT (pGT) since it acts as the reference to measure performance. We present two different scenarios for ρ. First, we propose

$$\rho_1 : \mathcal{C}^{H \times W} \to \mathbb{R}, \tag{1}$$

which represents the most common setup in autoencoder-based anomaly detection. Due to the generic nature of ρ_1, well-suited metrics for segmentation quality assessment can be used, such as the DSC or the Hausdorff Distance (HD). Secondly, we propose

$$\rho_2 : \mathcal{C}^{H \times W} \to \mathbb{R}^{H \times W}. \tag{2}$$

This function generates a visual map of the inconsistencies between the two masks. We use as $\rho_2(\cdot)$ a pixel-wise XOR operation between the segmentation mask $\widehat{X}$ and the pGT.

These two types of measures can be used jointly for performance assessment and model monitoring. Measures obtained from ρ_1-type functions can be paired with a threshold to flag poor segmentation results. The raised alert would then be used to take application-specific countermeasures as, for instance, a visual inspection of an inconsistency map generated by ρ_2.

Layer	Output Size	Parameters		
		Kernel	Stride	Padding
Input	256x256x4			
Block1	128x128x32			
		4x4	2	1
Block2	64x64x32			
		4x4	2	1
Block3	32x32x32			
		4x4	2	1
Block4	32x32x32			
		3x3	1	1
Block5	16x16x64			
		4x4	2	1
Block6	16x16x64			
		3x3	1	1
Block7	8x8x128			
		4x4	2	1
Block8	8x8x64			
		3x3	1	1
Block9	8x8x32			
		3x3	1	1
Conv2d	4x4x100			
		4x4	2	1

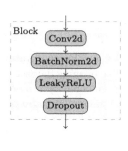

Fig. 2. Architecture of the encoding module. The decoder is built by reversing this structure and replacing convolutions with transposed convolutions.

Network Architecture. We use the CA architecture proposed in [4] as the backbone network (Fig. 2) with the following modifications. We use a latent space dimension to accommodate 100 feature maps of size 4×4. A softmax activation function is added to the last layer to normalize the output to a probability distribution over predicted output classes, as well as batch-normalization and dropout to each hidden layer. We use the loss function $\mathcal{L} = \mathcal{L}_{\mathrm{MSE}}(X, X') + \mathcal{L}_{\mathrm{GD}}(X, X')$, where $\mathcal{L}_{\mathrm{MSE}}$ is the mean squared error loss and $\mathcal{L}_{\mathrm{GD}}$ the generalized dice loss [16]. Trained over 500 epochs, for the first 10 epochs $\mathcal{L}_{\mathrm{GD}}$ is computed leaving aside the background class to avoid the convergence to a dummy blank solution. The network weights are set using a *He* normal initializer. The Adam optimizer is initialized with learning rate 2×10^{-4} and a weight decay of 1×10^{-5}. After every epoch, the model is evaluated on the validation set. The weights retrieving the lowest $\mathcal{L}$ value are stored for testing.

3   Experiments and Results

Section 3.1 describes the datasets used, the setup and implementation of the experiments. Experimental results are then presented in Sect. 3.2.

3.1 Experimental Setup

Data. We used data from the MICCAI 2017 Automatic Cardiac Diagnosis Challenge (ACDC) [5]. The dataset consists of an annotated set with 100 short-axis cine magnetic resonance (MR) images, at end-diastole (ED) and end-systole (ES), with corresponding labels for the left ventricle (LV), right ventricle (RV), and myocardium (MYO). The set was split into training and validation subsets using an 80:20 ratio. The challenge also provides a testing set with 50 cases, with no ground truth publicly available. To have uniform image sizes, these were placed in the middle of a 256×256 black square. Those exceeding this size were center cropped.

Setup. We trained the monitoring framework using the ground truth masks from the ACDC training set and used it to assess the performance of five methods participating in the ACDC Challenge [3,6,7,17,19] and an additional state-of-the-art cardiac segmentation model [2]. We trained five models [2,3,6,7,19] using the challenge's full training set (MR images and masks) and then segmented the ACDC test images. For the remaining method [17], we obtained the segmentation masks directly from the participating team.

The segmentations from every method were fed to the monitoring framework. The resulting pGTs were used to compute ρ_1-type measures (Def. 1), the DSC and the HD, and a ρ_2-type measure, an inconsistency map (Def. 2). We also computed pseudo DSC/HD using the RCA [18]. The ACDC challenge platform estimates different performance measures (DSC, HD, and other clinical measures) on the testing set upon submission of the segmentation results. We uploaded the masks from every model to obtain real DSC and HD. To differentiate the real measures computed by the platform from our estimates, we denote the latter ones pDSC and pHD. In our experiments, we set pHD > 50 or pDSC < 0.5 to flag a segmentation as suspicious and pHD = pDSC = 0 to raise an erroneous segmentation alert flag.

Implementation. We implemented our framework in PyTorch. All the cardiac segmentation models used the available implementations, except for [17] where we had the segmentation masks. The RCA was implemented following the guidelines in [18] using a previously validated atlas propagation heart segmentation framework [22]. All experiments ran on Amazon Web Services with a Tesla T4 GPU. To encourage reproducibility, our code and experiments are publicly available from a Github repository[1].

3.2 Results

Figure 3 presents scatter plots of the real DSC and HD from the ACDC platform and the pDSC and pHD obtained with our framework and the RCA [13,18]. We

---

[1] https://github.com/robustml-eurecom/quality_control_CMR.

present results for LV, RV, and MYO over all generated segmentation masks, and report the Pearson correlation coefficient r.

The results show a high positive correlation between real scores and our estimations. Our framework consistently outperforms RCA. For both RCA and our proposed approach, the real and pseudo HDs show stronger correlations than the DSC. This can be explained by the higher sensitivity of the HD to segmentation errors. Instead, the DSC shows little variability in the presence of minor segmentation, suggesting that both methods have difficulties in modeling small deviations from the reference ground truth data, i.e. very high-quality segmentations.

Table 1 simulates the ACDC Challenge results using real HD and pHD for every model to determine if our framework is a reliable means to rank the performance of the different cardiac segmentation methods. We do not use the pDSC for the ranking, as it reported a lower Pearson correlation coefficient r in our first experiment (Fig. 3).

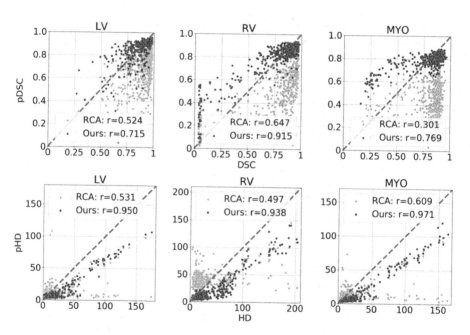

Fig. 3. DSC vs, pDSC (top) and HD vs. pHD (bottom) for our framework (blue dots) and RCA (yellow dots) on the left ventricle (LV), right ventricle (RV) and myocardium (MYO). (Color figure online)

We report results for LV, RV, and MYO in ED and ES and compare them against the RCA. The ranking quality is assessed using Spearman's rank correlation coefficient r_s between the real and the pseudo measures. The r_s assesses if there is a monotonic relationship between both measures, i.e. it allows to determine if the pseudo measure is a valid criterion to rank the different methods. For

Table 1. ACDC Challenge simulation with 6 models using the real HD (GT) and the pHD obtained with our framework (Ours) and the RCA. HD and pHD reported in mm. The Spearman's rank correlation coefficient r_s measures the ranking accuracy (the closer to 1.0 the better). pHD scores using RCA are excluded for [6], where the registration step failed.

ED									
	LV			RV			MYO		
Model	GT	**Ours**	RCA	GT	Ours	RCA	GT	**Ours**	RCA
Bai [2]	39.01	23.38	15.55	50.21	31.82	56.22	47.10	28.46	20.42
Baumgartner [3]	7.14	3.87	9.30	14.00	7.72	37.63	9.49	4.43	10.52
Isensee [6]	7.01	3.88	–	11.40	7.82	–	8.44	4.38	–
Khened [7]	16.81	6.39	10.58	13.25	6.87	39.01	16.09	6.08	11.22
Tziritas [17]	8.90	4.69	8.92	21.02	9.86	41.10	12.59	4.58	10.65
Yang [19]	16.95	5.29	12.96	86.08	47.24	44.75	31.93	16.39	15.12
r_s	–	0.90	0.90	–	1.00	0.80	–	1.00	1.00
ES									
	LV			RV			MYO		
Model	GT	**Ours**	RCA	GT	**Ours**	RCA	GT	**Ours**	RCA
Bai [2]	50.53	29.56	20.01	52.73	31.40	53.68	52.72	31.05	26.60
Baumgartner [3]	10.51	4.41	9.56	16.32	7.10	35.50	12.47	4.77	9.33
Isensee [6]	7.97	4.07	–	12.07	6.99	–	7.95	4.27	–
Khened [7]	20.14	6.96	11.72	14.71	7.07	35.65	16.77	6.03	10.36
Tziritas [17]	11.57	5.00	10.46	25.70	9.61	36.51	14.78	5.59	10.60
Yang [19]	19.13	6.11	11.78	80.42	33.21	40.68	32.54	16.98	13.68
r_s	–	1.00	0.90	–	1.00	0.80	–	1.00	0.90

fairness in the comparison, the ranking does not include one method [6], where RCA failed. In five out of six cases, we were able to perfectly reproduce the real ranking ($r_s = 1.0$). In the remaining case, the left ventricle in end-diastole (ED), there is only one difference between the real and our pseudo ranking ($r_s = 0.9$), where the 3rd and 4th places were swapped. The positive high r_s scores obtained by our framework confirm that it is a reliable mean for method ranking.

Through the use of alert flags we were able to detect 16 cases for which the challenge platform had reported NaN values indicating errors in the submitted results. Fifteen cases were flagged as erroneous (pHD = pDSC = 0) and one as suspicious (pHD > 50) by our framework. Figure 4 middle and bottom row illustrates two of these cases. The middle row presents a segmentation flagged as erroneous, where the inconsistency map confirms that the left ventricle has not been segmented. The bottom row shows the case of a segmentation flagged as suspicious, where the left ventricle's pHD is high (pHD = 104.93), although the pDSC = 0.814 is within normal range. The inconsistency map confirms the clear segmentation error.

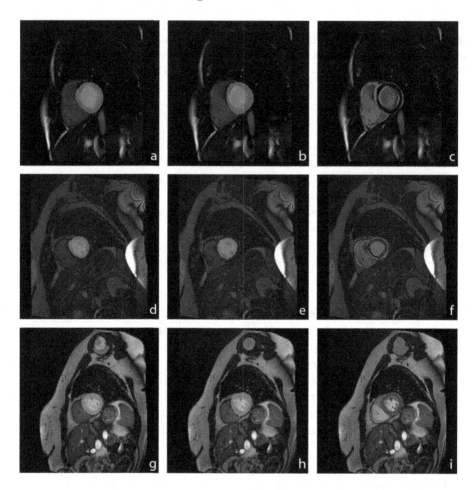

Fig. 4. Segmentation masks (a, d and g), pGTs (b, e and h), and inconsistency maps (c, f and i) for **first row:** successful segmentation, according to the alert thresholds; **middle row:** a segmentation mask flagged as erroneous with pHD = pDSC = 0 in the LV; and **bottom row:** a segmentation mask flagged as suspicious with pHD = 104.93 for the LV (pDSC = 0.814). The inconsistency maps confirm the errors.

Additionally, Fig. 4 top row illustrates an example of a segmentation mask that has not been flagged, i.e. a segmentation result considered good by the ρ_1-type metrics of our framework (pHD $<$ 50 and pDSC $>$ 0.5 for all LV, RV and MYO). The inconsistency map (Fig. 4c) flags pixels in the edges of all the structures as suspicious. While it is clear from the images that the segmentation mask for the RV (pDSC = 0.75, DSC = 0.89) has some errors in the edges, which require a revision, the flagged pixels in the LV and MYO are more difficult to assess. In such a case, a more informative ρ_2-type measure than the selected XOR is desirable.

4   Discussion and Conclusions

We presented a novel learning framework to monitor the performance of cardiac image segmentation models in the absence of ground truth. Our framework addresses the limitations of previous learning-based approaches [8,11,12,14] thanks to its formulation under an anomaly detection paradigm that allows training without requiring quality scores labels. Our results show a good correlation between real performance measures and those estimated with the pseudo ground truth (pGT), making it a reliable alternative when there is no reference to assess a model. Compared with state-of-the-art RCA [13,18], our method avoids the use of image registration which makes it more robust, scalable, and considerably faster ($\sim$ 20 min RCA vs. $\sim$ 0.2 s ours, per case). CAs allow for fast inference which conforms to real-time use, thus permitting a quick quality assignment, for example, in a clinical setting. All these characteristics make the proposed framework a viable option for quality control and system monitoring in clinical setups and large population studies.

A current limitation of the proposed framework is that it is less reliable when assessing high-quality segmentations. A simple and practical way to address this could be to increase the alert thresholds (lower pHDs, higher pDSCs), making them more sensitive to small segmentation errors, and to develop more informative ρ_2-type measures, similar to those proposed by uncertainty quantification methods [11]. However, we consider that a more principled approach should be favored by investigating mechanisms to model these smaller deviations from the reference.

Another straightforward extension of this work could be to embed the proposed framework within the segmenter network, as a way to perform quality control during training. This idea has been previously explored by different AE-based segmentation methods [9,10,20]. However, the results reported in [10] suggest that this does not fully solve the problem of unexpected erroneous and anatomically impossible results, originally reported in [5]. Therefore, we consider that separate and modular monitoring frameworks should be favored over end-to-end solutions and future works should focus on improving the limitations of the current quality control techniques, in particular, their poor sensitivity to smaller errors.

Acknowledgments. The authors would like to thank Christian Baumgartner, Elios Grinias, Jelmer M. Wolterink, and Clement Zotti for their help in the reproduction of the ACDC Challenge rankings by sharing their code or results and, overall, through their valuable advice.

References

1. Audibert, J., Michiardi, P., Guyard, F., Marti, S., Zuluaga, M.A.: USAD: unsupervised anomaly detection on multivariate time series. In: Proceedings of the 26th ACM SIGKDD International Conference on Knowledge Discovery & Data Mining, pp. 3395–3404 (2020)

2. Bai, W., et al.: Automated cardiovascular magnetic resonance image analysis with fully convolutional networks. J. Cardiovasc. Magn. Reson. **20**(1), 65 (2018)
3. Baumgartner, C.F., Koch, L.M., Pollefeys, M., Konukoglu, E.: An exploration of 2D and 3D deep learning techniques for cardiac MR image segmentation. In: Pop, M., et al. (eds.) STACOM 2017. LNCS, vol. 10663, pp. 111–119. Springer, Cham (2018). https://doi.org/10.1007/978-3-319-75541-0_12
4. Bergmann, P., Löwe, S., Fauser, M., Sattlegger, D., Steger, C.: Improving unsupervised defect segmentation by applying structural similarity to autoencoders. In: 14th International Joint Conference on Computer Vision, Imaging and Computer Graphics Theory and Applications, pp. 372–380 (2019)
5. Bernard, O., et al.: Deep learning techniques for automatic MRI cardiac multi-structures segmentation and diagnosis: is the problem solved? IEEE Trans. Med. Imaging **37**(11), 2514–2525 (2018)
6. Isensee, F., Jaeger, P.F., Full, P.M., Wolf, I., Engelhardt, S., Maier-Hein, K.H.: Automatic cardiac disease assessment on cine-MRI via time-series segmentation and domain specific features. In: Pop, M., et al. (eds.) STACOM 2017. LNCS, vol. 10663, pp. 120–129. Springer, Cham (2018). https://doi.org/10.1007/978-3-319-75541-0_13
7. Khened, M., Alex, V., Krishnamurthi, G.: Densely connected fully convolutional network for short-axis cardiac cine MR image segmentation and heart diagnosis using random forest. In: Pop, M., et al. (eds.) STACOM 2017. LNCS, vol. 10663, pp. 140–151. Springer, Cham (2018). https://doi.org/10.1007/978-3-319-75541-0_15
8. Kohlberger, T., Singh, V., Alvino, C., Bahlmann, C., Grady, L.: Evaluating segmentation error without ground truth. In: Ayache, N., Delingette, H., Golland, P., Mori, K. (eds.) MICCAI 2012, Part I. LNCS, vol. 7510, pp. 528–536. Springer, Heidelberg (2012). https://doi.org/10.1007/978-3-642-33415-3_65
9. Oktay, O., et al.: Anatomically constrained neural networks (ACNNs): application to cardiac image enhancement and segmentation. IEEE Trans. Med. Imaging **37**(2), 384–395 (2017)
10. Painchaud, N., Skandarani, Y., Judge, T., Bernard, O., Lalande, A., Jodoin, P.M.: Cardiac segmentation with strong anatomical guarantees. IEEE Trans. Med. Imaging **39**(11), 3703–3713 (2020)
11. Puyol-Antón, E., et al.: Automated quantification of myocardial tissue characteristics from native t1 mapping using neural networks with uncertainty-based quality-control. J. Cardiovasc. Magn. Reson. **22**(1), 1–15 (2020)
12. Robinson, R., et al.: Real-time prediction of segmentation quality. In: Frangi, A.F., Schnabel, J.A., Davatzikos, C., Alberola-López, C., Fichtinger, G. (eds.) MICCAI 2018, Part IV. LNCS, vol. 11073, pp. 578–585. Springer, Cham (2018). https://doi.org/10.1007/978-3-030-00937-3_66
13. Robinson, R., et al.: Automated quality control in image segmentation: application to the UK biobank cardiovascular magnetic resonance imaging study. J. Cardiovasc. Magn. Reson. **21**(1), 1–14 (2019)
14. Ruijsink, B., et al.: Fully automated, quality-controlled cardiac analysis from CMR. JACC Cardiovasc. Imaging **13**(3), 684–695 (2020)
15. Sculley, D., et al.: Hidden technical debt in machine learning systems. Adv. Neural Inf. Process. Syst. **28**, 2503–2511 (2015)
16. Sudre, C.H., Li, W., Vercauteren, T., Ourselin, S., Jorge Cardoso, M.: Generalised dice overlap as a deep learning loss function for highly unbalanced segmentations. In: Cardoso, M.J., et al. (eds.) DLMIA/ML-CDS -2017. LNCS, vol. 10553, pp. 240–248. Springer, Cham (2017). https://doi.org/10.1007/978-3-319-67558-9_28

17. Grinias, E., Tziritas, G.: Fast fully-automatic cardiac segmentation in MRI using MRF model optimization, substructures tracking and B-spline smoothing. In: Pop, M., et al. (eds.) STACOM 2017. LNCS, vol. 10663, pp. 91–100. Springer, Cham (2018). https://doi.org/10.1007/978-3-319-75541-0_10

18. Valindria, V.V., et al.: Reverse classification accuracy: predicting segmentation performance in the absence of ground truth. IEEE Trans. Med. Imaging **36**(8), 1597–1606 (2017)

19. Yang, X., Bian, C., Yu, L., Ni, D., Heng, P.-A.: Class-balanced deep neural network for automatic ventricular structure segmentation. In: Pop, M., et al. (eds.) STACOM 2017. LNCS, vol. 10663, pp. 152–160. Springer, Cham (2018). https://doi.org/10.1007/978-3-319-75541-0_16

20. Yue, Q., Luo, X., Ye, Q., Xu, L., Zhuang, X.: Cardiac segmentation from LGE MRI using deep neural network incorporating shape and spatial priors. In: Shen, D., et al. (eds.) MICCAI 2019, Part II. LNCS, vol. 11765, pp. 559–567. Springer, Cham (2019). https://doi.org/10.1007/978-3-030-32245-8_62

21. Zuluaga, M.A., Burgos, N., Mendelson, A.F., Taylor, A.M., Ourselin, S.: Voxelwise atlas rating for computer assisted diagnosis: application to congenital heart diseases of the great arteries. Med. Image Anal. **26**(1), 185–194 (2015)

22. Zuluaga, M.A., Cardoso, M.J., Modat, M., Ourselin, S.: Multi-atlas propagation whole heart segmentation from MRI and CTA using a local normalised correlation coefficient criterion. In: Ourselin, S., Rueckert, D., Smith, N. (eds.) FIMH 2013. LNCS, vol. 7945, pp. 174–181. Springer, Heidelberg (2013). https://doi.org/10.1007/978-3-642-38899-6_21

Domain Adaptation for Automatic Aorta Segmentation of 4D Flow Magnetic Resonance Imaging Data from Multiple Vendor Scanners

Jordina Aviles[1], Gonzalo D. Maso Talou[2], Oscar Camara[1(✉)],
Marcos Mejía Córdova[1], Xabier Morales Ferez[1], Daniel Romero[1],
Edward Ferdian[3], Kathleen Gilbert[2], Ayah Elsayed[3], Alistair A. Young[3,4],
Lydia Dux-Santoy[5,6], Aroa Ruiz-Munoz[5,6], Gisela Teixido-Tura[5,6],
Jose Rodriguez-Palomares[5,6], and Andrea Guala[5,6]

[1] Physense, BCN Medtech, Department of Information and Communications
Technologies, Universitat Pompeu Fabra, Barcelona, Spain
oscar.camara@upf.edu
[2] Auckland Bioengineering Institute, Auckland, New Zealand
[3] Faculty of Medical and Health Sciences, The University of Auckland,
Auckland, New Zealand
[4] Department of Biomedical Engineering, King's College London, London, UK
[5] Vall d'Hebron Institute of Research, Barcelona, Spain
[6] Department of Cardiology, Hospital Vall d'Hebron Universitat
Autònoma de Barcelona, Barcelona, Spain

Abstract. The lack of standardized pipelines for image processing has prevented the application of deep learning (DL) techniques for the segmentation of the aorta in phase-contrast enhanced magnetic resonance angiography (PC-MRA). Furthermore, large, well-curated and annotated datasets, which are needed to create DL-based models able to generalize, are rare. We present the adaptation of the popular nnU-net DL framework to automatically segment the aorta in 4D flow MRI-derived angiograms. The resulting segmentations in a large database (> 300 cases) with normal cases and examples of different pathologies of the aorta provided from a single centre were excellent after post-processing (Dice score of 0.944). Subsequently, we explored the generalisation of the trained network in a small dataset of images (around 20 cases) acquired in a different hospital with another scanner. Without domain adaptation, only with a model trained with the large dataset, the obtained results were substantially worst than with adding a few cases of the small dataset (Dice scores of 0.61 vs 0.86, respectively). The obtained results created good quality segmentations of the aorta in 4D flow MRI, which can later be post-processed to assess blood flow patterns, similarly than with manual annotations. However, advanced domain adaptation schemes are very important in 4D flow MRI due to the large differences in image characteristics between different vendor scanners available in multiple centers.

© Springer Nature Switzerland AG 2021
D. B. Ennis et al. (Eds.): FIMH 2021, LNCS 12738, pp. 112–121, 2021.
https://doi.org/10.1007/978-3-030-78710-3_12

Keywords: Aortic segmentation · Deep learning · nnU-net · 4D flow magnetic resonance imaging

1 Introduction

There is a wide spectrum of pathological conditions affecting blood vessels, but in the case of aortic diseases the most common are: aortic aneurysms, aortic valve disease, acute aortic syndromes including aortic rupture, atherosclerotic affections beyond others, as well as genetic diseases.

In this regard, 3D time-resolved phase-contrast magnetic resonance imaging (MRI), more colloquially known as 4D flow MRI, allows for the characterization of 3D haemodynamics throughout the whole cardiac cycle, allowing retrospective quantification of flow measurements such as peak velocities and flow direction, thereby helping to understand blood flow changes caused by disease [1]. In order to complete such analysis, a careful and precise segmentation of the anatomy of interest is indispensable, allowing the estimation of relevant clinical parameters [2].

Until recently, manual segmentation used to be the standard approach for aortic segmentation in 4D flow MRI. However, it is a tedious and time-consuming task, where expert annotators are needed and it is subject to large intra- and inter-subject variability. In this sense, convolutional neural networks (CNN) have completely revolutionized medical segmentation. Among the wide range of architectures that have emerged in recent years, the nnU-Net [3] has become widely popular by managing to streamline the tedious process of network fine-tuning, enabling the automatic adaptation of the segmentation pipeline to arbitrary datasets.

However, despite this unparalleled success, most models still have difficulties generalizing well when presented with data from different vendors or healthcare institutions. In this regard, domain adaptation (DA), enables to learn transferable features for adapting models from a source domain to a different target domain, avoiding the time-consuming process of creating and optimizing models from scratch or using un-optimized open source models for your objective. DA techniques have been applied in several medical imaging segmentation and classification approaches with promising results [4]. Hence, the aim of this study was to apply a simple domain adaptation approach to automatically perform aortic segmentation with 4D flow MRI data coming from different clinical sites and scanners.

2 Materials and Methods

2.1 Population Datasets

Data from two distinct centers was considered. One cohort was provided by the Centre for Advanced MRI - CAMRI (University of Auckland, Auckland, New Zealand), consisting of 22 cases (16 healthy, 4 with aortic regurgitation and 2 with left ventricular hypertrophy); a larger cohort was provided by the Hospital

Vall d'Hebron (Barcelona, Spain), consisting in 352 cases (VH), including healthy and a large variety of aortic disease cases: 44 healthy, 154 bicuspic aortic valve, 71 Marfan syndrome, 5 Ehlers-Danlons syndrome, 14 Loeys-Dietz syndrome, and 65 aneurysm and tricuspid aortic valve.

The acquisition of the Vall d'Hebron (VH) dataset was performed on a 1.5 T scanner (Signa, General Electric Healthcare, Waukesha, Wisconsin, USA) with a velocity encoding (VENC) of 200 cm/s, an spatial resolution of 2.5 × 2.5 × 2.5 mm^3, flip angle of 8°, repetition time (TR) of 4.2–6.4 ms and echo time (TE) of 1.9–3.7 ms. Endovenous contrast was not given. The acquisition volume was set to include the entire thoracic aorta. Acquisitions were made with an eight-channel cardiac coil (HD Cardiac, GE Healthcare) using the following parameters: velocity encoding (VENC) of 200 cm/s, field of view (FOV) 400 × 400 × 400 mm, scan matrix 160 × 160 × 160 (voxel size of 2.5 × 2.5 × 2.5 mm), flip angle 8°, repetition time (TR) 4.2–6.4 ms and echo time (TE) 1.9–3.7 ms. VENC was increased in case of aortic valve stenosis to avoid aliasing.

Similarly, the data from the Centre for Advanced MRI (CAMRI) Dataset was acquired through a 1.5 T scanner (Siemens Avanto Siemens Healthcare, Erlangen, Germany) without the use of contrast. The acquisition volume was set to include the whole heart, the aorta and the main pulmonary artery. Common acquisition parameters included: VENC of 150 cm/s (except for Aortic Regurgitation patients, where VENC was determined using VENC scout and tipically ranged from 150 to 300 cm/s), spatial resolution of 2.4 × 2.4 × 2.4 mm, a flip angle of 7° and TR/TE of 38.8/2.3 ms.

The VH dataset was manually annotated by three experts during the expand of three years. Data from the CAMRI was labeled through semi-automatic segmentation leveraging region-growing techniques from 3D Slicer[1].

2.2   nn-Unet Architecture and Experimental Setup

Among all choices provided by the nnU-Net, we opted for the 3D full resolution U-net architecture; due to the available computational resources the training of the U-Net cascade was not feasible. We employed two distinct architectures for each training dataset. The VH architecture, which will be further mentioned as Large Dataset (LD) network, used a patch size of 128 × 128 × 128 pixels with a batch size of 2. Six resolution stages were used together with instance normalization and leaky ReLU activations. We employed an alternative architecture for the CAMRI training data, which will be further mentioned as Small Dataset (SD) network. It employed a patch size of 48 × 128 × 160 with a batch size of 2 due to the small target voxel spacing (1 × 1 × 1 mm) and the median image shape in the cohort. The architecture consisted on 5 resolution stages, each one consisting of two computational blocks (each composed by convolution-instance normalization-leaky ReLU).

The VH dataset was splitted randomly into 70% for training and 30% for testing. The training set was further divided by the network into 80% training

---

[1] https://www.slicer.org/.

and 20% validation. Data from CAMRI was also as randomly divided into 70% for training and 30% for testing. The training set was further divided by the network into 80% training and 20% validation.

A 5-fold-cross validation was employed during training. The model was trained from scratch in the ABI High Performance Computing cluster with a NVIDIA GeForce v100 w/32 GB. Networks were trained for a maximum of 1000 epochs using stochastic gradient descent with nesterov momentum (0.99) and an initial learning rate of 0.01 with a decay of $(1 - epoch/epoch_{max})^{0.9}$. The loss function chosen was the sum of cross-entropy and Dice loss. Data augmentation was performed (rotation, scaling and Gaussian noise) only on the CAMRI dataset, due to its small sample size and with the aim of increasing variability in the training dataset.

The CNN performance was assessed with several metrics: Dice Score (DS), Haussdorf distance (HD), Jaccard (J) and Average Symmetrical Surface Distance (ASSD). For the evaluation of common inconsistencies in the predicted segmentations, all cases with a DS below 0.85 were visually assessed to evaluate possible post-processing strategies.

2.3 Domain Adaptation

To enable the comparison between the different networks in a different data domain (e.g., CAMRI dataset with LD network), a new network architecture, which will be further mentioned as Modified Small Dataset (MSD) network, with the same characteristics as the LD network, was created and trained for the CAMRI dataset. Cross-predictions were obtained, i.e., inference for the VH dataset was done using the modified CAMRI model, and CAMRI predictions were obtained using the VH model. An additional step was taken to process the CAMRI dataset by applying a fine-tuning approach to the LD network by adding 5 random cases of the CAMRI dataset and retrain for additional 50 epochs.

2.4 Post-processing

The VH dataset consisted on both expert manual segmentations and anatomical landmarks, each corresponding to a different anatomical structure of the aorta: sinotubular junction, pulmonary artery bifurcation, first subaortic trunk, third subaortic trunk and descendent aorta at the diaphragm level.

Parts of the left ventricle were included in the provided segmentations, facilitating the extraction of the aorta centreline. However, manual annotations were performed without an standardised protocol, thus compromising the CNN predictions near the LV. Furthermore, this area was not required for aorta segmentation. To ensure a more robust prediction, a semi-automatic post-processing approach was performed to exclude LV points from the VH testing dataset binary masks. An schematic description of the developed pipeline performed is depicted in Fig. 1. The pipeline required the introduction of the binary masks and anatomical landmarks. First, a three-dimensional (3D) model in .vtk format (The Visualization Toolkit [5]) was automatically extracted for each patient using the provided binary masks. Secondly, the centreline of the 3D model was extracted by computing a line between the first and last anatomical landmarks, corresponding to the sinotubular junction and descending aorta at the diaphragm level, respectively, using the Vascular Modelling Toolkit (VMTK) [6] Python's library. Once the centreline was obtained, a direction vector (i.e., vector with the direction of the centreline) was computed using the first anatomical landmark (sinotubular junction) and a consecutive point in the centreline. Then, a 3D cylinder was created with its top cap centered at the sinotubular landmark and oriented by the computed direction vector. Cylinder diameter and height were the same for all processed cases. To eliminate LV points, the intersection between the 3D aortic model and the cylinder was then subtracted. However, due to the fixed height and diameter for all cases, some LV points could be missed. Therefore, those areas were further processed by keeping the largest connected component of the segmentation. Finally, the output of the post-processing was a modified binary mask with only aortic points.

Post-processing methodology was applied to the VH testing set (both ground-truth and predicted masks) to assess changes in score values when LV was not included. Despite being a nearly fully automatic process, 4 out of 30 cases needed manual adjustment of the process. Due to this lack of full-automatisation of the pipeline (VTK Polydata to NIFTI step), the post-processing step was not applied to VH testing set.

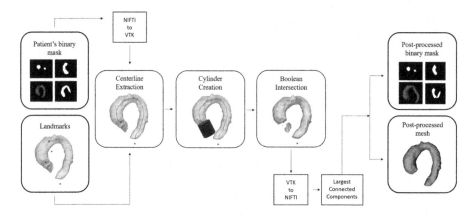

Fig. 1. Processing pipeline designed to modify from the Vall d'Hebron dataset.

2.5   Experiments

A total of five different experiments were performed:

1. Experiment 1 - LD network applied to VH testing set;
2. Experiment 2 - SD network applied to CAMRI testing set;

3. Experiment 3 - LD network applied to CAMRI dataset;
4. Experiment 4 - SD network applied to VH dataset;
5. Experiment 5 - LD network fine-tuned with 5 random CAMRI cases applied to CAMRI dataset.

3   Results

Table 1 summarises the average values for the accuracy metrics obtained in all the experiments designed in our study. The best performance of the aorta segmentation in the VH testing set by LD network was obtained on the last fold of the five-fold cross-validation. The best average results for fold 4 were a DS of 0.901, J of 0.822, ASSD of 1.49 and HD of 5.78. Twelve cases had a DS lower than 0.85, all in patients with associated pathologies: Marfan syndrome ($N = 5$), Bicuspid aortic valve ($N = 6$), Ehlers-Danlons syndrome ($N = 1$). Three examples of ground-truth (green), automated (red) segmentations for subjects with worst DS (0.684), intermediate DS (0.798) and best DS (0.969) are depicted in Fig. 2.

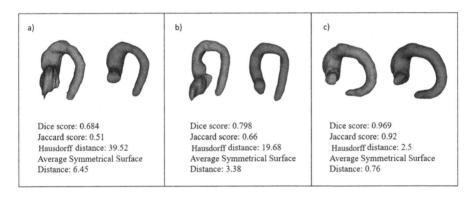

a)

Dice score: 0.684
Jaccard score: 0.51
Hausdorff distance: 39.52
Average Symmetrical Surface
Distance: 6.45

b)

Dice score: 0.798
Jaccard score: 0.66
Hausdorff distance: 19.68
Average Symmetrical Surface
Distance: 3.38

c)

Dice score: 0.969
Jaccard score: 0.92
Hausdorff distance: 2.5
Average Symmetrical Surface
Distance: 0.76

Fig. 2. Representative 3D segmentations of thoracic aorta for VH testing dataset. Ground-truth is shown in green and automatic segmentation is shown in red. a) Segmentation with worst Dice Score (DS); b) Segmentation with a below average DS; c) Segmentation with best DS. (Color figure online)

The best performance of the aorta segmentation in the VH testing set given by the LD network was obtained on the second fold of the five-fold cross-validation. Best segmentation metrics had average values for DS of 0.906, J of 0.837, ASSD of 1.44 and HD of 6.32. Two cases showed a DS lower than 0.85, being one healthy control and one with Marfan syndrome (Table 1).

The best performance of the aorta segmentation in the CAMRI testing set was achieved with the SD network on the second fold of the five-fold cross-validation. Best segmentation metrics had average values of a DS of 0.899, J of 0.818, ASSD of 0.61 and HD of 2.05, where no cases showed a DS lower than 0.85.

Table 1. Experiment results. Average metrics across five folds. DS: Dice Score; J: Jaccard similarity; ASSD: Average Symmetric Surface Distance; HD: Hausdorff Distance.

	DS	J	ASSD	HD
Experiment 1	0.921	0.8561	1.218	4.647
Experiment 2	0.8952	0.812	0.598	2.054
Experiment 3	0.6126	0.467	10.224	49.826
Experiment 4	0.3754	0.2746	61.926	133.38
Experiment 5	0.856	0.7508	2.336	9.74

Three examples of ground-truth (green) and automated (red) segmentations for subjects with worst DS (0.859), intermediate DS (0.900) and best DS (0.927) are depicted in Fig. 3.

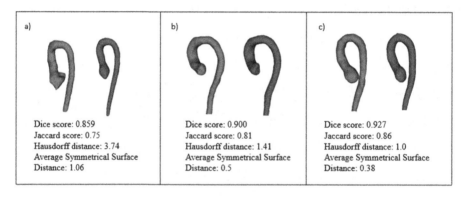

Fig. 3. Representative 3D segmentations of thoracic aorta in the CAMRI testing dataset. Ground-truth is shown in green and automatic segmentation is shown in red. a) Segmentation with worst Dice Score (DS); b) Segmentation with a below average DS; c) Segmentation with best DS. (Color figure online)

Most noticeable inconsistencies between ground-truth (GT) and automatic segmentations in the VH dataset were found near the aortic root. GT segmentations included part of the LV. Therefore, predicted segmentations failed on the segmentation of those areas. Fold 1 outperformed the others, with a DS of 0.944, J of 0.895, ASSD of 0.87 and HD of 2.80. No case showed a DS lower than 0.85. Two examples of non-modified ground-truth (green) together with modified ground-truth (blue) and automated (purple) for subjects with worst DS (0.895) and best DS (0.975) are depicted in Fig. 4.

The performance of the MSD network applied to the VH dataset was really poor in all cross-validation folds, with a maximum DS of 0.41. Similarly, results obtained on CAMRI dataset using LD network had a maximum DS of 0.70.

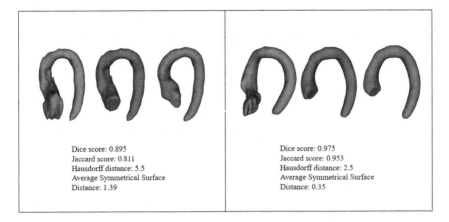

Dice score: 0.895
Jaccard score: 0.811
Hausdorff distance: 5.5
Average Symmetrical Surface
Distance: 1.39

Dice score: 0.975
Jaccard score: 0.953
Hausdorff distance: 2.5
Average Symmetrical Surface
Distance: 0.35

Fig. 4. Representative 3D postprocessed segmentations of thoracic aorta for VH dataset. Non-modified ground-truth is shown in green, modified ground-truth in blue and automatic modified 3D aorta segmentation is shown in red. a) Segmentation with worst Dice Score (DS) value; b) Segmentation with best DS value. (Color figure online)

4   Discussion

The nnU-net network showed good performance across a wide range of demographic and clinical characteristics. Several semi-automatic or automatic methods for aorta segmentation have been proposed lately [2,7,8]. Berhane et al. [2] developed a fully automated segmentation method based on a modification of a 3D U-Net with denseblocks, achieving a DS of 0.951. On their data, almost all patient studies included the use of a contrast agent. A major difference to the previously mentioned work is that we obtained a comparable performance with a DS of 0.944 for VH dataset after post-processing without the need of contrast in the ground-truth images.

Using the knowledge of pre-existing anatomical landmarks together with 4D flow MR images and aortic segmentations, we were able to create a semi-automatic post-processing that eliminated the LV points that were not relevant for aorta evaluation. Elimination of LV points resulted in an increase of approximately 0.038 in the best case, which gave more accurate results for aortic segmentation.

Most current publications on the field of 4D-flow MRI aorta segmentation use data coming from one center alone, with the same acquisition machine and protocols, resulting in selection bias. In order to explore the generalisation of our trained networks, we applied them to a cohort from a different centers, acquisition machines and protocols. As expected, the VH network was able to generalise in a better way, achieving a performance of an average DS of 0.921. The network training dataset was relatively small, thus leading to a more specific network. Specific networks tend to derive features that are not able to generalise. As expected, performance using the CAMRI network on the VH dataset was poor, with an average DS of 0.3754. To improve the generalisation of the VH network, a fine-tuning DA approach was applied in two different ways. As we could anticipate, performance improved when DA was applied, with an increase in DS of 0.167 in the best fold for the first configuration, and of 0.168 for the second one. DA offered similar results for the CAMRI dataset compared to the CAMRI trained network, with a small difference of 0.03 in DS, suggesting that similar performance can be obtained with less annotated cases from a second dataset using DA. Finally, pre-processing also turned to be a key step to improve accuracy.

5   Conclusions

The applicability of the nnU-Net framework for fast and automated 3D aortic segmentation in 4D flow MRI datasets with different characteristics has been demonstrated. To the best of our knowledge, we managed to obtain accurate contrast free 4D flow MRI automatic segmentations, with performance comparable to that of contrast-enhanced state of the art methods which could reduced contrast injection in future acquisition protocols. The DA approach improved the generalisation capabilities of the VH network, offering similar results when compared to the CAMRI trained network with the CAMRI dataset. An appealing advantage of DA is that only a small fraction of the CAMRI dataset required manual segmentations, diminishing human intervention in the overall segmentation process. Further work should investigate federated and other Domain Adaptation approaches to improve generalisation capabilities of neural networks.

References

1. Markl, M., Kilner, P.J., Ebbers, T.: Comprehensive 4D velocity mapping of the heart and great vessels by cardiovascular magnetic resonance. J. Cardiovasc. Magn. Reson. **13**(1), 7 (2011). https://doi.org/10.1186/1532-429X-13-7
2. Berhane, H., Scott, M., Elbaz, M., et al.: Fully automated 3D aortic segmentation of 4D flow MRI for hemodynamic analysis using deep learning. Magn. Reson. Med. **84**(4), 2204–2218 (2020)
3. Isensee, F., Jaeger, P.F., Kohl, S.A., Petersen, J., Maier-Hein, K.H.: nnU-Net: a selfconm guring method for deep learning-based biomedical image segmentation. Nat. Methods **18**(2), 203–211 (2021)
4. Oliveira, H., Dos Santos, J.: Deep transfer learning for segmentation of anatomical structures in chest radiographs. In: Proceedings - 31st Conference on Graphics, Patterns and Images, SIBGRAPI 2018, pp. 204–211. Institute of Electrical and Electronics Engineers Inc. (Jan 2019)
5. Schroeder, W., Ken, M., Lorensen, B.: The Visualization Toolkit (VTK), 4th edn. Kitware, Clifton Park (2006). https://www.vtk.org
6. Antiga, L., Piccinelli, M., Botti, L., Ene-Iordache, B., Remuzzi, A., Steinman, D.A.: An image-based modeling framework for patient-specific computational hemodynamics. Med. Biol. Eng. Comput. **46**(11), 1097–1112 (2008). https://doi.org/10.1007/s11517-008-0420-1

7. Herment, A., Kachenoura, N., Lefort, M., et al.: Automated segmentation of the aorta from phase contrast MR images: validation against expert tracing in healthy volunteers and in patients with a dilated aorta. J. Magn. Reson. Imaging **31**(4), 881–888 (2010)

8. Bustamante, M., Gupta, V., Forsberg, D., Carlhäll, C.J., Engvall, J., Ebbers, T.: Automated multi-atlas segmentation of cardiac 4D flow MRI. Med. Image Anal. **49**, 128–140 (2018)

A Multi-step Machine Learning Approach for Short Axis MR Images Segmentation

Andre Von Zuben$^{(\boxtimes)}$, Kylie Heckman , Felipe A. C. Viana ,
and Luigi E. Perotti

Department of Mechanical and Aerospace Engineering, University of Central Florida,
Orlando, FL 32816, USA
avzuben@knights.ucf.edu

Abstract. Segmentation of cardiac magnetic resonance (cMR) images is
often the first step necessary to compute common diagnostic biomarkers,
such as myocardial mass and left ventricle (LV) ejection fraction. Often
image segmentation and analysis require significant, time-consuming user
input. Machine learning has been increasingly adopted to automatically
segment medical images to lessen the burden on image segmentation
and image analysis for model construction and validation. In this work
we present a multi-step machine learning approach to segment short axis
cMR images based on a heart locator and the weighted average of 2D
and 2D++ UNets. The presence of a heart locator led to more accurate results and allowed to increase the neural network training batch
size. Finally, the obtained segmentations are post-processed using spline
interpolation and the Loop scheme to generate left ventricular endocardial and epicardial surfaces at the end of diastole and end of systole.

Keywords: Image segmentation · Machine learning · Heart locator ·
Weighted average · Left ventricular surfaces

1   Introduction

Cardiac magnetic resonance imaging (cMRI) is a common imaging modality used
to investigate cardiac function and dysfunction. Starting from long and short axis
images of the ventricles, several patient-specific biomarkers—e.g., ejection fraction (EF), left ventricular (LV) and right ventricular (RV) mass, myocardial wall
thickness—can be computed. A first common processing step necessary to analyze cMRI data and extract these biomarkers, consists in segmenting the images
to isolate the left and right ventricles' myocardial tissue and cavities. Manual
data segmentation is very time consuming as it can easily involve thousands of
images when several locations along the ventricular long axis, different cardiac
phases from end diastole to end systole, and multiple patients are considered.

Supported by the University of Central Florida, Mechanical and Aerospace Engineering
Department.

D. B. Ennis et al. (Eds.): FIMH 2021, LNCS 12738, pp. 122–133, 2021.
https://doi.org/10.1007/978-3-030-78710-3_13

In recent years, many studies have applied deep learning to cMRI segmentation [4]. However, semi-automatic segmentation is still a daily practice due to the lack of accuracy of fully automatic cardiac segmentation methods leading to time-consuming tasks [3].

In this work, we propose a framework to automatically segment the LV and RV based on a heart locator, a weighted average of two convolutional neural networks, and a multi-step post-processing algorithm to improve the final image segmentation. The proposed framework is tested and compared using the short axis cMRI database provided with the 2017 MICCAI Automated Cardiac Diagnosis Challenge (ACDC) [10]. The final short axis segmentations will be combined to define – by spline and Loop interpolation – the endo and epicardial LV surfaces at end diastole and end systole. In the future, these 3D surfaces will be the base to generate automatically LV anatomical models.

2 Method

2.1 Imaging Data

The ACDC dataset [10] includes short-axis, standard SSFP, cine-MRI of 150 patients. As described in [3], short axis images were acquired in each subject using either a 1.5 T (Siemens Aera) or a 3 T (Siemens Trio Tim) scanner. Slice thickness varied from 5 mm to 10 mm (including a 5 mm gap in some cases) and in plane isotropic spatial resolution varied from 1.34 mm × 1.34 mm to 1.68 mm × 1.68 mm. Ground truth segmentations of LV myocardium (LVM), LV and RV cavities (LVC and RVC, respectively) are provided within the ACDC dataset at end diastole (ED) and end systole (ES).

2.2 Image Preprocessing

Before the images are analyzed, the dataset is split into 100 training and 50 test patients. Furthermore, the images' signal intensity is normalized to overcome considerable contrast variation amongst images. This is achieved by: 1) normalizing the central 98% of the pixelwise intensity values between −1 and 1; and 2) truncating the tails of the intensity distribution at −1 and 1. In addition, all images are re-sampled (using bi-cubic interpolation and the scikit-image processing Python toolbox) to obtain a uniform 1.25 mm × 1.25 mm in plane resolution across all patients.

2.3 Convolutional Neural Networks

Convolutional neural networks, typically UNet-like architectures, have been particularly successful in performing medical image segmentation [14]. Recently, the UNet++ architecture has been proposed as a more powerful network for medical image segmentation [19]. In this new approach, the encoder and decoder sub-networks are connected through nested, dense skip pathways that aim at reducing the semantic gap between the feature maps of the encoder and decoder. This work utilizes three different models based on the UNet and UNet++ architectures shown in Fig. 1: a heart locator network (HL), a 2D network (2DN) – following the work of [7] – and a 2D++ network (2DN++). The complete prediction framework is shown in Fig. 2 and in Fig. 7 in the Appendix using a representative midventricular slice.

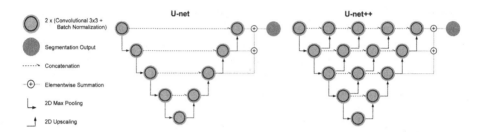

Fig. 1. Left: UNet architecture where several convolutional blocks in the encoder part of the network are followed by the same number of deconvolutional blocks in the decoder portion. In the UNet architecture, several connections between blocks on the same level are skipped. Right: UNet++ architecture containing additional convolutional layers between the skipped connections.

The heart locator network (HL) is a UNet that receives a 352×352 pixels image as input and outputs each pixel's probability to be part of the heart. This network's batch size is 10 and its loss function is the summation of a Dice similarity coefficient (DSC) loss and a binary cross-entropy loss. The pixels containing the heart with the largest/smallest horizontal and vertical locations are selected, their midpoint is computed, and the 144×144 size region centered at the midpoint is extracted. The cropped 144×144 image size was chosen after the in plane resampling described above (so that all images have the same in plane resolution) and based on the difference between the HL central point prediction and the ground truth central point. In order to guarantee that LV and RV are contained in the cropped image, the 144×144 image contains an additional buffer of 8 pixels in all directions. For every patient and cardiac phase, the full short axis stack is passed to the HL so that a single central point per patient/cardiac phase is identified as the average of the central point locations on each slice. Consequently, the cropped images maintain their original alignment along the longitudinal axis as they are passed to the segmentation networks (Fig. 2).

The segmentation 2D network (2DN) is also a UNet that receives as input 144×144 images processed with the heart locator network and outputs the probability of four different classes for each pixel: background, LVM, LVC, and RVC. Given the reduced image size, compared to the original size used in the HL, the 2DN batch size can be increased to 32. The loss function is the summation of DSC and categorical cross-entropy loss.

The segmentation 2D++ network (2DN++) is based on a UNet++ architecture with the same inputs, outputs, batch size, and configuration of loss

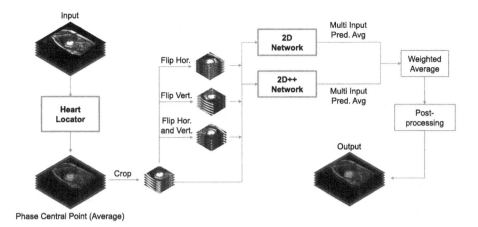

Fig. 2. A stack of images from the same patient and cardiac phase is passed as input to the HL, which outputs 144×144 cropped images containing the heart. Flipped inputs are generated and passed, together with the original input, to the 2DN and 2DN++ to classify each pixel as LVM, LVC, and RVC. The multi-input prediction average is computed from each network, and a weighted image average is calculated. Finally, the segmented images are post-processed to eliminate the smaller disconnected regions.

functions as the 2DN. While the 2DN uses skipped connections to help in the image reconstruction based on features from different levels, in the 2DN++ nested, dense skip pathways aim at decreasing the semantic gap between the feature maps of the encoder and decoder.

2.4   Training

A 5-fold cross-validation strategy was adopted for training. The data was split into five groups, each containing the images for 20 patients: four groups are used for training and the remaining one for validation. Therefore, five training models were created for each architecture. Each training consisted in 300 epochs with a learning rate of $5e-4 \cdot 0.985^{\text{epochs}}$. Data augmentation was applied to increase the diversity of the dataset and to prevent overfitting. A random combination of rotation, scaling, horizontal flip, vertical flip, and elastic transformation [16] was applied as described in [7]. Data augmentation was performed online for the heart locator (a different view of each input image was generated for each sample during each training step) and offline for the 2DN and 2DN++ segmentation networks (ten different images were generated from each image of the original dataset to be used during training). Additionally, to account for the HL uncertainty and variability in identifying the central point, an online central point random translation in X and Y (drawn from a uniform distribution between -10 and 10 pixels) was included in each epoch before cropping the images.

Regarding model parameters and training time, the HL and 2DN UNet architectures have more than 17M parameters each, and the UNet++ has more than

20 M parameters. Training of the HL required ≈ 10 h, whereas the segmentation nets required ≈ 30 h each. All networks were trained on a GPU Nvidia Tesla P100 of 16 GB.

2.5 Weighted Average and Image Correction

Instead of making a single prediction for each input image, different input image views are generated applying horizontal and vertical flips. These images are passed to the segmentation network together with the original input image. The resulting segmentations are then averaged to increase the model robustness. Single shots predictions can lead to incorrect LVM segmentations, especially near the RV insertion points, while a multi-input strategy removes incorrect LVM segments at the RV/LV boundaries as seen, for example, in Fig. 3. Subsequently, the output of the 2DN and 2DN++ models are linearly combined with weights that are inversely proportional to each model's mean loss as described in [17].

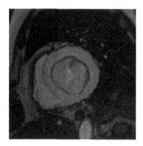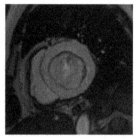

Fig. 3. Single prediction result leading to incorrect LVM (green) segmentation near the lower RV insertion point (left) compared to the corresponding multi-input prediction where horizontally, vertically, and both horizontally and vertically flipped images are used as input alongside the original image (right). (Color figure online)

Finally, an algorithm is applied to remove prediction errors. As LVM, LVC, and RVC are each continuously connected regions, if multiple regions with the same label are detected, the components with the smallest areas or volumes are removed from the classifier. For each cardiac phase, starting from the middle slice, the algorithm is applied iteratively to each slice separately and to each slice together with the remaining slices to verify that LVM, LVC, and RVC are continuously connected across slices.

2.6 Spline Interpolation and LV Surface Generation

The outlines of the LVM segmentation can be represented using a spline interpolant. This step provides a geometric description that can be translated into a model of the LV surfaces and eliminate any remaining non-physiological jaggedness. For both endocardial and epicardial outlines, a spline with 20 equally spaced (18° apart) control points is adopted. First, the endocardial and epi-cardial contours of the LVM are identified and represented using a refined spline with N control points, where N is the number of voxels along the endocardial and epicardial contours. In this first step, no details are lost. Subsequently, an arbitrary point is marked as #0, and other 19 points, 18° apart, are taken to calculate the first discrete spline. After storing the results, the following point becomes point #0, and the process repeats $N/20$ times until the first point is again point #0. Finally, the average of the generated discrete splines with 20 control points is calculated. Different numbers of control points were tested and 20 control points yielded the best results in terms of smoothing the endocar-dial and epicardial contours without losing important anatomical details and decreasing the DSC (Fig. 4).

In order to obtain the endocardial and epicardial 3D surfaces, the gaps between 2D endo and epicardial splines along the longitudinal axis are linearly interpolated and subsequently passed through a Loop filter [1].

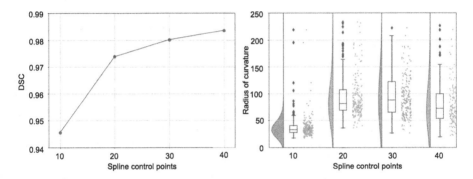

Fig. 4. Number of spline control points versus DSC (left). Pointwise endocardial and epicardial radius of curvature versus spline control points (right). Splines used to gen-erate the data shown in this figure were computed on ground truth segmentations.

3   Results

Figure 5 illustrates the image segmentations obtained with the 2DN, 2DN++, and combined framework for mid-ventricular short axis images in four patients chosen to represent different levels of complexity due to low contrast between the myocardium and the background, the presence of a highly corrugated endo-cardial trabecular structure, and/or a more complex RV anatomy.

We report the results of our framework both during cross-validation (when the available patient data is subdivided into training and validation sets - see Method section) and against the testing dataset only available through the MIC-CAI challenge website and not directly available to users.

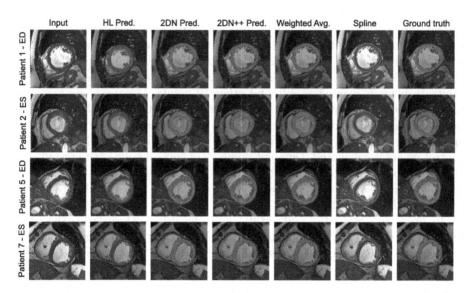

Fig. 5. Ground truth and segmentation results obtained with the proposed framework: LVM is green, LVC is blue, and RVC is red. Note that the entire segmentation computed with the HL is represented in red as no distinction is made between LVM, LVC, and RVC in this case. (Color figure online)

Table 1 shows the cross-validation method comparison. The DSC and the Hausdorff surface distance (d_H) are computed for LVM, LVC, RVC, and overall between the ground truth and the segmentations obtained using the 2DN, 2DN++, and weighted average framework. We also report the DSC and d_H between the HL and the ground truth. In this latter comparison, LVM, LVC, and RVC pixels in the ground truth segmentations are combined and classified as part of the heart.

The DSC is defined as $2 \cdot (|V_{\text{prediction}} \cap V_{\text{reference}}|)/(|V_{\text{prediction}}| \cup |V_{\text{reference}}|)$ and is a measure of overlap between the predicted segmentation $V_{\text{prediction}}$ and the corresponding ground truth segmentation $V_{\text{reference}}$. The DSC can assume values between 0 (no overlap) and 1 (full overlap). The d_H assesses the local maximum distance between the two surfaces $S_{\text{prediction}}$ and $S_{\text{reference}}$ and is reported here in millimeters.

Before generating the LV endocardial and epicardial surfaces at end diastole and end systole, we compared our framework with models that analyzed the same dataset used in this work as part of the 2017 MICCAI ACDC [3] (see Table 2). Finally, Fig. 6 shows the generation of 2D splines based on the segmentation of short axis images for a representative patient and the corresponding endocardial and epicardial surfaces at end diastole and end systole.

Table 1. Cross-validation methods comparison: bolded DSC and dH correspond to closest match with ground truth.

Method	End diastole							
	LVC		RVC		LVM		Total	
	DSC	d_H	DSC	d_H	DSC	d_H	DSC	d_H
HL	N/A	N/A	N/A	N/A	N/A	N/A	0.911	**15.27**
2DN without HL	0.932	12.21	0.856	23.07	0.777	18.01	0.921	20.37
2DN	**0.950**	**10.05**	0.903	17.62	0.833	14.25	**0.948**	15.85
2DN++	0.947	10.22	0.901	17.50	0.831	14.29	0.946	15.95
Weighted Avg. 2DN & 2DN++	0.949	**10.05**	**0.904**	**17.41**	**0.835**	**14.20**	**0.948**	15.36
Method	End systole							
HL	N/A	N/A	N/A	N/A	N/A	N/A	0.854	18.10
2DN without HL	0.867	14.83	0.722	27.08	0.790	18.94	0.871	19.92
2DN	**0.881**	**12.64**	**0.784**	22.12	**0.850**	**14.63**	**0.911**	16.98
2DN++	0.877	12.92	0.780	22.23	0.846	15.70	0.909	**16.73**
Weighted Avg. 2DN & 2DN++	**0.881**	12.85	**0.784**	**21.89**	**0.850**	15.12	**0.911**	16.78

Table 2. Comparison among our framework with weighted averages and other frameworks reported in [3]. "Ours" refers to the weighted average presented in this study.

Study	End diastole						End systole					
	LVC		RVC		LVM		LVC		RVC		LVM	
	DSC	d_H	DSC	d_H	DSC	d_H	DSC	d_H	DSC	d_H	DSC	d_H
Isensee et al. [7]	0.968	7.4	0.946	10.1	0.902	8.7	0.931	6.9	0.899	12.2	0.919	8.7
Baumgartner et al. [2]	0.963	6.5	0.932	12.7	0.892	8.7	0.911	9.2	0.883	14.7	0.901	10.6
Ours - cross-valid	0.949	10.0	0.904	17.4	0.835	14.2	0.881	12.8	0.784	21.9	0.850	15.1
Ours - test data	0.946	11.7	0.898	29.1	0.823	13.0	0.870	13.7	0.798	35.4	0.845	14.8
Tziritas-Grinias [6]	0.948	8.9	0.863	21.0	0.794	12.6	0.865	11.6	0.743	25.7	0.801	14.8
Yang et al. [18]	0.864	47.9	0.789	30.3	N/A	N/A	0.775	53.1	0.770	31.1	N/A	N/A

4   Discussion

We have presented a multi-step machine learning based approach to segment short axis MR images that can be subsequently used to construct endocardial and epicardial LV surfaces at different cardiac phases. The use of the heart locator reduces the amount of noise in the input of the segmentation model and the memory needed for its training, therefore enabling a larger image batch size.

Combining predictions from different UNet models is also reported in [7,8]. However, previous approaches employ a simple arithmetic average and do not include weights that are inversely proportional to the models' mean loss [17]. Similarly, literature shows that using dedicated models to work as locators of the regions of interest is a recurrent approach. The interested reader is referred to [9,12,15] for further details. In a way, even the recently proposed approach based on cascaded UNets [8,13], where the first segmentation output is used as an additional input for the following network instead of cropping the images, can be interpreted as a segmentation done in multiple stages. Alternatively, there are approaches that use the direct cropping of the image to reduce the input size of the segmentation networks [5]. For example, this is the case when a localization network outputs a bounding box enclosing the LVM in the ED frame. These networks do not incorporate a decoder sub-network, as the bounding-box is obtained directly from the encoder sub-network. Moreover, instead of using a model, [11] simply examines the voxels' intensity through time using Fourier analysis and Hough transform to find the region of interest. Similarly to cascaded UNets, in this work, we used a UNet as the first model that generates a mask as an output. This takes advantage of the UNet's ability to process MRI data. However, we use it as a locator model so that the subsequent image segmentation is concentrated in a specific region of the original image. One benefit of this strategy is that it reduces the model size of the UNets handling image segmentation.

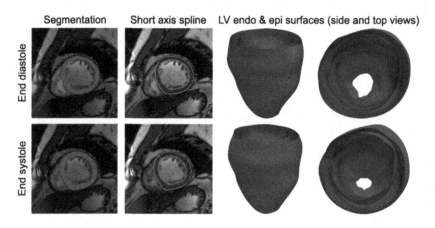

Fig. 6. Predicted segmentation and their correspondent LV myocardium contour generated by splines and Loop scheme.

From the analysis of the different architectures (Table 1), the framework combining the 2DN and 2DN++ with a weighted average lead, in most cases, to increased DSC and lower d_H. Although these improvements are small, this ensemble approach combining 2DN and 2DN++ acts as an insurance that can guarantee optimal solutions.

Table 2 shows that the proposed framework leads to results similar to the best models participating in the MICCAI 2017 challenge in terms of DSC, as reported in [3]. Although the d_H measures are comparable to the ones obtained by other frameworks presented in [3], further improvements regarding this measure are necessary. Furthermore, the proposed cross-validation results are in agreement with the results obtained using the testing dataset. This confirms that our framework has not been overfitted to the available training set.

Although the obtained segmentations reasonably agree with the ground truth, the proposed framework presents several limitations. A careful analysis of

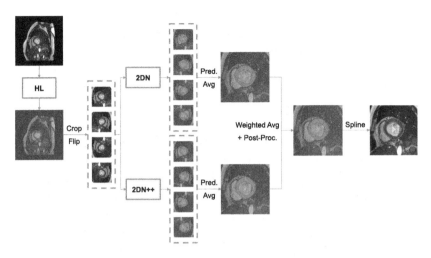

Fig. 7. Complete prediction framework shown with a representative midventricular slice image to highlight the role of different parts of the model.

the results in Fig. 5 shows that the model may: 1) classify small regions at the RV insertion points as LVM instead of RV myocardium (last row in Fig. 5); 2) the RVC may be slightly over (last row in Fig. 5) or under (third row in Fig. 5) predicted with respect to the given ground truth. Another limitation of the current framework becomes evident when basal short axis slices including the valves' structure are passed to the network. In this case, the model wrongly predicted LVM, LVC, and RVC. In future work, the HL will be enhanced to detect the valves' structure and overcome this limitation. Finally, in the current form, the spline and LV surface calculations are implemented as a post-processing step. The creation of the spline and/or a method to smooth the endo and epicardial outlines can be included as a constraint directly in the network.

In summary, we have presented a multi-step framework based on a heart locator and a weighted average of 2D networks to segment 2D cMRI data and generate by spline and Loop subdivision endo and epicardial LV surfaces. In the future, these will be the base to automatically create LV anatomical models.

Appendix

A complete prediction from the framework shown in Fig. 2 is illustrated in Fig. 7 using a representative midventricular slice. This example highlights the role of the weighted average approach as an insurance policy, as well as the spline as a way to further smooth the final LV contour.

References

1. Ayachit, U.: The ParaView Guide: A Parallel Visualization Application. Kitware (2015)

2. Baumgartner, C.F., Koch, L.M., Pollefeys, M., Konukoglu, E.: An exploration of 2D and 3D deep learning techniques for cardiac MR image segmentation. In: Pop, M., et al. (eds.) STACOM 2017. LNCS, vol. 10663, pp. 111–119. Springer, Cham (2018). https://doi.org/10.1007/978-3-319-75541-0_12

3. Bernard, O., et al.: Deep learning techniques for automatic MRI cardiac multi-structures segmentation and diagnosis: is the problem solved? IEEE Trans. Med. Imaging **37**(11), 2514–2525 (2018)

4. Chen, C., et al.: Deep learning for cardiac image segmentation: a review. Front. Cardiovasc. Med. **7**, 25 (2020). https://doi.org/10.3389/fcvm.2020.00025. https://www.frontiersin.org/article/10.3389/fcvm.2020.00025

5. Ferdian, E., et al.: Fully automated myocardial strain estimation from cardiovascular MRI–tagged images using a deep learning framework in the UK Biobank. Radiol. Cardiothorac. Imaging **2**(1), e190032 (2020). https://doi.org/10.1148/ryct.2020190032

6. Grinias, E., Tziritas, G.: Fast fully-automatic cardiac segmentation in MRI using MRF model optimization, substructures tracking and B-spline smoothing. In: Pop, M., et al. (eds.) STACOM 2017. LNCS, vol. 10663, pp. 91–100. Springer, Cham (2018). https://doi.org/10.1007/978-3-319-75541-0_10

7. Isensee, F., Jaeger, P.F., Full, P.M., Wolf, I., Engelhardt, S., Maier-Hein, K.H.: Automatic cardiac disease assessment on cine-MRI via time-series segmentation and domain specific features. In: Pop, M., et al. (eds.) STACOM 2017. LNCS, vol. 10663, pp. 120–129. Springer, Cham (2018). https://doi.org/10.1007/978-3-319-75541-0_13

8. Isensee, F., Jaeger, P.F., Kohl, S.A.A., Petersen, J., Maier-Hein, K.H.: nnU-Net: a self-configuring method for deep learning-based biomedical image segmentation. Nat. Methods (2020). https://doi.org/10.1038/s41592-020-01008-z

9. Islam, M., Vibashan, V.S., Jose, V.J.M., Wijethilake, N., Utkarsh, U., Ren, H.: Brain tumor segmentation and survival prediction using 3D attention UNet. In: Crimi, A., Bakas, S. (eds.) BrainLes 2019, Part I. LNCS, vol. 11992, pp. 262–272. Springer, Cham (2020). https://doi.org/10.1007/978-3-030-46640-4_25

10. Jodoin, P.M., Lalande, A., Bernard, O., Humbert, O., Zotti, C., Cervenansky, F.: Automated cardiac diagnosis challenge. https://acdc.creatis.insa-lyon.fr/

11. Khened, M., Kollerathu, V.A., Krishnamurthi, G.: Fully convolutional multi-scale residual DenseNets for cardiac segmentation and automated cardiac diagnosis using ensemble of classifiers. Med. Image Anal. **51**, 21–45 (2019). https://doi.org/10.1016/j.media.2018.10.004. https://www.sciencedirect.com/science/article/pii/S136184151830848X

12. Li, S., Zhang, J., Ruan, C., Zhang, Y.: Multi-stage attention-Unet for wireless capsule endoscopy image bleeding area segmentation. In: 2019 IEEE International Conference on Bioinformatics and Biomedicine (BIBM), pp. 818–825 (2019). https://doi.org/10.1109/BIBM47256.2019.8983292

13. Li, S., Chen, Y., Yang, S., Luo, W.: Cascade dense-Unet for prostate segmentation in MR images. In: Huang, D.-S., Bevilacqua, V., Premaratne, P. (eds.) ICIC 2019, Part I. LNCS, vol. 11643, pp. 481–490. Springer, Cham (2019). https://doi.org/10.1007/978-3-030-26763-6_46

14. Litjens, G., et al.: A survey on deep learning in medical image analysis. Med. Image Anal. **42**, 60–88 (2017). https://doi.org/10.1016/j.media.2017.07.005

15. Oktay, O., et al.: Attention U-Net: learning where to look for the pancreas (2018)

16. Simard, P.Y., Steinkraus, D., Platt, J.C.: Best practices for convolutional neural networks applied to visual document analysis. In: ICDAR, vol. 3 (2003)

17. Viana, F.A.C., Haftka, R.T., Steffen, V.: Multiple surrogates: how cross-validation errors can help us to obtain the best predictor. Struct. Multi. Optim. **39**(4), 439–457 (2009). https://doi.org/10.1007/s00158-008-0338-0
18. Yang, X., Bian, C., Yu, L., Ni, D., Heng, P.-A.: Class-balanced deep neural network for automatic ventricular structure segmentation. In: Pop, M., et al. (eds.) STACOM 2017. LNCS, vol. 10663, pp. 152–160. Springer, Cham (2018). https://doi.org/10.1007/978-3-319-75541-0_16
19. Zhou, Z., Rahman Siddiquee, M.M., Tajbakhsh, N., Liang, J.: UNet++: a nested U-Net architecture for medical image segmentation. In: Stoyanov, D., et al. (eds.) DLMIA/ML-CDS -2018. LNCS, vol. 11045, pp. 3–11. Springer, Cham (2018). https://doi.org/10.1007/978-3-030-00889-5_1

Cardiac Microstructure: Measures and Models

Diffusion Biomarkers in Chronic Myocardial Infarction

Tanjib Rahman[1]($\boxtimes$) , Kévin Moulin[2] , Daniel B. Ennis[2] ,
and Luigi E. Perotti[1]

[1] Department of Mechanical and Aerospace Engineering,
University of Central Florida, Orlando, FL 32816, USA
tanjib@knights.ucf.edu
[2] Department of Radiology, Stanford University, Stanford, CA 94305, USA

Abstract. Cardiac diffusion tensor magnetic resonance imaging (cDTI) allows estimating the aggregate cardiomyocyte architecture in healthy subjects and its remodeling as a result of cardiac disease. In this study, cDTI was used to quantify microstructural changes occurring in swine (N = 7) six to ten weeks after myocardial infarction. Each heart was extracted and imaged *ex vivo* with 1 mm isotropic spatial resolution. Microstructural changes were quantified in the border zone and infarct region by comparing diffusion tensor invariants – fractional anisotropy (FA), mode, and mean diffusivity (MD) – radial diffusivity, and diffusion tensor eigenvalues with the corresponding values in the remote myocardium. MD and radial diffusivity increased in the infarct and border regions with respect to the remote myocardium ($p < 0.01$). In contrast, FA and mode decreased in the infarct and border regions ($p < 0.01$). Diffusion tensor eigenvalues also increased in the infarct and border regions, with a larger increase in the secondary and tertiary eigenvalues.

Keywords: Cardiac diffusion tensor imaging · Myocardial infarction · Cardiac microstructure · Radial diffusivity

1 Introduction

Amongst existing imaging modalities, MRI and in particular T1 mapping and late gadolinium enhancement (LGE) are the most commonly used imaging techniques to detect and quantify infarct extension and location. Although recent studies have proposed the use of native T1 mapping to image myocardial infarction (MI), commonly used LGE and T1 mapping techniques rely on the use of contrast agents, which are not recommended for many patients with renal insufficiency. Furthermore, LGE and T1 mapping techniques do not provide

This work was supported by NIH/NHLBI K25-HL135408 to LEP, AHA 20POST352 10644 to KM, and the University of Central Florida. The content is solely the responsibility of the authors and does not necessarily represent the official views of the Funders.

direct information regarding the microstructural reorganization occurring in the infarcted myocardium.

Using cardiac diffusion tensor magnetic resonance imaging (cDTI), a non-contrast imaging technique, the myocardium microstructural organization can be inferred from the self-diffusion of water molecules. In particular, diffusion tensor invariants can be used to quantify tissue changes occurring as a result of MI. For example, mean diffusivity (MD) represents the overall magnitude of diffusivity within the tissue and it has been shown to increase in infarcted myocardium [4,13,14]. Similarly, the fractional anisotropy (FA) represents the magnitude of anisotropy of the myocardial microenvironment and it decreases in infarcted myocardium due to cellular death and the loss of cellular structure.

The objective of this study was to further investigate changes in diffusion tensor invariants in remote, border, and infarcted myocardium, focusing on the analysis of primary, secondary, and tertiary eigenvalues and radial diffusivity.

2   Methods

2.1   Experimental and Imaging Procedures

Female Yorkshire swine (N = 7) were considered in this study. Animal care and handling during the experimental procedures followed protocol # 2015–124 approved by the UCLA Institutional Animal Care and Use Committee.

As described in [5], under general anesthesia, a standard Seldinger technique was adopted to access the femoral artery. Subsequently, under X-ray fluoroscopy, a 7 French (Fr) balloon wedge-pressure catheter (110 cm, Teleflex Arrow) was advanced over a metal guidewire to the left aortic sinus. At this location, a micro guidewire (Hi-Torque, Balance Middleweight Universal Guide Wire, 0.014 inches, Abbot Vascular) was used to sub-select a branch of the left anterior descending (LAD) or left circumflex (LCx) artery. Specifically, referring to the subjects number used in Figs. 4 and 5, the LAD was sub-selected in subjects 3, 5, and 7, while the LCx was sub-selected in subjects 1, 2, 4, and 6. Subsequently, a balloon catheter (MINI TREK Coronary Dilatation Catheter, 1.5 mm diameter, Abbot Vascular) was inflated in the selected coronary branch and 2.5–3.0 ml of microspheres (Polybead, Polystyrene 90 micron from Polysciences Inc) were injected. The balloon catheter was left inflated for one minute past microspheres' injection to avoid microspheres backflow and subsequently deflated and removed. In one subject, myocardial infarction was caused by a thrombus formed in the proximal LAD and microspheres were not injected. A comprehensive MRI exam was conducted six to ten weeks after infarct induction to allow enough time for scar formation.

At the end of the MRI exam, a double dose of a gadolinium-based contrast agent was injected (0.6 ml/kg gadopentetate dimeglumine diluted in 10 ml of saline). Ten minutes past the injection of the gadolinium-based contrast agent, the subjects were euthanized under general anesthesia by injecting intravenously a veterinary grade euthanasia solution at 0.1 ml/lb of body weight (Euthasol®, Virbac). Post euthanasia, the heart was extracted, rinsed, and prepared for *ex vivo* imaging, including insertion in a 3D-printed mold to preserve the left and right ventricular anatomy as described in [3]. The 3D printed molds were based on the mid-diastasis ventricular configuration. A knee coil was used for imaging and the hearts were submerged in a perfluoropolyether solution (Fomblin, Solvay).

Ex vivo imaging was conducted in a 3T MRI scanner (Prisma, Siemens). Imaging protocols included: T1-weighted GRE (TE/TR = 3.15 ms/12 ms; flip angle = 25°; field of view = 160 mm × 160 mm; spatial resolution = 1.0 × 1.0 × 1.0 mm^3, N_{avg} = 6), T2-weighted SE (TE/TR = 89 ms/15460 ms; flip angle= 180°; field of view = 192 mm × 192 mm; spatial resolution = 1.0 × 1.0 × 1.0 mm^3, N_{avg} = 8), and cDTI (TE/TR = 62 ms/15560 ms; field of view = 150 mm×150 mm; spatial resolution = 1.0 × 1.0 × 1.0 mm^3; b-values = 0 s/mm^2 and 1000 s/mm^2; N_{dir} = 30; N_{avg} = 5). In average, *ex-vivo* MRI exams started 2.5 hours post euthanasia. The cDTI sequence was a readout segmented sequence [10] with a twice refocused spin-echo encoding for correction of eddy currents [11].

2.2 Diffusion Tensor Reconstruction and Image Labeling

Diffusion tensors were reconstructed from the diffusion weighted images using an in-house Matlab [6] code (github.com/KMoulin/DiffusionRecon). Subsequently, tensor invariants and eigenvalues were calculated at each voxel location.

T1-weighted images were used to segment the myocardium into remote, border, and infarct zones. For this purpose, five short axis slices were selected from apex to base in each subject. Per each subject, small regions of interest (ROIs) were outlined on the remote myocardium of each slice. The mean signal intensity (μ_{remote}) and standard deviation (σ_{remote}) of the selected remote myocardium across slices were computed for each heart. All voxels with a signal intensity SI less than two standard deviations above the mean signal intensity of the remote myocardium (i.e., SI $\leq \mu_{remote} + 2\sigma_{remote}$) were labeled as remote zone. Small ROIs were then drawn in the brightest regions of the otherwise unlabeled myocardium to determine the mean signal intensity of the infarcted myocardium ($\mu_{infarct}$). Following the segmentation method used by Schelbert et al. [12], a halfway value was calculated by averaging the mean of the remote myocardium and the mean of the infarcted myocardium. Voxels with signal intensity greater than two standard deviations above the mean of the remote myocardium, but less than the halfway value (i.e., $\mu_{remote} + 2\sigma_{remote} < \text{SI} < \frac{1}{2}(\mu_{remote} + \mu_{infarct})$) were labeled as border zone. The remaining voxels with signal intensity above the halfway value (SI $\geq \frac{1}{2}(\mu_{infarct} + \mu_{remote})$) were labeled as infarct.

2.3 Data Analysis

In order to compare tensor invariants and eigenvalues across remote, border, and infarct zones, 100 voxels per zone were randomly chosen from the previously labeled regions on the T1-weighted images for each subject. The labeled ROIs were then rigidly registered to the corresponding cDTI images, with the goal of computing FA, MD, mode, eigenvalues (e_1, e_2, and e_3) and radial diffusivity (Rd) – defined as the mean of e_2 and e_3 – associated with the remote, border, and infarct regions. Since a mold was inserted to preserve the heart configuration at mid-diastasis during the *ex vivo* image acquisition, a rigid registration was sufficient to register ROIs in the T1-weighted images to the corresponding cDTI regions.

Data was visualized using maps on representative slices as well as box and rain-cloud plots in order to compare the distributions of tensor invariants and eigenvalues across remote, border, and infarct regions. T1- and T2-weighted images were normalized by the standard deviation of the noise per slice.

All infarct, border, and remote data were screened using a non-parametric Kruskal-Wallis test with Bonferroni *ad hoc* adjustment. Pairwise non-parametric Mann-Whitney U test was also performed on the infarct-remote, border-remote, and infarct-border regions. $p < 0.01$ was considered significant.

3   Results

Figure 1 illustrates five representative T1-weighted and T2-weighted short axis slices from an infarcted heart along with their corresponding $b = 0\,\mathrm{s/mm^2}$ and $b = 1000\,\mathrm{s/mm^2}$ images.

Figure 2 illustrates five representative T1-weighted short-axis slices from the same infarcted heart, along with the corresponding FA, MD, e_1, e_2, e_3, and Rd maps. The border and infarct regions are identifiable due to their hyperenhanced nature in the T1-weighted images. In the corresponding quantitative maps, the infarct and border regions are marked by a higher MD, e_1, e_2, e_3, Rd and a lower FA compared to the remote myocardium.

Figure 3 quantitatively illustrates the differences in MD, FA, Mode, e_1, e_2, e_3, Rd, T1, and T2-weighted normalized signals in the infarct, border, and remote regions. The upper and lower edges of each boxplot indicate the 25th (Q1) and 75th (Q3) percentiles and the top and bottom whiskers indicate 1.5 times the interquartile range measured from Q1 and Q3. Overall mean, median, and IQR across subjects are further summarized in Table 1.

Figure 4 illustrates the percent difference – per subject and overall – observed in median values for border and infarct regions with respect to the remote myocardium. Amongst diffusion tensor invariants, eigenvalues, and Rd the largest percent differences are recorded for e_2 (e_2 increases by 36% and 115% from remote to border and infarct zones, respectively), e_3 (e_3 increases by 40% and 155% from remote to border and infarct zones, respectively), and Rd (Rd increases by 37% and 128% from remote to border and infarct zones, respectively). Mean diffusivity (MD) and radial diffusivity (Rd) per subject are also reported in Fig. 5 (ref. Appendix).

In all statistical analysis (ref. Sect. 2.3), the computed p values were less than 0.01, rejecting the null hypothesis that observations were part of the same distributions.

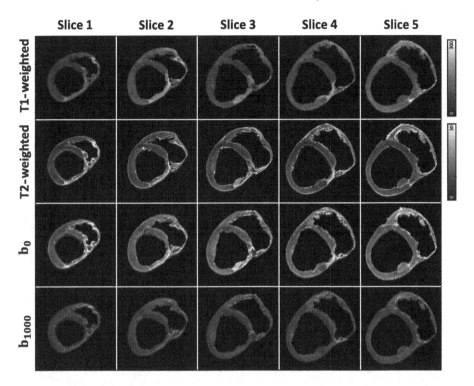

Fig. 1. T1-weighted, T2-weighted, $b = 0\,\mathrm{s/mm}^2$, and $b = 1000\,\mathrm{s/mm}^2$ images for representative short-axis slices in an infarcted subject (subject number 3 in the current study).

4   Discussion

In this study, *ex vivo* cDTI data acquired six to ten weeks after myocardial infarction was used to quantify differences in tensor invariants and eigenvalues in infarct, border, and remote myocardial regions. Rd, e_2, and e_3 present the largest increase in the infarct and border zones compared to the remote myocardium.

Across all subjects, FA and Mode decreased in the infarct and border regions, while MD, eigenvalues, and Rd increased. In all cases, the percent increase or decrease was higher in the infarct region than in the border zone. These trends concur with previous studies [2,4,7]. After chronic myocardial infarction, it has been histologically validated that the infarct region is largely comprised of replacement fibrosis [2] causing an overall increase in the extracellular volume ratio [9]. This increase in extracellular volume ratio likely drives the observed increase in MD and eigenvalues and decrease in FA found in the infarct region.

Indeed, a higher extracellular volume fraction accords with less hindered water self-diffusion owing to fewer cardiomyocyte diffusive barriers compared to healthy myocardium. In fact, water molecules in healthy myocardium are

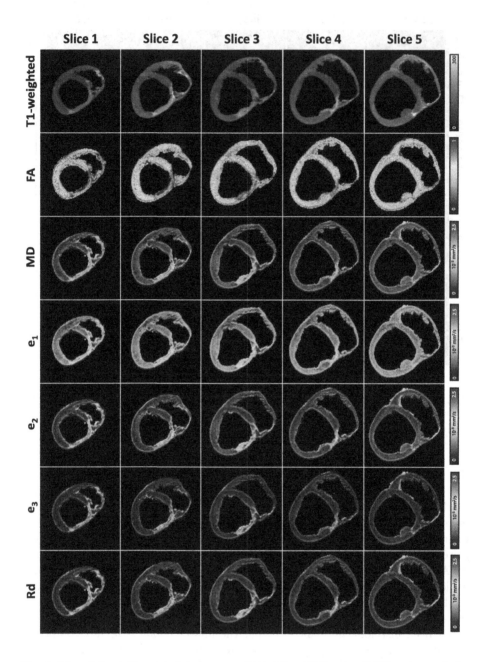

Fig. 2. T1-weighted, Fractional Anisotropy (FA), Mean Diffusivity (MD), primary (e_1), secondary (e_2), and tertiary (e_3) eigenvalue maps for representative short-axis slices. Infarct and border regions are visible from diffusion tensor invariants, eigenvalues, and Rd maps. Images are shown for subject 3 in the current study.

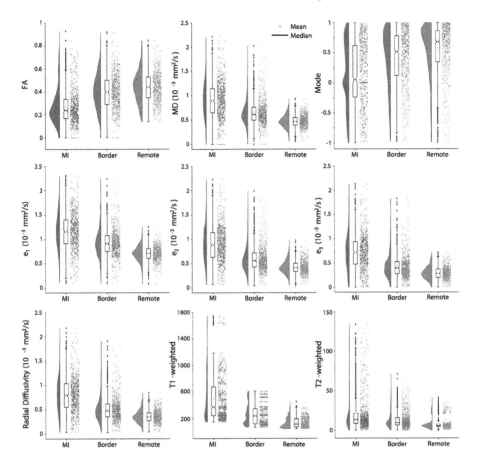

Fig. 3. Rain-cloud plots [1] overlaid to the corresponding box plots for FA, MD, Mode, eigenvalues e_1, e_2, and e_3, radial diffusivity (Rd), normalized T1- and T2-weighted signal intensities in the infarct, border, and remote myocardial regions.

more constrained in the radial direction, therefore the increase in self-diffusion in infarcted myocardium is noticeably larger in the radial direction.

This agrees with the markedly higher e_2 (associated to the sheetlets' direction) and e_3 (associated with the direction normal to the sheetlets) values in the infarct region compared to e_1 values (associated with the aggregate cardiomyocyte preferential direction). Similarly, although less significant, an unequal increase in e_1, e_2, and e_3 with respect to the remote myocardium is visible in the border region, which is likely a mixture of scar tissue and viable myocardium.

Radial diffusivity is a concise measure representative of the changes in the sheetlet and cross fiber directions, where, as discussed above, the largest differences due to an increase in extracellular volume ratio are expected. Accordingly, together with e_2 and e_3, Rd increased the most in the infarct and border regions compared to the remote myocardium.

Table 1. Overall mean, median, and Q1–Q3 computed over all subjects for FA, MD, Mode, eigenvalues, Rd, T1- and T2-weighted normalized signals.

	FA			MD $(1e - 3\,mm^2/s)$			Mode		
	Infarct	Border	Remote	Infarct	Border	Remote	Infarct	Border	Remote
Mean	0.27	0.40	0.44	0.91	0.65	0.47	0.14	0.40	0.54
Median	0.24	0.40	0.44	0.90	0.61	0.46	0.05	0.52	0.67
Q1, Q3	0.17, 0.34	0.29, 0.50	0.35, 0.53	0.65, 1.15	0.49, 0.76	0.39, 0.55	−0.24, 0.62	0.12, 0.78	0.34, 0.86
	e_1 $(1e - 3\,mm^2/s)$			e_2 $(1e - 3\,mm^2/s)$			e_3 $(1e\text{-}3\,mm^2/s)$		
	Infarct	Border	Remote	Infarct	Border	Remote	Infarct	Border	Remote
Mean	1.16	0.93	0.71	0.90	0.60	0.42	0.72	0.44	0.29
Median	1.16	0.91	0.70	0.88	0.55	0.41	0.72	0.39	0.28
Q1, Q3	0.91, 1.40	0.75, 1.10	0.61, 0.80	0.63, 1.14	0.43, 0.71	0.33, 0.50	0.48, 0.93	0.29, 0.53	0.20, 0.37
	Rd $(1e\text{-}3\,mm^2/s)$			T1-weighted			T2-weighted		
	Infarct	Border	Remote	Infarct	Border	Remote	Infarct	Border	Remote
Mean	0.81	0.52	0.36	491	253	138	21.1	13.0	8.2
Median	0.79	0.47	0.35	374	235	112	13.1	9.4	5.6
Q1, Q3	0.55, 1.04	0.35, 0.62	0.27, 0.43	243, 678	117, 346	72, 191	8.4, 21.2	6.5, 15.6	4.6, 7.3

The diffusion tensor eigenvalues and invariants reported in the current study are in good agreement with previously computed values. For example, the mean e_1, e_2, and e_3 values computed in Pashakhanloo et al. [8] compared to the ones computed in the current study (values computed in the current study are reported in parenthesis) were, respectively: $1.125(1.16)e - 3\,mm^2/s$, $0.923(0.90)e - 3$ mm^2/s, and $0.725(0.72)e - 3\,mm^2/s$ in the infarct region; 0.901 $(0.71)e - 3\,mm^2/s$, 0.625 $(0.42)e - 3\,mm^2/s$, and $0.495(0.29)e - 3\,mm^2/s$ in the remote myocardium.

Pashakhanloo et al. [8] reported mean MD and FA in the infarcted tissue to be $0.924e - 3\,mm^2/s$ and 0.23 respectively. These values are similar to mean MD and FA values computed in this study and equal to $0.91e - 3\,mm^2/s$ and 0.27, respectively. However, in the non-infarct regions, Pashakhanloo et al. [8] have reported mean MD and FA to be $0.674e - 3\,mm^2/s$ and 0.31, respectively, whereas mean MD and FA values computed in the current study were $0.47e - 3\,mm^2/s$ and 0.44, respectively. This larger discrepancy may be due to the fact that in [8] the myocardium has been subdivided in remote and infarct regions only, as opposite to the current study where a border region has also been considered. A higher MD and lower FA in the remote myocardium may be a consequence of including part of the border region in the remote myocardium in studies considering only two regions. Differences in the precise tissue handling methods and duration from explant to imaging may also play a role.

Kung et al. [4] computed median FA values equal to 0.464, 0.417, and 0.330 in the remote, border, and infarct regions, respectively, whereas the corresponding values computed in the current study are 0.44, 0.40, and 0.24 respectively.

This study also presents limitations. First, only seven subjects were included. Although the data across all subjects is consistent, future studies should include a larger cohort. Moreover, all data were acquired *ex vivo*. This allowed acquiring high resolution and high SNR data without motion artifacts. Further studies using

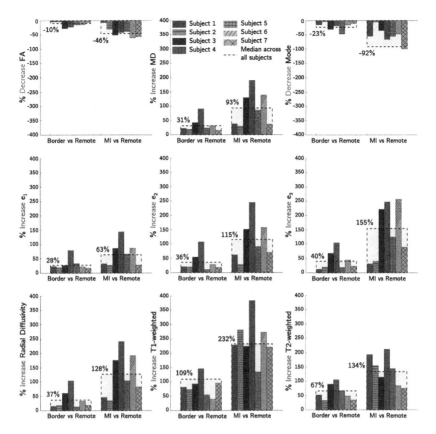

Fig. 4. Bar plots illustrating percent differences in median FA, MD, Mode, e_1, e_2, e_3, Rd, T1-weighted, and T2-weighted normalized signal intensities in the border and infarct regions with respect to the remote myocardium. The dotted line indicates the median percentage change when data from all subjects are considered together. The percent differences reported directly on the plots are computed between remote-border and remote-infarct overall medians.

in vivo data at a lower resolution and SNR are necessary. Additional work is also needed to compare how border and infarct zones detected with cDTI data alone compare to LGE data. Owing to its relatively large changes, Rd is as a key candidate to determine if infarct and border zones are adequately detected from cDTI data. Finally, the size of the infarct varied significantly across subjects, likely leading to variations in the border zone and extent of the microstructural changes. Although controlling the infarct size is challenging, a larger sample size will enable better understanding the effect of infarct size on the measured cDTI quantities.

In conclusion, diffusion tensor invariants, eigenvalues, and Rd are promising biomarkers that can be acquired without the use of a contrast agent and can help detecting and quantifying the microstructural changes occurring in infarct and border regions.

Appendix

Mean diffusivity (MD) and Radial diffusivity (Rd) data are reported separately for each subject in Fig. 5. As also evident in Fig. 4, there is a significant variability across subjects, in part as a function of infarct size.

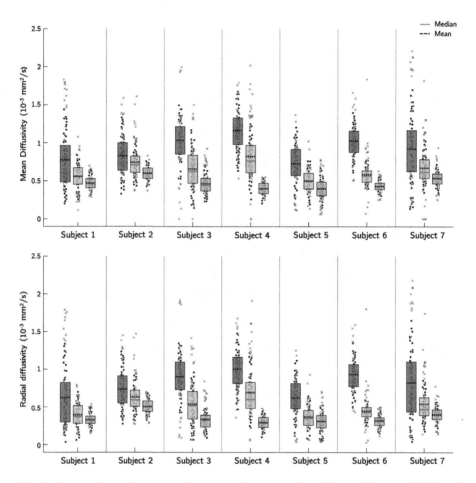

Fig. 5. Grouped box plots overlaid on top of corresponding Mean diffusivity (MD) and Radial diffusivity (Rd) data points (cross and circular markers) in the infarct (left, purple), border (center, yellow), and remote (right, blue) myocardial regions for each subject. Cross markers represent data points 1.5 IQR above Q3 or below Q1. (Color figure online)

References

1. Allen, M., Poggiali, D., Whitaker, K., Marshall, T.R., Kievit, R.A.: Raincloud plots: a multi-platform tool for robust data visualization. Wellcome Open Res. **4**, 63 (2019)
2. Chen, J., et al.: Remodeling of cardiac fiber structure after infarction in rats quantified with diffusion tensor MRI. Am. J. Physiol. Heart Circ. Physiol. **285**(3), H946–H954 (2003)
3. Cork, T.E., Perotti, L.E., Verzhbinsky, I.A., Loecher, M., Ennis, D.B.: High-resolution *Ex Vivo* microstructural MRI after restoring ventricular geometry via 3D printing. In: Coudière, Y., Ozenne, V., Vigmond, E., Zemzemi, N. (eds.) FIMH 2019. LNCS, vol. 11504, pp. 177–186. Springer, Cham (2019). https://doi.org/10.1007/978-3-030-21949-9_20
4. Kung, G.L., et al.: Microstructural infarct border zone remodeling in the post-infarct swine heart measured by diffusion tensor MRI. Front. Physiol. **9**, 826 (2018)
5. Li, X., et al.: Real-time 3T MRI-guided cardiovascular catheterization in a porcine model using a glass-fiber epoxy-based guidewire. PloS one **15**(2), e0229711 (2020)
6. MATLAB: version 9.7.0 (R2019b). The MathWorks Inc., Natick, MA (2019)
7. Moulin, K., et al.: MRI of reperfused acute myocardial infarction edema: ADC quantification versus T1 and T2 mapping. Radiology **295**(3), 542–549 (2020)
8. Pashakhanloo, F., et al.: Submillimeter diffusion tensor imaging and late gadolinium enhancement cardiovascular magnetic resonance of chronic myocardial infarction. J. Cardiovasc. Magn. Reson. **19**(1), 9 (2017). https://doi.org/10.1186/s12968-016-0317-3
9. Pop, M., et al.: Quantification of fibrosis in infarcted swine hearts by ex vivo late gadolinium-enhancement and diffusion-weighted MRI methods. Phys. Med. Biol. **58**(15), 5009 (2013)
10. Porter, D.A., Heidemann, R.M.: High resolution diffusion-weighted imaging using readout-segmented echo-planar imaging, parallel imaging and a two-dimensional navigator-based reacquisition. Magn. Reson. Med. Off. J. Int. Soc. Magn. Reson. Med. **62**(2), 468–475 (2009)
11. Reese, T.G., Heid, O., Weisskoff, R., Wedeen, V.: Reduction of eddy-current-induced distortion in diffusion MRI using a twice-refocused spin echo. Magn. Reson. Med. Off. J. Int. Soc. Magn. Reson. Med. **49**(1), 177–182 (2003)
12. Schelbert, E.B., et al.: Late gadolinium-enhancement cardiac magnetic resonance identifies postinfarction myocardial fibrosis and the border zone at the near cellular level in ex vivo rat heart. Circ. Cardiovas. Imaging **3**(6), 743–752 (2010)
13. Wu, E.X., et al.: MR diffusion tensor imaging study of postinfarct myocardium structural remodeling in a porcine model. Magn. Reson. Med. **58**(4), 687–695 (2007)
14. Wu, M.T., et al.: Sequential changes of myocardial microstructure in patients post-myocardial infarction by diffusion-tensor cardiac MR: correlation with left ventricular structure and function. Circ. Cardiovas. Imaging **2**(1), 32–40 (2009)

Spatially Constrained Deep Learning Approach for Myocardial T1 Mapping

María A. Iglesias[3], Oscar Camara[3], Marta Sitges[2], and Gaspar Delso[1,2(✉)]

[1] GE Healthcare, Barcelona, Spain
[2] Hospital Clínic de Barcelona, Barcelona, Spain
delso@clinic.cat
[3] Physense, Department of Information and Communication Technologies,
Universitat Pompeu Fabra, Barcelona, Spain

Abstract. Parametric cardiac magnetic resonance techniques, such as
T1 mapping with MOLLI sequences, enable quantitative imaging of tissue properties, which can be a powerful tool in the diagnosis and prognosis of different cardiovascular conditions. Conventional parameter estimation methods are often based on pixel-wise curve fitting, ignoring
spatial information. In this study, an automatic pipeline based on a spatially constrained deep learning algorithm is presented, to compute the
myocardial T1 values from MOLLI sequences, within clinically acceptable computation times. The proposed algorithm is based on the Deep-
BLESS architecture, modified to incorporate local spatial information
and regularization. The model was trained on a large database of clinical MOLLI cases (from 186 patients), showing promising preliminary
results, obtaining T1 maps faster and more robust to noise.

Keywords: T1 mapping · MOLLI series · Convolutional Neuronal
Networks · Regularization

1 Introduction

Cardiovascular magnetic resonance (CMR) provides insights into cardiac morphology, myocardial architecture, blood flow, tissue perfusion and function with
high diagnostic accuracy and without needing ionizing radiation [1]. Quantitative relaxation mapping techniques deliver objective diagnostic criteria based on
non-invasive characterization of tissue physical parameters [2,3]. Such is the case
of myocardial T1 mapping, which allows the detection of heart muscle abnormalities and is a proven indicator of several myopathies related to changes of
extracellular composition (e.g., edema, fibrosis) as well as fat, iron and amyloid
content [2–8].

The standard approach is based on estimating the spin-lattice relaxation
time (T1) with Modified Look-Locker Inversion recovery (MOLLI) sequences
[9], which rely on multiple Look-Locker inversion pulses over several heartbeats
to acquire a series of images with varying inversion times (TI). This allows the
estimation of T1 maps of the myocardium within one breath-hold.

© Springer Nature Switzerland AG 2021
D. B. Ennis et al. (Eds.): FIMH 2021, LNCS 12738, pp. 148–158, 2021.
https://doi.org/10.1007/978-3-030-78710-3_15

Traditional cardiac T1 mapping with MOLLI sequences is based on pixel-by-pixel estimation performed with exponential curve fitting [7], typically with non-linear least square method or Levenberg-Marquardt method. This approach is time-consuming and neglects spatial information, since neighboring pixel intensities are generally not considered, due to the additional computational cost of incorporating a regularization term in the optimization. For this reason, regularized T1 mapping methods are not currently applied in clinical practice.

Due to the frequent presence of artifacts in breath-hold cardiac imaging (e.g., respiratory motion, triggering accuracy, etc.) it would be desirable to include spatial information in the T1 map calculation, also in a time-efficient manner compatible with the clinical workflow. Hence, our main objective was to introduce spatial regularization without compromising the computational time of T1 map estimation. While many studies can be found in the literature using Deep Learning methods (DL) for CMR motion correction, artifact reduction or accelerated reconstruction [10,11], we are not aware of any prior work demonstrating clinically-usable regularized DL-based T1 mapping.

Although DeepBLESS model proposed by Shao, J. et al. [12] did not considered spatial regularization, it showed promising results in pixel-wise T1 estimation, however its training was mainly based on synthetic data using Block Equation and phantom studies since it only included 8 real cases of healthy volunteers, hence its training on bigger data set of real cases should be performed to evaluate the model. Also, a study adding regularization with U-net model for T1 estimation was presented by Jeelani, H., et al. [13], however timestamps are not included, making the model not clinically-usable.

The approach presented in our study consists of an automated heart detection algorithm, followed by regularized T1 map computation with a convolutional neural network. The proposed method has been validated on a large clinical database of CMR cases (186 patients).

2 Methods

2.1 Dataset

A cohort of 186 patients referred for a clinically-indicated cardiac scan (either with cardiomyopathies as patients with risk factors for atrial fibrillation), was included in this study. Demographic data of the patients is shown in Table 1. Scans were performed on a 3.0T GE Signa Àrchitect at Hospital Clínic de Barcelona (Spain). The acquisition protocol included one or more MOLLI sequences, with the following target parameters: 2D balanced steady-state free precession (bSSFP), 160×148, phase field-of-view 0.8–1.0, 1.4×1.4 mm, slice thickness 8 mm, echo time 1.4 ms, repetition time 3.0 ms, flip angle 35 deg, 1 average, bandwidth 100 kHz, $2\times$ ASSET acceleration.

Table 1. Demographic data of the cohort of 186 patients used.

Demographic data	
Sex	115 male, 71 female
Weight	75±15
Age	55±17

The acquisition scheme used was 5(3 s)3 (5 heartbeats, 3 s pause and 3 more heartbeats), with a Body48 radiofrequency coil mode (i.e., anterior and posterior arrays, 48 cm coverage), which is the predefined combination of elements of both acquisition antennas. Motion-corrected reconstructions and T1 maps by exponential fitting were exported for each case.

2.2 Identification of Myocardial Tissue for Training

The correct prediction of myocardial T1 values has been prioritized during the training of the DL model. For this purpose, prior to training, the myocardium was first automatically identified in each of the MOLLI series. The whole heart was extracted with a U-net convolutional neural network (CNN), as in [14], hence only myocardial tissue and blood pool are included in our mask.

All algorithms evaluated in the present study were implemented in Python 3.8, using the Tensorflow (Google LLC, CA) version 2.0.0 and Keras [15] version 2.3.1 libraries. The U-net architecture was based on the one described by Ronneberger et al. [16]. For the training of the segmentation task, binary cross-entropy was selected as loss function, with ADAM [17] as optimizer. As pre-processing, images were normalized based on the soft tissue histogram peak. Manually segmented MOLLI frames (n = 1137) were used to train the model (90% train, 10% test). Data augmentation (i.e., shift, rotation, zoom) was carried out on-the-fly. A post-processing step based on connected component analysis was applied to eliminate regions outside the heart. The evaluation of the results included Dice scores, specificity, sensitivity, precision and Hausdorff distance. Qualitative validation of the obtained masks was performed.

With the purpose of reducing even more the computational time required for myocardial T1 mapping, the heart mask CNN algorithm was not only used with the objective to prioritize the myocardial and blood pool T1 values for training the model, but also used to extract the cardiac bounding box.

2.3 Spatially Constrained Deep Learning T1 Mapping

To implement the T1 map calculation, the DeepBLESS [14] architecture was selected, based on a literature review since it has shown promising pixel-wise estimation performance. The method was modified to incorporate information of the local spatial neighborhood of each temporal series. The basic architecture of the model is shown in Fig. 1. It takes as input a 1D signal with 2 channels: the first channel encodes the acquisition timestamps of the MOLLI samples; the second channel encodes the temporal series of intensity values at a given pixel. The data is then processed by 13 layers, including convolutional, residual, and dense networks.

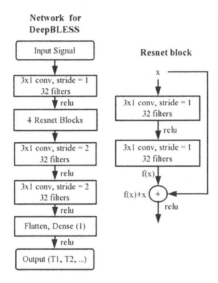

Fig. 1. DeepBLESS architecture used for pixel-wise T1 value prediction.

In its constrained form, an additional input vector was included, encoding the information about the local spatial neighborhood of the pixel being processed. A basic embodiment of this concept was evaluated, where the vector's elements represented the median value of the 8 neighbor's intensities at each timeframe. The input is extended beyond boundaries by reflecting about the edge of the last pixel to obtain the median value there.

The model was pre-trained with synthetic data, using samples generated by the Bloch Equation of an inversion recovery acquisition. The timestamp series for the simulations were extracted from the clinical MOLLI database described. As target output, the corresponding T1 maps obtained with the motion correction reconstruction algorithm of the scan and exponential fitting were used.

Once the training converged with synthetic data, re-training was performed using the database of clinical MOLLI series. The cardiac cavity was automatically extracted from each series with the CNN, as previously described. To ensure that myocardial tissue pixels were prioritized during training, intensity normalization and heuristic thresholding were applied to reduce blood pool and residual background sampling. Pixels belonging to the heart mask were assigned a pre-defined probability of still being included in the training despite not belonging to the intensity and T1 ranges expected for myocardial tissue.

Finally, regarding the training of the regularized model, synthetic exponentials based on real parameters and timestamps from the dataset are created, keeping the same procedure of myocardial prioritization described. Afterwards, noise was added to these exponentials. The additive noise was uniformly distributed in a range of 1% of the maximum signal level. Additionally, multiplicative impulse noise was applied with a 5% probability and [0.9, 1.1] uniform distribution. This extra noise aims to model movement artifacts, commonly present in real data.

For T1 mapping estimation, the Mean Square Error (MSE) was used as loss function with ADAM as optimization method. For training, 90% of the dataset was used, the remaining 10% was used as test set. A comparison between the predicted values and the curve-fitting values used as ground truth was carried out to confirm the algorithm performance.

A comparison of the average time required to obtain the T1 map per case with the DL algorithm versus with exponential fitting based on non-linear least square method was done, with and without the use of the bounding box extraction based on the heart mask, to prove the contribution of the DL itself in the speed up of the computation.

3   Results

Concerning the results of the U-Net CNN-based heart identification task, it took on average 0.13 s to infer the segmentation of an image (including the post-processing step), when run with NVIDIA GTX 1050 GPU. The evaluation metrics obtained for the heart mask segmentation are shown in Table 2. For a qualitative assessment of the obtained segmentation results, Fig. 2 shows representative inferred heart masks.

As for the results obtained with the DeepBLESS-based algorithm, a comparison between the spatially constrained DL T1-mapping method and the standard algorithm based on curve fitting was carried out.

Table 2. Evaluation metrics of the obtained whole heart segmentations.

	Mean Value	Standard deviation
DICE	0.93	0.04
Hausdorff (mm)	3.4	0.6
Sensitivity (%)	92.1	6.6
Specificity (%)	99.6	0.4
Precision (%)	95.2	6.0

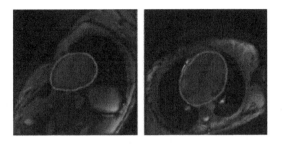

Fig. 2. Sample of the inferred heart masks obtained with U-Net.

The average times required for computing the T1 values with the DL-based method and with curve-fitting based on non-linear least square method, running both methods with NVIDIA GTX 1050 GPU, are presented in Table 3. The computational time has been measured having as input the whole image as with only the bounding box containing the region of interest.

Table 3. Average computational times required per case with the regularized DL and curve-fitting (non-linear least squares method) methods, with and without the inferred heart bounding box.

	Regularized DL	Curve-fitting
With heart bounding box	3.6 s	6.0 s
Without heart bounding box	10.0 s	55.2 s

Regarding the effect of the noise added to corrupt the synthetic exponentials, the Mean Signal-to-Noise Ratio (SNR) value of the original exponential values used for training was 3.29. As for synthetic exponentials, the mean SNR decreased from 3.15 to 3.08 after corrupting them with noise.

A summary of the training results of the regularized model is shown in Fig. 3 with scatter and Bland-Altman plots. The mean of the absolute difference between the DL-based and curve-fitting T1 values estimation was 13.2 ± 15.2. Regarding the statistical significance (assessed with t-test), the t-value was 0.83 and the corresponding p-value is 0.41. The Root Mean Square Error (RMSE) was 20.2 while the Pearson's correlation coefficient was 0.9969. Figure 4 depicts an example of the T1 maps obtained with the DL-based algorithm without regularization, its corresponding ground truth (e.g., curve-fitting algorithm), and the difference between them.

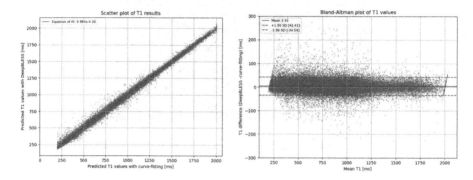

Fig. 3. Left: predicted T1 values with the regularized deep learning implementation vs the reference prediction obtained with pixel-wise curve fitting. Right: Bland-Altman plot of the same results.

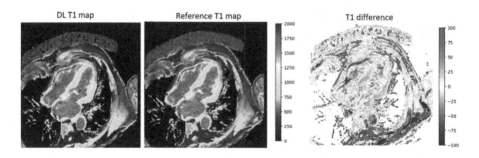

Fig. 4. Example of the reference and deep learning without regularization DeepBLESS T1 maps obtained on clinical data, along with the corresponding color-coded difference maps. (Color figure online)

Regarding the resulting T1 maps of the model adding regularization, Fig. 6 illustrates an example of the T1 maps obtained with the proposed DL-based

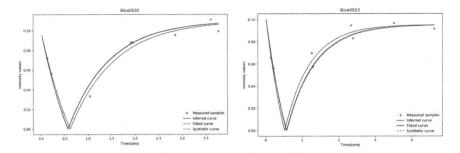

Fig. 5. Example of intensity exponential values prediction of pixels from the myocardium. Showing the normalized input intensity values (corrupted exponential with noise) as a cross, the inferred curve with DL in blue, the exponential curve obtained with curve-fitting prediction in red, and in green the original synthetic exponential without noise. (Color figure online)

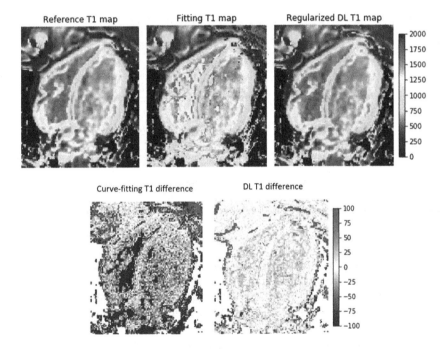

Fig. 6. Example of the reference and calculated T1 maps on synthetic data corrupted with random and impulse noise, with the regularized DL algorithm and with standard curve-fitting. Color-coded maps of the difference with respect to the reference are provided in each case.

algorithm, with and without regularization (curve-fitting method), and its corresponding ground truth. For a clearer comparison, the heart bounding box has also been inferred for the curve-fitting estimation.

Knowing the T1 values of surrounding heart structures could be interesting from a clinical perspective, it has been also considered the algorithm performance in the prediction of those structures taking as input the whole image, which is illustrated in Fig. 7.

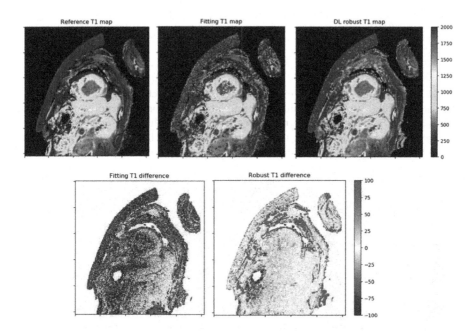

Fig. 7. Example with whole image as input of the reference and calculated T1 maps on synthetic data corrupted with random and impulse noise described. Color-coded difference maps with respect to the reference are provided in each case.

4   Discussion

The traditional T1-mapping estimation techniques from MOLLI sequences still could be further improved. One approach is by making it faster through its computation with deep learning. Furthermore, in the traditional curve-fitting pixel-by-pixel approach, spatial relationships are being ignored despite there is correlation between tissue properties at neighboring pixels.

In the present study, the main objective has been fulfilled since the implemented algorithm results achieve a performance similar to curve-fitting results, but in a faster way as shown in Table 3, in which the DL implementation is almost twice as fast even without the use of the bounding box. The interest in speeding up the computation time of algorithms increases with the new 3D acquisition sequences with free breathing, which cope with more data making them slower due to computational burden. In Fig. 5 can be observed that curve-fitting tries to adapt to noisy data, whereas regularized DL gets closer to the synthetic curve, as if it is more robust to noise.

Regarding the difference image between predictions in Fig. 7 and 6, it can be observed that the algorithm is focused on myocardial T1 value prediction, since the background was erroneously predicted for not being included during network training. Also, it can be appreciated more issues predicting blood pool than myocardial tissue, since although some blood pool pixels were included for training, myocardial tissue was prioritized. Consequently, blood pool T1 values estimation could be further enhanced modifying the training to improve Extracelullar Volume (ECV) mapping.

Furthermore, in order to focus on relevant myocardial tissue, pixels belonging to the intensity and T1 range values expected for myocardial tissue were considered for training. As a consequence, the present algorithm is more focused on healthy tissue estimation rather than pathological one. A future approach could be to extract the blood pool in an alternative way, not based on T1 and intensity values, to increase the presence of pathological pixels in the training. Also, concerning the training, the chosen hyperparameters could be further optimized.

Having demonstrated that the DL implementation was more time-efficient and robust to noise than traditional curve-fitting, the next step would be to explore which is the optimal methodology for introducing spatial regularization including more pathological cases for the training and improving the blood pool T1 values estimation. Due to the robustness of the algorithm, the training could be modified to include images with lower resolution in order to make the model able to cope with images with worse quality.

Finally, concerning the myocardial tissue identification for the training of the neural network, it can be concluded from the metrics that the segmentation goal has been successfully achieved. However, although the heart masks obtained were good enough for the proposed problem, they could be further improved by training separately pre- and post-contrast images, since most of the segmentations that obtained poor Dice scores are post-contrast images.

5   Conclusions

In the present article a spatially constrained deep learning algorithm was implemented to estimate T1 values over the cardiac region. The obtained results were compared with the traditional exponential fitting method. The results suggest a successful implementation of DL to predict T1 values in a faster way and including local spatial information.

References

1. von Knobelsdorff-Brenkenhoff, F., Schulz-Menger, J.: Role of cardiovascular magnetic resonance in the guidelines of the European Society of Cardiology. J. Cardiovasc. Magn. Reson. **18**(1), 6 (2015)
2. Matsumoto, S., et al.: Myocardial T1 values in healthy volunteers measured with saturation method using adaptive recovery times for T1 mapping (SMART1Map) at 1.5 T and 3 T. Heart Vessels **34**(11), 1889–1894 (2019)

3. Burkhardt, B.E.U., Menghini, C., Rücker, B., Kellenberger, C.J., Valsangiacomo Buechel, E.R.: Normal myocardial T1 values in Children using Single-point saturation recovery and Modified Look-Locker Inversion Recovery (MOLLI). Int. Soc. Magn. Reson. Med. **51**(3), 897–903 (2020)
4. Kellman, P., Hansen, M.S.: T1-mapping in the heart: accuracy and precision. J. Cardiovasc. Magn. Reson. **16**(1), 2 (2014)
5. Jellis, C.L., Kwon, D.H.: Myocardial T1 mapping: modalities and clinical applications. Cardiovasc. Diagn. Therapy **4**(2), 126–137 (2014)
6. Radenkovic, D., Weingärtner, S., Ricketts, L., Moon, J.C., Captur, G.: T1 mapping in cardiac MRI. Heart Fail. Rev. **22**(4), 415–430 (2017)
7. Taylor, A.J., Salerno, M., Dharmakumar, R., Jerosch-Herold, M.: T1 mapping: basic techniques and clinical applications. JACC Cardiovasc. Imaging **9**(1), 67–81 (2016)
8. Puntmann, V.O., Peker, E., Chandrashekhar, Y., Nagel, E.: T1 mapping in characterizing myocardial disease: a comprehensive review. Circ. Res. **119**(2), 277–299 (2016)
9. Messroghli, D.R., Radjenovic, A., Kozerke, S., Higgins, D.M., Sivananthan, M.U., Ridgway, J.P.: Modified look-locker inversion recovery (MOLLI) for high-resolution T1 mapping of the heart. Magn. Reson. Med. **52**(1), 141–6 (2004)
10. Zhang, Q., et al.: Deep learning with attention supervision for automated motion artefact detection in quality control of cardiac T1-mapping. Artif. Intell. Med. **110** (2020)
11. Leiner, T., et al.: Machine learning in cardiovascular magnetic resonance: basic concepts and applications. J. Cardiovasc. Magn. Reson. **21**(1), 61 (2019)
12. Jeelani, H., Yang Y., Zhou, R., Kramer, C., Salerno, M., Weller, D.: A myocardial t1-mapping framework with recurrent and U-Net convolutional neural networks. In: IEEE 17th International Symposium on Biomedical Imaging (ISBI) (2020)
13. Shao, J., Ghodrati, V., Nguyen, K., Hu, P.: Fast and accurate calculation of myocardial T1 and T2 values using deep learning Bloch equation simulations (DeepBLESS). Magn. Reson. Med. **84**(5), 2831–2845 (2020)
14. Iglesias, M.A., et al.: Improving T1 mapping robustness by automatic segmentation of myocardial tissue in MOLLI series. ISMRM 2021 (accepted)
15. Chollet, F., et al.: Keras (2015)
16. Ronneberger, O., Fischer, P., Brox, T.: U-net: convolutional networks for biomedical image segmentation. In: International Conference on Medical image computing and computer-assisted intervention, pp. 234–241 (2015)
17. Kingma, D.P., Ba, J.: Adam: a method for stochastic optimization. In: 3rd International Conference for Learning Representations, San Diego, CA, USA (2015)

A Methodology for Accessing the Local Arrangement of the Sheetlets that Make up the Extracellular Heart Tissue

Shunli Wang[1,2], François Varray[1(✉)], Feng Yuan[2], and Isabelle E. Magnin[1]

[1] Univ Lyon, INSA -Lyon, Université Claude Bernard Lyon 1, CNRS, Inserm, CREATIS UMR 5220, U1206, 69621 Villeurbanne, France
`francois.varray@creatis.insa-lyon.fr`
[2] Instrumentation Science and Engineering, Harbin Institute of Technology (HIT), Harbin 150001, China

Abstract. The sheetlet angle is a biomarker that can describe the local arrangement of the sheetlets making the heart wall tissue. It could be of major interest for the analysis of cardiac function. In this preliminary study, we use the *skeleton method* to measure the local sheetlet angle in two human left ventricular transparietal tissue samples. The samples were imaged using synchrotron X-rays phase contrast micro-tomography and reconstructed in 3D with isotropic voxels of 3.5 μm edges. We extract the skeleton from each sheetlet. Next, we scan the skeleton, voxel by voxel, performing principal component analysis (PCA) in a sliding cubic working window (WW) of 112 μm edges. The tertiary eigenvector of the PCA provides the sheetlet angle, step by step, along the skeleton. We show that the proposed methodology is able to provide the transmural distribution of sheetlets angle at a scale of 112 μm and locally to investigate the spatial evolution of the angle along the sheetlets. We compare the results obtained at a larger scale with the Fourier-based method.

Keywords: Left ventricular wall · Laminar sheetlets · Sheetlet angle · Surface skeleton · Synchrotron radiation phase contrast micro-tomography

1 Introduction

Ventricular myocytes are embedded in an extracellular collagen matrix (ECM) within a complex network. The endomysium tightly couples the myocytes into discrete laminar "sheetlets" about two-to-four cells thick while the perimysium appears as "cleavage planes (CP)" located between the adjacent sheetlets [14]. Sheetlets roughly stack from apex to base [9]. This laminar structure is crucial to the cardiac electro-physiology [5] and mechanical properties [1].

Supported by China Scholarship Council (CSC, No. 201806120173) and Metislab, CNRS LIA n°1124, INSA Lyon.

D. B. Ennis et al. (Eds.): FIMH 2021, LNCS 12738, pp. 159–167, 2021.
https://doi.org/10.1007/978-3-030-78710-3_16

Diffusion tensor magnetic resonance imaging (DT-MRI) is an alternative method to measure the sheetlet angle in a non-destructive way [2,4,7]. DT-MRI can measure the sheetlet angle throughout the entire heart *in vivo* or *ex vivo* [12]. However, its low spatial resolution (about $2 \times 2 \times 2\,\mathrm{mm}^3$ in clinics and $60 \times 60 \times 60\,\mu\mathrm{m}^3$ in lab) further limits its application by not allowing, for example, the detection of two populations of sheetlets with two locally distinct angular arrangements.

More recently, the 3D Synchrotron X-rays Phase Contrast Tomography (SR-PCT) was used to investigate the cardiac tissue architecture [16]. SR-PCT can image tissue samples of several cubic centimeters and provides histologic-like contrast of reconstructed volumes (cm^3) [8]. In our group, we have an SR-PCT dataset of human LV wall [11,16,17]. The samples were imaged at the European Synchrotron Radiation Facility located in Grenoble and reconstructed with an isotropic voxel of $3.5\,\mu\mathrm{m}$ edges. Using this dataset, Mirea et al. proposed a Fourier-based method to calculate the average normal direction of sheetlets inside cubic sub-volumes of edges dimensions equal to $0.9\,\mathrm{mm}$ [11]. The large size of the sub-volumes shows results in the DT-MRI resolution range, which is not suitable for analyzing the local layout of the sheetlets, especially for evaluating the evolution of the sheetlet angle along a single sheetlet. To succeed, we propose to use the skeleton-based method [17]. The skeleton-based method calculates the sheetlet angle using a working window centered on a voxel belonging to the skeleton of the sheetlet. In this way, we can study the relationship between tissue architecture and sheetlet arrangement. The method makes it possible to measure different characteristics of the tissue as a function of the eigenvector considered.

The paper is organized as follows. First, we give a brief presentation of the SR-PCT dataset. Then, we show how we adapt the skeleton-based method to calculate the local sheetlets angles and highlight results according to the scale. A discussion followed by a short conclusion end the paper.

2 Material

The SR-PCT reconstructed volumes of two LV free wall samples used in this paper (respectively named LV_post_S5_1 and LV_post_S1_2) have been deeply described in [16]. In short, the samples were extracted from heart slices (about $5\,\mathrm{mm}$ thick) of two males aged 32 and 35 years old, who had no heart disease history and died with an unexpected death. The samples cross the LV wall from endocardium to epicardium, revealing the transparietal micro-architecture. The spatial resolution of the reconstructed volumes is isotropic with an edge voxel equal to $3.5\,\mu\mathrm{m}$. Myocytes and laminar structures are visible at this resolution. The size of the cardiac myocytes is approximately $10\,\mu\mathrm{m}$ in diameter and $100\,\mu\mathrm{m}$ in length [6]. The average thicknesses of sheetlets and CPs, are about $180\,\mu\mathrm{m}$ and $30\,\mu\mathrm{m}$ [11] respectively.

We manually select a transmural parallelepipedal sub-volume from the reconstructed volume of each sample to avoid the influence of edge distortion. The edges of each block are parallel to the axes in the local cylindrical (r, c, l) coordinate system, where r, c and l respectively correspond to the radial, circumferential, and longitudinal directions. We note that the myocytes are not always

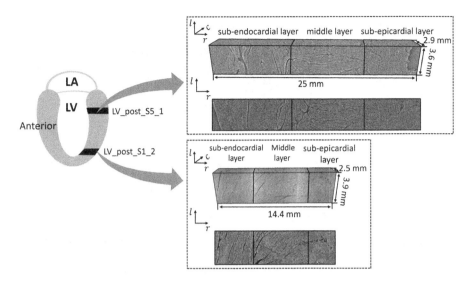

Fig. 1. Reconstructed SR-PCT volumes for samples LV_post_S5_1 and LV_post_S1_2: (up) 3D display, (down) middle rOl plane. LA: left atrium; LV: left ventricle.

organized into sheetlets throughout a sample, as seen in Fig. 1. We manually segment each sample into three layers that are the sub-endocardial, middle and sub-epicardial layers. We observe that the CPs are regular and relatively parallel in the middle layer, so we focus our study on this region-of-interest (ROI).

3 Data Processing Strategy

3.1 The Sheetlet Angle Measured Using the Skeleton Method

Extraction of the Surface Skeleton of the Sheetlets. We extract the surface skeleton of the sheetlets contained in SR-PCT reconstructed volumes [17]. The main processing steps are the following:

(1) Eliminate the slow variations in background intensity of the reconstructed volumes, induced by the acquisition procedure by filtering the low-frequencies with a centered Gaussian filter $G(0, \sigma)$ such that:

$$\hat{D} = D - D * G(0, 44) \tag{1}$$

where $\hat{D}$ denotes the "flattened" data, $*$ the convolution operator and σ the variance given in *voxels*. The kernel of the Gaussian filter is defined on a compact cubic support with edges 177 voxels long. The numerical values, chosen for the edges of the support and for σ, are large enough to keep the structures of interest [10].

(2) Separate ECM from myocytes using Otsu algorithm [13].

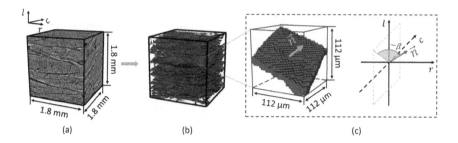

Fig. 2. Main steps for the calculation of the sheetlet angle β using the skeleton method. (a) SR-PCT reconstructed ROI (b) Skeletons of the sheetlets. (c) Sheetlet angle β calculated in a sliding cubic working window (WW) of edges 32 voxels (i.e. 112 μm).

(3) Extract the main CPs by filtering the ECM with a 2D connex filter, successively in each orthogonal plane, to eliminate barbs and small size branching structures.

(4) Create the *distance map* by giving to each voxel of a sheetlet, a value equal to the chamfer $(3, 4, 5)$ distance which separates it from the nearest CP [3].

(5) Select the voxels of maximal chamfer value in the *distance map*. They correspond to centers of balls of maximal size contained in a sheetlet and compose its surface skeleton [15]. Repeat the process for all the sheetlets.

Measurement of the Local Sheetlet Angle. The strategy, used to measure the angle of every sheetlet, consists in extracting the skeleton (Fig. 2b) of each sheetlet from the "flattened" SR-PCT data (Fig. 2a). Then, each voxel of the skeletons is processed using a sliding cubic working window (WW) whoses edges are 112 μm long (i.e. 32 voxels). The processing consists in a principle component analysis (PCA) followed by the selection of the eigenvector corresponding to the tertiary eigenvalue. Indeed, this vector is the mean normal direction $\overrightarrow{n}$ of the skeleton at the WW scale. The sheetlet angle β (Fig. 2c) is defined between $\overrightarrow{n}$ and the local cOl plane.

3.2   The Sheetlet Angle Measured Using the Fourier-Based Method

In previous work [11], Mirea *et al.* proposed a method, that we call the "Fourier-based method", to measure the mean normal direction of CPs in SR-PCT reconstructed volumes. The strategy contains three main steps:

(1) Divide the SR-PCT reconstructed volume into sub-volumes of edge size equals 0.9 mm (i.e. 256 voxels). Perform a fast Fourier transform (FFT) of each sub-volume with an overlapping of 50% between adjacent sub-volumes.

(2) Keep only 4% of the highest frequency components of the FFT spectrum (this step aims at selecting the most significant components of the tissue).

(3) Perform a principle component analysis (PCA) of the filtered spectrum. The primary eigenvalue is the eigenvector $(\overrightarrow{n})$ corresponding to the mean normal

direction of CPs in each sub-volume. The angle between $\overrightarrow{n}$ and the local cOl plane is the sheetlet angle β.

3.3   Angle β : Comparison Between the Skeleton and Fourier-Based Methods

The sheetlet angle β is computed at two different scales by the two methods. The skeleton method provides results at the higher scale of 112 μm compared to the 896 μm of the Fourier based method. To compare the two techniques, we average the values, obtained at a high scale by the skeleton method, in overlapping sub-volumes of edge size equal to 256 voxels identical to the Fourier based method. We consider sub-volumes which overlap at 75% instead of 50%.

4   Results

The 3D distribution and histogram of the sheetlets angle β calculated in the SR-PCT reconstructed ROI (Fig. 2a) are displayed in Fig. 3. Figure 3a depicts the distribution of β obtained by the skeleton-based method. We note the regular profile of the distribution showing a quasi symmetrical shape with most values located in the range $[-20°, +20°]$. On the contrary, the range of values of β obtained from the Fourier-based method Fig. 3c, is much narrower. The values range from $[-5°, +4°]$. This is due to the lower scale of observation. We confirm that hypothesis by comparing the results given by the skeleton method and transposed to the same scale. We note that the values of β decrease to a similar range as shown in Fig. 3b. The shapes of the distributions, at lower scale, are similar between the two methods. The difference in mean and standard deviation between the two techniques is close to 0°.

We now process the middle layer of the reconstructed volumes for samples LV_post_S5_1 and LV_post_S1_2 with the skeleton method. Figure 4 and Fig. 5 depict the distributions of β in the two volumes respectively. The display, in 2D planes is interesting: (1) In LV_post_S5_1 sample, we observe quasi-parallel lines with a β value around 0 (Fig. 4a, green). The arrangement is particularly regular. The global range of β is about $[-50°, +50°]$ with most values between $[-20°, +15°]$. The histogram, Fig. 4b, of the volume and middle layer are quite identical, showing the high regularity, layer after layer, of the 3D sheetlets structure. The profiles of the volume (Fig. 4c, red) and middle layer (Fig. 4c, blue) show the same behavior along the radial direction. The second one experiences oscillations at the scale of 112 μm. (2) In LV_post_S1_2 sample, we observe two populations of sheetlets (Fig. 5a, red and green) with angles respectively close to 60° and −5°. The global range of β is about $[-40°, +80°]$ with most values between $[-10°, +50°]$. In the middle layer, we note a strong discontinuity of the average profile of β along the radial direction (Fig. 5c, blue). It corresponds to the transition between the two populations of sheetlets of different orientations.

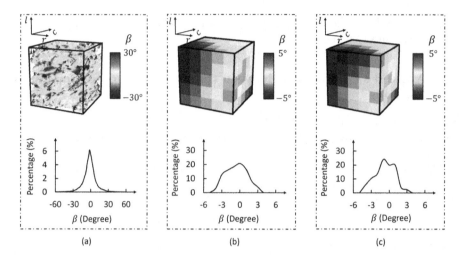

Fig. 3. Sheetlet angle β in the SR-PCT reconstructed ROI: (up) 3D distribution, (down) Histogram. (a) Skeleton-based method (WW edges equal 32 voxels). (b) β measured with the skeleton-based method and averaged in a window of edges 256 voxels. (c) Fourier-based method (WW edges equal 256 voxels).

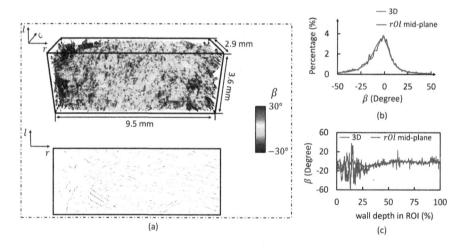

Fig. 4. Middle layer of sample LV_post_S5_1. Sheetlet angle β measured with the skeleton-based method: (a) 3D distribution and middle rOl plane, (b) Histogram, (c) Profile of the average angle β along the radial direction (r axis). (Color figure online)

5   Discussion

In this paper, we measured the sheetlet angle in SR-PCT reconstructed dataset of the human LV wall. Our skeleton-based method evaluates the local inclination of sheetlets' surface skeleton at a high scale. We compared the obtained sheetlet

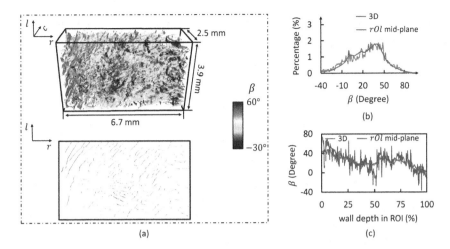

Fig. 5. Middle layer of sample LV_post_S1_2. Sheetlet angle β measured with the skeleton-based method: (a) 3D distribution and middle rOl plane, (b) Histogram, (c) Profile of the average angle β along the radial direction (r axis). (Color figure online)

angles, with those given by the Fourier-based method at a larger scale. The results, at same scale, were similar.

Concerning the methods, we note that the PCA performed to measure the normal direction $\overrightarrow{n}$ of local sheets in single sub-volumes is calculated on the sheet surface skeleton for the skeleton-based method and on the filtered spectrum of the local FFT in the Fourier-based method. The different targeted objects decide that the normal direction of local sheets corresponds respectively to the tertiary eigenvector and the primary eigenvector of the PCA results.

The main interest of the skeleton-based method, is its ability to measure the sheetlet angle at a scale around $100\,\mu m$, which corresponds to the size of tissue structures as sheetlets, CPs, bundles of myocytes. This method makes possible to analyze the evolution of such a biomarker along a sheetlet or through the LV wall, along the radial direction of the tissue. Though promising, the method has still some limitations: it is dependent on the quality of the CPs segmentation and rather time-consuming because not yet optimized. Indeed, it calculates the angle β at each voxel of the sheetlets' surface skeletons.

6   Conclusion

We could successfully measure the sheetlets angles of human LV tissue samples at the scale of myocytes aggregates. This allowed us to identify several populations with locally varying orientations and arrangement between the CPs. Future work will focus on processing more data to get accurate local measures and better understand the local mechanical behavior of the cardiac wall. Moreover, a multiscale strategy coupling the Fourier-based and skeleton methods should allow a global analysis of the orientation of myocytes aggregates from endocardium to epicardium. It would also increase the computation speed in view to provide a useful tool for clinicians.

References

1. Ashikaga, H., Covell, J.W., Omens, J.H.: Diastolic dysfunction in volume-overload hypertrophy is associated with abnormal shearing of myolaminar sheets. Am. J. Physiol.-Heart Circulatory Physiol. **288**(6), H2603–H2610 (2005)
2. Bernus, O., et al.: Comparison of diffusion tensor imaging by cardiovascular magnetic resonance and gadolinium enhanced 3D image intensity approaches to investigation of structural anisotropy in explanted rat hearts. J. Cardiovasc. Magn. Reson. **17**(1), 1–27 (2015)
3. Borgefors, G.: Distance transformations in arbitrary dimensions. Comput. Vis. Graph. Image Process. **27**(3), 321–345 (1984)
4. Hales, P.W., Schneider, J.E., Burton, R.A., Wright, B.J., Bollensdorff, C., Kohl, P.: Histo-anatomical structure of the living isolated rat heart in two contraction states assessed by diffusion tensor MRI. Prog. Biophys. Mol. Biol. **110**(2–3), 319–330 (2012)
5. Hooks, D.A., Trew, M.L., Caldwell, B.J., Sands, G.B., Legrice, I.J., Smaill, B.H.: Laminar arrangement of ventricular myocytes influences electrical behavior of the heart. Circulation Res. **101**(10), e103–e112 (2007)
6. Iaizzo, P.A.: Handbook of Cardiac Anatomy, Physiology, and Devices. Springer, Heidelberg (2009). https://doi.org/10.1007/978-1-60327-372-5
7. Ipp, H., Zemlin, A.: The paradox of the immune response in HIV infection: when inflammation becomes harmful. Clinica Chimica Acta **416**, 96–99 (2013)
8. Lang, S., et al.: Experimental comparison of grating-and propagation-based hard X-ray phase tomography of soft tissue. J. Appl. Phys. **116**(15), 154903-1–154903-12 (2014). https://doi.org/10.1063/1.4897225
9. LeGrice, I.J., Smaill, B., Chai, L., Edgar, S., Gavin, J., Hunter, P.J.: Laminar structure of the heart: ventricular myocyte arrangement and connective tissue architecture in the dog. Am. J. Physiol.-Heart Circulatory Physiol. **269**(2), H571–H582 (1995)
10. Mirea, I.: Analysis of the 3D microstructure of the human cardiac tissue using X-ray phase contrast micro-tomography. Ph.D. thesis, INSA Lyon (2017)
11. Mirea, I., Wang, L., Varray, F., Zhu, Y.M., Serrano, E.D., Magnin, I.E.: Statistical analysis of transmural laminar microarchitecture of the human left ventricle. In: 2016 IEEE 13th International Conference on Signal Processing (ICSP), pp. 53–56. IEEE (2016)
12. Nielles-Vallespin, S., et al.: Assessment of myocardial microstructural dynamics by in vivo diffusion tensor cardiac magnetic resonance. J. Am. Coll. Cardiol. **69**(6), 661–676 (2017)
13. Otsu, N.: A threshold selection method from gray-level histograms. IEEE Trans. Syst. Man Cybern. **9**(1), 62–66 (1979)
14. Pope, A.J., Sands, G.B., Smaill, B.H., LeGrice, I.J.: Three-dimensional transmural organization of perimysial collagen in the heart. Am. J. Physiol.-Heart Circulatory Physiol. **295**(3), H1243–H1252 (2008)
15. Pudney, C.: Distance-ordered homotopic thinning: a skeletonization algorithm for 3D digital images. Comput. Vis. Image Underst. **72**(3), 404–413 (1998)

16. Varray, F., Mirea, I., Langer, M., Peyrin, F., Fanton, L., Magnin, I.E.: Extraction of the 3D local orientation of myocytes in human cardiac tissue using X-ray phase-contrast micro-tomography and multi-scale analysis. Med. Image Anal. **38**, 117–132 (2017)
17. Wang, S., Mirea, I., Varray, F., Liu, W.-Y., Magnin, I.E.: Investigating the 3D local myocytes arrangement in the human LV mid-wall with the transverse angle. In: Coudière, Y., Ozenne, V., Vigmond, E., Zemzemi, N. (eds.) FIMH 2019. LNCS, vol. 11504, pp. 208–216. Springer, Cham (2019). https://doi.org/10.1007/978-3-030-21949-9_23

A High-Fidelity 3D Micromechanical Model of Ventricular Myocardium

David S. Li[1], Emilio A. Mendiola[1], Reza Avazmohammadi[2], Frank B. Sachse[3], and Michael S. Sacks[1(✉)]

[1] James T. Willerson Center for Cardiovascular Modeling and Simulation, Oden Institute for Computational Engineering and Sciences, Department of Biomedical Engineering, The University of Texas at Austin, Austin TX, USA
msacks@oden.utexas.edu
[2] Computational Cardiovascular Bioengineering Lab, Department of Biomedical Engineering, Texas A&M University, College Station, TX, USA
[3] Nora Eccles Harrison Cardiovascular Research and Training Institute, Department of Biomedical Engineering, The University of Utah, Salt Lake City, UT, USA

Abstract. Pulmonary arterial hypertension (PAH) imposes a pressure overload on the right ventricle (RV), leading to myofiber hypertrophy and remodeling of the extracellular collagen fiber network. While the macroscopic behavior of healthy and post-PAH RV free wall (RVFW) tissue has been studied previously, the mechanical microenvironment that drives remodeling events in the myofibers and the extracellular matrix (ECM) remains largely unexplored. We hypothesize that multiscale computational modeling of the heart, linking cellular-scale events to tissue-scale behavior, can improve our understanding of cardiac remodeling and better identify therapeutic targets. We have developed a high-fidelity microanatomically realistic model of ventricular myocardium, combining confocal microscopy techniques, soft tissue mechanics, and finite element modeling. We match our microanatomical model to the tissue-scale mechanical response of previous studies on biaxial properties of RVFW and examine the local myofiber-ECM interactions to study fiber-specific mechanics at the scale of individual myofibers. Through this approach, we determine that the interactions occurring at the tissue scale can be accounted for by accurately representing the geometry of the myofiber-collagen arrangement at the micro scale. Ultimately, models such as these can be used to link cellular-level adaptations with organ-level adaptations to lead to the development of patient-specific treatments for PAH.

Keywords: Soft tissue mechanics · Image based modeling · Finite element modeling

1 Introduction

Cardiac diseases like pulmonary arterial hypertension (PAH) lead to substantial adaptations in cardiac tissue structure and mechanical behavior. PAH imposes

© Springer Nature Switzerland AG 2021
D. B. Ennis et al. (Eds.): FIMH 2021, LNCS 12738, pp. 168–177, 2021.
https://doi.org/10.1007/978-3-030-78710-3_17

a chronic pressure overload on the right ventricle (RV) of the heart, which can cause the RV to initially thicken via hypertrophy of the myofibers to mitigate increased wall stress, but later dilate and lose contractile function, leading to RV failure. Previous work characterizing the complex changes in PAH has identified that the amount of RV remodeling resulting from pressure overload is one of the major predictors of poor prognosis in RV function [1,2]. As such, there is a need to quantify the mechanical state of the RV in order to better understand the factors influencing the onset, progression, and potential reversibility of remodeling in response to PAH through the development of computational models.

While the tissue-scale (~1-cm) structure and mechanics of RV myocardial tissue have been studied previously [3], knowledge of the mechanical contributions of and interactions between myofibers and the surrounding extracellular matrix (ECM) remains incomplete. Avazmohammadi et al. highlighted the importance of accounting for myofiber-collagen interactions in order to capture the full anisotropic mechanical behavior of the tissue [4]. This was accomplished through the addition of a myofiber-collagen interaction term in the strain energy density function based on the joint extension of myofibers and collagen fibers. Despite this important finding, the precise form of this coupling has yet to be characterized at the scale of individual myofibers, requiring more advanced histological approaches focused on the specific arrangement of myofibers and collagen in the myocardium.

To this end, multiscale computational modeling, particularly the development of models that are faithful to the microstructure of the myocardium, is important in order to gain insight into the cellular-level mechanical environment. Our specific objective was to determine to what extent the coupling behavior observed at the tissue scale could be accounted for by recapitulating the cellular-scale geometric arrangement of myofibers and ECM in the myocardium. For this first study, we based our model on a high-resolution 3D imaging dataset from rabbit myocardium, with the aim to match the results of our previous tissue-level model in planar biaxial deformations [4]. Then, based on the matched stress-strain responses, we quantify myofiber and ECM strain profiles under applied biaxial strain and further investigate the influence of myofiber-collagen mechanical interactions derived from the micro-scale deformations of myofibers and extracellular matrix.

2   Methods

2.1   Finite Element Mesh Construction

A high-fidelity representation of the myocardial tissue structure is essential in understanding cell-scale adaptations and linking them with remodeling events at the tissue and organ level. To achieve this, we developed a model using previously acquired 3D imaging datasets of ventricular myocardium from New Zealand white rabbits. Harvesting of the heart, sectioning, labeling, and imaging were performed at the University of Utah using previously developed methods [5, 6]. Briefly, myocardium samples of 5-mm diameter were obtained from the left ventricle and were then cryosectioned into slices with thickness of $100\,\mu$m. Slices were labeled and mounted compression-free by sealing within Fluoromount-G (Electron Microscopy Science, PA, USA). 3D image stacks were acquired via laser scanning confocal microscopy, with a tissue volume of $204 \times 204 \times 60\,\mu$m imaged. The segmentation and reconstruction of the 3D tissue geometry was performed using a semi-automatic approach based on watershed methods and histogram-based thresholding [5]. This method was used to segment myocytes, fibroblasts, coronary blood vessels, and extracellular space (Fig. 1). Manual smoothing and feature separation was performed using Seg3D (sci.utah.edu).

For simplicity in the development of the initial finite element (FE) model, myocytes were connected and combined into a "myofiber" phase, rather than individual myocytes, according to knowledge of the interconnected arrangement of myocytes in the myocardium. Coronary vessels, fibroblasts, and extracellular space were combined into an "extracellular matrix" or "collagen" phase in order to eliminate cavities within the model domain. This two-phase segmentation was cleaned with island removal and Gaussian smoothing and was then used to generate a volumetric mesh consisting of ~1.1 million linear tetrahedral elements within Simpleware ScanIP (Synopsys, CA, USA) (Fig. 1), with final dimensions of $204 \times 204 \times 40\,\mu$m. Image processing was performed such that the orientation of myocytes was aligned with the $\mathbf{e}_1$ axis in the image, and the cross-fiber direction with the $\mathbf{e}_2$ axis.

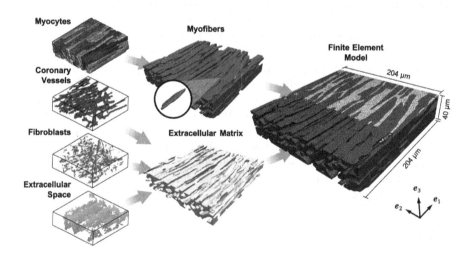

Fig. 1. Finite element model developed from 3D geometry. Left: Geometry of myocytes, coronary vessels, fibroblasts, and extracellular space [6]. Center: Myocytes joined into the myofiber phase (red), with representative myocyte highlighted. Coronary vessels, fibroblasts, and extracellular space joined into the extracellular matrix phase (gray). Right: Cross-section of FE model showing myofiber elements embedded in ECM elements. Coordinate axes $\{\mathbf{e}_1, \mathbf{e}_2, \mathbf{e}_3\}$ indicate the image and laboratory axes. (Color figure online)

2.2 Constitutive Modeling

We focused this study on passive myocardial mechanics at the cell and tissue scales. Thus, the mechanical properties of the myofiber and ECM phases were characterized with hyperelastic, nonlinear, anisotropic, incompressible material models adapted from our previous studies [3,4]. Deformations were described with the deformation gradient tensor $\mathbf{F}$, given by

$$
F_{ij} = \frac{\partial x_i}{\partial X_j} = \begin{bmatrix} \frac{\partial x_1}{\partial X_1} & \frac{\partial x_1}{\partial X_2} & \frac{\partial x_1}{\partial X_3} \\ \frac{\partial x_2}{\partial X_1} & \frac{\partial x_2}{\partial X_2} & \frac{\partial x_2}{\partial X_3} \\ \frac{\partial x_3}{\partial X_1} & \frac{\partial x_3}{\partial X_2} & \frac{\partial x_3}{\partial X_3} \end{bmatrix},
\tag{1}
$$

where $\mathbf{X}$ is a material point in the reference configuration moving to a new point $\mathbf{x}$ in the deformed configuration. From this we defined the right Cauchy-Green tensor $\mathbf{C} = \mathbf{F}^\mathrm{T}\mathbf{F}$ and the Green-Lagrange strain tensor $\mathbf{E} = (\mathbf{C} - \mathbf{I})/2$, where $\mathbf{I}$ is the identity tensor.

Myofibers. Since individual myofiber orientations could be directly obtained from the imaging dataset, we defined the myofiber direction $\mathbf{f}$ as the predominant direction of the long axes of myofibers in the mesh ($\mathbf{e}_1$), which was fairly constant throughout the geometry with a splay of 2.5°. We defined the cross-fiber (or sheet) direction $\mathbf{s}$ as the direction normal to the myofibers within the plane of the mesh, aligned with $\mathbf{e}_2$ (Fig. 1).

The myofiber strain energy ψ_myo was modeled using a modification of a well-established orthotropic constitutive form from Holzapfel and Ogden [7]. In particular, this form consisted of an isotropic matrix stiffened by mutually orthogonal fiber families along $\mathbf{f}$ and $\mathbf{s}$ within the layer, given by

$$
\begin{aligned}
\psi_\mathrm{myo} = {}& \frac{\mu_\mathrm{myo}}{2}(I_1 - 3) + \frac{a_\mathrm{f}}{2b_\mathrm{f}}\left\{\exp\left[b_\mathrm{f}(I_{4\mathrm{f}} - 1)^2\right] - 1\right\} \\
& + \frac{a_\mathrm{s}}{2b_\mathrm{s}}\left\{\exp\left[b_\mathrm{s}(I_{4\mathrm{s}} - 1)^2\right] - 1\right\},
\end{aligned}
\tag{2}
$$

where $I_1 = \mathrm{tr}\,\mathbf{C}$ is related to the isotropic stretch, and $I_{4\mathrm{f}} = \mathbf{f}{\cdot}\mathbf{C}\,\mathbf{f}$ and $I_{4\mathrm{s}} = \mathbf{s}{\cdot}\mathbf{C}\,\mathbf{s}$ are analogous to square of the stretch along the $\mathbf{f}$ and $\mathbf{s}$ directions, respectively. $\{\mu_\mathrm{myo},\, a_\mathrm{f}, b_\mathrm{f},\, a_\mathrm{s}, b_\mathrm{s}\}$ are the material parameters. Here, the exponential ground matrix term in [7] was replaced with a Neo-Hookean form, and the fiber-sheet interaction term was omitted with the aim of recovering this behavior instead from the interaction of discrete myofiber and ECM phases in the model geometry.

Extracellular Matrix. Individual collagen fibers, having diameters at the scale of less than 1 μm, were not visible in the image. In light of this, the ECM/collagen phase was modeled as an isotropic matrix reinforced by a distribution of undulated, linearly stiffening collagen fibers described by the orientation distribution function (ODF) Γ_θ and recruitment function Γ_s. The collagen fiber splay was represented with a von Mises distribution, given by

$$\Gamma_\theta(\theta) = \frac{\exp\left(\kappa \cos(\theta - \mu^c)\right)}{2\pi I_0(\kappa)}, \tag{3}$$

where $\theta \in [-\frac{\pi}{2}, \frac{\pi}{2}]$ is the in-plane angle used to define the collagen fiber orientation $\mathbf{n}$, $I_0(\kappa)$ is the modified Bessel function of order 0, and μ^c and $1/\kappa$ are related to μ and σ_θ^2 in the normal distribution, respectively, with σ_θ being the fiber splay. The recruitment function was given by the half-normal distribution

$$\Gamma_s(\lambda_s) = \begin{cases} \dfrac{\exp\left(-\frac{(\lambda_{\mathrm{ub}} - \lambda_s)^2}{2\sigma_s^2}\right)}{\sigma_s \sqrt{2\pi} \, \mathrm{erf}\left(\frac{(\lambda_{\mathrm{ub}} - 1)}{\sqrt{2}\sigma_s}\right)} & \text{for } \lambda_s \in (1, \lambda_{\mathrm{ub}}] \\ 0 & \text{otherwise} \end{cases}, \tag{4}$$

where λ_s is the fiber slack stretch at which collagen fibers begin contributing to the ECM stress, λ_{ub} is the upper-bound slack stretch of maximum recruitment, and σ_s controls the rate of recruitment. The final expression for the ECM constitutive form ψ_{ECM} was thus given by

$$\psi_{\mathrm{ECM}} = \frac{\mu_{\mathrm{col}}}{2}(I_1 - 3) + \frac{\eta_{\mathrm{col}}}{2} \int_{-\pi/2}^{\pi/2} \Gamma_\theta(\theta) \int_1^{\lambda_\theta} \Gamma_s(\lambda_s) \left(\frac{\lambda_\theta}{\lambda_s} - 1\right)^2 d\lambda_s \, d\theta, \tag{5}$$

where $\lambda_\theta = \sqrt{\mathbf{n} \cdot \mathbf{C}\mathbf{n}}$ is the component of the fiber stretch in the $\mathbf{n}$ direction. $\{\mu_{\mathrm{col}}, \eta_{\mathrm{col}}, \mu^c, \sigma_\theta, \lambda_{\mathrm{ub}}, \sigma_s\}$ are the material parameters. In order to match the tissue-scale collagen fiber recruitment, the fiber slack stretch was determined based on previous modeling results [4], which also indicated the collagen fiber distribution was aligned with the myofibers ($\mathbf{f}$) in the model ($\mu^c = 0°$).

2.3   FE Simulations

Single Layer Simulation. The structure-based model was originally developed to capture tissue-scale myofiber/collagen/interaction stresses with more dispersed fiber distributions. Thus, as a fitting target for the microanatomical model, we simulated a stress-strain response using the structure-based model (Eq. 27 in [4]) guided by the highly aligned structure of the microanatomical finite element geometry. Myofiber and ECM elements were assigned their corresponding fitted material models (Eqs. 2 and 5) in the FE geometry and assumed to be perfectly bonded. Deformations were applied to the boundary surfaces of the model in varying biaxial strain configurations $E_{11}:E_{22} = 0.30:0.30$, $0.30:0.15$, and $0.15:0.30$. The microanatomical constitutive parameters were then calibrated such that the second Piola-Kirchhoff stress response of both models matched. All simulations were performed on the Stampede2 supercomputer at the Texas Advanced Computing Center using the open-source software FEniCS (fenicsproject.org).

Linking to the Tissue Scale. In order to reproduce the fiber ensemble response associated with the higher-splay myofiber and collagen fiber structure of the full-thickness RVFW, rather than the highly aligned structure of the confocal microscopy dataset, the microanatomical results required a homogenization routine to link the cell-scale and tissue-scale mechanics. Using the histologically measured myofiber and collagen orientation distribution functions of previous studies (Fig. 7 in [4]), the micro-scale FE model stress-strain response was integrated over the tissue-scale orientation distribution to compute an effective dispersed stress as a function of transmural depth (from endo- to epicardium). The effective stress was then integrated over the transmural depth to predict the tissue-level stress.

3   Results

The stress-strain response of the microanatomical constitutive models successfully reproduced the tissue-scale structural model predictions (Fig. 2). Fitted parameters are listed in Table 1. Since the micro-level structure involved highly aligned myofibers oriented in the $\mathbf{e}_1$ direction, the model exhibited exponentially stiffening myofiber stress along $\mathbf{e}_1$ with negligible stress along $\mathbf{e}_2$. The ECM exhibited nonlinear stiffening during collagen fiber recruitment ($0.17 < E < 0.20$), after which fibers were fully recruited and the ECM response became linear, as demonstrated previously [4]. The overall response was governed by myofiber behavior in the low-strain regime ($E < 0.17$), whereas at higher strains, the ECM became the major contributor to the RVFW stress. This behavior was primarily observed in loading paths that stretched the collagen (along $\mathbf{e}_1$) beyond the slack stretch, as opposed to the $E_{11}{:}E_{22} = 0.15{:}0.30$ in which collagen recruitment was not significant.

Table 1. Constitutive parameters for micro-scale myofiber and ECM models fitted to tissue-scale structure-based model prediction [4].

Myofiber					ECM					
μ_{myo} (Pa)	a_f (kPa)	b_f	a_s (kPa)	b_s	μ_{col} (Pa)	η_{col} (MPa)	μ^c (deg)	σ_θ (deg)	λ_{ub}	σ_s
0.059	8.94	1.59	0.0813	1.29	0.050	1.91	0.0	11.0	1.205	0.022

At maximum deformation, the strain fields within the myofiber and ECM phases (Fig. 3) showed that although the distribution of myofiber and ECM strains along $\mathbf{e}_1$ and $\mathbf{e}_2$ followed the applied boundary strains in an average sense, there was a high level of heterogeneity throughout the volume (Fig. 4). Along $\mathbf{e}_1$, regions of ECM at the tips of myofibers exhibited decreases in E_{11}, but increases in areas between adjacent myofibers. An opposite trend was observed along $\mathbf{e}_2$, where significant regions of ECM located between the tips of adjacent myofibers showed pronounced increases in E_{22}. This was observed for all biaxial loading paths, although the amount of increase was not as prominent when collagen

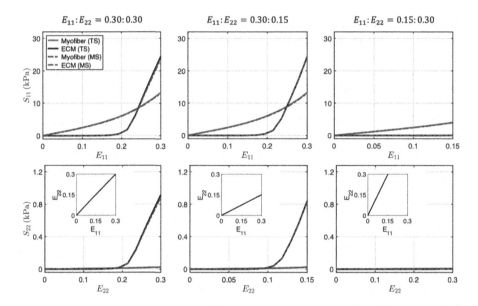

Fig. 2. Myofiber (red) and extracellular matrix fiber (black) stresses in response to biaxial strain deformations: $E_{11}:E_{22}$ = 0.30:0.30, 0.30:0.15, 0.15:0.30. Solid lines: predictions from tissue-scale structural model. Dotted lines: fitted response of microanatomical model. Abbreviations: tissue scale (TS), micro scale (MS). (Color figure online)

fibers were not being recruited (e.g., $E_{11}:E_{22}$ = 0.15:0.30). Substantial shear strains also developed in the model.

Using the structure-based model, the mechanical response of a single tissue-scale layer of RV myocardium was modeled by applying the histologically measured fiber splays for myofibers and collagen fibers rather than the attributes of the confocal microscopy dataset. The micro-scale model was homogenized over the orientation distribution function and offered a very close match (Fig. 5a). The full transmural homogenization also demonstrated the effect of transmurally varying fiber orientation through the generation of significant cross-fiber stress in both the myofiber and ECM phases (Fig. 5b). Homogenization over the fiber angle θ microanatomical model response was also found to generate this effect, though to a lesser extent. The combination of both increased splay at a given level along with transmural variation in preferred direction resulted in stress-strain behavior showing qualitative similarity to tissue-level studies [4].

4   Discussion

In this study, we developed a finite element model of ventricular myocardium based on a high resolution imaging dataset of a representative tissue element volume. We then used structurally motivated constitutive modeling approaches

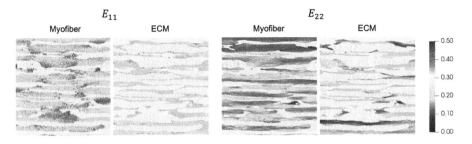

Fig. 3. Midplane cross-sections of E_{11} and E_{22} at maximum deformation for the myofiber and ECM phases. Myofibers and ECM are shown separately for clarity.

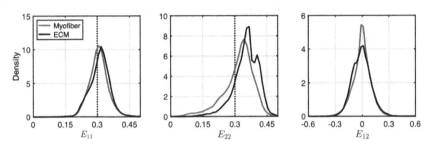

Fig. 4. Normalized histograms of E_{11} E_{22}, and E_{12} in myofiber (red) and ECM (black) elements at maximum applied strain for the equibiaxial loading path. Dotted lines indicate applied strain. (Color figure online)

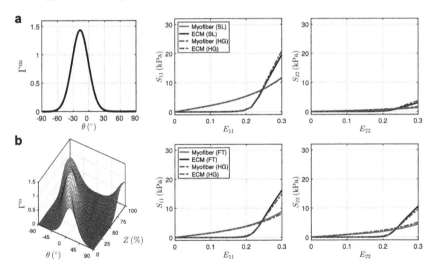

Fig. 5. (a) Homogenization of micro-scale stress-strain response using fiber orientation distribution (Γ^{m}) of a single RV layer. (b) Homogenization to full-thickness RVFW fiber structure over transmural depth Z. Solid lines: tissue-level model. Dotted lines: micro-level model homogenized with tissue-scale fiber splay and transmural variation. Abbreviations: single layer (SL), full-thickness (FT), homogenization (HG).

to both quantify the micro-scale strain and stress fields in the tissue as well as link the micro-scale mechanical response to previously established tissue-scale modeling studies. To the best knowledge of the authors, this study constitutes a major first step in the development of microanatomically-based computational models of cardiac tissue. Use of FE techniques to represent individual myofibers reveals the complexity of the cellar-scale mechanical environment under biaxial deformation, requiring the present approach.

The direction-dependent effects on local strain profiles found in the microanatomical simulation (Figs. 3 and 4) suggest that the complex cellular-scale arrangement of myofibers and collagen fibers in the myocardium indeed contribute to interactions that manifest at the tissue scale. As an example to highlight the localized effects brought on by accounting for discrete myofiber geometry, we conducted simulations of a simplified geometry based on an isolated myofiber from the microanatomical simulation in Fig. 6. At the myofiber-ECM interface, there are increased strains along the $\mathbf{e}_1$ direction in the myofibers paired with decreased strains in the ECM (compared to the applied boundary strain). Further away from the myofiber, the strain in the ECM returns to the applied strain. In contrast, in the $\mathbf{e}_2$ direction, the presence of the myofiber produces a band of increased strain along its long axis, and strains in the ECM are slightly below the applied strain further away from the myofiber. We theorize that these alterations in the local strain fields contribute to myofiber-collagen coupling stresses.

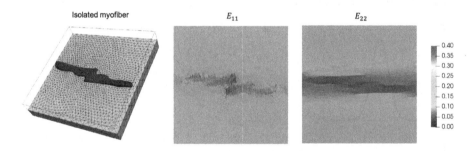

Fig. 6. Left: FE mesh of isolated myofiber (red) embedded in ECM (gray). Right: Midplane cross-sections of E_{11} and E_{22} at maximum equibiaxial deformation. (Color figure online)

Additionally, successful prediction of the tissue-scale stress-strain response based on the homogenization of the microanatomical model, for both a single layer and for the transmural thickness, provides another possible mechanism for the coupling interactions proposed in previous tissue-scale work. Accounting for discrete myofibers embedded in extracellular matrix, while very appearing highly aligned at the scale of individual myofibers, still gives rise to the dispersed stress-strain response observed at greater length scales. For the full-thickness homogenization, the transmural variation in orientation also contributes significantly to the overall stress response. It is clear that both effects are needed to account for the experimental observations; however, it must be noted that the transmural variation in fiber orientation is likely the dominant factor determining the mechanical anisotropy of the full-thickness tissue.

In summary, this model establishes a baseline set of mechanical identifiers at the micro scale that undergo alterations in PAH. Future work will be focused on construction of a microanatomical representation of post-PAH RVFW in order to perform direct comparisons in the local mechanics under biaxial deformation. Through this microanatomical approach, exploration of cellular-scale stress transfer between myocytes and their surroundings can be integrated into larger-scale models and used to further probe the effect of micro-level events on organ-level function. Ultimately, our microanatomical model will allow us to investigate fiber-specific remodeling of the cardiac microstructure in response to structural heart disease.

Acknowledgments. This work is supported by the National Institutes of Health (T32 EB007507, F31 HL139113 to D.S.L., K99 HL138288 to R.A.).

References

1. Bogaard, H., Abe, K., Vonk Noordegraaf, A., Voelkel, N.F.: The right ventricle under pressure: cellular and molecular mechanisms of right-heart failure in pulmonary hypertension. Chest **135**(3), 794–804 (2009)
2. McLaughlin, V.V., Shah, S.J., Souza, R., Humbert, M.: Management of pulmonary arterial hypertension. J. Am. Coll. Cardiol. **65**(18), 1976–1997 (2015)
3. Hill, M.R., Simon, M.A., Valdez-Jasso, D., Zhang, W., Champion, H.C., Sacks, M.S.: Structural and mechanical adaptations of right ventricle free wall myocardium to pressure overload. Ann. Biomed. Eng. **42**(12), 2451–2465 (2014)
4. Avazmohammadi, R., Hill, M.R., Simon, M.A., Zhang, W., Sacks, M.S.: A novel constitutive model for passive right ventricular myocardium: evidence for myofiber-collagen fiber mechanical coupling. Biomech. Model. Mechanobiol. **16**(2), 561–581 (2017)
5. Seidel, T., Draebing, T., Seemann, G., Sachse, F.B.: A semi-automatic approach for segmentation of three-dimensional microscopic image stacks of cardiac tissue. In: Ourselin, S., Rueckert, D., Smith, N. (eds.) FIMH 2013. LNCS, vol. 7945, pp. 300–307. Springer, Heidelberg (2013). https://doi.org/10.1007/978-3-642-38899-6_36
6. Stenzel, O.: The Physics of Thin Film Optical Spectra. SSSS, vol. 44, pp. 163–180. Springer, Cham (2016). https://doi.org/10.1007/978-3-319-21602-7_8
7. Holzapfel, G.A., Ogden, R.W.: Constitutive modelling of passive myocardium: a structurally based framework for material characterization. Philos. Trans. R. Soc. London A Math. Phys. Eng. Sci. **367**(1902), 3445–3475 (2009)

Quantitative Interpretation of Myocardial Fiber Structure in the Left and Right Ventricle of an Equine Heart Using Diffusion Tensor Cardiovascular Magnetic Resonance Imaging

Hilke C. H. Straatman[1](✉), Imke van der Schoor[1](✉), Martijn Froeling[2], Glenn Van Steenkiste[3], Robert J. Holtackers[4], and Tammo Delhaas[4]

[1] Eindhoven University of Technology, Eindhoven, The Netherlands
{h.c.h.straatman,i.v.d.schoor}@student.tue.nl
[2] Department of Radiology, University Medical Centre Utrecht,
Utrecht, The Netherlands
[3] Equine Cardioteam, Department of Large Animal Internal Medicine,
Faculty of Veterinary Medicine, Ghent University, Merelbeke, Belgium
[4] Maastricht University Medical Centre, Maastricht, The Netherlands

Abstract. In this study an equine heart of a 13-year-old Belgian Warmblood is scanned using diffusion tensor magnetic resonance imaging (DT-CMR) to perform quantitative analysis on myocyte orientation in the left and right ventricular free wall (LVFW and RVFW) as well as in the interventricular septum (IVS). Transmural helical angle profiles are obtained at basal, mid-ventricular and apical height. This equine heart with high transmural DT-CMR resolution revealed that the transmural helix angle range in the RVFW is approximately twice as large as the one in the LVFW. Moreover, the transmural helix angle distributions in the IVS could be separated into an RV and LV side with gradients in fiber angle orientation that were comparable to the RVFW and LVFW, respectively.

Keywords: DT-CMR · Helical angle · Right ventricular free wall · Septum

1 Introduction

Structural and mechanical properties of the myocardium determine ventricular function [13]. The myocardium composes a complex heterogeneous three-dimensional network of cardiomyocytes exhibiting both helical and transmural orientations. For the left ventricle (LV) it has been shown that the distribution of helical and transmural angles is strongly related to LV deformation in general and to LV rotation, torsion and stress distribution in particular [3,4,9,12,17]. Although it is evident that the anatomy and mural thickness of the right ventricle (RV) and LV are far from identical, a study by Omann et al. [16] has

© Springer Nature Switzerland AG 2021
D. B. Ennis et al. (Eds.): FIMH 2021, LNCS 12738, pp. 178–188, 2021.
https://doi.org/10.1007/978-3-030-78710-3_18

shown that the rearrangement of cardiomyocytes during the cardiac cycle also differs between both ventricles. This finding stresses the importance of a thorough understanding of the orientation of the cardiomyocytes in both LV and RV to understand cardiac function.

In the last decades, many attempts were made to quantify myofiber orientation in animal models using diffusion tensor cardiovascular magnetic resonance imaging (DT-CMR). It can be argued that DT-CMR as an imaging modality is mainly histopathologically validated in the LV as opposed to the RV [16]. Currently the most common voxel size for DT-CMR is $[400 \times 400 \times 800]\,\mu\mathrm{m}$ [1,16]. While this voxel size suffices for proper transmural resolution in the porcine thick-walled LV, the average thickness of 3–5 mm for the RV free wall results in a maximum of 12 transmural measurements, making interpretation of RV data difficult [1,2,5,9–11,18]. In vivo studies on the human heart are hampered even more due to its even lower resolution. Even if fiber structure of the thin-walled RV is visualized, it remains challenging to validate the results of such studies due to the lack of quantification [1,15].

In this study, the complex architectural myofiber structure is explored with DT-CMR using an equine heart for which a standard medical resolution results in a high transmural resolution due to the thick-walled structure of both ventricles. Myocyte orientation in the LV and RV free wall as well as in the interventricular septum at three different levels is quantified by its helical angle.

2   Method

2.1   MRI Protocol

A fresh equine heart (13 year old Belgian Warmblood mare) was obtained directly after euthanasia due to orthopedic reason. The heart was rinsed and thereafter filled with small plastic bags containing uncooked rice to main its shape and prevent susceptibility artifacts (Fig. 1). Within 12 h after extraction, the prepared heart was scanned on a clinical 3T MR system (Achieva CX; Philips Healthcare, Best, The Netherlands) using a 32 channel head coil for approxiamately 6 h. A multislice spin-echo single-shot EPI sequence was performed with the following parameters: echo time (TE) 104 ms, repetition time (TR) 40.8s, field of view $264 \times 264 \times 340.6$ mm, 1.5 mm isotropic resolution (acquired and reconstructed), 32 gradient directions ($b = 1500\,\mathrm{mm}^2/\mathrm{s}$) and 6 interleaved $b = 0\,\mathrm{mm}^2/\mathrm{s}$ volumes (Fig. 2A), SENSE factor 3.8 and 12 signal averages. Fat suppression was performed using spectral presaturation with inversion-recovery (SPIR) and gradient reversal. A total of 2 sets were scanned, each including 227 slices and both with the same b-value, a different direction and a duration of 2.5 h.

2.2   Processing

Data was processed using *QMRITools* for *Mathematica* [7]. It comprised PCA based denoising, affine registration for eddy current correction, and tensor fitting

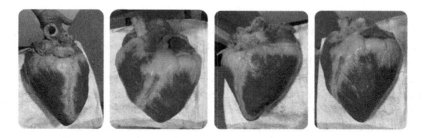

Fig. 1. Equine heart of a 13-year old Belgian Warmblood mare. Orientation going from left to right is as follows: Left, anterior, posterior, left-anterior oblique.

using iterative weighted linear least squares with outlier rejection. After processing, masks of the LV and RV were manually drawn using *ITKSnap* (Fig. 2**B**). Papillary muscles were excluded from the segmentation. In this study LV and RV local myocardial coordinate systems (LMCS) were obtained as described by Froeling et al. [8,14]. First, a second order polynomial was fitted through the centers of mass of the smallest enclosing disks in each mask to obtain the central axes of both the LV and RV (Fig. 2**C**). Next, for each voxel in the ventricular wall the minimum distance to the inner (set to zero (= 0)) and outer surface (set to one (= 1)) was determined. The 3D gradient of this wallmap defined the wall normal in each voxel. For each voxel the longitudinal axis of the LMCS was defined as the orthogonal projection of the central axis to the wall normal. Finally, the circumferential axis was defined as the cross product between the wall normal and the longitudinal axis. This resulted in orthogonal LMCS for LV and RV (Fig. 2**C**). Note that the method described above combines two commonly used approaches [6,9], i.e. defining the longitudinal vector by the ventricle barycenter, and the transmural vector directly by using the manually drawn epi- and endocardial boundaries. Cross-product of the wall normal vector and its perpendicular longitudinal vector yields the circumferential vector. The main advantage of this combination is that for each location the LMCS is defined based on the 3D geometry of the heart wall and therefore not dependent on the orientation of the 2D scanning plane.

2.3   Helix Angle Calculation

Data from diffusion-weighted images were used to compute the diffusion tensor for every voxel. Based on this tensor, three eigenvalues (λ_1, λ_2, λ_3) with corresponding eigenvectors (ϵ_1, ϵ_2, ϵ_3) were obtained, spanning an ellipsoid shape describing 3-dimensional diffusion [19]. For each voxel the primary eigenvector was projected on the plane spanned by the local longitudinal and circumferential axis, where the helix angle was defined as the elevation of this projection with respect to the circumferential axis.

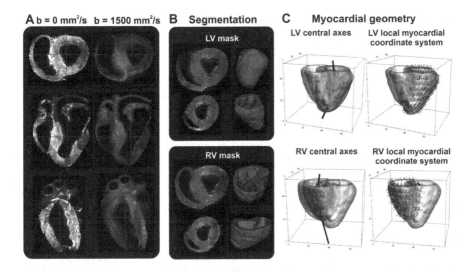

Fig. 2. Processing steps. **A**. DT-CMR images were obtained using different diffusion-weighted settings. **B**. Masks for the LV and RV were manually segmented in ITK-snap. **C**. A second order polynomial was fitted through the centers of mass of the smallest enclosing disks in each mask to obtain the central axes of both the LV and RV which were used as a basis to calculate the local myocardial coordinate systems.

2.4   Analysis

Transmural helix angle distributions were obtained at the basal (1), mid-ventricular (2), and apical (3) section. To optimize the intersection direction for both the LV and RV free wall, the tilted heart was rotated to align the septal region vertically. This resulted in intersections at approximately the same height for RV, septum and LV (Fig. 3A). At every location, three segments were taken in the LV and RV free wall as well as in the septal region (Fig. 3B). For each segment the helix angle was radially sampled. All transmural samples per segment were collected over 10 slices, corresponding to a thickness of 15 mm (1.5 mm per slice). To ensure physiological fiber orientations within the septal region, angles exceeding ±90° were forced to show a continuous profile. As fibers pointing up or down are identical, angles of 91° will be displayed as −89°. To ensure smooth angle maps, locations in which non-physiological jumps in angles were found, were corrected for by adding 180°. To obtain transmural profiles of the helix angle, the median values and 95% confidence intervals of all transmural samples per segment as a function of relative depth were calculated. The median values of the transmural profile were fitted using a linear function to determine the transmural range with a 95% confidence interval as well as the slope of the profiles, the latter with respect to both the absolute (degree/mm) and the relative (degree/%)wall thickness.

To determine the slope of the LV and RV free wall with respect to the absolute wall thickness, the transmural distance of inner to outer surface was obtained from the data, yielding the absolute change in helix angle in degrees per mm. For the slope with respect to the relative wall thickness, the transmural distance of inner to outer surface is set to 100%. Hence, these slopes give the relative change in helix angle in degrees per percent wall thickness.

For the septum, a slightly different approach was taken as the slopes were determined for the LV and RV side of the septum separately. While the helix angle distribution of the entire septal region is calculated with respect to the LMCS of the LV, a difference in slope is found at the right and left side of the septum. This transition in slope is found at approximately 2/3 of the septum, which means that 2/3 of the septal region could be attributed to the LV and the residual 1/3 to the RV. For each of these regions the absolute and relative slopes were determined by assuming that the entire septum would follow the helix angle distributions found at either the LV or RV side.

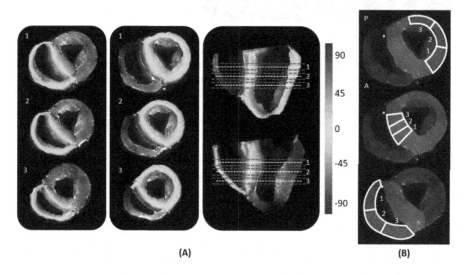

(A)   (B)

Fig. 3. A. Transmural helix angle distribution obtained at three locations for RV (left panel) and LV (middle panel) as expressed with regards to the respective ventricular Local Myocardial Coordinate System. Red dotted lines indicate the thickness of each region (15 mm). Red colors indicate maximum helix angle values (100°). Blue colors indicates minimum helix angle values (−100°). **B**. Schematic representation of the cardiac segments for LV free wall (upper), septum (middle) and RV free wall (lower). (Color figure online)

3 Results

Endo- to epicardial distributions of helix angles for various LV free wall segments and depths are shown in Fig. 4. All plots show an almost linear transmural helical angle profile with very narrow 95% interval ranges. Whereas endocardial helix angle varied from 0 to 50° between the 9 regions, the average relative slope of the linear approximation of the transmural distribution, estimated as $-0.6°/\%$, barely varied with values ranging from -0.4 to $-0.8°/\%$ (Table 1). The same consistency is found for the absolute slope estimated as $-3.4°/mm$, with values ranging from -1.9 to $-4.9°/mm$. The linear approximation showed that the endo- to epicardial range of the helical angle was $60 \pm 10°$.

Figure 5 shows endo- to epicardial distributions of helix angles for various RV free wall segments and depths. Not only the endo- to epicardial range in all RV free wall segments (estimated as $100 \pm 30°$ with a linear approximation) was larger than those of the LV free wall, but also the 95% interval range was larger throughout the whole RV free wall myocardium. The average values of the relative and absolute slopes of the linear approximation of the transmural helix angle profiles in the RV free wall ($-1.0°/\%$ and $-5.0°/mm$, respectively) were almost twice as high as in the LV free wall (Table 1).

Figure 6 shows transmural helix angle distributions for septal regions at 3 levels. Helix angles were estimated with respect to LV LMCS. Of note is that two areas seem to be present with two distinct slopes. At approximately 2/3 of the septum in transmural direction, as seen from LV endo, a clear transition in slope is found as it becomes steeper. The average relative and absolute slope of the linear approximation of the transmural helix angle profile at the LV side ($-0.8°/\%$ and $-2.9°/mm$, respectively) were comparable to the slopes as found in the LV free wall ($-0.6°/\%$ and $-3.4°/mm$). The average relative and absolute slopes of the helix angle profiles of the RV side of the septum ($-1.35°/\%$ and $-4.7°/mm$, respectively) were also comparable to the slopes as found in the RV free wall ($-1.0°/\%$ and $-5.0°/mm$) (Table 1).

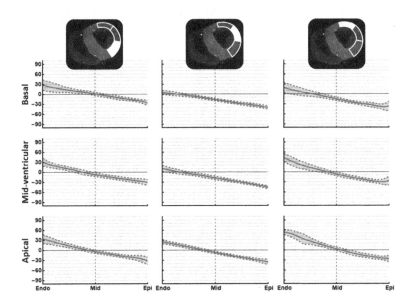

Fig. 4. Transmural helix angle distributions for the LV free wall segments at three depths (Fig. 3). Ranges indicate the 95% interval.

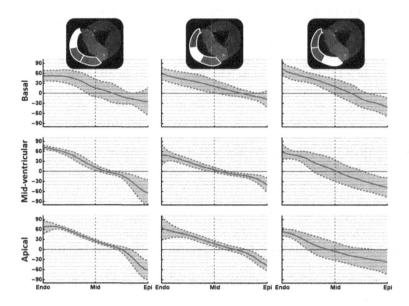

Fig. 5. Transmural helix angle distributions for the RV free wall segments at three depths (Fig. 3). Ranges indicate the 95% interval.

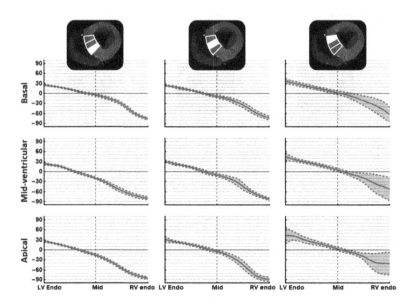

Fig. 6. Transmural helix angle distributions for the septum segments at three depths estimated with respect to LV LMCS (Fig. 3). Segments are ordered from anterior to posterior from left to right. Ranges indicate the 95% interval.

Table 1. Range of angles and slopes found by linearizing the transmural helix angle distribution for LV and RV free wall and left and right septal region (first 67% respectively last 33% in transmural direction as seen from LV endo), with a 95% confidence interval.

Left ventricle

Location	Segment 1				Segment 2				Segment 3			
	Range (deg.) [Endo, Epi]	95% int	Slope (deg./%)	Slope (deg./mm)	Range (deg.) [Endo, Epi]	95% int	Slope (deg./%)	Slope (deg./mm)	Range (deg.) [Endo, Epi]	95% int	Slope (deg./%)	Slope (deg./mm)
Basal	[26, −24]	±2	−0.5	−1.9	[5, −40]	±1	−0.4	−2.6	[12, −43]	±3	−0.6	−3.4
Mid	[24, −32]	±3	−0.6	−2.5	[9, −45]	±1	−0.5	−3.6	[35, −41]	±11	−0.8	−4.9
Apical	[27, −32]	±4	−0.6	−2.9	[24, −37]	±2	−0.6	−3.8	[49, −35]	±8	−0.8	−4.9

Right ventricle

Location	Segment 1				Segment 2				Segment 3			
	Range (deg.) [Endo, Epi]	95% int	Slope (deg./%)	Slope (deg./mm)	Range (deg.) [Endo, Epi]	95% int	Slope (deg./%)	Slope (deg./mm)	Range (deg.) [Endo, Epi]	95% int	Slope (deg./%)	Slope (deg./mm)
Basal	[68, −30]	±14	−1.0	−4.9	[50, −20]	±9	−0.7	−3.2	[71, −31]	±7	−1.0	−2.7
Mid	[84, −59]	±21	−1.4	−7.2	[49, −34]	±10	−0.8	−4.3	[55, −45]	±9	−1.0	−3.5
Apical	[87, −44]	±23	−1.3	−8.3	[69, −36]	±9	−1.0	−7.2	[45, −48]	±16	−0.9	−3.7

Left ventricular side of septum

Location	Segment 1				Segment 2				Segment 3			
	Range (deg.) [Endo, Epi]	95% int	Slope* (deg./%)	Slope (deg./mm)	Range (deg.) [Endo, Epi]	95% int	Slope* (deg./%)	Slope (deg./mm)	Range (deg.) [Endo, Epi]	95% int	Slope* (deg./%)	Slope (deg./mm)
Basal	[28, −17]	±3	−0.6	−1.9	[27, −21]	±2	−0.7	−1.9	[37, −9]	±3	−0.7	−2.1
Mid	[30, −40]	±7	−1.0	−3.5	[33, −24]	±3	−0.8	−2.7	[44, −8]	±4	−0.7	−2.6
Apical	[30, −34]	±5	−0.9	−4.4	[33, −20]	±6	−0.7	−3.4	[49, −12]	±4	−0.9	−3.4

Right ventricular side of septum

Location	Segment 1				Segment 2				Segment 3			
	Range (deg.) [Endo, Epi]	95% int	Slope* (deg./%)	Slope (deg./mm)	Range (deg.) [Endo, Epi]	95% int	Slope* (deg./%)	Slope (deg./mm)	Range (deg.) [Endo, Epi]	95% int	Slope* (deg./%)	Slope (deg./mm)
Basal	[−19, −78]	±8	−1.9	−5.6	[−24, −75]	±4	−1.5	−4.2	[−10, −47]	±4	−1.6	−4.9
Mid	[−50, −81]	±4	−0.9	−3.2	[−32, −86]	±5	−1.6	−5.2	[−5, −42]	±9	−1.2	−4.3
Apical	[−45, −89]	±7	−1.3	−6.4	[34, −100]	±16	−1.9	−8.8	[−20, −26]	±22	−0.1	−0.12

*Slope as if the entire septum would follow the helix angle distribution found at the left or right side of the septum. Note that the left side of the septum was defined as 67% of the septal region, while the right side of the septum was defined as the remaining 33%.

4   Discussion

The present study on the transmural distribution of myofiber helix angles in the equine heart shows that the endo- to epicardial ranges of the helix angles of the RV free wall are approximately twice as high as those of the LV free wall, being 60 and 120° respectively. Moreover, the septum showed different steepness of the helix angle profiles for the RV and the LV-side. When using the depth at which the slope of the helix angle profiles clearly showed a transition as an arbitrarily criterion to define the boundaries of the RV- and LV-side of the interventricular septum, the RV-side occupied approximately 33% of the total septum and had gradients in helix fiber angles that were comparable to those in the RV free wall, whereas the gradients in helix fiber angle at the LV-side of the septum were like those in the LV free wall. Though the ventricular free walls do not show transitions in the slope of the transmural helix angle profiles, the finding of a transition in slope of the septal transmural helix angle fiber profiles does not reveal anything about differences in anatomy or functionality of the RV- and LV-side of the septum. Finally it was shown that the helix angle distribution of the LV free wall and the septum were quite uniform with a very narrow 95% interval range, whereas the RV free wall showed less uniformity as indicated by the broad 95% interval ranges.

It is important to remark that the helix angle distribution found in the septal region depends on the definition of the LMCS. In this study the helix angle distribution of the septum is calculated with respect to the LV LMCS. The same analysis could also have been performed relative to the RV LMCS. Changing the local coordinate system does not impact the profile and total range of the helix angle distributions, but only the offset of the profiles will change. Therefore, the observations done in this study will remain valid independent of the chosen LMCS. On top of that, we found that the angle of the intersection plane with the central axis influences the helix angle distribution. Larger angles increase the cross-section and variations in helix angles distributions due to larger height differences. These findings stress the importance of reporting the LMCS used as well as the chosen plane of intersection.

The quite small myofiber helix angle range of the LV free wall (60°) in this equine heart is surprising, giving the generally reported +60 to −60° range in human, porcine, canine, rabbit and rat hearts. Though it might be argued that this small range is induced because the papillary muscles were omitted from the LV free wall masks with consequently less longitudinal fibers, it should be noted that segment 2 in the LV free wall, situated in between the two papillary muscles, shows endocardial fiber angles close to zero, indicating a true circumferential endocardial orientation. Even though measuring in diastole can not explain the low helix angle range completely, it does contribute. Additionally, these findings are shown in one equine heart only. Therefore, it is important to repeat this study in multiple hearts to ascertain whether the low helix angle range in the LV free wall of an equine heart is reproduced or that it is caused by a methodological error. If these findings can be corroborated in next studies, the currently existing hypothesis on necessity of a +60 to −60° helix angle range in a thick-walled ventricle to ensure homogeneous fiber shortening throughout the wall has to be reconsidered [3].

The remarkable finding that this equine heart exhibits twice as high helix angle ranges and/or gradients in the RV free wall and the RV side of the septum as compared with the LV free wall and the LV side of the septum has also not been shown in previous studies in other species. Because of the high transmural resolution with approximately 20 voxels across the RV free wall we are quite confident about the reliability of this first report. Further DT-CMR studies are needed to corroborate our findings with an increased sample size, whereas modeling studies are required to investigate the implications for RV and LV deformation.

References

1. Agger, P., et al.: Changes in overall ventricular myocardial architecture in the setting of a porcine animal model of right ventricular dilation. J. Cardiovasc. Magn. Reson. **19**(1), 1–16 (2017)
2. Agger, P., et al.: The myocardial architecture changes in persistent pulmonary hypertension of the newborn in an ovine animal model. Pediatric Res. **79**(4), 565–574 (2016)
3. Arts, T., Prinzen, F.W., Snoeckx, L., Rijcken, J.M., Reneman, R.S.: Adaptation of cardiac structure by mechanical feedback in the environment of the cell: a model study. Biophys. J. **66**(4), 953–961 (1994)
4. Bovendeerd, P., Arts, T., Huyghe, J., Van Campen, D., Reneman, R.: Dependence of local left ventricular wall mechanics on myocardial fiber orientation: a model study. J. Biomech. **25**(10), 1129–1140 (1992)
5. Chen, J., et al.: Regional ventricular wall thickening reflects changes in cardiac fiber and sheet structure during contraction: quantification with diffusion tensor MRI. Am. J. Physiol.-Heart Circulatory Physiol. **289**(5), H1898–H1907 (2005)
6. Chen, J., et al.: Remodeling of cardiac fiber structure after infarction in rats quantified with diffusion tensor MRI. Am. J. Physiol.-Heart Circulatory Physiol. **285**(3), H946–H954 (2003)
7. Froeling, M.: Qmrtools: a mathematica toolbox for quantitative MRI analysis. J. Open Source Softw. **4**(38), 1204 (2019)
8. Froeling, M., et al.: Comprehensive comparison of in- and ex-vivo whole heart fiber architecture: similar yet different. In: Proceedings 24th Scientific Meeting, International Society for Magnetic Resonance in Medicine, p. 3113 (2016)
9. Geerts, L., Bovendeerd, P., Nicolay, K., Arts, T.: Characterization of the normal cardiac myofiber field in goat measured with MR-diffusion tensor imaging. Am. J. Physiol.-Heart Circulatory Physiol. **283**(1), H139–H145 (2002)
10. Healy, L.J., Jiang, Y., Hsu, E.W.: Quantitative comparison of myocardial fiber structure between mice, rabbit, and sheep using diffusion tensor cardiovascular magnetic resonance. J. Cardiovasc. Magn. Reson. **13**(1), 1–8 (2011)
11. Jiang, Y., Pandya, K., Smithies, O., Hsu, E.W.: Three-dimensional diffusion tensor microscopy of fixed mouse hearts. Magn. Reson. Med. Off. J. Int. Soc. Magn. Reson. Med. **52**(3), 453–460 (2004)
12. Kroon, W., Delhaas, T., Bovendeerd, P., Arts, T.: Computational analysis of the myocardial structure: adaptation of cardiac myofiber orientations through deformation. Med. Image Anal. **13**(2), 346–353 (2009)
13. Liu, W., Wang, Z.: Current understanding of the biomechanics of ventricular tissues in heart failure. Bioengineering **7**(1), 2 (2020)

14. Mekkaoui, C.: Mapping cardiac tissue architecture systems and methods. US Patent 9,678,189, 13 June 2017

15. Nielsen, E., et al.: Normal right ventricular three-dimensional architecture, as assessed with diffusion tensor magnetic resonance imaging, is preserved during experimentally induced right ventricular hypertrophy. Anat. Rec. Adv. Integr. Anat. Evol. Biol. **292**(5), 640–651 (2009)

16. Omann, C., et al.: Resolving the natural myocardial remodelling brought upon by cardiac contraction; a porcine ex-vivo cardiovascular magnetic resonance study of the left and right ventricle. J. Cardiovasc. Magn. Reson. **21**(1), 1–19 (2019)

17. Rijcken, J., Bovendeerd, P., Schoofs, A., Van Campen, D., Arts, T.: Optimization of cardiac fiber orientation for homogeneous fiber strain during ejection. Ann. Biomed. Eng. **27**(3), 289–297 (1999)

18. Sosnovik, D.E., Wang, R., Dai, G., Reese, T.G., Wedeen, V.J.: Diffusion MR tractography of the heart. J. Cardiovasc. Magn. Reson. **11**(1), 1–15 (2009)

19. Watson, S.R., Dormer, J.D., Fei, B.: Imaging technologies for cardiac fiber and heart failure: a review. Heart Failure Rev. **23**(2), 273–289 (2018). https://doi.org/10.1007/s10741-018-9684-1

Analysis of Location-Dependent Cardiomyocyte Branching

Alexander J. Wilson[1]([✉]) [iD], Gregory B. Sands[2][iD], and Daniel B. Ennis[1][iD]

[1] Department of Radiology, Stanford University, Stanford, CA, USA
wilsonaj@stanford.edu
[2] Auckland Bioengineering Institute, University of Auckland,
Auckland, New Zealand

Abstract. Cardiomyocytes branch and interconnect with one another, providing important redundancy for propagation of electro-chemical signals. Despite this, cardiomyocyte branching structure remains poorly understood. Herein, myocardium from spontaneously hypertensive rats (SHR) was imaged using extended volume confocal microscopy. Samples from untreated SHRs ($n = 2$) were compared with SHRs undergoing ACE inhibitor treatment ($n = 2$). From these image-stacks the cardiomyocyte network (center-lines and branches) were manually tracked. The frequency of cardiomyocyte branching was calculated and these branching frequencies were compared according to spatial position within a myocardial sheetlet. ACE inhibitor treatment resulted in significantly reduced total cardiomyocyte branching compared with untreated SHR myocardium at 24 mo (0.49 ± 0.04 vs 1.07 ± 0.15 branches per $100\,\mu$m, $P = 0.020$). Cardiomyocytes on the sheetlet-surface branched more frequently within their respective cell-layer (0.59 ± 0.07 branches per $100\,\mu$m) compared with cardiomyocytes in the sheetlet interior (0.29 ± 0.12 branches per $100\,\mu$m). The cardiomyocytes in the sheetlet interior exhibited more frequent between-layer branching (0.56 ± 0.06 branches per $100\,\mu$m) compared to sheetlet-surface cardiomyocytes (0.17 ± 0.03 branches per $100\,\mu$m). The ratio of within-layer to between-layer branching was significantly greater at the surface layer compared with the interior layer (3.93 ± 1.06 vs 0.47 ± 0.16, $P = 0.018$). This proof-of-concept study demonstrates an approach to measuring branching cardiomyocyte networks and shows the spatial heterogeneity of cardiomyocyte branching.

Keywords: Cardiomyocyte · Network · Sheetlet · Confocal · Histology

1 Introduction

Cardiomyocytes form a complex, continuously branching syncytial mesh. The branching characteristics of this mesh-like network, however, remains incompletely characterized. Electron microscopy has previously been used to analyze

Supported by funding from the American Heart Association.

the complex branching structure of the cardiomyocyte network [4,5,9]. In these studies, cardiomyocyte branching frequencies were calculated along the fiber direction, and in the cross-fiber plane (sheetlet/sheetlet-normal plane). This branching analysis was performed using two dimensional images, and to our knowledge there has been no attempt to perform branching analysis in three dimensions.

The spontaneously hypertensive rat (SHR) is a well-studied model of chronic hypertensive heart disease. The SHR displays left ventricular hypertrophy as well as micro-structural remodeling and fibrosis. Previous studies by our group have shown that ACE inhibitor treatment, in addition to lowering blood pressure, alters collagen deposition and sheetlet organization [11]. However, it is not known how the cardiomyocyte network changes in disease, or the potential impact of treatment on this three dimensional network.

Herein, we characterized the branching structure of cardiomyocytes in three dimensions by measuring the branching frequency with regards to location within a sheetlet. From extended volume confocal microscopy [1,6–8,11] image volumes, a segment of a sheetlet was identified and the cardiomyocyte bodies and branches were manually tracked. We compared the branching of cardiomyocytes on the sheetlet surface with those in the sheetlet interior. We propose that the sheetlet surface acts as a topological barrier, constraining branching in the sheetlet-normal direction. Consequently, we hypothesized that cardiomyocytes on the sheetlet surface branch more in the sheetlet direction than the sheetlet-normal direction as compared with cardiomyocytes in the sheetlet interior.

1.1   Definition of Terms

Within the myocardium, three orthogonal directions are defined as the fiber direction ($\vec{f}$), which aligns with the long axis of cardiomyocytes, the sheetlet direction ($\vec{s}$), which is orthogonal to the fiber direction and lies within the plane of the sheetlet, and the sheetlet-normal direction ($\vec{n}$), orthogonal to the sheetlet plane.

Herein we refer to a cardiomyocyte *layer* as a subset of cardiomyocytes lying within the fiber/sheetlet plane. Layers are stacked in $\vec{n}$: the *surface layer* is adjacent to the sheetlet surface, the *intermediate layer* is the next layer towards the interior of the sheetlet, and the *interior layer* lies in the middle of the sheetlet (Fig. 1). Branches are defined as morphological features that allow the connection of one cardiomyocyte to more than two other cardiomyocytes [5,9]. Branches connecting cells in the same layer (within the fiber/sheetlet plane) are termed *within-layer* branches, while branching between cells in different layers (along $\vec{n}$) are termed *between-layer* branches.

2   Methods

This study was approved by the Animal Ethics Committee of the University of Auckland (Ref: 001119) and conforms to the Care and Use of Laboratory Animals Guide (NIH Publication No. 85–23). In this study we imaged and analyzed

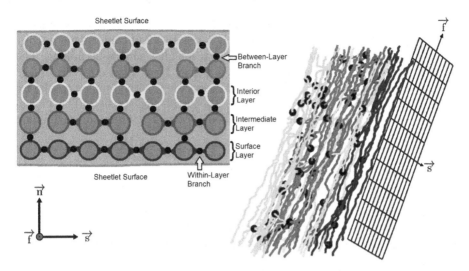

Fig. 1. Cardiomyocyte network analysis: A schematic cross-section of a sheetlet segment (top left) illustrates the cardiomyocyte layers, as well as the within-layer and between-layer branching types. The orientation of the schematic is indicated by the coordinate system (bottom left), which shows the fiber ($\vec{f}$), sheetlet ($\vec{s}$) and sheetlet-normal ($\vec{n}$) directions. A connected network (right) results from the cardiomyocyte tracking of the confocal image volumes. Colored lines represent the cardiomyocyte center-lines, and the black dots represent the cardiomyocyte branch-points. In order to differentiate between the different cardiomyocyte-layers, each cardiomyocyte-layer was colored differently. The surface layer is colored blue, the intermediate layer colored red, and the interior layer cardiomyocytes colored cyan. The grid showing the fiber-sheetlet plane indicates the orientation of the network. (Color figure online)

four rodent hearts from the SHR model of hypertensive heart disease, which are a subset of rats included in previously published studies [10–12]. At three months (mo) of age, two of the four SHRs were given ACE inhibitor (quinapril) treatment, with a dosage of 7.5 mg/kg/day delivered via their water source for the duration of their lives. The remaining two SHRs were not treated, and exhibited chronic hypertension for the duration of their lives.

2.1   Extended Volume Confocal Microscopy

Each rat was anaesthetized with isoflurane and culled via cervical dislocation. The chest cavity was opened and an intracardiac injection of heparin delivered. The heart was excised and perfused with St. Thomas Hospital cardioplegic solution using a Lagendorff rig. The heart was then stained via retrograde perfusion with picrosirius red, which stains for collagen, and fixed with Bouin's solution. After seven days in fixative, a transmural block from the left ventricular free wall was cut, dehydrated in a graded ethanol series, and embedded in resin.

The resin-embedded blocks were mounted on the precision three-axis stage of the extended volume confocal microscope. Extended-volume confocal microscopy [1,6–8,11] utilizes a repeated cycle of confocal imaging and tissue milling. The confocal microscope acquired images at $0.4\,\mu m$ or $1\,\mu m$ voxel size with image stacks acquired to a depth of 20–$30\,\mu m$. The sample was then milled to remove the top layer of tissue, leaving $5\,\mu m$ overlap for 3D registration. The image-mill cycle was repeated to build up extended 3D image volumes approximately $500\,\mu m \times 500\,\mu m \times 300\,\mu m$.

2.2 Cardiomyocyte Network Analysis

A region of approximately ten × five × three cardiomyocytes was selected, in the radial, longitudinal and circumferential directions respectively. An operator selected points corresponding with the center-line of the cardiomyocytes (Fig. 2) using the open-source software package ImageJ (1.53c, National Institutes of Health, Bethesda, MD, USA). Branch-points were identified where the cardiomyocyte bodies of adjacent cardiomyocytes merged, producing a connected network (Fig. 1). For each cardiomyocyte, the layer of the cardiomyocyte in the sheetlet-normal direction was recorded, and designated as *surface layer* for cardiomyocytes at the sheetlet boundary, *intermediate layer* for the cardiomyocyte layer immediately adjacent to the surface layer, and *interior layer* for the third cardiomyocyte layer.

The connections between cardiomyocytes was recorded, and the within-layer (fiber/sheetlet plane) and between-layer ($\vec{n}$) branching frequencies were calculated as follows: for each branch point, if both of the cells connected to it were

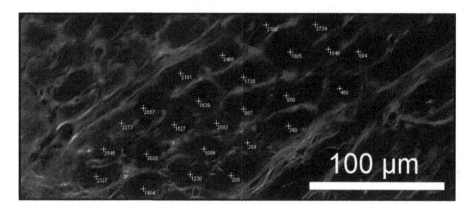

Fig. 2. Cardiomyocytes in cross-section: An example 2D image produced from the extended volume confocal microscopy. Cardiomyocytes were tracked from a sheetlet segment (approximately five by ten by three cardiomyocytes). The centerline of each cardiomyocyte was tracked, with branch-points identified where the cardiomyocyte bodies of adjacent cardiomyocytes merged. The yellow hash marks represent manual tracking used to identify each cardiomyocyte. The image is colored using the *glow* colormap, with collagen indicated by yellow/orange. (Color figure online)

in the same layer, 0.5 was added to the within-layer count for that layer for each cell. If the connected cells were in different layers, 0.5 was added to the between-layer count of both of the layers. These totals were then divided by the sum of lengths of all cells within each layer multiplied by 100 to give branching frequencies per 100 μm. Values are reported as mean ± standard error.

3   Results

The between-layer and within-layer cardiomyocyte branching frequencies showed differences according to their cardiomyocyte-layer (Fig. 3). Sheetlet-surface cardiomyocyte-layers branched within their own cardiomyocyte-layer (0.59 ± 0.07 branches per 100 μm) more frequently than did cardiomyocytes in the

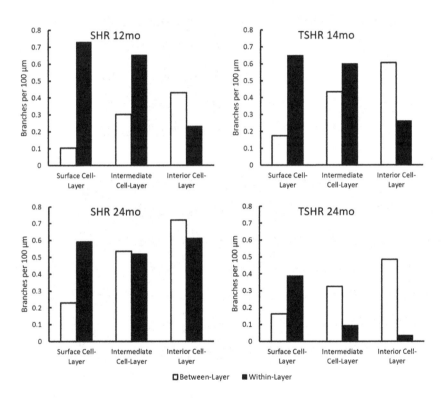

Fig. 3. Between-layer and within-layer branching frequency: Branching frequencies for the untreated-diseased spontaneously hypertensive rat (SHR) at 12 mo (top left) and 24 mo (bottom left), and the ACE inhibitor treated spontaneously hypertensive rat (TSHR) at 14 mo (top right) and 24 mo (bottom right). Bars represent mean between-layer and within-layer branching frequencies per 100 μm for the three cell-layers. While the number of within-layer branches decreases from the sheetlet surface layer to the sheetlet interior layer, the number of between-layer branches increases.

sheetlet interior (0.29 ± 0.12 branches per $100\,\mu$m). Cardiomyocytes on the sheetlet surface also branched less frequently with cardiomyocytes of a different cardiomyocyte-layer (0.17 ± 0.03 branches per $100\,\mu$m) compared with cardiomyocytes in the interior layer (0.56 ± 0.06 branches per $100\,\mu$m).

As reported in previous studies [11, 12], the untreated SHRs exhibited hypertension at both the 12 mo (122 mmHg $\pm$ 11) and 24 mo (142 mmHg $\pm$ 10) timepoints. The ACE inhibitor treatment curtailed the hypertension, with the 12 mo (83 mmHg $\pm$ 7) and 24 mo (69 mmHg $\pm$ 8) TSHRs exhibiting reduced mean arterial pressure compared with the untreated SHRs. The SHRs also exhibited cardiomyocyte hypertrophy, which was curtailed in the TSHRs (Fig. 4).

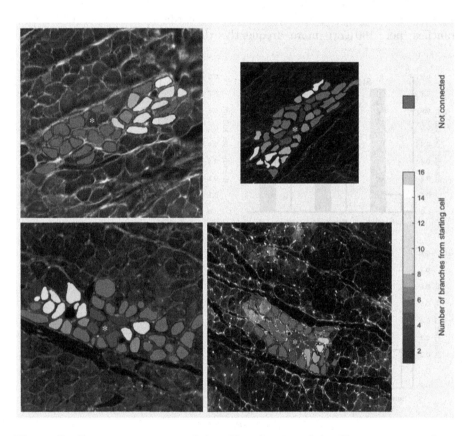

Fig. 4. Cardiomyocyte connectivity: Visualization of the 3D connectivity of cardiomyocytes from a 2D perspective. From the starting cardiomyocyte (*) the color map shows the number of branch-points that are traversed to reach each of the colored cells, with red cardiomyocytes not connected to the starting cell. Qualitative cardiomyocyte hypertrophy is evident from the greater cardiomyocyte cross-sectional area in the SHR samples (left column) compared with the TSHR samples (right column). Shown are confocal images of samples from 12mo SHR (top left, $350\,\mu$m $\times$ $350\,\mu$m), 14mo TSHR (top right, $220\,\mu$m $\times$ $200\,\mu$m), 24mo SHR (bottom left, $350\,\mu$m $\times$ $350\,\mu$m), 24mo TSHR (bottom right, $350\,\mu$m $\times$ $350\,\mu$m). (Color figure online)

The 3D branching connectivity was also visualized on a 2D slice (Fig. 4) which shows that cardiomyocytes that appear adjacent on a 2D slice may not be immediately connected within a 3D network. This visualization also showed the directionality of the local connectivity. The 12 mo SHR and 24 mo TSHR samples showed a predominant direction of local connectivity, which was not present in the 12 mo TSHR sample.

The ratio of within-layer to between-layer branching also varied by layer (Fig. 5, left). The surface-layer showed a consistently greater number of within-layer branches compared to between-layer branches (3.93 ± 1.06 within:between ratio). Conversely the interior cardiomyocyte-layer showed a predominance of between-layer branches (0.47 ± 0.16 within:between ratio). These ratios were significantly different between surface and interior layers ($P = 0.018$). Total cardiomyocyte branching was similar between 14 mo SHR and 14 mo TSHR samples (Fig. 5, right). However, by 24 mo the TSHR myocardium showed significantly lower total branching compared with the untreated SHR sample ($P = 0.020$).

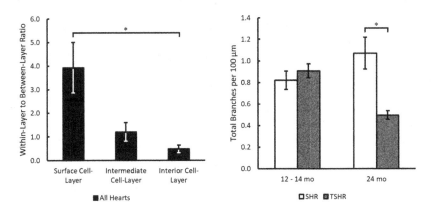

Fig. 5. Branching ratios and total branching: The ratio of within-layer to between-layer branching (left) shows a significantly greater proportion of within-layer branches at the surface cell-layer compared with the interior cell-layer ($P = 0.018$). Within-layer to between-layer ratios were averaged over all hearts ($n = 4$ for each cell-layer). A single t-test was performed to compare ratios between surface and interior cell-layers. The total number of branches per $100\,\mu\mathrm{m}$ (right) shows no difference between untreated and treated spontaneously hypertensive rat myocardium at 12–14 mo. At 24 mo, however, SHR myocardium had significantly greater total branches than treated SHR myocardium ($P = 0.020$). Total branching frequencies were calculated as the sum of the within-layer and between-layer branching frequencies, averaged over the three layers ($n = 3$ per myocardial sample). A single t-test was performed to compare total branching between 24 mo SHR and 24 mo TSHR samples.

4   Discussion

To our knowledge this work represents the first attempt to quantify cardiomyocyte branching characteristics in three dimensions. We found that cardiomyocyte branching is location-dependent. The number of within-layer branches decreased and the number of between-layer branches increased from the sheetlet surface layer to the interior layer.

4.1   Physiological Significance

Cardiomyocyte branching of the SHR, as well as the Wistar Kyoto (WKY) control rat has previously been investigated using high resolution electron microscopy [5]. Okabe *et al.* (1999) described different types of cardiomyocyte branching, particularly *series* branches along $\overrightarrow{f}$, and *lateral* branches in the cross-fiber plane (sheetlet/sheetlet-normal plane). In the present study, we were not able to distinguish these different branching types with confocal microscopy. However, the combined total of lateral and series branches in 14 mo SHR myocardium (0.97 ± 0.06) from Okabe *et al.* (1999) is similar to the total branching per $100\,\mu$m in our 12 mo SHR myocardium (0.82 ± 0.09).

Comparing healthy and diseased tissue, Okabe *et al.* (1999) found that total cardiomyocyte branching per cell was significantly greater in diseased myocardium compared with healthy myocardium (14 mo SHR $= 0.97 \pm 0.06$, 14 mo WKY $= 0.62 \pm 0.06$). In the present study, we found no difference in total branching between SHR and TSHR samples at 12–14 mo (12 mo SHR $= 0.82 \pm 0.09$, 14 mo TSHR $= 0.91 \pm 0.06$). However, by 24 mo the TSHR sample showed significantly lower branching than the SHR sample (24 mo SHR $= 1.07 \pm 0.15$, 24 mo TSHR $= 0.49 \pm 0.04$). With the assumption that 24 mo TSHR myocardium is healthy, this result aligns with the greater branching in diseased myocardium compared with healthy myocardium [5].

Considering the link between cardiac microstructure and function, previous work by our group has shown that at 14 mo SHR and TSHR have similar systolic performance, as measured by ejection fraction, but that 24 mo SHR have a significantly lower ejection fraction compared with the TSHRs and WKYs [11]. The differences in cardiomyocyte branching frequency align with the functional differences observed at the ventricular level.

4.2   Cardiomyocyte Branching and Myocardial Deformation

Cardiomyocytes are organized into a structural hierarchy, with many connected cardiomyocytes locally comprising a sheetlet, and the bulk myocardium being comprised of many sheetlets. Deformation of the myocardium during the cardiac cycle requires significant rearrangement of cardiomyocytes. These cardiomyocytes however are constrained by their intercalated discs, providing mechanical and electro-chemical coupling, as well as the extracellular matrix. The organizational rearrangement of individual cardiomyocytes during contraction has not been measured, however detailed measurement of cardiomyocyte network structure and subsequent analysis of this structure using computational modelling, may provide insight into the cardiomyocyte rearrangements that are possible for a given cardiomyocyte network.

In this study we confirm that sheetlet boundaries result in altered branching characteristics compared with those observed in the sheetlet interior. The sparsity of connections between sheetlets allows for greater mobility of sheetlets between one another, supporting the sheetlet sliding model of systolic mechanics [2]. The degree of connectivity within the sheetlet is evidence for restricted within-sheetlet rearrangement of cardiomyocytes.

4.3   Growth and Remodeling

Although cardiomyocyte shortening can be approximated as one-dimensional, cardiomyocyte branching must result in forces within the cardiomyocyte network that are off-axis with respect to the predominant long-axis direction of the cell. As a result, a branching cardiomyocyte is, in theory, less efficient at producing one-dimensional shortening in comparison to a cardiomyocyte that does not branch. However, the observation of extensive cardiomyocyte branching in the heart suggests that the sub-optimal performance at the cardiomyocyte-level is somehow more efficient for whole-organ performance.

Cardiomyocyte growth and remodelling is also known to be dependent upon the mechanics of the cardiomyocyte environment. Off-axis forces that result from branching may act as a growth and remodeling signal, which may explain the correlation between increased branching frequency and cardiomyocyte hypertrophy in SHRs [5]. As branching impacts cardiomyocyte forces and therefore cardiomyocyte growth signalling, our result suggests that cardiomyocyte growth signalling is location-dependent.

4.4   Limitations

As we were unable to observe intercalated discs in the microscopy volumes stained with picrosirius red, we are unable to ascertain if the branching patterns we observed are large branching myocytes, or multiple small connected cardiomyocytes. For simplicity we have chosen to use the term cardiomyocyte *branching* although perhaps a more apt term may be *morphological heterogeneity*. Our present study does not explore the process by which this branching structure is reached, whether individual cardiomyocytes themselves branch, or whether connections of linear cells form this branching structure as the cardiomyocytes fuse into multi-nucleated cells. The present analysis is also a description of this complex network at a fixed point in time. Due to the time and complexity of the image acquisition and analysis, only four myocardial samples were analysed in this study. In order to increase confidence in the between-sample comparisons, a greater sample size is required. We are presently improving the throughput of this workflow with the use of CUBIC tissue clearing technique [3] combined with Wheat Germ Agglutinin dying of the extracellular matrix.

5   Conclusion

Cardiomyocytes on the sheetlet surface branch within their own cell-layer more frequently, but between cell-layers less frequently compared with cardiomyocytes in the sheetlet interior. ACE inhibitor treatment resulted in significantly reduced total cardiomyocyte branching compared with untreated SHR myocardium at 24 mo. Sheetlet boundaries act as a topological constraint for cardiomyocyte branching, demonstrating that cardiomyocyte branching is location-dependent.

Acknowledgements. We thank Linley Nisbet for assistance with the animal experiments and Dane Gerneke for support with confocal imaging.

Funding Sources. This work was supported by funding from the American Heart Association (19IPLOI34760294 to DBE), the Marsden Fund administered by the Royal Society of New Zealand, and the Health Research Council of New Zealand.

Ethical Statement. This study was approved by the Animal Ethics Committee of the University of Auckland (Ref: 001119).

Conflict of Interest. None.

References

1. LeGrice, I.J., Pope, A.J., Sands, G.B., Whalley, G., Doughty, R.N., Smaill, B.H.: Progression of myocardial remodeling and mechanical dysfunction in the spontaneously hypertensive rat. Am. J. Physiol.-Heart Circulatory Physiol. **303**(11), H1353–H1365 (2012)
2. LeGrice, I.J., Takayama, Y., Covell, J.W.: Transverse shear along myocardial cleavage planes provides a mechanism for normal systolic wall thickening. Circ. Res. **77**(1), 182–193 (1995)
3. Nehrhoff, I., et al.: 3D imaging in cubic-cleared mouse heart tissue: going deeper. Biomed. Opt. Express **7**(9), 3716–3720 (2016)
4. Okabe, M., et al.: Backscattered electron imaging: a new method for the study of cardiomyocyte architecture using scanning electron microscopy. Cardiovasc. Pathol. **9**(2), 103–109 (2000)
5. Okabe, M., Kawamura, K., Terasaki, F., Hayashi, T.: Remodeling of cardiomyocytes and their branches in juvenile, adult, and senescent spontaneously hypertensive rats and Wistar Kyoto rats: comparative morphometric analyses by scanning electron microscopy. Heart Vessels **14**(1), 15–28 (1999)
6. Pope, A.J., Sands, G.B., Smaill, B.H., LeGrice, I.J.: Three-dimensional transmural organization of perimysial collagen in the heart. Am. J. Physiol.-Heart Circulatory Physiol. **295**(3), H1243–H1252 (2008)
7. Sands, G.B., Gerneke, D.A., Hooks, D.A., Green, C.R., Smaill, B.H., Legrice, I.J.: Automated imaging of extended tissue volumes using confocal microscopy. Microscopy Res. Tech. **67**(5), 227–239 (2005)
8. Sands, G.B., Smaill, B.H., LeGrice, I.J.: Virtual sectioning of cardiac tissue relative to fiber orientation. In: 2008 30th Annual International Conference of the IEEE Engineering in Medicine and Biology Society, pp. 226–229. IEEE (2008)

9. Sawada, K., Kawamura, K.: Architecture of myocardial cells in human cardiac ventricles with concentric and eccentric hypertrophy as demonstrated by quantitative scanning electron microscopy. Heart Vessels **6**(3), 129–142 (1991)
10. Wilson, A.J.: Biomechanics of cardiac remodelling in heart failure. Ph.D. thesis, University of Auckland (2017)
11. Wilson, A.J., et al.: Myocardial laminar organization is retained in angiotensin-converting enzyme inhibitor treated SHRs. Exp. Mech. **61**(1), 31–40 (2020). https://doi.org/10.1007/s11340-020-00622-4
12. Wilson, A.J., Wang, V.Y., Sands, G.B., Young, A.A., Nash, M.P., LeGrice, I.J.: Increased cardiac work provides a link between systemic hypertension and heart failure. Physiol. Rep. **5**(1), e13104 (2017)

Systematic Study of Joint Influence of Angular Resolution and Noise in Cardiac Diffusion Tensor Imaging

Yunlong He[1], Lihui Wang[2(✉)], Feng Yang[3], Yong Xia[4], Patrick Clarysse[1], and Yuemin Zhu[1(✉)]

[1] Univ Lyon, INSA Lyon, Université Claude Bernard Lyon 1, CNRS, Inserm, CREATIS UMR 5220, U1294, 69621 Lyon, France
`{Yunlong.He,patrick.clarysse,Yue-Min.Zhu}@creatis.insa-lyon.fr`
[2] Key Laboratory of Intelligent Medical Image Analysis and Precise Diagnosis of Guizhou Province, School of Computer Science and Technology, Guizhou University, Guiyang, China
[3] National Library of Medicine, National Institute of Health, 8600 Rockvill Pike, Bethesda, MD 20894, USA
`Feng.yang2@nih.gov`
[4] School of Computer Science, NPU, Xian, China

Abstract. Diffusion tensor imaging (DTI) is a promising imaging technique to non-invasively study diffusion properties and fiber structures of myocardial tissues. Previous studies have investigated the influence of noise or angular resolution independently on the estimation of diffusion tensors in DTI. However, the joint influence of these two factors in DTI remains unclear. In this paper, we propose to systematically study the joint influence of angular resolutions and noise levels on the estimation of diffusion tensors and tensor-derived fractional anisotropy (FA) and mean diffusivity (MD). The results showed that, as expected, given a certain noise level and sufficient acquisition time, the accuracy of diffusion tensor, FA and MD all increase as the angular resolution. Moreover, when the angular resolution reached a certain value, further increasing the number of angular resolutions has little effect on the estimation of diffusion tensor, FA and MD. Also, both the mean and variance of FA or MD decrease as the angular resolution increases. For an imposed acquisition time, increasing the angular resolution reduces SNR of DW images. When fixing SNR, higher angular resolution can be obtained at the expense of longer acquisition time. These findings suggest the necessity of an optimized trade-off when designing DTI protocols.

Keywords: Diffusion tensor magnetic resonance imaging · Angular resolution · Fractional anisotropy · Mean diffusivity

1 Introduction

Diffusion tensor imaging (DTI) refers to a magnetic resonance imaging (dMRI) technique that measures diffusion of water molecules within biological tissues

© Springer Nature Switzerland AG 2021
D. B. Ennis et al. (Eds.): FIMH 2021, LNCS 12738, pp. 200–210, 2021.
https://doi.org/10.1007/978-3-030-78710-3_20

using diffusion-weighted pulse sequences. DTI makes it possible to explore diffusion properties and fiber structures of tissues non-invasively compared to conventional imaging modalities [1]. In cardiac imaging, for example, DTI has been increasingly used to investigate myocardial microstructure changes related to many cardiac disorders such as hypertrophic cardiomyopathy [2], myocardial infarction [3], etc.

In DTI, water diffusion in tissues is described by diffusion tensors estimated from a set of diffusion-weighted (DW) images associated with noncollinear diffusion gradient directions. The number of gradient directions define the angular resolution. Specifically, it is possible to estimate the diffusion tensor D, a 3×3 symmetric positive definite matrix, according to the Stejskal-Tanner equation [4]: $S_i = S_0 \exp\left(-bg_i^T D g_i\right)$, for $i = 1, 2, \cdots n(n \geq 6)$, where n is the number of diffusion gradient directions, S_i is the DW signal intensity acquired in the i-th gradient direction g_i, and b is the diffusion weighting factor [5]. At a given voxel, the water diffusion properties are quantified by calculating the eigenvalues of the diffusion tensor and tensor-derived measures [6] such as fractional anisotropy (FA) and mean diffusivity (MD).

Accurate estimation of diffusion tensors is important for characterizing water diffusion and thus assessing the myocardial fiber structure. In practice, however, the estimation accuracy of diffusion tensors is influenced by several factors, e.g., image quality, estimation algorithm, tissue complexity, etc. Two factors, noise and angular resolution, have attracted much attention. The noise sources in DW images refer to random signals arising from the hardware (e.g., gradient-coil noise [7], field inhomogeneity [8], etc.) and inherent motions of the subject (e.g., heartbeat and respiration). These noises give rise to perturbed DW signals, and thus produce errors in the calculated diffusion tensor as well as its derived measures. The angular resolution in DTI determined by the number of diffusion gradient directions is used in acquiring DW images. It also influences the estimation of diffusion tensors.

The influence of noise or angular resolution in DTI have been independently investigated by researchers. For example, Pierpaoli et al. [9] show that the MD derived from diffusion tensors decreases as the noise level increases. Anderson in [10] demonstrated that higher noise levels increase the maximum eigenvalue of tensors, and thus result in a larger FA value. For the angular resolution, Papadakis et al. [11] compared various DTI angular sampling schemes, and show that the minimum number of gradient directions for accurate estimation of FA is around 18–21. Jones [12] found that at least 20 gradient directions are required for robust estimation of FA, and at least 30 directions are required for robust estimation of tensor orientation and MD. Despite these useful findings, it remains unclear how the angular resolution and noise jointly affect the diffusion tensor estimation in DTI.

In this paper, we systematically studied the joint influence of angular resolution and noise in cardiac DTI. Specifically, a set of DW images with different diffusion gradient directions are simulated from a simulated diffusion tensor field. Besides, real datasets of four human hearts are acquired with a large number of diffusion gradient directions. Then, different levels of Rician noise are added to both the simulated and real DW images to produce datasets with different signal-to-noise ratio (SNR). Different angular resolutions of data are achieved by varying the number of considered diffusion gradient directions for each SNR. Finally, the quality of diffusion tensors and tensor-derived FA, MD for different angular resolutions and different noise levels is quantitatively assessed. In addition, the relationship between angular resolution and SNR for a fixed acquisition time is also studied.

The current study has two main contributions. First, this is, to our knowledge, the first work to systematically investigate the joint influence of different angular resolutions and different levels of noise in cardiac DTI. Second, we quantitatively defined the minimum angular resolution required to obtain near-optimal diffusion tensors, FA and MD at a certain noise level, and also investigated the maximum angular resolution for keeping certain SNRs when imposing acquisition time.

2   Materials and Methods

2.1   Simulated Datasets

Simulated DW data were generated to analyze the influence of angular resolution and noise level on the estimation accuracy of diffusion tensors. The simulation strategy is similar to the method proposed in [13]. Specifically, a $20 \times 20 \times 20$ diffusion tensor field containing 9 homogeneous tensor regions with discontinuities of different amplitudes was created to represent 9 different structures of a human heart. Each z-slice of the tensor field is defined by

$$\begin{pmatrix} R_0 \ R_1 \ R_0 \ R_2 \\ R_0 \ R_3 \ R_0 \ R_4 \\ R_0 \ R_5 \ R_0 \ R_6 \\ R_0 \ R_7 \ R_0 \ R_8 \end{pmatrix} \tag{1}$$

where R_i represents a 5×5 homogeneous region containing 25 of the same diffusion tensors. Here we set these tensors by using the same coefficients as in [13], namely the eigenvalues of each tensor were $(2, 1, 1) \times 10^{-3}\,\mathrm{mm^2/s}$. The b-value was $700\,\mathrm{s/mm^2}$, which is the same value that is used in the real cardiac datasets. The FA and MD values for the simulated noisy DW images associated with 12 directions were FA $= 0.33 \pm 0.06$ and MD $= 1.3 \pm 0.11 \times 10^{-3}\,\mathrm{mm^2/s}$, which are typical for the left ventricular (LV) myocardium of *ex vivo* human hearts (according to the statistical results in [14]).

The simulated DW images were computed from simulated diffusion tensor field using Stejskal-Tanner equation with associated diffusion gradient directions. In order to analyze the effects of noise, different levels of Rician noise [15] were added to the ideal DW images with different standard deviation values. The noise levels of the simulated DW images were measured via SNR (as shown in Fig. 1). Different angular resolutions of DW images were achieved by sampling different

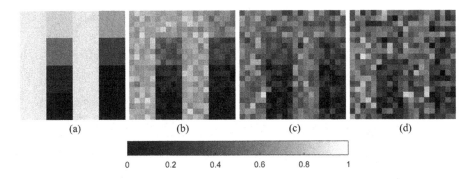

Fig. 1. An example of simulated DW images with: (a) no noise (ground-truth), noise levels with standard deviation (b) $\sigma = 0.02$ ($SNR = 23\,$dB), (c) $\sigma = 0.05$ ($SNR = 15\,$dB), and (d) $\sigma = 0.1$ ($SNR = 15\,$dB). The encoding gradient direction related to the DW image g_i is (0.894, 0 , 0.447).

number of associated diffusion gradient directions. The sampling method for diffusion gradient directions is detailed in Sect. 2.3. In this paper, the number of excitations used for signal averaging for each direction was set only when acquisition time is fixed; otherwise, the number was set to 1.

2.2   Real Datasets

DTI of *ex vivo* human hearts including two infants and two adults were performed. The infants datasets were acquired in clinical conditions with a Siemens 3T MRI Magnetom Verio. The imaging parameters are the following: $TE = 74\,$ms, $TR = 7900\,$ms, FOV $= 144 \times 144\,$mm^2, slice thickness=1.4 mm, in-plane resolution $= 2\,$mm, slice spacing $= 1.4\,$mm, slice duration $= 123.2\,$ms, number of slices $= 35$, slice size: 104×104 pixels, diffusion sensitivity b $= 700\,$s/mm^2, and number of gradient directions $= 192$, 64 or 12. In each direction, MRI scans were acquired 6 to 10 times for noise reduction. The adult datasets were acquired using Siemens 3T MRI Magnetom Prisma with following parameters: $TE = 71\,$ms, $TR = 9600\,$ms, FOV $= 177 \times 177\,$mm^2, slice thickness $= 1.5\,$mm, slice spacing $= 1.5\,$mm, slice duration $= 123.1\,$ms, number of slices $= 70$, slice size: 122×122 pixels, diffusion sensitivity b $= 700\,$s/mm^2, and number of gradient directions $= 192$. In each direction, MRI scans were performed 3 times for noise reduction. An example of real DW images from an infant and an adult is given in Fig. 2.

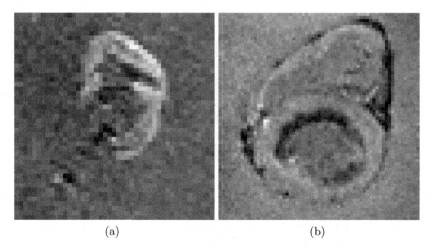

(a)   (b)

Fig. 2. An example of DW images from real datasets. (a) An DW image of infant. (b) An DW image of adult.

2.3 Gradient Direction Sampling

We considered two commonly used angular sampling schemes to generate spherically uniform distributed gradient directions for different angular resolutions of simulated datasets: Spherical Tessellation (ST) [16] and Electrostatic Energy Minimization (EEM) [17]. The ST employs spherical polyhedrons to generate specified numbers of directions geometrically uniformly distributed on a sphere. The EEM is capable of generating an arbitrary number of uniformly distributed directions by minimizing the electrostatic energy based on Coulomb's law. In order to systematically analyze the effects of various angular resolutions, we used EEM to generate different angular resolutions with uniformly-spaced numbers of diffusion gradient directions and one resolution with 6 directions (minimum number of gradient directions required for tensor estimation). For the real datasets, the popular spherical code sampling method [18] was used to obtain different angular resolutions from the real 192 directions. The directions for each subset are uniformly distributed on a sphere.

2.4 Evaluation

In the case of the simulated data, the SNR (dB) is defined by

$$SNR = 20 \cdot \log_{10} \left(\frac{V_n}{\frac{1}{N_v} \|V_n - V_f\|_2} \right) \tag{2}$$

where V_n denotes a discrete volume corrupted by Rician noise and V_f its noise-free representation. $\| \cdot \|$ is the standard Euclidean norm, and N_v is the number of voxels. In the case of real cardiac data, the SNR (dB) is defined by

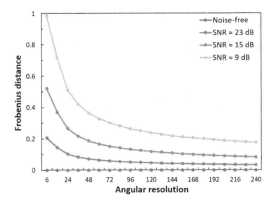

Fig. 3. Differences (Frobenius distance (10^{-3} mm^2/s)) between estimated and ground-truth tensor fields as a function of angular resolution and level of noise.

$$SNR = 20 \cdot \log_{10}\left(\frac{\mu_{rm}}{\sigma_{rm}}\right) \qquad (3)$$

where μ_{rm} and σ_{rm} are the mean and standard deviation (SD) of the voxel intensity values over the region of the myocardium.

We also evaluated the simulated and real data with different angular resolutions and different levels of noise by estimating diffusion tensors using standard least-squares estimation. The FA and MD were then derived from the diffusion tensors. For the simulated data, the mean Frobenius distance, root mean square errors (RMSE) of FA and MD (In the following sections, we call them FA or MD errors.) between ground-truth (the simulated tensor field) and the estimated diffusion tensors were computed. For the real data, the mean and standard deviation of FA and MD were computed.

3 Results

3.1 Results on Simulated Data

In Fig. 3, Frobenius distance curves are illustrated for different angular resolutions and different levels of noise. Here the used numbers of diffusion gradient directions were $N_d = \{6, 12, 24, 36, 48, \cdots, 240\}$. From this figure, we can clearly see that for each SNR, Frobenius distance decreases as the angular resolution increases, and that for each angular resolution, the distance decreases as the SNR increases. The vales of Frobenius distances for the maximum angular resolution (240) are smaller than 0.2 even for smaller SNR. Besides, the values of Frobenius distance for the dataset without noise (blue line) are always less than $= 1.2 \times 10^{-15}$ mm^2/s, which means that varying the angular resolution of noise-free dataset has no impact on the estimated diffusion tensors.

Figure 4 shows the curves of FA and MD errors as a function of the angular resolution and noise. The values of FA and MD were computed from the diffusion tensors (corresponding to the results in Fig. 3). From this figure, it can

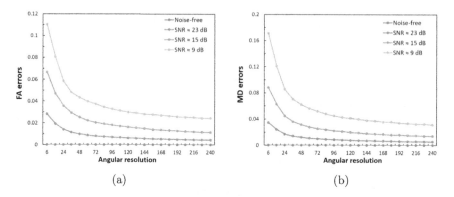

Fig. 4. Errors on tensor-derived measures for different angular resolution and level of noise. (a) FA errors. (b) MD errors. The unit of MD is $10^{-3}\,\mathrm{mm}^2/\mathrm{s}$

be observed that for all SNRs, the errors of FA or MD decrease as the angular resolution increases. Moreover, for each SNR, there is a tendency for both FA and MD errors to stabilize when the number of diffusion gradients (i.e. angular resolution) reached a certain number. This tendency can also be found in Fig. 3. To describe it clearly, we defined this number as the minimum angular resolution required for which the Frobenius distance, FA or MD errors come within 1% of their range (subtract the minimum value from the maximum value). For example, with this rule, the near-optimal diffusion tensors (the range of Frobenius distances smaller than 0.02) for SNR = 23 dB, 15 dB and 9 dB were achieved with 48, 120 and 156 diffusion gradient directions, respectively. The near-optimal FAs (the range of FA errors smaller than 0.006) for SNR = 23 dB, 15 dB and 9 dB were achieved with 48, 120 and 132 diffusion gradient directions, respectively, while for the near-optimal MDs (the range of MD errors smaller than 0.007), these numbers were 48, 108 and 124, respectively.

Fig. 5 gives the SNR curves as a function of the angular resolution for three acquisition times. To facilitate comparison, we assumed that the scan time for each simulated DW image was approximately 1s, and that the total acquisition time T is defined by the number N_d of diffusion gradient directions multiplied by the number N_e of excitations for each direction (used for signal averages): $T = N_d \times N_e$. Here $N_d = \{6, 12, 18, 24, 30, \cdots, 120\}$ was sampled for achieving different angular resolutions. N_e was set such that for a given angular resolution, we kept roughly the same acquisition time, e.g. $N_e = \{20, 10, 7, 5, 4, \cdots, 1\}$ was chosen for a acquisition time of approximately 40 minutes. From this figure, we can see that for each acquisition time, the SNR value decreases as the angular resolution increases. Moreover, longer acquisition time allows the use of higher angular resolution while keeping a certain value of SNR. For example, when SNR $\geq 25\,\mathrm{dB}$, the maximum available angular resolutions were 25, 48, and 78 for T $\approx 40\,\mathrm{min}$, 1 h 20 min, and 2 h, respectively.

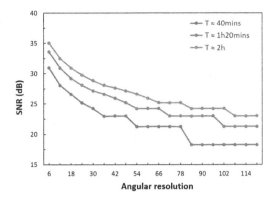

Fig. 5. SNR as a function of angular resolution for three fixed acquisition times. The values of SNR were computed using Eq. (2).

3.2   Results on Real Data

Figure 6 shows how the mean and standard deviation (SD) of FA and MD vary with the angular resolution and noise on real cardiac data. To evaluate the effect of noise, three additional datasets (SNR = 17 dB, 11 dB and 4 dB) were generated by adding different levels of Rician noise to the original DW images. The SNR was computed using Eq. (2), where V_n and V_f are the discrete volume corrupted with Rician noise and without noise, respectively. It can be seen that for all the noisy datasets with different levels of noise, the mean FA and MD value decreases as the angular resolution increases. Besides, the results also show that the MD value decreases as the noise level increases. This is consistent with the results in Pierpaoli et al.'s study [9]. In Table 1, the mean and SD of FA and MD were estimated for the four real datasets. For higher noise level (e.g., SNR = 4), the difference between original and noisy datasets decreases significantly as the angular resolution increases. In addition, when the number of gradient directions

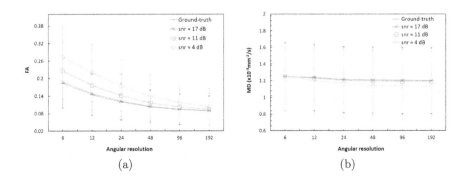

Fig. 6. Mean±SD of (a) FA and (b) MD values for different angular resolutions and different levels of noise on a real dataset of an infant heart.

are greater than 48, increasing the angular resolution has smaller effects on FA and MD, as also observed on simulated data.

Table 1. Mean±SD of FA and MD on four real datasets of *ex vivo* human hearts with different angular resolutions. The unit of MD is $10^{-3}\,\mathrm{mm}^2/\mathrm{s}$.

	Subject	Number of diffusion gradient directions (i.e., angular resolution)					
		6	12	24	48	96	192
FA	Infant # 1	0.18 ± 0.08	0.14 ± 0.07	0.12 ± 0.06	0.10 ± 0.06	0.09 ± 0.05	0.09 ± 0.05
	Infant # 2	0.16 ± 0.06	0.12 ± 0.04	0.11 ± 0.04	0.10 ± 0.04	0.10 ± 0.04	0.09 ± 0.04
	Adult # 1	0.22 ± 0.11	0.19 ± 0.11	0.17 ± 0.10	0.16 ± 0.10	0.15 ± 0.10	0.15 ± 0.10
	Adult # 2	0.17 ± 0.08	0.14 ± 0.07	0.13 ± 0.07	0.12 ± 0.06	0.12 ± 0.06	0.11 ± 0.05
MD	Infant # 1	1.31 ± 0.45	1.30 ± 0.45	1.27 ± 0.45	1.27 ± 0.45	1.27 ± 0.45	1.26 ± 0.45
	Infant # 2	1.42 ± 0.33	1.43 ± 0.33	1.42 ± 0.32	1.41 ± 0.32	1.41 ± 0.32	1.41 ± 0.32
	Adult # 1	0.91 ± 0.32	0.87 ± 0.33	0.86 ± 0.34	0.85 ± 0.34	0.84 ± 0.34	0.84 ± 0.34
	Adult # 2	0.90 ± 0.21	0.88 ± 0.22	0.88 ± 0.22	0.87 ± 0.22	0.87 ± 0.22	0.87 ± 0.22

Fig. 7 illustrates the variation of SNR as a function of the angular resolution on a real cardiac dataset (an infant heart) when the acquisition time is fixed to approximately 1 h 40 min. Different angular resolutions were achieved by sampling $N_d = \{32, 38, 44, 50, \cdots, 140\}$ from the original 192 diffusion gradient directions. $N_e = \{6, 5, 4, 3, \cdots, 1\}$ so that we kept a roughly equivalent acquisition time. The step shape occurs in Fig. 7 because for some continuous N_d, a same N_e was set for them in order to make sure that their total acquisition time were closer to 1 h 40 min. For example, for N_d from 62 to 92, the N_e was set to 2; but for $N_d \geq 98$, N_e was set to 1. It can be seen from Fig. 7 that for

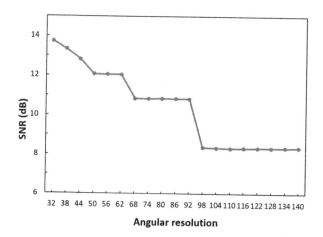

Fig. 7. SNR as a function of the angular resolution for fixed acquisition time on a real cardiac dataset. The values of SNR were computed using Eq. (3).

a fixed acquisition time, SNR decreases as angular resolution increases. Besides, the maximum angular resolution for SNR $\geq$ 12 dB is 62.

4   Conclusions and Discussions

This paper investigated the joint influence of angular resolutions and noise levels in cardiac DTI. The results on both synthetic and actual DW images showed that, as expected, given sufficient acquisition time and a certain noise level, the accuracy of diffusion tensors, FA and MD measurements increase as angular resolutions increase. Besides, continuing to increase the angular resolution beyond a certain value has little effect on the accuracy of diffusion tensor, FA and MD. For higher SNR (23 dB), the near-optimal diffusion tensor, FA and MD were obtained with 48 gradient directions. With lower SNR, the near-optimal diffusion tensor, FA and MD were obtained with, for example, 120 directions if SNR $=$ 15 dB and with 156 directions if SNR $=$ 9 dB. In addition, the results on real data show that both the mean and variance of FA or MD decrease as the angular resolution increases. When acquisition time is imposed, increasing angular resolution often comes at the cost of a reduced number of excitations used for signal averaging, thus reducing the SNR of DW images. Longer acquisition times allow the use of higher angular resolution while keeping a certain level of SNR. These findings suggest the necessity of an optimized trade-off when designing DTI protocols. In the future, we would like to conduct a similar study on *in vivo* DTI of the human heart.

Acknowledgements. This work was partly supported by the Program PHC-Cai Yuanpei 2018 (N° 41400TC), the LabEx PRIMES (Physics, Radiobiology, Imaging and Simulation), and the CNRS International Research Project METISLAB.

References

1. Tuch, D.S., Reese, T.G., Wiegell, M.R., Wedeen, V.J.: Diffusion MRI of complex neural architecture. Neuron **40**(5), 885–895 (2003)
2. Mekkaoui, C., et al.: Myocardial infarct delineation in vivo using diffusion tensor MRI and the tractographic propagation angle. J. Cardiovasc. Magn. Reson. **15**(1), 1–3 (2013)
3. Wu, M.-T., et al.: Diffusion tensor magnetic resonance imaging mapping the fiber architecture remodeling in human myocardium after infarction: correlation with viability and wall motion. Circulation **114**(10), 1036–1045 (2006)
4. Stejskal, E.O., Tanner, J.E.: Spin diffusion measurements: spin echoes in the presence of a time-dependent field gradient. J. Chem. Phys. **42**(1), 288–292 (1965)
5. Le Bihan, D.: IVIM method measures diffusion and perfusion (1990)
6. Taylor, W.D., Hsu, E., Krishnan, K.R.R., MacFall, J.R.: Diffusion tensor imaging: background, potential, and utility in psychiatric research. Biol. Psychiatry **55**(3), 201–207 (2004)
7. Hurwitz, R., Lane, S.R., Bell, R.A., Brant-Zawadzki, M.N.: Acoustic analysis of gradient-coil noise in MR imaging. Radiology **173**(2), 545–548 (1989)

8. Vovk, U., Pernus, F., Likar, B.: A review of methods for correction of intensity inhomogeneity in MRI. IEEE Trans. Med. Imaging **26**(3), 405–421 (2007)

9. Pierpaoli, C., Basser, P.J.: Toward a quantitative assessment of diffusion anisotropy. Magn. Reson. Med. **36**(6), 893–906 (1996)

10. Anderson, A.W.: Theoretical analysis of the effects of noise on diffusion tensor imaging. Magn. Reson. Med. Off. J. Int. Soc. Magn. Reson. Med. **46**(6), 1174–1188 (2001)

11. Papadakis, N.G., Murrills, C.D., Hall, L.D., Huang, C.L.-H., Carpenter, T.A.: Minimal gradient encoding for robust estimation of diffusion anisotropy. Magn. Reson. Imaging **18**(6), 671–679 (2000)

12. Jones, D.K.: The effect of gradient sampling schemes on measures derived from diffusion tensor MRI: a Monte Carlo study. Magn. Reson. Med. Off. J. Int. Soc. Magn. Reson. Med. **51**(4), 807–815 (2004)

13. Frindel, C., Robini, M., Croisille, P., Zhu, Y.-M.: Comparison of regularization methods for human cardiac diffusion tensor MRI. Med. Image Anal. **13**(3), 405–418 (2009)

14. Zhang, Y.-L., Liu, W.-Y., Magnin, I.E., Zhu, Y.-M.: Feature-preserving smoothing of diffusion weighted images using nonstationarity adaptive filtering. IEEE Trans. Biomed. Eng. **60**(6), 1693–1701 (2013)

15. Gudbjartsson, H., Patz, S.: The Rician distribution of noisy MRI data. Magn. Reson. Med. **34**(6), 910–914 (1995)

16. Teanby, N.A.: An icosahedron-based method for even binning of globally distributed remote sensing data. Comput. Geosci. **32**(9), 1442–1450 (2006)

17. Jones, D.K., Horsfield, M.A., Simmons, A.: Optimal strategies for measuring diffusion in anisotropic systems by magnetic resonance imaging. Magn. Reson. Med. Off. J. Int. Soc. Magn. Reson. Med. **42**(3), 515–525 (1999)

18. Cheng, J., Shen, D., Yap, P.-T., Basser, P.J.: Single-and multiple-shell uniform sampling schemes for diffusion MRI using spherical codes. IEEE Trans. Med. Imaging **37**(1), 185–199 (2017)

Novel Approaches to Measuring Heart Deformation

Arbitrary Point Tracking with Machine Learning to Measure Cardiac Strains in Tagged MRI

Michael Loecher[1](✉) ⓘ, Ariel J. Hannum[2] ⓘ, Luigi E. Perotti[3] ⓘ,
and Daniel B. Ennis[1] ⓘ

[1] Department of Radiology, Stanford University, Stanford, USA
mloecher@stanford.edu
[2] Department of Bioengineering, Stanford University, Stanford, USA
[3] Department of Mechanical and Aerospace Engineering, University of Central
Florida, Orlando, USA

Abstract. Cardiac tagged MR images allow for deformation fields to
be measured in the heart by tracking the motion of tag lines throughout the cardiac cycle. Machine learning (ML) algorithms enable accurate
and robust tracking of tag lines. Herein, the use of a massive synthetic
physics-driven training dataset with known ground truth was used to
train an ML network to enable tracking any number of points at arbitrary positions rather than anchored to the tag lines themselves. The tag
tracking and strain calculation methods were investigated in a computational deforming cardiac phantom with known (ground truth) strain
values. This enabled both tag tracking and strain accuracy to be characterized for a range of image acquisition and tag tracking parameters.
The methods were also tested on *in vivo* volunteer data. Median tracking
error was < 0.26 mm in the computational phantom, and strain measurements were improved *in vivo* when using the arbitrary point tracking for
a standard clinical protocol.

Keywords: Cardiac strains · Machine learning · MRI Tagging

1 Introduction

Measuring the regional function of the heart is a useful tool to aid in the diagnosis and treatment planning for several forms of heart disease. One tool to
measure the function of the heart is cine tagged MRI, which places tag lines
or grid patterns throughout the myocardium that are subsequently imaged as
they deform during the cardiac cycle. Tracking the tag lines enables computing
the myocardial deformation mapping, which can be used to calculate derived
parameters such as cardiac strains, or twist and torsion [4].

While the acquisition of tagged MRI images is widely available and straightforward, obtaining the deformation mapping from the images remains a difficult

© Springer Nature Switzerland AG 2021
D. B. Ennis et al. (Eds.): FIMH 2021, LNCS 12738, pp. 213–222, 2021.
https://doi.org/10.1007/978-3-030-78710-3_21

process with a variety of proposed methods. Tracking methods utilizing active contours, B-splines, or optical flow techniques have all been proposed to solve the problem [6]. Spectral methods such as HARP have also been proposed to produce displacement maps at the expense of image filtering [8]. Recently, machine learning (ML) has emerged as a promising method for measuring displacements from tagged images [3], wherein neural networks were used to process the tagged data. In our previous work, we demonstrated the achievable accuracy and precision when using physics-driven synthetic data to train a tag tracking neural network. This approach allowed training with known ground truth motion and did not require a large manually annotated *in vivo* datasets [7].

After obtaining the deformation mapping the myocardial strains can be calculated by looking at the changes in "lengths" along different directions with respect to the reference configuration [11]. When the tracked points are distant from one another, then radial basis function (RBF) interpolation can be combined with arbitrary point tracking. RBFs provide robustness with respect to different measurement positions and densities. Indeed, RBF has been used effectively in ultrasound [2] and tagged MRI studies [1]. Therefore, we incorporated RBF interpolation after tag tracking to investigate the impact of tag tracking point density on the strain measurements.

In this work, we have developed and trained a physics driven ML-based tag tracking method that allows selecting arbitrary tracking points. These points do not need to be anchored to the tag lines. This method increases the number of locations in the myocardium to be tracked compared to conventional methods. Subsequently, we investigated the effect of the arbitrary tracking point density and tag line spacing on the strain calculations to test if this enhanced tracking method can improve strain measurements.

2   Methods

2.1   Tag Tracking

In brief, our ML-based tag tracking approach used a Resnet-18 neural network [5]. Similar to previous work, this network was trained on a physics-driven synthetically generated dataset of tagged deforming objects. We've shown this approach leads to accurate tracking of tag intersections when used on *in vivo* data (positional error <1.0 mm *in vivo*) [7]. Herein, we generalize the method to track arbitrary points. The points to track were selected from the initial time-frame of a cine tagged MR image series, and fed into the neural network for patching. The output of the network was a vector of Lagrangian displacements of the tracked points for each time point.

The synthetic training dataset affords access to the ground truth motion needed for training the network to track arbitrary points in the image, rather than just along tag lines or at tag lines intersections. To explore this capability, the network was trained to track random points in the training images. After training the network, tag tracking was evaluated using both a computational deforming cardiac phantom embedded in an anatomical image and *in vivo* data

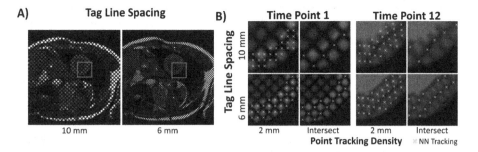

Fig. 1. A) Computational deforming phantom images with two different tag line spacings (10 mm and 6 mm). B) Cropped images (from red bounding box in panel A) showing the point tracking density in the reference configuration (time point 1) and at peak systole (time point 12). Green 'X' markers represent the tracking locations for 2 mm rectilinear grid tracking and tag intersection tracking. (Color figure online)

with three different tag tracking point density selections: tag intersections only, and tracking on a 2 mm and 1 mm rectilinear grid (Fig. 1).

2.2   Strain Calculation

RBFs were used to interpolate the deformation at a point $\mathbf{R}$ (defined in the reference configuration) from the tracked point locations according to:

$$s(\mathbf{R}) = \sum_{k=0}^{N} w_k \varphi (\mathbf{R} - \mathbf{P}_k) , \tag{1}$$

where N is the number of tracked points $\mathbf{P}_k$ defined in the reference configuration, w_k are the interpolation coefficients, and φ is the RBF function, which for our work is Gaussian:

$$\varphi (\mathbf{r}) = \exp \left(\frac{-\|\mathbf{r}\|^2}{\alpha^2} \right) , \tag{2}$$

where α is the shape parameter that defines the size of the Gaussian RBF. The deformation gradient tensor is calculated by taking the analytical derivative of $s(\mathbf{R})$ and it is then used to compute the Green-Lagrange strain tensor.

A first order polynomial fit was added to $s(\mathbf{R})$ to better account for translations, and Tikhonov regularization was used when solving for w_k to account for the presence of noise. Optimal values for RBF size (α) and regularization amount (λ) were determined using the computational phantom and searching for the parameters that produced the least discrepancy between computed and ground truth strain values.

2.3   Computational Deforming Cardiac Phantom

As a reference standard to test tracking accuracy, we used an computational deforming cardiac phantom generated to have prescribed strains [10,12]. The phantom was constructed with an axially symmetric heart-like geometry with initial ($t = 0$) inner radius equal to 25 mm and wall thickness equal to 12 mm. The radial, circumferential, and longitudinal displacement fields for the phantom were computed by solving an optimization problem to best match target peak systolic strains in the myofiber ($E_{ff} = -0.13$) [9], radial ($E_{rr} = 0.45$ to 0.30, endocardium to epicardium), circumferential ($E_{cc} = -0.20$ to -0.15, endocardium to epicardium), and longitudinal ($E_{\ell\ell} = -0.15$) directions.

The computational deforming cardiac phantom was then combined with a static body MR image and used to simulate a tagged MR image acquisition with a full Bloch simulation. To test the effects of tag line spacing on tracking and strain accuracy, the image simulation was performed with tag line spacing of 10 mm, 8 mm, 6 mm, and 4 mm. The tag line spacing in pixels was held constant by increasing the resolution (image resolutions = 1.75 mm, 1.40 mm, 1.05 mm, and 0.70 mm respectively). An example of the simulated images for 6 mm and 10 mm tag spacing are shown in Fig. 1, and the computational phantom strains at the peak systolic time-frame are shown in Fig. 2.

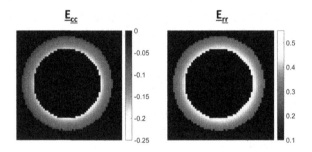

Fig. 2. Circumferential (E_{cc}) and radial (E_{rr}) strain maps from the computational phantom at peak systole used for reference validation.

2.4   *In Vivo* Data

In vivo images were acquired from healthy pediatric volunteers (N = 9, median age = 15 Y, with IRB approval and informed consent). Cine grid-tagged images were acquired with 110° total tagging flip angle for the tagging pulse and 8 mm grid spacing, 8 mm slice thickness, TE/TR = 2.5 ms/4.9 ms, flip angle = 10°, field of view = 260 mm × 320 mm, spatial resolution = 1.4 mm × 1.4 mm, 25 temporal time frames, 8–12 s breath hold. Three short axis slices at basal, middle, and apical locations were acquired for each subject for 27 total slices. The LV was manually segmented on each slice, and tag tracking and strain calculations were performed as above with intersection tracking, and 2 mm and 1 mm rectilinear grid tracking. Strain maps and values were compared across the subjects.

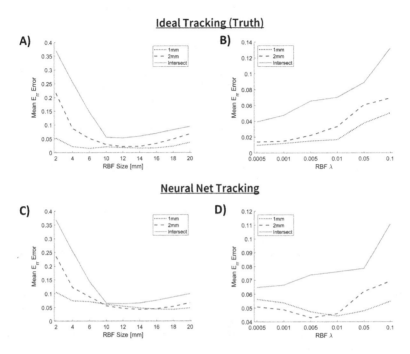

Fig. 3. A) Plot of mean E_{rr} error vs RBF size for ideal tracking locations (true displacements from computational phantom). B) Mean E_{rr} error vs RBF regularization for ideal tracking locations C) and D) show similar plots but for strain calculated with neural net tracked deformations rather than the true deformation. Error is the global mean difference in strain values across the LV area.

3 Results

Figure 3 shows the results of testing a range of RBF interpolation parameters for estimating strains. With ideal tracking locations, the lowest error was seen with an RBF window size of $10-14$ mm, and the lowest error corresponded to the lowest regularization value of 1e−4. Using the neural net tracked locations, a similar range was observed for the optimal RBF size, while a regularization parameter equal to 1e−3 produced the lowest overall error. In general, 1 mm point density led to the lowest errors in all comparisons, particularly with small RBF sizes. For the rest of this study, the RBF size was set to 14 mm and regularization was chosen equal to 1e−3.

Figure 4 shows the results of comparing the tracked displacements with the known ground truth motion from the computational deforming cardiac

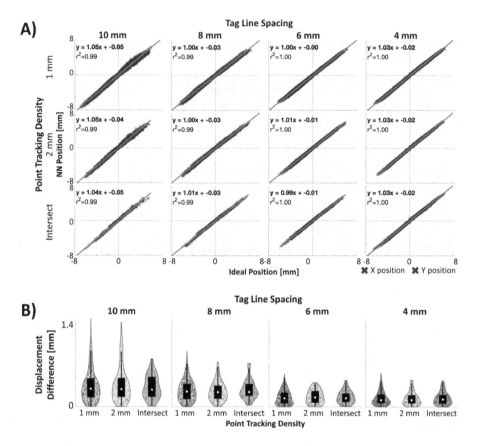

Fig. 4. A) Linear regressions of displacements for all tracked points with different line spacing and tracking point densities. x- and y-displacements are combined into an aggregate comparison of displacements for this analysis, and all time points are used. Best fit linear regressions and their equations are also displayed, where very good agreement is seen for all comparisons. B) Displacement error is also calculated for all points during peak systole and displayed as violin plots. The box and circle within the violin plots show IQR and median values. Good agreement is seen between tracked and true locations, with little difference in error between intersection tracking and arbitrary point tracking.

phantom. Linear regression shows very good agreement for all combinations of tag line spacing and tracking point density, with slopes between 0.99 and 1.05, and $r^2 >= 0.99$. Violin plots show the displacement error for each of the tag line spacing and point densities at peak systole, where the median error was $[0.26, 0.26, 0.25]$ mm, $[0.22, 0.22, 0.22]$ mm, $[0.13, 0.14, 0.13]$ mm, and $[0.10, 0.10, 0.11]$ mm for [1 mm, 2 mm, and intersection] tracking densities with 10 mm, 8 mm, 6 mm, and 4 mm tag line spacing, respectively. The median error and interquartile range did not change noticeably when tracking arbitrary points versus intersections, but there were slightly more outliers with 10 mm line spacing when tracking arbitrary points.

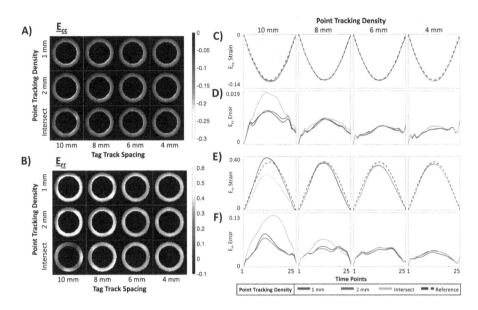

Fig. 5. A) E_{cc} and B) E_{rr} strain maps at peak systole for all combinations of tag line spacing and tracking point densities. C) Global E_{cc} and D) E_{rr} time curves and corresponding strain errors E), F) relative to the ground truth phantom values.

Strains were computed from the tracked points and the results are shown in Fig. 5. Strain maps demonstrated better qualitative agreement with the reference strain values (Fig. 2), particularly with 10mm tag line spacing and for E_{rr}. Line plots show the E_{cc} and E_{rr} curves over time, as well as the error over time relative to the known true strain in the phantom. The strain error decreased with smaller tag line spacing. Tracking more points with arbitrary point tracking improved strain quantification with 10 mm and 8 mm tag line spacing. For example, with 10 mm line spacing peak E_{cc} error reduced from 0.018 to 0.012, and peak E_{rr} error reduced from 0.12 to 0.07. Little difference was seen between strain calculated with 2 mm and 1 mm point tracking density.

Results from the *in vivo* analysis are shown in Fig. 6. Strain maps show qualitatively better spatial localization, similar to the phantom experiment. Global E_{cc} measurements were minimally different between the 1 mm and 2 mm tracking point densities ($<.01$), while E_{rr} increased with the 1 mm and 2 mm arbitrary point tracking with respect to intersection tracking (Mean value $= 0.208 \pm 0.077$ with intersection tracking, 0.230 ± 0.087, 0.231 ± 0.080 with 2 mm and 1 mm tracking point density). Line plots in Fig. 6 show similar global E_{cc} values with both methods, but increased E_{rr} with 2 mm tracking.

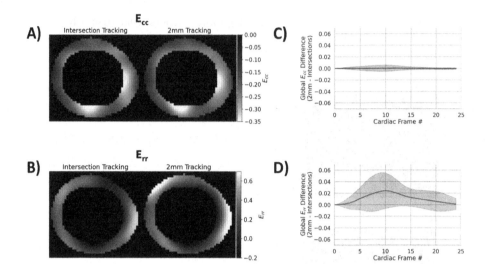

Fig. 6. A) E_{cc} and B) E_{rr} strain maps at peak systole for a representative *in vivo* dataset for intersection tracking and 2 mm tracking. Line plots show the difference in C) E_{cc} and D) E_{rr} between the two methods for all time points. Shaded errors in the line plots represent standard deviation across all slices measured.

4   Discussion

This study utilized a neural network in conjunction with synthetic data training to track arbitrary points in cine MRI tagged images of the heart. This allowed for tracking a large number of arbitrary points that are not necessarily tied to the tag lines intersections. In turn, this enabled the calculation of a more complete deformation map for analysis. When tested in a computational phantom, the tracking algorithm was shown to be very accurate, with no decrease in median error for tracking arbitrary points as compared to tracking tag line intersections. When strains were computed, the extra points enabled by arbitrary tracking improved measurements qualitatively and quantitatively for standard clinical protocol parameters. The strain improvement is reduced and eventually negligible as tag line spacing and resolution are improved beyond clinically realizable protocols.

Tracking point density was investigated by tracking points on a 2 mm and 1 mm rectilinear grid within a left ventricular mask. No improvement was seen when moving from 2 mm to 1 mm point density, showing that the spatial information at that level may be beyond the capabilities of the imaging and tracking method, or may be unnecessary given the measured deformation. RBF interpolation with a fixed shape parameter α was used for all point tracking densities as it was seen to give the best strain results. However the use of a higher tracking point density allows for smaller shape parameters to be used and may provide better spatial resolution of strain maps at the expense of accuracy.

One limitation of this work is that no regional pathologies were investigated to better understand spatial characteristics of the tracking method. Future work will aim to address this, with a non-symmetric computational deforming phantom and patients with grossly abnormal contraction patterns. However, computed transmural gradients of displacements and strains were better represented, demonstrating the potential for similar spatial representation improvements in these advanced cases. The synthetic data used to train the network contains a wide range of motions, with more variation and irregularities than seen in the computational phantom data presented here. It is therefore expected that the presented model will perform similarly well when applied to complex and abnormal cardiac motions.

Additionally, both the computational phantom and the synthetic data could be further improved. For instance, the computational phantom's cylindrical wall geometry can be modified to account for more realistic trabeculations. Through plane motion is also not simulated in either the training data or test data. While through plane motion cannot be reliably tracked with any 2D tagging technique, it is still important to understand the in-plane tracking performance of the method in the presence of through plane effects.

In vivo results showed differences consistent with the improvements seen in the computational phantom, suggesting that these results may also be more accurate. However, no ground truth strain measurement exists for comparison in these patients. Future work aims to compare to other tracking methods *in vivo*, as well as with other pulse sequence techniques for measuring strain with MRI.

In conclusion, a machine learning algorithm trained with synthetic data to track tags can track any point in a cardiac image, thereby enabling higher spatial resolution deformation mapping and more accurate strain measurements from tagged cardiac MRI.

Acknowledgments. This project was supported, in part, by NIH/NHLBI R01 HL131823, HL131975, and HL152256 to DBE.

References

1. Bistoquet, A., Oshinski, J., Škrinjar, O.: Myocardial deformation recovery from cine MRI using a nearly incompressible biventricular model. Med. Image Anal. **12**(1), 69–85 (2008)
2. Compas, C.B., et al.: Radial basis functions for combining shape and speckle tracking in 4D echocardiography. IEEE Trans. Med. Imaging **33**(6), 1275–1289 (2014)
3. Ferdian, E., et al.: Fully automated myocardial strain estimation from cardiovascular MRI-tagged images using a deep learning framework in the UK Biobank. Radiol. Cardiothorac. Imaging **2**(1), e190032 (2020)
4. Götte, M.J., et al.: Myocardial strain and torsion quantified by cardiovascular magnetic resonance tissue tagging: studies in normal and impaired left ventricular function. J. Am. Coll. Cardiol. **48**(10), 2002–2011 (2006)

5. He, K., Zhang, X., Ren, S., Sun, J.: Deep residual learning for image recognition. In: Proceedings of the IEEE Conference on Computer Vision and Pattern Recognition, pp. 770–778 (2016)
6. Ibrahim, E.S.H.: Myocardial tagging by cardiovascular magnetic resonance: evolution of techniques-pulse sequences, analysis algorithms, and applications. J. Cardiovasc. Magn. Reson. **13**(1), 1–40 (2011)
7. Loecher, M., Perotti, L.E., Ennis, D.B.: Using synthetic grid tagged images to train a neural network for tag tracking. In: Proceedings of the 24th Annual SCMR Meeting (2021)
8. Osman, N.F., Kerwin, W.S., McVeigh, E.R., Prince, J.L.: Cardiac motion tracking using CINE harmonic phase (HARP) magnetic resonance imaging. Magn. Reson. Med. Off. J. Int. Soc. Magn. Reson. Med. **42**(6), 1048–1060 (1999)
9. Perotti, L.E., Magrath, P., Verzhbinsky, I.A., Aliotta, E., Moulin, K., Ennis, D.B.: Microstructurally anchored cardiac kinematics by combining *In Vivo* DENSE MRI and cDTI. In: Pop, M., Wright, G.A. (eds.) FIMH 2017. LNCS, vol. 10263, pp. 381–391. Springer, Cham (2017). https://doi.org/10.1007/978-3-319-59448-4_36
10. Perotti, L.E., et al.: Estimating cardiomyofiber strain in vivo by solving a computational model. Med. Image Anal. **68**, 101932 (2020)
11. Scatteia, A., Baritussio, A., Bucciarelli-Ducci, C.: Strain imaging using cardiac magnetic resonance. Heart Fail. Rev. **22**(4), 465–476 (2017). https://doi.org/10.1007/s10741-017-9621-8
12. Verzhbinsky, I.A., Perotti, L.E., Moulin, K., Cork, T.E., Loecher, M., Ennis, D.B.: Estimating aggregate cardiomyocyte strain using in vivo diffusion and displacement encoded MRI. IEEE Trans. Med. Imaging **39**, 656–667 (2019)

Investigation of the Impact of Normalization on the Study of Interactions Between Myocardial Shape and Deformation

Maxime Di Folco[1]($\boxtimes$), Nicolas Guigui[2], Patrick Clarysse[1], Pamela Moceri[3,4], and Nicolas Duchateau[1]

[1] Univ Lyon, Université Claude Bernard Lyon 1, INSA-Lyon, CNRS, Inserm, CREATIS UMR 5220, U1294, 69621 Lyon, France
difolco@creatis.insa-lyon.fr
[2] Université Côte d'Azur, Inria, Epione, Sophia-Antipolis, France
[3] Centre Hospitalier Universitaire de Nice, Service de Cardiologie, Nice, France
[4] Université Côte d'Azur, Faculté de médecine, Nice, France

Abstract. Myocardial shape and deformation are two relevant descriptors for the study of cardiac function and can undergo strong interactions depending on diseases. Manifold learning provides low dimensional representations of these high-dimensional descriptors, but the choice of normalization can strongly affect the analysis. Besides, whether the shape normalization should include a scale factor is still an open question.

In this paper, we investigate the influence of normalization choices on the study of the interactions between cardiac shape and deformation using Multiple Manifold Learning, a dimensionality reduction method that considers inter- and intra-descriptors link between samples. By studying the main variations of two different shape normalizations (one including scaling, the other one not) we observed that the scaled normalization concentrates variations of a given physiological characteristic on only one mode. The influence of the associated choice of the deformation normalization was evaluated by quantifying differences between the estimated low-dimensional spaces (one for each choice against a combination of both), revealing potential analysis biases that may arise depending on such choices.

Keywords: Cardiac imaging · Manifold learning · Myocardial strain · Heart shape

1 Introduction

Medical imaging can provide high-dimensional descriptors of an organ's shape and function that are very complementary to the measurements used in clinical routine. However, identifying the appropriate descriptor for a particular diagnostic situation and finding an understandable representation of it is not an easy task due to the complexity of the anomalies that can co-exist.

© Springer Nature Switzerland AG 2021
D. B. Ennis et al. (Eds.): FIMH 2021, LNCS 12738, pp. 223–231, 2021.
https://doi.org/10.1007/978-3-030-78710-3_22

Myocardial shape, considered at End-Diastole (ED), and deformation are two relevant descriptors to assess the cardiac function, but these are often reduced to scalars in clinical routine (such as volumes or ejection fraction, respectively). However, these descriptors can undergo strong structural [1] or disease-related interactions [2,3].

Representation learning techniques can offer statistically-relevant and simplified views of a population or disease, but do not explicitly account for the interaction between descriptors. Within the field of manifold learning, Multiple Manifold Learning (MML) [4], generalized in [5] not only attempts to find a suitable latent space associated to a given descriptor, but also conditions the link between the representations associated to different descriptors.

Nonetheless, these methods still depend on the choice of descriptors and how these are normalized across a population. When analyzing shape features in a population, whether shape normalization should include scaling or volume invariance is still an open question. Besides, in computational anatomy, the transport of data from an individual's mesh to a reference template should be handled carefully [6], and the effect of scaling or volume differences on the transported data are still debated [7,8].

In this paper, we explore the influence of different normalization choices on the representation of myocardial shape and deformation while considering their interactions with non-linear manifold learning. We specifically examined a population of meshes representing the Right Ventricle (RV) from controls and patients with volume-overload, two conditions that affect differently shape and deformation depending on the pathology. We compared two different shape features and three strain normalization strategies, which correspond respectively to a rigid shape matching and no rescaling of the strain values, or finer shape changes with potential effect on the strain patterns.

2 Methods

2.1 Data and Descriptors

Data: We processed a database of 110 meshes of RV extracted from 3D echocardiography using the commercial software 4D-RV Function 2.0 (TomTec Imaging Systems GmbH, Germany). Point-to-point correspondences are ensured between the meshes, each containing 1587 cells and 822 points (valves excluded). The database consists of patients suffering congenital diseases with RV volume overload (Atrial Septal Defect (ASD) and Tetralogy of Fallot (ToF)), and 55 age- and sex-matched controls. In the following, we represent the myocardial shape and deformation using high-dimensional descriptors at each point of the RV mesh.

Shape Descriptors: In order to align the meshes spatially, a Procrustes analysis is applied with a rigid transformation (with no scale factor) and a similarity transformation (with a scale factor). This difference in normalization on the population will affect the shape and our analysis. Indeed, we characterize shape differences through a 3D vector between each point of a given individual mesh and its corresponding point on the reference. This reference was obtained by a Procrustes alignment only of the controls to encode shape normality. In the following, we denote this shape descriptor (of dimensions 822×3) either *rigid* or *similarity* depending in the type of Procrustes alignment used (Fig. 1a).

Deformation Descriptors: The deformation descriptor is the area strain corresponding to the relative area change of each mesh cell (in %) between ED and End-Systole (ES). It quantifies the local deformation of the RV surface with a high-dimensional descriptor of dimensions 822×1. We first computed this descriptor on the original population, it is denoted *original*.

We also applied a normalization procedure in order to disentangle this descriptor from the organ's shape. The normalization procedure relies on the parallel transport of deformations in the diffeomorphic registration setting, with a geometric representation of deformations as elements of the group of diffeomorphisms endowed with a right-invariant Riemannian metric [6,9]. In order to define a volume-invariant normalization, we used the metric proposed in [7]. It involves normalizing each shape so that its volume matches that of the reference shape. Two projection strategies were used, one by scaling around the barycentre of each shape, and the other one by following the flow of the gradient of the volume function. They are denoted respectively *scaling* and *gradient* (Fig. 1b). The former may be biased if the center of mass of the shapes is unnatural, while the latter depends on the choice of the embedding metric, in our case the Large Deformation Diffeometric Metric Mapping (LDDMM) metric, to define the gradient.

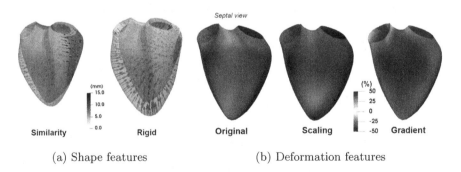

     (a) Shape features   (b) Deformation features

Fig. 1. (a) Distance vector field (mm) colored by magnitude between a sample and the reference mesh (in red) for *similarity* and *rigid* transformations. (b) Average area strain (%) for the *original*, *scaling* and *gradient* deformation features. (Color figure online)

2.2 Manifold Learning

Given K subjects and one high-dimensional descriptor for shape $\mathbf{X^S} = \left[\mathbf{x}_1^S, \cdots, \mathbf{x}_K^S\right]^T \in \mathbb{R}^{K \times M_S}$ and another one for deformation $\mathbf{X^D} = \left[\mathbf{x}_1^D, \cdots, \mathbf{x}_K^D\right]^T \in \mathbb{R}^{K \times M_D}$ (M_D and M_S are the number of input dimensions, here 822 and $2466 = 822 \times 3$), we jointly represented each descriptor and the interactions between their latent space using the MML algorithm. We first estimated the following affinity matrix:

$$\mathbb{W} = \begin{bmatrix} \mathbf{W}^S & \mu\mathbf{M} \\ \mu\mathbf{M}^T & \mathbf{W}^D \end{bmatrix} \tag{1}$$

$\mathbb{W}$ is composed of the sub-matrices $\mathbf{W}^S$ and $\mathbf{W}^D$ obtained by spectral embedding on the shape and deformation descriptors independently, which represent the intra-descriptor correspondences, and the matrix $\mathbf{M}$ (weighted by a scalar value $\mu > 0$) which encodes the inter-descriptor correspondences:

$$W_{ij}^S = \exp\left(-\|\mathbf{x_i}^S - \mathbf{x_j}^S\|^2/\sigma^2\right) \qquad M_{ij} = \frac{<\mathbf{w}_i^S, \mathbf{w}_j^D>}{\|\mathbf{w}_i^S\|\|\mathbf{w}_j^D\|}$$
$$W_{ij}^D = \exp\left(-\|\mathbf{x_i}^D - \mathbf{x_j}^D\|^2/\sigma^2\right)$$

where σ is the width of the kernel used for spectral embedding, determined as the average distance between a sample and its k_σ nearest neighbors, and $\mathbf{w}_i^S$ (resp. $\mathbf{w}_i^D$) is the i-th row of the affinity matrix $\mathbf{W}^S$ (resp. $\mathbf{W}^D$). The matrix $\mathbf{M}$ is sparsified afterwards by setting to zero each element that does not belong to the k_M closest neighbors of each sample (determined as the k_M highest values in each row of the affinity matrix).

MML aims at minimizing:

$$\Phi(\mathbf{Y}) = \sum_{i,j=1}^{K} \|\mathbf{y}_i^S - \mathbf{y}_j^S\|^2 W_{ij}^S + \sum_{i,j=1}^{K} \|\mathbf{y}_i^D - \mathbf{y}_j^D\|^2 W_{ij}^D + \mu \sum_{i,j=1}^{K} \|\mathbf{y}_i^S - \mathbf{y}_j^D\|^2 M_{ij}, \tag{2}$$

where $\mathbf{Y^S} = \left[\mathbf{y}_1^S, \cdots, \mathbf{y}_K^S\right]^T \in \mathbb{R}^{K \times N}$ and $\mathbf{Y^D} = \left[\mathbf{y}_1^D, \cdots, \mathbf{y}_K^D\right]^T \in \mathbb{R}^{K \times N}$ respectively stand for the latent spaces coordinates of the shape and deformation descriptors, N being the number of latent dimensions retained. They are obtained by computing the normalized graph Laplacian and then using Diffusion Maps as described in [10]. The first half of the eigenvectors of $\mathbb{W}$ encodes the data from the first descriptor (shape, K first eigenvectors), while the second half corresponds to the second descriptor (deformation, K last eigenvectors). Thus, MML provides one latent space for each descriptor, meaning that we have two low-dimensional latent spaces, one representing the shape and the other one the deformation.

3   Experiments and Results

The following section describes the application of MML to the entire population (except one control due to obvious mesh defects), with the hyperparameters:

$k_\sigma = 10$, $k_M = 10$ and $\mu = 1$. These parameters have been determined empirically from the energy defined in Eq. 2.

Figure 2 shows the obtained deformation latent spaces for the different combinations of shape and deformation descriptors. Note that for the *rigid-gradient* latent space, the second and third dimensions are displayed. Indeed, with spectral embedding several eigenvectors may encode the same spatial direction and make some dimensions redundant [11]. This is the case with the first and the second dimensions here.

ASD samples are more centered and close from each other using the *original* and *scaling* deformation descriptors compared to the *gradient* descriptors. The different populations seem more separated for these strategies especially the ToF and controls. However, the distribution of samples in the latent spaces is not enough to assess the difference between normalization strategies. In the next section, we explore the influence of the shape normalization on the main variations encoded in the latent spaces.

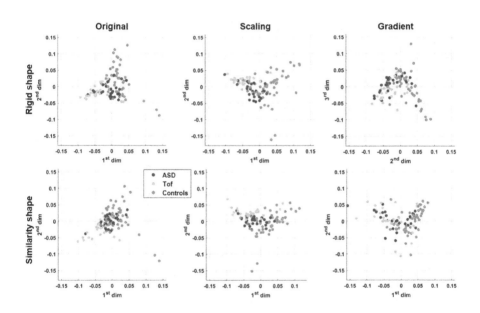

Fig. 2. Latent spaces encoding deformation, provided by MML for the 6 different normalization strategies.

3.1   Shape Descriptors

For the *similarity-original* and *rigid-original* latent spaces encoding shape, we reconstructed the main characteristics encoded from -2σ to 2σ along the first dimensions (σ being the standard deviation along the considered dimension), using multi-scale regression as described in [12]. Visualization consisted of the shapes obtained after adding the reconstructed vector fields to the reference shape. We then compared the different normalization strategies by defining some shape characteristics on the reconstructed meshes: the intra-valve distance, the width, the depth and the length of the meshes as depicted in Fig. 3a and also the volume. Each distance was computed between chosen points on the mesh for each measure, which are straightforward to obtain for different subjects as point-to-point correspondences are known.

Figure 3b shows the relative change of a given measure when moving between -2σ and $+2\sigma$ for the first three modes of variations. We can observe that the *similarity* strategy concentrates volume and depth variations on the first mode, while finer changes predominate on the next modes, such as length and valve distance on the second mode. For the *rigid* descriptor, substantial variations of volume, length and width happened on the first two modes. The depth and the valve distance variations are mainly on one mode (first and second respectively).

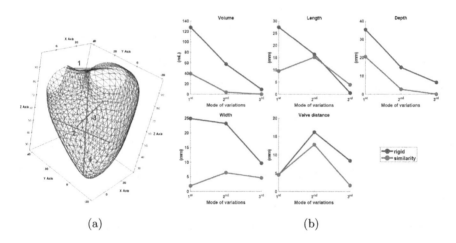

(a)   (b)

Fig. 3. Variations of 5 different shape characteristics defined on each mesh, when moving from -2σ to $+2\sigma$ along a given dimension of the latent space with rigid and similarity normalizations. The characteristics are defined in (a) with (1) valve distance, (2) RV width, (3) depth and (4) length.

3.2   Deformation Descriptors

In this section, we evaluated the influence of the normalization choice for the deformation descriptors. We first applied MML on two combinations of the deformation descriptors (*original/scaling* or *original/gradient*) and each shape normalization. These spaces are considered as reference and we evaluated the difference to the corresponding latent space displayed in Fig. 2. By comparing them to a combination of both descriptors equally weighted (to do so, we replaced the submatrix $\mathbf{W}^D$ by the arithmetic average of the associated affinity matrices), a predominant normalization may appear. Differences between the latent spaces were quantified by the point-to-point differences between samples. As these may be noisy, a 2D kernel regression was used on the first two dimensions of the latent spaces to smooth out the differences and better visualize trends for the whole population or specific subgroups. The regression involved a 18×18 grid equally distributed between -0.1 and 0.1 for the first two dimensions. To leverage the contribution of samples in less populated regions, we encoded the density around each sample as the opacity of arrows.

Figure 4a shows the results for the *original/scaling* combination. The *original* deformation vector field seems to rotate around the center for the latent space, for the two shape normalizations. Few changes are observed for the *scaling* deformation, even though the *rigid + scaling* exhibits a reduced rotation. Concerning the combination of *original/gradient* displayed in Fig. 4b, *rigid + original* and *rigid + gradient* show substantial changes that seem to be in opposite directions. *similarity + original* also exhibits a rotation and changes occur for *similarity + gradient* mainly in a low density zone.

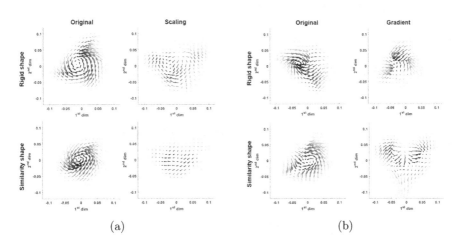

Fig. 4. Vector field encoding the local distances in the latent spaces between a combination of (a) *original/scaling* and (b) *original/gradient*. The opacity reflects to the estimated density at each point of the grid.

4   Discussion and Conclusion

In this paper, we investigated the influence of the normalization on the study of the interactions between the myocardial shape and the cardiac deformation using MML.

We evaluated the differences between two shape normalization strategies (*rigid* and *similarity* with a scaling factor), through the main modes of variations of the population. Several measurements on the meshes show that the variations are mainly concentrated on one mode for *similarity* (first mode for volume and depth, second mode for length and valve distance), unlike *rigid* where substantial variations jointly appear on several modes, except for the valve distance. Specific latent dimensions provide more understandable variations, which is an asset for the study of the population intra-variability.

Then, we assessed the influence of the normalization on the deformation feature. The two normalizations (*scaling* and *gradient*) predominate on *original* (a rotation appeared on every combination with no normalization). This rotation might correspond to the bias introduced by an improper pair of shape and deformation normalizations. Indeed, it can be expected that applying a volume-related normalization on only one of the features can introduce such a bias. The *similarity* descriptor has more affinity to the normalized deformation descriptor (less change in deformation field) compared to the *rigid*.

As we focus our work on the study of interactions between cardiac shape and deformation, we would like to prevent potential biases introduced by the normalization. These experiments provided insights into the influence of the shape normalization by itself, and the choices of pairs of shape and deformation normalizations. They revealed differences between the two strategies of normalization for the deformation feature. However the choice of an optimal strategy is likely to depend on the application and the associated data. Other state-of-the-art strategies to normalize the descriptors will be explored in future work.

Acknowledgements. The authors acknowledge the support from the French ANR (LABEX PRIMES of Université de Lyon [ANR-11-LABX-0063] within the program "Investissements d'Avenir" [ANR-11-IDEX-0007], and the JCJC project "MIC-MAC" [ANR-19-CE45-0005]), and the EEA doctoral school. This work was also partially funded by the ERC grant Nr. 786854 G-Statistics from the European Research Council under the European Union's Horizon 2020 research and innovation program, and by the French government through the 3IA Côte d'Azur Investments ANR-19-P3IA-0002 managed by the National Research Agency. The authors are grateful to the OPAL infrastructure from Université Côte d'Azur for providing resources and support.

References

1. Bijnens, B., Cikes, M., Butakoff, C., et al.: Myocardial motion and deformation: what does it tell us and how does it relate to function? Fetal Diagn. Ther. **32**, 5–16 (2012)
2. Moceri, P., Duchateau, N., Gillon, S., et al.: Three-dimensional right ventricular shape and strain in congenital heart disease patients with right ventricular chronic volume loading. Eur. Heart J. Cardiovasc. I maging (2020) (in press)
3. Sanz, J., Sánchez-Quintana, D., Bossone, E., et al.: Anatomy, function, and dysfunction of the right ventricle. J. Am. Coll. Cardiol. **73**, 1463–82 (2019)
4. Ham, J., Lee, D.D., Saul, L.K.: Semisupervised alignment of manifolds. In: Proceedings of the AISTATS 120, pp. 120–7 (2005)

5. Clough, J.R., Balfour, D., Cruz, G., et al.: Weighted manifold alignment using wave kernel signatures for aligning medical image datasets. IEEE Trans. Pattern Anal. Mach. Intell. **42**, 988–97 (2019)
6. Lorenzi, M., Pennec, X.: Efficient parallel transport of deformations in time series of images: from schild to pole ladder. J. Math. Imaging Vis. **50**, 5–17 (2014)
7. Niethammer, M., Vialard, F.: Riemannian metrics for statistics on shapes: parallel transport and scale invariance. In: Proceedings of the MFCA-MICCAI (2013)
8. Guigui, N., Moceri, P., Sermesant, M., et al.: Cardiac motion modeling with parallel transport and shape splines. In: Proceedings of the ISBI (2021) (in press)
9. Shapes and Diffeomorphisms. AMS, vol. 171. Springer, Heidelberg (2019). https://doi.org/10.1007/978-3-662-58496-5
10. Coifman, R.R., Lafon, S.: Diffusion map. Appl Comput. Harmon Anal **21**, 5–30 (2006)
11. Nadler, B., Lafon, S., Coifman, R., et al.: Diffusion Maps - a probabilistic interpretation for spectral embedding and clustering algorithms. In: Gorban, A.N., Kégl, B., Wunsch, D.C., Zinovyev, A.Y. (eds.) Principal Manifolds for Data Visualization and Dimension Reduction, pp. 238–260. Springer, Heidelberg (2008). https://doi.org/10.1007/978-3-540-73750-6_10
12. Duchateau, N., De Craene, M., Sitges, M., Caselles, V.: Adaptation of multiscale function extension to inexact matching: application to the mapping of individuals to a learnt manifold. In: Nielsen, F., Barbaresco, F. (eds.) GSI 2013. LNCS, vol. 8085, pp. 578–586. Springer, Heidelberg (2013). https://doi.org/10.1007/978-3-642-40020-9_64

Reproducibility of Left Ventricular CINE DENSE Strain in Pediatric Subjects with Duchenne Muscular Dystrophy

Zhan-Qiu Liu[1], Nyasha G. Maforo[2], Pierangelo Renella[2,3], Nancy Halnon[4], Holden H. Wu[2], and Daniel B. Ennis[1(✉)]

[1] Department of Radiology, Stanford University, Palo Alto, CA, USA
dbe@stanford.edu
[2] Department of Radiological Sciences, University of California, Los Angeles, CA, USA
[3] Department of Medicine (Pediatric Cardiology), Children's Hospital, Orange, CA, USA
[4] Department of Pediatrics, University of California, Los Angeles, CA, USA

Abstract. Cardiomyopathy is the leading cause of mortality in boys with Duchenne muscular dystrophy (DMD). Left ventricular (LV) peak mid-wall circumferential strain (E_{cc}) is a sensitive early biomarker for evaluating both the subtle and variable onset and the progression of cardiomyopathy in pediatric subjects with DMD. Cine Displacement Encoding with Stimulated Echoes (DENSE) has proven sensitive to changes in E_{cc}, but its reproducibility has not been reported in a pediatric cohort or a DMD cohort. The **objective** was to quantify the intra-observer repeatability, and intra-exam and inter-observer reproducibility of global and regional E_{cc} derived from cine DENSE in DMD patients (N = 10) and age- and sex-matched controls (N = 10). Global and regional E_{cc} measures were considered reproducible in the intra-exam, intra-observer, and inter-observer comparisons. Intra-observer repeatability was highest, followed by intra-exam reproducibility and then inter-observer reproducibility. The smallest detectable change in E_{cc} was 0.01 for the intra-observer comparison, which is below the previously reported yearly decrease of 0.013 ± 0.015 in E_{cc} in DMD patients.

Keywords: Duchenne muscular dystrophy · MRI · Strain · Reproducibility

1 Introduction

Duchenne muscular dystrophy (DMD) is a fatal inherited genetic disorder, affecting 2.63 to 11.66 per 10,000 male births for which cardiomyopathy is the leading cause of mortality. Reduced left ventricular (LV) ejection fraction (EF < 55%) is a widely used marker of cardiac function and outcomes, but the decline in LVEF is a relatively late finding among DMD patients. Late gadolinium enhancement (LGE) MRI is the gold standard for detecting focal myocardial fibrosis, but positive LGE is also a late finding in DMD. Owing to the subtle and highly variable onset and progression of cardiomyopathy in pediatric subjects with DMD, there is a clinical need for sensitive non-contrast CMR

© Springer Nature Switzerland AG 2021
D. B. Ennis et al. (Eds.): FIMH 2021, LNCS 12738, pp. 232–241, 2021.
https://doi.org/10.1007/978-3-030-78710-3_23

biomarkers prior to obvious systolic dysfunction in order to evaluate patient-specific treatment strategies of early cardiac dysfunction in DMD.

LV peak mid-wall circumferential strain (E_{cc}) using tagging has been identified as an early biomarker of dysfunction in DMD [1]. Alternatively, Cine Displacement Encoding with Stimulated Echoes (DENSE) has notable post-processing advantages over tagging and is also sensitive to changes in E_{cc}. E_{cc} derived from cine DENSE, however, has not been reported a pediatric cohort or a DMD cohort. Thus, the quantification of E_{cc} reproducibility using cine DENSE in a pediatric cohort and a DMD cohort is important for quantifying longitudinal disease progression. Consequently, our **objectives** were: 1) To evaluate the intra-observer repeatability, and intra-exam and inter-observer reproducibility of global and regional LV E_{cc} in DMD patients and healthy controls using cine DENSE; and 2) To quantify the corresponding smallest detectable change in E_{cc}.

2 Methods

2.1 Study Enrollment

LGE(-) DMD patients (N = 10, 15 ± 6.0 years old) and age- and sex-matched healthy controls (N = 10,16 ± 6.0 years old) were enrolled in an IRB-approved multi-center HIPAA compliant prospective study. Parental permission and statements of informed consent were obtained for each participant. DMD patients and healthy controls were recruited at one of two children's hospitals on a referral basis.

2.2 MR Imaging and Post-processing

All subjects underwent a cardiac MRI exam at 3T (Skyra, Siemens) using identical software, coils, and protocol. Single slice, LV short-axis datasets at the mid-ventricular level were acquired with navigator-gated free-breathing 2D cine DENSE [2] (2-point phase cycling, spatial resolution = $2.5 \times 2.5 \times 8$ mm^3, TE/TRes = 1.2/15, XY displacement encoding $k_e = 0.08$ cycles/mm, spirals = 10, Navg = 3, scan time ~2.5 min).

Mid-ventricular LV borders were manually traced over the entire cardiac cycle using the open-source DENSEanalysis and E_{cc} was evaluated using the approach outlined in Spottiswoode et al. [3]. Mid-ventricular myocardium was equally partitioned into three transmural layers (i.e., sub-endocardium, mid-myocardium, and sub-epicardium). Global mid-wall E_{cc} was averaged within the mid-myocardium. Regional mid-wall E_{cc} was also evaluated in anteroseptal, inferoseptal, inferior, inferolateral, anterolateral, and anterior wall segments.

2.3 Reproducibility and Statistics

Reproducibility. For the intra-exam reproducibility, each of the 20 subjects underwent consecutive DENSE acquisitions (i.e., Scan-1 and Scan-2) without repositioning during a single MRI exam. Post-processing was carried out by User-1. The post-processing analyses for all 20 subjects from the first DENSE acquisition (Scan-1) were repeated by User-2 for the inter-observer reproducibility. User-1 also repeated the post-processing analyses for Scan-1, with at least two weeks in between, for the intra-observer repeatability.

Reproducibility of a given metric was defined as the modified coefficient of variation (CoV) [4]. CoV $\leq$ 20% was considered reproducible. Absolute agreement between two observations of a given metric was defined as the intraclass correlation coefficient (ICC) [5]. ICC $\leq$ 40% was considered poor; ICC between 40% and 59% was considered fair; ICC between 60% and 74% was considered good; and ICC between 75% and 100% was considered excellent. The smallest detectable change (SDC) in a metric was quantified to provide a threshold above which a change in the metric is reliable at a 95% confidence interval [6].

Statistics. All statistical analyses were performed in MATLAB. A linear-regression model was used to identify the correlation between two observations of a given metric. The coefficient of determination R^2 and sum of squared error (SSE) were reported. For current observations of a given metric, an R^2 value between 70% and 100% was considered to be a "high correlation", while an R^2 value less than 40% was considered to be a "low correlation". Subsequently, the predicted residual error sum of squares (PRESS) was calculated by removing each pair of observations in turn from n pairs of observations, followed by refitting a linear-regression model using the remaining pairs of observations [7]. PRESS was defined as follows [8]:

$$\text{PRESS} = \sum\nolimits_{i=1}^{n} (y_i - f(x_i))^2 \qquad (1)$$

where y_i is the removed dependent observation; x_i is the removed independent observation; and $f(x_i)$ is the predicted value calculated from the refitted linear-regression model. Then, the predictive R^2 was defined as follows [9]:

$$\text{predictive } R^2 = \left(1 - \frac{\text{PRESS}}{\text{SST}}\right) \times 100 \qquad (2)$$

where SST is the sum of squared deviations of the measures from the mean measure. The predictive R^2 value indicates the probability of future pair of observations of a given metric being perfectly linearly correlated. A Bland-Altman analysis was performed using a non-parametric distribution assumption for a given metric to assess and visualize bias and 95% limits-of-agreement (LOA) between two observations of a metric. Data is reported as median and Interquartile range (IQR).

3 Results

3.1 Global and Regional E_{cc}

Global and Regional E_{cc} results are summarized in Table 1 – intra-exam reproducibility (first two rows); the inter-observer reproducibility data (rows one and three); and the intra-observer repeatability data (rows one and four). The values of predictive R^2 are summarized in Table 2 for peak global and regional circumferential strain.

Table 1. Peak mid-wall circumferential strain (E_{cc}) results. Data is reported as median (IQR).

Peak mid-wall E_{cc}	Antero-septal	Infero-septal	Inferior	Infero-lateral	Antero-lateral	Anterior	Global
Scan-1, User-1, Day-1	−0.15(0.03)	−0.13(0.04)	−0.17(0.03)	−0.21(0.04)	−0.21(0.03)	−0.18(0.05)	−0.17(0.02)
Scan-2, User-1, Day-1	−0.15(0.04)	−0.14(0.03)	−0.17(0.05)	−0.21(0.04)	−0.20(0.03)	−0.17(0.03)	−0.17(0.02)
Scan-1, User-2, Day-1	−0.20(0.05)	−0.15(0.04)	−0.15(0.05)	−0.18(0.05)	−0.21(0.07)	−0.21(0.03)	−0.18(0.02)
Scan-2, User-1, Day 2	−0.15(0.03)	−0.14(0.03)	−0.17(0.03)	−0.21(0.03)	−0.21(0.03)	−0.18(0.05)	−0.17(0.02)

Table 2. The predictive R^2 estimates the probability of future pair of observations of peak global or regional circumferential strain being perfectly linearly correlated

Predictive R^2 (%)	Intra-user observations	Intra-exam observations	Inter-user observations
Peak global circumferential strain	97	78	71
Peak regional circumferential strain	94	71	21

3.2 Intra-observer Repeatability

Table 3 summarizes CoV, ICC, and SDC values for the global and regional E_{cc} comparisons assessed by User-1 for intra-observer repeatability. Both the global and regional E_{cc} were highly reproducible with all CoVs $\leq$ 4% and exhibited excellent agreement between analyses with all ICCs $\geq$ 88%. The SDCs for global and regional E_{cc} ranged from 0.01 to 0.03.

Table 3. Intra-observer repeatability of peak global and regional circumferential strain (E_{cc}).

Intra-user reproducibility	Antero-septal	Infero-septal	Inferior	Infero-lateral	Antero-lateral	Anterior	Global
Coefficient of variation (%)	3	4	3	2	2	3	1
ICC (%)	95	88	96	97	97	95	99
Smallest detectable change	0.02	0.03	0.02	0.02	0.01	0.02	0.01

Figure 1A shows a high correlation between current intra-user observations of the global E_{cc} measures ($R^2 = 0.98$). And Fig. 2A also shows a high correlation between current intra-user observations of the regional E_{cc} measures ($R^2 = 0.94$). Table 2 shows a very high probability that future pair of observations (as assessed by User-1 with at least two weeks in between) will be perfectly linearly correlated (predictive $R^2 \geq 94\%$ for global and regional E_{cc} measures). Bland-Altman analysis shows excellent agreement between analyses with a mean difference and 95% LOA of 0.0002 ± 0.0142 for regional E_{cc}.

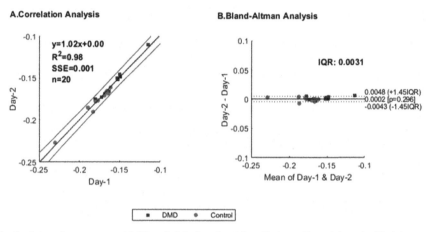

Fig. 1. Intra-observer repeatability of global peak mid-wall circumferential strain (E_{cc}) between scans in the pooled cohort of DMD patients and healthy controls.

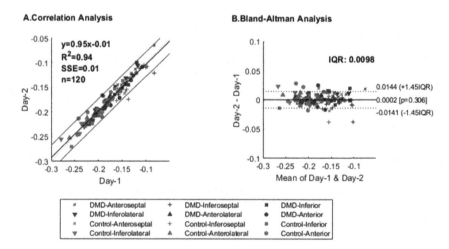

Fig. 2. Intra-observer repeatability of regional peak mid-wall circumferential strain (E_{cc}) between scans in the pooled cohort of DMD patients and healthy controls.

3.3 Intra-exam Reproducibility

The global E_{cc} measures ($R^2 = 0.82$, Fig. 3A) and regional E_{cc} measures ($R^2 = 0.75$, Fig. 4A) were highly correlated between consecutive scans when assessed by User-1. Table 2 shows it is likely that future pair of observations in two consecutive scans will be linearly correlated (predictive $R^2 \geq 71\%$ for global and regional E_{cc} measures). Bland-Altman analysis shows good agreement between scans with a mean difference and 95% LOA of 0.001 ± 0.016 for global E_{cc} (Fig. 3B) and 0.001 ± 0.029 for regional E_{cc} (Fig. 4B).

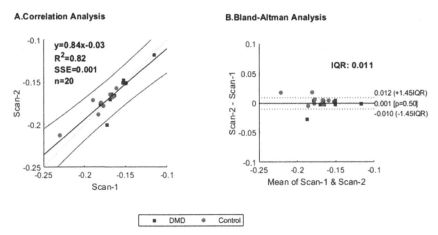

Fig. 3. Intra-exam reproducibility of peak mid-wall global circumferential strain (E_{cc}) between scans in the pooled cohort of DMD patients and healthy controls.

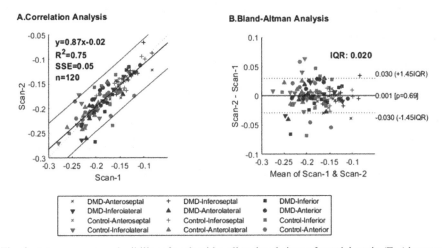

Fig. 4. Intra-exam reproducibility of peak mid-wall regional circumferential strain (E_{cc}) between scans in the pooled cohort of DMD patients and healthy controls.

Table 4 summarizes CoV, ICC, and SDC values for the comparisons of the global and regional E_{cc} between scans. Global and regional E_{cc} demonstrated excellent reproducibility (CoV: 2–8%). The inferior and anterior E_{cc} exhibited good agreement between scans (ICCs = 69% and 72%, respectively), while E_{cc} in the other regional segments and the global E_{cc} exhibited excellent agreement (ICC: 80–91%). The SDC was 0.02 for global E_{cc} while regional E_{cc} ranged from 0.03 to 0.06.

Table 4. Intra-exam reproducibility of peak global and regional circumferential strain (E_{cc}).

Intra-exam reproducibility	Antero-septal	Infero-septal	Inferior	Infero-lateral	Antero-lateral	Anterior	Global
Coefficient of variation (%)	5	6	8	4	4	8	2
ICC (%)	84	86	69	84	80	72	91
Smallest detectable change	0.03	0.03	0.06	0.04	0.03	0.05	0.02

3.4 Inter-observer Reproducibility

Table 5 summarizes CoV, ICC, and SDC values for the comparisons of the global and regional E_{cc} between users. The global E_{cc} was highly reproducible (CoV = 4%), while regional E_{cc} was less reproducible (CoV $\leq$ 20%). The global E_{cc} exhibited excellent agreement between users (ICC = 83%). However, regional E_{cc} exhibited poor agreement in anteroseptal and anterior segments (ICCs = 26% and 4%, respectively) and fair agreement in the other segments (ICC: 33–51%). The SDC was relatively low for global E_{cc} (SDC = 0.02), while the SDCs for regional E_{cc} were relatively high and more variable (SDCs: 0.06–0.10).

Table 5. Inter-observer reproducibility of peak global and regional circumferential strain.

Inter-user reproducibility	Antero-septal	Infero-septal	Inferior	Infero-lateral	Antero-lateral	Anterior	Global
Coefficient of variation (%)	20	12	14	12	10	15	4
ICC (%)	26	51	49	55	33	4	83
Smallest detectable change	0.10	0.06	0.07	0.07	0.07	0.09	0.03

The measures of global E_{cc} were highly correlated between users ($R^2 = 0.75$, Fig. 5A), while the regional E_{cc} measures were poorly correlated between users (R^2

= 0.23, Fig. 6A). Table 2 shows that there is a 71% probability that future pair of observations of global E_{cc} assessed by the two users will be perfectly linearly correlated, while the probability for regional E_{cc} measures drops sharply to 21%. The Bland-Altman analysis shows good agreement between users with a mean difference and 95% LOA of $-$0.007 ± 0.029 for global E_{cc} (Fig. 5B), while poor agreement with a mean difference and large 95% LOA of $-0.006 ± 0.088$ was indicated for regional E_{cc} (Fig. 6B).

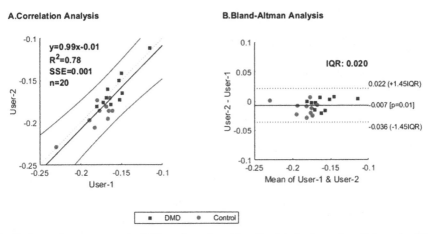

Fig. 5. Inter-observer reproducibility of peak mid-wall global circumferential strain (E_{cc}) between days in the pooled cohort of DMD patients and healthy controls.

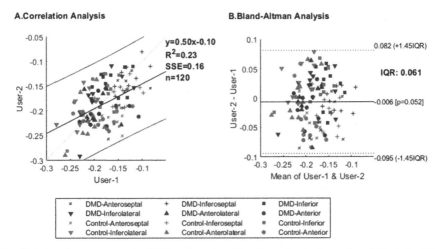

Fig. 6. Inter-observer reproducibility of peak mid-wall regional circumferential strain (E_{cc}) between days in the pooled cohort of DMD patients and healthy controls.

4 Discussion

To our knowledge, this is the first report to characterize the reproducibility of global and regional E_{cc} in a pediatric or DMD cohort using cine DENSE MRI.

Current intra-observer and intra-exam measures of global and regional E_{cc} were highly correlated ($R^2 \geq 70\%$). Between current inter-observer measures, a high correlation was found in the global E_{cc} measures ($R^2 = 0.75$), while poor correlation existed for the regional E_{cc} measures ($R^2 = 0.23$). Likewise, there is a probability larger than 71% that future pair of intra-observer and intra-exam measures of global and regional E_{cc}, and inter-observer measures of global E_{cc} will be perfectly linearly correlated (predictive $R^2 \geq 71\%$). However, there is a merely 21% probability that future pair of inter-observer measures of regional E_{cc} will be perfectly linearly correlated (predictive $R^2 = 21\%$).

Intra-observer repeatability was excellent for global and regional E_{cc} with all CoVs $\leq$ 4% and ICCs $\geq$ 75%, which agrees well with a previous report on healthy controls (CoV = 6%, ICC = 99% [10]).

For the intra-exam comparisons, the inferior and anterior E_{cc} were slightly less reproducible (CoVs = 8%, ICCs < 75%), while E_{cc} in the other regional segments and the global E_{cc} were highly reproducible (CoVs $\leq$ 6%, ICCs $\geq$ 75%).

For the inter-observer comparisons, global E_{cc} measures were highly reproducible (CoV = 4%, ICC $\geq$ 75%), which agrees well with a previous report on healthy controls (CoV = 4%, [11]). Regional E_{cc} demonstrated good reproducibility (CoVs: 10–15%) except in the anteroseptal segments (CoV = 20%), which was considered borderline reproducible.

Overall for global and regional E_{cc}, the intra-observer repeatability was found to be uniformly lower than the intra-exam reproducibility. And the intra-exam reproducibility was also uniformly lower than inter-observer reproducibility, which indicates that the user variability was higher than the scan variability in the study. Thus, the inter-observer reproducibility is the limiting factor in the overall reproducibility of E_{cc} measured by DENSE.

The intra-exam SDC of global E_{cc} was 0.020, which is below the inter-study SDC of mid-ventricular circumferential strain measured by MRI tagging (SDC = 0.027, [12]). Previous studies using MRI tagging suggest that peak systolic circumferential strain decreases uniformly among DMD patients at a rate of 0.013 ± 0.015 strain per year [13]. In this study, we can be 95% sure that a change in global E_{cc} exceeding 0.01 was not due to measurement error based on the intra-observer SDC. Thus, cine DENSE is capable of detecting subtle changes in global E_{cc} among DMD patients.

5 Conclusion

This study shows that cine DENSE CMR in healthy pediatric controls and DMD patients demonstrated excellent reproducibility with high ICCs (> 75%) and low CoVs (< 20%) for global midwall E_{cc}. Regional E_{cc} exhibited excellent intra-observer repeatability, good intra-exam reproducibility, and moderate inter-observer reproducibility. With low intra-observer and intra-exam SDCs, cine DENSE can be used to quantify progressive changes in E_{cc} in a longitudinal study for DMD patients.

Acknowledgements. We are especially grateful to our co-investigator and friend, Richard Patrick Magrath III who passed away in July of 2020. This project was supported by NIH HL131975 to DBE.

References

1. Hor, K.N., et al.: Circumferential strain analysis identifies strata of cardiomyopathy in Duchenne muscular dystrophy: a cardiac magnetic resonance tagging study. J. Am. Coll. Cardiol. **53**(14), 1204–1210 (2009). https://doi.org/10.1016/j.jacc.2008.12.032
2. Zhong, X., Spottiswoode, B.S., Meyer, C.H., Kramer, C.M., Epstein, F.H.: Imaging three-dimensional myocardial mechanics using navigator-gated volumetric spiral cine DENSE MRI. Magn. Reson. Med. **64**(4), 1089–1097 (2010). https://doi.org/10.1002/mrm.22503
3. Spottiswoode, B.S., et al.: Tracking myocardial motion from cine DENSE images using spatiotemporal phase unwrapping and temporal fitting. IEEE Trans. Med. Imaging **26**(1), 15–30 (2007). https://doi.org/10.1109/tmi.2006.884215
4. Bland, J.M., Altman, D.: Statistical methods for assessing agreement between two methods of clinical measurement. Lancet **327**(8476), 307–310 (1986)
5. McGraw, K.O., Wong, S.P.: Forming inferences about some intraclass correlation coefficients. Psychol. Methods **1**(1), 30 (1996)
6. Weir, J.P.: Quantifying test-retest reliability using the intraclass correlation coefficient and the SEM. J. Strength Cond. Res. **19**(1), 231 (2005)
7. Allen, D.M.: The prediction sum of squares as a criterion for selecting predictor variables. University of Kentucky (1971)
8. Tarpey, T.: A note on the prediction sum of squares statistic for restricted least squares. Am. Stat. **54**(2), 116–118 (2000)
9. Kotz, S., Johnson, H.L., Read, C.B.: Encyclopedia of statistical sciences (1982)
10. Lin, K., Meng, L., Collins, J.D., Chowdhary, V., Markl, M., Carr, J.C.: Reproducibility of cine displacement encoding with stimulated echoes (DENSE) in human subjects. Magn. Reson. Imaging **35**, 148–153 (2017). https://doi.org/10.1016/j.mri.2016.08.009
11. Suever, J.D., et al.: Right ventricular strain, torsion, and Dyssynchrony in healthy subjects using 3D spiral cine DENSE magnetic resonance imaging. IEEE Trans. Med. Imaging **36**(5), 1076–1085 (2017). https://doi.org/10.1109/TMI.2016.2646321
12. Donekal, S., et al.: Inter-study reproducibility of cardiovascular magnetic resonance tagging. J. Cardiovasc. Magn. Reson. **15**(1), 37 (2013)
13. Hagenbuch, S.C., et al.: Detection of progressive cardiac dysfunction by serial evaluation of circumferential strain in patients with Duchenne muscular dystrophy. Am. J. Cardiol. **105**(10), 1451–1455 (2010). https://doi.org/10.1016/j.amjcard.2009.12.070

M-SiSSR: Regional Endocardial Function Using Multilabel Simultaneous Subdivision Surface Registration

Davis M. Vigneault[1]([envelope]) [ID], Francisco Contijoch[2][ID], Christopher P. Bridge[3][ID],
Katherine Lowe[2], Chelsea Jan[2], and Elliot R. McVeigh[2][ID]

[1] Department of Internal Medicine, Scripps Mercy Hospital, San Diego, CA, USA
[2] Department of Bioengineering, UC San Diego, La Jolla, CA, USA
[3] Athinoula A. Martinos Center for Biomedical Imaging, Massachusetts General Hospital, Charlestown, MA, USA

Abstract. Quantification of regional cardiac function is a central goal of cardiology. Multiple methods, such as Coherent Point Drift (CPD) and Simultaneous Subdivision Surface Registration (SiSSR), have been used to register meshes to the endocardial surface. However, these methods do not distinguish between cardiac chambers during registration, and consequently the mesh may "slip" across the interface between two structures during contraction, resulting in inaccurate regional functional measurements. Here, we present Multilabel-SiSSR (M-SiSSR), a novel method for registering a "labeled" cardiac mesh (with each triangle assigned to a cardiac structure). We compare our results to the original, label-agnostic version of SiSSR and find both a visual and quantitative improvement in tracking of the mitral valve plane.

Keywords: Cardiac Computed Tomography (CCT) · Regional cardiac function · Convolutional Neural Networks (CNNs) · Subdivision surfaces · Mesh registration

1 Introduction

Fully-automated quantification of regional cardiac function is a central goal of cardiology. Multiple methods, such as Coherent Point Drift (CPD) [13] and Simultaneous Subdivision Surface Registration (SiSSR) [19], have been used to register meshes to the endocardial surface of the left ventricle (LV). However, there is increasing interest in quantifying function beyond the LV, including the left atrium (LA). Although these methods could in principle be extended, naïve application to a multi-chamber mesh does not actually incorporate the additional information which the labels provide into the registration. Consequently,

This work is supported by National Institutes of Health grants R01HL144678 and K01HL143113.

D. B. Ennis et al. (Eds.): FIMH 2021, LNCS 12738, pp. 242–252, 2021.
https://doi.org/10.1007/978-3-030-78710-3_24

the mesh may "slip" across the interface between two structures during contraction, resulting in inaccurate regional functional measurements. Here, we present Multilabel-SiSSR (M-SiSSR), a novel extension of the SiSSR method which may be used to register a "labeled" cardiac mesh (where each cell is labeled according to its corresponding structure). We compare our results to the original, label-agnostic version of SiSSR, and find both a visual and quantitative improvement in tracking of the mitral valve plane.

2   Methods

2.1   Dataset and Annotation

100 3D+Time CCT scans were collected retrospectively from three institutions as part of an IRB-approved study. Segmentation of the first acquired and end systolic (ES) frames was performed using ITK-Snap [20,21]. The segmentation was initialized using spheres with radius 5 throughout the anatomy of interest, and the segmentation was allowed to proceed until the structure was filled. Region competition force and smoothing (curvature) force were left at the default values of 1.0 and 0.2, respectively. The left and right inferior and superior pulmonary veins (PVs), LA, left atrial appendage (LAA), LV, left ventricular outflow tract (LVOT), and ascending aorta (AA) were separately labeled. Planes separating adjacent structures were chosen using the ITK-Snap scalpel tool.

2.2   Left Heart Segmentation

The V-Net [10], an extension of the U-Net [14] for 3D segmentation, was chosen as the basic network for prediction of a segmentation S_f from an input volume I_f. This architecture consists of a convolutional layer, a batch-normalization layer, and a rectified linear unit (ReLU) repeated in sequence. Every second cycle, a max-pooling step or an interpolation step is inserted, along the downsampling- and upsampling-paths, respectively (four downsampling/upsampling steps total in our implementation). With each downsampling and upsampling step, the length of the feature vector was doubled and halved, respectively.

Data augmentation consisted of random uniform translation (up to 10% of the image width), rotation (up to 15°), and scaling (up to 10%). All networks were trained using five-fold cross validation for 96 epochs, and all images from a single patient were partitioned into the same cross-validation fold. The learning rate was initialized to 10^{-3} and decreased by a factor of 10 every 24 epochs. The loss function was the categorical cross entropy between the output of the final softmax layer and the ground truth segmentation. The batch size, initial feature map depth, and spacing to which the input images were isotropically resampled were treated as hyperparameters.

2.3   Boundary Candidate Selection

The first partition (20 subjects, 40 image volumes) was used for all subsequent experiments. Given a segmentation $\mathcal{S}_f$, we calculated the segmentation surface using the Cuberille algorithm [5]. In order to appropriately limit the scope of this work, only the immediately relevant labels (LA, LV, and LVOT) were retained. The boundary candidates $\mathcal{C}_f$ were calculated as the centroids of the surface mesh cells.

2.4   Mesh Model Generation

Next we derived a template mesh $\mathcal{T}$ from the mid systolic segmentation $\mathcal{S}_{MS}$; the mid systolic phase was chosen so as to minimize the required displacement of the model during registration. The Cuberille algorithm was again used to derive an initial triangle mesh, and each cell was associated with the label of the adjacent foreground structure. Our application requires that the extracted surface be a water-tight 2D manifold, and so the Cuberille algorithm was preferred over the ubiquitous Marching Cubes algorithm [9] due to the latter's well-known topological inconsistencies [4,11,12]. The resulting mesh was then decimated to a target number of cells using successive edge contraction. Two strategies were compared: midpoint decimation (in which the contracted edge is placed at the geometric mean of the two vertices) and Lindstrom-Turk decimation (in which the new vertex is placed so as to preserve the volume of the original model, [7]).

We found that naïve application of either method to a multi-label mesh resulted in distorted boundaries between structures and poor approximation of the initial mesh. To mitigate this effect, edge contraction was handled differently depending on the relationship between the candidate edge and the label boundaries, according to the following rules:

- "Interior" edges (both vertices surrounded by cells of the same label) and "boundary" edges (flanked by cells of differing labels) were contracted as usual, using the respective decimation technique.
- "Hanging" edges (one vertex lying on a structure boundary, the other surrounded by cells entirely of the same label) were contracted to the boundary vertex.
- "Bridging" edges (flanked by cells of the same label, but both vertices lying on structure boundaries) were not candidates for contraction.

2.5   Subdivision Surface Evaluation

Following the notation given in [19], let $\boldsymbol{X} \in \mathbb{R}^{3 \times (C \times F)}$ be the matrix of control point coordinates defining the sequence of template meshes $\mathcal{T}_f$, where C is the number of control points and F is the number of cardiac phases. Displacements from the elements of $\boldsymbol{X}$ serve as the $3 \times C \times F$ parameters of the optimization. This control point matrix defines F triangular Loop subdivision surfaces, which we use to model the endocardial and endovascular surface. The position of a point $u \in \mathbb{R}^3$ on the surface of $\mathcal{T}_f$ may be calculated in terms of the matrix X, the frame f, the patch index i, and a set of parametric coordinates $t = (r, s)$, which are local to a given patch. We define such a function $\chi : t, X_{f,i} \to \mathbb{R}^3$ as follows:

$$\chi(t, X_{f,i}) = X_{f,i} \left(P_k \bar{A} A^{n-1} \right)^{\mathsf{T}} b\left(\hat{t}\right). \tag{1}$$

Here n is the number of required subdivisions, A is the *extended subdivision matrix*, $\bar{A}$ is the *bigger subdivision matrix*, P_k is the *picking matrix*, b is the column of basis functions, and $\hat{t}$ is a tuple representing the transformed patch coordinates. We refer the reader to [16,19] for precise definitions of these matrices.

2.6 Multilabel SiSSR (M-SiSSR)

Registration of the sequence of template meshes $\mathcal{T}$ to the boundary candidates $\mathcal{C}$ was performed using Levenberg-Marquardt least squares optimization. In the SiSSR algorithm, the primary residuals E_{cf} were the Cartesian components of the distance between the points sampled from the surface of $\mathcal{T}_f$ and the nearest point in the corresponding boundary candidate mesh $\mathcal{C}_f$:

$$E_{cf} = \sum_{f,i,t} \| u_{f,i,t} - \phi\left(u_{f,i,t}\right) \|^2. \tag{2}$$

Here, $\phi : \mathbb{R}^3 \to \mathbb{R}^3$ is the function which computes the nearest point in the boundary candidate mesh, and is represented internally as a vector of K_d trees. In M-SiSSR, we propose a modified cost function $E_{\hat{cf}}$ (relying on a modified function $\hat{\phi} : \mathbb{R}^3 \to \mathbb{R}^3$), which represents the nearest point *of the same label* in the corresponding boundary candidate mesh. Internally, this is represented as a vector *of vectors* of K_d trees (one for each label and cardiac phase).

Some amount of regularization is necessary to encourage biologically plausible surfaces and control point trajectories in time. In this work, we consider thin plate energy $E_{\hat{tp}}$ (which encourages surface smoothness in the second derivative), acceleration E_{ac} (which encourages smooth control point trajectories in time), and edge length E_{el} (which provides tension and discourages "foldover" artifacts), as defined in [19].

The overall loss is formulated as follows:

$$E = \min \left(\alpha_{\hat{cf}} E_{\hat{cf}} + \alpha_{\hat{tp}} E_{\hat{tp}} + \alpha_{ac} E_{ac} + \alpha_{el} E_{el} \right)$$

Here the coefficients $\alpha_{\hat{cf}}$, $\alpha_{\hat{tp}}$, α_{ac}, and α_{el} are scaling factors, which were treated as hyperparameters (defined as in [19]).

2.7 Implementation

The CNN segmentation architecture was implemented in Python 3.6.8 using the Keras 2.3.1 interface to Tensorflow 1.15.0 [1] in an Ubuntu 18.04 Docker

Table 1. *CNN segmentation performance.* Jaccard index for the LA, LV, and LVOT is reported as a function of image spacing, feature map depth, and batch size. The Jaccard index for the best performing network is bolded.

Spacing (mm)	Feature Map Depth	Batch Size	LV	LA	LVOT
1.0	8	1	71.9 ± 17.3	82.4 ± 12.0	63.6 ± 18.0
1.0	16	1	68.6 ± 21.7	74.7 ± 23.6	53.3 ± 26.7
2.0	8	14	75.3 ± 11.7	75.4 ± 9.8	0.4 ± 0.7
2.0	16	8	78.6 ± 10.0	84.1 ± 7.6	68.1 ± 8.8
2.0	32	4	**79.2 ± 11.3**	**86.1 ± 6.5**	**70.8 ± 9.7**
3.0	8	20	39.6 ± 24.0	42.2 ± 23.8	0.0 ± 0.0
3.0	16	12	68.2 ± 10.4	74.8 ± 6.3	42.9 ± 23.6
3.0	32	6	69.5 ± 9.6	76.1 ± 6.4	59.1 ± 9.4
3.0	64	3	69.6 ± 9.7	76.4 ± 6.4	59.7 ± 10.0

container. Models were trained on Amazon Web Services p2.xlarge instances with a single NVIDIA K80 GPU with 12 GB memory. Mesh visualization was performed using VTK [15]; mesh decimation was performed using CGAL [18]; mesh registration was implemented using the Ceres Solver [2]; and the remaining mesh processing was performed using ITK [6].

3 Results

3.1 CNN Segmentation

The CNN was trained for a range of spatial resolutions and feature map depths. The best-performing network for the LA, LV, and LVOT was that with images resampled to 2mm isotropic with 32 channels in the initial feature map and trained with a batch size of 4; the predictions of this network were used for all subsequent experiments (Table 1).

3.2 Template Mesh Generation

When generating the template mesh, we target a specific number of faces in the decimation algorithm. As the number of faces increases, the computational time of the mesh registration as well as the ability to represent fine anatomical detail increases, while the regional influence of a particular control point decreases. As expected, Jaccard index increased as a function of the target number of faces (Fig. 1) for both midpoint and Lindstrom-Turk decimation techniques.

Moreover, in terms of Jaccard index, the Lindstrom-Turk method outperformed the midpoint method for any fixed target number of faces. Note, for example, that even at 64 faces, the LVOT is preserved in the Lindstrom-Turk model (Fig. 2b), whereas it is entirely lost in the midpoint model (Fig. 2h).

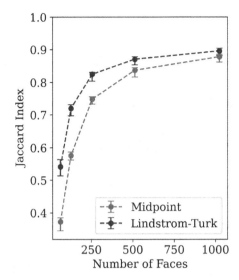

Fig. 1. *Template mesh resolution vs Jaccard Index.* Jaccard index is plotted as a function of the number of faces targeted in the mesh decimation algorithm (either Midpoint or Lindstrom-Turk). Note that there is a precipitous decline in accuracy below 256 faces, and that the Lindstrom-Turk method outperforms the midpoint method at all resolutions.

Note also that for models generated via the Lindstrom-Turk method, there was noticable loss of detail around certain features such as the cardiac apex below 256 faces, but conversely there was minimal improvement either in Jaccard index or in subjective anatomic detail above approximately 256 faces. For these reasons, the 256-face Lindstrom-Turk model was chosen as the starting point for registration. Representative models are shown in Fig. 2.

As discussed above, naïve application of either midpoint or Lindstrom-Turk decimation algorithm to a multi-label mesh without careful preservation of the edge resulted in unacceptable distortion of the structure boundary (Fig. 3c). However, modification of the decimation procedure to avoid large displacements in the boundary edges resulted in a marked visual improvement (Fig. 3a).

3.3   Determining Scaling Factors

The coefficients of the regularization terms were explored in a stepwise fashion (Fig. 4), using Jaccard index and triangle condition (measures of segmentation accuracy and mesh quality, respectively) as outcomes. The goal is to maximize Jaccard index while minimizing triangle condition. The coefficients chosen for the final registration algorithm were $E_{\hat{t}p} = 10^{-1}$, $E_{ac} = 10^{-1}$, and $E_{el} = 10^{-0.5}$.

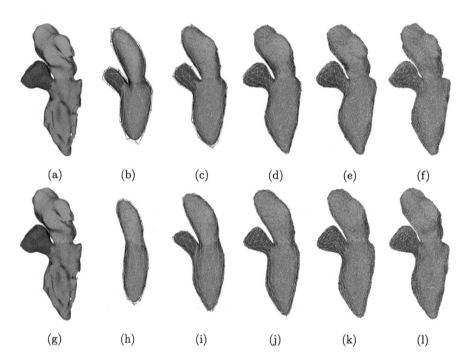

Fig. 2. *Template mesh resolution.* Representative template meshes $\mathcal{T}$ are shown at 64 (b, h), 128 (c, i), 256 (d, j), 512 (e, k), and 1024 (f, l) faces. The corresponding boundary candidate mesh $\mathcal{C}$ (a, g) is also shown for comparison. In this and subsequent figures, the LA is green, the LV is red, and the LVOT is blue. (Color figure online)

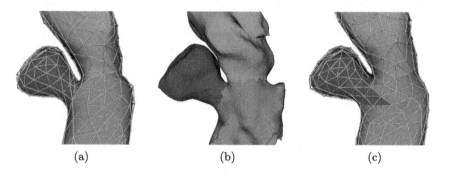

Fig. 3. *Edge preservation.* Template meshes $\mathcal{T}$ generated with (a) and without (c) edge preservation. The boundary candidates for the mid systolic frame are shown for comparison (b).

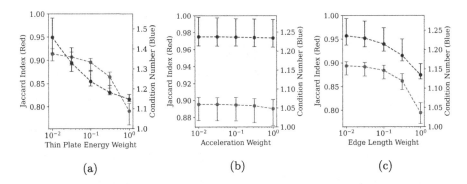

Fig. 4. *Hyperparameter search.* (a) Thin plate energy parameter search, with $\alpha_{\hat{c}f} = 1.0$. (b) Acceleration parameter search, with $\alpha_{\hat{c}f} = 1.0$ and $\alpha_{\hat{t}p} = 10^{-1}$. (c) Edge length parameter search, with $\alpha_{\hat{c}f} = 1.0$, $\alpha_{\hat{t}p} = 10^{-1}$, and $\alpha_{ac} = 10^{-1}$.

3.4 Comparison to Label-Agnostic Approach: M-SiSSR vs SiSSR

As a comparison, the registration was run using the same regularization weight, but without accounting for the labels (i.e., using the label-agnostic SiSSR algorithm). Representative results are shown in Fig. 5. In the meshes registered through using the M-SiSSR method (Fig. 5a, 5d), the boundary between the LA and LV tracks that of the boundary candidates (Fig. 5b, 5e), excursing downward during systole. By contrast, in the meshes registered through using the SiSSR method (Fig. 5c, 5f), the boundary stays stationary, lying below and above that of the boundary candidates in the ED and ES frames, respectively.

Additionally, the centroid of the mitral valve annulus (approximated by the geometric mean of all points lying on the boundary between the LV and LA) was calculated at end diastole and end systole for the boundary candidate meshes, the meshes registered using the M-SiSSR method, and those registered using the SiSSR method. The error (as judged by the Euclidean distance between the boundary candidate and registered mesh centroids) was significantly lower for the M-SiSSR meshes compared to the SiSSR meshes both at end diastole ($2.5\,\text{mm}$ vs $3.9\,\text{mm}$, $p < 10^{-2}$) and at end systole ($2.8\,\text{mm}$ vs $5.6\,\text{mm}$, $p < 10^{-7}$).

We also calculated SQUEEZ (a metric of regional cardiac function used in CCT, [13]) from both the SiSSR and M-SiSSR methods. At left ventricular end systole, we expect there to be a sharp transition in SQUEEZ between the contracting LV and the filling LA. And in fact we do see such a sharp transition at the mitral valve plane in the M-SiSSR meshes (Fig. 6a), whereas in the SiSSR method, there is a gradual transition that appears to underestimate both LV contraction and LA expansion.

4   Discussion

In this work, we present a fully-automated pipeline in which a CNN is used to segment the left heart and connected vasculature from a sequence of CCT

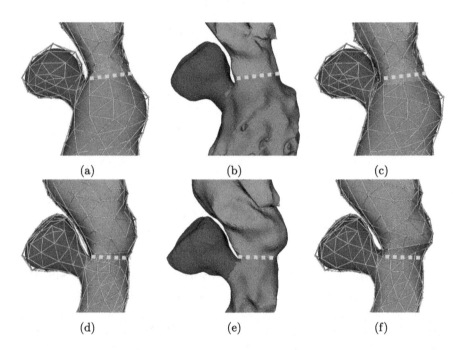

Fig. 5. *Registered meshes.* Registered meshes using label information (a, d) and ignoring label information (c, f), and boundary candidates for comparison (b, e). Both ED (top) and ES (bottom) frames are shown. The mitral annular plane of the boundary candidate mesh is approximated by a dashed yellow line and reproduced over the M-SiSSR and SiSSR meshes. Note that the M-SiSSR meshes more closely track the mitral annular plane. (Color figure online)

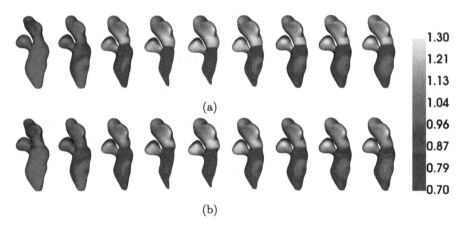

Fig. 6. *SQUEEZ.* Representative meshes registered using (a) and ignoring (b) label information and colored according to their SQUEEZ values.

volumes; a labeled Loop subdivision surface is generated from the mid-systolic frame; and this model surface is registered to the CNN segmentations across all phases simultaneously, accounting for the label in the registration. The result is a patient-specific model which is topologically consistent across all timepoints, and therefore may be used to calculate regional function.

The SiSSR algorithm makes use of data structures and techniques described by Loop [8], Stam [16], Cashman [3], and Stebbing [17]. Compared to the first iteration of the SiSSR algorithm [19], the present work has a number of notable advantages and improvements. In particular, experiments were performed using human subjects rather than canine hearts, and the number of subjects included was higher (from 13 cases to 20); the segmentations were derived using a neural network, rather than through semi-automated level set segmentation; the LA and LVOT were included in the registration, rather than the LV alone; and the algorithm was modified to incorporate the identity of the underlying structure into the registration.

Although the number of subjects is increased from [19], the absolute number is still small; future work should be expanded to include the remaining 80 cases in the dataset. The LV myocardium, papillary muscles, and right heart are all of great interest, but are not included here; these should all be explored in future work. And although here we demonstrated a quantitative improvement in the tracking of the mitral valve compared to the label-agnostic SiSSR algorithm in terms of the error between the mitral valve centroids, in future work, it would be useful to additionally report an error metric with more immediate clinical relevance, such as mitral annular plane systolic excursion (MAPSE).

We hope that this work represents a substantive contribution to the use of subdivision surfaces for modeling of cardiac structures, and to the clinically important goal of quantifying regional cardiac function from CCT images.

References

1. Abadi, M., et al.: TensorFlow: a system for large-scale machine learning (2016) https://arxiv.org/pdf/1605.08695.pdf
2. Agarwal, S., Mierle, K.: Ceres Solver (2010) http://ceres-solver.org
3. Cashman, T.J., Fitzgibbon, A.W.: What shape are dolphins? building 3D morphable models from 2D images. IEEE Trans. Pattern Anal. Mach. Intell. **35**(1), 232–244 (2013). https://doi.org/10.1109/TPAMI.2012.68
4. Chernyaev, E.: Marching cubes 33: construction of topologically correct Isosurfaces. In: GRAPHICON 1995, Saint-Petersburg, Saint-Petersburg, Russia (1995)
5. Herman, G.T., Liu, H.K.: Three-dimensional display of human organs from computed tomograms. Comput. Graph. Image Process.**9**(1), 1–21 (1979). https://doi.org/10.1016/0146-664X(79)90079-0. https://linkinghub.elsevier.com/retrieve/pii/0146664X79900790
6. Johnson, H.J., Ibanez, L., Mccormick, M.: The ITK Software Guide: Book 1, 4 edn. (2016)
7. Lindstrom, P., Turk, G.: Fast and memory efficient polygonal simplification. In: Proceedings of the IEEE Visualization Conference, pp. 279–286 (1998). https://doi.org/10.1109/visual.1998.745314

8. Loop, C.: Smooth subdivision surfaces based on triangles. Ph.D. thesis, University of Utah (1987) http://www.citeulike.org/group/5490/article/2864922

9. Lorensen, W.E., Cline, H.E.: Marching cubes: a high resolution 3D surface construction algorithm. ACM SIGGRAPH Comput. Graph. **21**(4), 163–169 (1987). https://doi.org/10.1145/37402.37422. http://portal.acm.org/citation.cfm?doid=37401.37422

10. Milletari, F., Navab, N., Ahmadi, S.A.: V-Net: fully convolutional neural networks for volumetric medical image segmentation. In: CVPR pp. 1–11 (2016). https://doi.org/10.1109/3DV.2016.79. http://arxiv.org/abs/1606.04797

11. Newman, T.S., Yi, H.: A survey of the marching cubes algorithm. Comput. Graph. **30**(5), 854–879 (2006). https://doi.org/10.1016/j.cag.2006.07.021

12. Nielson, G.M., Hamann, B.: The asymptotic decider: Resolving the ambiguity in marching cubes. In: Proceedings of the 2nd Conference on Visualization 1991, VIS 1991 (1991). https://doi.org/10.1109/visual.1991.175782

13. Pourmorteza, A., Schuleri, K.H., Herzka, D.a., Lardo, A.C., McVeigh, E.R.: A new method for cardiac computed tomography regional function assessment: Stretch quantifier for endocardial engraved zones (SQUEEZ). Circulation Cardiovascular Imaging **5**(2), 243–250 (2012). https://doi.org/10.1161/CIRCIMAGING.111.970061

14. Ronneberger, O., Fischer, P., Brox, T.: U-Net: Convolutional Networks for Biomedical Image Segmentation. In: Navab, N., Hornegger, J., Wells, W.M., Frangi, A.F. (eds.) MICCAI 2015. LNCS, vol. 9351, pp. 234–241. Springer, Cham (2015). https://doi.org/10.1007/978-3-319-24574-4_28

15. Schroeder, W., Martin, K., Lorensen, B.: The Visualization Toolkit. Kitware, 4 edn. (2006)

16. Stam, J.: Evaluation of Loop subdivision surfaces. In: SIGGRAPH, pp. 1–15 (1998)

17. Stebbing, R.: Model-Based Segmentation Methods for Analysis of 2D and 3D Ultrasound Images and Sequences. DPhil, University of Oxford (2014)

18. The CGAL Project: CGAL User and Reference Manual. CGAL Editorial Board, 5.2 edn. (2020). https://doc.cgal.org/5.2/Manual/packages.html

19. Vigneault, D.M., Pourmorteza, A., Thomas, M.L., Bluemke, D.A., Noble, J.A.: SiSSR: simultaneous subdivision surface registration for the quantification of cardiac function from computed tomography in canines. Med. Image Anal. **46**, 215–228 (2018). https://doi.org/10.1016/j.media.2018.03.009

20. Yushkevich, P.A., et al.: User-guided segmentation of multi-modality medical imaging datasets with ITK-SNAP. Neuroinformatics **17**(1), 83–102 (2018). https://doi.org/10.1007/s12021-018-9385-x

21. Yushkevich, P.A., et al.: User-guided 3D active contour segmentation of anatomical structures: significantly improved efficiency and reliability. Neuroimage **31**, 1116–1128 (2006). https://doi.org/10.1016/j.neuroimage.2006.01.015

CNN-Based Cardiac Motion Extraction to Generate Deformable Geometric Left Ventricle Myocardial Models from Cine MRI

Roshan Reddy Upendra[1(✉)], Brian Jamison Wentz[3,5], Richard Simon[2], Suzanne M. Shontz[3,4,5], and Cristian A. Linte[1,2]

[1] Center for Imaging Science, Rochester Institute of Technology, Rochester, NY, USA
ru6928@rit.edu
[2] Biomedical Engineering, Rochester Institute of Technology, Rochester, NY, USA
[3] Bioengineering Program, University of Kansas, Lawrence, KS, USA
[4] Electrical Engineering and Computer Science, University of Kansas, Lawrence, KS, USA
[5] Information and Telecommunication Center, University of Kansas, Lawrence, KS, USA

Abstract. Patient-specific left ventricle (LV) myocardial models have the potential to be used in a variety of clinical scenarios for improved diagnosis and treatment plans. Cine cardiac magnetic resonance (MR) imaging provides high resolution images to reconstruct patient-specific geometric models of the LV myocardium. With the advent of deep learning, accurate segmentation of cardiac chambers from cine cardiac MR images and unsupervised learning for image registration for cardiac motion estimation on a large number of image datasets is attainable. Here, we propose a deep leaning-based framework for the development of patient-specific geometric models of LV myocardium from cine cardiac MR images, using the Automated Cardiac Diagnosis Challenge (ACDC) dataset. We use the deformation field estimated from the VoxelMorph-based convolutional neural network (CNN) to propagate the isosurface mesh and volume mesh of the end-diastole (ED) frame to the subsequent frames of the cardiac cycle. We assess the CNN-based propagated models against segmented models at each cardiac phase, as well as models propagated using another traditional nonrigid image registration technique. Additionally, we generate dynamic LV myocardial volume meshes at all phases of the cardiac cycle using the log barrier-based mesh warping (LBWARP) method and compare them with the CNN-propagated volume meshes.

Keywords: Patient-specific modeling · Deep learning · Image registration · Cine Cardiac MRI · Mesh warping

1 Introduction

To reduce the morbidity and mortality associated with cardiovascular diseases (CVDs) [3], and to improve their treatment, it is crucial to detect and predict the

© Springer Nature Switzerland AG 2021
D. B. Ennis et al. (Eds.): FIMH 2021, LNCS 12738, pp. 253–263, 2021.
https://doi.org/10.1007/978-3-030-78710-3_25

progression of the diseases at an early stage. In a clinical set-up, population-based metrics, including measurements of cardiac wall motion, ventricular volumes, cardiac chamber flow patterns, etc., derived from cardiac imaging are used for diagnosis, prognosis and therapy planning.

In recent years, image-based computational models have been increasingly used to study ventricular mechanics associated with various cardiac conditions. A comprehensive review of patient-specific cardiovascular modeling and its applications is described in [17]. Cardiovascular patient-specific modeling includes a geometric representation of some or all cardiac chambers of the patient's anatomy and is derived from different imaging modalities [8].

The construction of patient-specific geometric models entails several steps: clinical imaging, segmentation and geometry reconstruction, and spatial discretization (i.e., mesh generation) [12]. For example, Bello et al. [2] presented a deep learning based framework for human survival prediction for patients diagnosed with pulmonary hypertension using cine cardiac MR images. Here, the authors employ a 4D spatio-temporal B-spline image registration method to estimate the deformation field at each voxel and at each timeframe. The estimated deformation field was used to propagate the ED surface mesh of the right ventricle (RV), reconstructed from the segmentation map, to the rest of the timeframes of a particular subject. Cardiac MRI is a current gold standard to assess global (ventricle volume and ejection fraction) and regional (kinematics and contractility) function of the heart under various diseases. In particular, cardiac MRI enables the generation of high quality myocardial models, which can, in turn, be used to identify reduced function.

In this work, we propose a deep learning-based pipeline to develop patient-specific geometric models of the LV myocardium from cine cardiac MR images (Fig. 1). These models may be used to conduct various simulations, such as assessing myocardial viability. In our previous work [19], we introduced a preliminary, proof of concept, CNN-based 4D deformable registration method for cardiac motion estimation from cine cardiac MR images, using the ACDC dataset [4]. Here, we demonstrate the use of the CNN-based 4D deformable registration technique to build dynamic patient-specific LV myocardial models across subjects with different pathologies, namely normal, dilated cardiomyopathy (DCM), hypertrophic cardiomyopathy (HCM) and subjects with prior myocardial infarctions (MINF). Following segmentation of the ED cardiac frame, we generate both isosurface and volume LV meshes, which we then propagate through the cardiac cycle using the CNN-based registration fields. In addition, we demonstrate the generation of dynamic LV volume meshes depicting the heart at various cardiac phases by warping a patient-specific ED volume mesh based on the registration-based propagated surface meshes, using the LBWARP method [15]. Lastly, we compare these meshes to those obtained by directly propagating the ED volume mesh using the CNN-based deformation fields.

2   Methodology

2.1   Cardiac MRI Data

We use the 2017 ACDC dataset that was acquired from real clinical exams. The dataset is composed of cine cardiac MR images from 150 subjects, divided into five equally-distributed subgroups: normal, MINF, DCM, HCM and abnormal RV. The MR image acquisitions were obtained using two different MR scanners of 1.5 T and 3.0 T magnetic strength. These series of short axis slices cover the LV from base to apex such that one image is captured every 5 mm to 10 mm with a spatial resolution of 1.37 mm^2/pixel to 1.68 mm^2/pixel.

2.2   Image Preprocessing

We first correct for the inherent slice misalignments that occur during the cine cardiac MR image acquisition. We train a modified version of the U-Net model [13] to segment the cardiac chambers, namely LV blood-pool, LV myocardium and RV blood-pool, from 2D cardiac MR images. We identify the LV blood-pool center, i.e., the centroid of the predicted segmentation mask and stack the 2D cardiac MR slices collinearly to obtain slice misalignment corrected 3D images [7,19].

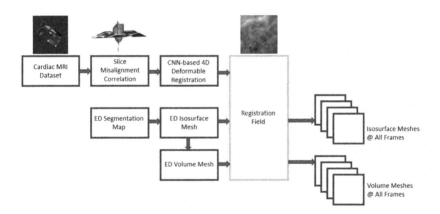

Fig. 1. Overview of the proposed CNN-based workflow to generate patient-specific LV myocardial geometric model.

2.3   Deformable Image Registration

CNN-Based Image Registration. We leverage our 4D deformable registration method described in [19] which employs the VoxelMorph [1] framework to determine the optical flow representation between the slice misalignment corrected 3D images. The CNN is trained using the following loss function:

$$L = L_{\text{similarity}} + \lambda L_{\text{smooth}}, \tag{1}$$

where $L_{\text{similarity}}$ is the mean squared error (MSE) between the target frame and the warped frame, L_{smooth} is the smoothing loss function to spatially smooth the registration field, and λ is the regularization parameter, which is set to 10^{-3} in our experiments. Inspired by Zhu et al. [20], we use the Laplacian operator in the smoothing loss function as it considers both global and local properties of the objective function $y = x^2$ instead of the traditional gradient operator which considers only the local properties of the function $y = x^2$. A detailed comparison of both these smoothing loss functions with respect to cardiac motion estimation from cine MR images is found in [19].

The 4D cine cardiac MRI datasets are composed of 28 to 40 3D image frames that cover the complete cardiac cycle. For this discussion, we shall refer to the 3D images as $I_{ED}, I_{ED+1},...,I_{ED+N_T-1}$ where I_{ED} is the ED image frame, and N_T is the total number of 3D images. We employ the fixed reference frame registration method, wherein the task is to find an optical flow representation between the image pairs $\{(I_{ED}, I_{ED+t})\}_{t=1,2,3,...,N_T-1}$.

During training, we use 110 of the total 150 MR image dataset for training, 10 for validation and the remaining 30 for testing. The CNN for cardiac motion estimation is trained using an Adam optimizer with a learning rate of 10^{-4}, halved at every 10th epoch for 50 epochs. Both the U-Net model used for segmentation and slice misalignment correction, and the VoxelMorph network trained to estimate cardiac motion were trained on NVIDIA RTX 2080 Ti GPU.

Conventional Image Registration. We compare the performance of the VoxelMorph framework with that of the B-spline free form deformation (FFD) nonrigid image registration algorithm [14]. This iterative intensity-based image registration method was implemented using SimpleElastix [9,11], which enables a variety of image-registration algorithms in different programming languages. The FFD algorithm was set to use the adaptive stochastic gradient descent method as the optimizer, MSE as the similarity measure and binding energy as the regularization function. The FFD-based image registration was optimized in 500 iterations, while sampling 2048 random points per iteration, on an Intel(R) Core(TM) i9-9900K CPU.

2.4   Mesh Generation and Propagation

We use the manual segmentation map of the ED frame to generate isosurface meshes. The slice thickness of each MRI image slice is 5 mm to 10 mm; however, in order to obtain good quality meshes, the segmentation maps were resampled to a slice thickness of 1 mm. We use the Lewiner marching cubes [10] algorithm to generate the meshes from the resampled segmentation maps of the ED frames, on an Intel(R) Core(TM) i9-9900K CPU, and then simplification techniques, such as r-refinement and edge collapse, were performed using MeshLab 2020.07 [5]. The simplification techniques are repeated multiple times to reduce the number of vertices until the mesh has been fully decimated while preserving the anatomical integrity and aspect ratio of the isosurface meshes.

Volume meshes of the initial surface meshes at the ED phases for four patients with various heart conditions were generated based on the decimated patient-specific surface meshes using Tetgen 1.6 [16]. In particular, a constrained Delaunay mesh generation algorithm was used to generate tetrahedral meshes based on the triangulated surface meshes. Steiner points were added within the boundary of the surface mesh so that the tetrahedra maintained a radius-edge ratio of 1.01 and a maximum volume of 9 mm^3 as needed for generation of valid meshes [16]. Volume mesh quality improvement was performed using the feasible Newton method in Mesquite [18]. This method iteratively minimizes the quadratic approximation of a nonlinear function and converges linearly toward a local minimum while performing an Armijo line search to ensure feasibility of the elements; feasibility in this case refers to a valid, non-inverted element. The volume mesh converged to the highest quality indicated by the minimum average scaled Jacobian of the elements in the mesh.

To demonstrate the VoxelMorph-based motion extraction and propagation to build patient-specific LV myocardial models, we generate two sets of volume meshes at each cardiac frame for each patient in each pathology group (Fig. 2).

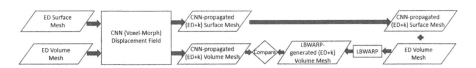

Fig. 2. Pipeline to generate dynamic volume meshes (at cardiac frames (ED + k)) by direct CNN-based propagation, as well as volume mesh warping based on dynamic boundary meshes.

The first set is produced by propagating the volume meshes at the ED frame to all the subsequent frames of the cardiac cycle using the deformation field estimated by the VoxelMorph-based registration method. For the second set, the ED volume mesh generated with Tetgen was used to generate the volume meshes corresponding to the other cardiac phases. We employed the LBWARP method [15] to deform the ED volume mesh onto the target surface mesh for the new cardiac phase (Fig. 3). The method computes new positions for the interior vertices in the ED volume mesh, while maintaining the mesh topology and point-to-point correspondence [15]. The simplification of the ED isosurface meshes, generation of the ED volume meshes and generation of the volume meshes corresponding to the other cardiac phases using the LBWARP method were performed on a machine equipped with AMD FX(tm)-6300 Six-Cores processor and a NVIDIA GeForce GTX 1050 Ti graphics card.

Briefly, LBWARP first calculates a set of local weights for each interior vertex in the initial ED volume mesh based on the relative inverse distances from each of its neighbors. The projected Newton method is used to solve the strictly convex optimization problems. For each set of local weights, a sparse system of linear equations is formed specifying the representation of each interior vertex in terms of its neighbors. Next, the boundary vertices in the ED surface mesh are mapped onto the new surface boundary. Finally, the interior vertices in the ED volume mesh are repositioned by solving the original system of linear equations with new right-hand side vectors to reflect the updated positions of the boundary nodes, while maintaining edge connectivity and point-to-point correspondence, and ultimately yielding the volume meshes that correspond to each new cardiac phase.

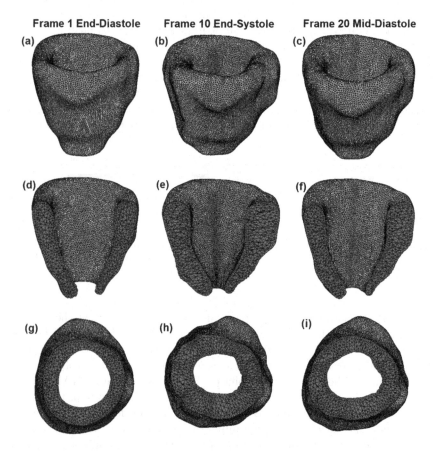

Fig. 3. Warped volume meshes for a patient with a healthy heart generated using LBWARP at three cardiac phases (a) end-diastole; (b) end-systole; and (c) mid-diastole; (d–f) Long axis cutaway view of volume meshes at the three cardiac phases, respectively; (g–i) short-axis cutaway view of volume meshes at the three cardiac phases, respectively.

3   Results and Discussion

To evaluate the registration performance, the LV isosurface (generated from the ED image segmentation map) is propagated to all the subsequent cardiac frames using the deformation field estimated by FFD and VoxelMorph. We then compare these isosurfaces to those directly generated by segmenting all cardiac image frames using a modified U-Net model [13] (Sect. 2.2), which we refer to as the "silver standard".

Table 1 summarizes the performance of the FFD and VoxelMorph registration by assessing the Dice score and mean absolute distance (MAD) between the propagated and directly segmented (i.e., "silver standard") isosurfaces.

Figure 4 illustrates the distance between the three sets of isosurfaces (segmented, CNN-propagated and FFD-propagated) for one patient randomly selected from each pathology. The MAD between the surfaces is less than 2 mm at all frames, with the CNN-propagated isosurfaces being closest to the "silver standard" segmented surfaces. Figure 5 illustrates the model-to-model distance between the FFD-propagated and CNN-propagated isosurface meshes at end-systole (ES) and mid-diastole frames for subjects from all four pathologies.

Since the CNN-propagated isosurfaces are in closer agreement to the "silver standard" segmented surfaces when compared to the FFD-propagated isosurfaces, we use the CNN-propagation method to generate the volume meshes at each phase of the cardiac cycle. As mentioned in Sect. 2.4 and shown in Fig. 2, we generate two sets of volume meshes at each frame of the cardiac cycle. Figure 6 shows the mean node distance between the two sets of volume meshes across all cardiac frames for one subject in each of the four pathologies. Figure 6 also shows the mean node distance between the two sets of volume meshes at each frame of the cardiac cycle for the four subjects. It can be observed that the two sets of volume meshes are in close agreement with each other, exhibiting a mesh-to-mesh distance within 0.5 mm.

Table 1. Mean Dice score (%) and mean absolute distance (MAD) (mm) between FFD and segmentation (FFD-SEG), CNN and segmentation (CNN-SEG), and FFD and CNN (FFD-CNN) results. Statistically significant differences were evaluated using the t-test (* for $p < 0.1$ and ** for $p < 0.05$).

	Normal		MINF		DCM		HCM	
	Dice	MAD	Dice	MAD	Dice	MAD	Dice	MAD
FFD-Segmentation	74.80	1.53	77.69	1.09	80.41	0.91	77.39	1.97
CNN-Segmentation	80.41**	1.15	81.21*	0.87	83.39*	0.91	82.46*	1.09
FFD-CNN	77.81	1.13	82.12	0.75	81.67	0.97	77.34	1.77

We also briefly investigated the effect of using initial-to-final frame vs. adjacent frame-to-frame registration to extract the cardiac motion throughout the

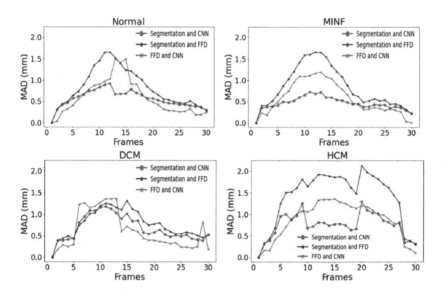

Fig. 4. MAD between FFD- and CNN-propagated, and segmented (i.e., "silver standard") isosurfaces at all cardiac frames for all patient pathologies.

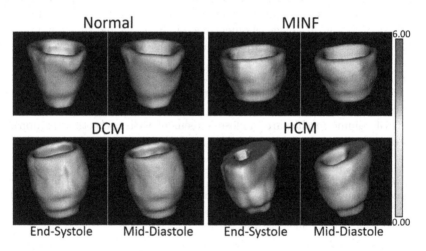

Fig. 5. Model-to-model distance between the isosurface meshes generated from FFD- and the CNN-propagation method for all patient pathologies at end-systole and mid-diastole frames.

cycle. Although the sequential registration method estimates smaller deformation between two consecutive, adjacent image frames compared to the larger deformations estimated by the initial-to-final frame registration, their concatenation across several frames accumulates considerable registration errors. As such, when using these concatenated registration-predicted deformation fields

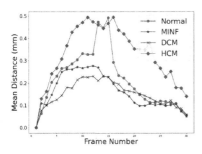 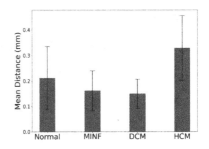

Fig. 6. Mean node-to-node distance at each cardiac frame between the CNN-propagated and LBWARP-generated volume meshes (left); mean (std-dev) node distance across all frames for each patient pathology (right).

to propagate the ED isosurfaces and volume meshes to the subsequent cardiac phases, the Dice score and MAD between the propagated and segmented geometries rapidly deteriorate, along with the quality of the propagated surface and volume meshes.

Following the generation of the dynamic, multi-phase meshes, we also assessed the quality of the ES meshes. One set of ES meshes was generated by propagating the ED mesh using the CNN-based extracted motion, while the other set of ES meshes was generated by warping the ED volume mesh based on the dynamic boundary meshes via the LBWARP approach. Unlike the starting ED phase meshes, the ES phase meshes contain some lower quality elements indicated by the lower minimum scaled Jacobian values, but are still suitable for use in simulations.

Moreover, although the proposed VoxelMorph-based cardiac motion extraction method can capture the frame-to-frame motion with sufficient accuracy, as shown in this work, our ongoing and future efforts are focused on further improving the algorithm by imposing diffeomorphic deformations [6]. This improvement will help maintain high mesh quality and prevent mesh tangling and element degeneration, especially for the systolic phases.

4   Conclusion

In this work, we show that the proposed deep learning framework can be used to build LV myocardial geometric models. The proposed framework is not limited to any pathology and can be extended to LV and RV blood-pool geometry.

Acknowledgments. This work was supported by grants from the National Science Foundation (Award No. OAC 1808530, OAC 1808553 & CCF 1717894) and the National Institutes of Health (Award No. R35GM128877).

References

1. Balakrishnan, G., Zhao, A., Sabuncu, M.R., Guttag, J., Dalca, A.V.: VoxelMorph: a learning framework for deformable medical image registration. IEEE Trans. Med. Imaging **38**(8), 1788–1800 (2019)
2. Bello, G.A., et al.: Deep-learning cardiac motion analysis for human survival prediction. Nature Mach. Intell. **1**(2), 95–104 (2019)
3. Benjamin, E.J., et al.: Heart disease and stroke statistics-2017 update: a report from the American heart association. Circulation **135**(10), e146–e603 (2017)
4. Bernard, O., et al.: Deep learning techniques for automatic MRI cardiac multi-structures segmentation and diagnosis: Is the problem solved? IEEE Trans. Med. Imaging **37**(11), 2514–2525 (2018)
5. Cignoni, P., et al.: Meshlab: an open-source mesh processing tool. In: Eurographics Italian Chapter Conference. vol. 2008, pp. 129–136. Salerno, Italy (2008)
6. Dalca, A.V., Balakrishnan, G., Guttag, J., Sabuncu, M.R.: Unsupervised learning of probabilistic diffeomorphic registration for images and surfaces. Med. Image Anal. **57**, 226–236 (2019)
7. Dangi, S., Linte, C.A., Yaniv, Z.: Cine cardiac MRI slice misalignment correction towards full 3D left ventricle segmentation. In: Medical Imaging 2018: Image-Guided Procedures, Robotic Interventions, and Modeling. vol. 10576, p. 1057607. International Society for Optics and Photonics (2018)
8. Gray, R.A., Pathmanathan, P.: Patient-specific cardiovascular computational modeling: diversity of personalization and challenges. J. Cardiovascular Transl. Res. **11**(2), 80–88 (2018)
9. Klein, S., Staring, M., Murphy, K., Viergever, M.A., Pluim, J.P.: Elastix: a toolbox for intensity-based medical image registration. IEEE Trans. Med. Imaging **29**(1), 196–205 (2009)
10. Lewiner, T., Lopes, H., Vieira, A.W., Tavares, G.: Efficient implementation of marching cubes' cases with topological guarantees. J. Graph. Tools **8**(2), 1–15 (2003)
11. Marstal, K., Berendsen, F., Staring, M., Klein, S.: SimpleElastix: a user-friendly, multi-lingual library for medical image registration. In: Proceedings of the IEEE Conference on Computer Vision and Pattern Recognition Workshops. pp. 134–142 (2016)
12. Morris, P.D., et al.: Computational fluid dynamics modelling in cardiovascular medicine. Heart **102**(1), 18–28 (2016)
13. Ronneberger, O., Fischer, P., Brox, T.: U-net: convolutional networks for biomedical image segmentation. In: Navab, N., Hornegger, J., Wells, W.M., Frangi, A.F. (eds.) MICCAI 2015. LNCS, vol. 9351, pp. 234–241. Springer, Cham (2015). https://doi.org/10.1007/978-3-319-24574-4_28
14. Rueckert, D., Sonoda, L.I., Hayes, C., Hill, D.L., Leach, M.O., Hawkes, D.J.: Non-rigid registration using free-form deformations: application to breast MR images. IEEE Trans. Med. Imaging **18**(8), 712–721 (1999)
15. Shontz, S.M., Vavasis, S.A.: A mesh warping algorithm based on weighted Laplacian smoothing. In: 12th International Meshing Roundtable, pp. 147–158 (2003)
16. Si, H.: Tetgen, a Delaunay-based quality tetrahedral mesh generator. ACM Trans. Math. Softw. (TOMS) **41**(2), 1–36 (2015)
17. Smith, N., et al.: Euheart: personalized and integrated cardiac care using patient-specific cardiovascular modelling. Interface Focus **1**(3), 349–364 (2011)

18. Trilinos Project Team, T.: The Trilinos Project Website (2020). https://trilinos.github.io. Accessed 12 Nov 2020
19. Upendra, R.R., Wentz, B.J., Shontz, S.M., Linte, C.A.: A convolutional neural network-based deformable image registration method for cardiac motion estimation from cine cardiac MR images. In: 2020 Computing in Cardiology, pp. 1–4. IEEE (2020)
20. Zhu, Y., Zhou Sr., Z., Liao Sr., G., Yuan, K.: New loss functions for medical image registration based on Voxelmorph. In: Medical Imaging 2020: Image Processing, vol. 11313, p. 113132E. International Society for Optics and Photonics (2020)

Multiscale Graph Convolutional Networks for Cardiac Motion Analysis

Ping Lu[1]([⊠]) [iD], Wenjia Bai[2,3] [iD], Daniel Rueckert[2] [iD], and J. Alison Noble[1] [iD]

[1] Institute of Biomedical Engineering, University of Oxford, Oxford, UK
ping.lu@eng.ox.ac.uk
[2] Department of Computing, Imperial College London, London, UK
[3] Department of Brain Sciences, Imperial College London, London, UK

Abstract. We propose a multiscale spatio-temporal graph convolutional network (MST-GCN) approach to learn the left ventricular (LV) motion patterns from cardiac MR image sequences. The MST-GCN follows an encoder-decoder framework. The encoder uses a sequence of multiscale graph computation units (MGCUs). The myocardial geometry is represented as a graph. The network models the internal relations of the graph nodes via feature extraction at different scales and fuses the feature across scales to form a global representation of the input cardiac motion. Based on this, the decoder employs a graph-based gated recurrent unit (G-GRU) to predict future cardiac motion. We show that the MST-GCN can automatically quantify the spatio-temporal patterns in cardiac MR that characterise cardiac motion. Experiments are performed on midventricular short-axis view cardiac MR image sequence from the UK Biobank dataset. We compare the performance of cardiac motion prediction of the proposed method with ten different architectures and parameter settings. Experiments show that the proposed method inputting node positions and node velocities with multiscale graphs achieves the best performance with a mean squared error of 0.25 pixel between the ground truth node locations and our prediction. We also show that the proposed method can estimate a number of motion-related metrics, including endocardial radii, thickness and strain which are useful for regional LV function assessment.

Keywords: Spatio-temporal graph convolutional networks · Cardiac MR · Motion analysis

1 Introduction

Cardiac motion analysis plays an important role in the diagnosis of heart conditions [7,11]. In diagnosis, displacement and strain are important biomarkers, which are sensitive to subtle changes in myocardial function and often indicate the early onset of cardiac disease [7]. Cardiac motion can be evaluated by tracking the contours of the endocardium and the epicardium from magnetic

© Springer Nature Switzerland AG 2021
D. B. Ennis et al. (Eds.): FIMH 2021, LNCS 12738, pp. 264–272, 2021.
https://doi.org/10.1007/978-3-030-78710-3_26

resonance imaging (MRI). The aim of cardiac motion analysis is to perform an accurate estimation of the motion trajectories for the myocardial contours.

In recent years, geometric deep learning-based methods have achieved promising results in medical image analysis, such as artery and vein classification [8] and landmark detection [3]. Inspired by these applications of spatio-temporal graph neural network (ST-GCN) [4,10], we propose to employ graph convolutional neural networks (GCN) to model cardiac motion in the geometry space of the GCN which can take advantage of a sparse representation of the cardiac myocadium using contours instead of images.

Our previous work [6] discussed different strategies for constructing cardiac structure graphs on a single scale. In this paper, we continue to explore relations between different parts of the endocardium and the epicardium, which convey essential information for cardiac motion. We represent myocardial geometry at different scales. At each scale, sample nodes on the left ventricular (LV) myocardial contour are connected as a graph. Then nodes across two scales are connected as a cross-scale graph, which is also named a bipartite graph. We propose a geometric deep learning-based architecture, named multiscale spatio-temporal graph convolutional networks (MST-GCN), with a self-supervised training strategy. This method predicts the future 2D LV cardiac motion given the previously observed motion trajectories. The cardiac motion is represented on a graph constructed from sample nodes on LV myocardial contours. The MST-GCN and a graph-based gated recurrent unit (G-GRU) are connected using an encoder-decoder framework. We investigate essential elements of the proposed architecture for modelling the spatio-temporal patterns of cardiac motion. We also analyse the regional function of the LV based on predicted cardiac motion trajectories.

Contributions. (1) We propose a multiscale geometric deep learning-based architecture for LV cardiac motion estimation. To our knowledge, this is the first method to explore the internal relations of the multiscale endo-epicardial geometry for feature extraction. (2) We evaluate the impact of different elements of the proposed architecture on the accuracy of the 2D LV cardiac motion prediction. (3) We demonstrate that multiscale spatio-temporal patterns achieve good performance for cardiac motion estimation and regional analysis of LV function.

2   Method

We investigate various orders of motion difference fed into the proposed architecture. We define the node locations as the 0-order difference and the node velocities as the 1-order difference [5]. The node velocities are the differences of node locations between the current and the immediately previous cardiac MR frame, without the division of the time between each frame (since the time between each frame is constant).

2.1 Multiscale Cardiac Graph Construction

Figure 1 illustrates the construction of a three scale cardiac graph. These endo-epicardial nodes are chosen by the left and right ventricle geometry. The detail of how to sample nodes is described in our previous work [6]. These selected node locations are the ground truth in our work. In order to explore the further detail of the beating heart described by endo-epicardial nodes, we design a multiscale cardiac graph based on multiscale endo-epicardial node components. According to the mid-slice 6-segments model of the 17-Segment AHA model, we use three cardiac scales: endo-epicardial nodes scale $S1$, two high-level part scale, $S2, S3$.

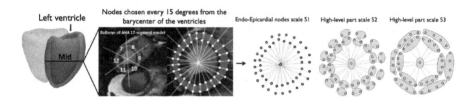

Fig. 1. *Overview of the proposed framework for the three scale cardiac structure graph construction in the mid-ventricular of short-axis view cardiac MR image sequences. At scale $S1$, the barycenter of the left ventricle (LV) and the 48 node locations from both the endocardium and epicardium are considered as cardiac structure graphs. At scale $S2$ and $S3$, 25 and 13 parts are considered respectively.*

2.2 Multiscale Graph Computation Unit (MGCU)

The aim of the multiscale graph computation unit (MGCU) is to extract and fuse features at different scales based on a multiscale graph. Each MGCU includes spatio-temporal graph convolution blocks and cross-scale fusion blocks. The spatio-temporal graph convolution block extracts features at each scale with a single-scale graph. The cross-scale fusion block interchanges features from one scale to another.

Spatio-Temporal Graph Convolution Block (ST-GCB). The ST-GCB includes a spatial graph convolution and then a temporal convolution, which extracts spatial and temporal features respectively. Let $A \in \{0,1\}^{N_S \times N_S}$ represent the adjacency matrix of the graph at the scale S, where N_S is the number of cardiac components. If the i-th and the j-th nodes are connected, $A_{i,j} = 1$. Otherwise, $A_{i,j} = 0$. Let the input feature at scale S be F_{in}, the spatial graph convolution is described as $F_{out} = \sum M \circ \widetilde{A} F_{in} W$. Here M is the edge weight matrix using the trainable weights, W is the feature importance matrix, $\widetilde{A} = D^{-\frac{1}{2}} A D^{-\frac{1}{2}}$ is the normalized adjacent matrix, $\circ$ represents the Hadamard product. A $1D$ temporal convolution is applied to extract features along time.

Cross-Scale Fusion Block (CS-FB). In order to obtain rich multiscale information, the SS-FB makes feature diffusion across different scales. It employs a bipartite graph [4] to deliver features from one scale graph to another. For instance, the features of a node at the segment 8 area in the high-level part scale $S2$ can offer the feature learning of a node at the segment 8 area in the endo-epicardial nodes scale $S1$. Here we describe CS-FB from $S1$ to $S2$ as an example.

Let $A_{S1S2} \in \{0,1\}^{N_{S1} \times N_{S2}}$ denote the cross-scale relations. Let $(A_{S1S2})_{k,i}$ denote the edge weight between the ith node and k part. Let the feature of the ith node be $F_{S1,i}$ and the feature of the kth part be $F_{S2,k}$. $h_{S1,i} = \theta_{S1}(F_{S1,i}), h_{S2,k} = \theta_{S2}(F_{S2,k}), (A_{S1S2})_{k,i} = softmax(h_{S2,k}^T, h_{S1,i})$, where θ_{S1} and θ_{S2} are implemented by MLP [4] and convolution [4]. The softmax function $softmax$ works on the result of the inner product matrix. After obtaining the adjacent matrix via inner product and softmax, we model the effects from the scale $S1$ to component in $S2$. The idea is to augment cardiac component features from the global related information. We obtain the edge weight from the inner product of two augmented features.

After that, we fuse the endo-epicardial node features in the scale $S1$ to the high-level part scale $S2$ with A_{S1S2}. The features F_{S2} at the scale $S2$ can be described as $A_{S1S2}F_{S1}W + X_{S2}$, where W is the trainable weights. The fused F_{S2} is used to the ST-GCB of the next MGCU in $S2$.

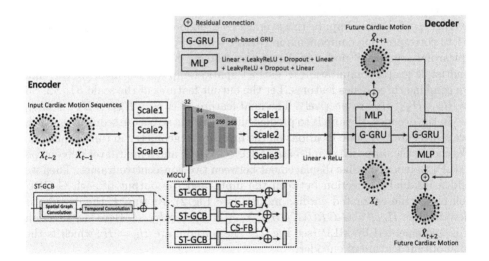

Fig. 2. *Network overview. The cardiac motion sequence is given as input to the MST-GCN encoder-decoder framework. In the encoder, each MGCU layer extracts spatio-temporal feature by multiscale graphs. The sum of the output of the decoder and the previous motion status represents the output future cardiac motion trajectory (predicted node locations), which is used in regional analysis of left ventricular function. $\oplus$ denotes the element-wise addition.*

2.3 Multiscale Spatio-Temporal Graph Convolutional Neural Network

Graph-Based Gated Recurrent Unit. The graph-based GRU (G-GRU) learns and updates hidden cardiac motion states with graph guidance. The graph is trained to regularise the states to generate a future cardiac structure. There are two inputs for the G-GRU: the initial state H_0 and the 2D cardiac structure-based feature $F_t \in \mathbb{R}^{m \times d}$ at time t. We define $A_h \in \mathbb{R}^{m \times m}$ as the adjacent matrix of the graph from time $t - 1, ..., t - n$. The G-GRU(F_t, H_t) is denoted as

$$z_t = \sigma(z_{in}(F_t) + z_{hid}(A_h H_t W_h)),$$
$$r_t = \sigma(r_{in}(F_t) + r_{hid}(A_h H_t W_h)),$$
$$g_t = tanh(g_{in}(F_t) + z_t \otimes g_{hid}(A_h H_t W_h)),$$
$$H_{t+1} = r_t \otimes H_t + (1 - r_t) \otimes g_t.$$

where the functions z_{in}, z_{hid}, r_{in}, r_{hid}, g_{in}, g_{hid} are linear transformations. The W_h is the trainable weights. A graph convolution is used on the hidden states H_t and generates the state for the next frame.

Encoder-Decoder Architecture. To initialize cardiac scales, we choose 2D average pooling to compute the average over endo-epicardial node clusters in $S1$ to corresponding components in coarser scales. For instance, we average two epicardial segment-8 nodes in $S1$ to the segment-8 part in $S2$. Then we use multiscale graph computation to extract spatio-temporal features. At the end, we combine three scales features. Let the output features of the scale $S1$, $S2$, $S3$ be H_{S1}, H_{S2}, H_{S3} respectively. The final feature H is $H = H_{S1} + H_{S2} + H_{S3}$.

In the decoder, the aim is to predict future cardiac structure sequences, which are represented by node locations on the myocardial geometry in time sequences. We choose the proposed graph-based GRU (G-GRU) and a multilayer perceptron (MLP) to model cardiac displacement between two consecutive frames. Then we add a residual connection between the input and the output of each G-GRU cell to get the estimated cardiac motion (see Fig. 2). At time t, the decoder is described as $H_{t+1} = G\text{-}GRU(\hat{X}_t, H_t), \hat{X}_{t+1} = \hat{X}_t + \tau(H_{t+1})$. Here the function τ is implemented by MLP (see Fig. 2). The initial state $H_0 = H$, which is the final output feature of encoder.

3 Experiments

3.1 Data Acquisition

In this study, we use 1611 short-axis view cardiac MR image sequences from the UK BioBank[1], acquired using a 1.5 Tesla scanner (Siemens Healthcare). A stack of short-axis images, around 12 slices, cover the entire left and right

---

[1] UK BioBank. https://www.ukbiobank.ac.uk/.

ventricles. In-plane resolution is 1.8×1.8 mm ^{2}, while the slice gap is 2.0 mm and the slice thickness is 8.0 mm. Each sequence contains 50 consecutive time frames per cardiac cycle. We randomly split image sequences into training/validation/testing with $1071/270/270$ subjects. We perform motion analysis on the 3 mid-ventricular slices.

3.2   Implementation Details

Pre-processing. The segmentation of the LV endo-epicardial borders and the RV was generated from using a publicly available FCN model [1] and used for node extraction. **Training.** The model is trained over 100 epochs via Adam with a learning rate 0.001 and a batch size of 1. In each training sample, we set the input difference operators length to 3 frames, and we predict future cardiac structure in 2 frames. The mean squared error (MSE) using the node locations is chosen as the evaluation metric. The proposed network was implemented using Python 3.7 with Pytorch. Experiments are run with computational hardware GeForce GTX 1080 Ti GPU 10 GB.

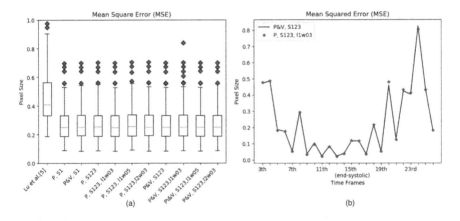

Fig. 3. (a) Using the proposed method - with ten different architectures and parameter settings - and the baseline, a comparison was made between the MSE on the predicted node locations on the endo-epicardial borders in the MRI sequences with the ground truth. The key elements of the architecture are inputting the 0-order difference features (P), 1-order difference features (V) in single scale 1 (S1), scale 1 to 3 (S123), fusion block in 1st MGCU layer (l1), fusion block in 1st and 2nd MGCU layers (l2), weight parameter 0.3 (w3) and 0.5 (w5). (b) Comparison of the mean squared error (MSE) for the best two performing architectures and parameter settings using the proposed method in each MRI frame. $P\&V, S123$ means inputting the 0-order and 1-order difference features using 3 scale graphs without the fusion block and $P, S123, l1w03$ means inputting the 0-order features using 3 scale graphs with the fusion block (hyperparameter 0.3) in 1st MGCU layer.

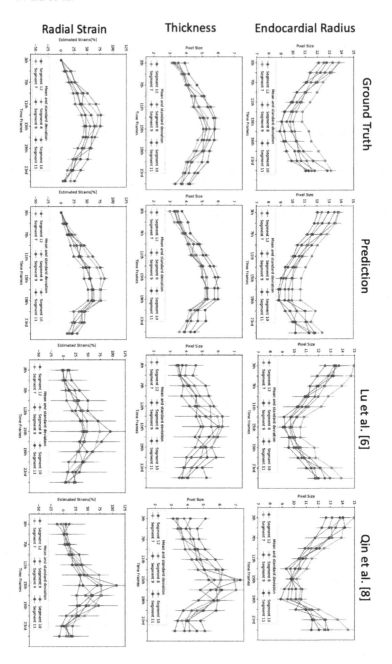

Fig. 4. Example results of the estimated endocardial radius (mean and standard deviation shown), thickness (mean and standard deviation shown) and radial strain (meanand standard deviation shown) for cardiac segments (7–12) plotted on frames 3, 5, 7, 9, 11, 13, 15, 17, 19, 21, 23, 25 over a cardiac cycle. Results (left to right) for ground truth, prediction, methods from Lu et al. [7] and Qin et al. [9] in a healthy volunteer.

3.3   Results

Quantitative Results. We compare the proposed method with ten different architectures and parameter settings with the baseline ST-GCN approach [6]. We also investigate key elements of the proposed architecture: effects of multiple scales, high-order motion difference, hyper-parameter in the fusion block balances the effect between the node-scale and other abstract scales. Based on the experiments on the validation data, we find that the best model is $P\&V, S123$ (inputting the 0-order and 1-order difference features using 3 scale graphs without the fusion block), which achieves a mean squared error of 0.251 pixels on MRI sequences. From Fig. 3(a), the multiscale architecture works better than the single architecture. The feature velocity improves performance. The fusion block between scale 1 and 2 works slightly better with the input 0th order difference feature node locations. Based on the performance with hyper-parameter 0.3 and 0.5 in the fusion block, the accuracy is not improved significantly. Figure 3(b) compares the MSE of the best two performing architectures and parameter settings using the proposed method in each MRI frame for an example subject. The MSE increases after the $19th$ frame. Later, we use the same example subject for left ventricular function evaluation.

Left Ventricular Function Evaluation. Based on the 17-Segment AHA model, the predicted nodes are classified into 6 segments [2]. Figure 4 shows an example of a time series of the endocardial radius, thickness, radial strain (Err) [7] in the six segments of myocardium from the ground truth of node locations, prediction and two other methods from Lu et al. [7] and Qin et al. [9] in a healthy volunteer. Compared to the ground truth, the prediction of the endocardial radius, the thickness and the Err has a similar plot shape.

4   Discussion and Conclusion

The focus of this paper is to investigate the effectiveness of graph convolutional networks for cardiac motion prediction. In clinical applications, motion trajectories may directly come from a motion tracking algorithm. Future work would extend the graph convolutional networks to the analysis of cardiac motion trajectories for disease diagnosis and motion-based biomarker discovery.

In this work, we propose a multiscale spatio-temporal graph convolutional network (MST-GCN) to characterise cardiac motion. We investigated the factors which can improve the accuracy of cardiac motion estimation. We found that the proposed method achieved a mean squared error of 0.25 pixel on MRI sequences. The accuracy is not significantly improved by a cross-scale fusion block. The proposed methods can characterise cardiac motion features for regional LV analysis.

Acknowledgements. This work is supported by SmartHeart. EPSRC grant EP/P001009/1. This study used data from the UK Biobank Resource under Application Number 40119. The authors declare no conflicts of interest related to this research study.

References

1. Bai, W., et al.: Automated cardiovascular magnetic resonance image analysis with fully convolutional networks. JCMR **20**(1), 65 (2018)
2. Cerqueira, M.D., et al.: Standardized myocardial segmentation and nomenclature for tomographic imaging of the heart: a statement for healthcare professionals from the cardiac imaging committee of the council on clinical cardiology of the american heart association. Circulation **105**(4), 539–542 (2002)
3. Lang, Y., et al.: Automatic localization of landmarks in craniomaxillofacial CBCT Images using a local attention-based graph convolution network. In: Martel, A.L., et al. (eds.) MICCAI 2020. LNCS, vol. 12264, pp. 817–826. Springer, Cham (2020). https://doi.org/10.1007/978-3-030-59719-1_79
4. Li, M., et al.: Dynamic multiscale graph neural networks for 3d skeleton based human motion prediction. In: Proceedings of the IEEE/CVF Conference on Computer Vision and Pattern Recognition, pp. 214–223 (2020)
5. Lu, P., et al: Dynamic spatio-temporal graph convolutional networks for cardiac motion analysis. In: ISBI 2021, pp. 122-125. IEEE (2021)
6. Lu, P., Bai, W., Rueckert, D., Noble, J.A.: Modelling cardiac motion via spatio-temporal graph convolutional networks to boost the diagnosis of heart conditions. In: Puyol Anton, E., Pop, M., Sermesant, M., Campello, V., Lalande, A., Lekadir, K., Suinesiaputra, A., Camara, O., Young, A. (eds.) STACOM 2020. LNCS, vol. 12592, pp. 56–65. Springer, Cham (2021). https://doi.org/10.1007/978-3-030-68107-4_6
7. Lu, P., Qiu, H., Qin, C., Bai, W., Rueckert, D., Noble, J.A.: Going deeper into cardiac motion analysis to model fine spatio-temporal features. In: Papież, B.W., Namburete, A.I.L., Yaqub, M., Noble, J.A. (eds.) MIUA 2020. CCIS, vol. 1248, pp. 294–306. Springer, Cham (2020). https://doi.org/10.1007/978-3-030-52791-4_23
8. Noh, K.J., Park, S.J., Lee, S.: Combining fundus images and fluorescein angiography for artery/vein classification using the hierarchical vessel graph network. In: Martel, A.L., et al. (eds.) MICCAI 2020. LNCS, vol. 12265, pp. 595–605. Springer, Cham (2020). https://doi.org/10.1007/978-3-030-59722-1_57
9. Qin, C., et al.: Joint learning of motion estimation and segmentation for cardiac MR image sequences. In: Frangi, A.F., Schnabel, J.A., Davatzikos, C., Alberola-López, C., Fichtinger, G. (eds.) MICCAI 2018. LNCS, vol. 11071, pp. 472–480. Springer, Cham (2018). https://doi.org/10.1007/978-3-030-00934-2_53
10. Yan, S., et al.: Spatial temporal graph convolutional networks for skeleton-based action recognition. In: Thirty-second AAAI conference on artificial intelligence (2018)
11. Zheng, Q., et al.: Explainable cardiac pathology classification on cine MRI with motion characterization by semi-supervised learning of apparent flow. Medical image analysis (2019)

An Image Registration Framework to Estimate 3D Myocardial Strains from Cine Cardiac MRI in Mice

Maziyar Keshavarzian[1], Elizabeth Fugate[2], Saurabh Chavan[3], Vy Chu[1], Mohammed Arif[4], Diana Lindquist[2], Sakthivel Sadayappan[4], and Reza Avazmohammadi[1,2(✉)]

[1] Department of Biomedical Engineering, Texas A&M University, College Station, TX 77843, USA
rezaavaz@tamu.edu

[2] Department of Radiology, University of Cincinnati, Cincinnati Children's Hospital Medical Center, Cincinnati, OH 45229, USA

[3] Department of Mechanical Engineering, Texas A&M University, College Station, TX 77843, USA

[4] Department of Internal Medicine, Division of Cardiovascular Health and Disease, Heart, Lung and Vascular Institute, University of Cincinnati, College of Medicine, Cincinnati, OH 45267, USA

Abstract. Accurate and efficient quantification of cardiac motion offers promising biomarkers for non-invasive diagnosis and prognosis of structural heart diseases. Cine cardiac magnetic resonance imaging remains one of the most advanced imaging tools to provide image acquisitions needed to assess and quantify in-vivo heart kinematics. The majority of cardiac motion studies are focused on human data, and there remains a need to develop and implement an image-registration pipeline to quantify full three-dimensional (3D) cardiac motion in mice where ideal image acquisition is challenged by the subject size and heart rate and the possibility of traditional tagged imaging is hampered. In this study, we used diffeomorphic image registration to estimate strains in the left ventricular wall in two wild-type mice and one diabetic mouse. Our pipeline resulted in a continuous and fully 3D strain map over one cardiac cycle. The estimation of 3D regional and transmural variations of strains is a critical step towards identifying mechanistic biomarkers for improved diagnosis and phenotyping of structural left heart diseases including heart failure with reduced or preserved ejection fraction.

Keywords: Cardiac magnetic resonance imaging · Left ventricle · Wall strain · Small animals

1 Introduction

The social and economic burden of cardiovascular diseases (CVDs), specifically acute cardiac events, has continued to increase over the past decade [3]. Accurate assessment of the cardiac contractile function is crucial to improving the

© Springer Nature Switzerland AG 2021
D. B. Ennis et al. (Eds.): FIMH 2021, LNCS 12738, pp. 273–284, 2021.
https://doi.org/10.1007/978-3-030-78710-3_27

morbidity and mortality associated with adverse cardiac events. While ejection fraction (EF) has been used to quantify the myocardial contractility and to classify the heart failures (HF), similar to most lumped parameters, it cannot provide etiological clues or information on the underlying mechanisms and ignores the regional variations in left ventricular diastolic and contractile behavior [2,9]. Considering the spatial heterogeneity of the pathologies which impair the left ventricle (LV) [1], the need to account for the local variations in the left heart motion in the diagnosis, studying, and treatment of these conditions still remains.

Myocardial strains, measured through non-invasive means, can be used to quantify cardiac diastolic and systolic functions. Over the past two decades, advances in cardiac magnetic resonance (cMR) tissue tagging, cine magnetic resonance imaging (cine-MRI), and speckle tracking echocardiography (STE) have paved the way for the non-invasive estimation of cardiac strains. While cMR tagging is accepted as the validated reference method for strain imaging [7,10,17,22], the need for specialized software, relatively complex protocols, and low temporal resolution have kept cMR tagging mostly a research tool for human and large animal data and not suitable for studying small animal models with high heart rate. STE, on the other hand, is widely accessible and easy to use. In 2D-STE, the cardiac strains are estimated by tracking the image texture and contours [8,14]. While this approach offers higher spatial and temporal resolutions, compared to cMR, reproducing the acquisition planes is difficult and it cannot capture out-of-plane motions. Although 3D echocardiography methods, such as 3D-STE, enable tracing speckles in all direction and therefore can address the latter issue [6], similar to tagged cMR, low spatial and temporal resolutions limit the accuracy of their predictions [13,21]. These issues become even more pronounced when it comes to studying LV dysfunction in small animal models. This gap has motivated interests in alternative techniques which are not limited by the complex imaging protocols, can capture out-of-plane motions, and offer reasonable spatial and temporal resolutions [15].

Cine-MRI, that provides detailed 3D anatomical images of the heart, offers great potential for the estimation of cardiac strains. Broadly speaking, one can divide the strain estimation methods which utilize cine-MRI into two categories: (a) mechanics-based nonlinear image registration algorithms, such as hyperelastic warping, which use finite element methods to simulate the myocardium and rely on assumptions about the passive and/or active mechanical properties of the tissue [19,20,23] and (b) deformable image registration methods that do not use finite element simulations [4,11,12]. In this study, our goal was to develop a MATLAB pipeline that enables the estimation of 3D cardiac strain using cardiac cine-MRI. We used the diffeomorphic elastic image registration algorithm, integrated with interpolation schemes, to estimate the displacement fields from short-axis (SA) cine-MRI scans and then calculated complete Green-Lagrange strain tensors at each point in the myocardial wall over one cardiac cycle. Importantly, our pipeline is not restricted to in-plane strain calculations which is a typical limitation with available commercial software using "feature tracking" technique. We were able to track out-of-plane movement of pixels within the stack of SA planes enabling the calculation of 3D strains at each point. We applied our pipeline to calculate the LV strains in two wild-type (WT) mice and one diabetic mouse exhibiting mild LV diastolic dysfunction (LVDD) and compared our predictions to those of two commercial software.

2   Materials and Methods

2.1   Image Acquisition

Short-axis cine-MRI scans of healthy WT female (n = 2, 21 weeks old, WT-1 and WT-2) and pathological male B6.BKS (D)-Lepr db/J (n=1, 14 weeks old, db-1) mice were acquired under anesthesia using a vertical wide-bore 9.4T Bruker Avance III HD scanner with a 36 mm proton volume coil (Bruker BioSpin MRI GmbH) for the WT and a 7T Biospec system (Bruker, Billerica, MA) ultizing a 72 mm ID linear volume transmit coil for db-1 mice. The imaging protocol was approved by the University of Cincinnati's Animal Care and Use Committee (UC IACUC Protocol 19-10-03-01 and 2018-0054). We used ECG signal to trigger image acquisition (R-R over 3 periods) and acquired a set of 10 short-axis scan with the following parameters: Fast Low Angle Shot, TE = 1.8 ms for the 9.4T and 1.6 ms for the 7T, TR = 9 ms, bit depth = 16, flip angle = 30 for the 9.4T and 20 for the 7T, bandwidth = 65789 Hz, resolution = 200 μm, slice thickness = 1 mm, matrix = 160×160, averages = 4. In the db-1 mouse, the plasma levels of insulin increases at 10 to 14 d and they become obese at 3 to 4 weeks of age [5]. They are expected to exhibit LVDD at 4 months of age and older.

2.2   Image Registration

The 3D image registration was performed using a custom MATLAB (Math-Works, MA, USA) library. The major steps in our pipeline included: (i) semi-automatically producing the region of interest (ROI) in each cine-MRI slice; the ROI comprises two contours defining the LV endocardial and epicardial borders, (ii) performing pre-processing that consists of cropping the stack of 2D images and linear resampling of the images, (iii) the 3D image registration over one complete cardiac cycle to calculate displacement field at each timepoint, and (iv) calculating strains from displacement field. The ROI borders were traced using a segmentation software.

We used the diffeomorphic demons (DD) registration algorithm to, successively, match the stream of images in 3D. In this algorithm, the object boundaries are considered as semi-permeable membranes and deformable grids in the fixed and the moving images, respectively [16,18]. The DD algorithm assumes that the pixel intensity of the moving image remains constant over time and identifies the transformation which aligns the moving image, $\mathcal{M}$, with the fixed image, $\mathcal{F}$, through minimizing the global energy function W, expressed as

$$W(\mathbf{c}, \mathbf{u}) = \frac{1}{\sigma_i^2}\|\mathcal{F} - \mathcal{M} \circ \mathbf{c}\|^2 + \frac{1}{\sigma_x^2}\|\mathbf{u} - \mathbf{c}\|^2 + \frac{1}{\sigma_T^2}\|\nabla\mathbf{u}\|^2 \qquad (1)$$

where σ_i and σ_x represent the image intensity noise and the spatial uncertainty on the correspondences respectively, σ_T controls the amount of the regularization, $\mathbf{u}$ is the spatial transformation, and $\mathbf{c}$ is the non-parametric spatial transformation [18]. Also, here, we used a diffusion-like regularization, and a histogram matching filter was used prior to the image registration step. The endocardial and epicardial contours were used to: (i) isolate the ROI after the estimation of the displacement fields and (ii) calculate the endocardial volume at each timepoint within the cardiac cycle.

2.3   Strain Calculations

We calculated the Green-Lagrange strain tensor, $\mathbf{E}$, from the deformation gradient tensor, $\mathbf{F}$, defined as

$$\mathbf{F} = \partial\mathbf{x}/\partial\mathbf{X} \tag{2}$$

and

$$\mathbf{E} = \frac{1}{2}(\mathbf{F}^T \cdot \mathbf{F} - \mathbf{I}) \tag{3}$$

where $\mathbf{x}$ is the reference frame, $\mathbf{X}$ is the target frame, and $\mathbf{I}$ is the identity tensor. The strain values were transferred from the Cartesian to the polar cylindrical coordinates ($\mathbf{E}_p = \mathbf{Q} \cdot \mathbf{E} \cdot \mathbf{Q}^T$, where $\mathbf{Q}$ is a rotation matrix) at each slice and the strain estimations were compared to values obtained from the strain calculation module commercially available in the Segment and CVi42 software packages.

3   Results

Assessments of LV chamber volume throughout a representative cardiac cycle, calculated by summing over the volume of all the intra-chamber pixels, is shown in Fig. 1. The volume was reduced by more than 50% at end-systole (ES) timepoint in all three mice.

Average sectional strains for three representative short-axis sections over one cardiac cycle is shown for WT-1 and WT-2 in Fig. 2. The sections were chosen from basal, mid, and apical regions and normalized average of radial and circumferential strains were calculated in each section. The maximums for the radial and circumstantially strains in both mice appeared to occur close to the global ES timepoint and nearly maintained their values across multiple SA planes from base to apex. Also, both radial and circumferential strains exhibited consistent patterns across the SA planes except for the circumferential strain in the apical plane in WT-2 where the maximum strain occurred twice, before and after the global ES timepoint.

Regional and transmural strain variations across an intra-myocardial long-axis cross-section are shown for WT-1 and db-1 in Fig. 3. The section is fixed in the space and the measurements for radial, circumferential, and longitudinal strains are given at three representative timepoints including ES. The interpolation scheme along the LV long axis used in our pipeline enabled the delivery of a continuous strain map across seven short-axis measurements originally used in

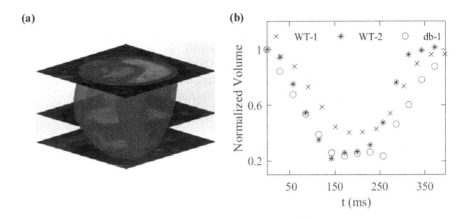

Fig. 1. (a) Biventricular reconstruction of a representative mouse heart using SA acquisitions. (b) Volume of the LV chamber drops by more than 50% at ES in WT-1, WT-2, and db-1 mice.

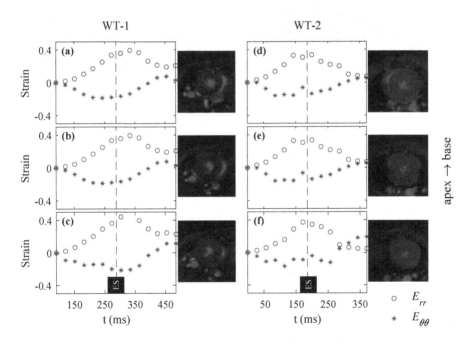

Fig. 2. Variation of radial and circumferential strains over one cardiac cycle in WT-1 and WT-2. The average radial (E_{rr}) and circumferential ($E_{\theta\theta}$) strains were calculated at basal (a and d), mid (b and e), and apical (c and f) left ventricular cross-sections for two WT mice.

the calculations. As depicted by representative LV short-axis slices, the chosen long-axis sections contain the basal and mid-wall regions and part of the apical region. As expected, spatial maps of strains qualitatively suggested that radial and circumferential strains exhibit notable tensile and compressive strains at ES timepoint, respectively, with radial positive strains being predominately present in the endocardial region. In contrast, the longitudinal strains exhibited stronger regional variations and tensile/compressive strain pattern in the given section with the basal regions remaining consistently in positive strains.

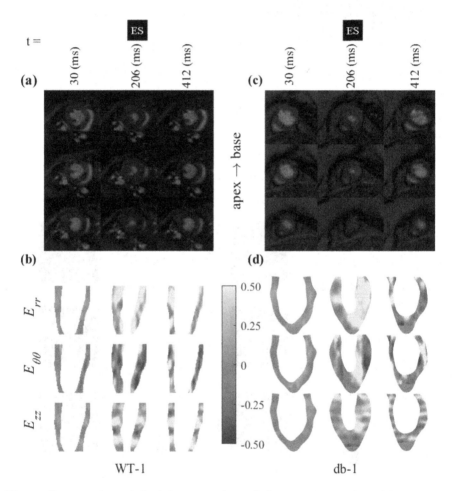

Fig. 3. Contour plots of the left ventricular radial, circumferential, and longitudinal strains in WT-1 (a, b) and db-1 (c, d) mice.

Regional variations in SA views were continuously quantified for all the strains (Fig. 4 shows the results for a chosen mid-section). The radial strain in the short-axis view showed the most pronounced variation followed by circumferential, and longitudinal strains. The regional variations at ES timepoint were not significant, leading to similar levels of strain for anterior, posterior, lateral, and septal regions *on average*. In contrast, all strains showed stronger regional variations prior and subsequent to ES timepoint. A significant wall thickening was observed as a synergistic result of large tensile and compressive strains in the radial and circumferential directions, respectively.

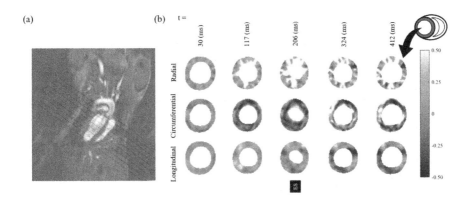

Fig. 4. (a) Short-axis slices shown in the long-axis view with oblique lines. The red arrows identify the location of strain contours shown in panel b. (b) Distribution of radial, circumferential, and longitudinal strains in the LV in WT-1. (Color figure online)

Two commercially available software that can calculate 2D strains from cine cardiac MRI are CVi42 and Segment. These software use feature tracking and can quantify strains from both clinical and pre-clinical images. We assessed our strain estimations against those calculated by both software (Fig. 5). We found an excellent agreement between our results for average radial and circumferential strains on a mid-wall short-axis section and corresponding results from Segment and CVi42 for WT-1 and db-1, respectively. An apparent discrepancy between our results and Segment estimation occurred in the radial strain following ES timepoint. Part of this discrepancy is likely due to the fact that the images representing one complete cardiac cycle are indeed averaged from different cycles, but at appropriate normalized timepoints, during data acquisition. Given that ventricular pressure waves slightly differ from one cycle to another, it is likely that the calculated strains from deformable image registration may not go back to zero at the end of the cycle. However, additional algorithms in commercial software, such as Segment and CVi42, are used to enforce equal strains at the beginning and end of each cycle.

Finally, using all the SA acquisitions, 3D strain maps were created for radial, circumferential, and longitudinal strains (Fig. 6 shows the results for WT-1 and db-1).

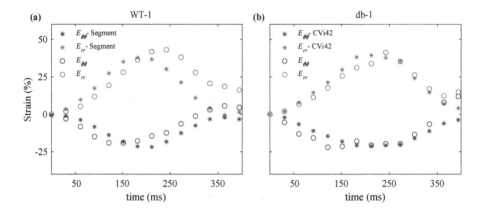

Fig. 5. Strain values obtained from the MATLAB pipeline match reasonably well with corresponding estimates from Segment and CVi42. Radial and circumferential strains in WT-1 (a) and db-1 (b) mice are compared with values obtained from two commercial software.

4   Discussion

Efficient non-invasive quantification of full 3D LV kinematics offers a comprehensive set of biomarkers that go beyond traditional measures of LV function and anatomy and can serve as insightful diagnostic and treatment-planning tools in the clinic. Surprisingly, despite significant advances in cine cardiac imaging, to the best of authors knowledge, an open-source, non-commercial package to quantify cardiac strains from un-tagged cine MRI remains elusive. The need to develop such a technology remains significant, especially for small animals where the acquisition of high quality imaging data is challenged by the subject size and heart rate. To summarize, in this work we proposed and implemented an image-registration pipeline to estimate full 3D cardiac strains from cine MRI in mice. Unlike some of existing advanced warping methods, the pipeline calculates the strains by processing raw image acquisitions without the need of finite-element simulations.

Using the deformable image registration framework, we provided 3D and continuous map of strain field in the LV myocardial wall. The generated deformation map allowed us to obtain strain at any given cross section and also with respect to an arbitrary coordinates. The results were presented in the clinically-relevant radial, circumferential, and longitudinal directions. Regional and trans-mural variations were evident and well-captured in the results. As expected, the radial and circumferential strains tended to be predominantly positive and negative, respectively, that consistently led to a pronounced wall thickening. Although not presented here, the shear strains including longitudinal-radial strains as well as principal strains can be easily obtained from the deformation map.

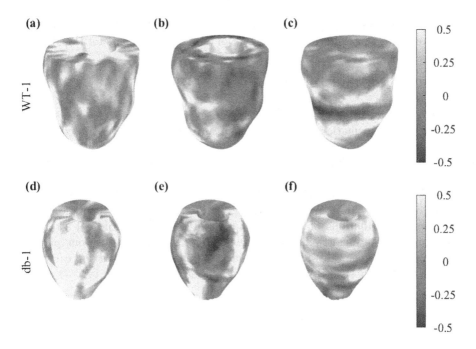

Fig. 6. 3D representation of radial (a and d), circumferential (b and e), and longitudinal (c and f) strains at ES in WT-1 and db-1 mice.

4.1 Overall Findings

4.2 Application to HFpEF Mice Model

Remodeling of the heart left ventricle is a key and common consequence in the setting of several structural heart diseases including myocardial infarction (MI) and heart failure with preserved ejection fraction (HFpEF). The remodeling events take place at multiple scales in the LV potentially altering tissue architecture and biomechanics collectively leading to functional adaptation of the LV. Therefore, traditional measures such as LV dilation and thickness used to characterize the LV remodeling provide limited information on cardiac performance. New biomechanics-driven biomarkers that reveal and quantify regional variations in LV remodeling are needed for early diagnosis and accurate prognosis of structural LV diseases.

In particular, we used our pipeline for a diabetic mouse with mild LVDD (db-1). Our results suggested a noticeable LV hypertrophy with slightly reduced circumferential strains in db-1. We plane to improve our understanding of how the LV adapts in HFpEF and quantify the regional variations in biomechanical measures of LVDD which is a key component in the progression of HFpEF. The pipeline in this work will allow us to study the tissue- and organ-level development of LVDD in murine models of HFpEF in a serial manner. Such explorations are needed to identify key LV kinematic metrics that can be used as clinical biomarkers for early diagnosis and phenotyping of HFpEF that tend to be an impeding factor in the development of efficient treatments for this disease.

5   Summary and Implications

Progressive myocardial remodeling events in several LV diseases lead to sizeable changes in the kinematics of the LV that, in turn, mount to LV dysfunction. The ability of non-invasive, high-fidelity, and 3D quantification of cardiac motion in small animals will severe as a significant tool to conduct longitudinal study of LV remodeling in mice. In this work, we developed and implemented a pipeline that uses diffeomorphic image registration technique to estimate 3D cardiac strains in the LV wall in WT mice. We presented and analyzed the regional and transmural distributions of radial, circumferential, and longitudinal strains in both long-axis and short-axis views. In future works, we will use our pipeline to identify LV kinematics-related biomarkers that can be used for early and image-based diagnosis of LV diastolic dysfunction.

6   Declaration of Competing Interest

Dr. Sadayappan provided consulting and collaborative research studies to the Leducq Foundation, Red Saree Inc., Greater Cincinnati Tamil Sangam, AstraZeneca, MyoKardia, Merck and Amgen, but such work is unrelated to the content of this manuscript. No other disclosures are reported.

Acknowledgements. This work was supported by the National Institutes of Health R00HL138288 to R.A. Dr. Sadayappan has received support from National Institutes of Health grants R01 HL130356, R01 HL105826, R01 AR078001, and R01 HL143490; American Heart Association 2019 Institutional Undergraduate Student (19UFEL34380251) and transformation (19TPA34830084) awards.

References

1. Abhayaratna, W.P., Marwick, T.H., Smith, W.T., Becker, N.G.: Characteristics of left ventricular diastolic dysfunction in the community: an echocardiographic survey. Heart **92**(9), 1259–1264 (2006)
2. Amzulescu, M.S., et al.: Myocardial strain imaging: review of general principles, validation, and sources of discrepancies. Eur. Heart J. Cardiovascular Imaging **20**(6), 605–619 (2019)
3. Benjamin, E.J., et al.: Heart disease and stroke statistics–2019 update: a report from the American heart association. Circulation **139**(10), e56–e528 (2019)
4. Bistoquet, A., Oshinski, J., Škrinjar, O.: Myocardial deformation recovery from cine MRI using a nearly incompressible biventricular model. Med. Image Anal. **12**(1), 69–85 (2008)

5. Coleman, D.L.: Obese and diabetes: two mutant genes causing diabetes-obesity syndromes in mice. Diabetologia **14**(3), 141–148 (1978)
6. De Craene, M., et al.: 3d strain assessment in ultrasound (straus): a synthetic comparison of five tracking methodologies. IEEE Trans. Med. Imaging **32**(9), 1632–1646 (2013)
7. Garot, J., et al.: Fast determination of regional myocardial strain fields from tagged cardiac images using harmonic phase MRI. Circulation **101**(9), 981–988 (2000)
8. Geyer, H., et al.: Assessment of myocardial mechanics using speckle tracking echocardiography: fundamentals and clinical applications. J. Am. Soc. Echocardiogr. **23**(4), 351–369 (2010)
9. Konstam, M.A., Abboud, F.M.: Ejection fraction: misunderstood and overrated (changing the paradigm in categorizing heart failure). Circulation **135**(8), 717–719 (2017)
10. Lima, J.A., et al.: Accurate systolic wall thickening by nuclear magnetic resonance imaging with tissue tagging: correlation with sonomicrometers in normal and ischemic myocardium. J. Am. Coll. Cardiol. **21**(7), 1741–1751 (1993)
11. Mansi, T., Pennec, X., Sermesant, M., Delingette, H., Ayache, N.: ilogdemons: a demons-based registration algorithm for tracking incompressible elastic biological tissues. Int. J. Comput. Vision **92**(1), 92–111 (2011)
12. Mansi, T., et al.: Physically-constrained diffeomorphic demons for the estimation of 3D myocardium strain from cine-MRI. In: Ayache, N., Delingette, H., Sermesant, M. (eds.) FIMH 2009. LNCS, vol. 5528, pp. 201–210. Springer, Heidelberg (2009). https://doi.org/10.1007/978-3-642-01932-6_22
13. Mor-Avi, V., et al.: Current and evolving echocardiographic techniques for the quantitative evaluation of cardiac mechanics: ASE/EAE consensus statement on methodology and indications endorsed by the Japanese society of echocardiography. Eur. J. Echocardiogr. **12**(3), 167–205 (2011)
14. Perk, G., Tunick, P.A., Kronzon, I.: Non-doppler two-dimensional strain imaging by echocardiography-from technical considerations to clinical applications. J. Am. Soc. Echocardiogr. **20**(3), 234–243 (2007)
15. Schuster, A., Hor, K.N., Kowallick, J.T., Beerbaum, P., Kutty, S.: Cardiovascular magnetic resonance myocardial feature tracking: concepts and clinical applications. Circulation: Cardiovascular Imaging **9**(4), e004077 (2016)
16. Thirion, J.P.: Image matching as a diffusion process: an analogy with maxwell's demons. Med. Image Anal. **2**(3), 243–260 (1998)
17. Thomas, D., et al.: Quantitative assessment of regional myocardial function in a rat model of myocardial infarction using tagged MRI. Magn. Reson. Mater. Phys., Biol. Med. **17**(3–6), 179–187 (2004)
18. Vercauteren, T., Pennec, X., Perchant, A., Ayache, N.: Diffeomorphic demons: efficient non-parametric image registration. Neuroimage **45**(1), S61–S72 (2009)
19. Veress, A.I., Phatak, N., Weiss, J.A.: Deformable image registration with hyperelastic warping. In: Suri, J.S., Wilson, D.L., Laxminarayan, S. (eds.) Handbook of Biomedical Image Analysis, pp. 487–533. Springer, Boston (2005). https://doi.org/10.1007/0-306-48608-3_12

20. Veress, A.I., et al.: Measuring regional changes in the diastolic deformation of the left ventricle of SHR rats using micropet technology and hyperelastic warping. Ann. Biomed. Eng. **36**(7), 1104–1117 (2008)
21. Voigt, J.U., et al.: Definitions for a common standard for 2D speckle tracking echocardiography: consensus document of the EACVI/ASE/industry task force to standardize deformation imaging. Eur. Heart J. Cardiovascular Imaging **16**(1), 1–11 (2015)
22. Zerhouni, E.A., Parish, D.M., Rogers, W.J., Yang, A., Shapiro, E.P.: Human heart: tagging with MR imaging-a method for noninvasive assessment of myocardial motion. Radiology **169**(1), 59–63 (1988)
23. Zou, H., et al.: Three-dimensional biventricular strains in pulmonary arterial hypertension patients using hyperelastic warping. Comput. Methods Programs Biomed. **189**, 105345 (2020)

Cardiac Mechanics: Measures and Models

Sensitivity of Myocardial Stiffness Estimates to Inter-observer Variability in LV Geometric Modelling

Abdallah I. Hasaballa[1]([envelope])[ID], Thiranja P. Babarenda Gamage[1],
Vicky Y. Wang[1,2], Debbie Zhao[1], Charlène A. Mauger[1,3], Kathleen Gilbert[1],
Zhinuo J. Wang[4], Bianca Freytag[1,5], Jie Jane Cao[6], Alistair A. Young[1,7],
and Martyn P. Nash[1,8]

[1] Auckland Bioengineering Institute, University of Auckland, Auckland, New Zealand
ahas804@aucklanduni.ac.nz
[2] Radiological Sciences Laboratory, Stanford University, Stanford, USA
[3] Department of Anatomy and Medical Imaging, University of Auckland,
Auckland, New Zealand
[4] Department of Computer Science, University of Oxford, Oxford, UK
[5] Creatis, CNRS UMR5220, INSERM U1206, Université Lyon 1, Lyon, France
[6] The Heart Center, St Francis Hospital, Roslyn, NY, USA
[7] Department of Biomedical Engineering, King's College London, London, UK
[8] Department of Engineering Science, University of Auckland,
Auckland, New Zealand

Abstract. Heart failure is known to be associated with substantial changes in the mechanical properties of the heart muscle. Biomechanical parameters, such as myocardial stiffness, have the potential to help clinicians diagnose and determine and monitor treatment options. The impact of inter-observer variability of geometric modelling on the estimation of passive myocardial stiffness has not yet been systematically investigated. We aimed to examine the sensitivity of myocardial stiffness estimates with respect to inter-observer geometric model variability. Twenty-four subjects (5 controls, 19 patients with heart failure) underwent left heart catheterisation and cardiovascular magnetic resonance (CMR) imaging. Three expert analysts independently constructed three-dimensional geometric models of the left ventricle (LV), which were used to estimate myocardial stiffness using finite element simulations that combined cine CMR data and LV pressure measurements. Bland-Altman analysis was used to assess the inter-observer effects on the reproducibility of myocardial stiffness. The inter-observer variations were ± 7.69 mL/m^2 and ± 9.78 g/m^2 for the LV end-diastolic volume and mass indices, respectively. Meanwhile, the variability ranged by up to ± 0.51 kPa for inter-observer analysis on the estimated intrinsic myocardial stiffness values. Findings from this pilot study highlight the importance of the accuracy of image-based geometric modelling when estimating myocardial stiffness.

A. I. Hasaballa and T. P. Babarenda Gamage—Joint first authorship.

© Springer Nature Switzerland AG 2021
D. B. Ennis et al. (Eds.): FIMH 2021, LNCS 12738, pp. 287–295, 2021.
https://doi.org/10.1007/978-3-030-78710-3_28

Keywords: Heart failure · Myocardial stiffness · Geometric modelling · Finite element analysis · Cine CMR · Inter-observer variability

1   Introduction

Heart failure (HF) is a complex clinical syndrome caused by either structural or functional impairment of the ventricles. Current clinical guidelines characterise HF patients based on the left ventricle (LV) ejection fraction (EF). Despite the increasing awareness of cardiac pathophysiology and advances in medical therapy for HF, most approved treatment plans for HF with reduced EF (HFrEF) have been demonstrated to be ineffective for HF with preserved EF (HFpEF), suggesting significant differences in the mechanism underlying the different forms of HF. Current clinical understanding of HF relies on simplistic chamber-level measurements and there is a need for more mechanistically driven approaches [1].

Biomechanical properties of the myocardium, such as passive tissue stiffness, can provide potential biomarkers for clinicians to better understand the underlying mechanisms of HF. There are, however, still considerable challenges in obtaining these parameters, which in turn limits the use of biomechanical modelling approaches in clinical practice. The estimation of subject-specific myocardial stiffness parameters is often achieved using an inverse finite element (FE) modelling approach [2]. The accuracy and reliability of the these FE frameworks depend on a number of key factors, including: (i) quantitative descriptions of LV geometry and tissue structure; (ii) kinematic boundary and loading conditions; (iii) constitutive equations that describe the mechanical properties of the myocardium; and (iv) the chosen objective function and optimisation strategy [3].

Geometric analyses of cardiovascular magnetic resonance (CMR) images are often performed by expert analysts. However, inter-observer variability is an important factor that can confound parameter estimates [4]. For myocardial stiffness to be used as a potential biomarker in clinical practice, it is important to examine the sensitivity and uncertainty of tissue stiffness estimates to variability in the geometric inputs. In this study, we used an existing FE modelling framework [5], which integrates cine CMR data with cardiac catheterisation pressure data to estimate LV myocardial stiffnesses for control and diseased hearts by matching model predicted global motion to that measured using cine CMR. The main goal was to investigate the sensitivity of tissue parameter estimates with respect to inter-observer variability in LV geometric model creation.

2   Methods

2.1   Overview

Subject-specific biomechanical models of the LV were used to estimate passive myocardial stiffness using an inverse FE modelling approach (Fig. 1) [5].

Briefly, the framework can be organised into three major components: input data, mechanics simulation, and parameter estimation. The input data (*blue block*) includes: LV geometric models constructed from *in vivo* cine CMR data throughout the cardiac cycle, with endocardial and epicardial surface data points quantifying LV wall motion; and LV pressure data throughout the cardiac cycle measured using interventional cardiac catheterisation. The mechanics simulation (*red block*) takes the haemodynamic measurements (pressure data between diastasis (DS) and end-diastolic (ED)) and geometric model at diastasis (derived from cine CMR data), along with a mamalian microstructural field, to simulate the passive mechanical function. The simulation generates model predictions of LV deformation during diastolic filling. The parameter estimation process (*green block*) involves evaluation of the difference between model predicted motion and that derived from cine CMR images, and minimises this difference by tuning a myocardial stiffness parameter (C) of the constitutive model.

The parameter estimation framework (as shown in Fig. 1) rests on the optimisation of a single constitutive parameter (C) to minimise an objective function. The objective function is defined as the error between model predictions at every frame of interest (during passive inflation) and corresponding frames of a geometric model derived from medical images. Datapoints were generated on the epicardial and endocardial surfaces of the image derived geometric model. The objective function was computed using the mean squared error of the closest projection of the epicardial and endocardial surface points onto the corresponding surfaces from the simulated mechanics model. Following the parameter estimation process, the identifiability of the passive stiffness parameter (C) was assessed based on a method introduced in [7]. Briefly, the approach introduces a threshold value for the estimated myocardial stiffness beyond which the model's deformation sensitivity fails to reach the resolution of the detectable wall motion.

2.2   Patient Data

Invasive haemodynamic measurements and *in vivo* cine CMR datasets for 5 control subjects (3 males) and 19 HF patients (10 males) were analysed in this study. HF patients were categorised into HFrEF (n = 12, 7 males) and HFpEF (n = 7, 3 males) groups. All participants provided written informed consent, and the study protocols were approved by the St Francis Hospital Internal Review Board (No. 07–014). For more details on the imaging and catheterisation protocols, and selection criteria of HF patients, see [5].

2.3   Cine CMR Acquisition and Processing

In vivo cine CMR datasets were acquired with a voxel size of $1.25\,\text{mm} \times 1.25\,\text{mm}$ and slice thickness of $6\,\text{mm}$. For analysis of *in vivo* cine CMR images, three-dimensional (3D) LV geometric models were created using Cardiac Image Modeller software (CIM version 8.1, University of Auckland, New Zealand), as described previously [6,8]. In order to investigate the inter-observer variability of the LV geometric models, three expert analysts independently examined the same dataset, which resulted in the generation of $24 \times 3 = 72$ subject-specific

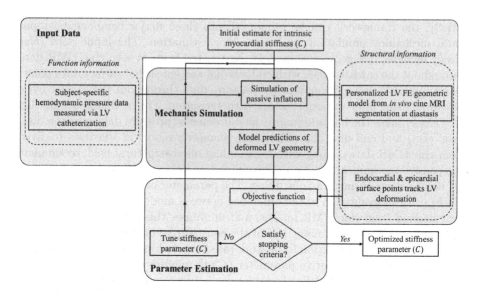

Fig. 1. Flowchart summarising the estimation process for subject-specific passive myocardial stiffness.

kinematic LV models. Geometric parameters such as LV end-diastolic volume (LVEDV) and ventricular wall mass were calculated using the CIM software. The selected end-diastolic, end-systolic and diastasis image frames were identical among the three observers.

2.4   Cardiac Catheterisation and Haemodynamic Pressure Processing

All participants underwent left heart catheterisation within five hours of imaging. For each participant, LV pressures from multiple cycles were beat-averaged to produce a single representative cycle. Beat-to-beat variation was present in the raw pressure data. A detailed description of the temporal registration between the cine CMR and catheterisation data can be found in [5]. Briefly, the end-diastolic time point was defined by the R wave of ECG, while diastasis, representing the state at which the pressure change was minimal, was identified by the inflection point of the LV pressure trace between minimum pressure and pressure at ED. We quantified the standard deviation of the ED pressures across the cycles for each participant as a measure of beat-to-beat variation and found that the standard deviations had an average across all participants of 0.38 kPa (IQR:[0.18 0.59] kPa).

2.5   Passive LV Mechanics Modelling and Tissue Stiffness Estimation

The cine CMR and cardiac catheterisation data were integrated using an inverse FE framework based on finite deformation hyperelasticity to estimate passive myocardial stiffness for each participant. The LV geometry at diastasis was chosen as the reference configuration for biomechanical analysis. The myocyte fibre arrangement of the LV was described by a typical mammalian distribution (see [5] for details).

The passive mechanical response of the myocardium was modelled using a transversely isotropic hyperelastic constitutive model [9], whereby the strain-energy density function (W) had the form:

$$W = \frac{C}{2}(e^Q - 1),$$
$$Q = b_1 E_{ff}^2 + b_2(E_{cc}^2 + E_{rr}^2 + E_{cr}^2 + E_{rc}^2) + b_3(E_{rf}^2 + E_{fr}^2 + E_{fc}^2 + E_{cf}^2) \tag{1}$$

where $E_{ij}, i, j \in \{f, c, r\}$ is the Green-Lagrange strain tensor with indices f, c and r denoting the myocyte fibre, cross-fibre and radial directions, respectively. The parameter C scales the overall stiffness of the myocardium, whereas the b parameters control the degree of tissue anisotropy and nonlinearity. For this study, these parameters were fixed to $(b_1 = 8.61, b_2 = 3.67, b_3 = 25.77)$, in accordance to the findings of a previous study [10]. To provide a subject-specific index of passive myocardial stiffness, the C parameter was estimated using a nonlinear optimisation procedure by minimising the discrepancy between model predictions and the displacements of the LV endocardial and epicardial surfaces measured in the images.

2.6   Statistical Analysis

All parameters are presented as mean $\pm$ standard deviation (SD) unless otherwise specified. The intra-class correlation coefficient (ICC, two-way mixed effects, absolute agreement, average measures) was used to measure inter-observer reliability and determine the degree of correspondence and agreement among observers. Meanwhile, the inter-observer variability was assessed using a combined Bland-Altman analysis, where symmetric 95% limits of agreement (± 1.96 SD) were calculated using the differences from each of the three observers to the mean of the measurements for each subject. All statistical tests were carried out using IBM-SPSS 22 (SPSS Institute, Chicago, IL, USA).

3   Results and Discussion

3.1   Inter-observer Variability in LV Geometry

The LVEDV index (LVEDVI: LVEDV/body surface area) and LV mass index (LVMI: LVM/body surface area) were used in the comparison of LV geometries between observers. Figure 2 shows Bland-Altman plots for the inter-observer variability of the LVEDVI and LVMI with symmetric limits of agreement of ± 7.69 mL/m^2 (± 1.96 SD) and ± 9.78 g/m^2 (± 1.96 SD), respectively. The ICC scores calculated for the LVEDVI between observers I-II, I-III, and II-III were 0.995, 0.995 and 0.994, respectively, and the ICC scores for LVMI were 0.991, 0.968 and 0.983, respectively. These results show that there were no statistically significant differences in the measurements among all observers. Such findings were consistent with studies [11,12] on inter-observer variability of CMR measurements, showing that the variability ranged by up to 13% for inter-observer analysis. It should be noted that higher LVMIs were associated with lower LVEDVIs for observer I and vice versa for observer III. This could be explained by the fact that the endocardial contours for observer I were smaller compared to observer III, while the epicardial contours were comparatively similar across all three observers.

To assess inter-observer variability with respect to regional LV geometry, mean absolute surface distances and masses for the corresponding geometric models were evaluated with respect to the American Heart Association (AHA) 17-segment model [12] at diastasis (Fig. 3).

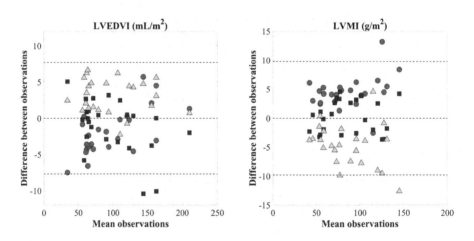

Fig. 2. Bland-Altman plots of inter-observer variability of observer I (●), II (■) and III (▲) for LVEDVI (*left*) and LVMI (*right*). Means and 95% limits of agreement are indicated using dotted lines. (Color figure online)

3.2 Identifiability of Stiffness Estimates

Stiffness parameters could not be estimated for 11 of the 72 biomechanical models due to issues regarding the lack of parameter identifiability, assessed using the threshold methodology introduced in [7]. For the remaining cases, there was a graded variation in the identifiability of the estimated parameters at the optimal solution (i.e., there was variability in the curvature of the objective function,

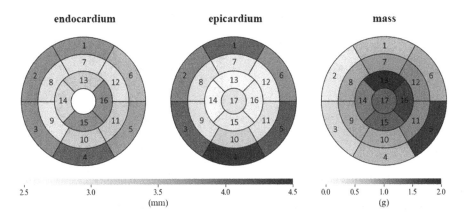

Fig. 3. Mean regional surface distances (mm) and absolute mass differences (g) at diastasis between corresponding geometric models, averaged between observers I-II, I-III and II-III. Numbers represent American Heart Association (AHA) segments [12].

quantified using the Hessian matrix at the optimal solution; data not shown). In general, smaller optimal stiffness values had objective functions with greater curvatures at the optimal estimates.

3.3 Inter-observer Variability in Stiffness Estimates

Figure 4a shows the individual stiffness estimates (C) for each of the three observers plotted against the mean values. In general, there was no indication of systematic bias, and the inter-observer variation in stiffness was lower for smaller mean stiffness values. Figure 4b shows a Bland-Altman plot comparing the estimated myocardial stiffness among the three observers with symmetric limits of agreement of ±0.51 kPa (±1.96 SD). When comparing myocardial stiffness parameters derived using models from each observer, the ICC scores for observers I-II, I-III, and II-III were 0.978 (n = 21), 0.980 (n = 18) and 0.977 (n = 18), respectively. In all but one case, we noted a general trend that inter-observer variations in stiffness were smaller for cases with lower mean stiffness values. In some cases, there were substantial differences (of up to 1.1 kPa) in stiffness parameters derived using models created by different observers for a given subject. We also noted that the comparatively lower stiffness values derived using models from observer I (red circles in Fig. 4b) were generally associated with comparatively higher LVMIs (Fig. 2), and vice versa for observer III (green triangles in Fig. 4b and Fig. 2).

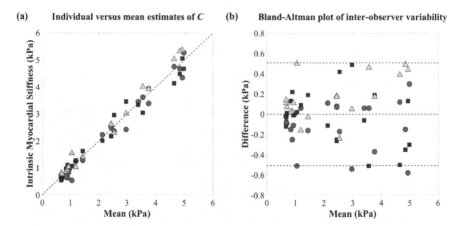

Fig. 4. Inter-observer variability in stiffness estimates of observer I (●), II (■) and III (▲): (a) individual versus mean estimates of C and (b) Bland-Altman plot. The equality line, mean and 95% limits of agreement are indicated using dotted lines. (Color figure online)

3.4 Study Limitations

There are a number of limitations to this study. First, diastasis was assumed to be a load-free state, which neglects effects of any non-zero residual pressures and delayed relaxation. The constitutive equation (Eq. 1) used in this study treats the myocardium as a transversely isotropic material, and we assumed that the microstructural orientations and the degree of anisotropy (controlled by the b parameters) were the same across all participants. Finally, several of the stiffness parameter estimation issues were related to challenges with image segmentation and construction of models suitable for biomechanical analyses [13]. Addressing these issues and analysing a larger number of cases are needed in order to investigate the correlation between parameter sensitivity and geometric variability.

4   Conclusion

This study examined the effects of inter-observer variation of LV geometric modelling on the estimation of passive myocardial stiffness. Our results show that inter-observer geometric variations can give rise to reasonably substantial differences in passive myocardial stiffness estimates in some cases, and relatively minor difference in other cases. This pilot study reinforces the need for objective and reliable cardiac segmentation and model building when analysing mechanical properties of the heart.

Acknowledgements. We gratefully acknowledge the financial support from the Health Research Council of New Zealand (grant 17/608).

References

1. Mangion, K., Gao, H., Husmeier, D., Luo, X., Berry, C.: Advances in computational modelling for personalised medicine after myocardial infarction. Heart **104**, 550–557 (2018)
2. Wang, V., Nielsen, P., Nash, M.: Image-based predictive modeling of heart mechanics. Ann. Rev. Biomed. Eng. **17**, 351–383 (2015)
3. Rumindo, G.K., Ohayon, J., Croisille, P., Clarysse, P.: In vivo estimation of normal left ventricular stiffness and contractility based on routine cine MR acquisition. Med. Eng. Phys. **85**, 16–26 (2020)
4. Feisst, A., et al.: Influence of observer experience on cardiac magnetic resonance strain measurements using feature tracking and conventional tagging. IJC Heart Vasculature **18**, 46–51 (2018)
5. Wang, Z.J., Wang, V.Y., Bradley, C.P., Nash, M.P., Young, A.A., Cao, J.J.: Left ventricular diastolic myocardial stiffness and end-diastolic myofibre stress in human heart failure using personalised biomechanical analysis. J. Cardiovas. Transl. Res. **11**, 346–356 (2018)
6. Young, A.A., Cowan, B.R., Thrupp, S.F., Hedley, W.J., Dell'Italia, L.: Left ventricular mass and volume: fast calculation with guide-point modeling on MR images. Radiology **216**, 597–602 (2000)
7. Wang, Z.: Left ventricular mechanics in human heart failure. Ph.D. thesis, The University of Auckland (2018)
8. Wang, V.Y., Lam, H., Ennis, D.B., Cowan, B.R., Young, A.A., Nash, M.P.: Modelling passive diastolic mechanics with quantitative MRI of cardiac structure and function. Med. Image Anal. **13**, 773–784 (2009)
9. Guccione, J.M., Costa, K.D., McCulloch, A.D.: Finite element stress analysis of left ventricular mechanics in the beating dog heart. J. Biomech. **28**, 1167–1177 (1995)
10. Wang, V.Y., Ennis, D.B., Cowan, B.R., Young, A.A., Nash, M.P.: Myocardial contractility and regional work throughout the cardiac cycle using FEM and MRI. In: Camara, O., Konukoglu, E., Pop, M., Rhode, K., Sermesant, M., Young, A. (eds.) STACOM 2011. LNCS, vol. 7085, pp. 149–159. Springer, Heidelberg (2012). https://doi.org/10.1007/978-3-642-28326-0_15
11. Luijnenburg, S.E., Robbers-Visser, D., Moelker, A., Vliegen, H.W., Mulder, B.J., Helbing, W.A.: Intra-observer and inter-observer variability of biventricular function, volumes and mass in patients with congenital heart disease measured by CMR imaging. Int. J. Cardiovascular Imaging **26**, 57–64 (2010)
12. Buechel, E.R.V., et al.: Remodelling of the right ventricle after early pulmonary valve replacement in children with repaired tetralogy of Fallot: assessment by cardiovascular magnetic resonance. Eur. Heart J. **26**, 2721–2727 (2005)
13. Xi, J., Lamata, P., Niederer, S., Land, S., Shi, W., Zhuang, X., Ourselin, S., Duckett, S.G., Shetty, A.K., Rinaldi, C.A.: The estimation of patient-specific cardiac diastolic functions from clinical measurements. Med. image Anal. **17**, 133–146 (2013)

A Computational Approach on Sensitivity of Left Ventricular Wall Strains to Fiber Orientation

L. Barbarotta🆔 and Peter H. M. Bovendeerd(✉)

Eindhoven University of Technology, Eindhoven, The Netherlands
{l.barbarotta,p.h.m.bovendeerd}@tue.nl

Abstract. In this work, we use a Finite Element model of the left ventricular (LV) mechanics to assess the sensitivity of strains to fiber orientation, modeled using both helix and transverse angles using a 5-parameter rule-based model. The ranges for the five parameters represent the variability within the human population, as inferred from DTI measurements found in literature. End-systolic Green-Lagrange strains with respect of the end-diastolic state were expressed according to the local circumferential (c), longitudinal (l) and transmural (t) direction. The results show that the transverse angle is more influential than the helix angle and that the strain components most affected by fiber orientation are E_{ct} (38 ± 18%), and E_{cl} (25 ± 8%). The most influential parameters are found to be t_u^0, describing the longitudinal offset of the transverse angle, and h_v^0, describing the transmural offset of the helix angle. Instead, h_v^1, describing the transmural slope of the helix angle, resulted to be the least influential parameter.

1 Introduction

Computational personalized cardiac medicine is based on data assimilation techniques to personalize both pathology-related and influential model parameters identifiable from data available in the clinic, such as strain measurements. In models of cardiac mechanics, fiber orientation is both an important determinant of tissue stress and strain [9,13] and a major source of uncertainty. Based on histological measurements, Streeter et al. [15] described fiber orientation through the helix angle, defining the base-to-apex component of the fiber vector (see Fig. 1a), and the transverse angle, describing the endocardium-to-epicardium component of the fiber vector. Findings in histological studies were later confirmed in studies using Diffusion Tensor Imaging (DTI) [8]. Lombaert et al. [10] built an atlas of myofiber orientation from a population of 10 ex-vivo healthy human hearts and provided an estimation of the distributions of the helix and the transverse angle within the human population. Typically, the measurement error of these angles is about 10° as measured by [8] and compatible with the one-standard deviation envelope reported in [10].

© Springer Nature Switzerland AG 2021
D. B. Ennis et al. (Eds.): FIMH 2021, LNCS 12738, pp. 296–304, 2021.
https://doi.org/10.1007/978-3-030-78710-3_29

Given the limited accuracy with which fiber orientation can be determined, it is important to quantify how the output of computer models of the mechanics of the heart is affected by this uncertainty. Most studies focused on sensitivity to fiber orientation during passive filling [12,16]. Geerts et al. [9] and Pluijmert et al. [13] addressed the sensitivity of end-systolic quantities to fiber orientation. Both studies suggest that fiber orientation plays a major role in the mechanics of the LV during the active phase. However, these studies used a local sensitivity approach without a proper quantification of the sensitivity. In fact, only perturbations of parameters around a single parameter combination point are considered, few parameter combinations are included, and only the average effect of the parameter perturbations on the output is computed. More recently, Campos et al. performed a sensitivity analysis of left ventricular function during the cardiac cycle [7]. They concluded that ventricular torsion is sensitive to fiber orientation, in line with earlier findings on torsion [1,2]. The study uses a global sensitivity approach, and considers sensitivity to a range of input parameters. However, it has a limited description of the fiber field and uses only one deformation parameter among the quantities of interest.

Considering the use of models in personalized medicine, it is important to consider myocardial strain as output quantity of interest: it is likely to contain important clinical information, as it is described by six components that vary both in time and location in the cardiac wall. Also, developments in MR tagging and ultrasound hold promise that strain can be obtained from the patient with increasing detail and accuracy. Thus, opportunities for successful model-based clinical decision support are increasing.

Therefore goal of this study is to systematically quantify the sensitivity of strain distributions to variation of the fiber field within the variability observed in the human population and to identify the most influential parameters in a five-parameter rule-based description of the fiber field. We focus on the spatial distribution of strain and assess strain at one moment in time only, the moment of end ejection.

2 Materials and Methods

2.1 Mathematical Model

We use a realistic geometry representing the average geometry from the data set we used in [4], that was derived from [11]. This geometry and its fiber field (Fig. 1c) are obtained by transforming a template geometry and its fiber field (Fig. 1b). Following the procedure in [4], we deflated the end-diastolic configuration until cavity volume was reduced to 40% of the end-diastolic volume. Fiber vectors were reoriented according to the deformation gradient tensor involved in the deflation process.

The orientation of fibers inside the template geometry is described by the helix angle (α_h) and the transverse angle (α_t), where α_h defines the component in the longitudinal-circumferential plane of the fiber unit vector, e_f, and α_t defines the component in the circumferential-transmural plane (Fig. 1a).

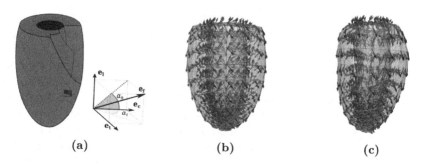

Fig. 1. Representation of the fiber orientation, e_f, in terms of helix, α_h, and transverse angle, α_t, (a). View of meshes and the central fiber orientation for the template geometry (b), and for the reconstructed unloaded geometry (c).

The functional description of α_h and α_t characterizes the fiber orientation inside the myocardium as follows

$$\alpha_h(u,v) = \left(h_v^0 + h_v^1 v\right)(1 + h_u^1 u), \tag{1}$$
$$\alpha_t(u,v) = \left(1 - v^2\right)\left(t_u^0 + t_u^1 u\right), \tag{2}$$

in which v denotes a normalized transmural coordinate, ranging from -1 at the endocardial surface to $+1$ at the epicardial surface. The normalized longitudinal coordinate u ranges from -1 at the apex, through 0 at the equator until $+0.5$ at the basal plane [4]. Here, for α_h, h_v^0 describes the transmural offset, h_v^1 describes the transmural slope, and h_u^1 describes the longitudinal variability. For α_t, t_u^0 describes the longitudinal offset ($t_u^0 = 0$ means that α_t is nil at equator), and t_u^1 describes the longitudinal slope of α_h.

Central values for α_h were obtained projecting the helix angle in Bovendeerd et al. [5] on the model in (1). Similarly, central values for α_t were obtained projecting the transverse angle reported in Geerts et al. [8] on the model in (2). For each parameter, we set a range around its central value to describe the interval in which the parameter will be varied during the sensitivity analysis.

We inferred the parameters ranges mainly from two studies: from Lombaert et al. [10], who used DTI to measure fiber fields in 10 ex-vivo healthy human hearts; and from Geerts et al. [8], who used DTI to measure fiber orientation in a population of 5 ex-vivo healthy goat hearts. These value are reported in Table 1.

Table 1. Central parameter combination $\pm$ parameter ranges.

α_h			α_t	
h_v^0	h_v^1	h_u^1	t_u^0	t_u^1
$20.43° \pm 8°$	$60° \pm 4°$	0 ± 0.25	$0° \pm 6°$	$9.96° \pm 8.67°$

The myocardium is described as a hyperelastic fiber-reinforced transversely isotropic active stress material [5]. Left ventricular wall mechanics was computed by solving for equilibrium between forces related to active stress, passive stress and cavity pressure as explained in [4].

2.2   FE Formulation and Simulations Performed

The LV was passively filled until a fixed preload of 1.5 kPa, entered the iso-volumic contraction phase during which active stress was generated and pressure increased, and ejected at a constant afterload level of 12 kPa. End-systolic Green-Lagrange strains with respect of the end-diastolic configuration were computed and expressed with respect to the wall-bound basis vectors e_c, e_l, and e_t (the circumferential, the longitudinal and the transmural directions, respectively Fig. 1a).

We used the same Finite Element implementation that we employed in [4]. The spatial discretization of the problem was set as a trade-off between accuracy and computational cost, leading to 5760 elements and 24519 degrees of freedom. Details about this choice can be found in [3]. The solver used our implementation of the Newton-Raphson algorithm combined with the LU decomposition implemented in the MUMPS library.

2.3   Sensitivity Analysis Approach

As in [4], we applied the elementary effects method [14]. This method relies on the definition of an elementary effect in a model with n parameters $p_1, \ldots, p_n$

$$EE_i\left(\boldsymbol{x}\right) = \frac{y\left(\boldsymbol{x}; p_1, \ldots, p_i + \Delta \cdot \Delta_{p_i}, \ldots, p_n\right) - y\left(\boldsymbol{x}; p_1, \ldots, p_i, \ldots, p_n\right)}{\Delta}, \quad (3)$$

where y is the Quantity Of Interest (QOI), $\boldsymbol{x}$ is the spatial coordinate, parameter p_i is perturbed by $\Delta \cdot \Delta_{p_i}$, with $\Delta \in [-1, 1]$ and Δ_{p_i} being the length of the interval in which the $i - th$ parameter is varied. The p_i parameters are the 5 fiber parameters ($n = 5$), each spanning a range of 1 SD around the mean. Using a sampling strategy that optimizes the spread of trajectories (see [14]), we generated 500 trajectories in a 4 level mesh built over the parameter space. The QOIs are the six end-systolic strain components, referred to the state at end-diastole. Following [14], three sensitivity indices can be computed as statistics over the trajectories: μ_i, the average of EE_i; μ_i^*, the average of $|EE_i|$, and σ_i, the standard deviation of EE_i. μ_i gives an overall estimation of the effect that a perturbation on parameter p_i has on the QOI, μ_i^* is used to detect whether variations on p_i are non-influential, and σ_i summarizes the ensemble effect of a parameter given by possible nonlinearities and interactions with other parameters. Note that, $EE_i\left(\boldsymbol{x}\right)$ and the sensitivity indices $\mu_i\left(\boldsymbol{x}\right)$, $\mu_i^*\left(\boldsymbol{x}\right)$, and $\sigma_i\left(\boldsymbol{x}\right)$ are scalar fields computed per-node ($\boldsymbol{x}$) of the FE discretization. Differently from [4], these indices are normalized as follows to be better interpreted as a percentage of a reference strain

$$\hat{\mu}_{i,\alpha\beta}(\boldsymbol{x}) = \frac{\mu_{i,\alpha\beta}(\boldsymbol{x})\,\Delta}{\|E_{\alpha\beta}^{ref}(\boldsymbol{x})\|_\infty}, \quad \hat{\mu}_{i,\alpha\beta}^*(\boldsymbol{x}) = \frac{\mu_{i,\alpha\beta}^*(\boldsymbol{x})\,\Delta}{\|E_{\alpha\beta}^{ref}(\boldsymbol{x})\|_\infty}, \quad \hat{\sigma}_{i,\alpha\beta}(\boldsymbol{x}) = \frac{\sigma_{i,\alpha\beta}(\boldsymbol{x})\,\Delta}{\|E_{\alpha\beta}^{ref}(\boldsymbol{x})\|_\infty}, \quad (4)$$

where the pair $\alpha\beta$ determines the strain component, and $E_{\alpha\beta}^{ref}$ is the strain computed in the central combination of the five fiber parameters.

Uncertainty is quantified by means of the Coefficient of Variation (CV) for each strain component and for each node $\boldsymbol{x}$. The CV is defined as the standard deviation of the strain component computed over all the M simulated fiber configurations normalized with respect of the norm of the average respective strain.

$$E_{\alpha\beta}^{avg}(\boldsymbol{x}) = \frac{1}{M}\sum_{j=1}^{M} E_{\alpha\beta}(\boldsymbol{x}; \boldsymbol{p}^j), \quad CV_{\alpha\beta}^{avg}(\boldsymbol{x}) = \sqrt{\frac{\sum_{j=1}^{M}\left(E_{\alpha\beta}(\boldsymbol{x}; \boldsymbol{p}^j) - E_{\alpha\beta}^{avg}(\boldsymbol{x})\right)^2}{(M-1)\|E_{\alpha\beta}^{avg}(\boldsymbol{x})\|_\infty^2}}, \quad (5)$$

where $\boldsymbol{p}^j$ represent the parameter combination associated with the $j-th$ simulated node in the parameter space.

3   Results

Figure 2 shows the histogram of the spatial distribution of strains CV. The norms used in its definitions were $\|E_{cc}^{avg}\|_\infty = 0.28$, $\|E_{ll}^{avg}\|_\infty = 0.26$, $\|E_{tt}^{avg}\|_\infty = 1.32$, $\|E_{cl}^{avg}\|_\infty = 0.19$, $\|E_{ct}^{avg}\|_\infty = 0.60$, and $\|E_{lt}^{avg}\|_\infty = 0.54$. For the circumferential, the longitudinal and the transmural strains, the average and the standard deviation are about $15\% \pm 7\%$ for E_{cc}, $22\% \pm 8\%$ for E_{ll}, and $17\% \pm 7\%$ for E_{tt}, respectively (see also blue bars in Fig. 4). For the circumferential-longitudinal, the circumferential-transmural and the longitudinal-transmural shear strains, the average and the standard deviation are $25\% \pm 8\%$ for E_{cl}, $38\% \pm 18\%$ for E_{ct}, and $21\% \pm 9\%$ for E_{lt}.

Figure 3 shows the mean over the nodes of the three sensitivity indices, $\hat{\mu}$, $\hat{\mu}^*$, and $\hat{\sigma}$, for the six end-systolic strains. Considering the mean $\hat{\mu}^*$, t_u^0 is the most influential. Variation of t_u^0 affects 6 out 6 strain components beyond 10% of which E_{ct} and E_{lt} beyond 20%. h_v^0 is the second most influential. Variation of h_v^0 affects 4 out of 6 strain components beyond 10%. h_u^1, t_u^1, and h_v^1 are the least influential. Variations of h_u^1 and t_u^1 affect only one strain component above 10% (E_{ll} and E_{ct}, respectively). Variation of h_v^1 does not affect any of the strain components above 10%.

Figure 3 also shows that the mean $\hat{\mu}$ is much smaller than the mean $\hat{\mu}^*$ for most of the parameter-strain combinations, indicating a symmetric variation of the strain. Exceptions to this come from parameter t_u^0 that averagely decreases E_{ct} by 35% of the reference value and from parameter h_v^0 that on average increases E_{cl} by 18% of the reference value. The mean $\hat{\sigma}$ ranges between 50% and 100% of the mean $\hat{\mu}^*$ for most of the parameter-strain combinations. The combinations of parameter-strain component for which the ratio between the mean $\hat{\sigma}$ and the mean $\hat{\mu}^*$ is below 50% are: h_v^0 for E_{cc} and E_{cl}, t_u^0 for E_{ct} and

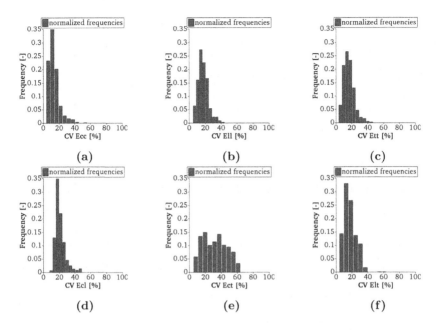

Fig. 2. Histogram of the coefficient of variation of the spatial distribution of strains. In the top row, from left to right the circumferential, longitudinal, and transmural strains (normal strains). In the bottom row, from left to right the circumferential-longitudinal, circumferential-transmural, and longitudinal-transmural strains (shear strains).

E_{lt}, and h_v^1, and h_u^1 for E_{ll}. For these latter combinations, the level of interactions is more limited.

We note that the bar plots in Fig. 3 are not directly comparable to those in [4] because of the different normalization used for the sensitivity indices. A direct comparison can be found in [3].

4   Discussion

Our results indicate that end-systolic shear strains are more sensitive to fiber orientation than normal strains. Among the shear strains, E_{ct} and E_{cl} are the most affected (Fig. 2).

The sensitivity indices indicate that none of the parameters shows a total lack of influence on end-systolic strains, but they allow to prioritize parameters by their effect on strains. t_u^0, determining the longitudinal position at which α_t is zero, and h_v^0, determining α_h at midwall at the equator, are the most influential, affecting 6 and 4 out of the 6 strain components beyond 10%, respectively. t_u^1, determining the base-to-apex range of α_t at midwall, and h_u^1, determining the longitudinal variation of α_h, are less influential, since affect only a single strain component beyond 10% (E_{ct}, and E_{ll}, respectively). h_v^1, determining the

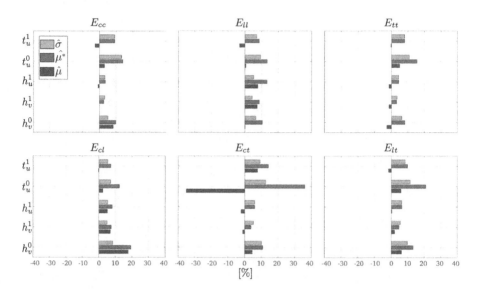

Fig. 3. Mean value of the spatial distribution of the sensitivity indices $\hat{\mu}$, $\hat{\mu}^*$, and $\hat{\sigma}$.

transmural slope of α_h, does not affect any of the strains beyond 10% and it is the least influential parameter among those considered. The overall high values of $\hat{\sigma}$ highlight the presence of interactions between parameters, which must be addressed with further investigations (see [3]).

The mechanism through which the most influential parameters h_v^0 and t_u^0 affect shear strains E_{cl} and E_{ct} has been explained before [1,2,6]. E_{cl} is related to LV twist, the difference in rotation of base and apex about the LV long axis, and LV torsion, twist per unit of LV length. It is the nett effect of the shortening action during systole of oblique fibers in the subendocardial layers, with a helix angle around $+45°$, and in the subepicardial layers, with a helix angle around $-45°$, during systole. Subendocardial fibers tend to rotate the apex in clockwise direction with respect to the base, when observing the LV in an apical view. Subepicardial fibers tend to rotate the apex in the opposite direction. The transmural offset parameter for the helix angle, h_v^0, affects the balance between oblique subendocardial and epicardial fibers, and hence it affects torsion and E_{cl}. The same mechanism also affects transmural shear strain E_{ct}, related to endo-to-epi differences in rotation. In addition this strain component is determined by the transmission of force in between the subendocardial and subepicardial layers, which is affected by the transmural component of the fiber vector α_t. The sign of α_t determines the sign of the relative rotation between subendocardial and subepicardial layers [5]. As the parameter t_u^0 determines the longitudinal level in the LV wall where α_t equals zero, eventually it also affects the spatial distribution of E_{ct}.

In our study, we represented strain components with respect to a wall-bound coordinate system $\{e_t, e_c, e_l\}$, as shown in Fig. 1. In clinical measurements,

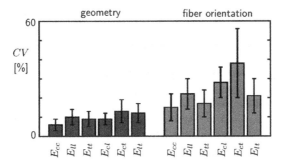

Fig. 4. Mean of the spatial distribution of CV (bars) ± SD (error bars) for geometry (red, results from [4]) and fiber orientation (blue). (Color figure online)

strains are often represented with respect to a cylindrical basis $\{e_r, e_c, e_z\}$, because they are derived from short-axis images. These bases coincide at the equatorial level, but they become increasingly different with increasing distance from the equator. We prefer the wall-bound system, since there the strain component E_{tt} is associated always with local wall thickening or thinning. In contrast, the cylindrical component E_{rr} can be related to change in wall thickness only at the equatorial level; outside this level it contains a combination of longitudinal and transmural strains.

Besides fiber orientation, also geometry is an important input of patient-specific models of cardiac mechanics. In a previous study, we assessed sensitivity of wall strain to geometry [4]. In Fig. 4 we combine the results. It can be observed that the uncertainty propagated from fiber orientation to end-systolic strain is at least twofold that propagated from the geometry. Current models tend to focus on patient-specific geometry and often use a generic rule-based fiber field, typically modeled using h_v^1 only and neglecting the transverse angle, the angle that was found to cause a large part of the uncertainty in our study. Our results suggest that a precise prediction of end-systolic strains may be hindered by the lack of information concerning fiber orientation, even when using a personalized geometry.

5   Conclusions

In this work we assessed the sensitivity of LV end-systolic wall strains to LV fiber orientation varied according to ranges inferred from data found in literature. We concluded that the strain components most affected by fiber orientation are two shear strains: the circumferential-transmural shear strain E_{ct} and the circumferential-longitudinal shear strain E_{cl} We found t_u^0, describing the longitudinal offset of the transverse angle, and h_v^0, describing the transmural offset of the helix angle, to be the most influential among all those considered while h_v^1, describing the transmural slope of the helix angle, was the least influential. Meaningful interactions between parameters are present and need to be investigated further.

References

1. Arts, T., Meerbaum, S., Reneman, R.S., Corday, E.: Torsion of the left ventricle during the ejection phase in the intact dog. Cardiovas. Res. **18**(3), 183–193 (1984)
2. Arts, T., Veenstra, P.C., Reneman, R.S.: Epicardial deformation and left ventricular wall mechanisms during ejection in the dog. Am. J. Physiol. Heart Circ. Physiol. **243**(3), H379–H390 (1982)
3. Barbarotta, L.: Towards computer assisted cardiac medicine: sensitivity analysis and data assimilation in models of left ventricular mechanics. Ph.D. thesis, Department of Biomedical Engineering, Eindhoven University of Technology, Eindhoven, The Netherlands (2021)
4. Barbarotta, L., Bovendeerd, P.: A computational approach on sensitivity of left ventricular wall strains to geometry. In: Coudière, Y., Ozenne, V., Vigmond, E., Zemzemi, N. (eds.) FIMH 2019. LNCS, vol. 11504, pp. 240–248. Springer, Cham (2019). https://doi.org/10.1007/978-3-030-21949-9_26
5. Bovendeerd, P.H.M., Kroon, W., Delhaas, T.: Determinants of left ventricular shear strain. Am. J. Physiol. Heart Circ. Physiol. **297**(3), H1058–H1068 (2009)
6. Bovendeerd, P., Huyghe, J., Arts, T., van Campen, D., Reneman, R.: Influence of endocardial-epicardial crossover of muscle fibers on left ventricular wall mechanics. J. Biomech. **27**(7), 941–951 (1994)
7. Campos, J.O., Sundnes, J., Dos Santos, R.W., Rocha, B.M.: Uncertainty quantification and sensitivity analysis of left ventricular function during the full cardiac cycle. Phil. Trans. R. Soc. A **378**(2173), 20190381 (2020)
8. Geerts, L., Bovendeerd, P.H.M., Nicolay, K., Arts, T.: Characterization of the normal cardiac myofiber field in goat measured with MR-diffusion tensor imaging. Am. J. Physiol. Heart Circ. Physiol. **283**(1), H139–H145 (2002)
9. Geerts, L., Kerckhoffs, R., Bovendeerd, P., Arts, T.: Towards patient specific models of cardiac mechanics: a sensitivity study. In: Magnin, I.E., Montagnat, J., Clarysse, P., Nenonen, J., Katila, T. (eds.) FIMH 2003. LNCS, vol. 2674, pp. 81–90. Springer, Heidelberg (2003). https://doi.org/10.1007/3-540-44883-7_9
10. Lombaert, H., et al.: Human atlas of the cardiac fiber architecture: study on a healthy population. IEEE Trans. Med. Imaging **31**(7), 1436–1447 (2012)
11. Medrano-Gracia, P., et al.: Left ventricular shape variation in asymptomatic populations: the multi-ethnic study of atherosclerosis. J. Cardiovasc. Magn. Reson. **16**(1), 56 (2014)
12. Nikou, A., Gorman, R.C., Wenk, J.F.: Sensitivity of left ventricular mechanics to myofiber architecture: a finite element study. Proc. Inst. Mech. Eng. [H] **230**(6), 594–598 (2016)
13. Pluijmert, M., Delhaas, T., de la Parra, A.F., Kroon, W., Prinzen, F.W., Bovendeerd, P.H.M.: Determinants of biventricular cardiac function: a mathematical model study on geometry and myofiber orientation. Biomech. Model. Mechanobiol. **16**(2), 721–729 (2016). https://doi.org/10.1007/s10237-016-0825-y
14. Saltelli, A.: Global Sensitivity Analysis: The Primer. Wiley, Hoboken (2008)
15. Streeter, D.D., Jr., Spotnitz, H.M., Patel, D.P., Ross, J., Jr., Sonnenblick, E.H.: Fiber orientation in the canine left ventricle during diastole and systole. Circ. Res. **24**(3), 339–347 (1969)
16. Wang, H., et al.: Structure-based finite strain modelling of the human left ventricle in diastole. Int. J. Numer. Meth. Biomed. Eng. **29**(1), 83–103 (2013)

A Framework for Evaluating Myocardial Stiffness Using 3D-Printed Heart Phantoms

Fikunwa O. Kolawole[1,2](✉) (iD), Mathias Peirlinck[2] (iD), Tyler E. Cork[1,3],
Vicky Y. Wang[1], Seraina A. Dual[1], Marc E. Levenston[1,2], Ellen Kuhl[2],
and Daniel B. Ennis[1,4] (iD)

[1] Department of Radiology, Stanford University, Stanford, CA 94305, USA
fikunwa@stanford.edu
[2] Department of Mechanical Engineering, Stanford University, Stanford, CA 94305, USA
[3] Department of Bioengineering, Stanford University, Stanford, CA 94305, USA
[4] Division of Radiology, VA Palo Alto Health Care System, Palo Alto, CA 94304, USA

Abstract. MRI-driven computational modeling is increasingly used to simulate *in vivo* cardiac mechanical behavior and estimate subject-specific myocardial stiffness. However, *in vivo* validation of these estimates is exceedingly difficult due to the lack of a known ground-truth *in vivo* myocardial stiffness. We have developed 3D-printed heart phantoms of known myocardium-mimicking stiffness and MRI relaxation properties and incorporated the heart phantoms within a highly controlled MRI-compatible setup to simulate *in vivo* diastolic filling. The setup enables the acquisition of experimental data needed to evaluate myocardial stiffness using computational constitutive modeling: phantom geometry, loading pressures, boundary conditions, and filling strains. The pressure-volume relationship obtained from the phantom setup was used to calibrate an *in silico* model of the heart phantom undergoing simulated diastolic filling. The model estimated stiffness was compared to a ground-truth stiffness obtained from uniaxial tensile testing. Ultimately, the setup is designed to enable extensive validation of MRI and FEM-based myocardial stiffness estimation frameworks.

Keywords: Computational modeling · 3D printed heart phantom · *In vitro* MRI · Material stiffness estimation

1 Introduction

Heart failure (HF) is a condition in which the heart is unable to meet the metabolic demands of the body. The US public health burden of heart failure is expected to grow significantly in the next decade with the prevalence projected to reach 8 million by 2030 [1]. As HF often occurs because of deteriorating cardiac function due to persistent remodeling, pathophysiological cardiac remodeling has been identified as a therapeutic target in HF [2]. Thus, understanding the various mechanisms and manifestations of remodeling is fundamental for formulating appropriate clinical intervention [3]. A significant consequence of cardiac remodeling is changes to the passive (diastolic) stiffness

© Springer Nature Switzerland AG 2021
D. B. Ennis et al. (Eds.): FIMH 2021, LNCS 12738, pp. 305–314, 2021.
https://doi.org/10.1007/978-3-030-78710-3_30

of ventricular myocardium. Measuring *in vivo* passive myocardial stiffness requires a comprehensive evaluation of the mechanical behavior (stress-strain) in diastole.

Passive stiffness of the left ventricle (LV) is commonly inferred from the LV end diastolic pressure-volume relationship (EDPVR) [4]. However, EDPVR only provides a global estimation of the apparent LV chamber stiffness. It is therefore inappropriate to infer intrinsic myocardial mechanical behavior from EDPVR. Cardiac Magnetic Resonance Elastography (MRE) has been used for direct measurement of myocardial shear stiffness [5], but its implementation requires assumptions that make solutions challenging for *in vivo* cardiac physiology.

Computational modeling using MRI data and Finite Element Modeling (FEM) enables the estimation of *in vivo* myocardial passive stiffness [6]. Provided ventricular geometry, microstructural organization, boundary conditions, kinematics from MRI, and endocardial filling pressures from catheterization, continuum balance laws and optimization techniques can be leveraged in FEM to inversely obtain the parameters of the constitutive model governing the material's mechanical behavior. Currently, however, validation of MRI and FEM-based myocardial stiffness estimation frameworks is exceedingly difficult *in vivo* as ground-truth *in vivo* myocardial stiffness remains elusive.

We have developed a 3D-printed heart phantom with a myocardium-mimicking material of known stiffness and MRI properties. The phantom was incorporated within an MRI compatible *in vitro* diastolic filling setup. We estimated stiffness of the heart phantom using an MRI-derived computational model and compared with ground-truth measures obtained from uniaxial tensile testing.

2 Methods

2.1 Phantom Development and Material Characterization

High-resolution T1-weighted images from a healthy *ex vivo* porcine heart (restored to *in vivo* mid-diastasis geometry) [7] were used to generate a 3D geometric heart model. The heart phantoms were subsequently manufactured using a lost-wax casting technique which has been used to manufacture anatomically detailed heart phantoms [8]. The epicardial surface of the 3D geometric heart model was used to create a negative epicardial mold. The ventricular blood pool segmentations were used to create LV and right ventricle (RV) blood pool casts. The epicardial mold and blood pool casts were converted into stereolithography files and 3D printed (Ultimaker 3 Extended) using tough polylactic acid and water-soluble polyvinyl alcohol, respectively. The heart phantom was cast using a silicone elastomer blend containing mass ratio of Sylgard 184:527 (Dow Corning) of 1:4 (Fig. 1). The suitability of different Sylgard blends was assessed, and the tissue-mimicking material was chosen based on mechanical and MR relaxation properties. Heart phantoms were cast by curing the Sylgard blend in the 3D printed mold for 48 h at room temperature. Ventricular basal ports and an apical anchor were added to the heart phantom to facilitate loading and motion stabilization.

Uniaxial tensile testing samples of the Sylgard blend were produced in parallel with the phantom development. Samples were punched out (ASTM cutting die A) from cured sheets (thickness 3.13 ± 0.11 mm) and three samples were mechanically tested according to the ASTM D412 standard using an Instron 5848 Microtester (100N load cell). Samples

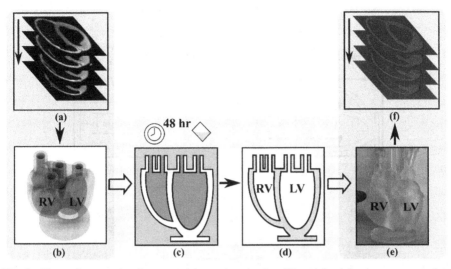

Fig. 1. Heart phantom development. (a) *ex vivo* porcine T1-weighted images segmented to develop a geometric model. (b) model fitted with ports and apical anchor. (c) Sylgard blend cured in mold for 48 h. (d) epicardial mold (pink) removed and ventricular blood-pool casts (yellow) dissolved in water. (e) subject-specific heart phantom with ports. (f) phantom 3D-SPGR images. (Color figure online)

were mounted by first clamping the specimen to the upper grip, zeroing the load cell, then clamping the specimen to the lower grip while taking care to avoid load application. The extensometer was then clamped to the samples with the gauge length set at 50.8 mm. The test was performed at ambient conditions with a strain rate of 500 mm/min. The Elastic Modulus was evaluated from a linear regression of the stress-stretch relation at stretches from 1.0 to 1.2. MRI relaxation properties were measured using T1-mapping (MOLLI 5-3-3, spatial resolution $1.00 \times 1.00 \times 5.0$ mm^3) and T2-mapping (T2-prep FLASH; flip angle 12°; spatial resolution, $1.00 \times 1.00 \times 5.0$ mm^3).

2.2 *In Vitro* Diastolic Filling Setup

The final heart phantom was embedded in a flow loop (Fig. 2) controlled by an MRI-compatible linear motion stage (MR-1A-XRV2, Vital Biomedical Technologies). To simulate *in vivo* diastolic LV filling, the loop was designed to deliver a cardiac-like late-diastolic filling cycle in the LV (sinusoidal flow: 13 mL/cycle mean, 1 s period).

Each ventricle was connected in a closed loop and the fluid volume in each loop was fixed. The motion stage was used to deliver the sinusoidal flow into the LV through a fluid filled syringe within the LV loop. The RV was kept at a constant volume throughout LV filling. LV pressure was acquired continuously in PowerLab (ADInstruments) using MRI-compatible pressure transducers (Micro-Tip SPR 350S, Millar). LV filling volume was recorded from the motion stage and synchronized with pressure measurements. The phantom was fixed at the apex and basal ports.

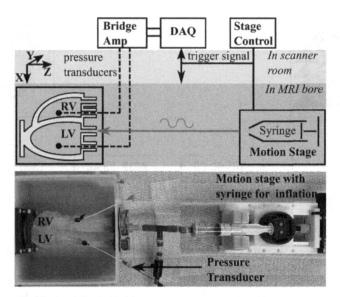

Fig. 2. Experimental setup. Schematic (top) and picture (bottom)

2.3 Image Acquisition and Processing

All *in vitro* images were collected on a 3T MRI (Skyra, Siemens) using a 32-channel chest and spine coil. In a static equilibrium phase, a 3D spoiled gradient recalled echo (3D SPGR; TE/TR = 2.17/5.5 ms; FA = 20°; isotropic 1.00 mm³; Static equilibrium phase) was performed over the entire volume of the phantom.

The 3D phantom geometric fidelity was evaluated using *in vitro* and *ex vivo* images semi-automatically segmented to extract binary masks of the myocardium, LV blood pool, and RV blood pool using Otsu's method and manual clean up (MITK Workbench). *In vitro* myocardium masks were registered to *ex vivo* myocardium using an automated 3D rigid regular-step gradient descent algorithm (Matlab, Mathworks). This transformation was applied to all *in vitro* binary masks. Volumetric accuracy and dice similarity coefficient (DSC) of binary masks were assessed by comparing the volume of the myocardium, LV blood pool, and RV blood pool from the *in vitro* images to *ex vivo* images.

2.4 *In Silico* Modeling and Stiffness Quantification

Starting from the 3D geometrical model (Sect. 2.1), a volumetric quadratic tetrahedral mesh with an average edge size of 1.5 mm was constructed using the 3-Matic meshing software suite (Materialise). For computational efficiency, the constrained apical anchor was removed from the mesh. The resulting mesh consisted of 118,337 nodes and 61,272 tet10 elements, summing up to 355,011 degrees of freedom to be solved in the finite element analysis software suite Abaqus (Dassault Systemes, Simulia Corp). Previous studies on mechanical characterization of Sylgard 184 and 527 indicate that the materials are nearly incompressible, or weakly compressible ($v = 0.495$) [9]. Thus, we assumed near incompressibility in the Sylgard blend. Given that we are working in a lower strain regime (strains under 40%), the Sylgard silicone elastomer blend was assumed to be linearly elastic [10] (consistent with our benchtop data). In accordance with the experimental setup, the computational heart phantom was kinematically constrained by constraining the apical bottom surface and the top surfaces of the ventricular basal ports. The port openings were virtually closed off to create enclosed fluid cavities, allowing an efficient computation of the temporal pressure and volume evolutions in the left and right ventricular 'blood' pools. Using a volume-driven boundary condition on the left ventricular fluid cavities, the deformation of the phantom following the sinusoidal inflow and outflow was virtually simulated. Similarly, for the RV fluid cavity a volume-driven boundary condition enforced the RV volume to remain constant, in accordance with experimental conditions. To include large-displacement effects, our simulation was set up to account for geometric nonlinearity.

The resulting pressure field was used to compute the elastic stiffness of the 3D casted Sylgard blend. More specifically, we found the Young's Modulus (E) using Abaqus as the forward solver wrapped inside Python's Nelder-Mead optimization algorithm [11, 12]. Starting from an initial value (E = 150 kPa), we computed the mismatch between the experimentally measured and simulated LV pressure evolution, and iteratively updated the Young's Modulus until this error was minimized.

3   Results

3.1   Phantom Geometric Accuracy

Analyses of the 3D *in vitro* heart phantom and *ex vivo* porcine subject MR images showed that the phantom development procedure adequately reproduces the porcine subject heart geometry. The *in vitro* and *ex vivo* images were all well registered.

In general, qualitatively, images of slices of the heart phantom and porcine subject heart were well matched. Qualitative comparison of the porcine subject and heart phantom images at a midventricular slice is shown (Fig. 3). Table 1 reports quantitative measures of the geometric fidelity of the heart phantom compared to the porcine subject heart. DSC is reported for the LV blood pool, RV blood pool, and myocardium. Geometric fidelity was also assessed by comparing the LV blood pool, RV blood pool, and myocardium volumes between the phantom and the swine subject.

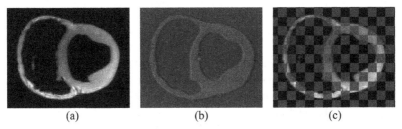

(a)   (b)   (c)

Fig. 3. Short-axis slice. (a) T1-weighted *ex vivo* porcine subject (b) 3D SPGR *in vitro* heart phantom (c) qualitative checkerboard of *ex vivo* and *in vitro*

Table 1. Geometry comparison between *in vitro* (phantom) and *ex vivo* (porcine subject)

Tissue	DSC	*In vitro* volume	*Ex vivo* volume	Volume error
LV blood pool	0.92	45.5 mL	43.9 mL	3.6%
RV blood pool	0.95	67.5 mL	68.9 mL	−2.0%
Myocardium	0.89	90.8 mm^3	93.0 mm^3	−2.4%

3.2 Uniaxial Tensile Testing

The Cauchy stress versus the stretch ratio for a representative sample of the Sylgard blend used to cast the heart phantom is shown in Fig. 4a. The elastic modulus was calculated for the three samples by fitting a linear model to the stress-stretch data for stretches between 1.0 and 1.2 and estimated to be 235 ± 6 kPa (mean ± SD). T1 and T2 times were 959.5 ± 5.0 ms and 313.8 ± 12.3 ms, respectively.

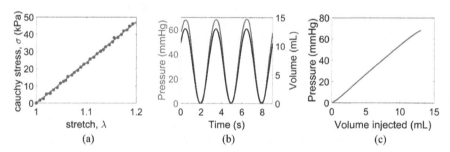

Fig. 4. (a) Cauchy stress vs stretch. (b) Pressure & volume vs time (c) *P-V* experimental

3.3 Pressure-Volume

Recorded pressures were sinusoidal (mean 34 mmHg, min. 0 mmHg, max. 68 mmHg) as were the volumes (peak 13 mL; mean 6.5 mL) (Fig. 4). The LV pressure-volume (*P-V*) relation was obtained after the loading and unloading cycles were steady and repeatable. The loading *P-V* curve was used for the *in silico* stiffness calibration.

3.4 *In Silico* Stiffness Calibration

Figure 5a shows the computed deformation and the spatial variation in maximum principal stretch state at the maximal loaded volume during the sinusoidal loading protocol. The maximal stretch in the phantom is located at the endocardial LV surface, reaching maximum principal stretches up to 1.2. The *in silico* stiffness calibration is shown in Fig. 5b. Starting from an initial elastic modulus of 150 kPa, the elastic modulus was iteratively updated, minimizing the pressure differences between the computational (line plot) and the experimental (x-markers) results. The line plot represents the converged forward simulation from which an elastic modulus of 328 kPa was obtained.

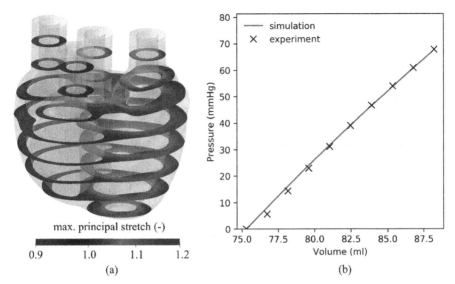

Fig. 5. (a) Computed phantom deformation at maximal inflation. Short-axis slices depict the spatially varying stretch of the Sylgard blend. (b) Calibrated elastic modulus of 328 kPa.

4   Discussion

To validate our MRI and FEM-based stiffness estimation framework, we have developed subject-specific heart phantoms with passive ventricular myocardium-mimicking mechanical and MRI relaxation properties. The heart phantom adequately replicates the porcine subject geometry as shown by the reported high DSC between registered 3D SPGR *in vitro* (heart phantom) and the *ex vivo* (porcine subject, geometrically restored to *in vivo* mid-diastasis) images.

The properties of the material used to construct the phantom were comparable to myocardium in some respects, but lacked some key features. Most notably, unlike myocardium, which is anisotropic and nonlinear, the phantom material is isotropic and linear in the strain range considered. These fundamental differences make it difficult to directly compare mechanical properties of the phantom material to *in vivo* myocardium. By comparison, Sommer *et al.* in their work on *ex vivo* mechanical characterization of myocardium obtained maximum Cauchy stresses in the fiber direction of about 8 kPa at stretch of 1.1 [13]. Our material, on the other hand, exhibited Cauchy stress of about 22 kPa at the same stretch. Although Sylgard blends with more Sylgard527 are softer, this blend was chosen for its far superior workability. Additionally, the T1 relaxation time of the Sylgard blend was identified as 959.9 ms at 3T compared with reported myocardium T1 times of 1158.7 ms at 3T [14]. The T2 relaxation time, which is less important for our study was however, significantly higher than that of myocardium at 313.8 ms compared with 45.1 ms [14].

Although we have yet to replicate *in vivo* myocardial anisotropy and nonlinearity in the heart phantom, our main objective, which is to validate MRI and FEM-based methods for estimating myocardial passive stiffness, does not require the phantom material properties to match perfectly with the properties of passive myocardium. This project aims to understand the ability of an MRI and FEM-based framework to reproduce accurate and precise estimates of the stiffness of soft heart phantoms under loading conditions similar to *in vivo* hearts in diastole. Necessary for the project are accurate quantification of ground-truth phantom material properties, and appropriate reproduction of experimental loading and boundary conditions in the computational model of simulated inflation. Validating the stiffness estimation framework in a controlled system, with isotropic, linear, heart shaped phantoms will build confidence in stiffness estimates of more complex materials like *in vivo* myocardium.

We estimated the phantom stiffness in an *in silico* model by matching the simulated compliance-pressure curve to the *in vitro* experimental *P-V* relation obtained from the heart phantom diastolic filling setup. The simulated solution converged to an elastic modulus of 328 kPa compared with the phantom material stiffness ground-truth of 235 ± 6 kPa obtained through tensile testing. A good match between experimental and simulated pressure-volume points was recorded, with some minor mismatch in the low-pressure regime which might suggest that the experimental pressure volume points do not form a perfect linear relationship. At this point, we saw no reason to assign additional weight to the pressure-volume points in the low-pressure regime to improve the fit. Palchesko *et al.* characterized blends of Sylgard 184 and 527 [15]. For our blend, their experimentally derived relationship between mass composition of a blend and its elastic modulus gives a stiffness of 215 kPa, comparing favorably with results from our tensile testing (235 kPa). This, and our measurement data, suggests that our stiffness simulation is overestimating the material stiffness.

The discrepancy between simulated and benchtop estimate of phantom mechanical properties could be due to inaccuracies in the *P-V* used for stiffness calibration. The connection between the syringe and the phantom has inherent compliance that is not currently modelled by our simulation. Hence, it is possible the experimental pressure-volume relationship overestimates the volume injected into the LV. Future work aims to eliminate this effect by directly obtaining the ventricular kinematics and volumes from MRI images, as opposed to obtaining the filling volumes from the inflation unit. Future studies will also use MRI tagging or displacement encoding to reconstruct local phantom displacements during the filling cycle. The filling pressures and kinematics will be used for *in silico* stiffness calibration, providing more constraints for stiffness estimation, thus improving the accuracy of simulated phantom stiffness.

It is also possible that the discrepancy is due to inaccurate replication of the effect of the RV in the simulation. In the simulation, the RV is modeled as a constant volume. The RV however is not closed off directly at the basal ports. Hence, it is possible that due to the compliance of the tubing connected to the phantom, some fluid volume escapes into the tubing from the RV during LV inflation. To more accurately model RV effects, future work will acquire RV and LV pressures simultaneously during inflation and use the RV pressure data to model the effect of the RV at the septal wall.

The discrepancy between simulated and benchtop phantom material stiffness may also be indicative of variations in mechanical properties between the heart phantom and tensile testing specimen. Although the heart phantom and tensile testing specimen were developed in parallel from the same batch, their mechanical properties may differ due to different curing conditions. The mechanical properties of Sylgard are sensitive to curing temperature and aging time between manufacture and testing [16]. This effect may be accelerated in thinner samples, which may explain variability in mechanical properties between the tensile testing samples and the heart phantom. To verify ground-truth phantom stiffness, in the future, testing strips could be cut out directly from the heart phantom after MRI and pressure data have been obtained.

In conclusion, we have developed an experimental setup for validating MRI and FEM-based myocardial passive stiffness in 3D-printed heart phantoms. The setup enables acquisition of MRI and pressure data necessary to estimate phantom material properties using an MRI and FEM-based constitutive modeling framework. The phantom development procedure effectively reproduces the subject geometry. The experimental setup will be refined to minimize discrepancy between the phantom mechanical properties and those of the tensile testing strips used to quantify ground-truth stiffness. The phantom stiffness can be tuned, and different subject-specific phantoms can be developed, enabling extensive quantification of the accuracy and repeatability of stiffness estimates obtained through our MRI and FEM-based stiffness estimation framework.

Acknowledgements. This work was supported by NIH R01 HL131823 to DBE.

References

1. Heidenreich, P.A., et al.: Forecasting the impact of heart failure in the united states. Circ. Heart Fail. **6**, 606–619 (2013)
2. Sakata, Y., Ohtani, T., Takeda, Y., Yamamoto, K., Mano, T.: Left ventricular stiffening as therapeutic target for heart failure with preserved ejection fraction. Circ. J. **77**, 886–892 (2013)
3. Peirlinck, M., et al.: Using machine learning to characterize heart failure across the scales. Biomech. Model. Mechanobiol. **18**, 1987–2001 (2019)
4. Burkhoff, D., Mirsky, I., Suga, H.: Assessment of systolic and diastolic ventricular properties via pressure-volume analysis: a guide for clinical, translational, and basic researchers. Am. J. Physiol. Heart Circ. Physiol. **289**, H501–512 (2005)
5. Khan, S., Fakhouri, F., Majeed, W., Kolipaka, A.: Cardiovascular magnetic resonance elastography: a review. NMR Biomed. **31**, (2018)
6. Wang, V.Y., Nielsen, P.M.F., Nash, M.P.: Image-based predictive modeling of heart mechanics. Annu. Rev. Biomed. Eng. **17**, 351–383 (2015)
7. Cork, T.E., Perotti, L.E., Verzhbinsky, I.A., Loecher, M., Ennis, D.B.: In: Coudière, Y., Ozenne, V., Vigmond, E., Zemzemi, N. (eds.) Functional Imaging and Modeling of the Heart 2019. LNCS, vol. 11504, pp. 177–186. Springer, Cham (2019)
8. Dual, S.A., et al.: Ultrasonic sensor concept to fit a ventricular assist device cannula evaluated using geometrically accurate heart phantoms. Artif. Organs **43**, 467–477 (2019)
9. Schneider, F., Draheim, J., Kamberger, R., Wallrabe, U.: Process and material properties of polydimethylsiloxane (PDMS) for optical MEMS. Sens. Actuators A Phys. **151**, 95–99 (2009)
10. Johnston, I.D., McCluskey, D.K., Tan, C.K.L., Tracey, M.C.: Mechanical characterization of bulk Sylgard 184 for microfluidics and microengineering. J. Micromech. Microeng. **24**, (2014)

11. Peirlinck, M., et al.: Kinematic boundary conditions substantially impact in silico ventricular function. Int. J. Numer. Methods Biomed. Eng. **35**, (2019)

12. Jones, E., Oliphant, T., Peterson, P.: SciPy: open source scientific tools for python (2001)

13. Sommer, G., et al.: Biomechanical properties and microstructure of human ventricular myocardium. Acta Biomater. **24**, 172–192 (2015)

14. von Knobelsdorff-Brenkenhoff, F., et al.: Myocardial T1 and T2 mapping at 3 T: reference values, influencing factors and implications. J. Cardiovasc. Magn. Reson. **15**, 53 (2013)

15. Palchesko, R.N., Zhang, L., Sun, Y., Feinberg, A.W.: Development of polydimethylsiloxane substrates with tunable elastic modulus to study cell mechanobiology in muscle and nerve. PLoS ONE **7**, (2012)

16. Hopf, R., Bernardi, L., Menze, J., Zündel, M., Mazza, E., Ehret, A.E.: Experimental and theoretical analyses of the age-dependent large-strain behavior of Sylgard 184 (10:1) silicone elastomer. J. Mech. Behav. Biomed. Mater. **60**, 425–437 (2016)

Modeling Patient-Specific Periaortic Interactions with Static and Dynamic Structures Using a Moving Heterogeneous Elastic Foundation Boundary Condition

Johane Bracamonte[1] ⓘ, John S. Wilson[2], and Joao S. Soares[1](✉) ⓘ

[1] Department of Mechanical and Nuclear Engineering, Virginia Commonwealth University, Richmond, VA, USA
jsoares@vcu.edu
[2] Department of Biomedical Engineering and Pauley Heart Center, Virginia Commonwealth University, Richmond, VA, USA

Abstract. Perivascular support and tethering are likely relevant factors in vascular mechanics and one of the possible causes of deformational heterogeneity of the aortic wall. Besides these effects, the thoracic aorta interacts with the heart and other moving tissues and organs. We propose a generalized approach to model the effect of aortic interactions with static and dynamic perivascular structures. Periaortic interactions are modeled as a heterogeneous Elastic Foundation Boundary Condition (EFBC). This is implemented in the Finite Element model as a collection of unidimensional springs attached to the adventitial surface and a movable opposite end. An optimization algorithm iterates over the material constants and EFBC parameters to fit the simulated nodal displacements or the aortic wall to patient-specific DENSE MRI-derived displacements. We hypothesize that the adventitial load distribution that replicates the *in vivo* motion and deformation of the aorta is representative of the actual periaortic interactions. We study 3 aortic locations: the distal aortic arch, the descending thoracic aorta, and the infrarenal abdominal aorta. Our method reproduced the *in vivo* DENSE MRI-derived displacements with a median error below 30% of the pixel-size resolution (1.2–1.6 mm). The resulting average adventitial load is circumferentially and axially heterogeneous and ranged between 30 and 60% of the luminal pressure-pulse depending on the local nature of the periaortic interaction. Adequate modeling of periaortic interactions may bring a better understanding of its role in the normal and pathological function of the aorta *in vivo*.

Keywords: Periaortic interactions · Image-based modeling · Inverse FEM · DENSE MRI

1 Introduction

The *in vivo* deformation of the aortic wall is circumferentially heterogeneous and regionally dependent along its axis as has been shown by different image-based kinematic

© Springer Nature Switzerland AG 2021
D. B. Ennis et al. (Eds.): FIMH 2021, LNCS 12738, pp. 315–327, 2021.
https://doi.org/10.1007/978-3-030-78710-3_31

analyses [1, 2]. This heterogeneous deformation is likely influenced by the inflation of the aorta against the materially heterogeneous perivascular structures. Thus, perivascular support and tethering may play an important role on the *in vivo* function of the aorta. Nevertheless, these effects are often neglected by stiffness indices used in clinical practice [3] and by many analytical and numerical studies [4].

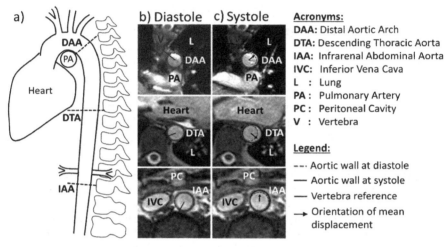

Fig. 1. a) Axial locations along the aorta under study. Aorta and periaortic structures over SSFP MRI b) at diastole c) at systole.

Previous FEM models have explored the effects of heterogeneous perivascular structures on aortic mechanics by attaching the adventitial surface to discrete computational domains of different properties [2, 5]. We recently proposed a method to model the effect of perivascular structures with a static elastic foundation boundary condition (EFBC) with heterogeneous stiffness. This allowed the patient-specific estimation of aortic wall properties and stiffness distribution of the periaortic structures at the infrarenal abdominal region [6]. However, all these approaches assume that the periaortic structures are static, and thus they fail to predict the mechanics of the thoracic aorta where the moving heart and the pulmonary artery (PA) interact with the aorta (Fig. 1). In this work, we develop a generalized model to account for the heterogeneous effect of interactions of the aortic wall with static and dynamic perivascular structures. To determine the stiffness of the moving periaortic structures, quantification of regional displacement fields from both the aortic wall and the moving structure are required, which is technically challenging. Instead, we aim here to model the interaction of the aortic wall with the perivascular structures through the means of the adventitial force distribution, and we estimate it such that the heterogeneous deformation of the aorta captured with MRI is achieved. For this, we prescribe a moving EFBC at the adventitial surface of the aortic wall as a collection of moving linear springs with uniform effective stiffness and heterogeneous spring elongation. We hypothesize that an optimal solution that mimics the *in vivo* deformation of the aorta will reproduce the adventitial loads equivalent to the aortic interactions with static and dynamic perivascular space.

2 Method

2.1 Patient-Specific Anatomical Models and Discretization

This study was performed using data from previously acquired cine steady-state free precession (SSFP) imaging and 2D displacement encoding with stimulated echoes (DENSE) MRI acquired on 27 healthy volunteers. At least one of three axial locations along the aorta were analyzed in each volunteer, the distal aortic arch (DAA), descending thoracic aorta (DTA) or infrarenal abdominal aorta (IAA) (Fig. 1a). SSFP images were acquired during a single breath-hold; DENSE MRI imaging was acquired during free-breathing with respiratory navigation. Full details of the MRI acquisitions can be found in Wilson et al. [7]. Diastolic images were segmented and smoothed to obtain anatomic models of the cross-section of the aortic wall. The patient-specific models were discretized into quadratic tetrahedral meshes. DENSE MRI derived displacements were estimated as previously described [6, 7]. A coordinate-based multiquadric interpolation was carried to estimate the target displacement of each FEM mesh node from the DENSE-MRI pixel-wise displacement data within a radius of three times the pixel-size, producing a smooth target displacement field.

2.2 Material Model

We assumed the aortic wall to be a homogeneous single layer with Fung's orthotropic pseudo-elastic constitutive equation as the material model, although other material models could be used to similar effect. Assuming orthotropic symmetry and negligible contributions of shear components in the axial direction, the resulting constitutive equation is given by

$$\Psi = \frac{c}{2}\left(e^{Q} - 1\right) + \frac{b}{2}(\ln J)^2 \tag{1}$$

$$Q = a_{rrrr}E_{rr}E_{rr} + a_{\theta\theta\theta\theta}E_{\theta\theta}E_{\theta\theta} + a_{rr\theta\theta}E_{rr}E_{\theta\theta} \tag{2}$$

Material parameters c and b with units of stress are related respectively to the stiffness and the bulk modulus. Non-dimensional constants a_{ijkl} are anisotropy parameters [8]. Nearly incompressible behavior was imposed by enforcing $b \gg c$. To enforce thermodynamic consistency and local convexity of the material model [9], material parameters were constrained by the following relations:

$$c > 0;\ a_{ijkl} > 0;\ a_{iiii} > a_{iijj};\ a_{iiii}a_{jjjj} > a_{ijkl}^2 \tag{3}$$

2.3 Boundary Conditions and Loads

An EFBC was applied to the outer wall as a distribution of radially-oriented linear springs. The constrained expansion process from diastole to systole was simulated in two quasi-static time-steps. In the first step, the spring ends were fixed and a uniform pressure load at the lumen was prescribed. Since patient-specific blood pressures were not recorded in the prior imaging dataset we set $p_l = 40$ mmHg as a representative magnitude for the pressure pulse increment from diastole to systole in the central arterial vasculature. In this step, the aorta expands into an approximately regular-shaped cylinder while compressing all the springs (Fig. 2a). In the second step, the distal ends of the springs were displaced on the radial direction to fit the aortic wall to its target shape and position; moving away from the centroid to relax the springs and decrease adventitial load, or moving in and pushing against the adventitial surface to increase the load (Fig. 2b). As a result, the moving EFBC can be regionally adjusted to reproduce the adventitial load distribution that deforms the anatomic model into the measured *in vivo* irregular geometry. Given the linearity of the springs, adjusting for their stiffness and displacement simultaneously is redundant. We fixed the uniform EFBC spring stiffness and used the heterogeneous spring-end displacement distribution as the fitting variable. Stiffness of each individual spring was calculated as a function of spring density such that the effective stiffness per unit length (κ) was proportional to the material constant c. The heterogeneous radial displacement (d_n) was set as a piece-wise constant function of a finite number of regions (n) depending on the circumferential position with smooth cubic transitions. The angles ϕ_n, representing the boundaries of each region, were measured from a line projected from the luminal centroid to the nearest point of the vertebra, to reduce dependence on minor differences in slice orientation and patient anatomy (Fig. 2). In this work we assumed the diastolic configuration as a stress-free reference.

Analytical analyses of pressurized vessels have shown a direct relation of the transmural pressure difference to the intramural stress distribution given by [10]

$$p_l - p_a = \int_{r_l}^{r_a} \left(\frac{\sigma_{rr} - \sigma_{\theta\theta}}{r} \right) dr, \tag{4}$$

and here, we report results as the ratio of the adventitial load per unit area to the prescribed luminal pressure pulse (p_a/p_l).

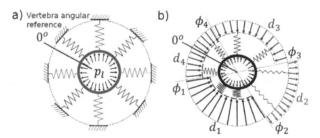

Fig. 2. Boundary conditions. a) Step 1: uniform luminal pressure and static EFBC. b) Step 2: uniform luminal pressure and moving EFBC.

2.4   Optimization Algorithm

Pixel-wise displacement information from DENSE-MRI was interpolated to the nodes of the FEM mesh as target data. An optimization algorithm, using the *Trust Region*

Reflective Method from *Python Scipy* library, was employed to iterate over material model constants (c, a_{rrrr}, $a_{\theta\theta\theta\theta}$, $a_{rr\theta\theta}$) and EFBC parameters (d_n, ϕ_n) to fit the simulated nodal displacement to the targeted displacement field. The optimization algorithm was set to minimize the squared nodal displacement error, defined as the magnitude of the vector difference of target to simulated displacement. Because the overall size of the aorta is different at each location and varies from patient to patient, nodal displacement error was normalized to the luminal radius at systole. The algorithm employed *FEBio 2.9* to solve the FEM forward problem for each iteration until a minimal error was found [11]. Each quasi-static step for the forward FEM problem was discretized in a minimum of 100 sub-steps. The convergence criteria for each simulation sub-step was reached when the iteration-to-iteration normalized error for the displacement and energy solutions dropped below a tolerance of 10^{-3}.

2.5 Parameter Sensitivity, Reproducibility, and Statistical Analysis

To better understand the relationship between predefined spring-stiffness, fitted spring displacement, and resulting adventitial load, we performed a model sensitivity analysis for the uniform EFBC stiffness per unit length (κ) and number of moving elastic regions (n). A preliminary screening revealed that the model is numerically unstable outside the range $10^{-7} \leq \kappa/c \leq 10^{-4}$. To explore the parameter sensitivity within the range of numerical stability, we run a total of 54 permutations of two levels of n and five levels of κ/c, distributed among nine cases, three from each aortic location. We conducted a standard Analysis of Variance (ANOVA) to stablish any significance influence of the fixed model parameters on estimations of material properties, nodal displacement error and adventitial load. Interobserver variability was explored by having a second independent observer segmenting the MRI images and generating the FEM mesh from other 3 random selected cases from each location group with $\kappa/c = 10^{-5}$ and $n = 8$. A paired Student t test was carried on the 9 cases to estimate the observer influence on geometric characteristics, material parameters, nodal displacement error and adventitial load. Statistical significance was set at 5% ($p^* = 0.05$).

3 Results

3.1 Parameter Sensitivity and Reproducibility

The proposed model requires setting two user-defined parameters, the number of moving elastic regions (n) and the relative stiffness of the elastic boundary (κ/c). The former discretizes the circumferential heterogeneity of the boundary condition and the latter modulates the ratio of boundary displacement d_n to adventitial load p_a. As expected, the optimization algorithm was able to minimize the normalized average or median errors for all cases and parameter combinations (Fig. 3a & Fig. 4a). However, we also analyzed the parameter effect on maximum error (excluding within-case outliers).

Increasing n from 4 to 8 significantly reduced the maximum error ($p^* = 0.04$) with no significant influence on any material properties or average adventitial load. Further increase of n can potentially improve the numerical fit although penalizing the computational expense by the increment of parameters to be optimized by a factor of 2.

The ANOVA revealed no significant differences on nodal displacement error or material properties within the range $10^{-6} \leq \kappa/c \leq 10^{-5}$ (Fig. 4b), suggesting that the model does not depend on κ/c in this range (Fig. 4c). The cases studied outside this range produced an increase of maximum displacement error and changes on the estimations of c and p_a/p_l, which suggest the existence of an optimal κ/c value. For illustrative purposes, best fit quadratic models with their respective confidence interval are shown in Fig. 4. When κ/c is increased to 10^{-4} springs are too stiff and the fitted end-displacements are amplified to large unphysiological sharp point-wise loads and deformations on the model, resulting in an EFBC that is locally unstable and prone to noise. On the other extreme, when $\kappa/c = 10^{-7}$, springs are too soft, rendering the EFBC ineffective, approximating the forward problem to a traction-free boundary condition, and resulting in cylinders of regular cross sections (Fig. 4a). Based on these results we selected the combination of $n = 8$ and $\kappa/c = 10^{-6}$ for the rest of the calculations, as a trade of in between reduction of nodal displacement error, numerical stability, and smoothness of the solution. The paired Student t analysis revealed no significant pairwise differences between observers' measurements of geometric characteristics, resulting material parameters, or average adventitial load (Table 1).

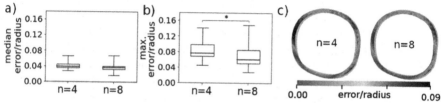

Fig. 3. Effect of number of elastic regions n with $\kappa/c = 10^{-6}$ on a) normalized median displacement error, b) maximum displacement error (without in-case outliers), c) nodal displacement error distribution.

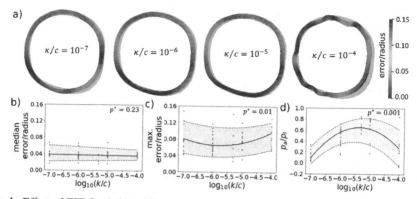

Fig. 4. Effect of EFBC relative stiffness (κ/c) on results a) Distribution of nodal displacement error for $n = 8$, b) median nodal displacement error, c) maximum nodal displacement error, and d) normalized adventitial load. Solid black lines represent a best fit model for the median value and the shaded area is the interval of confidence between the 5% and 95% limits.

3.2 Nodal Displacement Error

As a measure of the accuracy of the method, the displacement error for each node was calculated as the magnitude of the vector difference of the target displacement and the optimal solution displacement. The median error was below 0.4 mm for all locations which is approximately 30% of the average pixel size. This indicates reasonable success in reproducing the systolic geometry for different combinations of periaortic interactions (Fig. 5a). The magnitude of the error increased with the diameter and displacement magnitude; thus, arterial sections with smaller diameters and bulk displacements (such as the IAA) were associated with smaller errors (Fig. 5).

Table 1. Statistics of pairwise interobserver differences from 9 random cases.

Parameter	Interobserver pairwise difference (observer 1 – observer 2)			Difference confidence interval	
	Mean	St. Dev.	p value	Lower 95%	Upper 95%
Average luminal radius at diastole [mm]	0.5	0.36	0.21	– 0.33	1.35
$c[kPa]$	6.46	3.62	0.12	– 1.90	14.82
a_{rrrr}	1.13	1.01	0.29	– 1.20	3.46
$a_{\theta\theta\theta\theta}$	1.13	1.11	0.34	– 1.42	3.69
$a_{rr\theta\theta}$	1.46	1.18	0.25	– 1.24	4.17
Median normalized nodal displacement error	0.021	0.016	0.23	– 0.059	0.016
p_a/p_l	– 0.069	0.05	0.21	– 0.19	0.05

3.3 Displacements and Strains

The proposed method was successful in reproducing the qualitative features of the aortic wall displacement reported by DENSE MRI based kinematic analyses [1, 9]. In the IAA, the displacement is minimum along the posterior wall adjacent to the vertebra, but maximum on the opposite anterior wall facing the peritoneal cavity, with increased circumferential stretch in the lateral walls (Fig. 6c & 7c). For the DTA, results showed two different modes of net displacement. For most patients younger than 30 years of age, net deformation was directed towards the left posterior lung space and away from the heart's left ventricle. For these cases, the anterior wall adjacent to the left atrium of the heart is displaced inwardly toward the centroid, producing an overall rigid body displacement to the posterior left side (Fig. 6b).

The minimal radial displacement is near the vertebra; however, in contrast to the IAA, the circumferential displacement and strain peak near this location. For some other young patients and all older adults, maximum radial displacement is anterior towards the left ventricle and minimum at the vicinity of the vertebra (Fig. 7b). Displacements at the

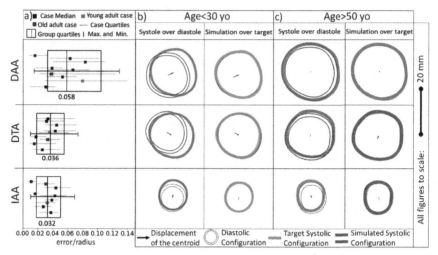

Fig. 5. a) Case specific and location group statistics for $\kappa/c = 10^{-6}$ and $n = 8$. Black markers: case-specific median nodal displacement. Gray whiskers: case-specific nodal displacement error quartiles. Green marker: median displacement error for representative old adult case. Blue marker: median nodal displacement error for representative young adult case Red box-plot: median, quartiles, and maximum and minimum nodal displacement error without outliers. Left column: target systolic (gray shade) over diastolic (black contour line) configuration. Right column: optimal solution (color shade) over target systolic (gray shade) configuration for b) representative young adult case of study, and c) representative old adult case of study. (Color figure online)

DAA were consistent among all patients and were on average oriented toward the left lung (Fig. 6a &7a). Radial displacement was negligible close to the vertebra and the PA and peaked adjacent to the lung. Regarding the circumferential strain fields, two groups could be differentiated depending on age. For patients >50 years of age, displacements and strains were smaller than for younger patients, with the former demonstrating peak strain between the vertebra and the wall adjacent to the PA, and the latter revealing peak strain along the greater curvature of the aorta [7].

3.4   Moving EFBC and Adventitial Loads

The circumferential distribution of EFBC displacements d_n showed good agreement with the location of anatomical features and their corresponding periaortic interactions. The EFBC moved in towards the wall to prevent radial displacements in the vicinity of stiff structures such as the vertebra in the $\phi_8 - \phi_1$ region for all cases (Figs. 6 & 7). Springs also moved inward to reproduce the interaction to other blood vessels depending on local mechanics, e.g. the displacements for the PA-DTA interaction (range $\phi_2 - \phi_4$ in Figs. 6a & 7a) were found larger than for the IVC-IAA interaction (range $\phi_6 - \phi_7$ in Figs. 6c & 7c).

Relatively soft tissues were modeled by either negligible or outward EFBC displacements (d_n), which allowed radial displacement of the aortic wall towards such regions. This was the case for the vicinity of the lungs at the DAA (range $\phi_4 - \phi_7$ in Figs. 6a

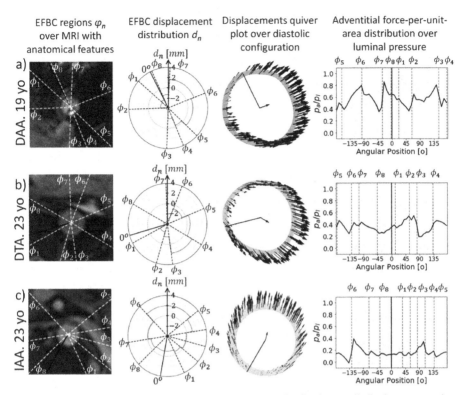

Fig. 6. EFBC locations over MRI, EFBC displacement distribution, wall displacement quiver plot, and adventitial pressure for a) DAA of a 19 yo, b) DTA of a 23 yo and c) IAA of a 23 yo.

& 7a) and DTA (range $\phi_3 - \phi_5$ in Figs. 6b & 7b), and the region corresponding to the peritoneal cavity at the IAA (range $\phi_3 - \phi_6$ in Figs. 6c & 7c). The two displacement modes at the DTA were reproduced by converging to different displacements for the regions adjacent to the heart (range $\phi_5 - \phi_8$ in Figs. 6b & 7b). For young adults, the displacement in this region was large and inward, producing a pushing effect that leads to the reported rigid body displacement to the posterior left side (Fig. 6b). For older adults, this region of the EFBC moved outward, reducing the adventitial resistance and leading to the displacement of the aortic wall in the anterior direction (Fig. 7b).

The distribution of p_a/p_l was consistent with the interactions of the aortic wall and the perivascular space. Adventitial force was higher where moving tissues displace the aorta (such as the heart and the PA) and rigid structures resist the radial displacement of the aortic wall. Lower p_a/p_l values were found around soft tissues where fitted spring-end displacement were either away from the aortic wall or close to zero, such as the lungs or the peritoneal cavity which offer little resistance to aortic wall displacement. Sharp changes in adventitial loads where found in transition regions between soft and rigid or moving perivascular structures. Results showed the adventitial load was larger on average at the DAA ($p_a/p_l \approx 0.6$) than at the DTA or IAA ($p_a/p_l \approx 0.3$).

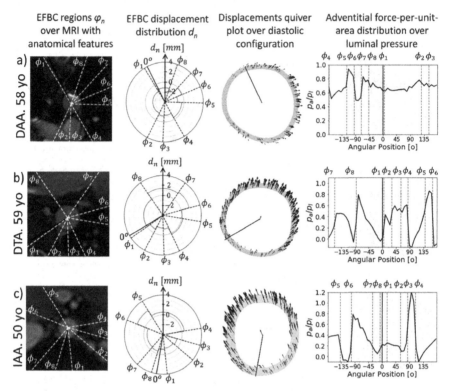

Fig. 7. EFBC locations over MRI, EFBC displacement distribution, wall displacement quiver plot, and adventitial pressure for a) DAA of a 58 yo, b) DTA of a 59 yo and c) IAA of a 50 yo.

3.5 Aortic Wall Properties

Significant proportional correlation of c to patient's age ($p^* = 0.04$) was found, which agrees with the expected increase of artery stiffness with age (Fig. 8a) [10]. No significant correlation was found between mechanical properties and location along the aorta. The circumferential anisotropic term ($a_{\theta\theta\theta\theta}$) was on average 60% larger than the radial term (Fig. 8b). The estimated material properties were in the same order of magnitude as those reported from experimental studies on excised human aortas [12].

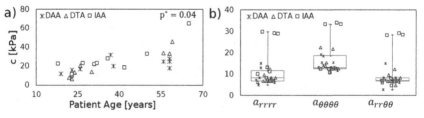

Fig. 8. Optimal material parameters: (a) c as a function of age, (b) anisotropic constants a_{ijkl}, the red box-plot show the median, quartiles and maximum and minim values without outliers. (Color figure online)

4 Discussion

In contrast to the heterogeneous *in vivo* deformation of the aortic wall, many *in vitro* tests on human and animal cadaveric aortas report deformations into cylindrical shapes of regular cross section under uniform pressurization [12, 13]. This supports the hypothesis that the naturally occurring material heterogeneities of healthy aortas may not be sufficient to produce measurable heterogeneous deformations, which in turn could be driven by *in vivo* interactions with periaortic structures. To further explore this hypothesis, the adventitial boundary was prescribed with moving EFBC implemented as a collection of springs of uniform stiffness with distal ends moving either inwards or outwards in the radial direction. This self-balancing system against the luminal pressure allows for the heterogeneous deformation and bulk displacement of the aortic wall model to match observed kinematics from patient-specific imaging. The EFBC parameters themselves pose no physiological meaning, but the resulting adventitial load distribution that reproduces the *in vivo* deformation of the aorta could be representative of the mechanical periaortic interactions. The model is able to replicate the aortic heterogeneous displacement to an error below 0.4 mm. Furthermore, it converges to an adventitial load distribution that agrees with the location of anatomical features and is consistent among individuals of similar characteristics at a given aortic location.

Notably, the model was not found sensitive to the user-defined parameters $(n, \kappa/c)$ within the range of parameters that produces reasonable results. The inverse FEM approach described herein, prescribes the luminal load as a boundary condition and the aortic wall deformation as the target optimization function. From a thermodynamic standpoint, both the pressure-pulse energy input and aortic wall deformation were prescribed as inputs for each problem. The optimization algorithm finds the remaining energy input from the adventitia to induce the heterogeneous deformation of the aortic wall at the same time that fits the material parameters to reconcile deformational stored energy with the imposed load. Since the adventitial load is a direct result of the overall energy/load balance from diastole to systole, it is only marginally affected by the selection of the material model and reference configuration.

In this work, we selected the widely known Fung orthotropic model as a relatively simple and computational robust alternative to account for material anisotropy. We also assume the aortic wall as a single layer of homogeneous properties, and the diastolic configuration as a reference stress-free configuration. Although these simplifications should not affect the boundary force balance that leads to the target systolic deformation, they influence the accuracy of the estimations of the material properties, and consequently, the estimation of strain/stress distributions. The model may be improved by the use of microstructural based material models, accounting for layer and circumferential heterogeneity of the aortic wall, and incorporating the effect of residual stress and prestrain. This work was also limited by the available prior data from the anonymized volunteers, for which we had to assume the same representative pressure-pulse magnitude for all cases. The use of patient and location specific arterial pressures could improve the accuracy of the estimation of material properties and stress, and could boost the capacity of this method to improve patient-specific diagnostics.

5 Conclusions

Modelling the effect of periaortic interactions successfully reproduced the *in vivo* heterogeneous deformation and rigid body translations of the aortic wall previously reported by image-based kinematic analyses using DENSE MRI. In this work, we presented a computationally inexpensive method that successfully modelled periaortic interactions to static and dynamic structures with moving EFBCs. Our method provides patient-specific magnitudes for the average adventitial load, which was estimated between 30 and 60% of the luminal pulse pressure depending on patient characteristics and the axial location. An adequate modeling of periaortic interactions and adventitial loads can bring a better understanding of the *in vivo* function of the aortic wall, the normal and pathological growth and remodeling of aortic tissue, and potentially the role of these periaortic interaction and loads on important aortic pathologies such as aortic dissection and aneurysmal rupture.

References

1. Wilson, J.S., Zhong, X., Hair, J., Robert Taylor, W., Oshinski, J.N.: In vivo quantification of regional circumferential green strain in the thoracic and abdominal aorta by two-dimensional spiral cine DENSE MRI. J. Biomech. Eng. **141**(6) (2019). https://doi.org/10.1115/1.4040910
2. Petterson, N.J., van Disseldorp, E.M.J., van Sambeek, M.R.H.M., van de Vosse, F.N., Lopata, R.G.P.: Including surrounding tissue improves ultrasound-based 3D mechanical characterization of abdominal aortic aneurysms. J. Biomech. **85**, 126–133 (2019). https://doi.org/10.1016/j.jbiomech.2019.01.024
3. Chirinos, J.A.: Arterial stiffness: basic concepts and measurement techniques. J. Cardiovasc. Transl. Res. **5**(3), 243–255 (2012). https://doi.org/10.1007/s12265-012-9359-6
4. Ferruzzi, J., Di Achille, P., Tellides, G., Humphrey, J.D.: Combining in vivo and in vitro biomechanical data reveals key roles of perivascular tethering in central artery function. PLoS ONE **13**(9), (2018). https://doi.org/10.1371/journal.pone.0201379
5. Kim, J., Peruski, B., Hunley, C., Kwon, S., Baek, S.: Influence of surrounding tissues on biomechanics of aortic wall. Int. J. Exp. Comput. Biomech. **2**(2), 105 (2013). https://doi.org/10.1504/ijecb.2013.056516
6. Bracamonte, J.H., Wilson, J.S., Soares, J.S.: Assessing patient-specific mechanical properties of aortic wall and peri-aortic structures from in vivo dense MR imaging using an inverse finite element method and elastic foundation boundary conditions. J. Biomech. Eng. **142**(12) (2020). https://doi.org/10.1115/1.4047721
7. Wilson, J.S., Taylor, W.R., Oshinski, J.: Assessment of the regional distribution of normalized circumferential strain in the thoracic and abdominal aorta using DENSE cardiovascular magnetic resonance. J. Cardiovasc. Magn. Reson. **21**(1), 59 (2019). https://doi.org/10.1186/s12968-019-0565-0
8. Mihai, L.A., Goriely, A.: How to characterize a nonlinear elastic material? A review on nonlinear constitutive parameters in isotropic finite elasticity. Proc. R. Soc. A Proc. Math. Phys. Eng. Sci. **473**(2207) (2017). https://doi.org/10.1098/rspa.2017.0607
9. Holzapfel, G.A., Gasser, T.C., Ogden, R.W.: A new constitutive framework for arterial wall mechanics and a comparative study of material models. J. Elast. **61**(1–3), 1–48 (2000). https://doi.org/10.1023/A:1010835316564
10. Roccabianca, S., Figueroa, C.A., Tellides, G., Humphrey, J.D.: Quantification of regional differences in aortic stiffness in the aging human. J. Mech. Behav. Biomed. Mater. **29**, 618–634 (2014). https://doi.org/10.1016/j.jmbbm.2013.01.026

11. Maas, S.A., Ellis, B.J., Ateshian, G.A., Weiss, J.A.: FEBio: finite elements for biomechanics. J. Biomech. Eng. **134**(1) (2012). https://doi.org/10.1115/1.4005694
12. Labrosse, M.R., Gerson, E.R., Veinot, J.P., Beller, C.J.: Mechanical characterization of human aortas from pressurization testing and a paradigm shift for circumferential residual stress. J. Mech. Behav. Biomed. Mater. **17**, 44–55 (2013). https://doi.org/10.1016/j.jmbbm.2012.08.004
13. Kim, J., Baek, S.: Circumferential variations of mechanical behavior of the porcine thoracic aorta during the inflation test. J. Biomech. **44**(10), 1941–1947 (2011). https://doi.org/10.1016/j.jbiomech.2011.04.022

An Exploratory Assessment of Focused Septal Growth in Hypertrophic Cardiomyopathy

Sandra P. Hager[1](✉), Will Zhang[2], Renee M. Miller[1], Jack Lee[1],
and David A. Nordsletten[1,2]

[1] School of Biomedical Engineering and Imaging Sciences,
King's College London, London, UK
sandra.p.hager@kcl.ac.uk
[2] Department of Biomedical Engineering and Cardiac Surgery,
University of Michigan, Ann Arbor, USA

Abstract. Growth and Remodelling (G&R) processes are typical responses to changes in the heart's loading conditions. The most frequent types of growth in the left ventricle (LV) are thought to involve growth parallel to (eccentric) or perpendicular to (concentric) the fiber direction. However, hypertrophic cardiomyopathy (HCM), a genetic mutation of the sarcomeric proteins, exhibits heterogeneous patterns of growth and fiber disarray despite the absence of clear changes in loading conditions. Previous studies have predicted cardiac growth due to increased overload in the heart [7,12,23] as well as modelled inverse G&R post-treatment [1,14]. Since observed growth patterns in HCM are more complex than standard models of hypertrophy in the heart, fewer studies focus on the geometric changes in this pathological case. By adapting established kinematic growth tensors for the standard types of hypertrophy in an isotropic and orthotropic material model, the paper aims to identify different factors which contribute to the heterogeneous growth patterns observed in HCM. Consequently, it was possible to distinguish that fiber disarray alone does not appear to induce the typical phenotypes of HCM. Instead, it appears that an underlying trigger for growth in HCM might be a consequence of factors stimulating isotropic growth (e.g., inflammation). Additionally, morphological changes in the septal region resulted in higher amounts of incompatibility, evidenced by increased residual stresses in the grown region.

Keywords: Growth and Remodelling · Hypertrophic cardiomyopathy · Computational modelling

1 Introduction

Nearly 1 in 200 people are affected by hypertrophic cardiomyopathy (HCM) [25], a genetic disease impacting the key sarcomeric proteins in cardiomyocytes. The histological phenotype of HCM includes hypertrophy, reorientation of the myocytes [18] and interstitial fibrosis (Fig. 1). Inflammation, as a response to fibrosis, is an

© Springer Nature Switzerland AG 2021
D. B. Ennis et al. (Eds.): FIMH 2021, LNCS 12738, pp. 328–339, 2021.
https://doi.org/10.1007/978-3-030-78710-3_32

Regional disarray   Focal fibrosis pattern
Healthy   Hypertrophic   Healthy   Hypertrophic

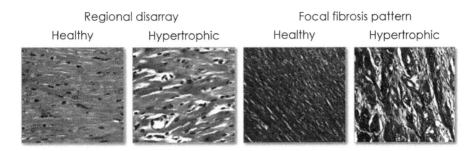

Fig. 1. Histological images of healthy and hypertrophic myocardial tissue (adapted with permission from [28], Copyright (2005) National Academy of Sciences, U.S.A.) visualizing the regional disarray (left) with H&E staining. Blue: nucleus, red: cytoplasm, varying red: collagen fibers and ECM. The right describes changes due to focal fibrosis with Masson's trichrome staining. Blue: cytoplasm in necrotic myocardium, red: cytoplasm in viable myocardium. (Color figure online)

accepted indicator of the severity in HCM cases [6]. The rationale for the disorganization of cardiac muscle cells in the ventricular septum is still not fully understood. However, septal disarray can be seen in 94% of patients with HCM [18].

On the whole-organ level, regional growth is observed despite no noticeable change in loading conditions [3,13,16,19,20]. Growth patterns in HCM are heterogeneous amongst patients, but around a third of patients have hypertrophy in the basal interventricular septal region [17] and in later stages develop an LV outflow obstruction (LVOTO) accompanied by systolic anterior motion of the mitral valve leaflet. A commonly observed HCM growth pattern involves hypertrophy in the antero-septal region of the left ventricular wall [3,26] which is quantified with a thickness of 13–15 mm in mild versions of the disease and ≥30 mm in severe cases [20]. Besides the noticeable changes in geometry, HCM patients are at higher risk of cardiac arrhythmias, sudden cardiac death (SCD) and heart failure [20].

Recent papers support utilizing the kinematic growth approaches to computationally model and predict hypertrophy due to deviation from the heart's homeostatic state. For a general overview of the phenomenological growth laws, the interested reader is referred to [27]. The challenge of most growth laws rests in the definition of the growth trigger that varies between mimicking sarcomergenesis inside the cells [7] to electromechanical changes in behaviour due to concentric hypertrophy [4].

This paper aims to investigate growth in HCM by examining drivers of septal thickening using computational modelling. Test cases are generated in order to evaluate possible mechanisms in an isotropic and orthotropic material model which lead to pathophysiological changes, such as remodelling of the fiber orientation, inflammation and different growth responses in a predefined septal region. Across the test cases, asymmetric septal growth might be triggered by an underlying isotropic growth stimulus that also results in enhanced residual stress in the septal area.

2   Methods

2.1   General Kinematic Framework for Finite Growth

The problem setup is based on a mechanical problem, accounting for the linear momentum balance and mass conservation equation, such that $(\mathbf{u}, p) \in \mathcal{U} \setminus \mathcal{Q}$ and $p \in \mathcal{P}$,

$$\int_{\Omega_X} \mathbf{P}(\mathbf{F}, p) : \nabla_X \mathbf{w} + q(J - 1) \mathrm{d}X = 0, \ \forall \ \mathbf{w} \in \mathcal{U} \setminus \mathcal{Q} \text{ and } q \in \mathcal{P}, \qquad (1)$$

where $\mathcal{U}, \mathcal{P}$ denote appropriate function spaces for the test functions $\mathbf{w}$ (displacement $\mathcal{U}$) and q (pressure p) and $\mathcal{Q}$ denotes the space of orthonormal rotations and translations. The problem is set up to solve for the physical displacement $\mathbf{u}$ and the hydrostatic pressure p. The system is defined over the initially ungrown reference domain Ω_X. $\mathbf{P}$ represents the first Piola-Kirchhoff stress tensor and is dependent on the deformation gradient $\mathbf{F}$ and the pressure in the continuum body. Subscript X of ∇ denotes that derivatives of the gradient are defined with respect to the material coordinates. J denotes the determinant of the deformation gradient $\mathbf{F}$ and provides information on the volume change due to growth and the subsequent elastic deformation from a continuity constraint.

2.2   Constitutive Equation for Growth

The kinematic growth approach represents one of the two traditional theories for computational modelling of growth, the other being the constrained mixture theory [11,24]. This postulate provides an easy parameterization of the material and the usage of phenomenological rate equations to interpret the growth process. An accepted belief is the existence of a stress-free intermediate grown reference configuration, which was mathematically confirmed by [8].

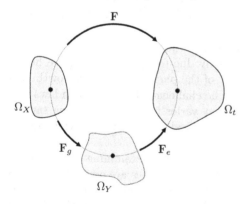

Based on this assumption, a multiplicative decomposition of the deformation gradient $\mathbf{F}$ into its elastic and inelastic parts (Fig. 2) leads to

$$\mathbf{F} = \mathbf{F}_e \cdot \mathbf{F}_g. \qquad (2)$$

Fig. 2. Schematics of multiplicative decomposition of $\mathbf{F}$ into an elastic ($\mathbf{F}_e$) and inelastic deformation gradient ($\mathbf{F}_g$).

Here, $\mathbf{F}_g$ denotes the growth from an initial traction-free configuration ($\Omega_X \subset \mathbb{R}^3$) to an intermediate local stress-free state ($\Omega_Y \subset \mathbb{R}^3$). The elastic part ($\mathbf{F}_e$) of the deformation gradient maps the intermediate state to the physical configuration and ensures continuity of the body. By enforcing a continuity constraint on the body, residual stress may be introduced. This stress links with another important G&R parameter which is the degree of incompatibility. Quantifying the degree of incompatibility of a growth process, where the measured geometric growth fails to reflect the kinematic growth tensors applied growth field, provides an opportunity to understand growth induced residual stress in the heart. From a mathematical perspective a compatible kinematic growth tensor can be rewritten as the gradient of the inelastic mapping, while, in the incompatible case, a direct mapping through a tensor gets lost. The proposed definition for compatibility is only true in the grown unloaded state and is based on the assumption that a compatible growth deformation holds

$$\mathbf{F}_e \cdot \mathbf{F}_g = \mathbf{I}. \tag{3}$$

Deviations from the introduced compatibility (Eq. 3) are defined by the degree of incompatibility in the region of interest. Since the elastic deformation is the sole contributor to the mechanical problem, Eq. 1 can be mapped from the reference domain to the intermediate state. Additionally, the deformation gradient is substituted with its multiplicatively-decomposed parts.

In the heart, two growth modes can be identified, concentric and eccentric hypertrophy. Concentric growth behaviour is often seen as a typical response to an elevated afterload in the heart, whereas eccentric hypertrophy aims to compensate for a volume overload inside the ventricle. These growth modes can be generally regarded as

$$\mathbf{F}_g = \vartheta^{\mathrm{f}} \mathbf{f}_0 \otimes \mathbf{f}_0 + \vartheta^{\mathrm{s}} \mathbf{s}_0 \otimes \mathbf{s}_0 + \vartheta^{\mathrm{n}} \mathbf{n}_0 \otimes \mathbf{n}_0, \tag{4}$$

where $\mathbf{f}_0$ defines the orientation of the myocyte and is defined using a rule-based method [5], Fig. 3 left. The sheet vector $\mathbf{s}_0$ describes the alignment of the cell bundles, and $\mathbf{n}_0$ is the sheet plane normal direction; ϑ^{i} with $\mathrm{i} = f, s, n$ represents the growth factor in each individual direction. Concentric hypertrophy leads to an increase in wall thickness, which can be defined by setting the growth multiplier in the myocytes' radial direction $\vartheta^{\mathrm{s,n}} > 1$ and keeping $\vartheta^{\mathrm{f}} = 1$. In contrast, eccentric growth causes growth along the axial direction of the fibers with $\vartheta^{\mathrm{f}} > 1$ and the axial components remaining 1. The value for the individual growth multiplier can be defined using ordinary differential equations as growth laws. However, for this study, they were set to be constant value fields and applied iteratively to converge.

2.3   Material Law, Boundary Conditions, and Growth Problem

Hypertrophy was analyzed and investigated at the final growth state in a patient-specific biventricular mesh. The geometry was created from neural network-generated segmentations of SSFP images from a patient with hypertrophic cardiomyopathy. From these segmentations, a biventricular model template was fit to the segmented contours of the left and right ventricles as well as valve landmarks [22]. To generate local muscle fiber directions, a rule-based fiber field was created using the Laplace-Dirichlet method, proposed by both [5] and [2]. Fiber angles varied from $-60°$ to $60°$ and $-25°$ to $90°$ from the epicardium to endocardium in the LV and RV, respectively. Fiber angles at the valve annuli were determined based on high-resolution DTI measurements from ex-vivo porcine hearts. Across the septum, fibers transitioned between $60°$ at the LV endocardium to $0°$ at the midwall to $90°$ at the right ventricle (RV) endocardium.

The presented simulations compare a Neo-Hookean material model (with a bulk modulus $\mu = 30$[kPa]) with the Holzapfel-Ogden cardiac model [10] (with $a = 2, b = 5, a_f = 10, b_f = 7.5, a_s = 0, b_s = 0, a_{fs} = 0, b_{fs} = 0$). Parameters a are stress-like parameters in [kPa] while b are unitless scalar values. The utilized cardiac law takes into consideration the increased stiffness along the fiber direction ($a_f, b_f > 0$) and assumes no enhanced stiffness rates along the radial direction of the myocytes ($a_s, b_s, a_{fs}, b_{fs} = 0$). Both of the material laws are assumed to be nearly incompressible. The following test cases modify the global growth tensors of concentric and eccentric hypertrophy with the aim of recreating growth patterns observed in HCM. We note that the growth simulated is driven by $\vartheta^{\parallel}$ and $\vartheta^{\perp}$ in these test cases and are not optimized to simulate patient data. Instead, these parameters are chosen and pushed to examine the influence of these mechanisms to meet the current objective of recreating an asymmetrical growth pattern in the septal area of a biventricular geometry.

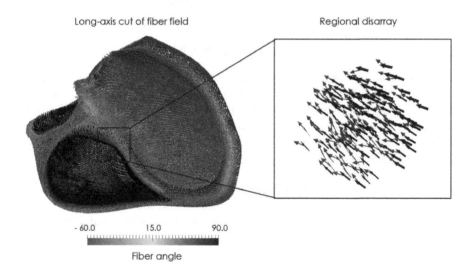

Long-axis cut of fiber field   Regional disarray

- 60.0   15.0   90.0

Fiber angle

Fig. 3. Left: Fiber angle of the field with respect to the local circumferential direction. Right: Regional disarray of the myocytes in the septal region. Red arrows: fibers are aligned in the septum; blue arrows: disarray of the fibers. (Color figure online)

Simulations of the test cases were run both with and without regional disarray introduced in the interventricular septum. The remodelling of the fiber orientation was achieved by adding stochasticity to the mean fiber $\mathbf{f}_0$ and sheet direction $\mathbf{s}_0$, which leads to a modification of the growth tensor in the defined septal region (Fig. 3). In this region, a new mean fiber, $\mathbf{f}_0'$ was defined whereby

$$\mathbf{f}_0'(\mathbf{X}) = \mathbf{f}_0(\mathbf{X}) + aw(\mathbf{X})\mathbf{b}(\mathbf{X}). \tag{5}$$

Here $w : \Omega_X \to \mathbb{R}$ is a weighting function $w(\mathbf{X}) \in [0,1]$ defining the region of disarray, $a \in [0,1]$ defining the degree of disarray, and $\mathbf{b} : \Omega_X \to \mathbb{R}^3$ is a vector where $\mathbf{b}(\mathbf{X})$ is a random vector on the unit sphere. Equation 5 provides a general formulation for adding stochasticity to the fiber field in the septal region (and similarly for the sheet vector, with the normal vector generated as the cross product of the two). Note that the vectors from Eq. 5 are subsequently normalized so that unit vectors are utilized in the growth and constitutive models.

To examine different potential explanations for the localized growth observed in HCM, three test cases are considered. **Test case 1** concentrates on the influence of disarray in the septal area to recreate a septal growth pattern seen in HCM. For this, a homogeneous global growth tensor (eccentric or concentric) is applied to the biventricular geometry. **Test case 2** explores the impact of spatially heterogeneous growth tensors with a variation of the growth multipliers in the LV septal region. The theory behind this test case is that the septum's diseased area reacts with higher amounts of growth than the remote area to global growth drivers. For this study, the first test case was extended doubling the growth factor in the septal region only. Finally, **test case 3** investigates the influence of a spatially heterogeneous growth tensor, particularly with an isotropic regional growth tensor in the septum. The theory is that a consequence of factors in the tissue may drive growth in equal amounts in all microstructural directions (e.g., inflammation). Computationally this behaviour is imitated by an adaption of Eq. 4, where ϑ^i with $i = f, s, n$ is defined in all directions equally in the septal region and using an eccentric or concentric growth stimulus outside of the septal area.

3   Results and Discussion

Figure 4 illustrates some of the permutations of growth applied in a biventricular geometry with a Neo-Hookean material. Each test contains data from eccentric and concentric growth modes and further compares the simulations' displacement magnitudes. Table 1 includes the values for the wall thickness in the septal region after growth was applied and the percentage change relative to the initial wall thickness (7.74 [mm]) for the isotropic Neo-Hookean and the orthotropic Holzapfel-Ogden material law. The reference values refer to continuous growth modes applied to the geometry and are used to identify the difference between HCM-triggered growth with disarray and growth due to increased loading.

The left column in Fig. 4 visualizes the first test case, with a spatially homogeneous growth tensor (concentric or eccentric) applied over the entire geometry. The significant aspect of this simulation is the introduced disarray in the predefined septal region. Neither the eccentric nor the concentric growth type showed clear manifestation of septal growth pattern. Table 1 additionally reveals that

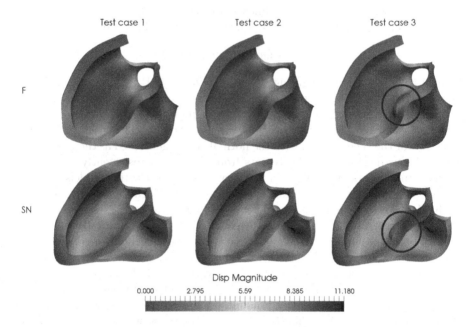

Fig. 4. Comparison of the displacement magnitude (mm) between three separate test cases and the applied growth types utilizing a Neo-Hookean material law. Yellow circles highlight a septal growth pattern in the LV. (Color figure online)

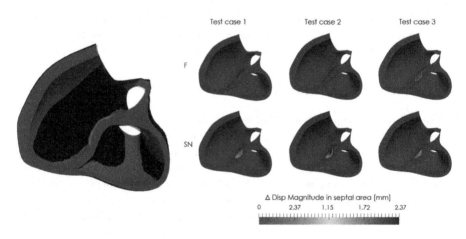

Fig. 5. Left: Long axis cut of eccentric growth in test case 3 to compare Neo-Hookean and Holzapfel-Ogden material law. Dark red: Neo-Hookean geometry, grey-semitransparent: Holzapfel-Ogden body. Right: Illustrates the difference of the displacement magnitude between the two constitutive laws in the septal region. (Color figure online)

Table 1. Septal wall thickness t (mm) after growth and change as a percentage t% of the original thickness for the reference simulations and the defined test cases with disarray and isotropic and cardiac material law, where SN and F represent concentric and eccentric growth respectively.

	Reference		Test case 1		Test case 2		Test case 3	
	SN	F	SN	F	SN	F	SN	F
t	9.59	8.61	9.66	8.50	11.19	9.00	12.56	11.65
t%	123.86	111.17	124.68	109.73	144.45	116.22	162.19	150.39

Holzapfel-Ogden

	Reference		Test case 1		Test case 2		Test case 3	
	SN	F	SN	F	SN	F	SN	F
t	9.78	8.57	9.79	8.46	11.25	8.97	12.34	10.67
t%	126.25	110.68	126.44	109.26	145.27	115.83	159.39	137.97

eccentric growth (F) in this test results in dilation of the septal wall compared to the reference simulation. Our results in this case do not support the theory that regional hypertrophy in HCM is caused by regional disarray alone.

Simulations with disarray and a spatially heterogeneous growth tensor with higher amounts of growth in the septal region (middle column Fig. 4) failed to reproduce a septal growth pattern. However, looking at this test case in Table 1, the value for the wall thickness in the SN growth case shows a difference of around 1 mm compared to the reference growth case.

In comparison to the first two test cases, the right panel in Fig. 4 displays the application of a heterogeneous growth tensor with isotropic growth in the septal area. With concentric and eccentric growth, the geometry exhibited growth patterns typical to the ones seen in septal HCM. Across the different test cases the difference between the two laws showed minimal deviation of the septal thickness except in test case 3, eccentric growth with disarray exhibited a difference of around 1 [mm] (Table 1). Figure 5 illustrates the difference of displacement between the isotropic Neo-Hookean and the orthotropic Holzapfel-Ogden material law in the growth simulation test case 3, eccentric growth. The left image visualizes the Neo-Hookean's (dark red) and Holzapfel-Ogden (grey-semitransparent) grown state compared to each other. The figure shows that both material laws are somewhat prone to buckle into the LV instead of solely increasing the mass in the septal region. Yet, it appears that the cardiac material tends to buckle more to hold Eq. 1. The right-hand side illustrates the difference in displacement magnitude between material laws for the individual test cases and growth modes focusing in the septal region. In the eccentric case of test case 3, a continuous difference of the displacement magnitude over the septal area can be observed. In contrast, across the test cases, the concentric growth modes seem to exhibit higher differences towards the base. Figure 6 illustrates a concentric growth tensor with isotropic growth in the septal region, equivalent

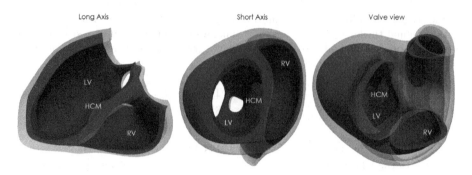

Fig. 6. Concentric growth tensor with isotropic growth in the septal region of the LV, viewed from different orientations. Blue: reference geometry, grey-semi-transparent: grown body. (Color figure online)

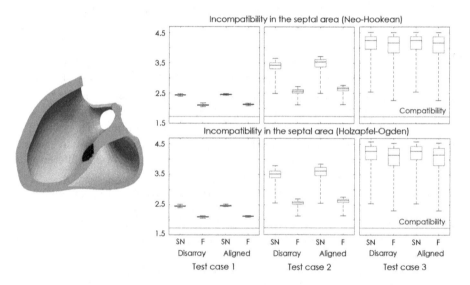

Fig. 7. Degree of incompatibility in the septal region (red area in the left image) over all test cases utilizing Neo-Hookean and Holzapfel-Ogden material law, defined with the Frobenius norm of $\mathbf{F}_e\mathbf{F}_g$ while the degree of compatibility is defined as the Frobenius norm of Eq. 3. (Color figure online)

to the test case in the bottom right corner in Fig. 4. This image indicates that with higher amounts of growth in the septal area, the septum's hypertrophic tissue starts to develop an LVOTO.

Figure 7 summarizes the degree of incompatibility in the septal region of the heart in the various test cases with and without regional disarray and enables an objective way to compare the generated residual stresses across simulations and material laws. Noticeable is that disarray in the heart's septal region has little impact on the degree of incompatibility. Given these simulation results, incompatibility is dependent on the type of growth tensor applied to the geometry. The highest incompatibility was seen in the third test case, indicating a high level of residual stress generated in the septal region by growth.

One of the limitations of this study is due to the simplification of the growth law. Utilizing a constant growth law to model G&R in HCM disregards growth drivers such as pressure or strain in the tissue that might also introduce hypertrophic patterns. Physiological growth laws will be implemented in subsequent studies.

4   Conclusion

With homogeneous growth, disarray alone within the septal region did in neither of the material laws result in septal growth patters observed in HCM patients. In contrast, a regional isotropic growth in the septal area did result in regional hypertrophy. This observation supports the idea that hypertrophy in the septum may not be driven by the fiber structures disruption but rather by focal fibrosis and inflammation.

Acknowledgements. Authors would like to acknowledge funding from Engineering and Physical Sciences Research Council (EP/R003866/1). This work was also supported by the Wellcome ESPRC Centre for Medical Engineering at King's College London (WT203148/Z/16/Z) and the British Heart Foundation (TG/17/3/33406).

References

1. Arumugam, J., Mojumder, J., Kassab, G., Lee, L.C.: Model of anisotropic reverse cardiac growth in mechanical dyssynchrony. Sci. Rep. **9**(1), 1–12 (2019)
2. Bayer, J.D., Blake, R.C., Plank, G., Trayanova, N.A.: A novel rule-based algorithm for assigning myocardial fiber orientation to computational heart models. Ann. Biomed. Eng. **40**(10), 2243–2254 (2012)
3. Davies, M.J., McKenna, W.J.: Hypertrophic cardiomyopathy–pathology and pathogenesis. Histopathology **26**(6), 493–500 (1995)
4. Del Bianco, F., Franzone, P.C., Scacchi, S., Fassina, L.: Electromechanical effects of concentric hypertrophy on the left ventricle: a simulation study. Comput. Biol. Med. **99**, 236–256 (2018)
5. Doste, R., et al.: A rule-based method to model myocardial fiber orientation in cardiac biventricular geometries with outflow tracts. Int. J. Numer. Meth. Biomed. Eng. **35**(4), e3185 (2019)
6. Fang, L., Ellims, A.H., Beale, A.L., Taylor, A.J., Murphy, A., Dart, A.M.: Systemic inflammation is associated with myocardial fibrosis, diastolic dysfunction, and cardiac hypertrophy in patients with hypertrophic cardiomyopathy. Am. J. Transl. Res. **9**(11), 5063–5073 (2017)
7. Göktepe, S., Abilez, O.J., Parker, K.K., Kuhl, E.: A multiscale model for eccentric and concentric cardiac growth through sarcomerogenesis. J. Theor. Biol. **265**(3), 433–442 (2010)

.

8. Goodbrake, C., Goriely, A., Yavari, A.: The mathematical foundations of anelasticity: Existence of smooth global intermediate configurations. Proc. R. Soc. A **477**(2245), 20200462 (2021)
9. Hadjicharalambous, M., Lee, J., Smith, N.P., Nordsletten, D.A.: A displacement-based finite element formulation for incompressible and nearly-incompressible cardiac mechanics. Comput. Method. Appl. M. **274**, 213–236 (2014)
10. Holzapfel, G.A., Ogden, R.W.: Constitutive modelling of passive myocardium: a structurally based framework for material characterization. Philos. Trans. A Math. Phys. Eng. Sci. **367**(1902), 3445–3475 (2009)
11. Humphrey, J.D., Rajagopal, K.R.: A constrained mixture model for growth and remodeling of soft tissues. Math. Mod. Meth. Appl. Sci. **12**(3), 407–430 (2002)
12. Kerckhoffs, R.C., Omens, J.H., McCulloch, A.D.: A single strain-based growth law predicts concentric and eccentric cardiac growth during pressure and volume overload. Mech. Res. Commun. **42**, 40–50 (2012)
13. Klues, H.G., Schiffers, A., Maron, B.J.: Phenotypic spectrum and patterns of left ventricular hypertrophy in hypertrophic cardiomyopathy: morphologic observations and significance as assessed by two-dimensional echocardiography in 600 patients. J. Am. Coll. **26**(7), 1699–1708 (1995)
14. Lee, L.C., Genet, M., Acevedo-Bolton, G., Ordovas, K., Guccione, J.M., Kuhl, E.: A computational model that predicts reverse growth in response to mechanical unloading. Biomech. Model. Mechanobiol. **14**(2), 217–229 (2014). https://doi.org/10.1007/s10237-014-0598-0
15. Lee, J., et al.: Multiphysics computational modeling in CHeart. SIAM J. Comput. **38**(3), C150–C178 (2016)
16. Liew, A.C., Vassiliou, V.S., Cooper, R., Raphael, C.E.: Hypertrophic cardiomyopathy–past, present and future. Clin. Med. **6**(12), 118 (2017)
17. Marian, A.J., Braunwald, E.: Hypertrophic cardiomyopathy: genetics, pathogenesis, clinical manifestations, diagnosis, and therapy. Circ. Res. **121**(7), 749–770 (2017)
18. Maron, B.J., Roberts, W.C.: Quantitative analysis of cardiac muscle cell disorganization in the ventricular septum of patients with hypertrophic cardiomyopathy. Circulation **59**(4), 689–706 (1979)
19. Maron, B.J., Epstein, S.E.: Hypertrophic cardiomyopathy: a discussion of nomenclature. Amer. J. Cardiol. **43**(6), 1242–1244 (1979)
20. Maron, B.J.: Hypertrophic cardiomyopathy: a systematic review. JAMA **287**(10), 1308–1320 (2002)
21. MATLAB. 9.9.0.1524771 (R2020b). Natick, Massachusetts: The MathWorks Inc. (2020)
22. Mauger, C., et al.: An iterative diffeomorphic algorithm for registration of subdivision surfaces: application to congenital heart disease. Ann. Int. Conf. IEEE Eng. Med. Biol. Soc. **2018**, 596–599 (2018)
23. Peirlinck, M., et al.: Using machine learning to characterize heart failure across the scales. Biomech. Model. Mechanobiol. **18**(6), 1987–2001 (2019). https://doi.org/10.1007/s10237-019-01190-w
24. Rodriguez, E.K., Hoger, A., McCulloch, A.: Stress-dependent finite growth in soft elastic tissues. J. Biomech. **27**(4), 455–467 (1994)
25. Semsarian, C., Ingles, J., Maron, M.S., Maron, B.J.: New perspectives on the prevalence of hypertrophic cardiomyopathy. J. Am. Coll. Cardiol. **65**(12), 1249–1254 (2015)
26. Teare, D.: Asymmetrical hypertrophy of the heart in young adults. Brit. Heart J. **20**(1), 1–8 (1958)

27. Witzenburg, C.M., Holmes, J.W.: A comparison of phenomenologic growth laws for myocardial hypertrophy. J. Elast. **129**(1), 257–281 (2017)
28. Wolf, C.M., et al.: Somatic events modify hypertrophic cardiomyopathy pathology and link hypertrophy to arrhythmia. P. Natl. Acad. Sci. USA **102**(50), 18123–18128 (2005)

Parameter Estimation in a Rule-Based Fiber Orientation Model from End Systolic Strains Using the Reduced Order Unscented Kalman Filter

Luca Barbarotta[ID] and Peter H. M. Bovendeerd[✉]

Eindhoven University of Technology, Eindhoven, The Netherlands
{l.barbarotta,p.h.m.bovendeerd}@tue.nl

Abstract. Fiber orientation is a major factor in the determination of end-systolic strains within models of cardiac mechanics. Unfortunately, direct patient-specific acquisition of fiber orientation is not readily available nowadays in the clinic. As an alternative, we propose to use the Reduced Order Unscented Kalman Filter to estimate rule-based fiber orientation parameters from end-systolic wall strains that can be obtained using more traditional imaging methodologies. We address the estimation of fiber orientation in the physiological left ventricle, where end-systolic strains were generated in-silico using a 12-parameter rule-based fiber model. The estimation process focused on the determination of the three most influential parameters of an imperfect 5-parameter rule-based fiber model. Our results show that these three fiber parameters can be estimated within an average deviation of 6° from a combination of three end-systolic strains even when the initial guess for each estimated parameter was set 10° away from the ground truth value.

1 Introduction

Patient-specific models of cardiac electromechanics might constitute a precious tool for assisting cardiologists during decision making. Patient-specific models are obtained by tailoring a generic model including patient data. When patient data are missing, because of the lack of reliable measurement techniques, they must be estimated indirectly using data assimilation techniques.

In models of cardiac mechanics, the orientation of fibers in the myocardium plays an important role in determining cardiac function [5,11]. Myofibers follow a complex path and constitute a major direction of anisotropy during myocardial contraction. Myofiber orientation can be measured in-vivo using Diffusion Tensor MRI (DT-MRI). However, the long acquisition time of such methodology prevents its use in clinical practice. Moreover, even when patient-specific fiber information is available, accuracy issues related with DT-MRI acquisition introduce uncertainty in the orientation of about 10° [13].

Given the sensitivity of cardiac wall mechanics to fiber orientation and the issues related to measuring this orientation in the individual patient in the clinic,

© Springer Nature Switzerland AG 2021
D. B. Ennis et al. (Eds.): FIMH 2021, LNCS 12738, pp. 340–350, 2021.
https://doi.org/10.1007/978-3-030-78710-3_33

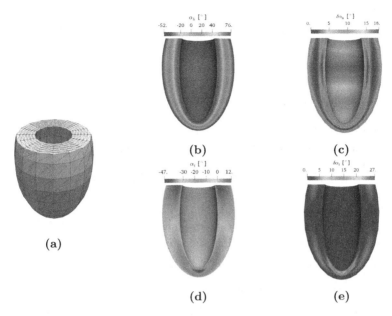

Fig. 1. View of the mesh (left), and spatial distribution of the helix angle α_h and the transverse angle α_t for the 12-parameter fiber model (center) and deviation from this 12-parameter model in the limited 5-parameter model (right).

alternative solutions are required to arrive at a patient-specific fiber field and hence to improve the predictive capabilities of patient-specific models. A possible solution might come from the application of data assimilation techniques. Nagler et al. [10] applied the Reduced Order Unscented Kalman Filter (ROUKF) to estimate the parameters of a rule-based fiber model from noisy DT-MRI measurements. However, this method still relies on the availability of a DT-MRI measurement of the patient. Instead, in this study, we investigate the possibility of computing patient-specific fiber orientation using data assimilation on strain data that could be measured in the clinic [6,8]. To this purpose, we use the ROUKF [9] to inversely estimate parameters of a rule-based fiber model using end-systolic strain data.

2 Materials and Methods

2.1 Model of the Mechanics of the LV

The LV geometry was described using an ellipsoidal shape with cavity volume 44 mL and wall volume of 136 mL, see Fig. 1a. The geometry is characterized by a cavity with a long axis of 70 mm, and is truncated at 24 mm above the equator to form a planar base. The orientation of fibers inside the myocardium is described by the helix angle α_h and the transverse angle α_t, where α_h defines the component in the longitudinal-circumferential plane of the fiber unit vector,

e_f, and α_t defines the component in the circumferential-transmural plane. Angles α_h and α_t were set according to a 12-parameter rule-based model, see Fig. 1b and d [4].

The myocardium is described as a hyperelastic fiber-reinforced transversely isotropic active stress material [4]. The active stress model introduces a stress along fiber that depends on sarcomere length, sarcomere shortening speed, and activation time. Left ventricular wall mechanics was computed by solving for equilibrium between forces related to active stress, passive stress and cavity pressure as explained in [2]. The lumped-parameter model presented in [4] is coupled to the LV mechanics model to provide physiological preload and afterload.

The resulting model is implemented in a Finite Element package based on FEniCS [7] presented in [1], where it has been validated and numerically verified. In balancing computational time and accuracy a discretization was performed using linear Lagrangian polynomials on the mesh in Fig. 1a, resulting in 5, 760 elements and 3, 381 degrees of freedom. This choice was dictated by the large number of simulations required by the estimation process, which required p+1 forward simulations (an entire cardiac cycle) at each iteration of the estimation process, being p the number of estimated parameters. The use of linear Lagrangian polynomials in our model allows to achieve consistent predictions of most of end-systolic strains, with E_{ct} being the strain presenting the largest error [1]. Conservation of mass of the myocardial tissue was affected by the interpolation scheme, but increasing the bulk modulus allowed to prevent excessive loss of mass during systole (<0.5%) without having locking phenomena. The solver used our implementation of the Newton-Raphson algorithm combined with the LU decomposition implemented in the MUMPS library. The average number of Newton iterations per solution of the nonlinear system was 2.71 and each linear system took approximately 0.03s to solve.

2.2   Reduced Order Unscented Kalman Filter

We implemented the Reduced Order Unscented Kalman Filter presented by Moireau et al. [9] to estimate parameters in a 5-parameter fiber orientation model, introduced later. The ROUKF algorithm requires three stages: a sampling stage, a prediction stage, and a correction stage. In the sampling stage, the so-called sigma-points are generated, representing the estimated average and covariance of the parameters involved in the estimation. In the prediction stage, the mechanics model predicts the states corresponding to the sigma-points. In this work, the state is the deformed configuration of the LV. At the correction stage, the average and the covariance of state and parameters are corrected using the difference between the measurements and the observations derived from the prediction stage. These three stages are repeated sequentially until convergence. For a more in depth description of the algorithm, refer to Moireau et al. [9].

The prediction stage relies on the simulation of a cardiac cycle until the end systolic configuration. Given $\boldsymbol{\theta}_n$, the vector of fiber parameters of the LV mechanics model to be estimated, the end systolic configuration $\boldsymbol{x}_{n+1}$ is achieved by means of the following nonlinear operator

$$x_{n+1} = \mathcal{A}_{n+1|n}\left(x_0, \theta_n\right),$$ (1)

where x_0 is the unloaded configuration and $\mathcal{A}_{n+1|n}$ is nonlinear operator that solves the circulation-mechanics coupled problem until the end systolic configuration at time $n + 1$ given the parameters θ_n. This operator $\mathcal{A}_{n+1|n}$ is defined as follows

$$\mathcal{A}_{n+1|n}\left(x_0, \theta_n\right) = \begin{cases} \text{(solve until end systole)} \\ \begin{cases} \mathcal{C}\left(p_{lv}\right) \to V_{lv}, & \text{circulation} \\ \mathcal{M}\left(u; p_{lv}, \theta_n\right) = 0 \to V_c, & \text{wall mechanics} \\ \text{until } V_c = V_{lv}, \end{cases} \\ x_{n+1} = x_0 + u\left(x_0; \theta_n\right), \end{cases}$$ (2)

where $\mathcal{C}$ is the nonlinear operator that represents the circulation model, $\mathcal{M}$ is the nonlinear operator that represents the mechanics model, and V_c and V_{lv} are the two cavity volumes computed within the fixed-point iteration scheme for the explicit coupling (see [1]). From the end-systolic configuration x_{n+1}, the observation z_{n+1} is derived, that contains the end systolic Green-Lagrange strains with respect to the end diastolic configuration, computed using the nonlinear observation operator $\mathcal{H}$. Given a set of pairs of normal vectors $(e_{i_1}, e_{j_1}) \dots (e_{i_n}, e_{j_n})$, with $i_1, \dots, i_n$ and $j_1, \dots, j_n$ in $\{c, l, t\}$ (c circumferential, l longitudinal, and t transmural), it reads

$$z_{n+1} = \mathcal{H}\left(x_{n+1}\right) + v, \quad \text{with:}$$ (3)

$$\mathcal{H}\left(x_{n+1}\right) = \left[\mathbf{E}^{ED \to ES}\left(x_{n+1}\right) e_{i_1} \cdot e_{j_1}, \dots, \mathbf{E}^{ED \to ES}\left(x_{n+1}\right) e_{i_n} \cdot e_{j_n}\right]^T,$$ (4)

$$\mathbf{E}^{ED \to ES}\left(x_{n+1}\right) = \frac{1}{2}\left[\nabla_0 x_{ED,n+1}^{-T} \nabla_0 x_{n+1}^T \nabla_0 x_{n+1} \nabla_0 x_{ED,n+1}^{-1} - \mathbf{I}\right],$$ (5)

where $x_{ED,n+1}$ is the end diastolic configuration, v is a Gaussian random variable with covariance matrix $\mathbf{W}$ that describes the measurements and the discretization errors. The parameter estimation is achieved by solving a sequence of application of $\mathcal{A}_{n+1|n}$.

2.3 Simulation Settings

In the estimation procedure the fiber orientation is modeled using the following simplified 5-parameters model

$$\alpha_h\left(u, v\right) = \left(h_v^0 + h_v^1 v\right)\left(1 + h_u^1 u\right),$$ (6)

$$\alpha_t\left(u, v\right) = \left(1 - v^2\right)\left(t_u^0 + t_u^1 u\right),$$ (7)

in which v denotes a normalized transmural coordinate, ranging from -1 at the endocardial surface to $+1$ at the epicardial surface. The normalized longitudinal

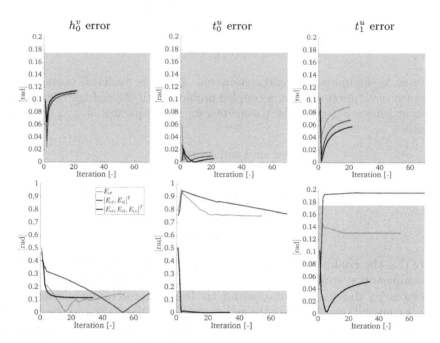

Fig. 2. Evolution of parameter estimation errors in experiment 1 (top row) and experiment 2 (bottom row), using 1 (E_{ct}, orange), 2 ($[E_{ct}, E_{cl}]$, red), or 3 ($[E_{ct}, E_{cl}, E_{cc}]$, black) strain components. The gray area indicates the accuracy of direct DT-MRI assessment. Note the increased strain range in the bottom left and middle figures. (Color figure online)

coordinate u ranges from -1 at the apex, through 0 at the equator until $+0.5$ at the basal plane [2]. Here, for α_h, h_0^v describes the transmural offset, h_v^1 describes the transmural slope, and h_u^1 describes the longitudinal variability. For α_t, t_u^0 describes the longitudinal offset ($t_u^0 = 0$ means that α_t is nil at equator), and t_u^1 describes the longitudinal slope of α_h.

Among the 5 parameters, we include in the estimation process only the three most influential parameters emerged from our previous sensitivity analysis in [3], which are h_0^v, t_0^u, t_1^u. Three combinations of the three observed end systolic strains are considered that were found to be most sensitive to variations in the fiber field in [3]: $z = [E_{ct}]^T$, $z = [E_{ct}, E_{cl}]^T$, $z = [E_{ct}, E_{cl}, E_{cc}]^T$.

We make use of the estimation of the variation of end systolic strains induced by perturbation of fiber orientation to model the covariance matrix of the observation error $\mathbf{W}_n$. This covariance matrix is assumed to be diagonal (spatially independent noise) with diagonal values being equal to the mean coefficient of variation computed from the sensitivity analysis multiplied by the infinity-norm of the measured strain component. These mean coefficients of variation for the three strains considered are: $E_{cc} = 0.15$, $E_{cl} = 0.25$, and $E_{ct} = 0.38$.

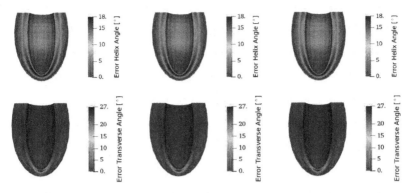

Fig. 3. Experiment 1: spatial distribution of the error between estimated and ground truth (12-parameter model) for helix (top row) and transverse angles (bottom row). Error in the helix and transverse angle when observing: E_{ct} (left column), $[E_{ct}, E_{cl}]^T$ (center), and $[E_{ct}, E_{cl}, E_{cc}]^T$.

Observations are generated using the 12-parameters fiber model proposed by Bovendeerd et al. [4]. Since we use different fiber models for generating the observations and for the estimation process, the definition of ground truth values is not straightforward. We projected the 12-parameter fiber model onto the 5-parameter model using the $L^2([-1, 1] \times [-1, 0.5])$ integral scalar product. This projection leads to the following ground truth values: $h_0^v = 0.3565$ rad, $h_1^v = -1.0471$ rad, $h_1^u = 0$ rad, $t_0^u = 0$ rad, and $t_1^u = 0.5164$ rad. The deviation from the 12-parameter model of the helix and the transverse angles generated by these parameters are shown in Fig. 1c and e. Observations generated using the 12-parameter fiber model result in end systolic strains with the following infinity-norm

$$\|E_{cc}\|_\infty = 0.2771, \quad \|E_{cl}\|_\infty = 0.0540, \quad \|E_{ct}\|_\infty = 0.1684. \quad (8)$$

Two estimation experiments were performed. In experiment 1, the initial values for the parameters were set to the ground truth values and the estimation was stopped once the absolute variation of all the estimated parameters was below a threshold value of 1×10^{-5} rad. In experiment 2, the estimation was started with values set at $10°$ away from the ground truth, resulting in the following initial values: $h_v^0 = 0.175$, $t_u^0 = -0.175$, and $t_u^1 = 0.34$. The estimation was stopped at a threshold value of 1×10^{-3} rad.

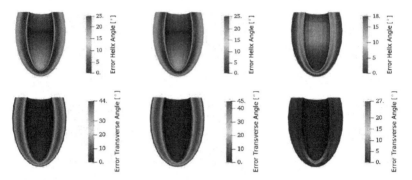

Fig. 4. Experiment 2: spatial distribution of the error between estimated and ground truth (12-parameter model) helix (top row) and transverse angles (bottom row). Error in the helix and transverse angle when observing: E_{ct} (left column), $[E_{ct}, E_{cl}]^T$ (center), and $[E_{ct}, E_{cl}, E_{cc}]^T$.

3   Results

3.1   Experiment 1

The top row in Fig. 2 shows the evolution of parameter errors in estimation experiments for the three combinations of end systolic strains, starting out from ground truth parameter values. All the parameter estimations converged within a maximum deviation of 0.11 rad (6°) from their projected truth values in about 20 iterations. Parameter errors generally decrease with increasing number of observed strain components. For t_u^0 it decreases from 0.016 rad to 0.006 rad, for t_u^1 it decreases from 0.088 rad to 0.057 rad, but for h_v^0, it remains unaffected at about 0.11 rad. The joint estimation of the three parameters leads to an average difference between the ground truth and the estimated fiber vectors of 6.5°, 6.3°, and 6.5° when observing E_{ct}, $[E_{ct}, E_{cl}]^T$, and $[E_{ct}, E_{cl}, E_{cc}]^T$, respectively. The error between the helix and transverse angles obtained using the estimated parameters and those obtained using the ground truth 12-parameter model are shown in Fig. 3. Note the spatial correlation with the deviation between the 5-parameter model and the 12-parameter model in Fig. 1c and e.

The error between observed strains in the 12-parameter model and the estimated strains in the 5-parameter model can be seen in Fig. 5. The error decreases from 0.027 through 0.018 until 0.015 when increasing the number of strain components in the estimation procedure (E_{ct}, $[E_{ct}, E_{cl}]^T$, and $[E_{ct}, E_{cl}, E_{cc}]^T$, respectively). All the three strain combinations present a sharp decrease of the observation error within the first 5 iterations. Note that the ground truth in Fig. 2 is defined by settings of the 5 parameters resulting from the projection of the 12-parameter model, while in Fig. 5 the ground truth is the observation of strains obtained with the 12-parameters model.

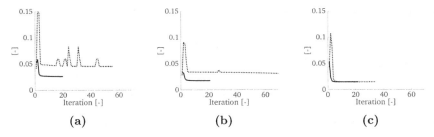

(a)   (b)   (c)

Fig. 5. Average 1-norm error of the observed strains, collected in the quantity z, for experiment 1 (solid line) and experiment 2 (dotted line). Results were obtained observing E_{ct} (left panel), observing $[E_{ct}, E_{cl}]^T$ (central panel), and observing $[E_{ct}, E_{cl}, E_{cc}]^T$ (right panel).

3.2   Experiment 2

The bottom row in Fig. 2 shows the evolution of parameter errors in experiment 2, starting out from initial parameter settings deviating from their ground truth values by 10°. All the parameter estimations converged. However, only when observing $[E_{ct}, E_{cl}, E_{cc}]^T$ all the three parameters converged to the same estimate as obtained in experiment 1. The error between the helix and transverse angles obtained using the estimated parameters and those obtained using the ground truth 12-parameter model are shown in Fig. 4. As expected, in that case the error between observed and estimated strains was equal to that in experiment 1, see Fig. 5. The increase in estimation error when using only E_{ct} or $[E_{ct}, E_{cl}]$, shown in Fig. 2, is reflected in an increase in the strain error, shown in Fig. 5.

4   Discussion

We estimated 3 fiber parameters in a 5-parameter model from strains generated with a 12-parameter fiber model. When initial values were set to ground truth values (experiment 1) convergence was obtained with an error between ground truth and estimated fiber vector of about 6°, which is on the order of the error estimated for DTI measurements, even when using only E_{ct} to drive the estimation. When initial values were set at one standard deviation away from these values (experiment 2), at least three strain components were needed to yield a similar error of 6°. The difference in the results achieved in the two experiments can be motivated by considering that the 5-parameter model is an approximation of the more detailed 12-parameter model around the ground truth values. When initializing the estimation farther from these values, the prediction of strains in the two models deviates and more strain components may be required to capture similarities in the produced strain fields. Estimations that worsen the initial set of fiber parameter values can be detected by an increase in the observation error as compared to the initial one (dashed lines in Fig. 5a and b). These estimations converged far from the ground truth because for certain sigma-points the resulting fiber field did not allow the LV to reach the ejection phase, thus affecting the prediction of strains and the capability of the algorithm to properly estimate fiber parameters. Our analysis can be improved by improving the treatment of failed simulations within the estimation process and by improving our modeling of the covariance matrix $\mathbf{W}$, which represents the measurement error of strains. In real applications, this matrix represents the many sources of error affecting the strain measurement in patients. This error depends on the measurement methodology, might not be uniform in time and space, and it might affect strain components differently.

Our results show that estimating personalized fiber parameters in the physiological LV using end-systolic strain components that can be measured in the clinic is viable, without the need of direct fiber measurements that are difficult to achieve and have accuracy issues. Different imaging modalities allows the computation of strains from images, such as ultrasound [8], MRI tagging [14], DENSE [6], and Cine-MRI [12]. These methodologies vary in terms of spatial and temporal resolution and strain components that can be measured. Concerning this, in our analysis we use shear strains such as E_{ct} and E_{cl} that can be more difficult to obtain for some of the aforementioned methodologies. We chose those strain components because they emerged to be highly sensitive to fiber orientation from our previous sensitivity analysis [3]. It would be interesting to repeat our analysis using strain components that are more easily obtainable with all imaging methodologies, such as normal strains E_{cc}, E_{ll}, and E_{tt}. The proposed methodology is promising, but a few challenges remain to be addressed. We tested this methodology in a purely synthetic experiment, assuming perfect knowledge on other model parameters such as the material properties of the myocardium. We assumed a symmetric fiber field as ground truth, although in reality the fiber field varies circumferentially, thus requiring to estimate more parameters. Moreover, in the pathological case similar results may be more difficult to achieve, mainly for two reasons. First, in the pathological case we would like to characterize the pathology in terms of contractility and tissue stiffness, which would have to be estimated simultaneously. Second, pathologies may very likely affect the spatial distribution of strains, thus making more difficult to estimate fiber orientation from strains. The current methodology might be used as a calibration stage before applying data assimilation techniques to extract information about patient pathology from images and measurements. It requires one measurement of the observed strain components at end systole only, leaving other available information for the estimation of the pathology.

5   Conclusions

In this work we propose a methodology based on the Reduced Order Unscented Kalman Filter for the patient-specific estimation of parameters in a rule-base fiber model. The algorithm is based on the observation of end systolic strains, generated in-silico with a 12-parameter fiber model. Estimates of the three most influential parameters in an imperfect 5-parameter fiber model could be obtained with an average deviation of about 6° between ground truth and estimated fiber orientation, when using three end-systolic strain components, even when initial estimates deviated from their ground truth values by 1 SD. The proposed methodology does have the potential to estimate patient-specific fiber better than direct DT-MRI acquisition.

References

1. Barbarotta, L.: Towards computer assisted cardiac medicine: sensitivity analysis and data assimilation in models of left ventricular mechanics. PhD thesis, Department of Biomedical Engineering, Eindhoven University of Technology, Eindhoven, The Netherlands (2021)
2. Barbarotta, L., Bovendeerd, P.: A computational approach on sensitivity of left ventricular wall strains to geometry. In: Coudière, Y., Ozenne, V., Vigmond, E., Zemzemi, N. (eds.) FIMH 2019. LNCS, vol. 11504, pp. 240–248. Springer, Cham (2019). https://doi.org/10.1007/978-3-030-21949-9_26
3. Barbarotta, L., Bovendeerd, P.: A computational approach on sensitivity of left ventricular wall strains to fiber orientation. In: D. B. Ennis et al. (eds.) FIMH 2021. LNCS, vol. 12738 , pp. 1–9. Springer, Heidelberg (2021). https://doi.org/10.1007/978-3-030-78710-3_29
4. Bovendeerd, P., Kroon, W., Delhaas, T.: Determinants of left ventricular shear strain. Am. J. Physiol. Heart Circ. Physiol. **297**(3), H1058–H1068 (2009)
5. Geerts, L., Kerckhoffs, R., Bovendeerd, P., Arts, T.: Towards patient specific models of cardiac mechanics: a sensitivity study. In: Magnin, I.E., Montagnat, J., Clarysse, P., Nenonen, J., Katila, T. (eds.) FIMH 2003. LNCS, vol. 2674, pp. 81–90. Springer, Heidelberg (2003). https://doi.org/10.1007/3-540-44883-7_9
6. Hess, A.T., Zhong, X., Spottiswoode, B.S., Epstein, F.H., Meintjes, E.M.: Myocardial 3d strain calculation by combining cine displacement encoding with stimulated echoes (DENSE) and cine strain encoding (SENC) imaging. Magn. Reson. Med. **62**(1), 77–84 (2009)
7. Logg, A., Mardal, K.A., Wells, G.: Automated Solution of Differential Equations by the Finite Element Method. The FEniCS Book, vol. 84. Springer, Heidelberg (2012). https://doi.org/10.1007/978-3-642-23099-8
8. Lopata, R.G., Nillesen, M.M., Thijssen, J.M., Kapusta, L., de Korte, C.L.: Three-dimensional cardiac strain imaging in healthy children using RF-data. Ultrasound Med. Biol. **37**(9), 1399–1408 (2011)
9. Moireau, P., Chapelle, D.: Reduced-order unscented Kalman filtering with application to parameter identification in large-dimensional systems. ESAIM Control Optim. Calc. Var. **17**(2), 380–405 (2011)
10. Nagler, A., Bertoglio, C., Gee, M., Wall, W.: Personalization of cardiac fiber orientations from image data using the unscented Kalman filter. In: Ourselin, S., Rueckert, D., Smith, N. (eds.) FIMH 2013. LNCS, vol. 7945, pp. 132–140. Springer, Heidelberg (2013). https://doi.org/10.1007/978-3-642-38899-6_16
11. Pluijmert, M., Delhaas, T., de la Parra, A.F., Kroon, W., Prinzen, F.W., Bovendeerd, P.H.M.: Determinants of biventricular cardiac function: a mathematical model study on geometry and myofiber orientation. Biomech. Model. Mechanobiol. **16**(2), 721–729 (2016). https://doi.org/10.1007/s10237-016-0825-y
12. Puyol-Antón, E., et al.: Fully automated myocardial strain estimation from cine MRI using convolutional neural networks. In: 2018 IEEE 15th International Symposium on Biomedical Imaging, ISBI 2018, pp. 1139–1143. IEEE (2018)

13. Scollan, D.F., Holmes, A., Winslow, R., Forder, J.: Histological validation of myocardial microstructure obtained from diffusion tensor magnetic resonance imaging. Am. J. Physiol. Heart Circ. Physiol. **275**(6), H2308–H2318 (1998)
14. Yeon, S.B., et al.: Validation of *in vivo* myocardial strain measurement by magnetic resonance tagging with sonomicrometry. J. Am. Coll. Cardiol. **38**(2), 555–561 (2001)

Effects of Fibre Orientation on Electrocardiographic and Mechanical Functions in a Computational Human Biventricular Model

Lei Wang[1]($\boxtimes$), Zhinuo J. Wang[1], Ruben Doste[1], Alfonso Santiago[2], Xin Zhou[1], Adria Quintanas[2], Mariano Vazquez[2], and Blanca Rodriguez[1]

[1] Department of Computer Science, University of Oxford, Oxford, UK
lei.wang@cs.ox.ac.uk
[2] Barcelona Supercomputing Centre (BSC), Barcelona, Spain

Abstract. The helix orientated fibres in the ventricular wall modulate the cardiac electromechanical functions. Experimental data of the helix angle through the ventricular wall have been reported from histological and image-based methods, exhibiting large variability. It is, however, still unclear how this variability influences electrocardiographic characteristics and mechanical functions of human hearts, as characterized through computer simulations. This paper investigates the effects of the range and transmural gradient of the helix angle on electro-cardiogram, pressure-volume loops, circumferential contraction, wall thickening, longitudinal shortening and twist, by using state-of-the-art computational human biventricular modelling and simulation. Five models of the helix angle are considered based on *in vivo* diffusion tensor magnetic resonance imaging data. We found that both electrocardiographic and mechanical biomarkers are influenced by these two factors, through the mechanism of regulating the proportion of circumferentially-orientated fibres. With the increase in this proportion, the T-wave amplitude decreases, circumferential contraction and twist increase while longitudinal shortening decreases.

Keywords: Fibre orientation · Electromechanical modelling and simulation · Helix angle range · ECG · Pressure volume loops

1 Introduction

The heart is an electrically-driven pump that pushes blood to circulate through the body and lungs by periodic contraction and relaxation of the myocardium. The electrome-chanical function of the cardiac ventricles depends on the underlying microstructure of myocyte aggregation [1]. Based on histological analysis, the myocardium has been structurally modelled as a helix aggregation of the rod-shaped myocyte within extracel-lular matrix [2], with anisotropy in the electric propagation, active tension generation and resistance to passive deformation [3]. Light-microscopy has also shown the helical

© Springer Nature Switzerland AG 2021
D. B. Ennis et al. (Eds.): FIMH 2021, LNCS 12738, pp. 351–361, 2021.
https://doi.org/10.1007/978-3-030-78710-3_34

structure in the canine heart [4], and the transmural variation in the helix angle has been commonly reported in many species, including porcine [5] and human [6].

The *helix angle* is commonly defined as the angle between fibre orientation and the ventricular circumferential direction. The reported range of helix angle from *in vivo* diffusion tensor magnetic resonance imaging (DTMRI) of the human ventricles varies from study to study: $30°\sim-40°$ [7], $55°\sim-30°$ [8], $60°\sim-60°$ [9] and $90°\sim-90°$ [10]. The range from DTMRI data was found lower than that from the histological data (see Fig. 3 in [11]). Some possible reasons for this are the different heart states and a noisy DTMRI signal at the endocardium and epicardium [11], which are just the locations for measuring the range of helix angle.

The transmural gradient of the helix angle also varies from linear to nonlinear. A nonlinear transmural variation of $90°\sim-80°$ from the endocardium to the epicardium (see Fig. 4 in [4]) was reported from the histological analysis of a canine heart. The proportion of *circumferentially* $(0 \pm 22.5°)$ to *longitudinally orientated fibres* $(\pm 67.5$ to $\pm 90°)$ is approximately 10:1 [4]. A similar nonlinearity is observed in the histological analysis of a human heart with a smaller transmural range of helix angles, about $50°\sim-80°$ (see Fig. 16 in [6]). The transmural gradient of helix angle has been found to influence contraction [12] and diastolic filling [13] through computation models of rat left ventricle. It is however still unclear how the transmural gradient of the helix angle modulates the electromechanical function of human ventricles.

This paper investigates the effects of fibre orientation on clinical electrocardiographic and mechanical biomarkers using high-performance computer (HPC) simulations with a multiscale model of the human biventricular electromechanical function. Specifically, we evaluate the roles of range and transmural gradient of the helix angle on the simulated electrocardiogram (ECG), pressure-volume (PV) loops, left ventricular ejection fraction (LVEF), circumferential contraction, wall thickening, longitudinal shortening and twist. Unravelling the impact of fibre orientation on clinical markers through computational modelling and simulation is important to aid in data interpretation, and facilitate advancements in image acquisition and analysis to characterise human ventricular fibre architecture.

2 Method

2.1 Image-Based Human Biventricular Multiscale Model of Healthy Electromechanical Function

The human image-based biventricular electromechanical modelling and simulation framework developed, calibrated and evaluated in [14] was used as the basis of this study, see Fig. 1. Briefly, electrical propagation was simulated using the monodomain equation with the latest ToR-ORd model [15] for the human-based cellular membrane kinetics. The myocyte active contractile force generation was modelled using the human-based active tension of Land *et al.* model [16], coupled to the ToR-ORd as in [17]. The passive mechanics was based on the Holzapfel-Ogden model [18].

The human ventricular electromechanical model was implemented as a strongly-coupled system in the HPC numerical software, Alya [19], and simulations were conducted in the supercomputer Piz Daint in the Swiss National Supercomputing Centre.

The simulation per heartbeat, for a mesh with over 1.1 M linear tetrahedral elements, takes about 1.7 h on 10 nodes. The physiological conduction velocities of 67 cm/s, 30 cm/s, and 17 cm/s along the fibre, sheet, and sheet-normal directions were achieved by calibrating the diffusivities for this mesh in [14]. The same mesh was used throughout the study for different cases with variations of fibre orientations, and the mesh resolution should have little effects on the results since the conduction velocity has been calibrated.

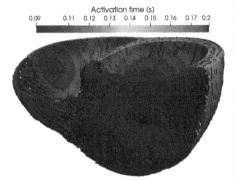

Geometry at diastasis	
LV-volume	103.3 mL
RV-volume	89.7 mL
Boundary conditions	
Endocardium	Pressure determined by two-element Windkessel flow model
Epicardium	Constrained by the elastic spring
Base	Fixed
Stimulus	
The current of 0.2 mA is applied at seven root patches for 1ms per 1 second.	

Fig. 1. Geometry of the human biventricular model with fibre structure and activation time map. Mechanical boundary conditions are applied to the endocardium, epicardium and base for a realistic deformation. Electrical propagation is elicited by stimulation on the entire endocardial surface originating from seven root nodes, as in [14], to mimic Purkinje-like activation.

2.2 Five Models of Helix Angle Variation

To evaluate the role of variability in fibre orientation in modulating ventricular electrocardiographic and mechanical biomarkers, five models of the helix angle in Table 1 were incorporated into the human biventricular model described in [14].

In Cases 1–4, the helix angle range varies based on the report in [7–10], with an assumed linear transmural gradient. However, with this linear variation, the histogram of the helix angle is inconsistent with that from the measurement in [10], which is a bell-shaped curve (see Fig. 3 in [10]). Instead, a nonlinear variation can produce this histogram, see Fig. 2 for Cases 4 and 5. Comparison of results from Case 5 to 4 can show the effects of nonlinear transmural gradient in the helix angle.

2.3 Quantification of Clinical Electrocardiographic and Mechanical Biomarkers

The simulated ECGs were computed by Alya on the fly using the pseudo-ECG method (spatial integral of the gradient of membrane potentials) at each time step [20, 21]. The electrocardiographic biomarkers, QRS duration, QT interval, T-wave amplitude, were quantified from the simulated ECGs. The start time of Q-wave and end times of S-, T-waves were identified through changes in the gradient of ECG signal.

The mechanical biomarkers, end-diastolic volume (EDV), end-systolic volume (ESV), ejection fraction (EF), ejection pressure, were quantified from the computed

Table 1. Five models of helix angle variation considered in the study, parameterized by the helix angle (α) range from endocardium to epicardium (second column), their transmural gradient through the left ventricular wall (as linear or nonlinear, third and fourth column), and proportion of circumferentially-oriented fibres with helix angle between $\pm 22.5°$ across the myocardium (fifth column). Variable t indicates a transmural position in the wall, and it is the percent of the distance between the position and endocardium out of the wall thickness.

Case no.	Helix angle range: endo ($t = 0$) ~ epi ($t = 100$)	Transmural gradient	Helix angles, α (°) $0 < t < 100$	Proportion of fibres with $-22.5° < \alpha < 22.5°$
1	30°~−40° [7]	Linear	$-0.7t + 30$	64%
2	55°~−30° [8]	Linear	$-0.85t + 55$	53%
3	60°~−60° [9]	Linear	$-1.2(t - 50)$	37%
4	90°~−90° [10]	Linear	$-1.8(t - 50)$	25%
5	90°~−90° [10]	Nonlinear	$-0.0005(t - 50)^3$ $-0.658(t - 50)$	48%

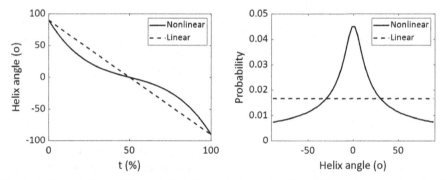

Fig. 2. Left: the helix angle through the left ventricular wall in Case 4 (linear transmural gradient) and Case 5 (nonlinear transmural gradient). Right: Histogram of helix angle values in Case 4 and Case 5, in which the histogram of the nonlinear model is consistent with that from the *in vivo* measurement of the human heart in [10].

PV loops. In each time step, the mean of each normal strain component in the circumferential-transmural-longitudinal coordinate system over all mesh nodes was computed as

$$mean\ E_{xx} = \frac{\sum_{i=1}^{N} E_{xx}^{(i)}}{N},$$

where, $xx = \{cc, tt, ll\}$ indicates the circumferential, transmural and longitudinal normal strain components, N is the number of mesh nodes, and $E_{xx}^{(i)}$ is the normal strain component of the node with index i along x direction. For any node on the endocardium, the transmural direction was determined by the vector to the node on the epicardium such that their distance is the shortest. The longitudinal direction was parallel to the apex-base direction, and the circumferential direction was orthogonal to both transmural and longitudinal directions. The circumferential contraction was *indicated* by the mean of circumferential strain, the longitudinal shortening by the mean of longitudinal strain, and the wall thickening by the mean of transmural strain, as they are naturally correlated. For example, when comparing results between different cases, a larger circumferential strain indicates larger circumferential expansion, and a lower circumferential strain or larger negative circumferential strain indicates larger circumferential contraction.

The twist angle is the relative rotational angle of the apex to the base. As shown in Fig. 3, the rotational angle of the apex was quantified by selecting all of the mesh nodes of the LV wall in a slice near the apex and parallel to the basal plane, and the average of the rotational angles over the selected nodes was computed and reported for each time step. Since the basal plane was fixed in our computation model [14], the twist angle was equal to this averaged rotational angle.

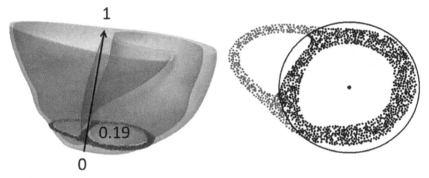

Fig. 3. Twist angle was computed by selecting a slice close to the apex (19% of the apex-base distance) parallel to the basal plane (left), and computing the rotational angles on the black-coloured nodes in the LV wall for each time step (right).

3   Results

The variation of electrocardiographic and mechanical biomarkers with altering the range and the helix angle transmural gradient were extracted from simulation results for the Cases 1–5 described in Table 1.

The comparison of mechanical effects between Cases 1–4 with various helix angle ranges is shown in Fig. 4(a)–(c). The PV loops show that the ejection pressure increases from Case 1 to 4, with increase in helix angle range, indicating a positive correlation between the ejection pressure and helix angle range. The LVESV in Case 3 is smaller than in other cases, so with a similar LVEDV for Cases 1–4, the LVEF in Case 3 is larger than in other cases. This is due to an intermediate helix angle range in Case 3 and a more detailed discussion on this point is provided in the next section.

Figure 4(d)–(g) shows effects of the helix angle range on simulated mean of circumferential, transmural and longitudinal strain, and twist angle of the apex against time.

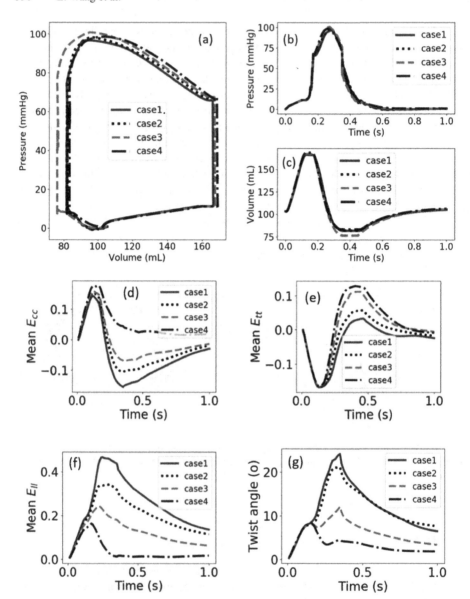

Fig. 4. Effects of the helix angle range on simulated mechanical properties for Cases 1–4: (a) PV loops of LV, (b) left ventricular pressure against time, (c) left ventricular volume against time, (d) mean of all nodal strain in circumferential (E_{cc}), (e) transmural (E_{tt}), (f) longitudinal (E_{ll}) directions, and (g) twist angle of the apex.

As seen in Fig. 4(b)–(c), the ventricles were sequentially inflated subject to a ramping ventricular pressure, activated, contracted and relaxed in the simulation.

In the first inflation phase, the circumferential diameter and longitudinal length of the ventricles were increased, meaning positive circumferential, E_{cc} in Fig. 4(d), and longitudinal strain, E_{ll} in Fig. 4(f), and the wall thickness was decreased, meaning negative transmural strain, E_{tt} in Fig. 4(e). The longitudinal strain, E_{ll}, was increased during the first 0.1 s due to inflation of the ventricles subject to pressure. Figure 4(f) shows that the longitudinal strain in Case 4 is the lowest. This is because the higher helix angle range than in other cases produces a larger proportion of longitudinally-orientated fibres, providing larger resistance to longitudinal extension than in other cases.

In the systolic phase, the circumferential contraction in Case 1 is the highest, as indicated by the largest negative value of E_{cc} at about t = 0.4 s in Fig. 4(d). The amplitude of circumferential contraction decreases from Cases 1 to 4, as indicated by the increase (less negative values) in E_{cc} at about t = 0.4 s. However, the wall thickening, indicated by E_{tt}, and longitudinal shortening, indicated by the negative values of E_{ll}, increases from Case 1 to 4. The twist angle decreases from Case 1 to 4, indicating that increase in the proportion of circumferentially-orientated fibres (Table 1, fifth column) can lead to increase in ventricular twist.

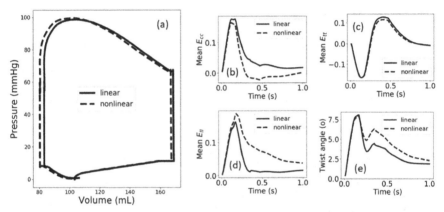

Fig. 5. Effects of linear (Case 4) versus nonlinear (Case 5) transmural gradient in helix angle on simulated mechanical properties: (a) PV loops of LV, (b) mean of all nodal strain in circumferential (E_{cc}), (c) transmural (E_{tt}), and (d) longitudinal (E_{ll}) directions, and (e) apical twist angle.

The effects of linear versus nonlinear transmural variation in helix angle on mechanical function are shown in Fig. 5 through a comparison between Case 4 and Case 5, respectively. The LVEF increases slightly by 2% with nonlinearity in Case 5 versus Case 4 in Fig. 5(a). The mean strain components in Fig. 5 (b)–(e) show that the deformation in diastolic phase before t = 0.2 s is similar, but it is different in the following systolic phase. Figure 5(b) shows that the circumferential strain E_{cc} at around t = 0.4 s is lower in Case 5 than in Case 4, so does the circumferentially expansion. While the effects on the longitudinal strain, E_{ll} in Fig. 5(d), are the opposite. Equivalently, the circumferential contraction in the nonlinear case is larger than in the linear one, while the longitudinal shortening is smaller. The nonlinear transmural variation also increases the circumferential strain rate, indicated by the shorter time over which E_{cc} increases and then decreases for case 5 compared to case 4. However, it decreases the longitudinal strain rate, indicated by the longer time taken for E_{ll} to increase and then decrease see

Fig. 5(d). The twist angle in Fig. 5(e) is larger in the case with nonlinear transmural helical variation than the one with linear.

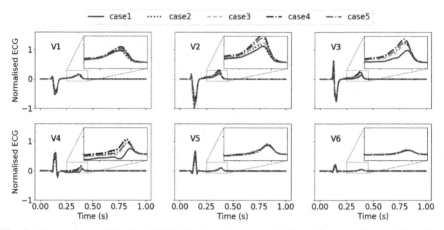

Fig. 6. Comparison of simulated ECG signals in the precordial leads between Cases 1–5 to show the effects of the range and transmural gradient of the helix angle in Table 1. The range increases from 30°~−40° in Case 1 to 90°~−90° in Case 4 with a linear transmural gradient. Case 5 has the same range as Case 4 but with nonlinear transmural gradient, see. The T-waves were zoomed in to show the influence in their amplitudes.

Figure 6 shows the effects of the range and the helix angle transmural gradient on ECGs. In leads V2, V3 and V4, the T-wave onset occurs progressively later from Case 4 to Case 1, which is especially apparent in lead V4. This is consistent with the comparison of repolarisation times between Cases 1 and 4 in Fig. 7, which shows earlier repolarisation of the LV in Case 4 than in Case 1. This could be because there is a greater proportion of longitudinally-oriented fibres in Case 4 than in Case 1, which means the electronic

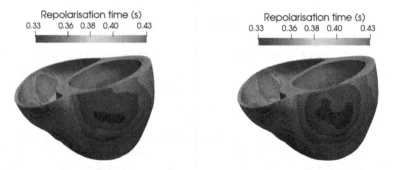

Fig. 7. Repolarisation times map in Case 1 (left) and Case 4 (right). Colour scale shows repolarization time at each ventricular location following propagation from endocardial activation. (Color figure online)

coupling is stronger in the apex-to-base direction, resulting in less heterogeneity in action potential duration and an earlier repolarisation overall. The other effect is that the T-wave amplitude increases with the helix angle range from Case 1 to 4 in leads V2 to V4 (see Fig. 6). This effect is likely due to differences in the distance between the precordial ECG leads and the systolic deformed ventricles. With increasing transmural fibre angle range, there is decreasing circumferential contraction, thus bringing the anterior portion of the LV in closer proximity to the precordial electrodes, resulting in a larger T-wave amplitude. A similar effect applies to the comparison between Cases 4 and 5, where the introduction of the transmural nonlinearity in Case 5 leads to greater circumferential contraction, resulting in decreased T-wave amplitude.

4 Discussion

In this study, we quantified the effects of the range and transmural gradient of the helix angle, on clinical electrocardiographic and mechanical biomarkers using simulations with a multiscale model of the human biventricular electromechanical function. We found that both electrocardiographic and mechanical biomarkers are influenced by these two helix angle properties, through regulating the proportion of circumferentially-orientated fibres. With an increase in this proportion, the circumferential contraction and twist increase while longitudinal shortening and T-wave amplitude decrease.

The comparison between Cases 1–4 shows that the largest LVEF is achieved in Case 3. This is attributed to the intermediate helix angle range. The lower range leads to an increase in the proportion of circumferentially orientated fibres, see Table 1. Case 1 with the lowest range has the most circumferentially-orientated fibres than other cases, leading to the circumferentially-dominated contraction. While Case 4 with the largest range has the most longitudinally-orientated fibres compared to other cases, leading to the longitudinally-dominated contraction. Case 3 with the intermediate helix angle range could have the optimal composition of circumferential and longitudinal contraction and consequently have the largest LVEF.

The nonlinearity in the transmural variation of the helix angle leads to a slight increase in LVEF of 2%, and also in twist angle (Case 5 versus Case 4 in Fig. 5). This can be explained by the increase in the proportion of circumferentially-orientated fibres as a result of this nonlinearity, see. The positive correlation between the proportion of circumferentially-orientated fibres and twist angle is consistent with analyses of results for Cases 1–4, and further prove that a large proportion of circumferentially orientated fibres plays an important role in ventricular twist. In Fig. 5(e), the twist angle at end of contraction, $t = 0.35$ s, increased by 41% in Case 5 to that in Case 4. The increase in circumferential strain rate and decrease in longitudinal strain rate is also attributed to the increase in the proportion of circumferentially orientated fibres in Case 5.

The results in Fig. 6 showed that the amplitude of T-wave increased with increasing the range of helix angle in Cases 1–4. This was due to a decrease in the proportion of circumferentially-orientated fibres. This correlation is also observed when comparing the ECG of Case 5 to Case 4 in Fig. 6: the amplitude of T-wave was lower in Case 5 than in Case 4. For Case 5, the proportion of circumferentially orientated fibres is larger than for Case 4, see the fifth column of Table 1.

The negative correlation between the proportion of circumferentially-orientated fibres and T-wave amplitude can be explained by the modulation of the distance between deformed ventricles and torso. As seen in the equation for computing the ECG potential or wave amplitude in (1) of [20], a shorter distance between the ventricles and torso will result in a larger amplitude. From Fig. 4(d), Case 4 had the least circumferential contraction or the largest diameters of ventricular cavities, leading to a shorter distance between ventricles and torso, and thus to the location of the ECG leads. Therefore, Case 4 has larger T-wave amplitude than in other cases.

5 Conclusion

A simulation study was conducted using a computational human biventricular model to quantify the effects of the range and transmural gradient of the helix angle on clinical electrocardiographic and mechanical biomarkers, using five models of the helix angle based on *in vivo* DTMRI data of the human heart. We found that both electrocardiographic and mechanical biomarkers are influenced by these two factors, through regulating the proportion of circumferentially-orientated fibres. With an increase in this proportion, the T-wave amplitude slightly decreases, circumferential contraction and twist increase while longitudinal shortening decreases. Thus, an optimal balance between circumferentially and longitudinally oriented fibres is crucial for the effective contraction of the heart. These findings have implications for modelling and simulation studies, and also for potential DTMRI data interpretation and to guide advancements in image acquisition and analysis to characterise human ventricular fibre architecture.

Acknowledgement. This work was funded by a Wellcome Trust Fellowship in Basic Biomedical Sciences to B.R. (214290/Z/18/Z), the CompBioMed2 Centre of Excellence in Computational Biomedicine (European Commission Horizon 2020 research and innovation programme, grant agreements No. 823712). The authors gratefully acknowledge PRACE for awarding access to SuperMUC-NG at Leibniz Supercomputing Centre of the Bavarian Academy of Sciences, Germany through project ref. 2017174226 and Piz Daint at the Swiss National Supercomputing Centre, Switzerland through an ICEI-PRACE project (icp005).

References

1. Hoffman, J.I.E.: Will the real ventricular architecture please stand up? Physiol. Rep. **5**, 1–13 (2017)
2. Sengupta, P.P., et al.: Left ventricular form and function revisited: applied translational science to cardiovascular ultrasound imaging. J. Am. Soc. Echocardiogr. **20**, 539–551 (2007)
3. Levrero-Florencio, F., et al.: Sensitivity analysis of a strongly-coupled human-based electromechanical cardiac model: effect of mechanical parameters on physiologically relevant biomarkers. Comput. Methods Appl. Mech. Eng. **361**, 112762 (2020)
4. Streeter, D.D., Spotnitz, H.M., Patel, D.P., Ross, J., Sonnenblick, E.H.: Fiber orientation in the canine left ventricle during diastole and systole. Circ. Res. **24**, 339–347 (1969)
5. Anderson, R.H., Smerup, M., Sanchez-Quintana, D., Loukas, M., Lunkenheimer, P.P.: The three-dimensional arrangement of the myocytes in the ventricular walls. Clin. Anat. **22**, 64–76 (2009)

6. Greenbaum, R.A., Ho, S.Y., Gibson, D.G., Becker, A.E., Anderson, R.H.: Left ventricular fibre architecture in man. Br. Heart J. **45**, 248–263 (1981)
7. Stoeck, C.T., et al.: Dual-phase cardiac diffusion tensor imaging with strain correction. PLoS ONE **9**, 1–12 (2014)
8. Toussaint, N., Stoeck, C.T., Schaeffter, T., Kozerke, S., Sermesant, M., Batchelor, P.G.: In vivo human cardiac fibre architecture estimation using shape-based diffusion tensor processing. Med. Image Anal. **17**, 1243–1255 (2013)
9. Nielles-Vallespin, S., et al.: In vivo diffusion tensor MRI of the human heart: reproducibility of breath-hold and navigator-based approaches. Magn. Reson. Med. **70**, 454–465 (2013)
10. Tseng, W.Y.I., Reese, T.G., Weisskoff, R.M., Brady, T.J., Wedeen, V.J.: Myocardial fiber shortening in humans: initial results of MR imaging. Radiology **216**, 128–139 (2000)
11. Wang, V.Y., et al.: Image-based investigation of human in vivo myofibre strain. IEEE Trans. Med. Imaging. **35**, 2486–2496 (2016)
12. Carapella, V., et al.: Quantitative study of the effect of tissue microstructure on contraction in a computational model of rat left ventricle. PLoS ONE **9**, 1–12 (2014)
13. Holmes, J.W.: Determinants of left ventricular shape change during filling. J. Biomech. Eng. **126**, 98–103 (2004)
14. Wang, Z.J., et al.: Human biventricular electromechanical simulations on the progression of electrocardiographic and mechanical abnormalities in post-myocardial infarction. EP Europace. **23**, i143–i152 (2021)
15. Tomek, J., et al.: Development, calibration, and validation of a novel human ventricular myocyte model in health, disease, and drug block. Elife **8**, e48890 (2019)
16. Land, S., Park-Holohan, S.-J., Smith, N.P., dos Remedios, C.G., Kentish, J.C., Niederer, S.A.: A model of cardiac contraction based on novel measurements of tension development in human cardiomyocytes. J. Mol. Cell. Cardiol. **106**, 68–83 (2017)
17. Margara, F., et al.: In-silico human electro-mechanical ventricular modelling and simulation for drug-induced pro-arrhythmia and inotropic risk assessment. Prog. Biophys. Mol. Biol. **159**, 58–74 (2020)
18. Holzapfel, G.A., Ogden, R.W.: Constitutive modelling of passive myocardium: a structurally based framework for material characterization. Philos. Trans. R. Soc. A: Math. Phys. Eng. Sci. **367**, 3445–3475 (2009)
19. Santiago, A., et al.: Fully coupled fluid-electro-mechanical model of the human heart for supercomputers. Int. J. Numer. Methods Biomed. Eng. **34**, e3140 (2018)
20. Gima, K., Rudy, Y.: Ionic current basis of electrocardiographic waveforms: a model study. Circ. Res. **90**, 889–896 (2002)
21. Mincholé, A., Zacur, E., Ariga, R., Grau, V., Rodriguez, B.: MRI-based computational torso/biventricular multiscale models to investigate the impact of anatomical variability on the ECG QRS complex. Front. Physiol. **10**, 1103 (2019)

Model-Assisted Time-Synchronization of Cardiac MR Image and Catheter Pressure Data

Maria Gusseva[1,2]⬤, Joshua S. Greer[3], Daniel A. Castellanos[4]⬤,
Mohamed Abdelghafar Hussein[3,5]⬤, Gerald Greil[3],
Surendranath R. Veeram Reddy[3]⬤, Tarique Hussain[3]⬤, Dominique
Chapelle[1,2], and Radomír Chabiniok[1,2,3(✉)]⬤

[1] Inria, Palaiseau, France
[2] Laboratoire de Mécanique des Solides (LMS), Ecole Polytechnique/CNRS/Institut Polytechnique de Paris, Palaiseau, France
[3] Division of Pediatric Cardiology, Department of Pediatrics, UT Southwestern Medical Center, Dallas, TX, USA
radomir.chabiniok@utsouthwestern.edu
[4] Department of Cardiology, Boston Children's Hospital, Boston, MA, USA
[5] Division of Pediatric Cardiology, Department of Pediatrics, Kafrelsheikh University, Kafr Elsheikh, Egypt

Abstract. When combining cardiovascular magnetic resonance imaging (CMR) with pressure catheter measurements, the acquired image and pressure data need to be synchronized in time. The time offset between the image and pressure data depends on a number of factors, such as the type and settings of the MR sequence, duration and shape of QRS complex or the type of catheter, and cannot be typically estimated beforehand. In the present work we propose using a biophysical heart model to synchronize the left ventricular (LV) pressure and volume (P-V) data. Ten patients, who underwent CMR and LV catheterization, were included. A biophysical model of reduced geometrical complexity with physiologically substantiated timing of each phase of the cardiac cycle was first adjusted to individual patients using basic morphological and functional indicators. The pressure and volume waveforms simulated by the patient-specific models were then used as templates to detect the time offset between the acquired ventricular pressure and volume waveforms. Time-varying ventricular elastance was derived from clinical data both as originally acquired as well as when time-synchronized, and normalized with respect to end-systolic time and maximum elastance value $(E_{orig}^N(t), E_{t-syn}^N(t),$ respectively). $E_{t-syn}^N(t)$ was significantly closer to the experimentally obtained $E_{exp}^N(t)$ published in the literature (p < 0.05, L^2 norm). The work concludes that the model-driven time-synchronization of P-V data obtained by catheter measurement and CMR allows to generate high quality P-V loops, which can then be used for clinical interpretation.

Keywords: Interventional cardiovascular magnetic resonance imaging · Time-synchronization of clinical data · Cardiovascular

© Springer Nature Switzerland AG 2021
D. B. Ennis et al. (Eds.): FIMH 2021, LNCS 12738, pp. 362–372, 2021.
https://doi.org/10.1007/978-3-030-78710-3_35

modeling · Pressure-volume loops · Personalized medicine · Translational research

1 Introduction

The analysis of intraventricular pressure-volume (P-V) loops contributes to the detailed assessment of the heart function and can be employed when planning a complex intervention. Combining the catheter pressure measurement with cardiovascular magnetic resonance imaging (CMR) yields rich morphological and functional datasets of the current state of patients' pathology. However, the recorded ventricular pressures and volumes are often not precisely synchronous, even during a simultaneous acquisition, which affects the shape of the P-V loop (see Fig. 1). There are three principal reasons for this P-V dyssynchrony:

1. Data of cine MRI are obtained over a number of cardiac cycles and the reconstructed time-volume plot therefore represents an average cycle during the acquisition.
2. The detection of R-peak in the strong magnetic field of MRI is known to be challenging, particularly in patients with a pathological QRS complex in whom the S wave or R' wave (e.g., in some right bundle branch block patients) may be detected instead. This leads into the time offset up to the QRS duration.
3. The fluid-filled catheter is known to record the pressure changes with some delay, which varies depending e.g., on the catheter size.

The time offset caused by these points varies among the patients and practically cannot be predicted beforehand. In this work we propose to create biophysical models of a reduced complexity using the patient-specific information of the maximum and minimum ventricular volume and pressure provided by the CMR and catheter data. These models will then be used as a template for the time-synchronization of the measured P-V data (see the pipeline depicted in Fig. 2). Qualitative assessment will be carried out by plotting the P-V loops using the original and the time-synchronized pressure and volume waveforms. The quantitative comparison will be performed by comparing the generated ventricular time-varying elastance functions with the experimentally obtained time-varying elastance published by Suga et al. [15]. Furthermore, some functional characteristics of the cardiovascular system not directly visible in the data can be obtained thanks to using a patient-specific biophysical model.

2 Methods

2.1 Data

Datasets of left ventricles of ten patients with repaired tetralogy of Fallot (rTOF) were included in the study. The data collections were performed under the ethical approvals of the Institutional Review Board of UT Southwestern Medical

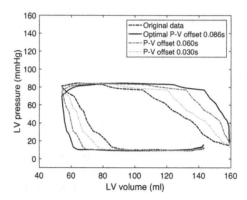

Fig. 1. Illustration of the pressure-volume (P-V) offsets on the shape of the P-V loop in the measured data.

Center Dallas (STU-2020-0023). The IRB waived the need for a consent to use the anonymized retrospective data. Cine MRI data with retrospective ECG gating (with parallel factor SENSE = 2, temporal resolution ∼30 ms) covering both ventricles and phase-contrast flow through the aortic valve were acquired. The ventricular volumes from cine MRI were obtained using the CVI42 software (Circle Cardiovascular Imaging Inc., Calgary, Canada) combined with the motion tracking algorithm of [4]. The catherization was performed in a separate session after CMR. A left-heart cathether was advanced into aorta and left ventricle, where the pressures were recorded during a breath-hold.

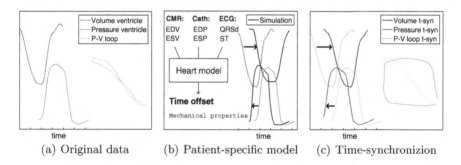

(a) Original data   (b) Patient-specific model   (c) Time-synchronizion

Fig. 2. Time-synchronization pipeline. (a) Original pressure-volume (P-V) data; (b) Patient-specific time offset detected by the model (black arrows); (c) Time offset detected in (b) applied on original P-V data in (a) to obtain time-synchronized (t-syn) P-V data (solid lines). ED/ESV: end-diastolic/-systolic volumes (ventricle); ED/ESP: end-diastolic/-systolic pressures (ventricle); ECG: electrocardiogram timings.

2.2   Biophysical Heart Model

A reduced-complexity biomechanical model of left ventricle (LV) and systemic circulation was employed [1]. The schematics of the model is depicted in Fig. 3. The geometry and kinematics of the ventricle are reduced to a sphere with an inner radius R and a wall thickness d, while the constitutive mechanical laws are preserved as in the full 3D heart model [3,12]. The mechanical behavior of the myocardium is described by a rheological model of Hill-Maxwell type [5] that contains an active contractile component (actin-myosin complex) and a visco-elastic component (collagen-rich elements in the surrounding connective tissue). The active component is represented within Huxley's sliding filament theory [6,7]. All myosin filaments in the sarcomere unit are considered as a series of springs, with active stress (τ_c) and active stiffness (k_c) produced for sarcomere extension (e_{fib}) given by, see [3]:

$$
\begin{cases}
\dot{k}_c = -(|u| + \alpha|\dot{e}_{fib}|)k_c + n_0(e_{fib})k_0|u|_+ \\
\dot{\tau}_c = -(|u| + \alpha|\dot{e}_{fib}|)\tau_c + \dot{e}_{fib}k_c + n_0(e_{fib})\sigma_0|u|_+,
\end{cases}
\tag{1}
$$

where $u(t)$ is an electrical activation function representing intracellular calcium kinetics that induces contraction (when $u > 0$) or relaxation ($u < 0$) of the myocardium (Fig. 3b). The parameter α is a bridge destruction rate upon rapid change in length in the sarcomere [3]. The parameter σ_0 is the active stress developed by the sarcomere during systole under optimal extension e_{fib}, and $n_0(e_{fib}) \in [0,1]$ represents the Frank-Starling law [3]. The parameter σ_0 will be further referred to as the contractility.

The model was turned into patient-specific regime according to the sequential calibration procedure described in [10]. Briefly, the 2-stage Windkessel model was adjusted by imposing the flow measured by phase-contrast MRI aiming to match the maximum and minimum aortic pressure and the pressure at the dicrotic notch. Then, the LV wall thickness and the ventricular volume were prescribed according to the cine MRI. The model stress-free configuration was assumed as in the experimental data of [8]. The ventricular preload was prescribed from the measured end-diastolic ventricular pressure. Passive myocardial properties were calibrated to match the measured end-diastolic volume under imposed preload. Myocardial contractility was adjusted to match the measured stroke volume. Patient-specific parameters adjusted during this process, therefore were in the Windkessel model: distal and proximal resistances of the circulation, distal capacitance of the circulation; in the heart model: myocardial stiffness and contractility. Ventricular stroke work was calculated as the area encompassed within the model-derived P-V loop.

2.3   Time-Synchronization

Thanks to the biophysical and physiological character of the model, the timing of the cardiac phases was considered as a reference. The simulated waveforms were therefore used as templates to time-synchronize the original data to find an

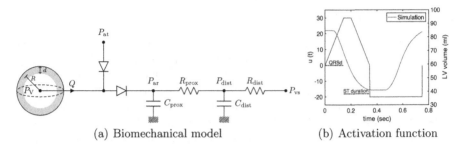

(a) Biomechanical model   (b) Activation function

Fig. 3. (a) Coupling of the heart with atrioventricular and arterial valves via system of diodes and with circulation system represented by a two-stage Windkessel model. $P_{at}, P_V P_{ar}, P_{dist}, P_{vs}$ stand for pressures in left atrium, left ventricle, aorta, distal systemic circulation and venous system; $R_{prox}, C_{prox}, R_{dist}, C_{dist}$ are proximal and distal resistances and capacitances in Windkessel model; (b) Electrical activation function $u(t)$ (black line) with imposed timings of duration of QRS and ST segment from electrocardiogram measurements.

optimal time offset t_{offset} between the measured data and the model, minimizing the criterion:

$$\min_{t_{\text{offset}}} \int_0^{t_{\text{ES}}} (f(t) - g(t - t_{\text{offset}}))^2 dt, \tag{2}$$

where $f(t)$ represents the simulated LV pressure or volume; $g(t)$ represents the measured ventricular pressure (volume); t_{ES} is the time of end-systole in the data.

The P-V loops were constructed by using both original and time-synchronized waveforms. Time-varying elastance was computed for the original and time-synchronized data as:

$$E_{\text{orig/t-syn}}(t) = \frac{P_{\text{orig/t-syn}}(t)}{V_{\text{orig/t-syn}}(t)}, \tag{3}$$

where $P_{\text{orig/t-syn}}(t)$, $V_{\text{orig/t-syn}}(t)$ are LV pressures and volumes as originally measured/time-synchronized. The time-varying elastances were then double-normalized in time (with respect to the end-systolic time t_{ES}) and with respect to the maximal elastance value, as in Suga et al. [15]:

$$E^N(\bar{t}) = \frac{E(\bar{t} \cdot t_{\text{ES}})}{\max(E)}. \tag{4}$$

2.4   Statistical Analysis

The distances between the normalized time-varying elastance obtained from the original, time-synchronized data, respectively (E^N_{orig}, $E^N_{\text{t-syn}}$, respectively) and the experimentally obtained E^N_{exp} (by Suga et al. [15]) were compared using the L^2 norm over the whole cardiac cycle, i.e. relative distance RD is given by

$$\mathrm{RD}(E^N_{\mathrm{orig/t\text{-}syn}}, E^N_{\mathrm{exp}}) = \frac{||E^N_{\mathrm{orig/t\text{-}syn}} - E^N_{\mathrm{exp}}||_{L^2}}{||E^N_{\mathrm{exp}}||_{L^2}}. \tag{5}$$

Wilcoxon signed-rank tests at $p < 0.05$ were conducted to assess the difference between $\mathrm{RD}(E^N_{\mathrm{orig}}, E^N_{\mathrm{exp}})$ and $\mathrm{RD}(E^N_{\mathrm{t\text{-}syn}}, E^N_{\mathrm{exp}})$.

3 Results

Figure 4 shows an example of the patient-specific model and the model-based time-synchronization for Patient #6. Figure 5 displays P-V loops for selected patients, while using original or time-synchronized pressure and volume waveforms. Figure 6(a) shows for patient #6 $E^N_{\mathrm{orig/t\text{-}syn}}(\bar{t})$ (original and time-synchronized data, respectively), in comparison with the experimentally obtained $E^N_{\mathrm{exp}}(\bar{t})$. Median $\mathrm{RD}(E^N_{\mathrm{orig}}, E^N_{\mathrm{exp}})$ and $\mathrm{RD}(E^N_{\mathrm{t\text{-}syn}}, E^N_{\mathrm{exp}})$ (original and time-synchronized data for all patients) were 0.16 and 0.03, respectively, as shown in Fig. 6(b). According to the Wilcoxon signed-rank test, $E^N_{t-syn}(\bar{t})$ was significantly closer to $E^N_{\mathrm{exp}}(\bar{t})$ ($p < 0.05$). Table 1 shows the time offsets for all patients detected by the proposed procedure between original and time-synchronized data. Table 1 (right columns) shows examples of patient-specific quantities derived by the model characterizing the functional state of the heart and circulation system.

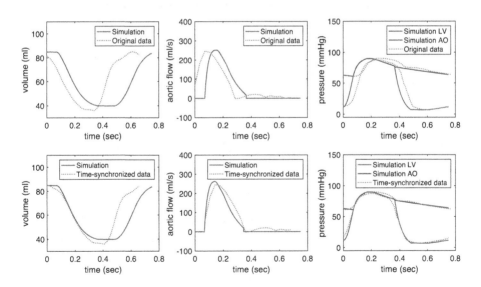

Fig. 4. Example of data-model coupling for Patient #6. Top: model calibration versus original (non-synchronized) data; bottom: model calibration versus time-synchronized data.

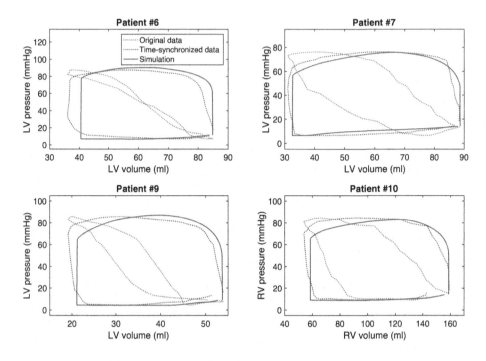

Fig. 5. Pressure-volume loops for selected patients.

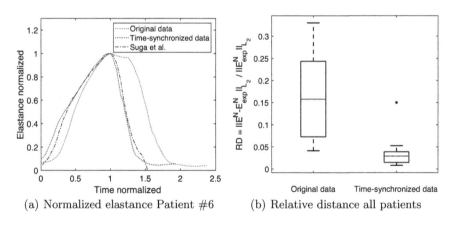

(a) Normalized elastance Patient #6   (b) Relative distance all patients

Fig. 6. (a) Normalized elastance for a selected patient versus elastance in [15]; (b) Boxplots of relative distance (RD) of normalized elastance between original/time-synchronized data and the elastance in [15] for all patients. Central line inside the boxes indicates the median, and the bottom and top edges of the boxes show the 25th and 75th percentiles, respectively. The star indicates significant difference at $p < 0.05$.

Table 1. Model-derived time offsets and biophysical parameters. QRSd: QRS duration; MRI: magnetic resonance imaging; Cath: catherization; R_{dist}: distal resistance of the circulation.

	Heart rate		Time offsets	Mechanical indicators		
QRSd	MRI	Cath	Pressure-volume	Contractility	Stroke work	R_{dist}
s	bpm	bpm	s	kPa	mJ	$\times 10^8 Pa \times s/m^3$
Pt #1	0.089	70	74	0.081	95	698	1.05
Pt #2	0.165	83	61	0.142	85	853	0.52
Pt #3	0.129	88	87	0.139	86	367	1.55
Pt #4	0.133	81	59	0.098	100	893	0.85
Pt #5	0.147	91	67	0.139	73	467	1.60
Pt #6	0.163	82	75	0.120	73	462	1.40
Pt #7	0.129	75	74	0.106	67	445	0.90
Pt #8	0.096	74	95	0.161	56	304	1.10
Pt #9	0.146	79	79	0.158	48	328	1.30
Pt #10	0.129	49	75	0.086	80	907	0.54

4   Discussion

In this work we developed a time-synchronization protocol for the ventricular pressure obtained by catheter measurement and volume from cine MRI using a patient-specific computational model. Figure 5 shows a qualitative improvement of the P-V loops after time-synchronizing the pressure and volume data. The box plot in Fig. 6 shows that in all 10 patients the shape of normalized time-varying elastance obtained from time-synchronized P-V data is substantially closer to the experimentally measured curve [15]. This suggests that the time-synchronization restores the time-varying elastance to give a physiologically more meaningful relationship. Even though the time-varying elastance obtained in [15] was obtained in dogs, some data obtained in other animal species (e.g., pigs [13]) or even from invasive measurements in human patients [9,14] suggest their compatibility with [15]. Therefore, in the present work we assumed the normalized time-varying elastance [15] to be a physiological representative of ventricular pressure-volume relationship. The time offsets detected by the model appeared to vary among the patients (Table 1) demonstrating that the same offset cannot be applied to all patients even though the same setup of MR acquisition was used. The offsets between P-V data were between 0.081 and 0.161 s, what is in the order of the QRS duration in all patients in our study. In addition to that in patients #2, #4, #5, #10 the difference between MRI and pressure heart rates was more that 30%. Therefore we hypothesize that when combining MRI and catherization techniques the offsets in the P-V data might be associated with several factors: wide QRS complex, difference in the heart rates, and an error in the order of temporal resolution of the MRI (0.020–0.030 s).

The model employed in this work was a biomechanical model incorporating physiology of muscle contraction built on various spatial scales. Of note, any other heart model with adequate timings of P-V change could be used as a

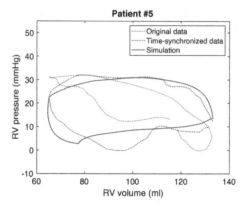

Fig. 7. Pressure-volume loops for right ventricle of patient #5 with 11% and 30% of tricuspid and pulmonary regurgitation, respectively.

template in the proposed time-synchronization method. The advantage of using our model [3] is that it can also account for the valvular mechanics (inflow and/or outflow valve regurgitation, see [2]) and hence can be used for time-synchronization of the ventricles with more complicated physiology. Figure 7 shows an example of model-assisted time-synchronization of the right ventricle of Patient #5 with 11% and 30% of tricuspid and pulmonary regurgitation, respectively, where the absence of isovolumic phases would have limited the ability to yield a physiological P-V loop without using a model as a resynchronization template. In addition, the advantage of using a truly biophysical model is that it provides, as a "side effect", additional patient-specific mechanical indicators of various compartments of the cardiovascular system (some examples are displayed in Table 1). Those may not be directly visible in the data but are clinically relevant. For instance, myocardial contractility characterizes the functionality of the ventricle, the stroke work can provide information about myocardial energetics, and the distal resistance of the circulation system is an important indicator of a functional state or cardiovascular health in a variety of pathologies. The proposed time-synchronization strategy is not limited neither to the CMR & catherization data acquisition protocol nor to the ventricular pressure & volume data itself, and could be utilized in the variety of patients' data. It can be used e.g., to time-synchronize data acquired in an interventional CMR suite (iCMR, [11]). Moreover, the model can substitute e.g., a missing part of the cardiac cycle in the volume waveform (typically, in highly accelerated cine MRI in prospective ECG trigger).

5   Conclusion

Time-synchronization of data using patient-specific biophysical modeling can be applied when combining various imaging techniques (echoardiography, dynamic

CT, or CMR) and/or when imaging is combined with catherization data. The proposed modeling framework is fast in computational time and can be employed in clinical settings allowing to plot high-quality P-V loops and extract some useful functional indices of the cardiovascular system.

Acknowledgments. The authors acknowledge the support of the Associated Team TOFMOD, created between Inria France and UT Southwestern Medical Center Dallas. The work was also funded in part by the W.B. & Ellen Gordon Stuart Trust, The Communities Foundation of Texas and by the Pogue Family Distinguished Chair (awarded to Dr F. Gerald Greil in February, 2015). In addition, we would like to acknowledge Dr Philippe Moireau, Inria research team MΞDISIM, for the development of the cardiac simulation software CardiacLab used in this work. Research reported in this publication was supported by Children's Health^{SM} but the content is solely the responsibility of the authors and does not necessarily represent the official views of Children's Health^{SM}.

References

1. Caruel, M., Chabiniok, R., Moireau, P., Lecarpentier, Y., Chapelle, D.: Dimensional reductions of a cardiac model for effective validation and calibration. Biomech. Model. Mechanobiol. **13**(4), 897–914 (2014)
2. Chabiniok, R., Moireau, P., Kiesewetter, C., Hussain, T., Razavi, R., Chapelle, D.: Assessment of atrioventricular valve regurgitation using biomechanical cardiac modeling. In: Pop, M., Wright, G.A. (eds.) FIMH 2017. LNCS, vol. 10263, pp. 401–411. Springer, Cham (2017). https://doi.org/10.1007/978-3-319-59448-4_38
3. Chapelle, D., Le. Tallec, P., Moireau, P., Sorine, M.: Energy-preserving muscle tissue model: formulation and compatible discretizations. Int. J. Multiscale Comput. Eng. **10**(2), 189–211 (2012)
4. Genet, M., Stoeck, C.T., Von Deuster, C., Lee, L.C., Kozerke, S.: Equilibrated warping: finite element image registration with finite strain equilibrium gap regularization. Med. Image Anal. **50**, 1–22 (2018)
5. Hill, A.V.: The heat of shortening and the dynamic constants of muscle. Proc. R. Soc. Lond. Ser. B Biol. Sci. **126**(843), 136–195 (1938)
6. Huxley, A.: Muscle structure and theories of contraction. Prog. Biophys. Biophys. Chem. **7**, 255–318 (1957). https://ci.nii.ac.jp/naid/10005175870/en/
7. Kimmig, F., Chapelle, D., Moireau, P.: Thermodynamic properties of muscle contraction models and associated discrete-time principles. Adv. Model. Simul. Eng. Sci. **6**(1), 1–36 (2019)
8. Klotz, S., et al.: Single-beat estimation of end-diastolic pressure-volume relationship: a novel method with potential for noninvasive application. Am. J. Physiol. Heart Circ. Physiol. **291**(1), H403–H412 (2006)
9. Le Gall, A., Vallée, F., Chapelle, D., Chabiniok, R.: Minimally-Invasive Estimation of Patient-Specific End-Systolic Elastance Using a Biomechanical Heart Model. In: Coudière, Y., Ozenne, V., Vigmond, E., Zemzemi, N. (eds.) FIMH 2019. LNCS, vol. 11504, pp. 266–275. Springer, Cham (2019). https://doi.org/10.1007/978-3-030-21949-9_29
10. Le Gall, A., et al.: Monitoring of cardiovascular physiology augmented by a patient-specific biomechanical model during general anesthesia. A proof of concept study. PLOS ONE **15**(5), e0232830 (2020)

11. Reddy, S.R.V., et al.: Invasive cardiovascular magnetic resonance (iCMR) for diagnostic right and left heart catheterization using an MR-conditional guidewire and passive visualization in congenital heart disease. J. Cardiovas. Magn. Reson. **22**(1), 1–11 (2020)
12. Sainte-Marie, J., Chapelle, D., Cimrman, R., Sorine, M.: Modeling and estimation of the cardiac electromechanical activity. Comput. Struct. **84**(28), 1743–1759 (2006)
13. Seemann, F., et al.: Noninvasive quantification of pressure-volume loops from brachial pressure and cardiovascular magnetic resonance. Circ. Cardiovasc. Imaging **12**(1), e008493 (2019)
14. Senzaki, H., Chen, C.H., Kass, D.A.: Single-beat estimation of end-systolic pressure-volume relation in humans: a new method with the potential for noninvasive application. Circulation **94**(10), 2497–2506 (1996)
15. Suga, H., Sagawa, K., Shoukas, A.A.: Load independence of the instantaneous pressure-volume ratio of the canine left ventricle and effects of epinephrine and heart rate on the ratio. Circ. Res. **32**(3), 314–322 (1973)

From Clinical Imaging to Patient-Specific Computational Model: Rapid Adaptation of the Living Heart Human Model to a Case of Aortic Stenosis

Andrew D. Wisneski[1], Salvatore Cutugno[2], Ashley Stroh[3], Salvatore Pasta[2], Jiang Yao[4], Vaikom S. Mahadevan[5], and Julius M. Guccione[1(✉)]

[1] Department of Surgery, University of California San Francisco, San Francisco, CA 94143, USA
julius.guccione@ucsf.edu

[2] Department of Engineering, Universita degli Studi Palermo, 90133 Palermo, Italy

[3] CATIA, Dassault Systemès, Wichita, KS 67208, USA

[4] Dassault Systemès, Johnston, RI 02919, USA

[5] Division of Cardiology, University of California San Francisco, San Francisco, CA 94143, USA

Abstract. Aortic stenosis (AS) is the most common acquired heart valve disease in the developed world. Traditional methods of grading AS have relied on the measurement of aortic valve area and transvalvular pressure gradient. Recent research has highlighted the existence of AS variants that do not meet classic criteria for severe AS such as low-flow, low-gradient AS. With the development of sophisticated multi-scale computational models, investigation into the left ventricular (LV) biomechanics of AS offers new insights into the pathophysiology that may guide treatment decisions surrounding AS. Building upon our prior study entailing LV-aortic coupling where AS conditions were applied to the idealized geometry of the Living Heart Human Model, we now describe the first patient-specific adaptation of the model to a case of low flow, low gradient AS. EKG-gated cardiac computed tomography images were segmented to provide surfaces to which the generic Living Heart model was adapted. The model was coupled to a lumped-parameter circulatory system; it was then calibrated to patient clinical data from echocardiography/cardiac catheterization with strong correlation (simulation versus clinical measurement): ascending aorta systolic pressure: 109 mmHg vs 116 mmHg, ascending aorta diastolic pressure 50 mmHg vs 45 mmHg, LV systolic pressure: 118 mmHg vs 128 mmHg, peak transvalvular gradient: 9 mmHg vs 12 mmHg, LV ejection fraction: 23% vs 25%. This work illustrates how the Living Heart Human Model geometry can be efficiently adapted to patient-specific parameters, enabling future biomechanics investigations into the LV dysfunction of AS.

Keywords: Aortic stenosis · Cardiac computed tomography · Patient-specific modeling

Electronic supplementary material The online version of this chapter (https://doi.org/10.1007/978-3-030-78710-3_36) contains supplementary material, which is available to authorized users.

D. B. Ennis et al. (Eds.): FIMH 2021, LNCS 12738, pp. 373–381, 2021.
https://doi.org/10.1007/978-3-030-78710-3_36

1 Introduction

The aortic valve is a tri-leaflet valve which enables blood to exit the left ventricle (LV) into the aorta and perfuse the rest of the body. Aortic stenosis (AS) is a progressive disease process by which calcium deposits on the valve leaflets prevent valve opening during systole, increasing the work the LV must perform to eject blood. Chronic increased afterload triggers LV remodeling and without treatment, AS will result in congestive heart failure with poor prognosis [1, 2]. AS remains the most common valvular heart disease in the world, and over 120,000 aortic valve replacement procedures are now performed annually in the United States [3]. While surgical aortic valve replacement was the original method of treating AS, the development of the minimally invasive transcatheter aortic valve replacement (TAVR) has revolutionized the treatment of this disease. Over recent years, the use of TAVR has continued to expand and its annual procedure volumes now supersede that of surgical aortic valve replacement.

With the growth of TAVR, there has been renewed study of AS and its pathophysiology. The classic means of grading AS severity as mild, moderate, and severe by measurable hemodynamic parameters such as estimated aortic valve area, transvalvular pressure gradient, or blood ejection jet velocity, remain widely used today. However, these are limited by being flow dependent posing challenges in diagnosis for patients that may have depressed cardiac function from multi-factorial reasons. Variants of AS have been identified that do not fit the classic hemodynamic definitions for severe AS but may warrant intervention for the patient's benefit. This includes low-flow, low-gradient (LFLG) AS, where the traditionally elevated trans-valvular pressure gradient is absent due to underlying LV disease [4, 5]. Clinicians must be cautious to not under estimate severity of AS, as subsets of patients with LFLG AS benefit from aortic valve replacement.

An improved understanding of AS pathophysiology and its impact on LV biomechanics should be pursued. AS should no longer be viewed as an isolated valve disease. With the advancement of computational modeling of the myocardium and the existence of multi-domain coupled models, we can now more readily study the biomechanical derangements associated with AS. The goal of treating AS is to prevent or halt irreversible myocardial damage from chronic afterload. Our prior work applied LV-aortic coupling principles to quantify myocardial stress associated with AS in the generic Living Heart Human Model (Simulia, Dassault Systèmes, Rhode Island, United States) [6]. Now, we present the model adapted to patient-specific geometry and describe the methods taking clinical imaging to arrive at a patient-specific adaption of the Living Heart Human Model. This work serves as a preliminary investigation for future biomechanical studies that will link LV function and biomechanics to AS severity and prognosis.

2 Methods

2.1 Patient Data and Image Acquisition

Diastolic phase images from an EKG-gated pre-procedural computed tomography (CT) scan were obtained for analysis (Fig. 1). CT scan slices were 0.625 mm contiguous axial images acquired on a GE Lightspeed VCT scanner (GE Healthcare, Chicago, Illinois,

United States). The patient was a 68-year-old man with LFLG AS scheduled to undergo TAVR, with reduced LV ejection fraction of 25–30% on echocardiography and a mean aortic transvalvular gradient (pressure drop) of 15 mmHg. The images were anonymized prior to analysis. This study was done in accordance with the standards of the University of California San Francisco's Committee for Human Research.

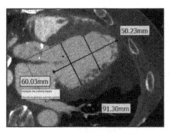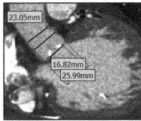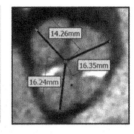

Fig. 1. Pre-processing of clinical computed tomography images prior to segmentation. Left panel: measurements of left ventricle chamber. Middle panel: Aortic root measurements. Right panel: the aortic cusps. A tri-leaflet aortic valve is clearly seen.

2.2   Segmentation of Cardiac CT Images

CT images were processed by Mimics (v.21, Materialise, Leuven, Belgium) to segment the heart, including the LV, left atrium (LA), and the aortic root, using different gray values and multiple masks (Fig. 2). Semi-automatic thresholding of the contrast-enhanced images was done to obtain the LA, aortic root, and the LV intraventricular chamber (i.e., blood volume). A Boolean mask difference was used to distinguish the LV myocardium from the LV chamber. Once the segmented regions were extracted, each mask was independently exported as stereolithographic (STL) files to use for adaptation of the Living Heart Human Model geometry.

2.3   The Living Heart Human Model

The Living Heart Human Model is a multi-domain anatomically comprehensive computational model coupled to a lumped parameter circulatory system that has been previously described [7–9]. The current iteration of the Living Heart Human Model utilizes second-order quadratic tetrahedral elements to eliminate volumetric locking, offering high quality mesh across a range of element sizes. Mesh size was selected based on convergence studies for the diastolic filling analyses, resulting in a model with approximately 100,000 elements. Cardiac fiber orientation followed a rule-based approach from $-60°$ from epicardium to $+60°$ at the endocardium.

The ventricular material model passive response employs the anisotropic hyperelastic formulation developed by Holzapfel and Ogden in which eight material parameters a, b, a_f, b_f, a_s, b_s, a_{fs}, b_{fs}, and four strain invariants $\left(\overline{I}_1, \overline{I}_{4f}, \overline{I}_{4s}, \overline{I}_{8fs}^2\right)$ define the strain energy potential (see Eq. 1), whereas the active myocardial tissue response is governed by a

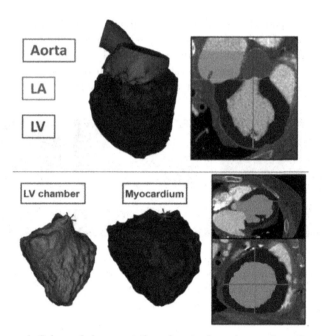

Fig. 2. Upper panel: Color coded segmentation of anatomic regions: red- aortic root, green- left atrium, blue- left ventricle. Lower panel: Differentiation of left ventricle chamber volume from myocardium. (Color figure online)

time-varying elastance model (see citations for further details of the myocardial model) [10–12].

$$\Psi_{dev} = \frac{a}{2b}\exp\left[b(I_1 - 3)\right] + \sum_{i=f,s}\frac{a_i}{2b_i}\left\{\exp\left[b_i\left((I_{4i} - 1)^2\right)\right] - 1\right\} + \frac{a_{fs}}{2b_{fs}}\left[\exp\left(b_{fs}I_{8fs}^2 - 1\right)\right] \quad (1)$$

The finite element model of the Living Heart Human Model for this study included the LV, a portion of the LA, the aortic root and ascending aorta. Given the clinical scenario being modeled does not entail primary right ventricular dysfunction, the right ventricle was excluded for computational efficiency. The structural finite element model was coupled to a lumped parameter model of the systemic and pulmonary circulatory system (see Fig. 4). The aortic valve resistance parameter was increased by a factor of approximately five above the baseline value ($5e-9$ MPa*s/mm^3) to create the appropriate degree of stenosis.

The 3DEXPERIENCE (Simulia, Dassault Systems, Johnston, RI, USA) platform "smart geometry" feature permitted rapid adaptation of the generic Living Heart ventricle geometry to the STL files of the patient-specific image segmentation (Fig. 3). Built-in parameterization for the mitral valve annulus and LV shape allowed for rapid model changes to roughly match the patient geometry. Manual morphing was then performed to fine-tune the geometry match. After these adjustments, the model is fully connected to the lumped parameter circulatory model and simulation ready. This process required 1.5 days to complete.

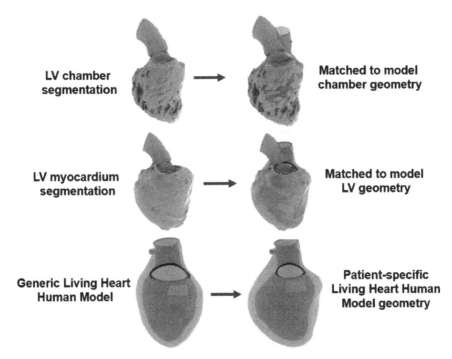

Fig. 3. Morphing Living Heart Human Model geometry to patient-specific geometry using STL files as constraints for Abaqus smart geometry.

2.4 Adaptation of Myocardial Material Model

The biaxial and triaxial experimental data published by Sommers et al. were used for initial calibration of the myocardial material properties [13]. Diastolic filling tests were used to augment the calibration of the eight material parameters: a, b, a_f, b_f, a_s, b_s, a_{fs}, b_{fs} describing the ventricular passive material properties based on the methods described in Klotz et al. [14]. In principle, the diastolic pressure-volume curve for a chamber can be estimated using a single measured volume at one pressure point. Best parameter fits were obtained while maintaining ratios between certain material parameters constant to preserve anisotropy. The parameters for the LV passive properties were as follows: $a = 7.89\mathrm{e}{-04}$, $b = 3.67\mathrm{e}{+0}$, $a_f = 3.70\mathrm{e}{-03}$, $b_f = 1.44\mathrm{e}{+01}$, $a_s = 2.11\mathrm{e}{-03}$, $b_s = 1.06\mathrm{e}{+01}$, $a_{fs} = 7.12\mathrm{e}{-07}$, $b_{fs} = 7.82\mathrm{e}{+0}$. Of note, different material parameters were used for the mitral valve annulus and aortic root/ascending aorta.

2.5 Cardiac Cycle Simulations

Cardiac cycles were simulated with Abaqus FEA (Simulia, Johnston, RI, United States) with the cardiac and aortic chamber pressures adjusted as close to the clinically obtained parameters of LV/aortic systolic and diastolic blood pressures, and LV ejection fraction (Fig. 4). A semi-automated machine-learning based workflow adjusts elastance of the chambers to provide as close agreement to the patient clinical data. Five cardiac cycles

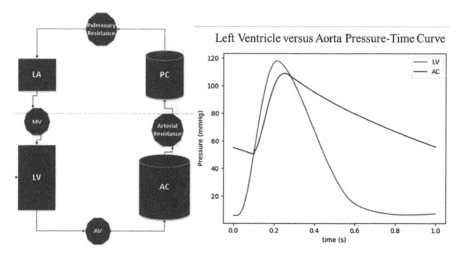

Fig. 4. Left panel: Diagram of the lumped parameter circulatory system coupled to the Living Heart Human Model. Parameters: aortic valve resistance (AV) $5e-9$ MPa*s/mm^3, arterial resistance $1.35e+02$ MPa*sec/mm^3, pulmonary vascular resistance $8e+0$ MPa*sec/mm^3, mitral valve resistance $2e+0$ MPa*sec/mm^3, arterial compliance (AC) $1.0e+07$ mm^3/MPa, pulmonary compliance (PC) $7.99e+06$ mm^3/MPa. Right panel: Pressure-time curve of the left ventricle (LV, red curve) and ascending aorta (AC, blue curve) demonstrating gradient from aortic stenosis, as the left ventricle's peak pressure exceeds the aortic peak systolic pressure. (Color figure online)

were considered adequate for the simulation to have achieved steady state, with <5% variation in chamber pressures thereafter.

3 Results

The results of major physiologic parameters from the patient compared to the tuned simulation are listed in Table 1. LV end diastolic volume was 266 ml and LV end systolic volume was 206 ml, corresponding to an LV ejection fraction of 23%. There was strong correlation with global variables of blood pressure and ejection fraction. Animations of the simulation are available (See Videos 1 and 2).

Table 1. Comparison of patient clinical variables with simulation results

Parameter	Clinical measurement	Simulation result	Error
LV ejection fraction	25%	23%	−8.0%
LV systolic pressure	128 mmHg	118 mmHg	−8.0%
LV diastolic pressure	12 mmHg	6 mmHg	−50%
Pulse pressure	71 mmHg	59 mmHg	−17%
Ascending aorta systolic blood pressure	116 mmHg	109 mmHg	−6%
Ascending aorta diastolic blood pressure	45 mmHg	50 mmHg	11%

4 Discussion

This simulation demonstrates preliminary feasibility that the Living Heart Human Model geometry can be efficiently adapted from clinical imaging to create patient-specific geometry for computational models. Clinical data to calibrate models can be readily obtained from echocardiography, cardiac CT or magnetic resonance imaging, and cardiac catheterization. Prior application of such methods for calibrating patient-specific myocardial material properties have been published [15, 16]. Cardiac-gated CT imaging is routinely obtained for any patient who may undergo a TAVR procedure. For research on the topic of patient-specific LV modeling in aortic stenosis patients, there is a vast amount of clinical data to be referenced to ensure fidelity of models and to link biomechanics data with clinical outcomes.

The methods used in this work will lay the foundation for future studies that will delve more extensively into the LV biomechanics derangements associated with AS. In our prior work, the generic Living Heart Human Model was subjected to varying degrees of AS and significant differences in global LV systolic stress were measured. Regional LV analysis by the American Heart Association of 17 standardized myocardial segments revealed a high degree of stress variability throughout the LV at systole (ranging from 6 kPa to 40 kPa) [6, 17]. Use of this standardized myocardial segmentation for regional biomechanical analysis could portray the unique biomechanical profile for a patient's LV, or serve as a telltale marker for populations of patients with similar states of disease severity. Further studies will need greater sample sizes of patients to characterize the link between LV biomechanics and AS severity. The longer-term goal of this project is to provide a meaningful tool to help clinicians decide whether to intervene for a patient with AS that does not fit the classic severe criteria. We believe LV biomechanics and computational modeling can play a role in detecting early onset myocardial dysfunction.

4.1 Limitations

There are several limitations of this simulation. For one, it does not incorporate directly measured patient-specific material properties, though computational methods incorporating clinical data can be highly accurate. Additionally, the experimental data used as the reference for passive myocardial material properties was derived from human heart tissue not necessarily representative of a patient with LV dysfunction in LFLG AS. The arrangement of myofiber orientation in the remodeled LV may differ from the applied rule-based method; the extent to which the rule-based approach requires modification in order to provide accurate results will require further investigation. Furthermore, this model excluded the right heart from the finite element component. While interventricular dependency may influence LV biomechanics in certain circumstances or with conditions that place strain on the right ventricle (i.e. pulmonary hypertension, tricuspid valve insufficiency), it is less crucial in isolated aortic stenosis. Afterall, treatment for AS with associated LV dysfunction not attributable to other cardiac disease is isolated aortic valve replacement.

This study's primary purpose was to demonstrate an efficient manner of adopting generic Living Heart Human Model geometry to a patient-specific case, and only major physiologic parameters were used to calibrate the model. Other parameters indicative of LV function should be incorporated in future iterations, such as segmental strain, LV end systolic/end diastolic volumes. Furthermore, matching the simulation to one specific set of physiologic variables has clinical limitations, as a patient's blood pressure and ejection fraction will naturally vary in daily life or depending on level of activity.

5 Conclusions

This study represents the first adaptation of the Living Heart Human Model to a patient-specific LV geometry for a case of AS. Strong correlation to clinical parameters was achieved, and this will serve as the foundation for future investigations on the impact of AS on LV biomechanics. Efficient methods to translate clinical imaging to patient-specific computational models will be needed to launch larger scale investigations elucidating the relationship between AS and the LV.

Acknowledgement. We thank Pamela Derish in the Department of Surgery, University of California, San Francisco, for assistance with proofreading the manuscript.

References

1. Lindman, B.R., Bonow, R.O., Otto, C.M.: Current management of calcific aortic stenosis. Circ. Res. **113**(2), 223–237 (2013). https://doi.org/10.1161/circresaha.111.300084
2. Miura, S., et al.: Causes of death and mortality and evaluation of prognostic factors in patients with severe aortic stenosis in an aging society. J. Cardiol. **65**(5), 353–359 (2015). https://doi.org/10.1016/j.jjcc.2015.02.011
3. Carroll, J.D., et al.: STS-ACC TVT registry of transcatheter aortic valve replacement. Ann. Thorac. Surg. **111**(2), 701–722 (2021). https://doi.org/10.1016/j.athoracsur.2020.09.002
4. Hachicha, Z., Dumesnil, J.G., Bogaty, P., Pibarot, P.: Paradoxical low-flow, low-gradient severe aortic stenosis despite preserved ejection fraction is associated with higher afterload and reduced survival. Circulation **115**(22), 2856–2864 (2007). https://doi.org/10.1161/circulationaha.106.668681
5. Pibarot, P., Dumesnil, J.G.: Low-flow, low-gradient aortic stenosis with normal and depressed left ventricular ejection fraction. J Am. Coll. Cardiol. **60**(19), 1845–1853 (2012). https://doi.org/10.1016/j.jacc.2012.06.051
6. Wisneski, A.D., et al.: Impact of aortic stenosis on myofiber stress: translational application of left ventricle-aortic coupling simulation. Front. Physiol. **11**, 1157 (2020). https://doi.org/10.3389/fphys.2020.574211
7. Baillargeon, B., Rebelo, N., Fox, D.D., Taylor, R.L., Kuhl, E.: The living heart project: a robust and integrative simulator for human heart function. Eur. J. Mech. A Solids **48**, 38–47 (2014). https://doi.org/10.1016/j.euromechsol.2014.04.001
8. Genet, M., et al.: Distribution of normal human left ventricular myofiber stress at end diastole and end systole: a target for in silico design of heart failure treatments. J. Appl. Physiol. **117**(2), 142–152 (2014). https://doi.org/10.1152/japplphysiol.00255.2014
9. Sack, K.L., Dabiri, Y., Franz, T., Solomon, S.D., Burkhoff, D., Guccione, J.M.: Investigating the role of interventricular interdependence in development of right heart dysfunction during LVAD support: a patient-specific methods-based approach. Front. Physiol. **9**, 520 (2018). https://doi.org/10.3389/fphys.2018.00520

10. Holzapfel, G.A., Ogden, R.W.: Constitutive modelling of passive myocardium: a structurally based framework for material characterization. Philos. Trans. A Math. Phys. Eng. Sci. **367**(1902), 3445–3475 (2009). https://doi.org/10.1098/rsta.2009.0091

11. Guccione, J.M., McCulloch, A.D.: Mechanics of active contraction in cardiac muscle: Part I-Constitutive relations for fiber stress that describe deactivation. J. Biomech. Eng. **115**(1), 72–81 (1993). https://doi.org/10.1115/1.2895473

12. Walker, J.C., et al.: MRI-based finite-element analysis of left ventricular aneurysm. Am. J. Physiol. Heart Circ. Physiol. **289**(2), H692–H700 (2005). https://doi.org/10.1152/ajpheart.01226.2004

13. Sommer, G., et al.: Biomechanical properties and microstructure of human ventricular myocardium. Acta Biomater. **24**, 172–192 (2015). https://doi.org/10.1016/j.actbio.2015.06.031

14. Klotz, S., et al.: Single-beat estimation of end-diastolic pressure-volume relationship: a novel method with potential for noninvasive application. Am. J. Physiol. Heart Circ. Physiol. **291**(1), H403–H412 (2006). https://doi.org/10.1152/ajpheart.01240.2005

15. Dabiri, Y., et al.: Method for calibration of left ventricle material properties using 3D echocardiography endocardial strains. J. Biomech. Eng. **141**(9), 0910071–09100710 (2019). https://doi.org/10.1115/1.4044215

16. Wenk, J.F., et al.: First evidence of depressed contractility in the border zone of a human myocardial infarction. Ann. Thorac. Surg. **93**(4), 1188–1193 (2012). https://doi.org/10.1016/j.athoracsur.2011.12.066

17. Cerqueira, M.D., et al.: Standardized myocardial segmentation and nomenclature for tomographic imaging of the heart: a statement for healthcare professionals from the Cardiac Imaging Committee of the Council on Clinical Cardiology of the American Heart Association. Circulation **105**(4), 539–542 (2002). https://doi.org/10.1161/hc0402.102975

Translational Cardiac Mechanics

Cardiac Support for the Right Ventricle: Effects of Timing on Hemodynamics-Biomechanics Tradeoff

Ileana Pirozzi[1]([⊠])[iD], Ali Kight[1][iD], Edgar Aranda-Michael[3][iD], Rohan Shad[4],
Yuanjia Zhu[1,4], Lewis K. Waldman[5], William Hiesinger[4], and Mark Cutkosky[2]

[1] Department of Bioengineering, Stanford University,
Stanford, CA 94305, USA
ipirozzi@stanford.edu
[2] Department of Mechanical Engineering, Stanford University,
Stanford, CA 94305, USA
[3] Department of Biomedical Engineering, Carnegie Mellon University, Pittsburgh,
PA 15289, USA
[4] Department of Cardiothoracic Surgery, Stanford University, Stanford,
CA 94305, USA
[5] Insilicomed, La Jolla, CA, USA

Abstract. A well-established treatment option for advanced heart failure is the implantation of a ventricular assist device (VAD) in the left heart. In over one quarter of patients, however, failure of the right ventricle (RV) occurs shortly after implantation, with a paucity of options for RV failure management in this clinical context. A possible treatment for RV failure is the application of regional mechanical support to the free surface of the RV. Here, we investigate the effect of this treatment using a multiscale finite element model. We discuss a trade-off between hemodynamic benefits and biomechanical effects of simulated interventions with respect to the complex dynamics of RV contraction. Specifically, we report on timing of support with respect to the cardiac cycle, duration of applied force, and force profile distribution. Insights from these preliminary studies can be informative in the rational design of RV-specific mechanical support solutions.

Keywords: Right heart failure · Biomechanical cardiac modeling · Medical device design

1 Introduction

For decades, the focus of heart failure treatment and interventions has largely been the left ventricle, with one of the most groundbreaking technologies added to the toolkit of heart failure being the Left Ventricular Assist Device (LVAD).

I. Pirozzi and A. Kight—Equally contributing authors.

© Springer Nature Switzerland AG 2021
D. B. Ennis et al. (Eds.): FIMH 2021, LNCS 12738, pp. 385–395, 2021.
https://doi.org/10.1007/978-3-030-78710-3_37

Yet, while LVADs have provided a glimmer of hope for patients with late-stage heart failure, right heart failure (RHF) following implantation remains a highly unpredictable, often fatal complication [1,2]. To date, up to 30–40% of LVAD patients develop RHF shortly after implantation, facing a paucity of clinical management options [1,3]. Following LVAD insertion, the RV is challenged in a number of ways. First, with near normalization of cardiac output, preload increases dramatically. Second, the LVAD causes unloading of the left ventricular cavity resulting in a interventricular septal shift and altered right ventricular (RV) geometry. Third, increases in pulmonary vascular resistance and pulmonary pressures encountered in the perioperative period further compromise the failing ventricle [4]. Such complications result in RV dysfunction. With an increasing number of patients undergoing LVAD implantation, the need for easily-deployed, safe, and effective devices for temporary support of the RV will continue to grow [5]. In order to design such therapies, the mechanisms of physiological (and pathophysiological) RV dynamics must be elucidated. External, non-blood contacting solutions provide several advantages [6]. Importantly, pulsatile flow augmentation (or assistance in pump function) can be achieved across the native cardiac vasculature, without the need to divert blood through artificial circuits, and thus not incurring the risk of thrombolytic events. To promote the development of alternative, external solutions supporting the RV free wall, we propose to characterize the effects of distributed compressive forces applied on the RV, with an eye on device-tissue interactions and effects on tissue strains and deformations. These considerations are generally difficult to evaluate due to the complex, multiscale nature of cardiac physiology. Additionally, the right ventricle presents more peculiar geometry and contraction patterns than the left ventricle, incurring additional complications. To meet these challenges, we employ a widely characterized multiscale finite element analysis software package (Continuity Pro, Insilicomed, La Jolla, CA) developed specifically for dynamic cardiac modeling and extensively used for medical device modeling. Using similar methods, recent interest in direct cardiac compression (DCC) has led to preliminary investigations to evaluate the effects of external forces on the cardiac surface to aid left ventricular (LV) performance [7]. Here, we highlight the need for investigations focused specifically on the right ventricle. Despite the existence of several other etiologies of RV failure, the clinical context of LVAD implantation provides an ideal scenario for such investigations, given that LV pump function is replaced by the LVAD. Specifically, any external device would provide a finite amount of energy to be applied to the ventricular surface. The goal of this study is to identify physically-interpretable conditions to apply such finite energy to maximise cardiac output and performance. To achieve long-term sustainable outcomes and promote myocardial health, ventricular deformations resulting from the application of such forces must also be minimized, in order to prevent the onset of adverse cardiac remodeling and ensure viability of this approach [8]. Such insights will aid the development of engineered solutions tailored to the RV complex shape, motion profiles, and function.

2   Methods

Stable beating-heart simulations were established using a multiscale finite element analysis software package for dynamic cardiac modeling (Continuity Pro, Insilicomed, La Jolla, CA) [7,9]. Continuity Pro (CPro) recapitulates patient-specific cardiac mechanics and electrophysiology by combining imaging-derived cardiac geometry and 3D fiber architecture, a dynamic model of calcium excitation-contraction coupling dynamics, and a lumped parameter circulatory model to simulate the function of the beating heart through both systemic and pulmonic vasculature. A transversely isotropic constitutive law, adapted from previous work was employed alongside a Fung-type hyperelastic model with exponential strain energy density function [10]. Interested readers can find in-depth description of the software and the patient-based model employed in these experiments in previous work [11].

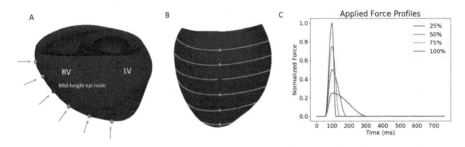

Fig. 1. Experimental model setup. A) Rendering of human biventricular model with applied nodal compressive forces. The mid-height endocardial node used for biomechanical analysis is shown in green. B) Rendering of the biventricular model displaying RV free wall. C) Time-varying force profiles of 25%, 50%, 75%, and 100% max force, over one cardiac cycle (Color figure online)

The human biventricular model employed in the following analyses is shown in Fig. 1. The model consists of 192 tri-cubic Hermite elements and 282 nodes, a refinement determined in previous work to obtain sufficient longitudinal resolution with respect to biomechanical metrics of interest [7]. The resulting refinement provides 6 longitudinal nodes at which compressive forces are applied during contraction (defined as $0 < t < 300\,\mathrm{ms}$, over a cycle duration of $750\,\mathrm{ms}$). Specifically, the coordinate system for the strain tensor was selected as the orthonormal basis of the local fiber coordinate system, with each fiber having (1) a direction along its own axis, (2) a direction perpendicular to the fiber and along the surface of the heart, and (3) a direction perpendicular to the fiber and directed radially towards the lumen. E11, E22 and E33 are thus the normal strains in the fiber, cross-fiber and radial directions respectively. The first principal strain is denoted as P3. Details and illustrations on the selected coordinate system are described in extensive detail in previous work [12].

In this study, Cpro was used to generate beating-heart simulations of a patient-specific model with anatomic dimensions congruent with a severe case of dilated cardiomyopathy and a LV ejection fraction of 16%, consistent with a clinical scenario benefiting from LVAD support. To simulate the effect of an external support device, compressive forces were applied to the LV and RV free wall epicardial surfaces using force boundary conditions established as a function of time throughout the cardiac cycle. Simulations were run at constant heart rate of 80 beats per minute and aortic valve opening defined as the end of isovolumetric contraction, IVC (t = 146 ms, where t = 0 is end-diastole).

A total of five beating-heart simulations were set up (see Fig. 1B): one baseline case where no external support is applied and four applied-force cases. The four active force simulations were prescribed time-varying force profiles applied to epicardial nodes and normal to the RV free wall surface throughout contraction. The maximum applied force to the epicardial RV is 0.4 N while the LV was kept at a constant 1 N throughout the simulations. The maximum force magnitude was informed by previous experimental work optimizing RV epicardial restraint pressure, with the goal of optimally reducing end-diastolic wall stress without impairing filling [13]. The integral of the force-time graph was kept constant at 31.25 N.s across all simulations in order to simulate a finite amount of energy available from a prospective device, and to investigate the effect of timing of the applied force on cardiac dynamics. To simulate the use of a device applied to the entire epicardial surface, 6 points along the longitude of the RV free wall were selected for force application (Fig. 1A). Force profiles were chosen based on reasonable expectations of physically realizable devices that could achieve similar force-time behavior. Peak forces were selected to be within physiological ranges. Particular consideration was given to the fact that higher RV peak pressure may result detrimental to pulmonary remodeling and have deleterious effects on pulmonary vasculature. Throughout the report, the four applied-force cases are referenced as a percentage of the maximum force case on the RV side: cases 25–100, corresponding to 25–100% of max force respectively. Time-varying force profiles for each simulated case are shown in Fig. 1C.

It is worth noting that, as this is an LV failure model, moderate compression was also applied to the LV and kept constant throughout the parametric simulations. The relevance of this simulated input to the physical case of LVAD support is discussed in later sections.

3   Results

Hemodynamic Analysis. The hemodynamic variables presented are end-systolic volume (ESV), end-diastolic volume (EDV), stroke volume (SV, defined as the difference between EDV and ESV), stroke work (SW, defined as the product of pressure and volume) and ejection fraction (EF, defined as SV/EDV). All RV hemodynamic metrics are reported in Fig. 2 as a percentage improvement of each applied-force case (25%, 50%, 75% and 100%) from the baseline case. The five aforementioned metrics were selected for their established relevance to cardiac performance. In patients with dilated cardiomyopathy, increases in EDV can

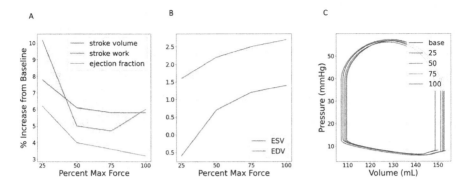

Fig. 2. Right ventricular (RV) hemodynamic variables as a function of percent max force. A) Stroke volume and stroke work, B) EDV and ESV. On average, hemodynamic benefit decreases as the percent max force increases. Note: Hemodynamic benefit is defined as minimized EDV and ESV, and maximized stroke work and stroke volume. C) Pressure-Volume loops for all cases

induce adverse tissue strains triggering a long-term process of ventricular remodeling and should thus be minimized [8]. Similarly, increases in ESV, for a given EDV, demonstrate decreased contractile function and are thus also undesirable [14]. SV, SW and EF on the other hand, are measures of cardiac output in each contraction, and increases are considered hemodynamically beneficial. Based on these insights, we note a clear trend of diminishing hemodynamic benefit across all hemodynamic parameters with increasing peak force. One interesting exception to this trend can be seen in Fig. 2A, where stroke work improvement does not monotonically decrease from 25 to 100% max force, but rather reaches a nadir at 75%. The trend is more pronounced in SW than SV due to the compounded contribution of both volume and pressure to the calculation of SW. Stroke volume is a particularly interesting metric as it is directly related to cardiac output, assuming constant heart rate. The aforementioned metrics can also be visualized through pressure-volume loops, illustrated in 2C for baseline and each applied-force case, While the primary focus of this study is on the RV, the authors also report improvements in EF in the LV, which is a critical measure of systolic function. The greatest improvement of left ventricular EF was 7.84%, corresponding to the 25% force case.

Biomechanical Analysis. Biomechanical analysis is performed on selected nodes at mid-ventricular height on the RV free wall surface, as this region undergoes substantial deformation during the cardiac cycle [15]. The model offers sufficient radial refinement to analyze transmural trends in mechanical behavior. Specifically we select three mid-height nodes for analysis: one epicardial, one in the midwall and one epicardial (see Fig. 3B inset). In Fig. 3 we report aggregate transmural trends in strain deviation from the baseline case (i.e. no external support applied). The maximum increase in strain is reported for each applied force case with respect to each of the three transmural nodes. We observe an inverse

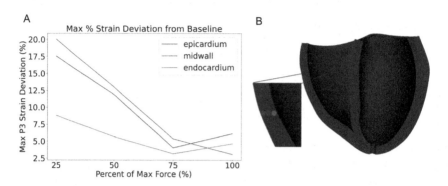

Fig. 3. Max Percent Strain Deviation from baseline case. A) deviation of principal strain between applied force trials and baselines case for transmural mid-height nodes; B) cross-sectional view of biventricular model; inset: location of epicardial (blue), midwall (orange) and endocardial nodes (green). Note: Strain deviation from baseline (healthy RV case) should be minimized (Color figure online)

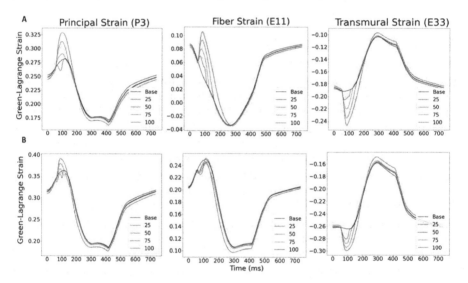

Fig. 4. Time-series analysis of Green-Lagrange Strain (P3, E11, E33) at mid-height nodes for one cardiac cycle where t = 0 ms = end diastole; t = 146 ms = aortic valve opening. BL = Baseline case. A) epicardial mid-height node, B) endocardial mid-height node

relationship between the applied load and the maximum principal strain, with a local minimum occurring at the 75% force case (Fig. 3). The midwall element experiences the greatest increase in strain (up to 20% from baseline), followed closely by the epicardial node, on which the force is applied. The endocardial node experiences the least increase in strain (max at 8% in the 25 case). Time-series data for the epicardial and endocardial mid-height nodes is presented in

Fig. 4. The first column of panels displays time-series data of the first principal strain (P3). The second and third columns display principal strain decomposition into fiber-directed components (E11) and transmurally-directed components (E33). The transmural strain (E33) is compressive, indicating a larger cavity volume, and becomes less compressive with the increasing RV load, accounting for the sharp increase in strain from baseline. Fiber strain (E11) exhibits a sharp increase peaking shortly around 150 ms, which coincides with the administration of the external force. Due to the incompressibility of cardiac tissue, the compressive strain in the transmural direction results in an tensile strain in the fiber direction.

We note that in the endocardium, P3 strain shows a much more pronounced E11 implication than the epicardium. This difference is likely due to the transmurally variable direction of cardiac fibers, as discussed in a later section. Moreover, the transmural strain (E33) is compressive, indicating a larger cavity volume, and becomes less compressive with the increasing RV load. Finally, the effects of the RV geometry are seen as the fiber strain (E11) plays little role in the endocardial node, a finding that differs from prior work examining the LV under compression [7]. This is likely attributable to the cresecent, asymmetric shape of the RV compared to the prolate shape of the LV.

Maximum deviations in P3 strain for each experimental case, occurring during mid-isovolumetric contraction (around 100 ms), are also graphically represented in Fig. 5 for the RV free wall view and the cross-sectional biventricular views. The observed trend extends transmurally throughout the RV free wall midsection and beyond the mid-height node used for analysis. The cross-sectional view (Fig. 5B) demonstrates two trends: (1) the RV geometrical conformation shifts from more enlarged at 25% force case, to more compressed at the 100% case, assuming a typical crescent-shape conformation; (2) the transmural strain concentration across the RV wall decreases as RV force is increased.

4   Discussion

The development of technologies for heart failure have been historically focused on the left ventricle. Mechanical interventions for left heart failure have benefited from ample research and development, whereas tailored support for the right heart remains lacking in comparison. Previous computational studies have investigated ideal epicardial pressure for direct cardiac compression devices, demonstrating that ventricle-specific requirements differ substantially between the left and right ventricles [9]. To our knowledge, we present the first study set out to quantitatively investigate the effect of externally applied forces in aiding ventricular contraction and describe the interplay between hemodynamic and biomechanical effects in the RV. In the context of a post-LVAD implantation RV failure, it is critical for RV mechanical support interventions to enhance hemodynamic function and ensure a comparable cardiac output to the supported LV. In this study, the physical rationale behind the experimental design was to test the effects of a prospective external device exerting a finite amount of energy on the

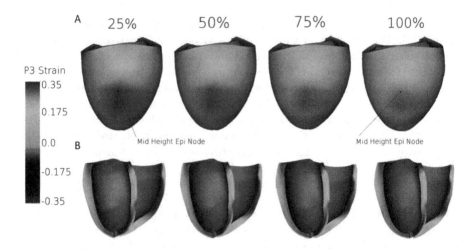

Fig. 5. Principal Strain Maps during Isovolumic Contraction (IVC) - A) RV Free Wall View. Maximum principal strain is plotted for each applied force case, 25%, 50%, 75% and 100%. Strain concentrations in the mid-ventricular free wall display a decreasing trend. B) Cross-sectional Biventricular View. Maximum principal strain is plotted for each applied force case, 25%, 50%, 75% and 100%.

RV free wall surface. Against the authors' initial intuition, lower peak applied force results in a larger stroke work than higher peak forces (10% increase from baseline compared to 5%, 4.5% and 6% for the four cases respectively), suggesting that force duration could be a more powerful predictor of hemodynamic benefit compared to peak magnitude.

Alterations to the complex biomechanics of myocardial tissue have been implicated as contributing factors to adverse ventricular remodeling, a mechanism universally proposed to underpin the pathophysiology of heart failure [16,17]. Therefore, it is imperative for interventions to be designed to minimize alterations to baseline regional biomechanics. Results of comparative biomechanical analysis of the four models tested demonstrate that the maximum strain deviation from baseline decreases as percent max force increases. This trend opposes the aforementioned hemodynamic observations, in that beneficial effects occur at higher percent max force and shorter force duration. Together, the contrasting effects imply a trade-off between hemodynamic benefit and biomechanical alterations with respect to force magnitude and timing. Increasing the duration of the applied force with lower magnitude seems to positively affect pump function but has a negative effect on tissue biomechanics. Surprisingly, deviation from baseline appears larger in the midwall as compared to the epicardial wall; an observation that can be attributed to the geometry of the RV and the myocardial fiber angle. In the midwall, the fibers are circumferentially aligned, meaning that the fiber strain, which is a major component of the principal strain, is closely related to hoop stress at this depth in the myocardium.

When force is applied, the hoop stress, and associated strain, decreases substantially.

One important observation is the seemingly incongruous finding that increasing force on the RV both increases the end diastolic volume (Fig. 2B) while also decreasing the maximum principal and fiber strain (Fig. 4). First, it may be counterintuitive that the application of epicardial forces results in increased volume parameters (ESV, EDV - see Fig. 2). However, it must be noted that this simulation provides transient support during contraction (not expansion), and the adaptive circulatory model in the ContinuityPro software responds by increasing systemic flow-back and thus increasing preload. Additionally, upon first inspection, it may seem logical that increasing EDV would cause an increased distention in the RV myocardium, resulting in increased strain. However, RV geometry plays a crucial, yet subtle, role. While the LV generally embodies a prolate shape with an axis-symmetric behavior, the RV presents and much more complex shape [18,19]. A possible interpretation is that, while the EDV is increasing as a whole in the RV, expansion is not uniform throughout the crescent shaped cavity, with some regions distending more then others (see Fig. 5). Upon inspection of different nodes in the model, such intuitions are confirmed by the evident disparity in regional deformations. Such findings are consistent with experimental studies attempting to characterize the RV's irregular motion, which previously concluded that the RV displays a clear pattern of regionally varying mechanical activation [20]. Specifically, the mid-height epicardial section where force is applied undergoes a decreased deformation with increasing force. While this specific finding is interesting in its own right, it also points to the more generalizable observation that these computational methods are necessary to address such discrepancies, especially ones which are intimately involved with patient specific geometry.

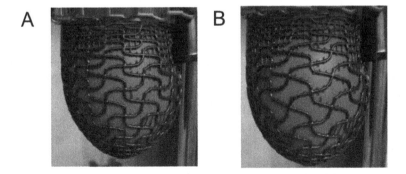

Fig. 6. Prototype of elastic metamaterial sleeve. A) Systolic conformation, B) Diastolic conformation

5   Conclusion and Implications for Device Design

In this study, we investigated the effects of external mechanical support on the RV to elucidate physically interpretable insights to inform the design of prospective devices. Studies leveraging modeling approaches have previously demonstrated that the interplay between hemodynamic and biomechanical effects of interventions can be surprising and counter-intuitive. For example, studies aimed at evaluating the effects of isotropic stiffening of newly infarcted cardiac tissue found no net hemodynamic benefit despite substantial alteration in mechanical properties - an effect that was later validated experimentally in a large animal model [21]. Similarly, it is worth noting that the results of the present study also contrast the authors' original predictions. It was initially hypothesized that a lower magnitude force delivered over a longer time period (i.e. the 25% case), would be biomechanically favorable and hemodynamically unfavorable compared to a higher, impulse-like force, the 100% case. Instead, the results show the opposite trend, evidencing the complex dynamics of cardiac function.

Nevertheless, these findings are useful to inform the design of physically realizable mechanical systems based on favorable force profiles characterized in this report. For example, elastic structures can be designed and tuned to display desired energy release dynamics, by modifying properties such as effective stiffness. Mechanical design coupled with additive manufacturing can be leveraged to yield smart mechanical metamaterials with effective stiffness properties that change as a function of time or deformation. An example of this approach is provided in Fig. 6, where early prototypes of an elastic, metamaterial support sleeve are shown. The two panels show the different configurations of the sleeve in systole (A) vs. diastole (B) in a benchtop setup. The passive support behavior was achieved using auxetic motifs, patterned onto a ventricular surface, to simulate ventricular mechanical behavior. Further investigation of material behavior and effective support provided by passive structures to ventricular tissue is warranted through %in-vitro and %in-vivo studies following initial computational findings.

Overall, we believe the elucidation and quantification of the trends exposed in this preliminary study can provide insight for the rational design of patient-specific mechanical support for LVAD-adjunct therapies aimed at supporting the right heart. Future investigations can build on this work by designing force profiles representative of physically realizable mechanical elements.

References

1. Argiriou, M., et al.: Right heart failure post left ventricular assist device implantation. J. Thorac. Dis. **6**(Suppl. 1), S52 (2014)
2. Lampert, B.C., Teuteberg, J.J.: Right ventricular failure after left ventricular assist devices. J. Heart Lung Transplant. **34**(9), 1123–1130 (2015)
3. Tang, P., et al.: Right ventricular failure following left ventricular assist device implantation is associated with a preoperative pro-inflammatory response. J. Cardiothorac. Surg. **14**, 80 (2019)

4. Kapur, N.K., et al.: Mechanical circulatory support devices for acute right ventricular failure. Circulation **136**(3), 314–326 (2017)
5. Sidney, S., Go, A.S., Jaffe, M.G., Solomon, M.D., Ambrosy, A.P., Rana, J.S.: Association between aging of the us population and heart disease mortality from 2011 to 2017. JAMA Cardiol. **4**(12), 1280–1286 (2019)
6. Horvath, M.A., et al.: An intracardiac soft robotic device for augmentation of blood ejection from the failing right ventricle. Ann. Biomed. Eng. **45**(9), 2222–2233 (2017)
7. Aranda-Michel, E., Waldman, L.K., Trumble, D.R.: Left ventricular simulation of cardiac compression: hemodynamics and regional mechanics. PLOS ONE **14**(10), e0224475 (2019)
8. Cokkinos, D.V., Belogianneas, C.: Left ventricular remodelling: a problem in search of solutions. Eur. Cardiol. Rev. **11**(1), 29 (2016)
9. Han, J., Kubala, M., Aranda-Michel, E., Trumble, D.R.: Ventricle-specific epicardial pressures as a means to optimize direct cardiac compression for circulatory support: a pilot study. PLOS ONE **14**(7), e0219162 (2019)
10. Guccione, J.M., McCulloch, A.D., Waldman, L.K.: Passive material properties of intact ventricular myocardium determined from a cylindrical model. J. Biomech. Eng. **113**(1), 42–55 (1991)
11. Kerckhoffs, R.C.P., Neal, M.L., Gu, Q., Bassingthwaighte, J.B., Omens, J.H., McCulloch, A.D.: Coupling of a 3d finite element model of cardiac ventricular mechanics to lumped systems models of the systemic and pulmonic circulation. Ann. Biomed. Eng. **35**(1), 1–18 (2007)
12. Costa, K.D., Hunter, P.J., Wayne, J.S., Waldman, L.K., Guccione, J.M., McCulloch, A.D.: A three-dimensional finite element method for large elastic deformations of ventricular myocardium: II-prolate spheroidal coordinates. J. Biomech. Eng. **118**(4), 464–472 (1996)
13. Cevasco, M., et al.: Right heart failure: an ischemic model and restraint therapy for treatment. Ann. Thorac. Surg. **97**(4), 1356–1363 (2014)
14. Kerkhof, P.L.M.: Characterizing heart failure in the ventricular volume domain. Clin. Med. Insights Cardiol. **9**, 11–31 (2015)
15. Truong, V.T., et al.: Cardiac magnetic resonance tissue tracking in right ventricle: feasibility and normal values. Magn. Reson. Imaging **38**, 189–195 (2017)
16. Mann, D.L., Bristow, M.R.: Mechanisms and models in heart failure: the biomechanical model and beyond. Circulation **111**(21), 2837–2849 (2005)
17. Kurrelmeyer, K., et al.: Cardiac remodeling as a consequence and cause of progressive heart failure. Clin. Cardiol. **21**(S1), 14–19 (1998)
18. Gripari, P.: Right ventricular dimensions and function: why do we need a more accurate and quantitative imaging? J. Cardiovasc. Echography **25**, 19–25 (2015)
19. Lang, R.M., et al.: Recommendations for cardiac chamber quantification by echocardiography in adults: an update from the American Society of Echocardiography and the European Association of Cardiovascular Imaging. J. Am. Soc. Echocardiogr. **28**, 1–39 (2015)
20. Auger, D.A., Zhong, X., Epstein, F.H., Spottiswoode, B.S.: Mapping right ventricular myocardial mechanics using 3d cine dense cardiovascular magnetic resonance. J. Cardiovasc. Magn. Reson. **14**(1), 4 (2012)
21. Fomovsky, G.M., Macadangdang, J.R., Ailawadi, G., Holmes, J.W.: Model-based design of mechanical therapies for myocardial infarction. J. Cardiovasc. Transl. Res. **4**(1), 82–91 (2011)

In Vivo Pressure-Volume Loops and Chamber Stiffness Estimation Using Real-Time 3D Echocardiography and Left Ventricular Catheterization – Application to Post-heart Transplant Patients

Bianca Freytag[1,2] , Vicky Y. Wang[1,3(✉)] , Debbie Zhao[1], Kathleen Gilbert[1],
Gina Quill[1], Abdallah I. Hasaballa[1], Thiranja P. Babarenda Gamage[1],
Robert N. Doughty[4,5], Malcolm E. Legget[4], Peter Ruygrok[4,5], Alistair A. Young[6,7],
and Martyn P. Nash[1,8]

[1] Auckland Bioengineering Institute, University of Auckland, Auckland, New Zealand
bianca.freytag@creatis.insa-lyon.fr, vicky.wang@auckland.ac.nz
[2] Creatis, CNRS UMR5220, INSERM U1206, Université Lyon 1, INSA Lyon, Lyon, France
[3] Department of Radiology, Stanford University, Palo Alto, CA, USA
[4] Department of Medicine, University of Auckland, Auckland, New Zealand
[5] Green Lane Cardiovascular Service, Auckland City Hospital, Auckland, New Zealand
[6] Department of Biomedical Engineering, King's College London, London, UK
[7] Department of Anatomy and Medical Imaging,
University of Auckland, Auckland, New Zealand
[8] Department of Engineering Science, University of Auckland, Auckland, New Zealand

Abstract. *In vivo* pressure-volume loops (PVLs) are the gold standard measurement to assess ventricular function. We developed a pipeline to integrate hemodynamic measurements with real-time three-dimensional (3D) echocardiographic data to construct *in vivo* PVLs for 25 post-heart transplant patients. We then evaluated left ventricular diastolic function for these patients by calculating chamber stiffness from a cubic polynomial fit of the diastolic pressure-volume relationships (PVR). We examined the ability of a well-established mathematical (Klotz) model to predict the patient-specific diastolic PVRs. We found that beat-to-beat variation in hemodynamic measurement was typical for this group of patients, which resulted in mean $\pm$ standard deviation end-diastolic chamber stiffness estimates of 0.75 ± 0.40 mmHg/ml. The cubic polynomial fits of the individual diastolic PVRs resulted in much smaller errors (0.25 ± 0.01 mmHg) compared to those associated with the Klotz predicted diastolic PVRs (4.0 ± 0.27 mmHg), which provided a poor representation of the *in vivo* diastolic PVRs. The proposed framework enables the temporal alignment between hemodynamic and 3D imaging data to produce *in vivo* PVLs that can be used not only to quantify global ventricular function, but also to estimate mechanical properties of the myocardium.

Keywords: In vivo pressure-volume loop · Real-time 3D echocardiography · Catheterization · Chamber stiffness · Heart transplant

© Springer Nature Switzerland AG 2021
D. B. Ennis et al. (Eds.): FIMH 2021, LNCS 12738, pp. 396–405, 2021.
https://doi.org/10.1007/978-3-030-78710-3_38

1 Introduction

Historically, analyses of the left ventricular (LV) pressure-volume loops (PVLs) have been considered the gold standard for assessment of cardiac function under both *in vivo* and *ex vivo* conditions [1]. PVLs reflect the most direct relationship between pressure and volume inside the LV and allow the derivation of other mechanical properties of the heart, such as chamber stiffness and end-systolic elastance [1]. Methods for measuring PVLs can be broadly categorized as: 1) simultaneous measurements of pressure and volume via a micromanometric conductance catheter [2]; and 2) image-based methods that combine catheterization measurements with cardiac geometric data derived from imaging, such as cardiac magnetic resonance (CMR) imaging or echocardiographic (echo) imaging [3, 4]. While conductance catheters allow for synchronized acquisitions of pressure and volume measurements, the accuracy of LV volume measurement remains limited and often involves a calibration against other types of imaging techniques such as ventriculography [5]. The advancement of non-invasive cardiac imaging has improved the accuracy and accessibility of LV volume quantification. However, accurate temporal registration between LV pressure and volume is not well developed. We have previously described a piecewise linear temporal scaling method based on cardiac events, identified on both invasive pressure traces and cine CMR images [3]. Time delays between catheterization and CMR can lead to discrepancies in haemodynamic states between the two data acquisitions. Real-time 3D-echocardiography (RT3DE) has evolved to a readily available and cost-effective modality for rapid LV assessment. Furthermore, scanner portability enables imaging to be performed immediately after catheterization, thereby minimizing time-dependent hemodynamic variability.

While chamber stiffness is straightforward to calculate, it is also load- and geometry-dependent, making it challenging to discern diastolic dysfunction at the myocardial tissue level. On the other hand, intrinsic myocardial tissue stiffness can only be estimated using inverse finite element modeling techniques. Personalized finite element modeling of ventricular mechanics generally requires patient-specific measurements of heart anatomy and motion across the cardiac cycle, hemodynamic loading conditions, and microstructural information. Estimation of the passive myocardial tissue stiffness can then be made by matching the model-predicted diastolic PVR with subject-specific PVR and/or global and regional deformations derived from imaging data. Although the diastolic PVR provides information about LV filling characteristics, personalized measurements of the PVR are not routinely available. Instead, an algebraic mathematical model, known as the Klotz curve [6], has been adopted in many studies to predict an individualized PVR based on a single set of pressure and volume estimates [7, 8]. However, to the authors' knowledge, there are no published reports that have compared the Klotz-predicted diastolic PVR with catheterization measurements.

In the present study, we propose a framework for temporally registering invasive LV and aortic pressures (LVP and AOP) acquired during left heart catheterization with RT3DE images to generate *in vivo* PVLs in a group of heart transplant (HTx) patients. We estimated chamber stiffness from the diastolic pressure-volume relationships (PVRs) for each patient and examined the subject-specific variability. Lastly, we investigated the predictive power of the Klotz model for the HTx patient cohort to examine its suitability to estimate an individualized single-beat diastolic PVRs.

2   Methods

Orthotopic heart transplantation patients (HTx) attending for routine coronary assessment were recruited for invasive hemodynamic measurement and RT3DE imaging. Ethical approval for the present study was granted by the Health and Disability Ethics Committee of New Zealand (17/NTB/46), and written informed consent was obtained from each participant. 49 post-HTx patients were prospectively enrolled and 25 cases were selected for analysis based on satisfactory apical echocardiographic windows for 3D geometric modeling and adequate hemodynamic data quality.

2.1   In Vivo Data Collection

LV Catheterization. A fluid-filled pigtail catheter (Impulse by Boston Scientific, Marlborough MA) was advanced into the LV through the aortic valve via radial access under X-ray guidance. Four continuous multi-cycle recordings (i.e., over 9–15 heartbeats) of LV cavity pressure were obtained using the Mac-Lab Hemodynamic Recording System (GE Healthcare, Chicago, IL, USA). The catheter was then withdrawn from the LV into the aorta, where AOPs were recorded for a similar period as the LVPs at the root of the aorta and aortic arch. An electrocardiogram (ECG) was simultaneously recorded during the entire procedure at a sample rate of 240 Hz.

RT3DE Imaging. Within an hour of catheterization, single-beat transthoracic RT3DE volumetric imaging of the LV was performed using a Siemens ACUSON SC2000 Ultrasound System with a 4Z1c transducer (Siemens Medical Solutions, CA, USA) from the apical window in a left lateral decubitus position. Imaging parameters were optimized for each patient to maximize the temporal resolution while maintaining an adequate spatial resolution (reconstructed to 1 mm^3 isotropic voxels in Cartesian space) for geometric analysis. This resulted in between 15–41 imaging frames per cardiac cycle across the study population.

2.2   Data Analysis

Hemodynamic Analysis. A piecewise linear temporal scaling method based on cardiac events of CMR images [3] was extended to handle RT3DE. LVPs and AOPs were processed using an in-house analysis tool written in Matlab R2020b (MathWorks Inc., Natick, MA, USA). Noise was removed interactively using a Fourier transform with a participant-specific low-pass filter with frequencies ranging from 10.3 Hz to 20 Hz. Breathing artifacts, which manifested as low-frequency shifts of the LVP traces across the cardiac cycles, were corrected with a high-pass filter with cut-off frequencies ranging up to 0.83 Hz. The difference in cut-off frequencies to remove breathing artifacts was mainly due to variations in patients' breathing motion. R-peaks were identified on the ECG traces and used to isolate the LVP and AOP traces from individual cycles. For the analyses, the following exclusion criteria were used: 1) no arrhythmia; 2) no visible pressure overshoot (e.g. air bubbles in the catheter can cause a positive LVP overshoot during ejection, and a LVP undershoot during relaxation [9]); and 3) acceptable noise.

For temporal alignment with RT3DE imaging data, we identified five cardiac events in the pressure traces (Fig. 1a) based on characteristics described in Table 1. Although identification of end-diastole (ED) and end-systole (ES) is well described in the literature, methods for identifying end of isovolumic contraction (eIVC), end of isovolumic relaxation (eIVR), and diastasis (DS) are sparsely discussed. DS is often assumed to correspond to the minimum LVP by commercial hemodynamic analysis software. However, neither pressure nor volume changes substantially at diastasis, thus the minimum pressure may not necessarily be the diastatic pressure, as rapid recoil during IVR can cause a significant drop of LVP due to a suction effect.

Geometric Modeling. Semi-automatic analysis of RT3DE volumetric imaging data was performed offline using EchoBuildR 3.5.0 prototype software (Siemens Medical Solutions, CA, USA) [10, 11]. In our imaging protocol, RT3DE from 4 cardiac cycles were typically acquired, but only the cycle with the best imaging quality was used for geometric modeling. Geometric LV models were manually constructed at ED and ES, followed by automatic tracking across intermediary frames to estimate LV volume and mass over the entire cardiac cycle. Image frames corresponding to the same five cardiac events identified for hemodynamic analysis were manually identified for each RT3DE dataset (Fig. 1b) using the methods described in Table 1.

Table 1. Characteristics of key cardiac events in LVP trace and RT3DE.

Cardiac event	Pressure trace	RT3DE
End-diastole (**ED**)	Rapid change in LVP slope occurs (maximum rate of change of acceleration)	R peak of the ECG and closure of the mitral valve
End of isovolumic contraction (**eIVC**)	LVP equal to minimum AOP, beyond which aortic valve opens	Sudden LV and/or opening of the aortic valve
End-systole (**ES**)	LVP equal to the AOP at dicrotic notch	Maximal LV contraction and/or aortic valve closure
End of isovolumic relaxation (**eIVR**)	Maximum change of rate of LVP (peak d^2P/dt^2)	Instance prior to the opening of the mitral valve
Diastasis (**DS**)	Inflection point of the LVP trace ($d^2P/dt^2 = 0$)	Plateau in volume curve or partial mitral valve closure

In Vivo PVL Generation. After identifying cardiac events, the individual pressure traces were divided into five segments: DS to ED, ED to eIVC, eIVC to ES, ES to eIVR, and eIVR to DS. Each segment of each loop was temporally scaled to match the duration of the respective echo segment and sampled at the echo imaging time points, which had coarser temporal resolution compared to the hemodynamic measurements, This resulted in multiple pressure values for each echo frame due to beat-to-beat variation in LVP. The temporally aligned pressure values were further averaged to find the beat-averaged LV PVL (Fig. 1c). The number of cycles used for beat averaging ranged between 5 to 14 cycles.

In Vivo Diastolic PVRs. To analyze diastolic PVRs, we isolated the portion of the PVL between DS and ED for each cardiac cycle. Quantification of chamber stiffness can be made by fitting a function to the diastolic PVRs, then evaluating the slope (dP/dV) of the fitted curve at LVEDV. A range of equations (e.g. exponential, polynomial, power laws) is summarized in [1], among which a mono-exponential with an offset is commonly adopted in the field. However, when we fitted the diastolic PVRs using the mono-exponential equation, we found that the fitted curves did not adequately represent the underlying data particularly at ED, which led to inaccurate estimation of the slope of the curve at LVEDV, and hence chamber stiffness. Instead, we fitted the diastolic PVRs using a cubic polynomial and obtained much more accurate fits to the data.

Next, we normalized the diastolic PVRs using maximum and minimum pressure and volume to examine whether the normalized curves conformed to one single relationship as proposed in [6]. Using the coefficients (A and B) published in their study, we estimated model-predicted diastolic PVRs using subject-specific EDV and EDP as input for the algorithm. To examine the accuracy of the Klotz model-predicted diastolic PVRs, we evaluated LV pressures at the LV cavity volumes estimated from RT3DE between DS and ED, and then calculated the root-mean-squared error (RMSE) between Klotz predicted LV pressures and the corresponding *in vivo* measurements.

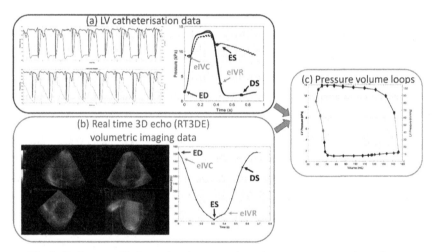

Fig. 1. Patient-specific input data for generation of *in vivo* pressure-volume loops.

3   Results

3.1   Demographics and Functional Indices

A summary of patient demographics and functional indices is shown in Table 2. The mean age was 54 years ($\pm$ 8) and 7 (28%) were women. The mean LV ejection fraction

(EF) was 58 ± 6% while only two patients had an EF below 50%, indicating majority of these patients have preserved EF.

Table 2. Patient demographics and LV functional indices (EF = ejection fraction).

Index (mean ± S.D.)	Male ($n = 18$)	Female ($n = 7$)	Total ($n = 25$)
Age (years)	54 ± 7	55 ± 9	54 ± 8
Height (cm)	178 ± 9	165 ± 3	174 ± 10
Weight (kg)	83 ± 11	69 ± 13	79 ± 13
Body surface area (m^2)	2.02 ± 0.17	1.77 ± 0.17	1.95 ± 0.20
EF (%)	56 ± 5	63 ± 3	58 ± 6

3.2 *In Vivo* LV PVLs and Diastolic Function

Multiple *in vivo* PVLs were constructed for each patient based on the number of selected cardiac cycles for hemodynamic analysis and a beat-averaged PVL (Fig. 2a) was also generated for each of the 25 post-HTx patients. With the exception of one case, the beat-averaged PVLs exhibited classically representative shape with distinct isovolumic contraction and isovolumic relaxation phases. Of the 25 patients, 2 patients showed a significantly higher LVESP than the others, while 6 patients had LVEDPs greater than 15 mmHg (2 kPa), indicating potential diastolic dysfunction.

The individual diastolic PVRs for all patients are shown in Fig. 2b, with beat-to-beat variation observed in most patients. For some cases, the variation manifested as an offset in LVP, whereas changes in the diastolic PVR slope were observed in other cases. The mean chamber stiffness across all patients was 0.75 ± 0.40 mmHg/ml (interquartile range (IQR): [0.51 0.86] mmHg/ml) and the mean beat-to-beat variation in chamber stiffness across all patients was 0.12 ± 0.06 mmHg/ml (IQR: [0.08 0.13] mmHg/ml), which were calculated using the chamber stiffness standard deviations across each set of cycles recorded for each participant.

3.3 Klotz Prediction of Diastolic PVR

The Klotz predicted diastolic PVRs for one representative case are shown in Fig. 2c along with the *in vivo* measurements. By definition, all Klotz predicted diastolic PVRs matched LVEDV and LVEDP, but the predicted LV pressures at other LV volumes were less accurate. The normalized diastolic PVRs for all 25 patients (Fig. 2d) did not conform to one single relationship as suggested in [6]. The nonlinearity of the diastolic PVRs differed across the HTx cohort. The average RMSE between *in vivo* diastolic PVRs and those predicted by the Klotz model was 4.0 ± 0.27 mmHg (IQR: [2.6 5.8] mmHg). In comparison, the average RMSE for the cubic polynomial fit was only 0.25 ± 0.01 mmHg (IQR: [0.13 0.28] mmHg).

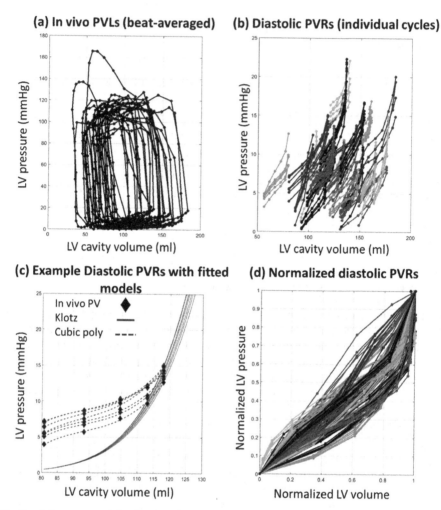

Fig. 2. a) Beat-averaged *in vivo* pressure-volume loops (PVLs) for all 25 post-HTx patients. b) Individual diastolic pressure-volume relationships (PVRs). c) Raw diastolic PVRs (diamonds) overlaid with cubic polynomial and Klotz fitted diastolic PVRs. d) Normalized diastolic PVRs.

4   Discussion

In the present study, a framework was developed to construct subject-specific PVLs using *in vivo* hemodynamic measurements and RT3DE data from the same subject, acquired within an hour of left heart catheterization. We proposed an algorithm to identify key cardiac events from the LVP and AOP traces as well as from the LV cavity volume-time curve quantified from RT3DE. Recognizing that a uniform temporal scaling based on the R-R interval difference was insufficient to align the pressure and volume data temporally, we implemented temporal scaling for each of the five cardiac phases individually. This scheme resulted in excellent *in vivo* PVLs with well-preserved isovolumic phases for the group of 25 HTx patients in comparison to previous studies [4, 5]. The constructed *in vivo* PVLs represent an efficient diagnostic tool for clinicians to gauge LV function for patients undergoing catheterization. It not only allows calculation of chamber stiffness, but it also enables estimation of indices that reflect cardiac energetics, such as stroke work and cardiac work, which are difficult to calculate without PVLs. For patients with cardiac interventions such as heart/lung transplantation, it is also a useful tool to monitor a patient's response to surgery.

Chamber stiffness is used widely to characterize ventricular diastolic function, and can be calculated directly from the diastolic PVR. It has been considered as an important indicator of diastolic dysfunction, which manifests as restrictive filling for HTx patients. In the present study, the chamber stiffness estimated for the HTx patients (0.75 ± 0.40 mmHg/ml) was larger than that reported for groups of control subjects with normal LV function (0.16 ± 0.11 mmHg/ml [2]; 0.09 [IQR:0.07,0.12] mmHg/ml [5]), and for patients with heart failure with preserved LV ejection fraction (0.24 [IQR:0.16,0.37] mmHg/ml) [5]. It is worth noting that the chamber stiffness derived in the present study was based on single beat diastolic PVRs, while the aforementioned studies transiently reduced the preload to measure EDPVRs, which may partially explain the differences in chamber stiffness estimates.

Direct measurement of LVP is not routinely accessible due to the invasive nature of the procedure. Consequently, several studies have used the Klotz model to estimate the diastolic PVR on a per-subject basis for the purpose of estimating myocardial tissue stiffness [7, 8]. Based on the *in vivo* diastolic PVRs in the present study, we found that the shape of the normalized diastolic PVRs was very different among the patients as opposed to a single relationship reported in [6]. The RMSE in the Klotz predicted diastolic PVRs ranged between 2.6 mmHg (25% quartile) and 5.8 mmHg (75% quartile), illustrating its inability to predict *in vivo* diastolic PVR accurately. These errors are comparable to the RMSE (2.79 ± 0.21 mmHg) previously reported in heart failure patients in [12], which showed poor prediction at low pressures (e.g. <10 mmHg). Such inaccuracies in the prediction of the diastolic PVR may confound estimates of passive myocardial stiffness.

The beat-to-beat variation observed in these patients is mostly due to the variability in hemodynamic measurements over several cardiac cycles because the volume estimate was derived from a single cardiac cycle. While this approach did not provide concurrent pressure and volume measurements, the volume measurement obtained from RT3DS is much more accurate. The quantification of beat-to-beat variation and reproducibility of volume from RT3DE was beyond the scope of the present study. However, previous experiments have demonstrated that RT3DE is able to provide low test-retest variation and high reproducibility of LV volumes [13, 14]. In addition, we ensured that the cavity volume used for generation of *in vivo* PVLs was derived from the cardiac cycle with the best imaging quality. Nevertheless, this framework can be readily extended to construct a family of *in vivo* PVLs based on multi-cycle pressure and volume data.

5   Conclusion

We proposed a framework to construct patient-specific *in vivo* PVLs from hemodynamic measurements obtained during left heart catheterization and cavity volume quantified from RT3DE using a temporal alignment scheme based on cardiac events. Application to patients post heart transplantation revealed beat-to-beat variation of hemodynamic state. Normalized diastolic PVRs showed varying degrees of nonlinearity among the patients, suggesting that the use of a simple algebraic mathematical model for the prediction of subject-specific diastolic PVR is insufficient.

Acknowledgement. We gratefully acknowledge the financial support from the Health Research Council of New Zealand (grant 17/608).

References

1. Burkhoff, D., Mirsky, I., Suga, H.: Assessment of systolic and diastolic ventricular properties via pressure-volume analysis: a guide for clinical, translational, and basic researchers. Am. J. Physiol.-Heart Circ. Physiol. **289**, H501–H512 (2005)
2. Zhang, W., Kovács, S.J.: The diastatic pressure-volume relationship is not the same as the end-diastolic pressure-volume relationship. Am. J. Physiol.-Heart Circ. Physiol. **294**, H2750–H2760 (2008)
3. Wang, Z.J., Wang, V.Y., Bradley, C.P., Nash, M.P., Young, A.A., Cao, J.J.: Left ventricular diastolic myocardial stiffness and end-diastolic myofibre stress in human heart failure using personalised biomechanical analysis. J. Cardiovasc. Transl. Res. **11**(4), 346–356 (2018). https://doi.org/10.1007/s12265-018-9816-y
4. Chowdhury, S.M., et al.: Echocardiographic detection of increased ventricular diastolic stiffness in pediatric heart transplant recipients: a pilot study. J. Am. Soc. Echocardiogr. **31**, 342-348.e341 (2018)
5. Westermann, D., et al.: Role of left ventricular stiffness in heart failure with normal ejection fraction. Circulation **117**, 2051–2060 (2008)
6. Klotz, S., et al.: Single-beat estimation of end-diastolic pressure-volume relationship: a novel method with potential for noninvasive application. Am. J. Physiol.-Heart Circ. Physiol. **291**, H403–H412 (2006)
7. Rumindo, G.K., Ohayon, J., Croisille, P., Clarysse, P.: In vivo estimation of normal left ventricular stiffness and contractility based on routine cine MR acquisition. Med. Eng. Phys. **85**, 16–26 (2020)
8. Genet, M., et al.: Distribution of normal human left ventricular myofiber stress at end diastole and end systole: a target for in silico design of heart failure treatments. J. Appl. Physiol. **117**, 142–152 (2014)
9. de Vecchi, A., et al.: Catheter-induced errors in pressure measurements in vessels: an in-vitro and numerical study. IEEE Trans. Biomed. Eng. **61**, 1844–1850 (2014)
10. Zheng, Y., Barbu, A., Georgescu, B., Scheuering, M., Comaniciu, D.: Four-chamber heart modeling and automatic segmentation for 3-D cardiac CT volumes using marginal space learning and steerable features. IEEE Trans. Med. Imaging **27**, 1668–1681 (2008)
11. Lin, Y., Georgescu, B., Yefeng, Z., Meer, P., Comaniciu, D.: 3D ultrasound tracking of the left ventricle using one-step forward prediction and data fusion of collaborative trackers. In: 2008 IEEE Conference on Computer Vision and Pattern Recognition, pp. 1–8 (2008)
12. ten Brinke, E.A., et al.: Single-beat estimation of the left ventricular end-diastolic pressure–volume relationship in patients with heart failure. Heart **96**, 213–219 (2010)

13. Thavendiranathan, P., et al.: Feasibility, accuracy, and reproducibility of real-time full-volume 3D transthoracic echocardiography to measure LV volumes and systolic function: a fully automated endocardial contouring algorithm in sinus rhythm and atrial fibrillation. JACC Cardiovas. Imaging **5**, 239–251 (2012)

14. Jenkins, C., Bricknell, K., Hanekom, L., Marwick, T.H.: Reproducibility and accuracy of echocardiographic measurements of left ventricular parameters using real-time three-dimensional echocardiography. J. Am. Coll. Cardiol. **44**, 878–886 (2004)

In Silico Mapping of the Omecamtiv Mecarbil Effects from the Sarcomere to the Whole-Heart and Back Again

Stefano Longobardi[1]($\boxtimes$) ⓘ, Anna Sher[2] ⓘ, and Steven A. Niederer[1] ⓘ

[1] Cardiac Electromechanics Research Group, School of Biomedical Engineering and Imaging Sciences, King's College London, London, UK
stefano.longobardi@kcl.ac.uk
[2] Pfizer Worldwide Research, Development and Medical, Cambridge, MA, USA

Abstract. Omecamtiv mecarbil (OM) is a cardiac myosin activator developed as a treatment of heart failure. OM acts on cross-bridge formation without disrupting intracellular calcium homeostasis. OM effects are extensively characterised both *in vitro* and *in vivo* yet how these mechanistically translate from the sarcomere to whole-heart function is not fully understood. We employed a 3D biventricular contraction model of a healthy rat heart that was fitted to anatomic, structural, and hemodynamic and volumetric functional data. The model incorporates pre-load, after-load, fibre orientation, passive material properties, anatomy, calcium transients, and thin and thick filament dynamics. We identified 4 sarcomere model parameters that reflect cross-bridge behavior. Gaussian process emulators (GPEs) were trained to map these parameters to pressure- and volume-based indexes of left ventricular (LV) function. We constrained the 4-parameter space using preclinical OM data, either (1) *in vivo* whole-heart hemodynamics data (using the Bayesian history matching technique), or (2) *in vitro* force-pCa measurements. The OM-compatible sarcomere parameter space from case (1) was used to directly calculate force-pCa curves, while the one resulting from case (2) was mapped to the LV indexes using the trained GPEs. We found that our mapping from LV features to force-pCa and vice versa was in agreement with experimental data. In addition, our simulations supported the latest evidence that OM indirectly alters thin filament calcium sensitivity. Our work demonstrates how quantitative mapping from cellular to whole-organ level can be used to improve our understanding of drug action mechanisms.

Keywords: Omecamtiv mecarbil · Gaussian process

1 Introduction

Heart failure (HF) is a leading cause of hospitalisation worldwide, with more than one million admissions annually in the US and Europe [2]. However, the treatment options are limited and, therefore, new pharmacotherapies are continuously

© Springer Nature Switzerland AG 2021
D. B. Ennis et al. (Eds.): FIMH 2021, LNCS 12738, pp. 406–415, 2021.
https://doi.org/10.1007/978-3-030-78710-3_39

sought. HF pathways can involve impaired cellular function and propagate up to the whole-organ dysfunction. Multi-scale contraction modelling represents a useful tool for understanding the underlying mechanisms and possibly identifying targets in these pathways.

Building on a previously developed multi-scale rat bi-ventricle mechanics modelling framework [8], we can investigate drug mechanisms of action and better understand the mechanistic processes linking cellular and whole-heart contraction. For this case study, we chose omecamtiv mecarbil (OM), a novel drug currently in Phase 3 clinical trial [13] for treating HF. OM is a selective allosteric cardiac myosin modulator. It increases the rate of cross-bridge cycling by accelerating phosphate release [9], without disrupting intracellular calcium dynamics [3]. Recently, OM was shown to enhance the duty ratio, resulting in increased calcium sensitivity and slowed force development [4,11].

This paper outlines our methodology of incorporating OM into our model, that consists of calibrating the cellular model using *in vitro* data from skinned cellular and trabecular preparations in OM-containing solutions [4,5,10], and validating the biventricular multi-scale contraction model using pressure-volume measurements from *in vivo* whole-heart studies in healthy animals with OM [1]. Our simulation results are consistent with the available experimental data on OM and support the hypothesis (e.g. [11]) that OM affects the thin filament.

2   Methods

2.1   Cellular Contraction Model

We employed the Land et al. [7] myocyte contraction model to simulate active tension generation at sarcomere level and isometric steady-state force-calcium relationship in the rat heart. This model describes the cooperative binding of calcium (Ca^{2+}) to the "C" binding site of troponin (TnC), which in turn causes unblocking of the actin sites for myosin cross-bridge cycling. Equations (1)–(2) summarise the two processes of Ca^{2+} to TnC binding and of cross-bridge formation, respectively:

$$\frac{dTRPN}{dt} = k_{on} \left(\frac{[Ca^{2+}]_i}{Ca_{T50}} \right)^{n_{trpn}} (1 - TRPN) - k_{off} TRPN \tag{1}$$

$$\frac{dXB}{dt} = k_{xb} \left[permtot(1 - XB) - \frac{1}{permtot} XB \right] \tag{2}$$

with

$$permtot := \sqrt{\left(\frac{TRPN}{TRPN_{50}} \right)^{n_{xb}}} \tag{3}$$

where TRPN and XB are the proportions of bound Ca^{2+}-TnC complexes and bound cross-bridges, respectively; $[Ca^{2+}]_i$ is a defined, representative calcium transient from healthy, 6 Hz-paced rat left ventricular myocytes at 37 °C; Ca_{T50}

is a phenomenological representation of the calcium sensitivity of troponin (Ca_{50}) dependence on sarcomere length:

$$Ca_{T50} := Ca_{50}[1 + \beta_1(\lambda - 1)] \tag{4}$$

where λ is the extension ratio along the fibre direction. In this work, we only considered the resting sarcomere configuration of $\lambda = 1$, i.e., $Ca_{T50} = Ca_{50}$. All the other parameters appearing in the presented equations are described in Table 1.

The steady-state solutions of Eq. (1)–(2) are:

$$TRPN = \frac{\left(\frac{[Ca^{2+}]_i}{Ca_{T50}}\right)^{n_{trpn}}}{\frac{k_{off}}{k_{on}} + \left(\frac{[Ca^{2+}]_i}{Ca_{T50}}\right)^{n_{trpn}}} \tag{5}$$

$$XB = \frac{(TRPN)^{n_{xb}}}{(TRPN_{50})^{n_{xb}} + (TRPN)^{n_{xb}}} \tag{6}$$

and the steady-state force F is expressed in terms of the fraction of bound cross-bridges XB, a maximal reference tension T_{ref} (a scaling factor), and of two components $\ell(\lambda)$ and $v(d\lambda/dt)$ representing length- and velocity-dependence respectively:

$$F = T_{ref} \cdot XB \cdot \ell(\lambda) \cdot v(d\lambda/dt) \tag{7}$$

The length- and velocity-dependence terms equal to 1 when, as in our case, $\lambda = 1$ [7]. Equation (7) can be re-written in a more classical form as

$$\frac{F}{F_0} = \frac{x^{h(x)}}{1 + x^{h(x)}} \tag{8}$$

where $F_0 := T_{ref}$, the Hill coefficient h is a function of x

$$h(x) := n_{xb} \left[n_{trpn} - \log_x (1 - TRPN_{50}(1 - x^{n_{trpn}}))\right] \tag{9}$$

and $x := [Ca^{2+}]_i/EC_{50}$, where the half-maximal effective concentration EC_{50} is given as a function of five model parameters:

$$EC_{50} := Ca_{T50} \left(\frac{k_{off}}{k_{on}} \frac{TRPN_{50}}{1 - TRPN_{50}}\right)^{1/n_{trpn}} \tag{10}$$

By construction, the Land et al. [7] model has a biphasic Hill coefficient. To simplify the analysis, we approximated the steady-state force by characterising F using a single Hill coefficient value h defined as the slope at half-maximal activation:

$$h := \lim_{x \to 1} h(x) = n_{xb} n_{trpn}(1 - TRPN_{50}) \tag{11}$$

As a result, the Hill coefficient does not depend on $[Ca^{2+}]_i$ and is a function of three model parameters. Using logarithmically-spaced calcium values Eq. (8) becomes

$$\frac{F}{F_0} = \frac{1}{1 + 10^{h(pCa_{50} - pCa)}} \tag{12}$$

where $pCa := -\log[Ca^{2+}]_i$ and $pCa_{50} := -\log EC_{50}$. Throughout this work, we will refer to Eq. (12) as the *force-pCa curve*.

Table 1. Land et al. [7] model parameters with Longobardi et al. [8] baseline values.

Parameter	Definition	Value
Ca_{50}	Ca^{2+} thin filament sensitivity	$2.1708\,\mu M^{1\,-\,1/n_{trpn}}$
β_1	Phenomenological tension length -dependence scaling factor	-1.5
n_{trpn}	Ca^{2+}-TnC binding degree of cooperativity	2.0
k_{on}	Binding rate of Ca^{2+} to TnC	$0.1\,\mu M^{-1}\,ms^{-1}$
k_{off}	Unbinding rate of Ca^{2+} from TnC	$0.0515\,ms^{-1}$
n_{xb}	Cross-bridge formation degree of cooperativity	5.0
k_{xb}	Cross-bridges cycling rate	$0.0172\,ms^{-1}$
$TRPN_{50}$	Fraction of Ca^{2+}-TnC bounds for half-maximal cross-bridges activation	0.35
T_{ref}	Maximal reference tension	$156.067\,kPa$

2.2 Sarcomere Parameter Space

We modelled the OM effect at the sarcomere level by altering parameters that are specifically responsible for cross-bridge dynamics. Parameters regulating Ca^{2+} binding to TnC, namely Ca_{50}, β_1, n_{trpn}, k_{on}, k_{off}, were kept fixed. As OM increases binding of cross-bridges, we considered the parameters n_{xb}, $TRPN_{50}$, k_{xb} and T_{ref} that represent cross-bridge binding cooperativity, the number of cross-bridges bound as Ca^{2+} binds to troponin, the rate of cross-bridge binding and the force generated by bound cross-bridges, respectively.

2.3 Mapping Sarcomere Parameters to Whole-Organ Function

To quantitatively link sarcomere properties to whole-organ function, we used a previously fitted 3D biventricular contraction model of a healthy rat heart [8]. The model simulation outputs of the LV contractile function can be described using 14 features (Table 2). We can thus treat the full model as a multi-scale map from sarcomere input parameters to LV output features.

2.4 Surrogate Mapping

To overcome the computational burden of simulating the full model, we replaced the multi-scale map with a probabilistic surrogate based on Gaussian process emulation (GPE) [8]. Briefly, we sampled 4096 points from a Latin hypercube design in the 4-dimensional sarcomere parameter space with initial ranges: $n_{xb} \in [0.90, 7.05]$, $TRPN_{50} \in [0.05, 0.50]$, $k_{xb} \in [0.0086, 0.0258]$ and $T_{ref} \in [109.25, 202.89]$. These were set as the union of a 50% perturbation around reference parameter values and the range of parameter values inferred from *in*

Table 2. Indexes of LV function.

Label	Units	Definition
EDV	µL	End-diastolic volume
ESV	µL	End-systolic volume
SV	µL	Stroke volume
EF	%	Ejection fraction
IVCT	ms	Isovolumetric contraction time
ET	ms	Ejection time
IVRT	ms	Isovolumetric relaxation time
Tdiast	ms	Diastolic time
ET/Tdiast	–	Systolic ejection time over diastiolic filling time
PeakP	kPa	Peak pressure
Tpeak	ms	Time to peak pressure
ESP	kPa	End-systolic pressure
dP/dt_{max}	$kPa\,ms^{-1}$	Maximum pressure rise rate
dP/dt_{min}	$kPa\,ms^{-1}$	Maximum pressure decay rate

vitro force-pCa data of skinned rat myocyte preparations [4,5,10] (see Sect. 3.2 for details). The full model was run at these points and the successfully completed simulations formed the training dataset (1189 points). Univariate GPEs were then trained with a 5-fold cross-validation to predict each of the 14 LV output features. The accuracy of each of the resulting 14 trained GPEs was evaluated using the R^2-score regression metric. The resulting mean cross-validation R^2 test score was >0.98 for all the GPEs.

3   Results

3.1   Inferring OM Effects on the Sarcomere from *in Vivo* Whole-Organ Hemodynamics data

We did not have *in vivo* measurements of OM effects on rat cardiac hemodynamics, so qualitative observations (Table 3) from a healthy pig study were used [1]. We aimed to match significant changes in hemodynamics from baseline after OM administration and keep all other features constant. Changes in parameters that could recover the desired hemodynamic changes were determined by using the available GPEs to predict LV features' values at $400,000$ input parameter points sampled using a Latin hypercube design over the training dataset's ranges. A single iteration of Bayesian history matching (HM) was then performed with an implausibility threshold set to 1.5 (see [8] for more details), that identified $3,469$ points (corresponding to 0.8672% of the initial space) as non-implausible for replicating the organ-scale effects of OM administration (Fig. 1). We predicted the intact steady-state force-pCa curves for each of these parameter sets by running the cellular contraction model and derived the pCa_{50} and h values from the curves. The median predicted changes in the intact force-pCa curves following OM administration were then compared with measured [4,5,10] changes in force-pCa curves from skinned rat preparations in OM-containing solutions (Fig. 2) and found to be in qualitative agreement.

Table 3. Experimental values to match. Values are given as percentage change from healthy, reference experimental mean values.

Label	Exp. mean	Exp. std	Label	Exp. mean	Exp. std
SV	100.00 %	0.00 %	Tdiast*	88.24 %	16.97 %
EDV*	87.14 %	17.14 %	ET*	116.16 %	9.61 %
ESV*	76.92 %	20.51 %	dP/dt^*_{max}	121.66 %	35.53 %
EF	100.00 %	0.00 %	dP/dt_{min}	100.00 %	0.00 %

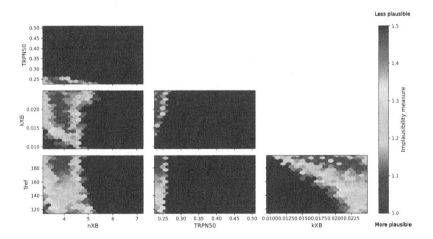

Fig. 1. First iteration of HM procedure. The full parameter space is constrained according to an implausibility criterion which evaluates how plausible is a point to yield model predictions that are matching experimental observations.

3.2 Inferring OM Effects on Whole-Organ Function from *in Vitro* Force-pCa Measurements

The effect of OM on Force-pCa measurements in healthy rats' skinned myocytes preparations were taken from the literature [4,5,10]. To map skinned experimental observations to intact/*in vivo* results we scaled our reference pCa_{50} and

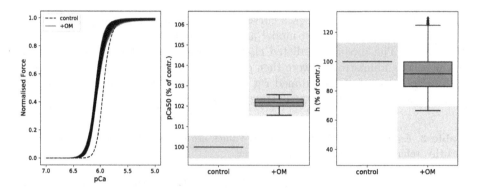

Fig. 2. Predicted *in silico* OM effects on the sarcomere as described by force-pCa curves calculated from the constrained, OM-compatible sarcomere parameter space and by the related percentage changes of pCa$_{50}$ and h values from control values. Experimental uncertainty ranges are displayed (the middle and far right panels) as shaded regions using percentage-from-control mean ± standard deviation values for both the healthy (gray) and +OM (orange) cases. (Color figure online)

h values by the experimentally observed percentage changes P_{shift} and P_{slope}, respectively, to estimate the *in vivo* change (Δ) of pCa$_{50}$ and h values due to OM. We indicate *in vivo* model parameters in the presece of OM with a superscript OM. We define and calculate values for α, $\beta \in \mathbb{R}$ such that the parameters $\mathrm{TRPN}_{50}^{\mathrm{OM}} := \alpha \cdot \mathrm{TRPN}_{50}$ and $n_{xb}^{\mathrm{OM}} := \beta \cdot n_{xb}$ achieved the desired change in the force-pCa curve encoded by a given combination of ΔpCa and Δh. To determine the plausible values of $\mathrm{TRPN}_{50}^{\mathrm{OM}}$ and n_{xb}^{OM}, we sampled $100,000$ points from a 2D Latin hypercube design using the experimentally observed ranges for $P_{shift} \in [0.0085, 0.0833]$ and $P_{slope} \in [-0.7282, -0.2470]$ and determined the corresponding pair of scaling coefficients (α, β) for each point to define a set of $100,000$ plausible values of $(n_{xb}^{\mathrm{OM}}, \mathrm{TRPN}_{50}^{\mathrm{OM}})$. Finally, we used the previously trained GPEs to predict EDV, ESV, ET/Tdiast and dP/dt$_{max}$ for the plausible values of $(n_{xb}^{\mathrm{OM}}, \mathrm{TRPN}_{50}^{\mathrm{OM}})$, while keeping the values of k_{xb} of T_{ref}, where we did not have data, fixed to control values. Figure 3 shows that the inferred median effects of these changes to pCa$_{50}$ and h due to OM on whole-organ function show a qualitative agreement in the direction of change for all the 4 LV features reported to be significantly altered by OM in the pig study [1].

4 Discussion

We have demonstrated that our virtual rat heart framework [8] can recapitulate the effects of OM on cardiac contraction and that multi-scale cardiac mechanics models can be used to infer the impact of drugs on cellular function from whole-organ observations. Calibration of model parameters to both cellular and whole-organ data on OM gives us additional insight into the mechanistic processes that link cellular and whole-heart contraction. Our results demonstrate that in order

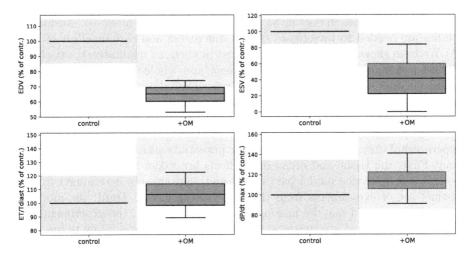

Fig. 3. Predicted *in silico* OM effects on whole-heart function as percentage changes of LV features' values from control values. Experimental uncertainty ranges are displayed as shaded regions using percentage-from-control mean ± standard deviation values for both the healthy (gray) and +OM (orange) cases. (Color figure online)

to reproduce available data, OM requires to alter the function of both thick and thin filaments. The OM effect on thick filament (the direct site of action of OM) is essential to reproduce OM effect on tension in whole-heart. Simulations show that the OM effect on force-pCa curves involve changes in thin filament function, namely by altering calcium myofilament sensitivity, which supports the recent hypothesis of the effect OM on thin filaments [11].

Limitations of this work include: (1) Both thin and thick filament dynamics are modeled using a simplified representation of the sarcomere, without a detailed mechanistic description of its components. Specifically, the cross-bridge kinetics is described by a two-state model where the strongly-/weakly-/un-bound states are collapsed into a single state. While more detailed models exist [6], their parameters are not necessarily constrained using experimental data due to difficulties in measuring subcellular processes. (2) As a result of (1), the OM effect is modelled using 2- to 4-parameters linked to the cross-bridge cycling but not directly incorporating the OM mechanism of action (which is to increase the rate of myosin-head attachment to actin [9]). (3) OM whole-organ data in healthy animals is limited. To the best of our knowledge, the *in vivo* pig hemodynamics data [1] used to constrain the model parameters is the only available non-human study of OM effect on LV function with therapeutic doses in healthy animals. (4) Differences across species and in acute vs chronic OM administration are not explored. For example, studies in healthy volunteers [12] suggest an increase in SV and EF following OM treatment as a result of the overall systolic function improvement. However, we only mapped LV features' values that were reported to significantly change from control after OM administration [1], thus our results excluded cases when change in SV and EF features was observed. Additional studies are needed to investigate species differences and chronic effect of OM. (5) We have shown that the median model predictions qualitatively match the observed changes due to OM at the tissue and whole-organ scale. However, the model does not quantitatively match experimental observations. Specifically, there are some non-implausible Hill coefficients in Fig. 2 that are increased and so do not qualitatively match the experimentally observed changes. Similarly, in some cases ET/Tdiast is predicted to decrease in Fig. 3, in contrast to the experimental observations. There are four potential contributors to this discrepancy. First, the tissue and organ drug effects are taken from different species. Second, the reference model parameters and anatomy are determined from a distinct set of experiments from the OM measurements. Third, the model is a simplification and may be missing some features. Forth, the experiments are performed in skinned preparations, while the model replicates intact tissue. This likely explains the discrepancy in predicted change in Hill coefficient, which is sensitive to the skinning process. Finally, we fitted the model to the data using an emulator, which itself has uncertainty, and this can increase the overall uncertainty. A self-consistent multi-scale data of the effects of OM on cell, tissue and organ scales would allow us to better identify and address the source of these discrepancies.

5 Conclusions

We demonstrated how multi-scale heart contraction models can aid our understanding of the mechanistic links between cellular and whole-organ function and can help in interpreting skinned experimental preparations in the context of *in vivo* whole-heart function.

References

1. Bakkehaug, J.P., et al.: Myosin activator Omecamtiv mecarbil increases myocardial oxygen consumption and impairs cardiac efficiency mediated by resting myosin ATPase activity. Circ. Hear. Fail. 8(4), 766–775 (2015). https://doi.org/10.1161/CIRCHEARTFAILURE.114.002152
2. Benjamin, E.J., et al.: Heart disease and stroke statistics - 2018 update: a report from the American Heart Association. Circulation 137(12), E67–E492 (2018). https://doi.org/10.1161/CIR.0000000000000558
3. Horváth, B., et al.: Frequency-dependent effects of Omecamtiv mecarbil on cell shortening of isolated canine ventricular cardiomyocytes. Naunyn. Schmiedebergs. Arch. Pharmacol. 390(12), 1239–1246 (2017). https://doi.org/10.1007/s00210-017-1422-z
4. Kampourakis, T., Zhang, X., Sun, Y.B., Irving, M.: Omecamtiv mercabil and blebbistatin modulate cardiac contractility by perturbing the regulatory state of the myosin filament. J. Physiol. 596(1), 31–46 (2018). https://doi.org/10.1113/JP275050

5. Kieu, T.T., Awinda, P.O., Tanner, B.C.: Omecamtiv Mecarbil slows myosin kinetics in skinned rat myocardium at physiological temperature. Biophys. J. **116**(11), 2149–2160 (2019). https://doi.org/10.1016/j.bpj.2019.04.020

6. Land, S., Niederer, S.A.: A spatially detailed model of isometric contraction based on competitive binding of troponin i explains cooperative interactions between tropomyosin and crossbridges. PLOS Comput. Biol. **11**(8) (2015). https://doi.org/10.1371/journal.pcbi.1004376

7. Land, S., et al.: An analysis of deformation-dependent electromechanical coupling in the mouse heart. J. Physiol. **590**(18), 4553–4569 (2012). https://doi.org/10.1113/jphysiol.2012.231928

8. Longobardi, S., et al.: Predicting left ventricular contractile function via Gaussian process emulation in aortic-banded rats. Philos. Trans. A. Math. Phys. Eng. Sci. **378**(2173), 20190334 (2020). https://doi.org/10.1098/rsta.2019.0334

9. Malik, F.I., et al.: Cardiac Myosin activation: a potential therapeutic approach for systolic heart failure. Science **331**(6023) (2011). https://doi.org/10.1126/science.1200113

10. Nagy, L., et al.: The novel cardiac myosin activator Omecamtiv mecarbil increases the calcium sensitivity of force production in isolated cardiomyocytes and skeletal muscle fibres of the rat. Br. J. Pharmacol. **172**(18), 4506–4518 (2015). https://doi.org/10.1111/bph.13235

11. Swenson, A.M., et al.: Omecamtiv mecarbil enhances the duty ratio of human β-cardiac myosin resulting in increased calcium sensitivity and slowed force development in cardiac muscle. J. Biol. Chem. **292**(9), 3768–3778 (2017). https://doi.org/10.1074/jbc.M116.748780

12. Teerlink, J.R., et al.: Dose-dependent augmentation of cardiac systolic function with the selective cardiac myosin activator, Omecamtiv mecarbil: a first-in-man study. Lancet **378**(9792), 667–675 (2011). https://doi.org/10.1016/S0140-6736(11)61219-1

13. Teerlink, J.R., et al.: Cardiac myosin activation with Omecamtiv mecarbil in systolic heart failure. N. Engl. J. Med. **384**(2) (2021). https://doi.org/10.1056/NEJMoa2025797

High-Speed Simulation of the 3D Behavior of Myocardium Using a Neural Network PDE Approach

Wenbo Zhang[1], David S. Li[1], Tan Bui-Thanh[2], and Michael S. Sacks[1(✉)]

[1] Willerson Center for Cardiovascular Modeling and Simulation, Oden Institute for Computational Engineering and Sciences, Department of Biomedical Engineering, University of Texas at Austin, Austin, TX 78712, USA
{wenbo,davidli,msacks}@oden.utexas.edu
[2] Oden Institute for Computational Engineering and Sciences, Department of Aerospace Engineering and Engineering Mechanics, University of Texas at Austin, Austin, TX 78712, USA
tanbui@oden.utexas.edu

Abstract. The full characterization of three-dimensional (3D) mechanical behaviour of myocardium is essential in understanding their function in health and disease. The hierarchical structure of myocardium results in their highly anisotropic mechanical behaviors, with the spatial variations in fiber structure giving rise to heterogeneity. The optimal set of loading paths has been used to estimate the constitutive parameters of myocardium using a novel numerical-experimental approach with full 3D kinematically controlled (triaxial) experiments [1,2]. Due to the natural variations in soft tissue structures, the mechanical behaviors of myocardium can vary dramatically within the same organ. To alleviate the associated computational costs for obtaining responses of myocardium under a range of loading conditions with a given realization of structure, we developed a neural network-based method integrated with finite elements. The boundary conditions were parameterized. The neural network generated a corresponding trial solution of the underling hyperelasticity problem for each boundary condition. Thus, the neural network approximated the parameter-to-state map. A physics-informed approach was used to train the neural network. Due to their learnability characteristics, the neural network was able to predict solutions for a range of boundary conditions with given individual specimen fiber structures. The neural network was validated with finite element solutions. This method will provide efficient and robust computational models for clinical evaluation to improve patient outcomes.

Keywords: Cardiac simulation · Myocardium · Machine learning

1 Introduction

The full characterization of three-dimensional (3D) mechanical behavior soft tissues, such as myocardium, is essential in simulating organ function in health and

© Springer Nature Switzerland AG 2021
D. B. Ennis et al. (Eds.): FIMH 2021, LNCS 12738, pp. 416–424, 2021.
https://doi.org/10.1007/978-3-030-78710-3_40

disease. The hierarchical structure of soft tissues dictates their highly anisotropic mechanical behaviors, with the spatial variations in fiber structure also giving rise to heterogeneity. To address these issues in a full 3D context, we have developed a novel numerical-experimental approach to determine the optimal model form and parameter estimation for continuum constitutive models of soft tissues, as applied to the myocardium [1, 2]. This approach utilized optimal experimental design of the full 3D kinematic (triaxial) experiments coupled to an inverse model that incorporated local fibrous structure to perform robust parameter estimation (Fig. 1). Due to the natural variations in soft tissue structures (Fig. 2), the mechanical behaviors of soft tissues can vary dramatically within the same organ. The set of optimal loading paths for a tissue specimen (Fig. 2) includes three pure shear and three simple shear loading conditions.

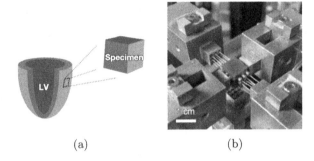

(a)   (b)

Fig. 1. Tissue specimen of left ventricular myocardium for triaxial mechanical testing [2]

It becomes prohibitive to obtain the responses of heart-specific models under varying parameters, including boundary conditions and fibrous structures, in translational clinical time frames. To shift the computational cost ahead of prediction time, neural network (NN) representations of the parameter-to-state map have been proposed as surrogates for parametric partial differential equations (PDEs) due to the representation power of neural networks. Data-driven approaches that require training datasets of finite element (FE) solutions were developed to train the neural network surrogate models. For example, derivative-informed projected neural networks were developed to improve generalization accuracy [3], and machine learning methods were used to enhance reduced order models [5]. To avoid generating training dataset by solving numerous linear or nonlinear FE equations, physics-informed approaches have been developed. The densely connected neural networks were utilized to approximate the solutions of the governing equations trained by minimizing the L_2 norm of the residuals and penalizing the violation of boundary conditions [4]. The convolutional neural networks with a finite difference method estimating spatial gradients [6] is limited to regular domains.

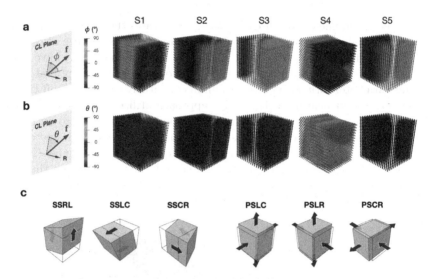

Fig. 2. Spatially varying fiber orientations on the cuboid domain for 4 different specimens, including (a) circumferential angle, and (b) out of plane angle. (c) Optimal loading paths for material parameter estimation including simple shear (SS) and pure shear paths in longitudinal (L), circumferential (C), and radial (R) directions. [2]

In-silico implementation of such complex 3D continuum soft tissue constitutive models to obtain the responses of varying boundary conditions and fibrous structures requires the solution of the associated hyperelasticity problem, which remains impractical in translational clinical time frames. To alleviate the associated substantial computational costs at the time of simulation, we have developed a neural network-based method that can simulate the 3D mechanical behavior of soft tissues. A physics-informed approach was employed to train the neural network (NN) surrogate model to give physically correct solution for a range of loading conditions by minimizing the potential energy without any training dataset generated by finite element (FE) solver. The FE discretization of the solution field is applicable to problems defined with complex geometry and boundary conditions such as ventricular simulations and it enables strong enforcement of the Dirichlet boundary conditions in a natural manner.

2   Methods

We aim to develop an efficient neural network representation of the solutions of parametric PDEs that describe heart-specific cardiac models. To this end, we streamline the neural network-based surrogates and finite elements in an end-to-end pipeline. The neural network generates the corresponding FE nodal values for a trial solution with a given realization of parameters. We construct the corresponding trial solution using finite element basis functions. The neural network-based surrogate is trained by minimizing the sum of the energy functional for a set of sampled parameters.

We denote the reference configuration $\mathbf{X}$, the current configuration $\mathbf{x}$, the displacement $\mathbf{u} = \mathbf{x} - \mathbf{X}$, the deformation gradient $\mathbf{F} = \nabla\mathbf{u}$, the right Cauchy-Green tensor $\mathbf{C} = \mathbf{F}^\mathrm{T}\mathbf{F}$. The unimodular right Cauchy-Green tensor is defined as $\bar{\mathbf{C}} = J^{-2/3}\mathbf{C}$, where $J = \det\mathbf{F}$. We modeled the myocardium as a hyperelastic nearly incompressible material using a recently developed strain invariant form as the strain energy function Ψ in the full 3D kinematic space [2]. The first isotropic invariant of $\bar{\mathbf{C}}$ is

$$\bar{I}_1 = \mathrm{Tr}(\bar{\mathbf{C}}). \tag{1}$$

Two pseudo-invariants of $\mathbf{C}$ for the squared fiber stretches in each fiber directions are defined using

$$I_{4f} = \mathbf{f}_0 \cdot \mathbf{C} \cdot \mathbf{f}_0, \ I_{4s} = \mathbf{s}_0 \cdot \mathbf{C} \cdot \mathbf{s}_0, \tag{2}$$

where $\mathbf{f}_0$ is the myofiber orientation of myocardium, $\mathbf{s}_0$ is the orthogonal direction within the tangent plane of the laminar sheet, and $\mathbf{n}_0$ is the normal direction to the sheet. Three pseudo-invariants of $\mathbf{C}$ for coupling effects are

$$I_{8fs} = \mathbf{f}_0 \cdot \mathbf{C} \cdot \mathbf{s}_0, \ I_{8fn} = \mathbf{f}_0 \cdot \mathbf{C} \cdot \mathbf{n}_0, \ I_{8sn} = \mathbf{s}_0 \cdot \mathbf{C} \cdot \mathbf{n}_0. \tag{3}$$

The constitutive model for myocardium consists of three exponentially stiffening terms, given by

$$\begin{aligned}
\Psi = {} & \frac{a}{2b}\left(\exp\left(b\left(\bar{I}_1 - 3\right)\right) - 1\right) \\
& + \sum_{i=\mathrm{f},\mathrm{s}} \frac{a_i}{2b_i}\left(\exp\left(b_i\left(I_{4i} - 1\right)^2\right) - 1\right) \\
& + \sum_{i,j=\mathrm{f},\mathrm{s},\mathrm{n} \ i\neq j} \frac{a_{ij}}{2b_{ij}}\left(\exp\left(b_{ij}I_{8ij}^2\right) - 1\right) \\
& + \kappa\left(J - 1 - \ln J\right)
\end{aligned} \tag{4}$$

where the first term is for isotropic extracellular matrix (ECM), the second term for fiber families within the sheet, the third term for coupling interactions, and the fourth term for enforcing incompressibility condition weakly.

The variational problem for hyperelasticity of myocardium can be describe solved by a minimization problem. Given the body force $\mathbf{f}$ and the traction $\mathbf{t}$ on the Neumann boundary, find the displacement $\mathbf{u}$ ($\mathbf{u} \in \mathcal{V}_0^h$) that minimize the energy functional (potential) Π. The displacement $\hat{\mathbf{u}}(\mathbf{x})$ is discretized using

$$\hat{\mathbf{u}}(\mathbf{x}) = \mathbf{U}\mathbf{N}(\mathbf{x}) \tag{5}$$

where $\mathbf{U} \in \mathbb{R}^{3\times d_u}$ is the nodal values for the displacement, and $\mathbf{N}$ represents trilinear basis functions for the Q_1 Lagrange element on a hexahedral mesh of d_u nodes. We choose the Q_1 element since it is sufficient for the present study. The present method is not limited to the element we use.

For each instance of varying loading condition parameterized by a d_m-D vector $\mathbf{M}^{(i)} \in \mathbb{R}^{d_m}$, a nonlinear FE equation drived from the stationary conditons for the potential minimization problem needs to be solve to obtain the FE solution $\mathbf{U}^{(i)} \in \mathbb{R}^{d_u}$. To avoid prohibitive computational cost for numerous evaluation the parameter-to-state map $\mathbf{U}(\mathbf{M})$, we use a neural network surrogate model $\hat{\mathbf{U}} = f_{\mathrm{NN}}(\mathbf{M}; \theta)$ to approximate the FE solutions for a range of $\mathbf{M}$ where θ parametrize the neural network.

The NN surrogate model $f_{\mathrm{NN}}(\mathbf{M}; \theta)$ is trained by the optimization problem

$$\min_{\theta} \quad \sum_{i=1}^{N} \Pi(\hat{\mathbf{u}}^{(i)}), \tag{6}$$

where the objective function is the aggregated potential on the sampled loading conditions $\mathbf{M}^{(i)}$. The nodal values for the corresponding trial solution $\hat{u}^{(i)}$ is $\hat{\mathbf{U}}^{(i)} = f_{\mathrm{NN}}(\mathbf{M}^{(i)})$. Then, the Dirichlet boundary condition is enforced strongly. The potential is

$$\Pi(\mathbf{u}) \overset{\text{def}}{=} \int_{\Omega} \Psi(\mathbf{u}) \mathrm{d}x - \int_{\Omega} \mathbf{f} \cdot \mathbf{u} \mathrm{d}x - \int_{\Gamma_N} \mathbf{t} \cdot \mathbf{u} \mathrm{d}s. \tag{7}$$

In the case of parameterized Dirichlet boundary conditions, $\mathbf{M}$ is a collection of prescribed nodal values for the fixed displacements. In the case of parameterized Neumann boundary conditions, $\mathbf{M}$ is a collection of prescribed nodal values for the loads.

In this work, we use fully connected network (FCN) which can be described as a sequence of composite functions of nonlinear functions and affine functions. For a FCN with L hidden layers, we have

$$f_{NN} = A_L \circ \phi \circ \cdots A_1 \circ \phi \circ A_0 \tag{8}$$

where A_i $(i = 0, \ldots, L)$ are affine functions, ϕ is a element-wise activation function. Using the FCN as an approximator of the parameter-to-state map, we can substitute the corresponding trial function into the potential.

To demonstrate the learnability of the neural network surrogate model, we considered the triaxial simulations parameterized by its Dirichlet boundary conditions on a cuboid domain (Fig. 3). The marked facets with aligned normal directions on two opposite sides of the cube has a single parameter dictates the magnitude of the fixed boundary condition. The Dirichlet boundary condition is $\mathbf{u} = \mathbf{n}M_i$ for the i-th pair of facets with normal directions $\mathbf{n} = \pm\mathbf{e}_i$ $(i = 1, \ldots, 3)$ where $\mathbf{e}_i$ is the Cartesian basis. Thus, the boundary conditions are parametrized by $\mathbf{M} \in \mathbb{R}^3$. We use low-discrepancy Halton sequence to sample N realizations of $\mathbf{M}$. The material parameters of the myocardium are listed in the Table 1. We trained the neural network for four different specimens to demonstrate the learnability of the neural networks.

Table 1. The material parameters for the myocardium specimens.

	a (Pa)	b	a_f (Pa)	b_f	a_s (Pa)	b_s	a_{fs} (Pa)	b_{fs}	a_{fn} (Pa)	b_{fn}	a_{sn} (Pa)	b_{sn}
S1	4.81	5.05	1175	0.276	1665	6.93	135.5	0.024	3440	2.01	3278	20.4
S2	0.100	0.346	2936	0.045	989.1	0.190	548.6	0.0020	384.0	1.47	6397	0.004
S3	6.07	4.90	2892	0.0300	178.2	0.0500	1553	0.78	4174	40.3	1791	44.6
S4	1.05	12.0	2964	3.10	496.0	0.0870	369.0	0.011	1712	68.8	547.7	1.17

Fig. 3. A cuboid domain with side lengths of 1 cm is discretized using hexahedron elements. Dirichlet boundary conditions are imposed on the yellow facets. (Color figure online)

3 Results

We consider a cuboid domain with side lengths of 1 cm. The domain is discretized by trilinear elements. The number of DOFs is $3 \times 9^3 = 2187$. The 2-th order Gaussian quadrature is used. There are six pads on each facet where we apply the Dirichlet boundary conditions. The training range for sampling all components of $\mathbf{M} = (M_1, M_2, M_3)$ is $[-0.2, 0.2]$ cm. We restrict $\mathbf{M}$ to be unimodular using $(1+M_1) \times (1+M_2) \times (1+M_3) = 1$ to respect the incompressibility condition. The training range is incrementally expanded in 8 steps. The number of $\mathbf{M}$ samples generated using Halton sequence is 400. The NN has 1 hidden layer of 10 neurons with hyperbolic tangent function as the activation function.

Fig. 4. Protocols for generating validation datasets.

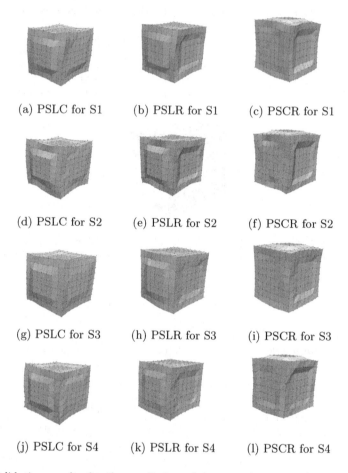

(a) PSLC for S1   (b) PSLR for S1   (c) PSCR for S1

(d) PSLC for S2   (e) PSLR for S2   (f) PSCR for S2

(g) PSLC for S3   (h) PSLR for S3   (i) PSCR for S3

(j) PSLC for S4   (k) PSLR for S4   (l) PSCR for S4

Fig. 5. Validation results for the prediction of the neural network (black wireframe) with the finite element solutions (red points) for pure shear (PS) deformations in longitudinal (L), circumferential (C), radial (R) directions for specimen S1–S4. (Color figure online)

To demonstrate the performance of the neural networks, we examine the relative L_2 error, defined as

$$e = \frac{\sum_i ||u_{NN}^{(i)} - u_{FE}^{(i)}||_2}{\sum_i ||u_{FE}^{(i)}||_2}. \tag{9}$$

The validation dataset includes three loading protocols: (1) $M_1 \in [0, 0.2]$ cm, $M_2 \in [-0.167, 0]$ cm, $M_3 = 0$; (2) $M_1 \in [0, 0.2]$ cm, $M_2 = 0$, $M_3 \in [-0.167, 0]$ cm; (3) $M_1 = 0$, $M_2 \in [0, 0.2]$ cm, $M_3 \in [-0.167, 0]$ cm as shown in Fig. 4. Each loading protocol has 10 uniformly spaced steps. The results obtained with present NN matched closely with the finite element solutions (Fig. 5). The relative L_2

errors for four different specimen with different set of material parameters are listed in Table 2. We also examine the average responses (Fig. 6) for the first loading protocol using the first specimen as an example. The neural network predictions and the finite element solutions have a good agreement.

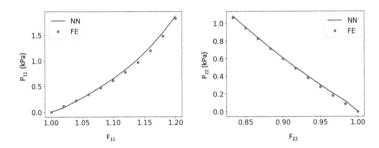

Fig. 6. The average first Piola-Kirchhoff stress on the boundary vs the deformation of the FE solutions and NN predictions.

Table 2. The relative L_2 error of neural network predictions using the corresponding FE solutions as ground truth on the validation datasets.

	S1	S2	S3	S4
e	6.2064%	6.9889%	6.9068%	7.6440%

The computation time for one prediction of the neural network surrogate model was 0.02236 s, while the FE solver takes 6.5762 s for assembly and solution for one step using a serial program, which is a speed-up of 294.1. The predictions of the neural network surrogate model and the FE solution were generated on using a Intel(R) Core(TM) i9-9920X on a System 76 Thelio Major computer. The average time needed to train the neural network surrogate model is 19 min 42.25 s on a NVIDIA(R) GeForce RTX 2080 Ti. The neural network surrogate model can give predictions in parallel while the FE solver needs a stepping scheme to incrementally obtain the solution for the fully loaded state which would multiply the cost with the number of steps. The number of steps is 10 in the present case.

4   Discussion

In this work, we have developed a high fidelity neural network surrogate model that is trained in a physics-informed approach to give a direct solution of 3D soft tissue hyperelasticity in-silico. The present method was found to be an order of 10^2 time faster than the equivalent FE model using the same mesh on the same machine. With the learnability of the neural networks, the architecture of the NNs can incorporate attributes such as spatially varying fiber structures.

By shifting the computation expense from FE solutions to NN training, the NN surrogate model can be used to give significantly fast atiredictions of complex 3D deformations in full kinematic space with given fiber structures by forward propagation in the neural network. More detailed studies on the error analysis of the NN surrogate model are reserved for the future. The future strategies for improving the accuracy of the NN surrogate models include efficient sampling method, scalable training algorithms, and advanced neural networks with improved representation power. One natural extension is to apply the present method to a ventricular model which has more complex geometry and boundary conditions. This method will pave the way for building an efficient template model of hearts with add-on heart-specific attributes, with neural network-based surrogates for fast predictions to evaluate the need to conduct high-fidelity simulations. The ultimate goal is to provide efficient and robust computational models for clinical evaluation to improve patient outcomes.

References

1. Avazmohammadi, R., et al.: An integrated inverse model-experimental approach to determine soft tissue three-dimensional constitutive parameters: application to post-infarcted myocardium. Biomech. Model. Mechanobiol. **17**(1), 31–53 (2017). https://doi.org/10.1007/s10237-017-0943-1
2. Li, D.S., Avazmohammadi, R., et al.: Insights into the passive mechanical behavior of left ventricular myocardium using a robust constitutive model based on full 3d kinematics **103**, 103508 (2020) https://doi.org/10.1016/j.jmbbm.2019.103508
3. O'Leary-Roseberry, T., Villa, U., Chen, P., Ghattas, O.: Derivative-informed projected neural networks for high-dimensional parametric maps governed by PDEs (2020)
4. Raissi, M., Perdikaris, P., Karniadakis, G.: Physics-informed neural networks: A deep learning framework for solving forward and inverse problems involving non-linear partial differential equations **378**, 686–707 (2019) https://doi.org/10.1016/j.jcp.2018.10.045
5. Sheriffdeen, S., Ragusa, J.C., Morel, J.E., Adams, M.L., Bui-Thanh, T.: Accelerating PDE-constrained inverse solutions with deep learning and reduced order models (2019)
6. Zhu, Y., Zabaras, N., Koutsourelakis, P.S., Perdikaris, P.: Physics-constrained deep learning for high-dimensional surrogate modeling and uncertainty quantification without labeled data. J. Comput. Phys. **394**, 56–81 (2019)

On the Interrelationship Between Left Ventricle Infarction Geometry and Ischemic Mitral Regurgitation Grade

Hao Liu[1], Harshita Narang[1], Robert Gorman[2], Joseph Gorman[2], and Michael S. Sacks[1(✉)]

[1] Willerson Center for Cardiovascular Modeling and Simulation, University of Texas at Austin, Austin, TX, USA
msacks@oden.utexas.edu
[2] Perelman School of Medicine, Gorman Cardiovascular Research Group, University of Pennsylvania, Philadelphia, PA, USA

Abstract. Ischemic mitral regurgitation (IMR) is manifested by the inability of the mitral valve (MV) to form a completed sealed shape, which is induced by rapidly impairing contractile function of acute myocardial infarction (MI). Mitral valve repair with undersized ring annuloplasty is currently the preferred treatment strategy for IMR. However, the overall persistence and recurrence rate of moderate or severe IMR within 12 months of surgery has been consistently reported as high, which is a direct consequence of adverse left ventricle (LV) remodeling after MI. In this study, we developed a detailed finite element model with coupled left ventricle-mitral valve structure including mitral valve leaflets, chordae tendineae (CT), papillary muscles, and myocardium. In addition, this model was consisted of high fidelity structure segmented from image data, a novel structural constitutive model of MV leaflets and mechanical properties of CT measured using an in-vitro mechanical testing in an integrated computational modeling framework. Discrepancy of strain mapping has been found between in-silico model and in-vivo strain analysis and including pre-strain of mitral valve leaflets in in-silico model was necessary to have more agreement with in-vivo data. Our findings suggests our LV-MV model is capable of predicting IMR results by shutting down regional contractility and pre-strain should be incorporated into future LV-MV model for more accuracy.

Keywords: Left ventricle-mitral valve structure · Mitral regurgitation · Finite element model

1 Introduction

The mitral valve (MV) regulates blood flow from left atrium (LA) into the left ventricle (LV). In situations where the MV fails to fully close, the resulting blood regurgitation into the left atrium causes pulmonary congestion, leading to heart

© Springer Nature Switzerland AG 2021
D. B. Ennis et al. (Eds.): FIMH 2021, LNCS 12738, pp. 425–434, 2021.
https://doi.org/10.1007/978-3-030-78710-3_41

failure and strokes. There are two main types of mitral valve regurgitation (MR): primary MR, caused by organic disease of one or more components of MV, and secondary or functional MR, which occurs secondary to LV dysfunction. One form of secondary MR is ischemic MR (IMR), which results from left ventricular myocardial infarciton (MI) [2]. Annuloplasty surgery, a common preferred treatment for patients with IMR, tightens the MV annulus using a ring. However, over a third of patients have consistently been reported to suffer persistence or recurrence of moderate to severe IMR within the first 12 months after annuloplasty repair surgery [10,17], a direct consequence of adverse LV remodeling after MI. In some patients, the major contributor to the pathogenesis of IMR is annular dilation and flattening, while in others the annular dilation is moderate and leaflet tethering is the primary pathological driver. Some previous studies involving finite element models of the MV have provided insights into IMR progression and treatment. However, the etiology of the disease is highly dependent on the evolution of LV geometry post MI, as supported by the pioneering work of Wenk et al. [22]. Therefore, in order to truly understand how the pathology of IMR affects its progression, such MV models must be integrated into a model of the LV as one functional unit.

The first integrated left ventricle-mitral valve computational model was developed in 2010, based on MRI data from a sheep with moderate ischemic MR [22]. This model has been extended to include mitral annuloplasty ring [23]. Subsequently, the living heart project model modeled the whole heart with four chambers and four valves [3], and was capable of capturing electrical potential, mechanical deformation, with good agreement with clinical observations. In a recent study on LV-MV model with fluid-structure interaction had been investigated [8]. However, detailed LV-MV interactions with anatomically and biomechanically accurate MV models, especially on the effects of infarct extent and anatomic location, have yet to be achieved.

The objective of this study was thus to investigate how regional loss of active contraction due to myocardial infarction and its relation to IMR grade. We firstly developed a pipeline to develop an anatomically faithful ovine LV-MV model, that including aspects such as MV pre-strain and a gradient of borderzone function. We then conducted a parametric study on how infarct extent and location and IMR severity. We then compared our results with published kinematic and IMR grade and demonstrated excellent results. In particular, the LV-MV model was able to predict severity of mitral valve regurgitation with specific infarct location. Prediction from posterobasal infarction in-silico model indicated largest regurgitation orifice area which was consistent from largest IMR grade reported from animal study. Our findings suggest our LV-MV in-silico model is capable of predicting similar IMR results, and that MV pre-strain should be incorporated into future LV-MV models to correctly simulate its behaviors.

2   Methods

2.1   Pipeline to Create the Left Ventricle-Mitral Valve Geometry

Geometry of left ventricle-mitral valve structure was created by integration of several key components of the structure: myocardium, mitral valve, papillary muscles, and chordal structures [Fig. 1]. Two ventricles in the ovine heart were segmented from MRI imaging data (details have been shown in previous study [21]) and some improvements have been made in the current model. The basal plane was assumed to be flat and mitral valve plane could not be identified from the previous geometry in the previous study while portion of atrium was segmented and included in the current version of geometry. Furthermore, the ovine heart was scanned under fixation by end-diastolic pressure after the animal was sacrificed. Under this circumstance, both structures of mitral valve and papillary muscles could not be clearly visualized and these geometry were different from in-vivo ones. Thus, we took advantage of real-time 3 dimensional echocardiography (rt-3DE) from other ovine heart to integrate these structures to our previous cardiac model. Furthermore, anatomically-equivalent structure of CT was created based on technique previously developed [12].

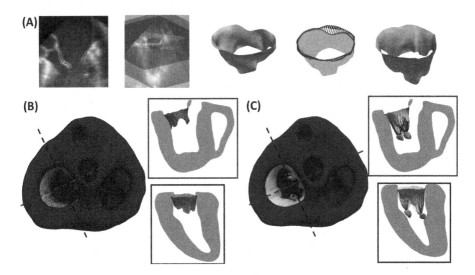

Fig. 1. (A) Segmentation of mitral valve from real-time 3d echo and morphing MV to fit LV; (B) integration of mitral valve into left ventricle; (C) integration of chordae and papillary muscles into left ventricle.

2.2   In-Silico LV-MV Model

The overall pipeline of creating LV-MV model has shown in Fig. 2. The first step was to develop an in-silico biventricular model based on experimental data collected from a single ovine heart at the Visible Heart Laboratory (VHL, University of Minnesota). Next, a finite element mesh was created based on the geometry segmented from MRI imaging data and local fiber directions obtained from diffusion tensor MRI imdaging data was aligned with finite element mesh. Then, fiber orientation from DTMRI data prescribed local coordinates of myocardium to represent anisotropic material law. Papillary muscles shared the same mechanical material response with myocardium and the local fiber directions followed the direction from base of PM to the tip of PM. The representative Cauchy stress for single element in FE mesh which consisted of passive response W^{vol}, active contraction S^{act} and volumetric response W^{vol} was shown as

$$T = -\frac{\partial W^{vol}}{\partial J} + \frac{1}{J}\tilde{F}\frac{\partial W^{dev}}{\partial \tilde{E}}\tilde{F}^T + \frac{1}{J}FS^{act}F^T \tag{1}$$

where $\mathbf{F}$ is the deformation gradient and the deviatoric deformation gradient is defined as $\tilde{\mathbf{F}} = J^{-1/3}\mathbf{F}$. Meanwhile, the Jacobian of the motion is defined as $J = \det \mathbf{F}$. Cavity pressure captured in vivo by catherization were treated as input kinetic boundary condition and cavity volume obtained from sonomicrometry study were used to fit the magnitude of active contraction forces for each time steps. Details could be found in previous publication [21].

Mitral valve closure was simulated under certain pressure on ventricular surface and value of pressure were calculated by subtraction of left ventricular pressure and end-diastolic pressure, which means pressure added on mitral valve leaflets was at 0 mmHg in end-diastolic state. The beginning of MV closure occurs right after end-diastolic state as pressure was applied on ventricular surface of MV. From previous findings of constitutive model of MV leaflets, accurate MV closure and strain analysis could be obtained with an incompressible nonlinear transversely isotropic material model [7,24]. The mechanical response of tissue is shown as

$$S = \frac{1}{\pi}\int_{-\pi/2}^{\pi/2} S_{ens}[E_{ens}(\theta)]\hat{\mathbf{n}}(\theta) \otimes \hat{\mathbf{n}}(\theta)d\theta + \mu_m(\mathbf{I} - C_{33}\mathbf{C}^{-1}) \tag{2}$$

where S is the second Piola-Kirchhoff stress tenor with contribution of collagen fibers oriented along $\hat{\mathbf{n}} = cos\theta\hat{\mathbf{c}} + cos\theta\hat{\mathbf{r}}$ and neo-Hookean ground matrix. Also, $S_{ens}(E_{end})$ is the PK2 stress in a unidirectional ensemble of collagen fibers where $E_{ens} = \hat{\mathbf{n}}E\hat{\mathbf{n}}$ and E is the Green-Lagrange strain tensor. Details of effective ensemble fiber response have been reported in the previous studies [5,14]. Further study on pre-strain in MV leaflets was performed by updating deformation matrix $F^{new} = F^{old}F^{pre-strain}$ and the extension of this equation was shown as

$$\begin{bmatrix} F_{11}^{new} & F_{12}^{new} \\ F_{21}^{new} & F_{22}^{new} \end{bmatrix} = \begin{bmatrix} F_{11}^{old} & F_{12}^{old} \\ F_{21}^{old} & F_{22}^{old} \end{bmatrix} \begin{bmatrix} \lambda_1 & 0 \\ 0 & \lambda_2 \end{bmatrix} \tag{3}$$

where λ_1 and λ_2 represent the amount of pre-stretch applied in circumferential and radial direction.

An incompressible isotropic hyperelastic material was intoduced to model the mechanical behavior of mitral valve chordae tendineae. The stress-strain relations for this model is to incorporate uniaxial strain without considering bending, which is shown in the following

$$S_{MVCT} = C_{10}[e^{C_{01}E_{11}} - 1] \tag{4}$$

where $\mathbf{E} = (\mathbf{C} - \mathbf{I})/2$ is the Green-Lagrange strain tensor and $E_{11} = \frac{1}{2}(F_{11}^2 - 1)$ is the uniaxial strain in the MVCT. The material constants C_{10} and C_{01} were chosen based on data reported by Lee et al. [14] and were set to be 0.1 MPa and 13.7 respectively for the present study. However, we did not assign individual pre-strain to each chordal tendineae because of we lack of data on closed state of mitral valve.

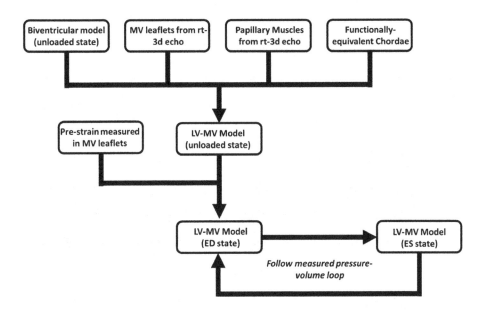

Fig. 2. Pipeline of creating LV-MV in-silico model

The finite element simulation was implemented using commercial software Abaqus (Dassault Systemes). All the previous material constitutive models were translated into fortran code in Abaqus subroutine VUMAT. The elements type for myocardium and papillary muscle was C3D10 (3d 10 nodes tetrahedron) and mitral valve leaflets were assigned as S3 element type (3-node shell). Truss element type with 2-node linear displacement (T3D2) was used for building chordal structure. Tie constriants were defined among two groups: myocardium with mitral valve and myocardium with papillary muscle. This constraint will force no relative motion between the surface connection. Also, self-contact constraint has been defined for mitral valve leaflets which uses a main-secondary kinematic scheme to avoid surface penetration. Finally, the simulation was performed by a Newton-Raphson quasistatic solver with explicit central difference integration.

2.3   Modeling Early Stage of Myocardial Infarction

During few minutes to hours post myocardial infarction, myocardium in infarct region lacks of oxygen supply which gradually adapts from active-force generating material to passive hyperelastic material. Multiple pieces of evidence [11,16,20] have shown that loss of contractility immediately occurs after MI. However, it is difficult to measure the quantity of loss of contractility. Thus, we assumed that the local contraction in infarct region reduced to zero right after MI. Borderzone, defined as a region with oxygen diffusion from healthy tissue to infarct tissue, tends to become hypocontractile and larger in size as myocardium keeps remodeling. In the current models, four infarct locations had been differentiated as anteroapical infarction, anterobasal infarction, posterobasal infarction, and laterobasal infarction [Shown in Fig. 3(A)] and we treated contractility reduced to 50% within borderzone. Infarcted tissue collected several weeks post-MI tends to be much stiffer than normal myocardium by triaxial mechanical test [15]. However, limited data has been reported in term of passive mechanical properties of infarcted tissue at immediate post-MI state and some of them are conflicted with each other [4,9,19]. Thus, we assumed that infarcted tissue in this study keeps the same passive mechanical properties as normal myocardium.

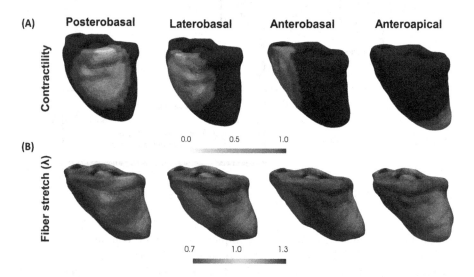

Fig. 3. (A) Different infarct locations defined in LV-MV model; (B) Fiber stretch of myocardium in end-systolic state

3   Results

The whole simulation of LV-MV model started from the unloaded state which was obtained by fitting Klotz's curve [13] and then followed left ventricular pressure-volume from experimental measurement (The whole cardiac cycle was calibrated to 1 s). The behavior of mitral valve was driven by boundary shared with left ventricle, pressure applied on ventricular surface, and chordal insertion connecting with papillary muscles. Mitral valve initiated closing at end-diastolic state as pressure applied on ventricular surface and closing period was lasting in 0.15 s. Then, the mitral valve remained fully-closed state for 0.45 s and began to open as pressure was removed from ventricular surface. Finally, the mitral valve was fully open and left ventricle was returned to relaxation state. Local cardiac contraction was shut down in the infarct tissue locations. From observations in Fig. 3(B), higher fiber stretch ($\lambda > 1$) was captured within infarct region which indicates extension of myocardium during end-systole.

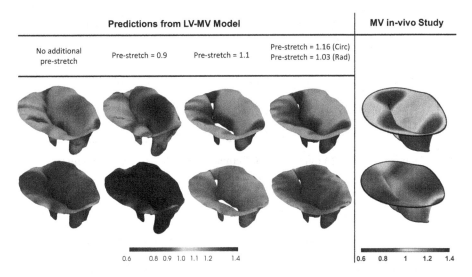

Fig. 4. Strain mapping (top row: circumferential; bottom row: radial) with additional pre-stretch applied on MV leaflets and strain mapping from in-vivo study

Comparing strain analysis from predictions in simulation [Fig. 4, first column] and in-vivo study [Fig. 4, last column] [18], some differences have been observed including less compression ($\lambda < 1$ and $\lambda = \sqrt{NFF^T N^T}$) along free edge in circumferential direction and larger fiber extension ($\lambda > 1$)in radial direction. Thus, we investigated the role of additional pre-strain in MV leaflets. With value of pre-strain obtained from experimental measurements [1], the strain mapping in second column had more agreements with strain mapping from in-vivo study than the one without pre-strain.

A critical aspect of the current study was to quantify the impact of specific infarct locations and extents on mitral valve function. Firstly, the gap between two Mv leaflets has been shown in infarction group where the enclosed area is known as mitral regurgitation orifice area [column 2–5 in Fig. 5]. Furthermore, the shape of MV annulus varies with different infarct locations due to tethering force from myocardium. However, the strain pattern of MV leaflets with and without infarction shows no significant difference. To quantity the severity of mitral regurgitation, regurgitant orifice area (ROA) was reported by measuring enclosed area. With comparisons, MV with largest ROA is induced by posterobasal infarction which reached an agreement from clinical study [6].

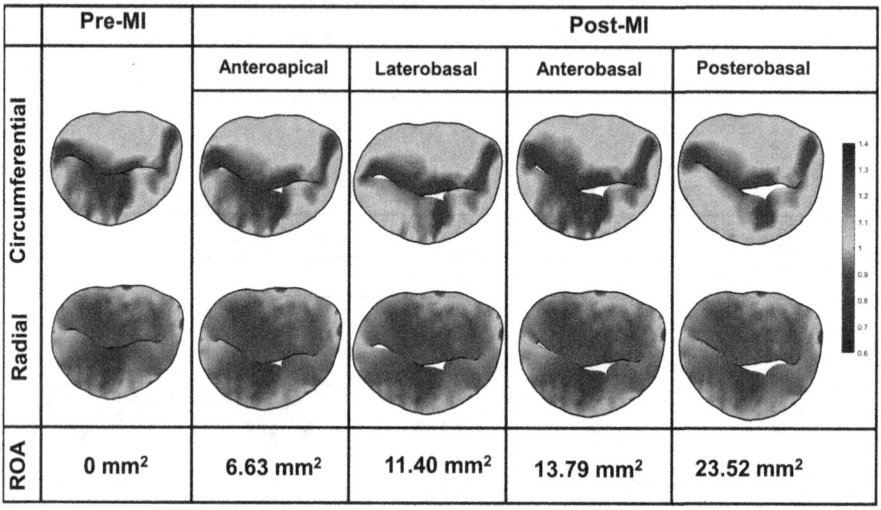

	Pre-MI	Post-MI			
		Anteroapical	Laterobasal	Anterobasal	Posterobasal
Circumferential					
Radial					
ROA	0 mm²	6.63 mm²	11.40 mm²	13.79 mm²	23.52 mm²

Fig. 5. Prediction of circumferential stretch, radial stretch and ROA from in-silico model for groups of different infarct locations

4   Discussion

The main goal of this study was to establish inner-relationship between ischemic mitral regurgitation and left ventricular infarction and evaluate the role of pre-strain on mitral valve strain mapping. We firstly developed a pipeline for creating LV-MV in-silico model with different critical component. Different factors resulting in IMR including loss of contractility and infarct location, have been investigated. Furthermore, the role of pre-strain in mitral valve leaflets was emphasized in computational model and more realistic pre-strain applied on MV leads to more agreements with in-vivo data. Moreover, we are using these approaches to inform novel therapies, such as the injection of hydrogel into LV wall to reduce the LV dilation and reduce the MR grade level. Another improvement of current LV-MV in-silico model is to incorporate growth and remodeling of myocardium and model myocardial infarction in the longer term. Second, plasticity of MV leaflets will be considered for long-term IMR which has been investigated in MV in-silico model. Finally, the success of this study could lead to develop further patient-specific finite element model. Differentiating with previous animal model, patient-specific LV-MV model will gain more insight into potential therapies instead of investigating mechanisms behind MI and IMR.

References

1. Amini, R., et al.: The posterior location of the dilator muscle induces anterior iris bowing during dilation, even in the absence of pupillary block. Invest Ophthalmol. Vis. Sci. **53**(3), 1188–1194 (2012). https://doi.org/10.1167/iovs.11-8408, http:// www.ncbi.nlm.nih.gov/pubmed/22281822

2. Atluri, P., et al.: Cardiac retransplantation is an efficacious therapy for primary cardiac allograft failure. J. Cardiothorac. Surg. **3**(1), 26 (2008). https://doi.org/ 10.1186/1749-8090-3-26

3. Baillargeon, B., Rebelo, N., Fox, D.D., Taylor, R.L., Kuhl, E.: The Living Heart Project: a robust and integrative simulator for human heart function. Eur. J. Mech. A. Solids **48**, 38–47 (2014). https://doi.org/10.1016/j.euromechsol.2014.04. 001, http://www.ncbi.nlm.nih.gov/pubmed/25267880

4. Connelly, C.M., McLaughlin, R., Vogel, W., Apstein, C.: Reversible and irreversible elongation of ischemic, infarcted, and healed myocardium in response to increases in preload and afterload. Circulation **84**(1), 387–399 (1991)

5. Drach, A., Khalighi, A.H., Sacks, M.S.: A comprehensive pipeline for multi-resolution modeling of the mitral valve: validation, computational efficiency, and predictive capability. Int. J. Numer. Methods Biomed. Eng. **34**(2), e2921 (2018)

6. Enomoto, Y., et al.: Surgical treatment of ischemic mitral regurgitation might not influence ventricular remodeling. J. Thorac. Cardiovasc. Surg. **129**(3), 504–511 (2005)

7. Fan, R., Sacks, M.S.: Simulation of planar soft tissues using a structural consti-tutive model: finite element implementation and validation. J. Biomech. **47**(9), 2043–2054 (2014)

8. Gao, H., Feng, L., Qi, N., Berry, C., Griffith, B.E., Luo, X.: A coupled mitral valve-left ventricle model with fluid-structure interaction. Med. Eng. Phys. **47**, 128–136 (2017)

9. Gupta, K.B., Ratcliffe, M.B., Fallert, M.A., Edmunds Jr., L.H., Bogen, D.K.: Changes in passive mechanical stiffness of myocardial tissue with aneurysm for-mation. Circulation **89**(5), 2315–2326 (1994)

10. Hung, J., et al.: Mechanism of recurrent ischemic mitral regurgitation after annu-loplasty: continued lv remodeling as a moving target. Circulation 110(11_suppl_1), II-85 (2004)

11. Jackson, B.M., et al.: Extension of borderzone myocardium in postinfarction dilated cardiomyopathy. J. Am. College Cardiol. **40**(6), 1160–1167 (2002)

12. Khalighi, A.H., Rego, B.V., Drach, A., Gorman, R.C., Gorman, J.H., Sacks, M.S.: Development of a functionally equivalent model of the mitral valve chor-dae tendineae through topology optimization. Ann. Biomed. Eng. **47**(1), 60–74 (2019)

13. Klotz, S., et al.: Single-beat estimation of end-diastolic pressure-volume relation-ship: a novel method with potential for noninvasive application. Am. J. Phys. Heart Circulatory Phys. **291**(1), H403–H412 (2006)

14. Lee, C.H., Rabbah, J.P., Yoganathan, A.P., Gorman, R.C., Gorman, J.H., Sacks, M.S.: On the effects of leaflet microstructure and constitutive model on the closing behavior of the mitral valve. Biomech. Model. Mechanobiology **14**(6), 1281–1302 (2015)

15. Li, D.S., et al.: Insights into the passive mechanical behavior of left ventricular myocardium using a robust constitutive model based on full 3d kinematics. J. Mech. Behav. Biomed. Mater. **103**, 103508 (2020)

16. Lindsey, M.L., et al.: Guidelines for experimental models of myocardial ischemia and infarction. Am. J. Phys. Heart Circulatory Phys. **314**(4), H812–H838 (2018)
17. McGee Jr., E.C., et al.: Recurrent mitral regurgitation after annuloplasty for functional ischemic mitral regurgitation. J. Thorac. Cardiovasc. Surg. **128**(6), 916–924 (2004)
18. Rego, B.V., et al.: Remodeling of the mitral valve: an integrated approach for predicting long-term outcomes in disease and repair. Ph.D. thesis (2019)
19. Sato, S., Ashraf, M., Millard, R., Fujiwara, H., Schwartz, A.: Connective tissue changes in early ischemia of porcine myocardium: an ultrastructural study. J. Mol. Cell. Cardiol. **15**(4), 261–275 (1983)
20. Shimkunas, R., et al.: Left ventricular myocardial contractility is depressed in the borderzone after posterolateral myocardial infarction. Ann. Thorac. Surg. **95**(5), 1619–1625 (2013)
21. Soares, J.S., Li, D.S., Lai, E., Gorman III, J.H., Gorman, R.C., Sacks, M.S.: Modeling of myocardium compressibility and its impact in computational simulations of the healthy and infarcted heart. In: Pop, M., Wright, G.A. (eds.) FIMH 2017. LNCS, vol. 10263, pp. 493–501. Springer, Cham (2017). https://doi.org/10.1007/978-3-319-59448-4_47
22. Wenk, J.F., et al.: First finite element model of the left ventricle with mitral valve: insights into ischemic mitral regurgitation. Ann. Thorac. Surg. **89**(5), 1546–1553 (2010). https://doi.org/10.1016/j.athoracsur.2010.02.036, http://www.ncbi.nlm.nih.gov/pubmed/20417775
23. Wong, V.M., et al.: The effect of mitral annuloplasty shape in ischemic mitral regurgitation: a finite element simulation. Ann. Thorac. Surg. **93**(3), 776–782 (2012)
24. Zhang, W., Ayoub, S., Liao, J., Sacks, M.S.: A meso-scale layer-specific structural constitutive model of the mitral heart valve leaflets. Acta Biomater. **32**, 238–255 (2016)

Cardiac Modeling for Multisystem Inflammatory Syndrome in Children (MIS-C, PIMS-TS)

Rebecca Waugh[1,2], Mohamed Abdelghafar Hussein[1,3], Jamie Weller[1], Kavita Sharma[1], Gerald Greil[1], Jeffrey Kahn[4], Tarique Hussain[1]🆔, and Radomír Chabiniok[1(✉)]🆔

[1] Division of Pediatric Cardiology, Department of Pediatrics, UT Southwestern Medical Center, Dallas, TX, USA
radomir.chabiniok@utsouthwestern.edu
[2] Natural Sciences and Mathematics, UT Dallas, Richardson, TX, USA
[3] Kafrelsheikh University, Kafr El-Sheikh, Egypt
[4] Division of Pediatric Infectious Disease, Department of Pediatrics, UT Southwestern Medical Center, Dallas, TX, USA

Abstract. Cardiovascular data of 8 patients with Multisystem Inflammatory Syndrome in Children (MIS-C), caused by an aberrant reaction of immune system to the SARS-CoV-2 coronarovirus, were retrospectively analyzed by using patient-specific biomechanical modeling. The first goal was to increase the understanding of the pathophysiology of the cardiovascular involvement in MIS-C, during the inpatient stay at Intensive Care Unit and after discharge from the hospital. Secondly, hypothetical action of various types of pharmacological therapy was tested in silico using the created patient-specific models, aiming to contribute into the optimal pharmacological management during the acute stage of MIS-C.

Keywords: Multisystem Inflammatory Syndrome in Children (MIS-C, PIMS-TS) · Heart failure · Circulatory shock · Cardiac modeling · Intensive Care Unit (ICU) monitoring · Echocardiography · Cardiovascular magnetic resonance imaging (CMR)

1 Introduction

Multisystem Inflammatory Syndrome in Children (MIS-C), also known as Pediatric Inflammatory Multisystem Syndrome-temporally associated with severe acute respiratory syndrome coronarovirus 2 SARS-CoV-2 (PIMS-TS), is a rare but serious condition appearing in children typically of the age 4–18 years, which was first described in April 2020 [17,23,25]. It is different to the world-spread coronavirus disease 2019 (Covid-19). The symptoms in Covid-19 are directly caused by SARS-Cov-2, they appear 2–14 days after the exposure to the coronavirus, the severity increases with the age of patient, while children have typically a mild (or even asymptomatic) course. MIS-C, on the contrary, appears

© Springer Nature Switzerland AG 2021
D. B. Ennis et al. (Eds.): FIMH 2021, LNCS 12738, pp. 435–446, 2021.
https://doi.org/10.1007/978-3-030-78710-3_42

in children 2–4 weeks after the exposure to the coronavirus and is caused by a delayed aberrant reaction of the immune system. As reported in the US, UK and Europe [4,8,27] (see also systematic review [2]), it affects various organs out of which cardiovascular system is the most common (80% of MIS-C patients). It often leads to a severe circulatory shock and heart failure (HF). Around 80% of all MIS-C patients require Intensive Care Unit (ICU) hospitalization for 5–7 days. The treatment is based on ceasing the unusual immune reaction typically by intravenous immunoglobulins (IVIG) or glucocorticoids. Once the immune reaction is under control, the heart function ameliorates.

By then, MIS-C is rapidly progresssive and life-threatening [9]. Therefore it is of utmost interest to diagnose MIS-C and start the IVIG therapy as soon as possible (every hour counts). Secondly, an optimal handling of the circulatory shock and heart failure at ICU would minimize the long-term consequences on the heart. Finally, after discharging from the hospital, the question about the actual long-term cardiovascular consequences remains and regular follow up exams are therefore indicated. The community of Functional Imaging and Modeling of Heart (FIMH) has the potential to contribute in all three stages – diagnosis, acute stage at ICU and long-term monitoring of cardiac consequences:

1. Currently, there exists no specific test for MIS-C and doctors of various disciplines are needed to support the diagnosis. A smart way of assembling pieces of information from a given patient (see Table 1), e.g. by using some artificial intelligence methods [19], could contribute to a faster diagnosis of MIS-C.
2. Monitoring of cardiovascular system at ICU could be augmented by some methods developed within the FIMH community – typically in cardiovascular modeling [6,11,12,15,24,26] or machine learning (see e.g. recent review [20] and references therein). This could contribute to a more targeted therapeutic management of HF and circulatory shock.
3. While echocardiography (ECHO) is used during the hospitalization, the regular follow up exams may combine ECHO and cardiovascular magnetic resonance (CMR). An objective assessment of heart in comparison to the previous exams (function and morphology, the latter typically represented by the size of each heart chamber and also e.g. size of coronary aneurysms, which develop in some MIS-C patients [14]) could benefit from, for example, robust image registration techniques [13,21,22,28] or multi-modality image analysis [16].

The aim of this study was to retrospectively analyze data from MIS-C patients to increase the insight into the pathophysiological basis of MIS-C in order to direct the treatment strategies. This paper shows our preliminary results using data during the ICU (i.e. Step 2. in the above list), while putting those into the context also with the follow up investigations 4–8 weeks (and later) after the discharge from the hospital.

Table 1. Example symptoms within Multisystem Inflammatory Syndrome in Children (MIS-C) according to the World Heatlh Organization (WHO, [1]).

System	Symptom
Global	Fever > 100.4 °F (38 °C) lasting > 3 d; age 0–19 years
Cardiac ECHO findings	Myocardial dysfunction; Pericarditis; Valvulitis; Coronary abnormalities
Cardiac lab findings	Elevated troponin; Elevated NT-proBNP
Circulation	Hypotension; Shock
Gastrointestinal	Diarrhea; Vomiting; Abdominal pain up to acute abdomen
Mucocutaneus	Rash; Nonpurulent conjcutivitis; Mucocutaneous inflammation
Evidence of coagulopathy	Elevated D-dimers; PT, APTT times;
Markers of inflammation	Elevated CRP; Procalcitonin; Erythrocyte sedimentation rate
Evidence of Covid-19	RT-PCR; Antigen test; Contact with patients with Covid-19

Legend: APTT, activated partial thromboplastin time; CRP, C-reactive protein; ECHO, echocardiography; NT-proBNP, N-terminal pro-B-type natriuretic peptide; PT, prothrombin time; RT-PCR, reverse transcriptase-polymerase chain reaction.

1.1 Hypothesis

The patients with MIS-C may develop HF and circulatory shock (low blood pressure, limited perfusion of organs). We hypothesize that synthesizing available measurements of cardiovascular system during monitoring at ICU (ECHO exam, blood pressure, heart rate) can contribute into distinguishing the cardiogenic component of the shock and the component given by the level of periphery vascular resistance. Secondly, the patient-specific models can behave pro-actively to contribute in optimizing the management of patients at ICU by predicting the effect of a given drug and its dose.

2 Methods

2.1 Data

Datasets of 8 patients with MIS-C were included in this retrospective study. The data collections were performed under the ethical approvals of Institutional Review Board of UT Southwestern Medical Center Dallas (STU-2020-0626). The IRB waived the need for a consent to use the anonymized retrospective data. Each patient underwent 1–4 ECHO exams during their inpatient stay (acute MIS-C) and a cardiovascular magnetic resonance scan (CMR) 1–2 months after discharging from the hospital. Additionally, follow up ECHO exams were performed in most patients. Periphery blood pressure (cuff measurement) was recorded at each exam. All image data were processed by a single observer to reduce the inter-study observer variance. The LV volume was assessed at the end-diastole (EDV) and end-systole (ESV) to compute stroke volume $SV = EDV - ESV$ and ejection fraction $EF = SV/EDV$. Flow through aortic valve was measured by Doppler (velocity time interval scaled by SV). LV myocardial mass was measured only in the CMR scan and assumed to be constant throughout all exams of a given patient.

2.2 Biomechanical Model

The synthesis of the patient's data was performed by patient-specific cardiac modeling which allows to access peripheral vascular resistance and myocardial contractility, while using patient's data acquired at each exam. A reduced-order biomechanical model of left ventricle (LV) and systemic circulation was employed [5]. In this model the passive myocardium is considered as transverse isotropic (with respect to the myocardial fiber direction). The fibers are isotropically distributed in the orthoradial plane through the wall thickness. The overall behavior then varies in the radial vs. orthoradial directions. While the geometry and kinematics were reduced to a sphere, all constitutive relations were preserved as in the full 3D model [7].

First, the model based on CMR data was built. The sequential calibration procedure described in [12] was used. The 2-stage Windkessel model was adjusted by imposing the aortic flow obtained from CMR and aiming to the simulated aortic pressure to match the measured systolic and diastolic pressure. Then, the LV wall thickness and the ventricular volume was prescribed according to the cine MRI. The model configuration at hypothetical zero filling pressure (reference volume V_{ref}) was assumed as in experimental data of Klotz et al. [10] $V_{\text{ref}} = EDV \cdot (0.6 - 0.006 \cdot EDP)$, EDV and EDV being end-diastolic volume (in ml) and pressure (in mmHg). As there was no sign of increased left atrial pressure in the final CMR exams, we assumed the filling (end-diastolic) pressure being 7 mmHg. Passive myocardial properties were calibrated in line with the measured end-diastolic volume. Myocardial contractility was adjusted to match the measured stroke volume.

The model specific to patient's CMR exam was then adjusted to each ECHO exam; while the geometrical properties (LV reference volume and myocardial mass) and passive tissue stiffness were kept as in CMR, all other properties were readjusted according to data of end-diastolic volume (adjustment of EDP), aortic flow (adjustment of the resistance and capacitance of the circulation), end-systolic volume (contractility adjusted according to stroke volume), and basic indices obtained from ECG (heart rate, PQ, QRS and ST interval durations (timing of electrical activation, in particular the action potential duration). We refer to [18] for details.

The patient-specific models provided the values of the proximal and distal resistances of the systemic circulation (R_p and R_d); and of myocardial contractility at each state. Furthermore, the models specific to the patients entering ICU were used to drive a hypothetical study of assessing the effect of a vasodilator (reducing periphery vascular resistance by 50 or 25%); increased ventricular

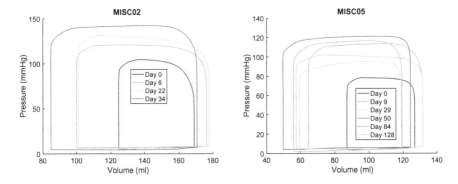

Fig. 1. Example of ventricular Pressure-Volume loops obtained by computational modeling in selected patients showing a picture of cardiovascular shock state at Day 0, and a progressive improvement and stabilization at after 3–4 weeks.

filling; a vasodilator with a mildly inotropic effect (in addition, increasing the contractility by 20%); vasoactive drug with a mild inotropic effect (increase of periphery vascular resistance by 60% and myocardial contractility by 14%, motivated by the effect of norepinephrine bolus studied in [12]); and finally a drug solely with an inotropic effect (increase of contractility by 30%).

3   Results

Table 2 summarizes the clinical measurements for each patient and each exam as well as the quantities derived by coupling the clinical data with the biomechanical model (contractility and periphery vascular resistance). Fig. 1 demonstrates LV pressure-volume (P-V) loops in selected patients during each exam. As can be seen, all patients have EF reduced below the normal value of 55% at the time of entering ICU. Figure 2 demonstrates the evolution of contractility values (normalized to the maximum value in a given patient) in each patient throughout the study.

Table 3 shows a hypothetical therapeutical effect of a vasodilator (possibly with a mild inotropic effect) applied on selected patients with the initially increased distal vascular resistance. Finally, in patient MISC05 (with EF exam 30% and blood pressure 78/49 mmHg during the initial ECHO1 exam), we demonstrated a hypothetical effect of the vasoactive drug norepinephrine and the effect of a drug with predominantly inotropic effect.

4   Discussion

In this paper we presented our preliminary results when applying cardiac modeling on patients with MIS-C. All patients had at acute stage increased cardiac enzymes (troponin, NT-proBNP), signifying the component of some myocardial

Table 2. Measurements and model-obtained physiological characteristics for each patient and clinical exam.

Patient	Exam	Day	HR (bpm)	BP (mmHg)	EF (%)	EDVi (ml/m^2)	R_p (MPa · s/m^3)	R_d (MPa · s/m^3)	Contractility (kPa)
MISC01	ECHO1	0	96	103/55	31	62	28	150	80
	ECHO2	7	92	123/68	60	75	12	90	123
	ECHO3	40	72	117/68	54	75	15	100	105
	CMR	152	64	91/61	56	78	10	110	77
MISC02	ECHO1	0	140	97/46	26	88	21	80	75
	ECHO3	6	110	115/73	33	91	15	80	103
	ECHO4	22	87	129/74	37	91	13	110	115
	CMR	34	76	142/86	51	86	16	100	130
MISC05	ECHO1	0	137	79/49	32	108	12.5	75	77
	ECHO3	9	61	115/59	51	115	17	140	125
	ECHO4	29	82	97/64	50	101	10	100	110
	ECHO5	50	105	123/88	59	111	10	95	142
	ECHO6	84	104	118/72	53	101	14	95	135
	CMR	128	80	102/61	57	116	8	95	127
MISC06	ECHO 1	0	118	106/66	43	64	28.5	180	84
	ECHO2	22	105	112/67	53	61	25	185	103
	ECHO3	44	84	133/61	60	62	27	190	104
	CMR	92	102	75/29	60	65	22	78	71
MISC12	ECHO1	0	136	101/49	44	52	24	115	87
	ECHO3	3	113	86/53	50	54	16	112	78
	ECHO5	34	121	118/67	56	55	22	119	115
	ECHO6	62	123	115/64	56	56	18	116	117
	CMR	117	100	117/66	63	56	26	200	110
MISC15	ECHO1	0	124	73/59	40	70	10	160	56
	ECHO2	3	85	115/76	53	83	23	212	99
	ECHO3	17	114	107/64	56	72	20.5	135	97
	ECHO4	49	104	107/71	55	85	16	141	107
	CMR	49	101	104/69	58	86	16.5	126	102
MISC19	ECHO1	0	83	100/66	46	74	27	220	91
	ECHO2	11	94	91/54	50	78	18.5	140	99
	ECHO3	46	93	109/57	53	79	24	146	118
	CMR	116	88	119/58	56	79	29	148	132
MISC21	ECHO1	0	151	85/53	38	58	28	179	64
	ECHO2	2	149	85/35	39	64	34	118	70
	ECHO3	16	96	110/77	43	69	30	290	92
	ECHO4	22	120	113/93	49	69	14.5	240	95
	ECHO5	45	108	107/72	43	69	24.5	198	99
	ECHO6	60	100	95/73	55	69	13	203	93
	CMR	60	110	110/82	53	71	14	210	108

Legend: BP, blood pressure; CMR, cardiovascular magnetic resonance imaging exam; Day, number of days after admission to Intensive Care Unit; ECHO, echocardiography exam; EDVi, LV end-diastolic volume indexed to body surface area; EF, ejection fraction of left ventricle (LV); HR, heart rate; R_p, R_d, proximal (distal) vascular resistance.

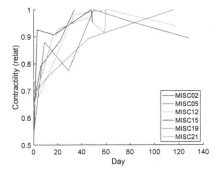 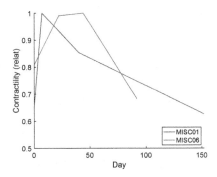

Fig. 2. Evolution of contractility (normalized in each patient) showing stabilization after ~30 d in majority (left), while a decrease of contractility in patients MISC01 and MISC06 (right).

Table 3. In silico trials (in italics) using models created for patients at the time of first ECHO at ICU.

Patient	Patient specific model/*In silico trial*	R_d *(Change)*	Contractility *(Change)*	EDP (mmHg)	BP (mmHg)	EF (%)
MISC01	ECHO1	150 MPa·s/m³	80 kPa	3	103/55	31
	Trial vasodil	× *0.50*	× *1.00*	*3*	*94/39*	*44*
	Trial vasodil	× *0.75*	× *1.00*	*3*	*99/47*	*35*
	Trial vasodil & Inotrope	× *0.75*	× *1.20*	*3*	*111/52*	*42*
MISC05	ECHO1	70 MPa·s/m³	75 kPa	7	78/49	35
	Trial norepinephrine	× *1.60*	× *1.14*	*7*	*94/62*	*26*
	Trial inotrope	× *1.00*	× *1.30*	*7*	*95/58*	*39*
MISC15	ECHO1	160 MPa·s/m³	56 kPa	7	73/59	40
	Trial vasodil	× *0.50*	× *1.00*	*7*	*64/45*	*57*
	Trial vasodil	× *0.50*	× *1.00*	*10*	*66/47*	*58*
	Trial vasodil	× *0.75*	× *1.00*	*7*	*71/54*	*47*
	Trial vasodil	× *0.75*	× *1.00*	*10*	*69/53*	*47*
	Trial vasodil & Inotrope	× *0.75*	× *1.20*	*7*	*77/58*	*55*
MISC19	ECHO1	220 MPa·s/m³	91 kPa	7	100/66	46
	Trial vasodil	× *0.50*	× *1.00*	*7*	*90/47*	*62*
	Trial vasodil	× *0.50*	× *1.00*	*10*	*93/49*	*62*
	Trial vasodil	× *0.75*	× *1.00*	*7*	*96/58*	*52*
	Trial vasodil	× *0.75*	× *1.00*	*10*	*99/60*	*52*
	Trial vasodil & Inotrope	× *0.75*	× *1.20*	*7*	*108/66*	*59*

Legend: BP, blood pressure; ECHO1, patient specific model based on the first echocardiography exam; EDP, end-diastolic pressure; EF, ejection fraction; inotrope, inotropic drug; R_d, distal vascular resistance; vasodil, vasodilation drug.

damage (such as due to myocarditis triggered by the immune reaction of MIS-C). Patient-specific modeling gave an insight into the level of periphery vascular resistance and heart contractility (i.e. the vascular vs. cardiac component of the shock state).

Five patients (MISC01, MISC06, MISC15, MISC19 and MISC21) had distal vascular resistance (R_d) initially increased above 150 MPa $\cdot$ s/m^3, while these patients have moderately or severely reduced EF. This might be caused by their heart not being able to cope with the increased vascular resistance (limitation in the myocardial contractility, causing a cardiogenic component of the shock). In Table 2, no obvious cut off value for the reduced contractility is seen. While the values of contractility represent the active stress developed by a sarcomere unit, the overall effect on the whole organ will also depend on the geometrical properties (volume and wall thickness of the LV).

Figure 2 plotted the evolution of contractility values in each patient throughout the study, while normalizing them to the maximum value in given patient. The left panel demonstrates that in majority of our patients the contractility increases by ~30–40% 2–4 weeks after the acute stage. This supports the hypothesis of a pathological decrease of heart contractility at the acute stage (in line with the increased cardiac enzymes). The values gradually improve, which supports the optimistic scenario that the heart is fully recovering within a month. The exceptions are shown in the right panel of Fig. 2 – patients MISC01 and MISC05. In both patients this was caused by a lower pressure in their final CMR exam. However, the stroke volume did not vary from previous time point, as can be appreciated in Fig. 3. A significantly decreased periphery vascular resistance during CMR of patient MISC05 can be explained by the CMR under general anesthesia and patient MISC01 will be followed further clinically.

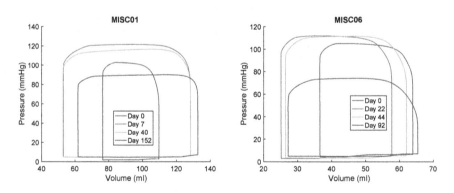

Fig. 3. Pressure-volume loops in patients MISC01 and MISC05: a low blood pressure and normal stroke volume explains the decreased contractility at this exam.

Our pilot study augments the physiological picture of the patients at ICU and their cardiovascular recovery after their discharge. It has a number of limitations, however. The model was used to monitor the patients only at the time of ECHO exams (which was every second day, at best). Implementing a true monitoring at bedside could increase our insight into the physiology of each patient. While the proposed modeling pipeline is simple and can be performed in clinical environment, the model simplifications brings a number of limitations.

For instance, the assumption of a non-changing passive material properties during the 2 months is not in line with the half-time of collagen fibers in myocardial tissue. However, the reduced information obtained from ECHO cannot be utilized to reliably modify the tissue passive properties, with respect to those adjusted according to the detailed CMR acquisition. The small sample of patients in this pilot study and using only retrospective data limits us from deriving definite conclusions, particularly in our in silico therapeutic management.

Even though some methods of artificial intelligence and machine learning could be used to enable a prompt diagnosis or predict long-term outcomes, in this work we focused on employing biophysical modeling. The advantage of modeling is that it has the potential to provide information about the reactivity of vascular resistance and level of myocardial contractility e.g. when using various heart failure drugs to support the cardiovascular conditions. This effort aims to provide advice for example about who needs inotropic support (e.g. epinephrine) and who would rather benefit from a vasoactive support (e.g. norepinephrine), or who is likely to be resilient to fluid resuscitation [17]. To address the latter, we would need to extend the model to a closed-loop circulation [3, 18].

By using the models specific to the patients at their entering ICU, we carried out a hypothetical study of using a vasodilation drug (for those with high periphery vascular resistance). As expected, we observed an increase of EF and a decrease of blood pressure. The latter could be counterproductive e.g. in perfusion of the myocardium or brain. A combined vasodilation & inotropic effect (such as e.g. for the phosphodiesterase 3 inhibitor milrirone) could be of a desired effect. Similarly, the effect of a preferential vasoactive drug (e.g. norepinephrine) in patients with a decreased periphery vascular resistance may lead to a decreased cardiac output (such as in silico presented in patient MISC05). A drug with mostly inotropic effect could be more efficient, in this case.

The action of every drug depends on a given patient. As an illustration, to predict the effect of norepinephrine we applied the average effect of norepinephrine as observed in a group of patients in the clinical modeling study by Le Gall et al. [12]. In this study, the periphery vascular resistance increased in average by 60% and myocardial contractility by 14% after administering a norepinephrine bolus. We are aware that this could only serve as an initial guess for a real-time ICU monitoring framework. By monitoring the patient and gathering data about the reactivity of patient's cardiovascular system during the first hours at ICU, such a monitoring system could train itself to perform predictions for the given patient in the following hours.

5   Conclusion

MIS-C is a rare, however, severe disease. The number of cases are proportional to the incidence of the Covid-19 disease in each wave of pandemic, while the peak of MIS-C cases is lagged by 2–4 week. With Covid-19 being still a severe problem worldwide, it is expected that we will be still encountering MIS-C cases at pediatric ICUs. The development of fast and effective diagnostic tools, a support at ICUs and efficient post-MIS-C follow up assessment could be very helpful. A number of already existing tools, of which some have been developed within the FIMH community, may be of direct benefit. Likewise, research focused on MIS-C may facilitate development of new tools applicable for other diseases. Some examples are ICU monitoring in complex cases (such as during cardiogenic shock of any origin) or an objective assessment of long-term disease progress during follow up exams – bread and butter in cardiology clinic.

Acknowledgment. The authors acknowledge the support of the Associated Team TOFMOD, created between Inria France and UT Southwestern Medical Center Dallas. The project was also funded in part from by the W.B. & Ellen Gordon Stuart Trust, The Communities Foundation of Texas and by the Pogue Family Distinguished Chair (award to Dr. F. Gerald Greil in February, 2015). In addition, we would like to acknowledge Dr. Philippe Moireau and Dr. Dominique Chapelle, Inria research team MΞDISIM, for the development of the cardiac simulation software CardiacLab used in this work. Research reported in this publication was supported by Children's Health[SM] but the content is solely the responsibility of the authors and does not necessarily represent the official views of Children's Health[SM].

References

1. Multisystem inflammatory syndrome in children and adolescents with COVID-19: scientific brief, 15 May 2020. Technical Report, World Health Organization (2020)
2. Ahmed, M., et al.: Multisystem inflammatory syndrome in children: a systematic review. EClinicalMedicine **26**, 100527 (2020)
3. Arts, T., Delhaas, T., Bovendeerd, P., Verbeek, X., Prinzen, F.W.: Adaptation to mechanical load determines shape and properties of heart and circulation: the CircAdapt model. Am. J. Phys. Heart Circulatory Phys. **288**(4), H1943–H1954 (2005)
4. Belhadjer, Z., et al.: Acute heart failure in multisystem inflammatory syndrome in children in the context of global SARS-CoV-2 pandemic. Circulation **142**(5), 429–436 (2020)
5. Caruel, M., Chabiniok, R., Moireau, P., Lecarpentier, Y., Chapelle, D.: Dimensional reductions of a cardiac model for effective validation and calibration. Biomech. Model. Mechanobiology **13**(4), 897–914 (2014)
6. Chabiniok, et al.: Multiphysics and multiscale modelling, data-model fusion and integration of organ physiology in the clinic: ventricular cardiac mechanics. Interface Focus **6**(2), 20150083 (2016)
7. Chapelle, D., Le. Tallec, P., Moireau, P., Sorine, M.: An energy-preserving muscle tissue model: formulation and compatible discretizations. Int. J. Multiscale Comput. Eng. **10**(2), 189–211 (2012)
8. Feldstein, L.R., et al.: Multisystem inflammatory syndrome in US children and adolescents. N. Engl. J. Med. **383**(4), 334–346 (2020)
9. Godfred-Cato, S., et al.: Covid-19-associated multisystem inflammatory syndrome in children-United States, march-july 2020. Morb. Mortal. Weekly Rep. **69**(32), 1074 (2020)

10. Klotz, S., et al.: Single-beat estimation of end-diastolic pressure-volume relationship: a novel method with potential for noninvasive application. Am. J. Phys. Heart Circulatory Phys. **291**(1), H403–H412 (2006)
11. Le Gall, A., Vallée, F., Chapelle, D., Chabiniok, R.: Minimally-invasive estimation of patient-specific end-systolic elastance using a biomechanical heart model. In: Coudière, Y., Ozenne, V., Vigmond, E., Zemzemi, N. (eds.) FIMH 2019. LNCS, vol. 11504, pp. 266–275. Springer, Cham (2019). https://doi.org/10.1007/978-3-030-21949-9_29
12. Le Gall, et al.: Monitoring of cardiovascular physiology augmented by a patient-specific biomechanical model during general anesthesia. a proof of concept study. PLoS ONE **15**(5), e0232830 (2020)
13. Lee, L.C., Genet, M.: Validation of equilibrated warping—image registration with mechanical regularization—on 3D ultrasound images. In: Coudière, Y., Ozenne, V., Vigmond, E., Zemzemi, N. (eds.) FIMH 2019. LNCS, vol. 11504, pp. 334–341. Springer, Cham (2019). https://doi.org/10.1007/978-3-030-21949-9_36
14. Loke, Y.H., Berul, C.I., Harahsheh, A.S.: Multisystem inflammatory syndrome in children: is there a linkage to Kawasaki disease? Trends in cardiovascular medicine (2020)
15. Niederer, S.A., Lumens, J., Trayanova, N.A.: Computational models in cardiology. Nat. Rev. Cardiology **16**(2), 100–111 (2019)
16. Pop, M., Sermesant, M., Camara, O., Zhuang, X., Li, S., Young, A., Mansi, T., Suinesiaputra, A. (eds.): STACOM 2019. LNCS, vol. 12009. Springer, Cham (2020). https://doi.org/10.1007/978-3-030-39074-7
17. Riphagen, S., Gomez, X., Gonzalez-Martinez, C., Wilkinson, N., Theocharis, P.: Hyperinflammatory shock in children during COVID-19 pandemic. Lancet **395**(10237), 1607–1608 (2020)
18. Ruijsink, et al.: Dobutamine stress testing in patients with Fontan circulation augmented by biomechanical modeling. PLoS ONE **15**(2), e0229015 (2020)
19. Seetharam, K., Shrestha, S., Sengupta, P.P.: Artificial intelligence in cardiovascular medicine. Curr. Treat. Options Cardiovasc. Med. **21**(5), 1–14 (2019)
20. Sermesant, M., Delingette, H., Cochet, H., Jaïs, P., Ayache, N.: Applications of artificial intelligence in cardiovascular imaging. Nat. Rev. Cardiol. 1–10 (2021, ahead of print). https://doi.org/10.1038/s41569-021-00527-2
21. Sotiras, A., Davatzikos, C., Paragios, N.: Deformable medical image registration: a survey. IEEE Trans. Med. Imaging **32**(7), 1153–1190 (2013)
22. Tobon-Gomez, C., et al.: Benchmarking framework for myocardial tracking and deformation algorithms: an open access database. Med. Image Anal. **17**(6), 632–648 (2013)
23. Verdoni, L., et al.: An outbreak of severe Kawasaki-like disease at the Italian epicentre of the SARS-CoV-2 epidemic: an observational cohort study. Lancet **395**(10239), 1771–1778 (2020)
24. Walmsley, J., van Everdingen, W., Cramer, M.J., Prinzen, F.W., Delhaas, T., Lumens, J.: Combining computer modelling and cardiac imaging to understand right ventricular pump function. Cardiovas. Res. **113**(12), 1486–1498 (2017)
25. Waltuch, T., et al..: Features of COVID-19 post-infectious cytokine release syndrome in children presenting to the emergency department. Am. J. Emerg. Med. **38**(10), 2246–e3 (2020)
26. Wang, V., Nielsen, P., Nash, M.: Image-based predictive modeling of heart mechanics. Ann. Rev. of Biomed. Eng. **17**(1), 351–383 (2015). https://doi.org/10.1146/annurev-bioeng-071114-040609

27. Whittaker, E., et al.: Clinical characteristics of 58 children with a pediatric inflammatory multisystem syndrome temporally associated with SARS-CoV-2. JAMA **324**(3), 259–269 (2020)
28. Xiong, Z., Zhang, Y.: A critical review of image registration methods. Int. J. Image Data Fusion **1**(2), 137–158 (2010). https://doi.org/10.1080/19479831003802790

Personal-by-Design: A 3D Electromechanical Model of the Heart Tailored for Personalisation

Gaëtan Desrues[1,2], Delphine Feuerstein[2], Thierry Legay[2], Serge Cazeau[2,3], and Maxime Sermesant[1(✉)]

[1] Inria, Universit Côte D'Azur and 3IA Côte D'Azur, Sophia Antipolis, France
maxime.sermesant@inria.fr
[2] Microport CRM, Clamart, France
[3] Hopital Saint-Joseph, Paris, France

Abstract. In this work we present a coupled electromechanical model of the heart for patient-specific simulations, and in particular cardiac resynchronisation therapy. To this end, we propose a fast fully autonomous and flexible pipeline to generate and optimise the data required to run the mechanical simulation. After the meshing of the biventricular segmentation image and the construction of the associated fibres arrangement, we compute the electrical potential propagation in the myocardial tissue from selected onset points on the endocardium. We generate a 12-lead electrocardiogram corresponding to the latter activation map by extrapolating the electrical potential on a virtual torso. This electrical activation is coupled to a mechanical model, featuring a small set of interpretable parameters. We also propose an efficient algorithm to optimise the model parameters, based on patient data. The whole pipeline including a cardiac cycle is computed in 30 min, enabling to use this digital twin for diagnosis and therapy planning.

Keywords: Personalisation · Digital twin · Cardiac electromechanical model · Electrophysiology · Electrocardiogram

1 Introduction

Building patient-specific 3D models of the heart can help improving diagnosis and therapy selection for various cardiac diseases. They can for instance help interventional cardiologists in choosing the best pacing method and anticipate the patient response based on the chosen pacing sites [12]. However, personalisation, i.e. adapting a generic cardiac model to a specific patient, is still challenging due to theoretical (identifiability) and practical (computation time) issues.

The model presented here combines different modelling approaches and tools that are well suited for personalisation, and results in a computational time compatible with clinical constraints. In this manuscript, we detail the different elements, how we can adjust their parameters, and present simulation results to evaluate their feasibility. The computational cost of each step is also detailed.

© Springer Nature Switzerland AG 2021
D. B. Ennis et al. (Eds.): FIMH 2021, LNCS 12738, pp. 447–457, 2021.
https://doi.org/10.1007/978-3-030-78710-3_43

This pipeline is based on various previous studies that have been combined to propose for the first time in the team a complete and fast method for running a personalised patient-specific simulation of a beating heart, including:

- a local anatomical correction tool,
- a realistic approach to simulate the Purkinje activation and electrical propagation,
- algorithms to quickly generate body surface electrocardiograms,
- a ready-to-use mechanical model,
- a sensitivity analysis to assess the model's observability and
- algorithms to personalise the models' parameters based on patient data.

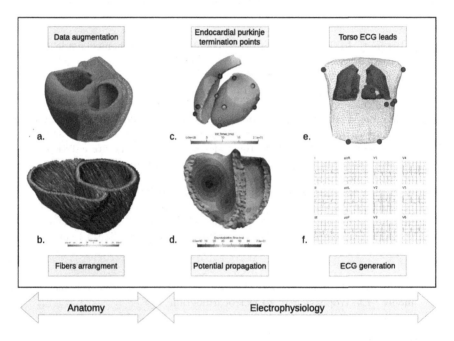

Fig. 1. Pipeline for patient-specific simulation. a. Initial and modified geometries. b. Fibre directions. c. Activation points and activation times on the endocardial surfaces. d. Activation map. e. Leads location on a virtual torso. f. 12-lead ECG. This pipeline generates all the data required to run the mechanical simulation

2   Anatomical Model

We suppose here that the patient biventricular myocardium was already segmented from 3D images. This is more and more available thanks to deep learning approaches. Note that the atria are not included in our 3D model, but their interactions with the ventricles are modelled in different ways: a delay is introduced in the electrical model to simulate the propagation of the electrical signal from the sino-atrial to the auriculo-ventricular node; the atrial pressure is modelled as a sigmoid curve; and the atrial contraction is taken into account in the boundary conditions, as a spring force on the basal tetrahedra.

2.1   Meshing and Labelling

The binary mask from the segmentation (Fig. 2a) is resampled into $1 \times 1 \times 1mm^3$ voxels and contains the labels for left and right ventricles. The labels for endocardium, myocardium and epicardium are built with a ray-tracing method, emerging from each ventricle barycentre. A tetrahedral mesh is built using the remeshing software MMG [6] (Fig. 2b), resulting in a mesh of approximately 90k tetrahedra and 20k vertices.

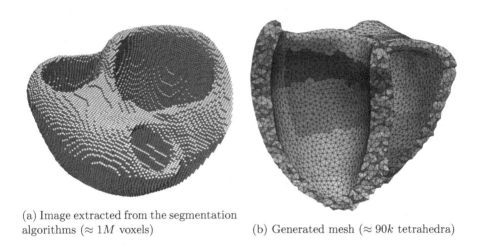

(a) Image extracted from the segmentation algorithms ($\approx 1M$ voxels)

(b) Generated mesh ($\approx 90k$ tetrahedra)

Fig. 2. Generated topologies

In addition to the generated mesh, tools were developed to virtually modify the anatomy, in order to increase the available number of healthy and pathological cases, and to correct any inaccuracies from the segmentation method. For instance, on Fig. 1a, the right ventricle volume has been reduced by 20%. Several pathological cases could be generated similarly from a healthy heart geometry, for example dilated or hypertrophic cardiomyopathies.

2.2   Cardiac Fibres

The spatial orientation of the muscle fibres plays a major role in the excitation and contraction of the heart [13]. We suppose that the fibre helix angle varies from $-\alpha$ to $+\alpha$ across the myocardial wall and remains constant throughout the cardiac cycle. We neglect the angle variations in any other direction.

To compute the fibres direction, we need to build the local coordinate system in each voxel. To this end, we first compute the tissue thickness using a simple diffusion equation on the endocardium and epicardium as in [14]. The resultant vector field gives the radial direction. The other directions are then easy to obtain. The fibre directions are given by the circumferential basis vector, rotated by an angle $\alpha_f \in [-\alpha, +\alpha]$ along the radial direction (Fig. 1b). In this framework, the angle variation is chosen to linearly increase from the endocardium ($\alpha_f = -80^o$) to the epicardium ($\alpha_f = 80^o$). Note that the septum is considered as left ventricle for the fibres generation.

2.3   AHA Regions

The 17 American Heart Association (AHA) segments are defined as a standard-ised segmentation of the left ventricle. We can extend this to 29 regions (Fig. 2b) for the whole myocardial tissue. These segments are mostly used in the mechan-ical simulation. They can also give a good insight on regional properties such as stress and strain, thus help in assessing intraventricular dyssynchrony.

To build these regions for each ventricle, we let an ellipsoid fit the ventricle mesh vertices and use the spherical coordinates to generate the regions on the ellipsoid. Finally, the segments are projected onto the tetrahedral mesh [3]. At the base of the mesh, a stiffer and less conductive region is introduced to model the valves fibrous tissue (beige region at the top of the mesh in Fig. 2b).

3   Electrophysiology Simulation

3.1   Electrical Activation

At each cardiac cycle, the mechanical contraction of the heart is driven by the electrical activation. The electrical signal emerging from the auriculo-ventricular node is conducted along the bundle of His and the Purkinje network to the endo-cardium (Fig. 1c). The numerous termination points of the Purkinje fibres on the endocardium are reduced to a dozen of points (Fig. 1d) and selected matching a real endocardial map (Fig. 3). To account for the fast potential propagation due to the Purkinje fibres, a thin endocardial layer is assigned a higher isotropic con-ductivity. These onset points are activated between 0 and $21ms$, $t = 0$ marking the beginning of the QRS complex. In the remaining tissue, the potential prop-agation is computed using a fast marching method and solving the anisotropic Eikonal equation [10] for each mesh vertex:

$$\sigma\sqrt{\nabla T^T D \nabla T} = 1$$

After a short period of time (duration of the QRS), all the myocardial cells are activated and contract until the repolarisation wave arrives. The action potential duration (APD) is linearly interpolated from the endocardium to the epicardium (shorter on the epicardium), using a wall depth map.

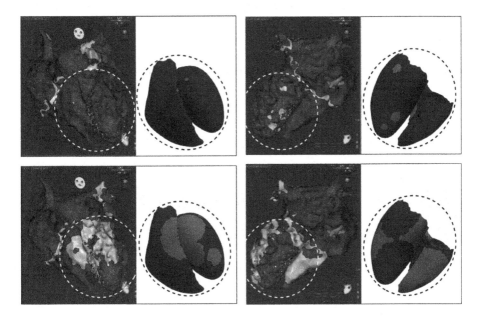

Fig. 3. Real endocardial activation map (courtesy of Dr. Mouhoub) and their computed homologous. The computed map (white background) shows both ventricles endocardium (no atria) and the fast isotropic propagation wave.

3.2   ECG Generation

The ECG records the bio-electric activity of the cardiac cells by measuring the electrical potential at the 6 precordial leads placed on the patient's torso and 3 (or 4) limb leads (Fig. 1e). In order to simulate this electrical activity, we consider each voxel of the image as a dipole of current density $\mathbf{j}_{eq} = -\sigma \nabla v$ where ∇v is the spacial gradient of the potential v [4]. By using the chain rule, we obtain:

$$\mathbf{j}_{eq} = -\sigma \frac{\partial v}{\partial T} \nabla T$$

where σ is the local conductivity, ∇T the gradient of the activation map and $\frac{\partial v}{\partial T}$ is given by solving the Mitchell-Schaeffer model [11] using a forward Euler scheme. We suppose here that the body is a homogeneous material with constant conductivity σ_T. The electrical potential at a distance r from the source is developed in [8] and is given by:

$$\Phi(r) = \frac{1}{4\pi\sigma_T} \int_V \mathbf{j}_{eq} \cdot \nabla \left(\frac{1}{r}\right) dV$$

We finally derive the potential contribution of each voxel with the Einthoven triangle to obtain the augmented leads and plot the electrocardiogram Fig. 1f.

4   Cardiac Mechanics

After the electrical depolarisation, the muscle cells in the heart release ions that lead to sarcomere shortening and active muscle contraction. These cardiomyocites are surrounded by an extracellular matrix that accounts for the passive material. We use here the Bestel-Clement-Sorine [1] model, further improved by [5], based on a multi-scale physiological description of the myocardial muscle function.

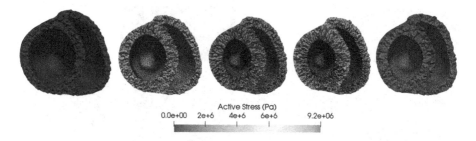

Fig. 4. Mesh deformation and active stress over a cardiac cycle

4.1   Bestel-Clement-Sorine Model

In this model, the heart is described as a passive isotropic Mooney-Rivlin material with 3 parameters, which accounts for the elasticity and friction in the cardiac extracellular matrix (mainly collagen) surrounding the fibres.

The electrical stimulation is derived from the Eikonal activation map computed beforehand and is coupled to the active orthotropic contraction part, which has 7 main parameters and accounts for the active stress (Fig. 4) along the cardiac fibres and the elasticity between sarcomeres and Z-discs. Note that combining the Mooney-Rivlin isotropic material with the elasticity of the Z-discs, the material is globally considered transversely isotropic.

The cardiac cycle is decomposed into four phases: filling, isovolumetric contraction, ejection and isovolumetric relaxation (Fig. 5 and Fig. 6). We couple a haemodynamic model implementing those phases and use the four-element Windkessel model to compute the arterial pressures.

Moreover, we performed a sensivity analysis (Table 1) on the mechanical model using a Latin Hypercube Sampling method, coupled with the computation of the Partial Ranked Correlation Coefficient [9] for a collection of relevant model outputs, selected from [2].

The four first parameters evaluated are the four elements of the Windkessel model and thus are not, or poorly, correlated with non-ejection indexes such as LPEI or the isovolumetric contraction time. Since they control the shape of the arterial and ventricular pressures during the ejection, a strong correlation on the ejection time and ejection fraction was expected. The closing time of the aortic and pulmonary valves is also depending on those pressures, thus the correlation

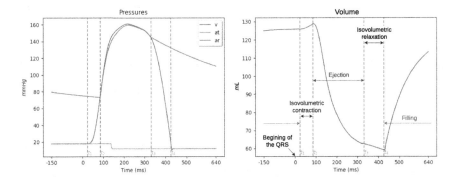

Fig. 5. Left ventricle pressure and volume curves extracted from the simulation

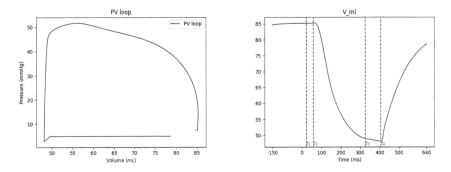

Fig. 6. PV loop and volume curves for right ventricle

observed on the isovolumetric relaxation time. We can notice that the relaxation rate, the myocardial passive stiffness and the fibres extremum angle do not seem to impact these model outputs. Finally, the contractility is strongly correlated with all the considered indexes.

5   Personalisation Method

This model can be customised with only a few set of parameters and is therefore adapted to personalisation. To this end, we use the Covariance Matrix Adaptation Evolution Strategy (CMA-ES), an evolutionary algorithm for difficult non-linear non-convex black-box optimisation problems [7]. The parameters are sampled according to a multi-variate normal distribution. The covariance matrix of this distribution is iteratively updated such as the new set of parameters minimises the cost function. The latter is built such that the output of the model with the current parameters fits the patient data. Moreover, this personalisation method is embarrassingly parallel, as all simulations are independent within an iteration.

Table 1. Results of the sensivity analysis. EF: Ejection Fraction, LVET: Left Ventricle Ejection Time, LPEI: Left ventricle Pre-Ejection Interval, SD: Systole Duration, Iso.CT: Isovolumetric Contraction Time, Iso.RT: Isovolumetric Relaxation Time, IVD: InterVentricular Delay. R_p: arterial peripheral resistance, Z_c: characteristic impedance of the artery, LL: total arterial inertance, C: total arterial compliance, K_{rs}: relaxation rate, K_{atp}: contraction rate, K: stiffness, σ: contractility, α: maximum fibres angle. Green: p-value<0.001, Yellow: p-value<0.01, Orange: p-value<0.1, Red: p-value>0.1.

	EF	LVET	LPEI	LPEI/LVET	SD	Iso.CT	Iso.RT	IVD
R_p	-0.42222	0.23128	0.02273	-0.11963	0.18141	0.02645	0.64766	-0.00766
Z_c	-0.03788	0.21657	0.00576	-0.10319	0.14087	0.00125	0.05784	-0.00157
LL	0.24441	0.58033	-0.00847	-0.36903	0.34962	-0.01144	-0.349	-0.00948
τ	0.32062	-0.29936	0.10974	0.304	-0.19011	0.10257	-0.48444	0.18499
K_{rs}	-0.02571	-0.03178	0.01345	0.01319	-0.02072	0.00205	-0.01778	0.02944
K_{atp}	0.07897	-0.02247	-0.4114	-0.39651	-0.25082	-0.41028	-0.00922	-0.30771
K	-0.02688	0.00409	0.03267	0.02106	0.00687	0.03839	0.02471	0.06666
σ	0.72809	-0.25892	-0.81655	-0.64763	-0.62813	-0.81896	-0.32104	-0.79371
α	-0.02548	-0.03125	0.02816	0.00826	-0.00697	0.01788	-0.00138	0.096

Preliminary results on the mechanical model are presented in Fig. 7. For these results, only the Windkessel and the contractility parameters have been included in the CMA-ES framework, thus spanning a five-dimensional parameter space.

5.1 Anatomy

Depending on the available patient data, it is possible to generate a mesh that meets several constraints: axis length, volumes, wall thickness, etc. If the anatomy of the heart is poorly defined, it is possible to adjust the geometry of mesh based on the patient's cardiomyopathy.

5.2 Electrophysiology

In order to match the QRS axis and the QRS duration of each patient, the following parameters are being optimised:

- depolarisation delay of the Purkinje breakthrough points (one parameter for all termination points or one per point),
- position on the endocardium of these points (two parameters per point) and
- endocardial and myocardial conductivities (two parameters).

5.3 Mechanics

All the indexes presented in Table 1 (columns of the table) will be used for the mechanical personalisation. Again, the following parameters (rows of the table) are being optimised:

- four elements of the Windkessel model. Note that the parameters for the right ventricle are adjusted according to those of the left ventricle and are not counted here (four parameters),
- contraction and relaxation parameters (two parameters),
- Mooney-Rivlin material parameters (three parameters) and
- active parameters, sarcomere contractility and viscosity, Z-discs elasticity (three parameters).

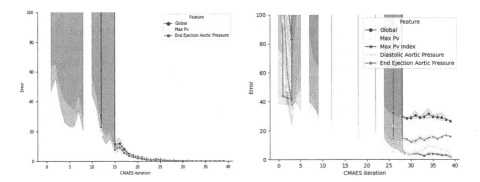

Fig. 7. Convergence of the CMA-ES features error (%) on a mechanical personalisation for two (left) and four (right) patient indexes

5.4 Discussion

Figure 7 presents some personalisation results regarding two and four patient indexes, on the mechanical model, optimised with five parameters. In both cases the algorithm converges toward the set of parameters that minimises the global error. With four indexes, the algorithm only achieved a global relative error of 25% compared to the user-defined target values. Either the global minimum of the error function was found, or the algorithm has fallen into a local minimum, and the global minimum has been excluded from the parameter space due to overly restrictive bounds.

This personalisation method has shown great promises in our preliminary tests, in terms of computational cost and global error. We are currently implementing the CMA-ES method for both the electrophysiological and mechanical frameworks on a computer cluster.

This algorithm converges in only a few iterations, for example 30 iterations if the population size is ten times the dimension of the parameter space. This space is spanned by a user-defined range of realistic values for each parameter.

With this configuration, the electrophysiological model can be personalised in less than two hours and the mechanical model in less than twelve hours on a computer cluster.

Table 2. Computation time of the different steps of the pipeline. Computed on an Intel Core i5-4430 CPU @ 3.00GHz machine.

Anatomical transformations	Meshing	Labelling	Fibres direction	Activation map	ECG generation	One cardiac cycle
10 s–1 min	30 s	1 min	15 s	30 s	1 min	20 min

6   Conclusion

In this manuscript, we have presented a fast (see Table 2) pipeline for patient-specific electromechanical simulations, using a model relying on a small set of interpretable parameters, allowing for extensive personalisation. This pipeline is built upon three submodels: an anatomical one, allowing local corrections, as well as mesh and fibres generation; an electrophysiological one, generating ECGs and simulating the complex activation and propagation of the electrical potential; and a mechanical submodel. We also lead a sensitivity analysis to assess the model's observability, and developed methods to efficiently fit the model output to patient data.

Such approach opens up possibilities in order to use cardiac modelling within clinical applications.

Acknowledgments. This work has been supported by the French government through the National Research Agency (ANR) Investments in the Future 3IA Côte d'Azur (ANR-19-P3IA-000) and by Microport CRM funding.

References

1. Bestel, J., Clément, F., Sorine, M.: A biomechanical model of muscle contraction. In: Niessen, W.J., Viergever, M.A. (eds.) MICCAI 2001. LNCS, vol. 2208, pp. 1159–1161. Springer, Heidelberg (2001). https://doi.org/10.1007/3-540-45468-3_143
2. Cazeau, S., Toulemont, M., Ritter, P., Reygner, J.: Statistical ranking of electromechanical dyssynchrony parameters for CRT. Open Heart 6(1), e000933 (2019)
3. Cedilnik, N., et al.: Fast personalized electrophysiological models from computed tomography images for ventricular tachycardia ablation planning. EP Europace 20(suppl_3), iii94–iii101 (2018)
4. Cedilnik, N., et al.: Fully automated electrophysiological model personalisation framework from CT imaging. FIMH 2019, 325–333 (2019)
5. Chapelle, D., Le Tallec, P., Moireau, P., Sorine, M.: An energy-preserving muscle tissue model: formulation and compatible discretizations. Int. J. Multiscale Comput. Eng. 10(2), 189–211 (2012)
6. Dapogny, C., Dobrzynski, C., Frey, P.: Three-dimensional adaptive domain remeshing, implicit domain meshing, and applications to free and moving boundary problems. J. Comput. Phys. 262, 358–378 (2014)
7. Hansen, N.: The CMA evolution strategy: a comparing review. In: Lozano, J.A., Larrañaga, P., Inza, I., Bengoetxea, E. (eds.) Towards a New Evolutionary Computation. Studies in Fuzziness and Soft Computing, vol. 192, pp. 75–102. Springer, Berlin https://doi.org/10.1007/3-540-32494-1_4

8. Malmivuo, J., Plonsey, R.: Bioelectromagnetism - Principles and Applications of Bioelectric and Biomagnetic Fields (10 1995)
9. Marino, S., Hogue, I.B., Ray, C.J., Kirschner, D.E.: A methodology for performing global uncertainty and sensitivity analysis in systems biology. J. Theor. Biol. **254**(1), 178–196 (2008)
10. Sermesant, M., et al.: An anisotropic multi-front fast marching method for realtime simulation of cardiac electrophysiology. In: Sachse, F.B., Seemann, G. (eds.) FIMH 2007. LNCS, vol. 4466, pp. 160–169. Springer, Heidelberg (2007). https://doi.org/10.1007/978-3-540-72907-5_17
11. Mitchell, C.: A two-current model for the dynamics of cardiac membrane. Bull. Math. Biol. **65**(5), 767–793 (2003)
12. Sermesant, M., et al.: Patient-specific electromechanical models of the heart for the prediction of pacing acute effects in CRT: a preliminary clinical validation. Med. Image Anal. **16**(1), 201–215 (2012)
13. Streeter, D.D., Spotnitz, H.M., Patel, D.P., Ross, J., Sonnenblick, E.H.: Fiber orientation in the canine left ventricle during diastole and systole. Circ. Res. **24**(3), 339–347 (1969)
14. Yezzi, A.J., Prince, J.L.: An eulerian pde approach for computing tissue thickness. IEEE Trans. Med. Imaging **22**(10), 1332–1339 (2003)

Modeling Electrophysiology, ECG, and Arrhythmia

Scar-Related Ventricular Arrhythmia Prediction from Imaging Using Explainable Deep Learning

Buntheng Ly[1], Sonny Finsterbach[2], Marta Nuñez-Garcia[3,4], Hubert Cochet[4,5], and Maxime Sermesant[1,4(✉)]

[1] Inria, Université Côte D'Azur, Epione Team, Sophia Antipolis, France
maxime.sermesant@inria.fr
[2] CHU Bordeaux, Bordeaux, France
[3] Université de Bordeaux, Bordeaux, France
[4] IHU Liryc, Electrophysiology and Heart Modeling Institute, fondation Bordeaux Université, Pessac, France
[5] Univ. Bordeaux, Inserm, CRCTB U1045, CHU Bordeaux, Bordeaux, France

Abstract. The aim of this study is to create an automatic framework for sustained ventricular arrhythmia (VA) prediction using cardiac computed tomography (CT) images. We built an image processing pipeline and a deep learning network to explore the relation between post-infarct left ventricular myocardium thickness and previous occurrence of VA. Our pipeline generated a 2D myocardium thickness map (TM) from the 3D imaging input. Our network consisted of a conditional variational autoencoder (CVAE) and a classifier model. The CVAE was used to compress the TM into a low dimensional latent space, then the classifier utilised the latent variables to predict between healthy and VA patient. We studied the network on a large clinical database of 504 healthy and 182 VA patients. Using our method, we achieved a mean classification accuracy of 75%±4 on the testing dataset, compared to 71%±4 from the classification using the classical left ventricular ejection fraction (LVEF).

Keywords: Conditional-VAE · Sustained ventricular arrhythmia · CT cardiac imaging · Myocardium thickness · Image classification

1 Introduction

VA is highly associated with sudden cardiac death (SCD), one of the major causes of death in the developed countries [10]. Increased risk of VA has been linked to the presence of left ventricular myocardial (LVMYO) scar. Although, the current gold standard for SCD risk stratification is the LVEF, most SCD events still occur in patients with higher LVEF than the cut-off value of 35%. Further evidence has shown LVMYO scar to be a better independent predictor of SCD than the LVEF. Nonetheless, the detection and characterisation of the scar region are still challenging, thus its application in clinical practice is limited.

© Springer Nature Switzerland AG 2021
D. B. Ennis et al. (Eds.): FIMH 2021, LNCS 12738, pp. 461–470, 2021.
https://doi.org/10.1007/978-3-030-78710-3_44

Recent studies have linked the potential of CT imaging in locating the LVMYO scar region through the assessment of wall thinning [9]. Moreover, the studies have also found VA substrate within or next to CT-defined scar region.

The advent of Deep Learning (DL) has given rise to several image classification networks of exceptional accuracy. Among the different network configurations, we investigate the embedding-classification architecture, which has been successfully implemented in classification tasks for medical imaging [1,2,6]. The two-part architecture includes an embedding model and a classification model. Multiple DL architectures have been used as the embedding network, such as the deep residual network [1], the convolutional autoencoder [6], and the variational autoencoder [2]. In our framework, we used the CVAE architecture [13], which allows a more robust reconstruction of the input. Moreover, it enables to analyse the explainability of the results thanks to its generative nature.

In this paper, we studied the correlation between the myocardial wall thinning and the history of VA. To this end, we constructed an automatic pipeline to generate 2D TM from the imaging input and the CVAE-Class network to embed and classify the 2D TM.

2 Methods

Our pipeline used 3D cardiac CT image as the main input. Myocardial masks were automatically segmented and were used for TM computation. The 3D TM was then mapped into a uniform circular 2D image, which ensures the shape consistency between input images. The 2D TM was then fed the CVAE-Class network, where it was projected into the latent variables, which were then used for VA classification.

2.1 Image Processing

Figure 1 outlines our automatic image processing pipeline. The main steps included the cardiac segmentation, the short-axis view (SAX) orientation, the TM calculation, the LV flattening, and the thickness value normalisation.

Image Segmentation. To start with, we generated the ventricular masks from cardiac CT using a pre-trained Dual-UNet segmentation network proposed in [4]. The required ventricle masks included the right ventricular epicardium (RVEPI), the left ventricular epicardium (LVEPI) and endocardium (LVENDO). Then, the LVMYO could be extracted from LVEPI and LVENDO.

Short-Axis Orientation. An automatically computed rigid transformation was applied to re-orient the ventricular masks in the standardised SAX view following the method presented in [11]. This step also included an isotropic resampling of the image and masks that along with the standardised SAX orientation highly facilitated tissue thickness calculation and the computation of the 2D TM map.

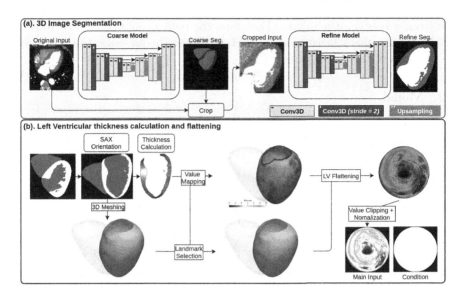

Fig. 1. Automatic pipeline to extract the 2D thickness map from 3D cardiac CT.

Thickness Map Calculation. The LVMYO thickness was calculated from the LVEPI and LVENDO masks using the Eulerian partial differential equation approach for tissue thickness calculation [14]. The python package implemented for [3] was made available by the author[1]. A 3D surface model of the LV was computed using the discrete marching cubes algorithm provided by the VTK package[2] and the thickness information was projected onto the mesh.

LV Flattening. To transform the TM from 3D to 2D, we used the LV quasi-conformal flattening algorithm[3] based on the left atrial flattening proposed by [12]. In their LV extension, the authors proposed the manual selection of three landmarks: one at the centre of the apex and two points at the base. In this work, thanks to the standardised representation of the ventricles in SAX view, the landmark selection step could be automated.

Value Clipping and Normalisation. To prepare the TM as input to the DL model, we mapped the 2D TM mesh onto a 2D array in 256×256 resolution. We clipped the maximum thickness value at $10\,\mathrm{mm}$, and divided the result by 10 to limit the values to $[0, 1]$. This pre-processing was done to generate a normalised map with increased separation between the regions with low value.

This processing pipeline reduced the input dimension from 3D into 2D, while still retained the 3D information in term of the wall thickness. Additionally, the uniform circular shape of the TM removed the variability of the LV shape in the different scan images, as the flattening landmarks ensured the rotation

---

[1] https://pypi.org/project/pyezzi/.
[2] https://vtk.org/.
[3] https://github.com/martanunez/LV_flattening.

consistency between the inputs. Therefore, the model could be trained to focus on the specific characteristics of the TM, including the presence, the extent and the location of the thin regions.

2.2 Model

The CVAE-Class network was a combination of a CVAE and a fully connected classifier model, as shown in Fig. 2. The network was built and trained using the Tensorflow 2.0 package[4].

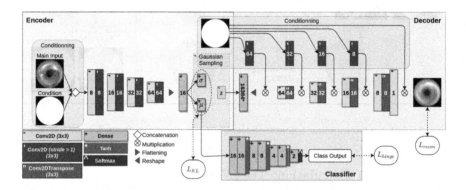

Fig. 2. CVAE-Class network architecture. Please refer to the colour or the symbol in the bottom-left corner to distinguish the layer types. The number in each layer indicates the number of output filters or units.

CVAE Model. The CVAE consisted of an encoder and a decoder branch. In the encoder branch, we used the concatenate layer to combine the main input image with the condition mask, following by a series of 1-stride and 2-stride convolution layers. The 2-stride layers progressively reduced the resolution from 256×256 to 16×16. At the bottleneck, we flattened the output features before applying the fully connected layers to generate the latent mean μ and log variance σ vector. Thereafter, we applied the Gaussian sampling method to generate the latent vector z. The latent dimension was set to 25.

In the decoder branch, we used a fully connected layer to increase the number of features of the latent vector z. Then, we reshaped the vector to the same shape as before the flattening layer. The decoder was then built using the sequences of regular and transpose convolution layer. We set the stride of the transpose layer to 2 to increase the resolution of the output. At each decoding level, we combined the condition mask and the main output using multiplication. A convolution layer with an adjusted number of stride was used to rectify the resolution difference. Finally, we multiplied the features with the conditional mask to get the final reconstruction output. We conditioned this VAE with the mask in order

---

[4] https://www.tensorflow.org/.

to prevent the network from using its parameters to correctly predict the circular shape of the map.

The CVAE was trained using the sum of the reconstruction loss (L_{recon}) and the Kullerback-Leibler divergence loss (L_{KL}), Eq. 1. The L_{recon} was calculated using the weighted root mean square error, defined in Eq. 2, where N is the number of pixels, $\hat{y}$ denotes the decoder reconstruction output, y the input TM and m is the binary mask. The L_{KL} was calculated between the latent distribution $\mathcal{N}(\mu, \sigma^2)$ and the normal distribution $\mathcal{N}(0, 1)$. The β value was set to 0.01.

$$L_{CVAE} = L_{recon} + \beta L_{KL} \tag{1}$$

$$L_{recon} = \sqrt{\frac{1}{N} \sum_i \sum_j (\hat{y}_{ij} - y_{ij})^2 \times (2m_{ij} - y_{ij})} \tag{2}$$

Classifier Model. The classifier model used the latent mean μ as input. The model was built using three blocks of two fully connected layers and a tanh activation layer. The numbers of units were set to 16, 8 and 4, respectively. To get the classification output (healthy and VA), we used a fully connected layer with 2 units followed by the softmax function.

The classifier was trained using the hinge loss, as defined in Eq. 3, where N denotes the number of classes, y denotes the ground truth and $\hat{y}$ denotes the classification score. Noted that the expected value of the ground truth for each class is between $[-1, 1]$, while the classification score is between $[0, 1]$.

$$L_{hinge} = \frac{1}{N} \sum_i^N \max\{1 - y_i * \hat{y}_i, 0\} \tag{3}$$

3   Experimental Setup

Dataset. A large series of CT scan datasets were retrospectively collected from Bordeaux University Hospital. The inclusion criteria were: history of myocardial infarction or ischaemic heart disease and available contrast-enhanced and ECG gated CT scan reconstructed in diastole. The medical records of these patients were reviewed to collect clinical characteristics, including demographics, the delay since myocardial infarction, the LVEF and the history of scar-related VA. Patients with the history of cardiac catheter ablation or ventricular surgery were not considered for inclusion. The LVEF was considered valid when measured with MRI or echocardiography closest to the time of CT within the limit of 1 year. VA events were qualified as past episodes of sustained ventricular tachycardia, ventricular fibrillation, aborted sudden cardiac arrest, or any appropriate therapy delivered by an implantable cardioverter defibrillator less then 1 year before the CT. VA occurring at the acute stage of the infarct (less than 1 month) were not considered as scar-related and were not analysed as such. Patients with incomplete medical records were also excluded from the analysis.

A total of 686 patients (age 73 ± 12, 83.4% men and 26.5% with VA) were included. We used the bootstrap method [8] to estimate the mean accuracy and the standard error of our model. To account for the computation time of the deep learning model, the number of resampling was set to 25, which has been shown to be sufficient in gauging the randomness caused by the sampling data [7]. For each resampling, we randomly selected 100 cases (50 healthy and 50 VA patients) to be used as testing dataset.

Training Setup. To balance between healthy and VA patients, we trained and validated the model with the same number of data from each class at each iteration.

The CVAE-Class networks were trained step-by-step, where we first trained the CVAE to adequately embed the input before optimising the classifier. Both models were optimised using Adam optimiser with the initial learning rate at $1e - 4$. The learning rate was reduced by half after 5 epochs with no validation improvement. The training was set to stop when the validation loss does not improve after 50 epochs. The model was rerun 5 times, afterward the model with the highest validation accuracy is selected.

Baseline Model. To compare our network performance, we also built a straightforward image classifier network using the architecture of the encoder and the classifier of the CVAE-Class network. Additionally, we run the test with a threshold model using the LVEF cut-off value. To adjust the value to our sample, we recalculated the best cut-off value to maximise VA classification accuracy in the training set, then applied the value to the testing data. We also trained a logistic regression model using the LVEF value.

4   Results

4.1   Accuracy

Table 1 shows the comparison of the classification accuracy of the CVAE-Class and the baseline models, at 25 resamplings. To adjust for the small resmapling size, the confidence intervals were calculated using the approximate bootstrap confidence interval method as proposed by [8]. The mean LVEF cut-off value was 39.4%.

The CVAE-Class model scored the highest accuracy, with an average of 75% with 4% standard deviation. The image classifier model achieved a mean accuracy of 73%. Lastly, the $LVEF_{LogReg}$ and $LVEF_{cut-off}$ model felt behind at 71% and 70%, respectively. The better performance of the CVAE-Class compared to straightforward image classifier model suggests that the latent variables generated by the CVAE network were relevant in improving the classification accuracy. Both models also showed better performance than the LVEF models, which further highlights the potential of the 2D TM in identifying patients with history of VA.

Table 1. The CVAE-Class and the baseline models classification accuracy.

Model	Mean accuracy	St. Deviation	90% CI
CVAE-Class	**0.75**	**0.04**	**(0.74, 0.77)**
Img-Class	0.73	0.04	(0.72, 0.75)
LVEF$_{LogReg}$	0.71	0.05	(0.70, 0.73)
LVEF$_{cut-off}$	0.70	0.04	(0.69, 0.72)

4.2 Explainability

We tested the explainability of the CVAE-Class model using the GradCam++ method proposed in [5], which computes the classification attention map using the prediction score. This could help the user in understanding what was used by the network for its prediction.

Figure 3 shows a series of attention maps including the true and false positive cases (Fig. 3.a and Fig. 3.b) and the true and false negative cases (Fig. 3.c and Fig. 3.d). As expected, we can see higher attention values in the thin regions, and higher positive prediction scores were placed on the cases with larger thin regions. In consequence, however, the model failed with the patient who presented wall thinning without VA history (Fig. 3.b) and vice versa when VA occurred in the absence of wall thinning (Fig. 3.d).

The generative nature of the model could also be used to explore the latent space and better understand the learned representation, as well as the differences between classes.

5 Discussion

Due to the compression of VAEs, the reconstructed TM did not contain some high frequency details, as shown in the Fig. 4.a. While this could be due to the decoder branch, it could also suggest that the latent vectors did not hold necessary information to decode the missing details. Although, our current framework focused on the classification and did not directly use the reconstructed output, ensuring the latent vector contains heterogeneity information could lead to better classification accuracy.

We analysed the latent variables of the test data using the partial least square regression, Fig. 4.b, where we could observe partial overlap between the two classes. Adding constraints to increase the separation of the latent variables could allow for a more robust classification model. Moreover, in a well separated latent space, we can exploit the latent variables to generate a TM of the different classes. This would allow us to display the personalised features that influence the classification.

As our current model was only trained with the TM input, its accuracy was ultimately tied to the cases where the VA was related to the wall thinning, as illustrated by the false predictions in Fig. 3. The subtle differences between the

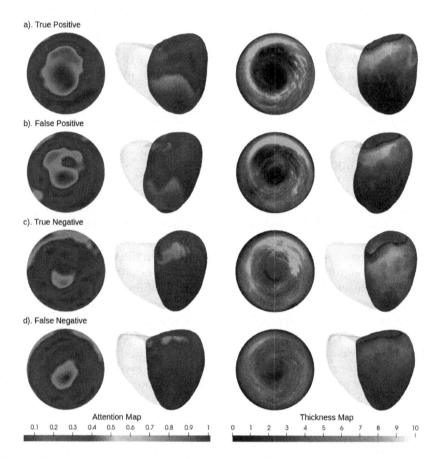

Fig. 3. The attention map and its corresponding thickness map. Higher gradients (Red) were placed on the thin regions of the thickness map, as expected, the scar was driving the decision of the model. (Color figure online)

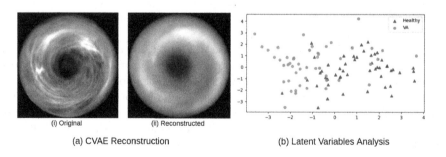

Fig. 4. Image reconstruction (a) and latent variables analysis (b).

TM may be the key to a more accurate classification, therefore it is crucial to improve the latent representation of the TM. On the other hand, integrating the clinical information into the current framework can be the direct answer to the single-input limitation.

6   Conclusion

We proposed an automatic pipeline for VA classification using 3D CT images. Using a 2D thickness map, our model showed a clear increase in classification accuracy as compared to the current gold standard classification using LVEF value. We also provided evidence that the CVAE network generated meaningful latent variables that improved the classification accuracy compared to a straightforward image classification network. In the future work, we wish to explore further into the latent variables enhancement and clinical information integration to increase classification accuracy, as well as to exploit the generative capability of the network.

Acknowledgement. Part of the authors' work has been supported by the French Government, through the National Research Agency (ANR) 3IA Côte d'Azur (ANR-19-P3IA-0002), IHU Liryc (ANR- 10-IAHU-04).

References

1. Abbet, C., Zlobec, I., Bozorgtabar, B., Thiran, J.-P.: Divide-and-rule: self-supervised learning for survival analysis in colorectal cancer. In: Martel, A.L., et al. (eds.) MICCAI 2020. LNCS, vol. 12265, pp. 480–489. Springer, Cham (2020). https://doi.org/10.1007/978-3-030-59722-1_46
2. Biffi, C., et al.: Learning interpretable anatomical features through deep generative models: application to cardiac remodeling. In: Frangi, A.F., Schnabel, J.A., Davatzikos, C., Alberola-López, C., Fichtinger, G. (eds.) MICCAI 2018. LNCS, vol. 11071, pp. 464–471. Springer, Cham (2018). https://doi.org/10.1007/978-3-030-00934-2_52
3. Cedilnik, N., et al: Fast personalized electrophysiological models from CT images for ventricular tachycardia ablation planning. EP-Europace **20**(3), iii94-iii101 (2018)
4. Cedilnik, N., Duchateau, J., Sacher, F., Jaïs, P., Cochet, H., Sermesant, M.: Fully automated electrophysiological model personalisation framework from CT Imaging. In: FIMH 2019–10th International Conference on Functional Imaging and Modeling of the Heart, pp. 325–333. Bordeaux, France (June 2019)
5. Chattopadhay, A., Sarkar, A., Howlader, P., Balasubramanian, V.N.: Grad-CAM++: Generalized gradient-based visual explanations for deep convolutional networks. In: Proceedings - 2018 IEEE Winter Conference on Applications of Computer Vision, WACV 2018, pp. 839–847 (2018)
6. Dercksen, K., Bulten, W., Litjens, G.: Dealing with label scarcity in computational pathology: A use case in prostate cancer classification. arXiv (2019)
7. Efron, B., Rogosa, D., Tibshirani, R.: Resampling Methods of Estimation, vol. 19. Elsevier, second edition edn. (2015)

8. Efron, B., Tibshirani, R.J.: An Introduction to the Bootstrap. In: An Introduction to the Bootstrap. CRC Monographs on Statistics and Applied Probability, CRC Press (1993)

9. Mahida, S., et al.: Cardiac imaging in patients with ventricular tachycardia. Circulation **136**(25), 2491–2507 (2017)

10. Nielsen, J.C., et al.: European Heart Rhythm Association (EHRA)/Heart Rhythm Society (HRS)/Asia Pacific Heart Rhythm Society (APHRS)/Latin American Heart Rhythm Society (LAHRS) expert consensus on risk assessment in cardiac arrhythmias: Use the right tool for the right outcome. Europace (2020)

11. Nuñez-Garcia, M., Cedilnik, N., Jia, S., Sermesant, M., Cochet, H.: Automatic multiplanar CT reformatting from trans-axial into left ventricle short-axis view. In: STACOM 2020–11th International Workshop on Statistical Atlases and Computational Models of the Heart. Lima, Peru (October 2020)

12. NuÃez-Garcia, M., et al.: Fast quasi-conformal regional flattening of the left atrium. IEEE Trans. Vis. Comput. Graph. **26**(8), 2591–2602 (2020)

13. Sohn, K., Yan, X., Lee, H.: Learning structured output representation using deep conditional generative models. In: Advances in Neural Information Processing Systems, pp. 3483–3491 (2015)

14. Yezzi, A.J., Prince, J.L.: An Eulerian PDE approach for computing tissue thickness. IEEE Trans. Med. Imaging **22**(10), 1332–1339 (2003)

Deep Adaptive Electrocardiographic Imaging with Generative Forward Model for Error Reduction

Maryam Toloubidokhti[1]([✉]), Prashnna K. Gyawali[1], Omar A. Gharbia[1], Xiajun Jiang[1], Jaume Coll Font[2], Jake A. Bergquist[3], Brian Zenger[3], Wilson W. Good[3], Dana H. Brooks[4], Rob S. MacLeod[3], and Linwei Wang[1]

[1] Rochester Institute of Technology, Rochester, NY, USA
mt6129@rit.edu
[2] Cardiovascular Bioengineering and Imaging Lab, Massachusetts General Hospital, and Harvard Medical School, Boston, MA, USA
[3] The University of Utah, Salt Lake City, UT, USA
[4] Northeastern University, Boston, MA, USA

Abstract. Accuracy of estimating the heart's electrical activity with Electrocardiographic Imaging (ECGI) is challenging due to using an error-prone physics-based model (forward model). While getting better results than the traditional numerical methods following the underlying physics, modern deep learning approaches ignore the physics behind the electrical propagation in the body and do not allow the use of patient-specific geometry. We introduce a deep-learning-based ECGI framework capable of understanding the underlying physics, aware of geometry, and adjustable to patient-specific data. Using a variational autoencoder (VAE), we uncover the forward model's parameter space, and when solving the inverse problem, these parameters will be optimized to reduce the errors in the forward model. In both simulation and real data experiments, we demonstrated the ability of the presented framework to provide accurate reconstruction of the heart's electrical potentials and localization of the earliest activation sites.

Keywords: Electrocardiographic imaging (ECGI) · Body surface potential (BSP) · Forward model · VAE-based generative model

1 Introduction

Electrocardiographic Imaging (ECGI) is a non-invasive procedure for estimating the heart's electrical activity from body surface potentials (BSPs) [1] and a model of the torso geometry. Traditionally, ECGI approaches start with a *forward* model that defines the relationship between the heart's electrical activity and the BSPs based on quasi-static electromagnetism [1]. Many studies, starting from [3], have shown that it is essential to customize this forward model to the heart and torso geometry specific to a subject. Using this physics-based subject-specific forward model, the inverse problem estimates the cardiac sources that

© Springer Nature Switzerland AG 2021
D. B. Ennis et al. (Eds.): FIMH 2021, LNCS 12738, pp. 471–481, 2021.
https://doi.org/10.1007/978-3-030-78710-3_45

best fit the BSP data with various optimization and regularization techniques [4–6]. While traditional ECGI approaches exploit the physics behind the problem, their success relies heavily on the accuracy of forward model. Due to the ill-posed nature of the inverse problem, small errors in the forward model can have magnified influence on the inverse solution. Unfortunately the ECGI forward model is inherently prone to errors in model parameters such as the shape and orientation of the heart and torso [3]. Using such a fixed froward model in the inverse problem does not allow us to correct errors in model parameters. On the other hand, it is time consuming to directly optimize these parameters during inverse estimation with typical forward model computation methods such as boundary element method (BEM). In [18] authors reparameterize the forward model and optimize a limited set of its underlying parameters, but generalizing this approach to account for other parameters is not clear.

In recent times, deep learning (DL)-based approaches are widely popular for modeling electrocardiography data [7–10]. Among them, DL-based ECGI techniques have gained traction due to the advantage of faster inference [10]. Like most existing DL models, these black-box inverse mappings learn from cardiac bioelectrical sources and BSP data pairs but cannot incorporate any physics knowledge. Furthermore, contrary to the knowledge that the forward and inverse relationship is specific to each individual, most of these approaches learn a single inverse mapping function that often fails to generalize to different geometries. In this paper, we address the critical gap between the traditional and DL-based approaches with a novel ECGI formulation, which we call Deep Adaptive ECGI (DA-ECGI) that allows the incorporation of *geometry-informed* forward-modeling physics while simultaneously allowing this *a priori* physics to be optimized to patient-specific data. DA-ECGI has two key ingredients:

- A deep generative model for forward modeling that approximates the physics behind the problem by exploiting available heart and torso geometry.
- An inverse estimation procedure that simultaneously optimizes the partially-known forward model's parameters and the unknown cardiac source of interest using the given BSP data.

Specifically, we learn to approximate ECGI forward model's generation using a variational autoencoder (VAE) [11]. The VAE generates a forward operator $\mathbf{H}$ dependent on two latent variables $\mathbf{g}$ and $\mathbf{z}$, where the former represents available partial knowledge about the heart-torso geometry while the latter represents unknown parameters. The trained generative model for $\mathbf{H}(\mathbf{g}, \mathbf{z})$ is then used in a regularization (here second-order Tikhonov) to reconstruct heart-surface potentials $\mathbf{u}$ from BSP, where $\mathbf{u}$ and the latent variable $\mathbf{z}$ are simultaneously optimized. As a proof of concept, we trained the VAE-based $\mathbf{H}(\mathbf{g}, \mathbf{z})$ on BEM forward operators generated by the SCIRun software [12], from a single torso tank and a heart with various rotations and translations. The VAE was trained to generate the forward operators from partial knowledge of the geometry; in particular we assumed it had accurate information about the rotation of the heart but no information about any translations. The trained $\mathbf{H}(\mathbf{g}, \mathbf{z})$ was applied to ECGI on BSP data and $\mathbf{z}$ was optimized to account for the translation of the heart and correct the errors in the forward operator and the resulting inverse solutions. We evaluated this method with real BSP data from varying pacing locations acquired from a different geometry. The results showed that DA-ECGI was able to correct errors in the forward model due to geometry errors and thereby deliver more accurate inverse solutions in terms of reconstruction of electrical propagation patterns and localization of the earliest activation sites.

2 Methodology

In Sect. 2.1 we first provide the necessary background information about the forward and inverse problem of ECGI, in Sect. 2.2 we explain the fundamentals of the variational autoencoder (VAE) and finally, in Sect. 2.3 and 2.4 we explain the components of DA-ECGI framework.

2.1 The Forward and Inverse Problems of ECGI

This paper considers the extracellular potential $\mathbf{u}$ on the joint epicardial and endocardial surfaces as the source model in the heart, whose relationship with the BSP follows the Laplace equation [1,2]. When solved numerically on a given heart-torso geometry, it gives rise to a forward matrix $\mathbf{H}$ that relates $\mathbf{u}$ and BSP as $\phi = \mathbf{Hu}$ where $\mathbf{\Phi}$ are the BSP measurements. A commonly-used second-order Tikhonov [13] approach estimates $\mathbf{u}$, by solving the following equation:

$$\mathbf{u} = \arg\min ||\mathbf{\Phi} - \mathbf{Hu}|| + \lambda||\mathbf{L}\mathbf{u}||; \tag{1}$$

where λ is a scalar, and $\mathbf{L}$ is a regularization matrix. The first term in Eq. 1 minimizes the error of fitting the measured BSP data, and the second term encourages a smooth solution if $\mathbf{L}$ is a spatial derivative estimator.

2.2 Variational Autoencoders (VAEs)

VAE [11] is a widely adopted deep generative model capable of learning the generation of observed distributions of data $\mathbf{x}$ from a compact set of latent variables $\mathbf{z}$. It comprises a neural decoder $p(\mathbf{x}|\mathbf{z})$ to approximate the likelihood of $\mathbf{x}$ given $\mathbf{z}$, and a neural encoder $q(\mathbf{z}|\mathbf{x})$ to approximate posterior density of $\mathbf{z}$ given $\mathbf{x}$. This encoding-decoding network is trained by maximizing the variational evidence lower bound (ELBO) of the marginal data likelihood:

$$\log p(\mathbf{x}) \geq \mathcal{L} = E_{\mathbf{z} \sim q(\mathbf{z}|\mathbf{x})} p(\mathbf{x}|\mathbf{z}) - D_{KL}(q(\mathbf{z}|\mathbf{x})||p(\mathbf{z})) \tag{2}$$

where the first term can be interpreted as data reconstruction, while the second penalty term constrains the approximated posterior density $q(\mathbf{z}|\mathbf{x})$ to be similar to a prior $p(\mathbf{z})$ (defined as isotropic Gaussian) by minimizing their Kullback-Leibler (KL) divergence. Once the VAE is trained, the decoder is able to generate new samples of $\mathbf{x}$ from samples of $\mathbf{z}$.

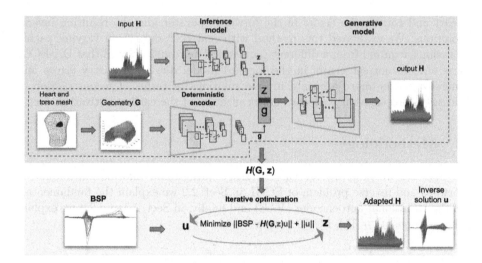

Fig. 1. Outline: forward model generator and parameter optimization are separated with the green and yellow box. The dashed box is the generative model. (Color figure online)

2.3   DA-ECGI Forward Model

DA-ECGI forward model consists of a deep generative model to define the forward model $\mathbf{H}$ as $\mathbf{H}(\mathbf{z}, \mathbf{g})$. The outline of DA-ECGI is shown in Fig. 1. The DA-ECGI forward model follows the structure of an independent conditional variational autoencoder ICVAE [14]. It encodes geometry information independent of the latent space of the VAE to keep the valid, partial geometric information $\mathbf{g}$, and encodes other parameters and the error attached to them in $\mathbf{z}$, such that they can be optimized when solving the inverse problem.

Generative Model : We assume a generative model as below,

$$p(\mathbf{h}, \mathbf{z}|\mathbf{g}) = p(\mathbf{h}|\mathbf{z}, \mathbf{g})p(\mathbf{z}) \tag{3}$$

This generative model consists of a deterministic encoder to encode geometry information $\mathbf{G}$ into a lower dimension representation $\mathbf{g}$, and a decoder to reconstruct forward model $\mathbf{H}$ given samples from latent distribution $p(\mathbf{z})$ (defined as isotropic Gaussian) and a known $\mathbf{g}$. Each subject's specific $G = (g_{ij})$ is modeled as the distance matrix between all heart and torso mesh nodes[1]. Since the highest values (or peaks) in $\mathbf{H}$ correspond to the closest heart-torso node pairs, we normalize $\mathbf{G}$ as $g_{ij} = \max{(G)} - g_{ij}$. Note that the deterministic encoder is used only to reduce the dimension of $\mathbf{G}$ to prevent the dominance of the geometry features $\mathbf{g}$ over $\mathbf{z}$ and allow the variational encoder to learn $q(\mathbf{z}|\mathbf{h})$ (defined in the next section). In this sequel, we refer to the $\mathbf{G}$ as geometry.

---

[1] For a homogeneous torso conductor model this captures all the information in the forward solution.

Inference Model : The inference model encodes the forward model **H** to the latent space **z** to approximate the intractable true posterior $p(\mathbf{z}|\mathbf{h})$ with $q(\mathbf{z}|\mathbf{h})$. Since **g** is separately encoded from **G**, the learned **z** is independent of **g**.

Variational Inference : The lower bound of the DA-ECGI forward model is:

$$\log p(\mathbf{h}|\mathbf{g}) \geq \mathcal{L} = E_{\mathbf{z} \sim q(\mathbf{z}|\mathbf{h})} p(\mathbf{h}|\mathbf{z}, \mathbf{g}) - \beta D_{KL}(q(\mathbf{z}|\mathbf{h}) || p(\mathbf{z})) \quad (4)$$

where $p(\mathbf{h}|\mathbf{z}, \mathbf{g})$ is the likelihood of output forward models given a specific geometry, considered as the reconstruction term of the loss function. Maximizing Eq. 4 forces the approximate posterior to match the true posterior as closely as possible. Thus, we can redefine the forward model as $H(z, g)$.

2.4   Optimizing the Forward Model

We consider an iterative optimization procedure after learning the forward model as $H(g, z)$. At each iteration, we search for **z** that minimizes $||\mathbf{\Phi} - H(\mathbf{z}, \mathbf{g})\mathbf{u}||$ and generate a new **H** corresponding to the optimized **z**, and use this adjusted **H** to find the inverse solution **u** for the whole heart beat. This iterates to continuously reduce the forward model's errors while estimating the inverse solution. We use the BOBYQA algorithm [15–17] to optimize z, and the second-order Tikhonov method to find the inverse solution.

We present the pseudo-code of our algorithm in Algorithm 1.. In order to balance the fitting error term and the regularization term in Eq. (1), we continuously anneal the regularization term λ over the iterations. Starting with a large value for λ and gradually decreasing it over iterations, prevents fitting the BSP data ($\mathbf{\Phi}$) in the initial iterations and allows the forward model to be adequately optimized. Optimization continues until the fitting error reaches the desired minimum value or if the maximum number of iterations is completed. In Algorithm 1. variable λ_0 is set to 0.1 and α to 0.8.

Algorithm 1. Forward-Model-Optimization

1: **function** INVERSEOPTIMIZE(maxIteration, fittingError, G, Φ)
2:     g = model.GeometryEncoder(G)
3:     $z = (z_{ij}), z_{ij} \sim \mathcal{N}(0, 1)$
4:     $\lambda \leftarrow \lambda_0, i \leftarrow 0, initH \leftarrow$ model.decode($concat(z, g)$)
5:     $u = TikhonovInverse(initH, \Phi, \lambda)$
6:     **while** $i < maxIteration$ or $fittingError > Error$ **do**
7:         $z =$ BOBYQA(objectiveFunction, g, z, u, Φ)
8:         $H =$ model.decode($concat(z, g)$)
9:         $\lambda = \alpha * \lambda$
10:         $u = TikhonovInverse(H, \Phi, \lambda)$
11:         $i = i + 1$
12:     **return** H, u
13: **function** OBJECTIVEFUNCTION(g, z, u, Φ)
14:     $f = ||model.decode(concat(z, z_g) * u - \Phi||$

---

3 Experiments and Results

We tested DA-ECGI with two different datasets: simulated data with controlled geometrical errors in the forward model and real data on a given heart-torso geometry with unknown errors in the forward model. We present training details and datasets descriptions in Sect. 3.1, and test results in Sect. 3.2 and 3.3.

3.1 Training and Testing the VAE-based Forward Modeling

Training: We trained the presented model on a synthetic dataset comprising 346 modes of variation applied on a pair of heart and torso volume of human data by rotating the base heart along the X-, Y- and Z-axes in the range of [–50°,20°], [–50°,20°], and [–60°,20°], respectively. We refer to this dataset as *rotated hearts*. We then translated the *rotated hearts* along the X-,Y- and Z-axis for [–30, 10] mm, resulting in what we refer to as *rotated+translated hearts*. To simulate the situation where geometry error is introduced into the forward model, we generated forward models corresponding to *rotated+translated hearts*, and then paired them with the geometries generated from *rotated hearts* creating a dataset of 834 pairs, which are split into training and test sets with an 80–20 ratio. The goal is to train the generative model such that using the partial knowledge about rotations of the heart, it can be used to estimate its unknown translation.

All networks are implemented in PyTorch and optimized with Adam optimizer. Both encoders and the decoder consist of 9 convolutional layers. We chose the batch size of 16, the learning rate of 1e–4, and the latent dimension of 16, representing an equal combination of output from the deterministic and the variational encoder. We used the mean-squared loss to calculate the reconstruction error, and β is annealed to balance the reconstruction and KL-term error. The final value for β is set to 0.001. The model is trained for 2200 epochs.

Testing with Synthetic Dataset: We first simulate the spatiotemporal propagation sequence of extracellular potential considering a combination of 5 different origins of activation on the heart using Aliev-Panfilov (AP) model [20]. We then randomly selected 15 forward models from the test-set corresponding to the *rotated+translated hearts* to generate BSPs reflecting effects of varying heart orientations. The total number of synthetic test data is 75. We should note that all BSPs are corrupted with 45 dB Gaussian noises before applying ECGI.

Testing with Real Dataset: We used a dataset consisting of a single canine heart suspended in a human-shaped torso tank geometry, with measurements of 247 electrodes placed on a sock surrounding the dog's beating heart and simultaneous BSP measurements of 192 electrodes on the tank. Experimental details are in [19]. We used recordings from five different pacing sites. Since this dataset's electrical potentials are recorded only on the epicardium, we first registered a geometry from our synthetic dataset to this geometry and used it to generate the forward model using a consistent source model.

3.2   Results – Synthetic Data Experiments

Fig. 2(a) shows an example forward operator **H**, visualized as a 3D image, generated from a *Rotated* geometry **g** and a random sample from $q(\mathbf{z}|\mathbf{h})$ which was used as the initial **H** in the optimization. Figure 2(b) is the final **H** after optimization and Fig. 2(c) is the ground-truth **H** generated from *Rotated+translated* geometry. Comparing (b) and (a) with (c) we can see that the final **H** spatial pattern is closer to the ground-truth than the initial **H**; note for example the relative dominance of more heart-torso node pairs in the original and final images, highlighted in the red ellipse.

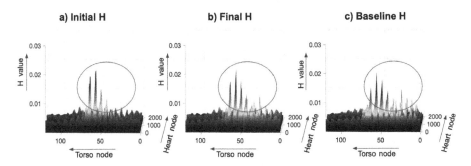

Fig. 2. (a) Initial forward operator H generated from given **g** and random sample of **z**, (b) Final H after optimization, (c) Ground-truth H. All matrices are visualized as 3D images

We compared our method's initial (generative model output) and final (forward model after optimizing for 60 iterations) solution with both the ground-truth heart potentials and the inverse solutions obtained from the ground-truth forward model, which we refer to as the *baseline solution*. The average distance of the origin of activation of final solutions to the ground-truth is 18.28 ± 9.61 mm and with the baseline solution is 11.60 ± 9.70 mm, while those for the initial solutions are 20.68 ± 11.23 mm and 17.98 ± 14.66 mm, respectively. In Fig. 3 we show that DA-ECGI achieved significantly higher spatial and temporal correlation ($p < 0.0029$) with the ground-truth than the initial solutions.

3.3   Results – Real Data Experiments

After generating the initial forward model with our generative model from the given geometry, assuming no rotation applied on the heart, we optimized the forward model's parameters for 80 iterations. Figure 4 compares the initial and the final forward model with the baseline forward model generated with SCIrun software. Figure 5 shows the worst and the best cases of electrical propagation of inverse solutions compared to the ground-truth at the time-points indicated by the colored bars on the electrograms on the right. The worst and the best

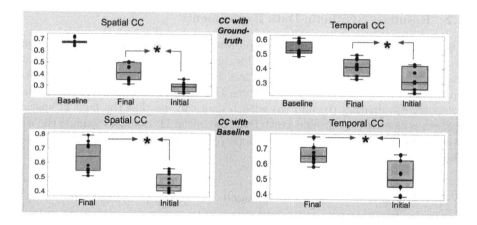

Fig. 3. Comparison of spatial and temporal correlation coefficients. The arrows and the red asterisk indicate that these distributions are significantly different.

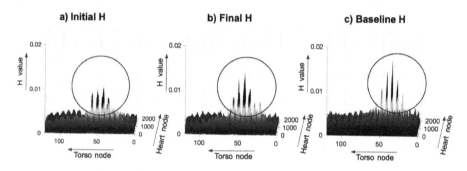

Fig. 4. Comparison of a) initial, b) final, and c) baseline forward models.

cases correspond to the solutions with the lowest and highest correlation with the ground-truth. The propagation pattern of the final solution is closer to the ground-truth than the initial solution for the best-case solution, whereas in the worst-case we do not observe many changes between the initial and final solution. In Fig. 5 all solutions are normalized before visualization. Figure 6 summarizes the spatial and temporal correlation coefficients of the ground-truth with our method's initial and final solutions. The correlation coefficients of all cases, including the worst-case, have improved after optimization. The average distance of the activation origin in the ground-truth and the final solutions is 29.28 ± 14.71 mm, and in the initial solutions is 38.58 ± 14.36 mm. Thus, optimizing the geometry errors led to more accurate localization.

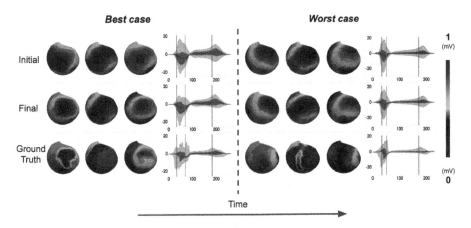

Fig. 5. Worst-case and best-case of electrical propagation. The vertical lines in the last column of each case marks the time stamps of the displayed potential distribution.

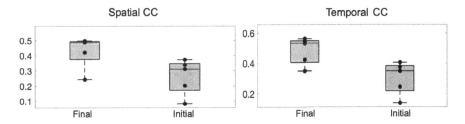

Fig. 6. Comparing spatial and temporal correlation coefficients with the ground-truth.

4   Discussion and Conclusion

This work presents DA-ECGI, a novel VAE-based ECGI framework for reducing errors in forward models. We defined a generative forward model to create a compact parameter space that can later be optimized to subject-specific BSP data and correct the errors in the forward model and the resulting inverse solutions. To our knowledge, this is the first deep-learning-based ECGI framework able to learn an adjustable forward model from geometry. While this initial study considered the second-order Tikhonov [13] formulation of ECGI and only geometric parameters as potential errors in the forward model, future work will explore generalizing the proposed approach to be used in other ECGI formulations and to consider other types of forward model errors.

Acknowledgments. This work is supported by the National Heart, Lung and Blood Institute of the National Institutes of Health (NIH) under Award Number R01HL145590 (Wang) and National Institute of General Medical Sciences of the National Institutes of Health under grant numbers P41 GM103545 and R24 GM136986 (MacLeod).

References

1. Gulrajani, R.: The forward and inverse problems of electrocardiography. IEEE Eng. Med. Biol. Mag. **17**(5), 84–101 (1998)
2. Plonsey, R., Fleming, D.G.: Bioelectric Phenomena. McGraw-Hill, New York (1989)
3. Messinger-Rapport, B.J., Rudy, Y.: The inverse problem in electrocardiography: a model study of the effects of geometry and conductivity parameters on the reconstruction of epicardial potentials. IEEE Trans. Biomed. Eng. **33**(7), 667–676 (1986)
4. Brooks, D., et al.: Inverse electrocardiography by simultaneous imposition of multiple constraints. IEEE Trans. Biomed. Eng. **46**(1), 3–18 (1999)
5. Serinagaoglu, Y., Brooks, D.H., MacLeod, R.S.: Improved performance of Bayesian solutions for inverse electrocardiography using multiple information sources. IEEE Trans. Biomed. Eng. **53**(10), 2024-2034 (2006)
6. Wang, L., et al.: Physiological-model-constrained noninvasive reconstruction of volumetric myocardial transmembrane potentials. IEEE Trans. Biomed. Eng. **57**(2), 296–315 (2010)
7. Gyawali, P.K., et al.: Sequential factorized autoencoder for localizing the origin of ventricular activation from 12-lead electrocardiograms. IEEE Trans. Biomed. Eng. **67**(5), 1505–1516 (2020)
8. Ghimire, S., Wang, L.: Deep generative model and analysis of cardiac transmembrane potential. In: Computing in Cardiology Conference (CinC) (2018)
9. Ghimire, S., et al.: Non-invasive reconstruction of transmural transmembrane potential with simultaneous estimation of prior model error. IEEE Trans. Med. Imaging **38**(11), 2582–2595 (2019)
10. Bacoyannis, T., Krebs, J., Cedilnik, N., Cochet, H., Sermesant, M.: Deep learning formulation of ECGI for data-driven integration of spatiotemporal correlations and imaging information. In: Coudière, Y., Ozenne, V., Vigmond, E., Zemzemi, N. (eds.) FIMH 2019. LNCS, vol. 11504, pp. 20–28. Springer, Cham (2019). https://doi.org/10.1007/978-3-030-21949-9_3
11. Kingma, D.P., Welling, M.: Auto-encoding variational Bayes. In: Proceedings of the International Conference on Learning Representations (ICLR) (2014)
12. Burton, B.M., et al.: A toolkit for forward/inverse problems in electrocardiography within the SCIRun problem-solving environment. In: Engineering in Medicine and Biology Society (2011)
13. Tikhonov, A.N., Arsenin, V.Y.: Solution of ill-posed problems. Mathematics of Computation (1977)
14. Pesteie, M., et al.: Adaptive augmentation of medical data using independently conditional variational auto-encoders. IEEE Trans. Med. Imaging **38**(12), 2807–2820 (2019)
15. Powell, M.J.D.: The BOBYQA algorithm for bound constrained optimization without derivative. Technical report, CMS University of Cambridge (2009)
16. Cartis, C., et al.: Improving the flexibility and robustness of model-based derivative-free optimization solvers. ACM Trans. Math. Softw. **45**, 3 (2019)
17. Cartis C., et al.: Escaping local minima with derivative-free methods: a numerical investigation. Technical report, University of Oxford (2018)
18. Coll-Font, J., Brooks, D.H.: Tracking the position of the heart from body surface potential maps and electrograms. Front. Physiol. **9**, 1727 (2018)

19. Shome, S., et al.: Ischemic preconditioning protects against arrhythmogenesis through maintenance of both active as well as passive electrical properties in ischemic canine hearts. J. Electrocardiol. **40**(6), 150–159 (2007)

20. Aliev, R.R., Panfilov, A.V.: A simple two-variable model of cardiac excitation. Chaos, Solitons Fractals **7**(3), 293–301 (1996)

EP-Net 2.0: Out-of-Domain Generalisation for Deep Learning Models of Cardiac Electrophysiology

Victoriya Kashtanova[1,2], Ibrahim Ayed[3,5], Nicolas Cedilnik[1],
Patrick Gallinari[3,4], and Maxime Sermesant[1,2(✉)]

[1] Inria, Université Côte d'Azur, Nice, France
maxime.sermesant@inria.fr
[2] 3IA Côte d'Azur, Sophia Antipolis, France
[3] Sorbonne University, LIP6, Paris, France
[4] Criteo AI Lab, Paris, France
[5] Theresis Lab, Thales, Paris, France

Abstract. Cardiac electrophysiology models achieved good progress in simulating cardiac electrical activity. However, it is still challenging to leverage clinical measurements due to the discrepancy between idealised models and patient-specific conditions. In the last few years, data-driven machine learning methods have been actively used to learn dynamics and physical model parameters from data. In this paper, we propose a principled deep learning approach to learn the cardiac electrophysiology dynamics from data in the presence of scars in the cardiac tissue slab. We demonstrate that this technique is indeed able to reproduce the transmembrane potential dynamics in situations close to the training context. We then focus on evaluating the ability of the trained networks to generalize outside their training domain. We show experimentally that our model is able to generalize to new conditions including more complex scar geometries, multiple signal onsets and various conduction velocities.

Keywords: Electrophysiology · Deep learning · Simulation

1 Introduction

Mathematical modelling of the heart has been an active research area for the last decades, and it is now more and more coupled with artificial intelligence approaches, see for instance [12]. Among the multi-physics phenomena involved in the cardiac function, cardiac electrophysiology models can accurately reproduce electrical behaviour of cardiac cells. However, it is still challenging to leverage clinical measurements due to the discrepancy between idealised models and patient-specific conditions. Machine learning (ML) approaches could help alleviate these difficulties.

The idea of leveraging ML methods in order to learn data-driven models of dynamical systems is not new: [14] gives a thorough introduction to the closely

© Springer Nature Switzerland AG 2021
D. B. Ennis et al. (Eds.): FIMH 2021, LNCS 12738, pp. 482–492, 2021.
https://doi.org/10.1007/978-3-030-78710-3_46

related field of Nonlinear System Identification while [6] gives an early example of such endeavours. More recently, those questions have seen a renewed interest with works such as [15,16,20] proposing to use Deep Neural Network models for solving differential equations while [1,19,22] use alternative statistical learning tools such as Gaussian Processes and sparse linear regressions to learn the explicit form of differential equations. In the last few years, Neural Networks have been increasingly used in order to learn dynamical models from data: [10,11] endow neural layers with additional structure, useful for learning PDEs while [2,5] use the adjoint method to learn differential equations parametrised with neural models and learn them in fully and partially observable settings. More generally, [21] propose a broader survey of ML in physics-based modeling.

We propose a framework for learning cardiac electrophysiology dynamics from data and we experimentally evaluate its ability to forecast cardiac dynamics on new conditions, unseen during training. Our models are trained and evaluated using data simulated from an electrophysiology model [13]. This is a classical experimental setting in the domain [7] which, although offering a simplification over real cardiac data, allows us to assess our framework using controlled conditions. This work builds on initial results [2,3] that were evaluated in an idealised setting with simple boundary conditions corresponding to a healthy slab of cardiac tissue with a uniform conductivity and a single onset of transmembrane potential.

(a)   (b)

Fig. 1. Example of transmembrane potential (yellow) propagation in the cardiac tissue slab in absence (a) and presence (b) of scar tissue, through successive time steps. (Color figure online)

Our extension here introduces more complex conditions. First, we consider diffusion in tissues with ischaemic (non conductive) regions denoted scars in the following. In clinical practice it is essential to be able to recognise and to estimate the impact of scars because they are the main cause of cardiac arrhythmias. For example, in Fig. 1 we can clearly see the changes in the dynamics of the depolarisation wave in the presence of scar tissue (black area). Second, we introduce in our simulations multiple onsets and colliding fronts, as it is a classical situation in cardiac electrophysiology. The focus of the paper is on the evaluation of the ability of our model to generalise to unseen conditions. The model is then trained on simulated data corresponding to relatively simple context (one type of scar, one front and set of several discrete conduction velocities) and its generalisation ability is evaluated on more challenging contexts like more complex scars, multiple fronts and any real conduction velocity sampled from a given interval. Figure 2 presents the general experimental setting used in the manuscript.

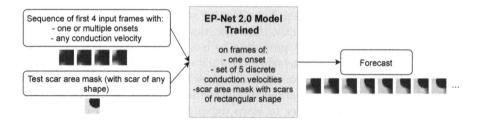

Fig. 2. General setting used throughout the manuscript. Once trained, the EP-Net 2.0 model takes as input a context consisting of a few (4 here) observations plus an indication of the scar area (left), and forecasts the depolarization wave dynamics (right).

2   Problem Formulation and Learning Framework

We used the Mitchell–Schaeffer model [13] for cardiac electrophysiology simulation. This two variables model has been successfully used in patient–specific modelling [18]. The variable v in Eq. 1 represents normalised ($v \in [0,1]$) dimensionless transmembrane potential while the "gating" variable h controls the repolarisation:

$$\partial_t v = div\,(\sigma \mathbf{I} \nabla v) + \frac{hv^2(1-v)}{\tau_{in}} - \frac{v}{\tau_{out}} + J_{stim}$$

$$\partial_t h = \begin{cases} \frac{1-h}{\tau_{open}} & \text{if } v < v_{gate} \\ \frac{-h}{\tau_{close}} & \text{if } v > v_{gate} \end{cases} \tag{1}$$

In practice, since h is a hidden variable, it is difficult to measure. Only the measurement of the potential v is available. Therefore, as in [3], we modify the system (1) by replacing variable h with an observation operator $\mathcal{H}$ which extracts the corresponding information from the current state X_t. This allows us to rewrite this model in a vector form:

$$\begin{cases} X_0 = g_\theta(V_{-k}) \\ \dfrac{dX_t}{dt} = F_\theta(X_t) \\ V_t = \mathcal{H}(X_t) \end{cases} \tag{2}$$

where $X = (V, H)^T$ is a spatio–temporal two–dimensional vector field over the domain $\Omega \subset \mathbb{R}^2$, g_θ and F_θ are parameterised functions which allow to model the ODE governing the dynamics of X and $V_{-k} = (V_{-k+1}, ..., V_0)$ is the sequence of past observations of transmembrane potential.

We then introduce a constraint corresponding to the presence of the scar:

$$\Omega_{scar} \subset \Omega \subset \mathbb{R}^2 : (X_t)_{\Omega_{scar}} \equiv 0 \tag{3}$$

Note that in our setting, scars are considered as binary masks for simplification.

In order to enforce the constraints (2) and (3), we compare the sequence of observations V_t generated by the parameterised model to data simulated from the actual equations and minimise the following loss:

$$\mathcal{L}(V, \widetilde{V}) = \mathcal{L}_{obs}(V, \widetilde{V}) + \lambda_{scar}\mathcal{L}_{scar}(\widetilde{V}), \tag{4}$$

$$\text{where: } \mathcal{L}_{obs}(V, \widetilde{V}) = \int_0^T \|V_t - \widetilde{V}_t\|^2 dt, \; \mathcal{L}_{scar}(\widetilde{V}) = \|\Omega_{scar} \odot \widetilde{V}_t\|^2$$

with $\odot$ the element-wise product and λ_{scar} a hyper-parameter used to balance the losses. We can then frame the statistical learning problem as:

$$\begin{aligned}
\underset{\theta}{\text{minimize}} \quad & \mathbb{E}_{V \in \text{Dataset}}\mathcal{L}(V, \mathcal{H}(X^\theta)) \\
\text{subject to} \quad & \frac{dX_t}{dt} = F_\theta(X_t), \\
& X_0 = g_\theta(V_{-k})
\end{aligned} \tag{5}$$

Learning Method. Operators F, g in problem (5) are implemented via Deep Neural Networks. We chose to use a ResNets [8] (illustrated in Fig. 3) to parameterise both F and g. Optimisation is performed via stochastic gradient descent, precisely ADAM algorithm [9], according to the following algorithm:

0. Randomly initialise θ (denoting the parameters of F and g);
1. Solve the forward state Eq. 2 to find X^θ with an explicit differentiable solver (Euler scheme);
2. Get the gradient of $\theta \to \mathbb{E}_{V \in \text{Dataset}} \left[\mathcal{J}(V, \mathcal{X}^\theta) \right]$ with automatic differentiation tools and update θ;
3. Repeat from step 1 until convergence.

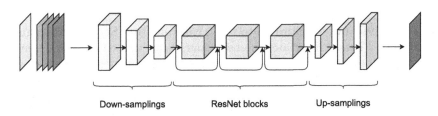

Down-samplings   ResNet blocks   Up-samplings

Fig. 3. The ResNet architecture used in EP-Net 2.0. It has 5 input frames (mask frame plus 4 frames of transmembrane potential) and 1 output frame of forecast.

3   Experiments

Data Collection. We generated 2D data frames using the Lattice Boltzmann method to solve the EP model [17] on a Cartesian grid. The Mitchell-Schaeffer model parameters are taken as in the original paper [13]: $\tau_{in} = 0.3$, $\tau_{out} = 6$, $\tau_{open} = 120$, $\tau_{close} = 150$, $v_{gate} = 0.13$. The computational domain represents a slab of cardiac tissue of size 24×24 mm^2 discretised with 1 mm^2 pixels. A stimulation current was applied for 10 ms to initiate the propagation in selected pixels (J_{stim}). We superposed a mask with randomly generated rectangular area (with random size and position) of excluded domain to simulate the cardiac scars. Training was performed with 5 different discrete conduction velocities, each corresponding to a given parameter of conductivity ($\sigma = 1, 2, 3, 4, 5$) in Eq. 1. The conductivity was applied uniformly on whole cardiac slab except scar area. One value of σ is used per simulation. The simulations were conducted for 30 ms, with a discrete time step of 0.1 ms, and stored every ms. Then random sequences of 10 data frames (one data sample) were extracted at different time points for training / validation data. Overall we have a database of 30000 training and 12000 validation samples.

Training Settings. Parameter λ_{scar} in loss (4) was set to 0.1 and a learning rate for ADAM optimiser was set to 10^{-3}. We use ResNet with 64 filters at the initial stage, three downsampling initial layers and three intermediary blocks and start with a reweighted orthogonal initialisation for its parameters. We also use exponential scheduled sampling [4] with parameter 0.9999 during training. We trained our EP-Net 2.0 model until full model convergence (about 5000 epochs). In each training (and validation) sequences of data frames we used the first 4 frames for initialisation and the rest (6 frames) to compute the losses.

3.1 Results

Tests were performed in two situations: first with scar and current distributions similar to the training set, second with different scars and initial current onsets distributions in order to test the model ability to generalise to new situations.

Testing Environment Similar to the Training One: Scars of Rectangular Shape. Figure 4 illustrates the behaviour of our trained EP-Net 2.0 model in test conditions similar to the training ones: rectangular scars with random size and position plus one onset only. The Fig. 4a shows the forecast over 9 time frames (9 ms) after assimilating the first 4 frames (not presented in the Fig. 4a). We observe very good agreement with the ground truth on this forecast, which represents an important part of cardiac dynamics within this virtual slab of tissue, from early depolarisation to full depolarisation. Figure 4b shows that EP-Net 2.0 model has a very good precision on depolarization during more than 50 ms, an equilibrium state for the model, but cannot predict a repolarisation (Fig. 4c). Quantitative results provided in the Table 1 for different forecasting horizons T (6, 12 and 24 ms) show excellent performance, while the training time horizon was only 6 ms.

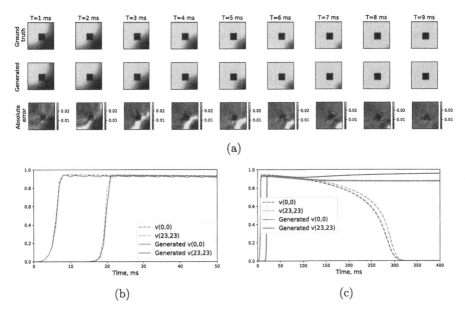

(a)

(b)   (c)

Fig. 4. (a) Results of trained EP-Net 2.0 model (9 ms of forecast, cardiac slab conductivity $\sigma = 2$). (b, c) Transmembrane potential graph at the leftmost upper point $(0, 0)$ and the rightmost bottom point $(23, 23)$ in the slab with different forecasting horizons.

Generalisation Ability of EP-Net 2.0: Scars of Various Shapes and Multiple Onsets. Our objective is to train models able to generalise to conditions outside the training environment. This is important since for example different patients will have different characteristics. In order to evaluate the capability of our model to generalise we performed two types of tests, one with scars with different shapes when training was performed only with rectangular shapes, and one with multiple onset when training considered only one onset.

As for the generalisation to different scar shapes, we evaluated our model with triangular, circular and complex scars (see Fig. 5). Table 1 shows that the model performs well on the different shapes. The errors are slightly larger than for the rectangular scars used for training, but remain low. They however increase for long term prediction (24 ms here). Figure 5 illustrates the behaviour of the model for typical test sequences.

We also performed tests with multiple onsets. The model shows good results for forecasting of multiple depolarisation waves on one cardiac slab tissue (Fig. 6), which is essential to capture correctly for ventricular tachycardia simulation. As one can see from the Table 2, relative mean-squared error is larger for multiple onsets than for one onset (as it was used for training) but still acceptable. However, this error does not increase proportionately to time of forecast (like in the Table 1), because the virtual slab reaches faster the full depolarisation with multiple onsets, an equilibrium state for EP-Net 2.0 model.

Table 1. Relative mean-squared error (MSE) of transmembrane potential forecasting in presence of scars of various forms for different forecasting horizons (cardiac slab conductivity $\sigma = 2$).

	MSE (6 ms)	MSE (12 ms)	MSE (24 ms)
Rectangular shape	$1.8 * 10^{-4}$	$4.45 * 10^{-4}$	$6,8 * 10^{-4}$
Triangular shape	$3.1 * 10^{-4}$	$8 * 10^{-4}$	$1.36 * 10^{-3}$
Circular shape	$2.7 * 10^{-4}$	$8.2 * 10^{-4}$	$3.4 * 10^{-3}$
Complex shape	$4.6 * 10^{-4}$	$1.9 * 10^{-3}$	$6.36 * 10^{-3}$

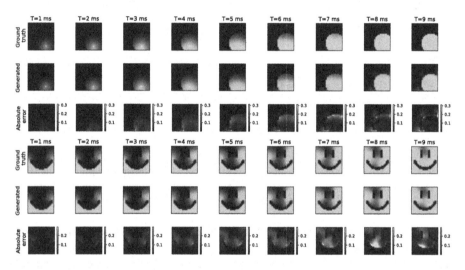

Fig. 5. Results of trained EP-Net 2.0 model on scar with circular (top three rows) and complex (bottom three rows) shape (9 ms of forecast, cardiac slab conductivity $\sigma = 2$).

Table 2. Relative mean-squared error (MSE) of potential forecasting in presence of multiple onsets and scar of rectangular form for different forecasting horizons (cardiac slab conductivity $\sigma = 2$).

	MSE (6 ms)	MSE (12 ms)	MSE (24 ms)
One onset	$1.8 * 10^{-4}$	$4.45 * 10^{-4}$	$6,8 * 10^{-4}$
Multiple onsets	$4.7 * 10^{-4}$	$5.8 * 10^{-4}$	$6.9 * 10^{-4}$

Generalisation Ability of EP-Net 2.0: Various Conduction Velocities.
To estimate the ability of EP-Net 2.0 model to learn the conduction velocity of cardiac tissue we performed tests with various cardiac slab conductivities (σ). The tests have been performed with sigma values used for training ($\sigma \in 1, 2, 3, 4, 5$), and sigma values uniformly sampled from the interval $[0.7, 6]$., i.e. outside the training set.

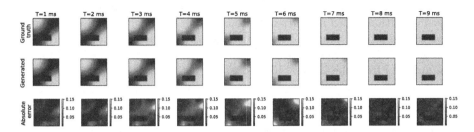

Fig. 6. Results of trained EP-Net 2.0 model with two stimulation currents applied on different pixels and at different times (9 ms of forecast, cardiac slab conductivity $\sigma = 2$).

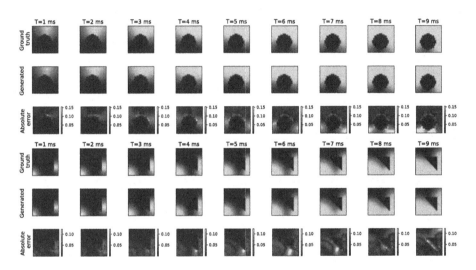

Fig. 7. Results of EP-Net 2.0 model on scar with circular shape and cardiac slab conductivity $\sigma = 3.8$ (top three rows), and on scar with triangular shape, two onsets and cardiac slab conductivity $\sigma = 1.5$ (bottom three rows).

Table 3. Relative mean-squared error (MSE) of potential forecasting in presence of various conduction velocities of cardiac slab and scar of rectangular form for different forecasting horizons.

	MSE (6 ms)	MSE (12 ms)	MSE (24 ms)
$\sigma = 0.7$	$4.65 * 10^{-4}$	$3.95 * 10^{-3}$	$1,9 * 10^{-2}$
$\sigma = 2$	$1.8 * 10^{-4}$	$4.45 * 10^{-4}$	$6,8 * 10^{-4}$
$\sigma = 2.5$	$3.5 * 10^{-4}$	$1.4 * 10^{-3}$	$1,6 * 10^{-4}$
$\sigma = 6$	$2 * 10^{-3}$	$4.7 * 10^{-3}$	$3 * 10^{-3}$

As shown in the Fig. 7, EP-Net 2.0 model keeps the capability to generalise to unseen conditions, such as scars of various shapes and multiple onsets. Quantitative results provided in the Table 3 show that model achieves a good precision in forecasting depolarisation waves in cardiac tissue slabs for any conductivity (real number) from the interval $[0.7, 6]$.

3.2 Limitations and Discussion

The Sect. 3.1 shows the ability of model to learn the local dynamics and to generalise to unseen conditions.

Although our approach can achieve compelling results in many cases, there are still limitations. For example, as shown in the Fig. 8, EP-Net 2.0 model does not work properly on thin scars (thickness less than 2 pixels) and produces additional transmembrane potential diffusion through the scar from generated noise. The current model has been trained only to model depolarization of the cardiac slab tissue and cannot predict its repolarisation (see Fig. 4c). This is left for future work.

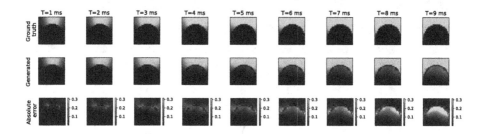

Fig. 8. Results of trained EP-Net 2.0 model on thin scar with circular shape.

4   Conclusion and Future Work

In this paper, we proposed the EP-Net 2.0 Deep Learning model to learn the cardiac electrophysiology dynamics in presence of complex initial boundary conditions (like scar area, multiple onsets and various conduction velocities). The obtained results show a great generalisation ability of this model to unseen conditions. Despite training on data with scars of only rectangular shape and one onset activation, EP-Net 2.0 model works on data with scars of any possible geometric shape and multiple onsets, even when normalised transmembrane potential stimulations were applied at different moments[1].

We believe that in future our approach can help upgrade and personalise mathematical model via additional data. However in clinical practice tissue properties are not binary and real data is always noised and sparse, future work includes looking into more complex formulations considering model parameter

---

[1] Visit our github page with trained EP-Net-2.0, for more detailed examples.

identification for continuously varying cardiac tissue properties and into possible strategies to complete real data via simulation data.

Acknowledgments. This work has been supported by the French government, through the 3IA Côte d'Azur Investments in the Future project managed by the National Research Agency (ANR) with the reference number ANR-19-P3IA-0002 and through the "Research and Teaching chairs in artificial intelligence (AI Chairs)" funding for DL4Clim project. The research leading to these results has also received European funding from the ERC starting grant ECSTATIC (715093). The authors are grateful to the OPAL infrastructure from Université Côte d'Azur for providing resources and support.

References

1. Alvarez, M.A., Luengo, D., Lawrence, N.D.: Linear latent force models using gaussian processes. IEEE Pattern Anal. Math. Intell. **35**(11), 2693–2705 (2013)
2. Ayed, I., de Bézenac, E., Pajot, A., Brajard, J., Gallinari, P.: Learning dynamical systems from partial observations. arXiv preprint:1902.11136 (2019)
3. Ayed, I., Cedilnik, N., Gallinari, P., Sermesant, M.: EP-net: learning cardiac electrophysiology models for physiology-based constraints in data-driven predictions. In: Coudière, Y., Ozenne, V., Vigmond, E., Zemzemi, N. (eds.) FIMH 2019. LNCS, vol. 11504, pp. 55–63. Springer, Cham (2019). https://doi.org/10.1007/978-3-030-21949-9_7
4. Bengio, S., Vinyals, O., Jaitly, N., Shazeer, N.: Scheduled sampling for sequence prediction with recurrent neural networks. arXiv preprint:1506.03099 (2015)
5. Chen, R.T.Q., Rubanova, Y., Bettencourt, J., Duvenaud, D.: Neural ordinary differential equations. In: Proceedings of Neural Information Processing Systems (2018)
6. Crutchfield, J.P., McNamara, B.: Equations of motion from a data series. Complex Syst. **1**(417–452), 121 (1987)
7. Fresca, S., Manzoni, A., Dedè, L., Quarteroni, A.: Deep learning-based reduced order models in cardiac electrophysiology. PLOS ONE **15**(10), e0239416 (2020)
8. He, K., Zhang, X., Ren, S., Sun, J.: Deep residual learning for image recognition. In: IEEE Conference CVPR, pp. 770–778 (2016)
9. Kingma, D.P., Ba, J.: Adam: a method for stochastic optimization. arXiv preprint:1412.6980 (2014)
10. Long, Z., Lu, Y.Y., Dong, B.: PDE-net 2.0: Learning PDEs from data with a numeric-symbolic hybrid deep network. J. Comput. Phys. **399**, 108925 (2019)
11. Long, Z., Lu, Y., Ma, X., Dong, B.: PDE-net: Learning PDEs from data. In: International Conference on ICML, pp. 3208–3216. PMLR (2018)
12. Mansi, T., Passerini, T., Comaniciu, D.: Artificial Intelligence for Computational Modeling of the Heart. Elsevier (2020)
13. Mitchell, C.C., Schaeffer, D.G.: A two-current model for the dynamics of cardiac membrane. Bull. Math. Biol. **65**(5), 767–793 (2003)
14. Nelles, O.: Nonlinear System Identification. Springer, Heidelberg (2001). https://doi.org/10.1007/978-3-662-04323-3
15. Raissi, M.: Deep hidden physics models: deep learning of nonlinear partial differential equations. J. Mach. Learn. Res. **19**(1), 932–955 (2018)

16. Raissi, M., Perdikaris, P., Karniadakis, G.E.: Machine learning of linear differential equations using gaussian processes. J. Comput. Phys. **348**, 683–693 (2017)
17. Rapaka, S., et al.: LBM-EP: Lattice-Boltzmann method for fast cardiac electrophysiology simulation from 3D images. In: Ayache, N., Delingette, H., Golland, P., Mori, K. (eds.) MICCAI 2012. LNCS, vol. 7511, pp. 33–40. Springer, Heidelberg (2012). https://doi.org/10.1007/978-3-642-33418-4_5
18. Relan, J., et al.: Coupled personalization of cardiac electrophysiology models for prediction of ischaemic ventricular tachycardia. Interface Focus **1**(3), 396–407 (2011)
19. Rudy, S.H., Brunton, S.L., Proctor, J.L., Kutz, J.N.: Data-driven discovery of partial differential equations. Sci. Adv. **3**(4), e1602614 (2017)
20. Sirignano, J., Spiliopoulos, K.: DGM: a deep learning algorithm for solving partial differential equations. J. Comput. Phys. **375**, 1339–1364 (2018)
21. Willard, J.D., Jia, X., Xu, S., Steinbach, M., Kumar, V.: Integrating physics-based modeling with machine learning: a survey. arXiv preprint:2003.04919 (2020)
22. Zhang, S., Lin, G.: Robust data-driven discovery of governing physical laws with error bars. Proc. R. Soc. Math. Phys. Eng. Sci. **474**(2217), 20180305 (2018)

Simultaneous Multi-heartbeat ECGI Solution with a Time-Varying Forward Model: A Joint Inverse Formulation

Jake A. Bergquist[1](✉)[iD], Jaume Coll-Font[2][iD], Brian Zenger[1][iD],
Lindsay C. Rupp[1][iD], Wilson W. Good[1][iD], Dana H. Brooks[3],
and Rob S. MacLeod[1][iD]

[1] Biomedical Engineering Department, University of Utah,
Salt Lake City, UT 84112, USA
jbergquist@sci.utah.edu
[2] Cardiovascular Bioengineering and Imaging Lab, Massachusetts General Hospital,
and Harvard Medical School, Boston, MA, USA
[3] Department of Electrical and Computer Engineering, Northeastern University,
Boston, MA, USA

Abstract. Electrocardiographic imaging (ECGI) is an effective tool for
noninvasive diagnosis of a range of cardiac dysfunctions. ECGI leverages
a model of how cardiac bioelectric sources appear on the torso surface
(the forward problem) and uses recorded body surface potential signals
to reconstruct the bioelectric source (the inverse problem). Solutions to
the inverse problem are sensitive to noise and variations in the body surface potential (BSP) recordings such as those caused by changes or errors
in cardiac position. Techniques such as signal averaging seek to improve
ECGI solutions by incorporating BSP signals from multiple heartbeats
into an averaged BSP with a higher SNR to use when estimating the
cardiac bioelectric source. However, signal averaging is limited when it
comes to addressing sources of BSP variability such as beat to beat
differences in the forward solution. We present a novel joint inverse formulation to solve for the cardiac source given multiple BSP recordings
and known changes in the forward solution, here changes in the heart
position. We report improved ECGI accuracy over signal averaging and
averaged individual inverse solutions using this joint inverse formulation
across multiple activation sequence types and regularization techniques
with measured canine data and simulated heart motion. Our joint inverse
formulation builds upon established techniques and consequently can
easily be applied with many existing regularization techniques, source
models, and forward problem formulations.

Keywords: Electrocardiographic imaging · Signal averaging · Inverse
problems

Supported by NIH NHLBI grant no. 1F30HL149327; NIH NIGMS Center for
Integrative Biomedical Computing (www.sci.utah.edu/cibc), NIH NIGMS grants
P41GM103545 and R24 GM136986; and the Nora Eccles Treadwell Foundation for
Cardiovascular Research.

© Springer Nature Switzerland AG 2021
D. B. Ennis et al. (Eds.): FIMH 2021, LNCS 12738, pp. 493–502, 2021.
https://doi.org/10.1007/978-3-030-78710-3_47

1   Introduction

Electrocardiographic imaging (ECGI) has emerged as an effective modality for noninvasive diagnosis of a range of cardiac electrical dysfunctions, *e.g.*, for guiding ablation procedures and for predicting characteristics of arrhythmic substrate, with novel implementations under continual development [8]. ECGI has also benefited from a wide variety of technical improvements [2,6,8]. A persistent challenge in ECGI is the high degree of variability in electrogram reconstruction accuracy, which limits clinical confidence and consequently utility [2,9]. In this study we pursued one opportunity to decrease this variability by leveraging the availability, in most practical settings, of recordings of multiple consecutive heartbeats.

ECGI is organized generally into a forward problem—estimation of the body surface potentials (BSPs) due to known cardiac sources given the geometry of the torso—and an inverse problem—the estimation of cardiac sources given the geometry and BSPs. Like all forms of ECGI, (we treat here the estimation of heart surface electrograms (EGMs) from BSPs), the main challenge is solving an ill-posed inverse problem, meaning that errors such as noise in the BSPs or uncertainty in the forward solution can have dramatic and unpredictable effects on the resulting EGM reconstructions [12]. Despite the many approaches that have been proposed to address this ill-posedness, including a variety of regularization approaches, source model formulations, and signal processing techniques, [1,3,12] there have been few studies of how best to use recordings of multiple consecutive heartbeats in ECGI. Signal averaging of the BSPs over time is one viable option *i.e.*, averaging the available heartbeats and then applying ECGI to the averaged signals. This method promises improved SNR and thus more robust ECGI solutions. However, there is a weakness in the basic assumption underlying standard signal averaging that the source electrograms and the forward model for each beat are the same and that differences in BSP signals are due only to additive noise that is uncorrelated across heartbeats; signal averaging approaches exclude the possible presence of correlated noise and ignore any model changes due to underlying physiological parameters such as cardiac position. An approach that could address changes such as varying cardiac position would be to solve individual inverse solutions for each heart beat, then average the resulting inverse estimations into a single estimate. This tehcnique however does not leaverage improving the BSP SNR before inverse estimation, and thus is open to weaknesses introduced by instability in the inverse estimation.

We propose here an alternate approach that extends signal averaging to take into account other sources of variability. In this extension we construct the ECGI inverse problem to directly take in all available BSP signals and forward solutions from multiple heartbeats and return a single set of EGMs as the solution. This formulation replaces the distinct steps of signal averaging followed by inverse solution computation with a single combined step that utilizes common information from multiple beats of BSPs to identify inverse solutions that match the population of input heartbeats. Additionally this approaches captures additional information such as beat to beat variations in the underlying forward model due to changes in cardiac position. In this study we describe and assess this formulation by comparing to solutions using signal averaging. We have evaluated its performance using a synthetic dataset in which beat to beat cardiac motion was simulated, and accurate cardiac positions were known. We found that our joint inverse method produced superior inverse reconstructions of EGMs across several activation sequences and regularization techniques in comparison to both signal averaging and individual inverse solutions.

2 Methods

2.1 Inverse Strategies

We implemented three approaches for computing inverse solutions: using our joint inverse method, using traditional signal averaging, and finally performing individual inverse solutions on a beat by beat basis then averaging the resulting reconstructions.

Joint Inverse Formulation. To estimate a single cardiac source from a series of BSPs we implemented what we refer to as a "joint inverse" formulation of ECGI. We start with a standard, regularized inverse solution based on extracellular potential sources (Φ_H^*) defined over the heart surface and a matrix (A) that contains the forward transfer coefficients to torso surface potentials (Φ_T), with regularization by some matrix R with some weight λ expressed as

$$\arg\min_{\Phi_H^*} ||\Phi_T - A\Phi_H^*||_2^2 + \lambda||R\Phi_H^*||_2^2. \tag{1}$$

To incorporate multiple BSP recordings and variations in the forward model, we replace Φ_T and A with the block matrices $\hat{\Phi}_T$ and $\hat{A}$ and write the system as

$$\hat{\Phi}_T = \begin{bmatrix} \phi_T^1 \\ \phi_T^2 \\ \phi_T^3 \\ ... \\ \phi_T^N \end{bmatrix}, \hat{A} = \begin{bmatrix} A(b_1) \\ A(b_2) \\ A(b_3) \\ ... \\ A(b_N) \end{bmatrix}. \tag{2}$$

The submatrices of $\hat{\Phi}_T$ are the individual Φ_T BSP recordings for the N recorded heart beats. The submatrices of $\hat{A}$ are the forward matrices computed as a function of the geometry during a specific beat b_n.

This expanded formulation allows for the incorporation of beat-specific parameters into the forward calculation, such as heart movement with respiration, and leads to the modified optimization expression

$$\arg\min_{\Phi_H^*} ||\hat{\Phi}_T - \hat{A}\Phi_H^*||_2^2 + \lambda||R\Phi_H^*||_2^2. \tag{3}$$

Signal Averaging: We compared the joint inverse formulation to a traditional signal averaging based approach. Heartbeats were aligned by QRS onset and length normalized by fitting a cubic spline to each time signal then sampling at 300 evenly spaced intervals before being arithmetically averaged per electrode into a single BSP signal. Regularization was applied in the same manner as for the joint inverse, described below.

Individual Inverse Average: As a final comparison we computed individual inverse solutions on a beat by beat basis and then averaged the resulting solutions into a single final EGM estimate. Like before the signals were length normalized. We then computed inverse solutions with regularization as described below. Finally the resulting inverse solutions for beat were averaged into a single EGM estimate.

Regularization: We implemented the joint inverse approach, the standard signal averaging approach, and the individual inverse solutions with three different regularization techniques. The first two were Tikhonov 0 order (T0), Tikhonov 2nd order (T2). In addition we computed truncated singular value decomposition (TSVD) solutions, which have a formulation somewhat different from the one given above but can be related formally using "filter factors" [11]. We compared results for all

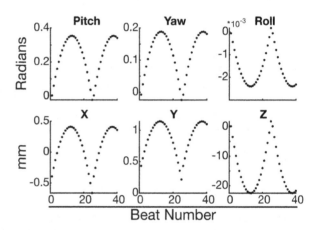

Fig. 1. Cardiac position parameters for 40 heartbeats. Pitch yaw and roll are defined (in radians) about the septal axis with pitch being rotation about the X axis, yaw rotation about the Y, and roll rotation about the septal axis. X, Y, and Z (mm) are defined as translations of the centroid of the cardiac geometry. The position 0,0,0,0,0,0 corresponds to the original registered position of the cardiac geometry.

these methods between joint and signal averaging approaches. For each technique, the L-curve method provided the regularization parameter (λ for T0 and T2, truncation level for TSVD). In the case of T2 regularization, the Laplacian matrix was badly conditioned and the singular value spectrum decayed so rapidly that the L curve lacked a sharp corner. To remedy this, we regularized the Laplacian by computing the SVD of the Laplacian, added a small constant (0.001) to the singular values that were less than 1e–3, and reconstructed the Laplacian from these modified SVD components. Forward calculations were performed using boundary element tools implemented in MATLAB within the SCIRun Forward-Inverse Toolkit [7].

2.2   Evaluation

Validation Data. We designed a synthetic dataset based on an experiment using a modified Langendorff preparation described in Bergquist *et al.* [6], but which included artificial variation of the position of the heart. Briefly, an isolated canine heart, suspended in a human-shaped torso tank, was instrumented with a 240-electrode ventricular epicardial sock. Recordings were performed during sinus rhythm (sinus) as well as during electrical pacing of the anterior left ventricular free wall (aVP) and the posterior left ventricular free wall (pVP). The recorded electrograms were processed using the open-source MATLAB-based PFEIFER software [13]. For each activation sequence (Sinus, aVP, pVP), 40 heartbeats were selected, totaling 120 heartbeats.

Post experiment MRI images of the heart were segmented and a cardiac mesh was generated using Seg3D (www.sci.utah.edu/cibc-software/seg3d.html) and Cleaver (www.sci.utah.edu/cibc-software/cleaver.html), open source segmentation and meshing tools. The epicardial sock geometry was registered along with the heart mesh into a surface mesh of the torso using GRÖMeR, a toolkit for registration of electrode arrays [5]. The epicardial sock collects signals only from the ventricular epicardium, resulting in an open surface mesh unsuitable for forward solution calculation by the boundary element method. To remedy this, a closed sock geometry was created by fitting an ellipsoid to the sock geometry and generating nodes to close the sock surface.

To create controlled cardiac motion in the data set, we parameterized the position and orientation of the heart geometry with six degrees of freedom as described by Coll-Font *et al.* [10]: the X, Y, and Z coordinates of the centroid and three rotations about a septal base-to-apex axis (pitch, yaw, and roll). We generated 40 cardiac positions by sampling a respiratory function fit to the position of the heart of a human subject during cine MRI (Fig. 1) [4,15]. Torso surface potentials at the 771 nodes of the tank geometry were calculated from the recorded EGMs of the moving heart using a potential-based boundary element forward solution. Gaussian white noise was added to the resulting BSP signals at an SNR of 30 dB. This resulted in 120 BSP heartbeats, 40 per activation sequence.

For each activation sequence type, we applied both the joint, signal averaged, and individual inverses as described above. In the case of the joint inverse and individual inverse, the forward matrix A was adjusted on a beat by beat basis according to the instantaneous heart position. In the case of signal averaging, we placed the heart in the first position, our nominal cardiac position.

Error Metrics: We assessed ECGI accuracy according to average spatial correlation (SC), average temporal correlation (TC), and root mean squared error (RMSE). Briefly, spatial correlation was calculated between measured and reconstructed epicardial potentials averaged across time throughout the QRST and

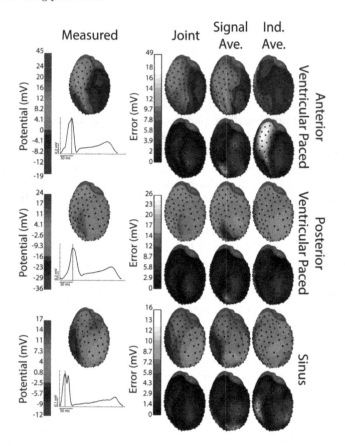

Fig. 2. Measured vs reconstructed epicardial potentials at the peak of the RMS QRS for tikhonov 0 order regularization. For each activation sequence, the top row shows the potential distribution while the second row shows the absolute difference between measured and reconstructed. The RMS electrogram shows the time int ant (red line) that is visualized for each activation sequence. (Color figure online)

temporal correlation between measured and reconstructed electrograms throughout the QRST averaged across all electrodes. RMSE was averaged over both over space and time. We also compared reconstructed and measured electrograms qualitatively. Analysis, and visualizations were prepared in MATLAB (www.mathworks.com) and the SCIRun problem solving environment (www.sci.utah.edu/cibc-software/scirun.html).

3   Results

Figure 2 shows an example of the epicardial potentials reconstructed by joint inverse, signal averaging, and averaged individual inverse solutions as compared to measured signals for Tikhonov 0 order regularization across all activation sequence types. At the peak of the QRS wave we observed shifts in key features of the potential distributions on the signal-averaged reconstructions that were not present in the joint inverse results. Severe shifts and errors are also present in the averaged individual inverse solutions. Additionally the signal averaging and averaged individual inverse frequently exhibited erroneously high maxima and minima potentials in the reconstruction compared to the joint inverse solutions. In the case of the anterior ventricular paced and sinus activation sequences the average individual inverse solution produced errors and potential maxima and minima that required clipping of the color map to not wash out the features of the other reconstructions.

Figure 3 shows comparisons of ECGI accuracy using the joint inverse, signal averaging, and individual inverse averaging across all three activation sequence types and regularization techniques. In all cases, the joint inverse provided superior ECGI reconstruction accuracy, with the aVP activation sequence showing the most pronounced improvement.

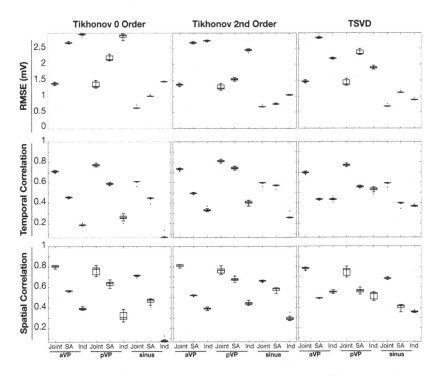

Fig. 3. Box and whisker plots or error metrics. Row 1 shows RMSE, row two shows spatial correlation, and row three shows temporal correlation. Left column shows Tikhonov zero order (T0), middle column 2nd order (T2), right column TSVD. Within each panel aVP, pVP, and sinus are shown left to right while for each comparison joint inverse results are on the left, signal averaging (SA) in the center, and averaged individual solution (Ind) on the right. Black dots indicate outliers.

4   Discussion

In this study we developed a method for including multiple beats in ECGI that extends traditional signal averaging approaches while incorporating beat-to-beat variation in the heart position (or more generally in forward solutions). We evaluated its performance against ECGI carried out after standard signal averaging of BSP signals as well as averaged individual inverse solutions across multiple heartbeats.

We found that the joint inverse produced superior solutions over signal averaging that ignored the changed in heart position /orientation on all observed metrics across all activation sequences and regularization techniques tested (Fig. 3). Additionally, when we computed individual inverse solutions that accounted for cardiac position and averaged these solutions we found that our joint inverse method was superior. Qualitatively, we saw shifts in the EGM potential distributions when using signal averaging or averaged individual inverse with respect to ground truth that were not present using the joint inverse method (Fig. 2). We theorize that improvement over signal averaging is due to the ability of the joint inverse to incorporate additional information, namely the correct beat-to-beat positions of the heart. The joint inverse method relaxes the assumption of a shared forward solution for each heartbeat, but rather allows each heart beat to be paired with an individual forward solution. This removes a source of error present in the signal averaging approach that would otherwise be difficult to address. Even in the averaged individual inverse solution, where the error caused by cardiac motion could be accounted for, we saw inferior performance when compared to joint inverse. We theorize that the improvement over averaged individual inverse solutions is due to simultaneous incorporation of multiple BSP signals, increasing the overall SNR of the input signal to the inverse problem in a manner similar to signal averaging. We also theorize that the joint inverse is a more well conditioned problem than any of the individual inverse solutions. The individual inverse solutions suffered from lower SNR due to a lack of averaging across the BSP signals, resulting in more unstable inverse solutions and thus poorer reconstruction quality even after averaging of these solutions.

This study assumed that per-beat cardiac positions were known and accurate, which is not currently common in clinical ECGI setups. However, it is feasible that cardiac positions could either be directly measured (via ultrasound or similar technology) or estimated on a beat to beat basis—an active area of research [4,10,14]. We also did not investigate the response of the joint inverse to errors in the cardiac position estimations, which we would expect to compromise accuracy. We plan to address the sensitivity of the joint inverse method to the accuracy of the geometric model in future studies.

This joint inverse method modifies existing mathematical formulations for solving the inverse problem (Eq. 1), while still allowing for established regularization methods, source models, and forward problem formulations to be implemented without any additional modification. This flexibility of implementation means that the joint inverse method can be readily applied to improve a wide variety of ECGI applications. Additional body surface signals are readily acquired with minimal additional overhead, meaning the improvements seen in this study may be readily available in other ECGI implementations. This study is limited by use of synthetic data, however the cardiac positions used were based on the realistic range of motion observed for a human heart during respiration. In the future, we plan to examine the response of the joint inverse to variation in the accuracy of the assumed cardiac positions, as well as the application of the joint inverse formulation using other source models, regularization techniques, and forward problem formulations. Future studies will also investigate the ability of the joint inverse method to improve inverse solutions in the context of sources of correlated noise.

References

1. Ahmad, G., Brooks, D., Jacobson, C., MacLeod, R.: Constraint evaluation in inverse electrocardiography using convex optimization. In: Proceedings of the IEEE Engineering in Medicine and Biology Society 17th Annual International Conference, pp. 209–210. IEEE Press (1995)
2. Bear, L.R., et al.: Advantages and pitfalls of noninvasive electrocardiographic imaging. J. Electrocardiol. **57**, S15–S20 (2019). https://doi.org/10.1016/j.jelectrocard.2019.08.007
3. Bear, L., et al.: Effects of ECG signal processing on the inverse problem of electrocardiography. IEEE Comput. Cardiol. **45**, 1–4 (2018)
4. Bergquist, J.A., et al.: Improving localization of cardiac geometry using ECGI. In: 2020 Computing in Cardiology, pp. 1–4 (2020). https://doi.org/10.22489/CinC.2020.273
5. Bergquist, J.A., Good, W.W., Zenger, B., Tate, J.D., MacLeod, R.S.: GRÖMeR: a pipeline for geodesic refinement of mesh registration. In: Coudière, Y., Ozenne, V., Vigmond, E., Zemzemi, N. (eds.) FIMH 2019. LNCS, vol. 11504, pp. 37–45. Springer, Cham (2019). https://doi.org/10.1007/978-3-030-21949-9_5
6. Bergquist, J.A., Zenger, B., Good, W.W., Rupp, L.C., Bear, L.R., MacLeod, R.S.: Novel experimental preparation to assess electrocardiographic imaging reconstruction techniques. In: 2020 Computing in Cardiology, pp. 1–4 (2020). https://doi.org/10.22489/CinC.2020.458
7. Burton, B., et al.: A toolkit for forward/inverse problems in electrocardiography within the SCIRun problem solving environment. In: Proceedings of the IEEE Engineering in Medicine and Biology Society 33rd Annual International Conference, pp. 1–4. IEEE (2011)
8. Cluitmans, M., et al.: Validation and opportunities of electrocardiographic imaging: from technical achievements to clinical applications. Front. Physiol. **9**, 1305 (2018)
9. Cluitmans, M., et al.: In vivo validation of electrocardiographic imaging. JACC Clin. Electrophysiol. **3**(3), 232–242 (2017)
10. Coll-Font, J., Brooks, D.H.: Tracking the position of the heart from body surface potential maps and electrograms. Front. Physiol. **9**, 1727 (2018)
11. Hansen, P.: Rank-deficient and discrete ill-posed problems: numerical aspects of linear inversion. Ph.D. thesis, Technical University of Denmark (1996)
12. Milanic, M., Jazbinsek, V.V., Macleod, R., Brooks, D., Hren, R.: Assessment of regularization techniques for electrocardiographic imaging. J. Electrocardiol. **47**(1), 20–28 (2014)

13. Rodenhauser, A., et al.: Pfeifer: preprocessing framework for electrograms intermittently fiducialized from experimental recordings. J. Open Source Softw. **3**, 472 (2018)
14. Rodrigo, M., et al.: Solving inaccuracies in anatomical models for electrocardiographic inverse problem resolution by maximizing reconstruction quality. IEEE Trans. Med. Imaging **37**(3), 733–740 (2018)
15. Swenson, D., Geneser, S., Stinstra, J., Kirby, R., MacLeod, R.: Cardiac position sensitivity study in the electrocardiographic forward problem using stochastic collocation and boundary element methods. Ann. Biomed. Eng. **39**(12), 2900–2910 (2011)

The Effect of Modeling Assumptions on the ECG in Monodomain and Bidomain Simulations

Dennis Ogiermann[1]([⊠])[ID], Daniel Balzani[1][ID], and Luigi E. Perotti[2][ID]

[1] Chair of Continuum Mechanics, Ruhr-University Bochum, Bochum, Germany
dennis.ogiermann@ruhr-uni-bochum.de
[2] Mechanical and Aerospace Engineering Department, University of Central Florida, Orlando, FL, USA

Abstract. Computing a physiologically accurate electrocardiogram (ECG) is one of the key outcomes of cardiac electrophysiology (EP) simulations. Indeed, the simulated ECG serves as a validation, may be the target for optimization in inverse EP problems, and in general allows to link simulation results to clinical ECG data. Several approaches are available to compute the ECG corresponding to an EP simulation. Lead field approaches are commonly used to compute ECGs from cardiac EP simulations using the Monodomain or Eikonal models. A coupled passive conductor model is instead common when the full Bidomain model is adopted. An approach based on solving an auxiliary Poisson problem propagating the activation field from the heart surface to the torso surface is also possible, although not commonly described in the literature. In this work, through a series of numerical experiments, we investigate the limits of validity of the different approaches to compute the ECG from simulations based on the Monodomain and Bidomain models. Significant discrepancies are observed between the common lead field and direct ECG approaches in most realistic cases – e.g., when conduction anisotropy is included – while the ECG computed via solution of an auxiliary Poisson problem is similar to the direct ECG approach. We conclude that either the direct ECG or Poisson approach should be adopted to improve the accuracy of the computed ECG.

Keywords: Electrocardiogram · Cardiac electrophysiology · Validation criteria · Monodomain model · Bidomain model

1 Introduction

Validation of mathematical models in cardiac electrophysiology is a challenging task, which gained significant attention over the last years. Indeed, satisfying verification and validation criteria is a critical step necessary to translate mathematical models and the respective numerical simulations to clinical practice. Once validated, EP simulations have a vast range of applications from personalized medicine to drug development. Our goal is to move one step further towards

© Springer Nature Switzerland AG 2021
D. B. Ennis et al. (Eds.): FIMH 2021, LNCS 12738, pp. 503–514, 2021.
https://doi.org/10.1007/978-3-030-78710-3_48

504 D. Ogiermann et al.

unlocking these applications by investigating commonly used methods to compute electrocardiograms (ECGs) in numerical studies. The ECG is one of the most widely used diagnostic tools in cardiology for its low cost, simplicity, and amount of information provided on cardiac function and disease. Given its wide adoption and significant relevance in the clinical practice, the ECG is one of the key validation criteria to be satisfied by an EP simulation, while simultaneously respecting physiological activation sequences [9].

The contribution of the current study towards reaching this milestone is to carefully review commonly adopted modeling assumptions to compute the ECG from EP simulations and to highlight discrepancies and their causes. Particular attention will be given to evaluating the lead field approach due to its common use in the literature. Preliminary analyses comparing the lead field approach and the direct ECG have been carried out in [12], while [3] investigated the effect on the ECG of different coupling models between the torso and the heart. In this work we continue this effort with an in depth analysis comparing the lead field approach and directly computed ECGs.

2 Methods

Cardiac electrophysiology and the ECG are commonly simulated using the Monodomain model together with the lead field approach (to recover the ECG) [14]. In comparison, fewer studies solve the Bidomain equations [16] coupled with a passive conductor model for the torso, where the extracellular potential is tied at the torso-heart interface. In this case the ECG is directly computed from the extracellular potential on the surface of the torso.

2.1 Direct ECG

The Bidomain model [16] coupled to a surrounding passive conductor can be stated in parabolic-elliptic form as:

$$\chi C_\mathrm{m} \frac{\partial \varphi_\mathrm{m}}{\partial t} = \nabla \cdot (\boldsymbol{\kappa}_\mathrm{i} \nabla \varphi_\mathrm{m}) + \nabla \cdot (\boldsymbol{\kappa}_\mathrm{i} \nabla \varphi_\mathrm{e}) - \chi I_\mathrm{ion}(\varphi_\mathrm{m}, \mathbf{s}) - \chi I_\mathrm{stim}(t) \quad \text{in } \Omega_\mathbb{H}$$

$$0 = \nabla \cdot (\boldsymbol{\kappa}_\mathrm{i} \nabla \varphi_\mathrm{m}) + \nabla \cdot (\boldsymbol{\kappa}_\mathrm{e} + \boldsymbol{\kappa}_\mathrm{i}) \nabla \varphi_\mathrm{e} \quad \text{in } \Omega_\mathbb{H}$$

$$\frac{\partial \mathbf{s}}{\partial t} = \mathbf{g}(\varphi_\mathrm{m}, \mathbf{s}) \quad \text{in } \Omega_\mathbb{H}$$

$$0 = \nabla \cdot (\kappa_\mathrm{b} \nabla \varphi_\mathrm{b}) \quad \text{in } \Omega_\mathbb{B}$$

with boundary conditions (BCs) at any time $t \in (0, T]$:

$$0 = (\boldsymbol{\kappa}_\mathrm{i} \nabla \varphi_\mathrm{i}) \cdot \mathbf{n} \quad \text{on } \partial \Omega_\mathbb{H} \qquad \varphi_\mathrm{e} = \varphi_\mathrm{b} \qquad \text{on } \partial \Omega_\mathbb{H} \cap \partial \Omega_\mathbb{B}$$

$$0 = (\kappa_\mathrm{b} \nabla \varphi_\mathrm{b}) \cdot \mathbf{n} \quad \text{on } \partial \Omega_\mathbb{B} \setminus \partial \Omega_\mathbb{H} \quad (\boldsymbol{\kappa}_\mathrm{e} \nabla \varphi_\mathrm{e}) \cdot \mathbf{n} = (\kappa_\mathrm{b} \nabla \varphi_\mathrm{b}) \cdot \mathbf{n} \quad \text{on } \partial \Omega_\mathbb{H} \cap \partial \Omega_\mathbb{B}$$

together with admissible initial conditions and a cellular ionic model. We denote with $\boldsymbol{\kappa}_\mathrm{i}$ and $\boldsymbol{\kappa}_\mathrm{e}$ the intracellular and extracellular conductivity tensor fields, with

κ_b the torso conductivity, with $\varphi_m, \varphi_e, \varphi_b$ the transmembrane, extracellular, and body potential fields, with χ the volume to membrane surface ratio, and with C_m the membrane capacitance. In this work we have used the Mahajan's model for rabbit ventricular cardiomyocytes [11] as adopted in [9] for $\mathbf{g}$ and I_{ion}. Note that this formulation yields semi-definite forms. Therefore an additional Dirichlet BC $\varphi_b = 0$ on a non-empty subset of $\partial\Omega_\mathbb{B} \setminus \partial\Omega_\mathbb{H}$ is introduced (this BC can be loosely interpreted as a grounding). In this model, under the implicit assumption of no mechanical deformation, the ECG is simply the evaluation of φ_b over time at specific locations in $\Omega_\mathbb{B}$. We will refer to this ECG as the *direct ECG*.

2.2 Poisson Reconstruction

More widely used than the Bidomain model is the Monodomain model. Here the assumption $\kappa_i = \lambda \kappa_e$ is added to the previous system of equations to collapse the first two equations into one. Although it allows to simplify significantly the model, this assumption is rather problematic as all experimental observations suggest that $\kappa_i \neq \lambda \kappa_e$, see, e.g., [8]. Furthermore, this simplification raises the question on how to enforce the boundary conditions on the extracellular potential, as they are known to affect the electrical wave propagation at the boundary. We adopt the most commonly used approach and set the flux of the transmural potential across the heart's boundary equal to zero with a suitable conductivity tensor $\overline{\kappa}$ representing the conductivity in the Monodomain model:

$$\chi C_m \frac{\partial \varphi_m}{\partial t} = \nabla \cdot (\overline{\kappa} \nabla \varphi_m) - \chi I_{ion}(\varphi_m, \mathbf{s}) - \chi I_{stim}(t) \quad \text{in } \Omega_\mathbb{H}$$

$$\frac{\partial \mathbf{s}}{\partial t} = \mathbf{g}(\varphi_m, \mathbf{s}) \quad \text{in } \Omega_\mathbb{H}$$

$$0 = (\overline{\kappa} \nabla \varphi_m) \cdot \mathbf{n} \quad \text{on } \partial\Omega_\mathbb{H}$$

With the solution of the Monodonain model we can reconstruct an approximate ECG as a postprocessing step by solving the following problem:

$$\nabla \cdot (\kappa_e + \kappa_i) \nabla \varphi_e = \underbrace{-\nabla \cdot (\kappa_i \nabla \varphi_m)}_{'=f'} \quad \text{in } \Omega_\mathbb{H}$$

$$\nabla \cdot (\kappa_b) \nabla \varphi_b = 0 \quad \text{in } \Omega_\mathbb{B}$$

$$0 = \varphi_b - \varphi_e \quad \text{on } \partial\Omega_\mathbb{H} \cap \partial\Omega_\mathbb{B}$$

Because of its familiar structure, we name this approach *Poisson reconstruction* throughout this work.

2.3 Pseudo-ECG

Most studies using the Monodomain or Eikonal model reconstruct the ECG with a lead field approach [14] because it is less expensive to compute and it does not require an additional mesh for the surrounding torso. The ECG computed using this approach is often refer to as *pseudo-ECG* and we adopt this convention here as well. A general integral form of the pseudo-ECG is given by:

$$\varphi_{\mathrm{b}}(\mathbf{x}, t) \mapsto \frac{1}{4\pi\kappa_{\mathrm{b}}} \int_{\Omega_{\mathbb{H}}} \frac{\nabla \cdot \boldsymbol{\kappa}_{\mathrm{i}}(\tilde{\mathbf{x}})\nabla\varphi_{\mathrm{m}}(\tilde{\mathbf{x}}, t)}{||\tilde{\mathbf{x}} - \mathbf{x}||_2} d\tilde{\mathbf{x}} \,, \tag{1}$$

where the original derivation assumes uniform, isotropic conductivity tensors. Moreover it is assumed that $\Omega_{\mathbb{B}}$ is a sphere with infinite radius and Dirichlet boundary condition $\varphi_{\mathrm{b}} = 0$ everywhere on the surface. Note that we introduced a modification to the formula, as we exchanged the commonly used bulk conductivity tensor with the intracellular conductivity tensor. This accounts for the intracellular anisotropic conductivity in the elliptic part of the Bidomain equation and is equivalent to 'f' in the Poisson reconstruction equations.

2.4 ECG Comparison: Similarity Measure

In order to quantitatively compare the morphology of computed ECGs, we adopt the PC^* correlation measure [12]. In contrast to the classic L_2 distance, the PC^* correlation measure is only sensitive to morphological differences – e.g., fractionations or spurious Q-waves – and not to differences in phase, amplitude, and duration.

2.5 Numerical Implementation

The Bidomain problem is solved by applying a Godunov operator splitting to separate the nonlinear cell model, containing the reaction term together with the system internal variables, from the linear system of partial differential equations (PDEs). The linear system of PDEs is solved using Rothe's method, where time is discretized via an implicit Euler scheme and space is discretized via the linear finite element method (see, e.g., [15]).

MFEM [1] is used as the numerical framework. The resulting linear system is solved via preconditioned conjugate gradient (PCG) method, where a block diagonal preconditioner combining HYPRE's [5] l1-scaled block SSOR for the first subsystem and BoomerAMG for the second subsystem is utilized. The discretization is finalized by applying an adaptive substepped explicit Euler scheme to the cell model as described in [10]. The Monodomain problem is solved following an analogous implementation, but using only a l1-scaled block SSOR preconditioner. The Poisson reconstruction is similarly implemented in MFEM and solved via PCG and BoomerAMG preconditioner. Default MFEM and HYPRE solver parameters have been adopted.

The used Monodomain and Bidomain solvers have been verified using the benchmark proposed in [13].

Equation 1 can be computed numerically using three approaches. A first possibility consists in using the *mass matrix method* (MMM) described in [4]. In this case, given φ_{m}, an auxiliary problem is solved for f:

$$f = \nabla \cdot \boldsymbol{\kappa}_{\mathrm{i}}(\tilde{\mathbf{x}})\nabla\varphi_{\mathrm{m}}(\tilde{\mathbf{x}}, t) \,,$$

and the pseudo-ECG is approximated by solving numerically

$$\varphi_{\mathrm{b}}(\mathbf{x}, t) \mapsto \frac{1}{4\pi\kappa_{\mathrm{b}}} \int_{\Omega_{\mathbb{H}}} \frac{f}{||\tilde{\mathbf{x}} - \mathbf{x}||_2} d\tilde{\mathbf{x}}.$$

The second approach is named *Gauss method* [6,9], since it relies on applying the Gauss theorem to Eq. 1. The boundary flux of the transmembrane potential is then set to zero (in agreement with the assumptions used in deriving the Monodomain model) leading to:

$$\varphi_{\mathrm{b}}(\mathbf{x}, t) \mapsto -\frac{1}{4\pi\kappa_{\mathrm{b}}} \int_{\Omega_{\mathbb{H}}} \kappa_{\mathrm{i}}(\tilde{\mathbf{x}}) \nabla \varphi_{\mathrm{m}}(\tilde{\mathbf{x}}, t) \cdot \frac{\tilde{\mathbf{x}} - \mathbf{x}}{||\tilde{\mathbf{x}} - \mathbf{x}||_2^3} d\tilde{\mathbf{x}},$$

which is then evaluated numerically given φ_{m}. This formulation can be interpreted as the projection of the electrical flux onto the scaled direction vectors pointing to the leads' positions [6].

A third possibility consists in evaluating directly Eq. 1, but is applicable only when φ_{m} is approximated using higher order interpolations (≥ 2).

3 Results

In this section we investigate the effect of the modeling assumptions on the ECG through numerical studies. We use the direct ECG as a reference, since the alternative approaches are derived from the full Bidomain model. We will analyze a simple "sphere in sphere" numerical setup and a biventricular model with a Purkinje network [9] embedded into a box torso to investigate the implications of alternative approaches to compute the ECG. In all numerical simulations a timestep equal to 0.01 ms is used, unless otherwise stated. Further we have set $C_m = 0.01\,\mu\mathrm{F}/\mathrm{mm}^2$ and $\chi = 140\,/\mathrm{mm}$ in all experiments as chosen in [13].

3.1 Sphere in Sphere Numerical Setup

The first series of numerical experiments is constructed to satisfy as close as possible the original assumptions from which the lead field approach is formulated. Subsequently, we drop these assumptions individually to investigate their effect on the ECG. With this goal in mind, the active myocardium is modeled with a sphere embedded in a larger spherical domain representing the surrounding tissue as illustrated in Fig. 1. In one numerical test, a region with a lower conductivity is also introduced in the outer sphere to conceptually mimic the presence of the lungs. In order to ensure convergence, linear hexahedral elements with an average edge size of 200 μm and 300 μm have been used to discretize the active cardiac domain and the torso domain, respectively.

In the first experiment, we set $\kappa_{\mathrm{b}} = 0.2\,\mathrm{mS}/\mathrm{mm}$, $\kappa_{\mathrm{i}} = \kappa_{\mathrm{e}} = 0.2\,\mathbf{I}\,\mathrm{mS}/\mathrm{mm}$, and a zero Dirichlet BC on the surface of the outer sphere. The radius R of the outer sphere is varied from 12.5 mm to 25 mm and 50 mm to analyze convergence of the Poisson reconstruction approach toward the pseudo-ECG. The pseudo-ECG is

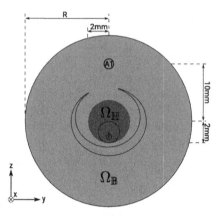

Fig. 1. "Sphere in sphere" numerical setup. An inner sphere centered at the origin with a radius of 2 mm models the heart. The inner sphere is embedded in a larger sphere, also centered at the origin and of radius R, which models the surrounding tissue. A spherical stimulus of 200 μA/mm is applied at [0, 0, −1] for the first 1 ms of the simulation. The electrical field φ_b is measured at [0, 0, 10] in all numerical experiments. In the second numerical experiment, a C-shaped subdomain constructed by the set difference between a sphere of radius 4 mm centered at [0,0,0] and a smaller sphere of radius 3.25 mm centered at [0.00, 0.25, 0.75] is inserted in $\Omega_{\mathbb{B}}$. (Note that the representation of the numerical setup is not to scale.)

computed using both the Gauss method and Mass Matrix method. As illustrated in Fig. 2 (left), the ECG computed using the Poisson approach converges toward the pseudo-ECG as the radius of the outer sphere increases, therefore satisfying a key assumption in the derivation of the pseudo-ECG equation (Eq. 1). In this case, the Gauss method and Mass Matrix method lead to the same computed ECG. Although the MMM and Gauss methods do not always lead to exactly the same quantitative result, they always agree qualitatively in our experiments. Hence all the conclusions presented here hold and for simplicity only the Gauss method is used to compute the pseudo-ECG in the following.

The second series of experiments utilizes the same setup except the Monodomain model is exchanged with the Bidomain model and the Poisson reconstruction is replaced with the direct ECG. Similarly to the first experiment, the direct ECG converges toward the pseudo-ECG for increasing R (Fig. 2, right). However, a difference between the pseudo-ECG computed from the Monodomain model and the direct ECG still exist even for R = 50.0 mm. This difference is due the boundary flux of φ_e across $\partial\Omega_{\mathbb{H}}$, which was assumed to be zero.

Next the effect on the ECG of inhomogeneous conductivities in the surrounding tissue is investigated. For the next two experiments κ_b is set equal to 0.01 mS/mm (Fig. 3, left) and 0.05 mS/mm (Fig. 3, right) within the C-shaped region (see Fig. 1). Inhomogeneous conductivities may be due to the presence of different tissues such as the lungs and bones in a torso model. In this case we observe that the magnitudes of the direct and Poisson reconstructed ECGs are lower than the pseudo-ECG magnitude (Fig. 3).

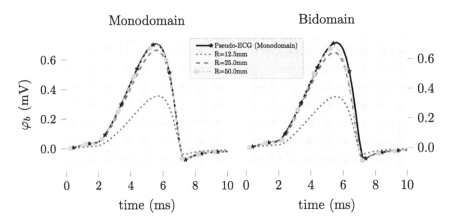

Fig. 2. Verification and validation of the pseudo-ECG formula with the Monodomain and Bidomain models using the setup from Fig. 1 with varying radius of the outer sphere and isotropic conductivity tensors $\kappa_i = \kappa_e = 0.2\,\mathbf{I}\,\mathrm{mS/mm}$ and $\kappa_b = 0.2\,\mathrm{mS/mm}$. Reference pseudo-ECGs were computed from the Monodomain solution. The Gauss and mass methods result in the same pseudo-ECG.

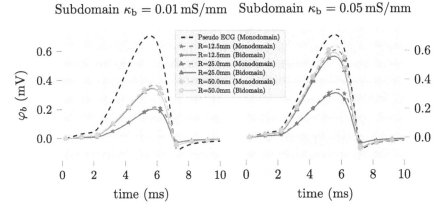

Fig. 3. ECG computed with inhomogeneous conductivities in $\Omega_{\mathbb{B}}$. All conductivities remain unchanged with respect to the previous experiment, except in the C-shaped subdomain (Fig. 1) where $\kappa_b = 0.01\,\mathrm{mS/mm}$ (left) and $\kappa_b = 0.05\,\mathrm{mS/mm}$ (right).

The subsequent set of experiments was designed to investigate the effect of the uniform and isotropic conductivity tensor assumption required to derive the pseudo-ECG approach, which is violated in realistic EP simulations. In all numerical experiments with an anisotropic conductivity tensor (uniform and non-uniform) the pseudo-ECG is not recovered for any sphere radius R with either the Poisson reconstruction or the direct ECG approach (Fig. 4).

In all numerical experiments involving the "sphere in sphere" numerical setup, the width of the ECG traces are nearly identical and the $PC*$

measures are all close to 1. The main differences are related to the magnitude of the pseudo-ECG compared to the ECG traces computed with the Poisson reconstruction or direct approach.

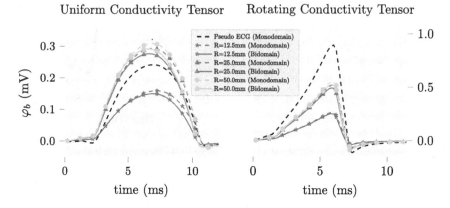

Fig. 4. ECG resulting from the setup illustrated in Fig. 1 with anisotropic uniform and non-uniform conductivity tensors. Solid lines represent Bidomain simulations and dashed lines Monodomain simulations. The eigenvalues of κ_i and κ_e are chosen as $[0.2, 0.1, 0.05]\,\mathrm{mS/mm}$ and $\kappa_b = 0.2\,\mathrm{mS/mm}$. In the uniform case (left) the eigenvectors are, in order, $[1, 0, 0]$, $[0, 1, 0]$, and $[0, 0, 1]$. In the non-uniform case, the second eigenvector is chosen along the radial direction, the third eigenvector along $[0,0,1]$, and the first eigenvector along the direction identified by the cross product of the second and third eigenvectors.

3.2 Biventricular Rabbit Model

Final experiments are conducted using a biventricular setup to investigate the effect of a more complex and realistic geometry together with more realistic boundary conditions. We have extended the rabbit biventricular model presented in [9] by embedding it in a box-shaped torso. Since we only investigate anterograde propagation from the Purkinje network in the healthy case, the activation sequence is precomputed via Dijkstra's algorithm as presented in [17]. Accordingly, a short stimulus of $250\,\mu\mathrm{A/mm}$ is applied for $4\,\mathrm{ms}$ to all nodes in a search radius equal to $300\,\mu\mathrm{m}$ from the activated Purkinje muscle junctions. Here we use a timestep of $0.05\,\mathrm{ms}$.

The ECG is computed using the Wilson leads located in the same positions as in [9] (see also Fig. 5). The ground electrode boundary condition is applied to the face closest to the subject right foot. The first experiment uses an isotropic conductivity tensor and focuses on the effect of boundary conditions and heart geometry. The second experiment includes an anisotropic conductivity tensor [7] in combination with the original experimentally measured myofiber, sheetlet, and normal directions to investigate the discrepancies in a real case scenario between pseudo-ECG and direct or Poisson-reconstructed ECGs. Direct ECG

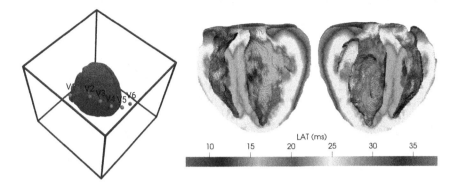

Fig. 5. Left: experimental setup with the heart embedded into a box torso of dimension 51.2 mm × 51.2 mm × 51.2 mm aligned with the biventricular model longitudinal axis. The Wilson leads' placement with respect to the biventricular model is also shown. Right: open biventricular view of the activation maps obtained in the Bidomain simulation with anisotropic conductivity tensors.

traces and ECG computed via Poisson reconstruction present the same overall features and progression. In contrast, the pseudo-ECG presents larger and lead-dependent discrepancies with respect to the direct ECG (Fig. 6). Results of the correlation analysis are reported in Table 1.

Table 1. Generalized Pearson correlation coefficients comparing the ECG computed in the isotropic and anisotropic biventricular simulations using the lead field method (pseudo-ECG), the Poisson reconstruction, and the direct approach.

		V1	V2	V3	V4	V5	V6
Isotropic	$PC^*_{\text{direct, pseudo}}$	0.838	0.930	0.941	0.921	0.907	0.932
	$PC^*_{\text{Poisson, pseudo}}$	0.910	0.943	0.936	0.931	0.919	0.940
	$PC^*_{\text{Poisson, direct}}$	0.947	0.931	0.930	0.945	0.961	0.979
Anisotropic	$PC^*_{\text{direct, pseudo}}$	−0.039	0.416	0.703	0.855	0.922	0.976
	$PC^*_{\text{Poisson, pseudo}}$	−0.247	0.244	0.772	0.940	0.963	0.976
	$PC^*_{\text{Poisson, direct}}$	0.989	0.985	0.989	0.995	0.996	0.997

4   Discussion

The first set of numerical experiments using the spherical setup confirmed that the widely used pseudo-ECG agrees with the Poisson reconstruction approach when an isotropic uniform conductivity tensor is adopted and for large enough surrounding domains. In this ideal case the ECG computed using the Monodomain model can be reconstructed using either approach. However, further experimentation on the idealized spherical setup, revealed that the hypothesis

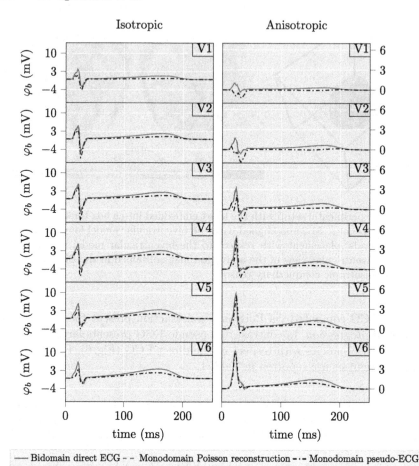

Fig. 6. Precordial ECG computed from EP simulations on a rabbit heart. Left: uniform, isotropic conductivity tensor $(0.2\,\mathbf{I}\,\mathrm{mS/mm})$. Right: full model as in [9] with Hooks' conductivities [7]. For the Monodomain simulation we have chosen $\overline{\kappa} = 0.5\,\kappa_\mathrm{i}$. Note that κ_b has been set equal to 0.2 mS/mm in these simulations.

of uniform isotropic conductivity tensor is essential for the pseudo-ECG to be accurate. As soon as an inhomogeneous surrounding domain or an anisotropic conductivity tensor – uniform or non-uniform over the inner sphere – is introduced, the pseudo-ECG approach leads to traces that only qualitatively agree with the direct ECG, i.e., their morphology is similar but their magnitude differs. In comparison, the Poisson reconstruction leads to results which remain quantitatively closer to the direct ECG.

Discrepancies between the pseudo-ECG and the direct ECG increase in a biventricular setup embedded in a homogeneous domain representing an idealized torso. In this case, the pseudo-ECG does not approximate the direct ECG qualitatively well in some of the Wilson leads, e.g., V1 and V2 in the anisotropic case as shown in Fig. 6 and by the PC^* measures in Table 1 being far from one. This discrepancy may be due to the complex geometry and strongly anisotropic conductivities in the biventricular simulations, which were not captured by the "sphere in sphere" experiment.

In all experiments, the ECG computed using the Poisson reconstruction approach is similar to the direct ECG. The remaining differences are likely due to the boundary flux of φ_e across $\partial\Omega_\mathbb{H}$, which was assumed to be zero in the Monodomain model. This missing boundary term can be resolved adopting the pseudo-Bidomain approach [2], which we plan to address in future work.

We conclude by noting some of the limitations of the current study. First, only a few cases are considered in the "sphere in sphere" experimental setup. These are likely not sufficient to pinpoint all major discrepancies between the pseudo-ECG and the direct or Poisson reconstructed ECGs. Second, we did not investigate the effect of different conductivity values on the ECGs computed with the presented methods and their similarity measures. Third, we emphasize that further refinement of the biventricular Bidomain model is needed to replicate a physiologically accurate ECG with the conductivity tensor employed in this study. In this context, the discrepancies between the Monodomain and Bidomain experiments in the biventricular setup could be due, at least in part, to the sub-optimal choice of the conductivity tensor for the Monodomain simulation. Finally, the current study was focused on the Wilson leads and did not analyze the possible discrepancies among methods in computing other electrodes' arrangements, such as the Einthoven leads' placement.

Based on the current results, we recommend that the Poisson reconstruction approach is selected in cardiac EP simulations based on the Monodomain model where obtaining a more accurate ECG is important. In all the numerical experiments reported herein, the computational cost associated with the Poisson reconstruction approach was only slightly higher than the cost required by the pseudo-ECG approach. Although no careful optimization of the numerical implementation has been carried out and the torso model employed here is small, these preliminary results suggest that the Poisson reconstruction is a viable solution to efficiently and accurately compute the ECG.

References

1. Anderson, R., et al.: MFEM: a modular finite element library. Comput. Math. Appl. **81**, 42–74 (2020)
2. Bishop, M.J., Plank, G.: Bidomain ECG simulations using an augmented monodomain model for the cardiac source. IEEE Trans. Biomed. Eng. **58**(8), 2297–2307 (2011)
3. Boulakia, M., Cazeau, S., Fernández, M.A., Gerbeau, J.F., Zemzemi, N.: Mathematical modeling of electrocardiograms: a numerical study. Ann. Biomed. Eng. **38**(3), 1071–1097 (2010)
4. Dupraz, M., Filippi, S., Gizzi, A., Quarteroni, A., Ruiz-Baier, R.: Finite element and finite volume-element simulation of pseudo-ECGs and cardiac alternans. Math. Methods Appl. Sci. **38**(6), 1046–1058 (2015)

5. Falgout, R.D., Yang, U.M.: *hypre*: a library of high performance preconditioners. In: Sloot, P.M.A., Hoekstra, A.G., Tan, C.J.K., Dongarra, J.J. (eds.) ICCS 2002. LNCS, vol. 2331, pp. 632–641. Springer, Heidelberg (2002). https://doi.org/10.1007/3-540-47789-6_66

6. Göktepe, S., Kuhl, E.: Computational modeling of cardiac electrophysiology: a novel finite element approach. Int. J. Numer. Methods Eng. **79**(2), 156–178 (2009)

7. Hooks, D.A., Trew, M.L., Caldwell, B.J., Sands, G.B., LeGrice, I.J., Smaill, B.H.: Laminar arrangement of ventricular myocytes influences electrical behavior of the heart. Circ. Res. **101**(10), e103–e112 (2007)

8. Johnston, B.M., Johnston, P.R.: Approaches for determining cardiac bidomain conductivity values: progress and challenges. Med. Biol. Eng. Comput. **58**(12), 2919–2935 (2020). https://doi.org/10.1007/s11517-020-02272-z

9. Krishnamoorthi, S., et al.: Simulation methods and validation criteria for modeling cardiac ventricular electrophysiology. PloS one **9**(12), e114494 (2014)

10. Krishnamoorthi, S., Sarkar, M., Klug, W.S.: Numerical quadrature and operator splitting in finite element methods for cardiac electrophysiology. Int. J. Numer. Methods Biomed. Eng. **29**(11), 1243–1266 (2013)

11. Mahajan, A., et al.: A rabbit ventricular action potential model replicating cardiac dynamics at rapid heart rates. Biophys. J. **94**(2), 392–410 (2008)

12. Mincholé, A., Zacur, E., Ariga, R., Grau, V., Rodriguez, B.: MRI-based computational torso/biventricular multiscale models to investigate the impact of anatomical variability on the ECG QRS complex. Front. Physiol. **10**, 1103 (2019)

13. Niederer, S.A., et al.: Verification of cardiac tissue electrophysiology simulators using an N-version benchmark. Philos. Transa. R. Soc. A Math. Phys. Eng. Sci. **369**(1954), 4331–4351 (2011)

14. Plonsey, R., Barr, R.C.: Bioelectricity: A Quantitative Approach. Springer, New York (2007). https://doi.org/10.1007/978-0-387-48865-3

15. Sundnes, J., Lines, G.T., Cai, X., Nielsen, B.F., Mardal, K.A., Tveito, A.: Computing the Electrical Activity in the Heart, vol. 1. Springer, Heidelberg (2007). https://doi.org/10.1007/3-540-33437-8

16. Tung, L.: A bi-domain model for describing ischemic myocardial D-C potentials. Ph.D. thesis, Massachusetts Institute of Technology (1978)

17. Wallman, M., Smith, N.P., Rodriguez, B.: A comparative study of graph-based, eikonal, and monodomain simulations for the estimation of cardiac activation times. IEEE Trans. Biomed. Eng. **59**(6), 1739–1748 (2012)

Uncertainty Quantification of the Effects of Segmentation Variability in ECGI

Jess D. Tate[1]([✉])[ID], Wilson W. Good[2][ID], Nejib Zemzemi[3][ID],
Machteld Boonstra[4][ID], Peter van Dam[4][ID], Dana H. Brooks[5], Akil Narayan[1][ID],
and Rob S. MacLeod[1][ID]

[1] University of Utah, Salt Lake City, USA
jess@sci.utah.edu
[2] Acutus Medical, INC., Carlsbad, CA, USA
[3] Inria Bordeaux Sud Ouest, Talence, France
[4] UMC Utrecht, Utrecht, The Netherlands
[5] Northeastern University, Boston, MA, USA

Abstract. Despite advances in many of the techniques used in Electrocardiographic Imaging (ECGI), uncertainty remains insufficiently quantified for many aspects of the pipeline. The effect of geometric uncertainty,
particularly due to segmentation variability, may be the least explored to
date. We use statistical shape modeling and uncertainty quantification
(UQ) to compute the effect of segmentation variability on ECGI solutions. The shape model was made with Shapeworks from nine segmentations of the same patient and incorporated into an ECGI pipeline. We
computed uncertainty of the pericardial potentials and local activation
times (LATs) using polynomial chaos expansion (PCE) implemented in
UncertainSCI. Uncertainty in pericardial potentials from segmentation
variation mirrored areas of high variability in the shape model, near the
base of the heart and the right ventricular outflow tract, and that ECGI
was less sensitive to uncertainty in the posterior region of the heart.
Subsequently LAT calculations could vary dramatically due to segmentation variability, with a standard deviation as high as 126ms, yet mainly
in regions with low conduction velocity. Our shape modeling and UQ
pipeline presented possible uncertainty in ECGI due to segmentation
variability and can be used by researchers to reduce said uncertainty or
mitigate its effects. The demonstrated use of statistical shape modeling
and UQ can also be extended to other types of modeling pipelines.

Keywords: Electrocardiographic imaging · Shape analysis ·
Uncertainty quantification · Activation times · Cardiac segmentation

Supported by the National Institutes of Health, P41GM103545, R24GM136986,
U24EB029012, U24EB029011, R01AR076120, and R01HL135568. Data used in this
study was made available by Drs. John Sapp and Milan Horáček and their research
collaboration with Dalhousie University. Thanks to Sophie Giffard-Roisin, Eric
Perez-Alday, Laura Bear, Beáta Ondrušová, Svehlikova, Machteld Boonstra, Martim
Kastelein, and Maryam Tolou for providing segmentations for this study.

© Springer Nature Switzerland AG 2021
D. B. Ennis et al. (Eds.): FIMH 2021, LNCS 12738, pp. 515–522, 2021.
https://doi.org/10.1007/978-3-030-78710-3_49

1 Introduction

Electrocardiographic Imaging (ECGI) has seen continued recent interest to non-invasively diagnose and guide treatment of cardiac arrhythmias and other abnormalities. ECGI estimates cardiac electrical activity from body surface potential recordings using a numerical model of a subject's thorax [2,13]. However, although ECGI solutions depend heavily on model parameters and assumptions, the impact of uncertainty in those models and assumption have not yet been carefully quantified. Understanding this impact of this uncertainty is critical for confident use in clinical settings. Specifically, one important source of uncertainty in ECGI comes from the segmentations of anatomical images required to build forward models, *i.e.*, estimates of expected surface measurements if the cardiac sources were known, that are in turn required in the "inverse procedures" of ECGI, We have previously demonstrated that segmentations can vary widely across ECGI implementations even for the same set of images, especially on the cardiac surface [11,20], and that changes in segmentation can alter ECGI solutions [19].

However, we have not yet actually quantified this segmentation-based uncertainty. Here we introduce a method to do so and report on the results. We use a mathematical technique for quantification of parameter uncertainty called Polynomial Chaos Expansion (PCE) [3,21–23]. Although PCE has been used previous in electrocardiographic forward models [7,9,10,16,18], employing it to quantify uncertainty due to segmentation requires parameterization of shape variability, because PCE depends on the availability of a relatively low-dimensional parameterization of the uncertain quantities. Thus we need an approach to shape models that can accurately capture geometric variability, yet can still be implemented efficiently in an uncertaintly quantification (UQ) pipeline we will refer to as ECGI-UQ. Advances in available tools for shape analysis [12] provide an opportunity to merge shape modeling with ECGI-UQ to systematically compute the effect of the segmentation variability on ECGI.

Specifically, we incorporate statistical shape modeling into UQ and use it to compute the effect of variations in segmentations on pericardial potentials obtained from ECGI as well as on the highly clinically relevant local activation times (LATs) computed from those potentials. We carried out this study in connection with colleagues in the Model Building workgroup of the Consortium for ECG Imaging (CEI, https://www.ecg-imaging.org) who generously provided a set of different segmentations of the same images from the same subject. Our pipeline was able to efficiently generate a parameterized shape model from that set of pericardial segmentations and use it with ECGI-UQ by combining the use of two open source tools: ShapeWorks [5] (https://www.sci.utah.edu/software/shapeworks.html) and a UQ tool called UncertainSCI [16] (https://www.sci.utah.edu/sci-software/simulation/uncertainsci.html). Our results showed that both the degree and location of variation in the ECGI solution corresponded to the degree and location of variation of the segmented cardiac surfaces.

2   Methods

To analyze the effect of cardiac segmentation variability on ECGI, we first computed a parameterized shape model across multiple segmented ventricular pericardium geometries and then applied it to an ECGI pipeline driven by our UQ
approach (ECGI-UQ).

2.1   Shape Model

CT scans from a single subject were segmented by nine CEI research groups. The
CT images, as well as the potential recordings used for ECGI, were collected as
described in by Sapp et al. [17] and are freely available for use via the EDGAR
database (http://edgar.sci.utah.edu) [1], a shared resource of the CEI. These
nine segmentations were then analyzed in ShapeWorks to generate a pericardial
shape model [20]. Specifically, shape analysis in ShapeWorks proceeds by finding
corresponding locations among points (512 in this study) distributed across all
of the segmentation surfaces. The points are initially placed randomly and then
moved to statistically corresponding locations using a particle optimizer that
minimizes the modes of variation for the cohort. Principle component analysis
is then applied to these optimized point sets to find a mean shape and the
modes of variation along with coefficients along each mode that approximate each
segmentation. We used the first four modes of variation to form our shape model
[20]. Thus the shape model that was passed to the ECGI-UQ pipeline consisted
of a vector of points representing the mean shape and four vectors indicating
the modes of variation. An approximation to each original segmentation can
then be found by translating the points from the mean shape using a linear
combination of the four vectors weighted by the coefficients computed for that
segmentation by projecting that nine shapes onto each of mode of variation.
The parameters of the shape model are then the scalar coefficients. ShapeWorks
also computes statistics with respect to each of the shape modes, which we use
to define parameter (coefficient) ranges in ECGI-UQ (Sect. 2.3). Figure 1 shows
the mean and two standard deviations of each of the four modes of variation
included in the pericardial shape model.

2.2   ECGI Pipeline

We used any parameterized segmentation given by the shape model (Sect. 2.1)
into an ECGI pipeline using the Forward/Inverse Toolkit in SCIRun [4,14,15]
(http://scirun.org). A pericardial surface mesh was created by triangulating the
points on a given shape model while maintaining local neighborhoods. Pericardial
surface potentials were computed from torso surface recordings using a boundary
element method (BEM) forward model and zero-order Tikhonov regularization.
Local activation times (LATs) were computed from the ECGI-estimated electrograms by finding the minimum temporal derivative at each mesh node [6]. The
root mean squared (RMS) potential over the pericardial surface as a function of
time was also computed. The computed pericardial potentials, LATs, and RMS

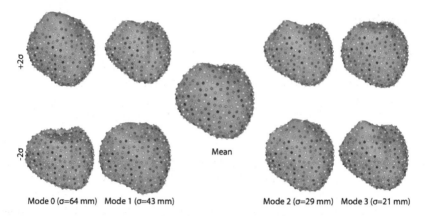

Fig. 1. Pericardial shape model based on nine segmentations of the same patient. The central image shows the mean shape while the other images show the shapes of the four dominant modes of variation (columns) computed at plus (top row) and minus (bottom row) two standard deviations along that mode. Correspondence points are shown with colors matching across segmentations and modes.

potentials were used as outputs of the ECGI pipeline in the UQ analysis. We computed the uncertainty of ECGI solutions for four activation profiles: sinus, LV paced, apically paced, and RV paced.

2.3   Uncertainty Quantification

We quantified the uncertainty of the ECGI solution resulting from shape variability using PCE in UncertainSCI, where the random parameters in the ECGI-UQ analysis were variations along the four principal component directions. UncertainSCI estimates uncertainty in the ECGI pipeline output due to variability in these parameters using a parsimonious experimental design in parameter space. With this ensemble of ECGI solutions, a multivariate polynomial function is constructed to estimate parametric variability and can be used to map specified distributions on the parameters to distribution of the pipeline outputs [3,16]. UncertainSCI employs a Weighted Approximate Fekete Points (WAFP) strategy, a special kind of D-quasi-optimal design [3]. This PCE emulator is used to estimate statistics of the distribution of the pipeline outputs based on the input parameter distributions [3,16]. For this study, we specified independent uniform distributions along each of the shape axes with bounds of ± 125, 85, 60, and 40 mm, corresponding to approximately two standard deviations the modes, cf. Fig. 1. The total degree polynomial order was set to five. UncertainSCI provided statistics for the predicted distributions of pericardial potentials, LATs, and pericardial RMS potentials.

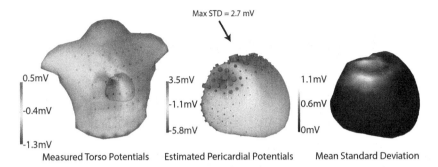

Measured Torso Potentials   Estimated Pericardial Potentials   Mean Standard Deviation

Fig. 2. Spatial distribution of uncertainty from segmentation variability. Left, center: Torso and pericardial potentials are shown near the peak of the QRS wave. Color shows potentials while the sizes of the cylinders on the surfaces represent relative standard deviation at each location, Right: Spatial distribution of the mean standard deviation, *i.e.*, square root of the average variance, over the cardiac cycle. (Color figure online)

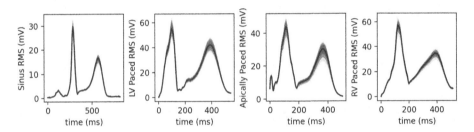

Fig. 3. Computed uncertainty of reconstructed RMS cardiac potentials for four activation patterns as indicated. Values shown are in mv as a function of time in ms. The blue line indicates the mean RMS potential and the red shading indicates the 8 quantile regions. (Color figure online)

3 Results

We compared the statistics of the results of our ECGI-UQ pipeline with to the statistics of the shape model itself. In general, the pericardial potential uncertainty, as shown by both standard deviation and quantile range, was correlated to amplitude of the median estimated potentials both spatially (Fig. 2) and temporally (Fig. 3). The uncertainty in pericardial potential was greater near the RV outflow tract and the base of the heart and was largely localized to anterior regions, as demonstrated by the image of the temporal mean of the standard deviation (square root of the average variance) of predicted potentials. The anterior areas of greater uncertainty in the pericardial potential roughly correlate to regions of high shape variability [20] (Fig. 1).

The uncertainty of the computed LATs due to segmentation variability showed broad areas of low uncertainty punctuated by smaller high uncertainty areas, as seen in the estimated quantile range (Fig. 4). The ECGI-UQ pipeline predicted a maximum standard deviations resulting from segmentation

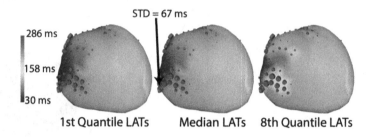

Fig. 4. Spatial distribution of LAT uncertainty illustrated by the median (center) and the minimum (left), and maximum (right) over 8 quantiles. Cylinder sizes represent relative standard deviation of the LAT at each location.

uncertainty of 67 ms for sinus activation, 66 ms for LV pacing, 111 ms for apical pacing, and 127 ms for RV pacing. The maximum predicted interquartile range was 45 ms for sinus, 64 ms for LV pacing, 103 ms for apical pacing, and 117 ms for RV pacing. Regions of high variability in LATs were not geometrically consistent, but varied with activation profile. High uncertainty was often located in regions of low conduction speeds, as estimated by the gradient in the mean activation times, and was not necessarily coincident with areas of highest shape or computed potential variability. Highly variable LAT areas corresponded with the presence of computed electrograms containing QRS fractionation and other abnormal morphologies, which also make LAT determination more challenging.

4   Discussion

Including shape modeling and UQ into an ECGI pipeline allowed us to quantify probabilisitcally the possible effects of segmentation variability on computed pericardial potentials and LATs. The uncertainties predicted by our pipeline indicate that segmentation variability could dramatically alter pericardial potentials or LATs in certain regions of the heart, while the specific activation pattern could also affect which regions are most sensitive. We found that the uncertainty of pericardial potential reconstructions roughly correlate to the segmentation variability except in the posterior region of the heart. This finding is consistent with our previous results with a simpler analysis [19]. Gander et al. [8] reported similar levels of variability in the uncertainty of pericardial potentials due to changes in heart shape over the cardiac cycle.

The predicted uncertainty of the LATs demonstrated the possibility that segmentation variability could alter secondary predictions of ECGI. The predicted standard deviation of the LAT, as high as 127 ms, indicate that segmentation variability could cause some early activating regions to be classified as late activating and vice versa (Fig. 4). The potential adverse effect of such an error may be mitigated because high variability regions are spatially limited. Areas of high predicted variability tend to co-occur with low conduction speed, but neighboring high conduction speed areas have low variability, causing a regionally limited shift in the activation pattern in most cases. Additionally, areas of high variability did not occur near the centers of early or late activating regions, making it unlikely to adversely affect results for relatively simple activation patterns such as paced beats. However, more complex activation profiles, such as ventricular tachycardia, will likely have more areas of high gradient LATs and may thus be more impacted by segmentation error. Furthermore, since some arrhythmia substrates, such as fibrosis, have with lower conduction speeds, substrate identification could be affected by segmentation variability. Poor signal quality recordings may also be more sensitive to segmentation variability, although these effects might be mitigated with improvements in signal processing and LAT detection. Because of the many nuances involved with computing LATs, more analysis is needed to fully explore the effect of segmentation variability on LAT prediction.

In summary, our methodology of incorporating ShapeWorks and Uncertain-SCI to model uncertainty in ECGI resulting from shape variability could be similarly applied to other applications. This method requires availability of a dataset large enough to sufficiently characterize shape variability. ShapeWorks and SCIRun have tools to facilitate the remaining tasks. We note that Uncertain-SCI is designed to interfaces with most modeling software through python [16], so we anticipate adopting it to other pipelines will be relatively straightforward.

References

1. Aras, K., et al.: Experimental data and geometric analysis repository: EDGAR. J. Electrocardiol. **48**(6), 975–981 (2015)
2. Barr, R., Ramsey, M., Spach, M.: Relating epicardial to body surface potential distributions by means of transfer coefficients based on geometry measurements. IEEE Trans. Biomed. Eng. **24**, 1–11 (1977)
3. Burk, K.M., Narayan, A., Orr, J.A.: Efficient sampling for polynomial chaos-based uncertainty quantification and sensitivity analysis using weighted approximate fekete points. Int. J. Numer. Methods Biomed. Eng. **36**(11), e3395 (2020). https://doi.org/10.1002/cnm.3395
4. Burton, B., et al.: A toolkit for forward/inverse problems in electrocardiography within the SCIRun problem solving environment. In: Proceedings of the IEEE Engineering in Medicine and Biology Society 33rd Annual International Conference, pp. 1–4. IEEE (2011)
5. Cates, J., Meyer, M., Fletcher, P., Witaker, R.: Entropy-based particle systems for shape correspondence. In: Workshop on Mathematical Foundations of Computational Anatomy, MICCAI 2006, pp. 90–99 (October 2006). http://www.sci.utah.edu/publications/cates06/Cates-miccai2006.pdf
6. Erem, B., Brooks, D., van Dam, P., Stinstra, J., MacLeod, R.: Spatiotemporal estimation of activation times of fractionated ECGS on complex heart surfaces. In: Proceedings of the IEEE Engineering in Medicine and Biology Society 33rd Annual International Conference 2011, pp. 5884–5887 (2011)
7. Fikal, N., Aboulaich, R., El Guarmah, E., Zemzemi, N.: Propagation of two independent sources of uncertainty in the electrocardiography imaging inverse solution. Math. Model. Nat. Phenom. **14**(2), 206 (2019). https://doi.org/10.1051/mmnp/2018065

8. Gander, L., Krause, R., Multerer, M., Pezzuto, S.: Space-time shape uncertainties in the forward and inverse problem of electrocardiography (2020)
9. Geneser, S., MacLeod, R., Kirby, R.: Application of stochastic finite element methods to study the sensitivity of ECG forward modeling to organ conductivity. IEEE Trans. Biomed. Eng. **55**(1), 31–40 (2008)
10. Geneser, S., Xiu, D., Kirby, R., Sachse, F.: Stochastic Markovian modeling of electrophysiology of ion channels: reconstruction of standard deviations in macroscopic currents. J. Theor. Biol. **245**(4), 627–637 (2007)
11. Ghimire, S., et al.: Overcoming barriers to quantification and comparison of electrocardiographic imaging methods: a community- based approach. In: Computing in Cardiology Conference (CinC), 2017, vol. 44, pp. 1–4 (2017)
12. Goparaju, A., et al.: Benchmarking off-the-shelf statistical shape modeling tools in clinical applications (2020)
13. Gulrajani, R.: The forward and inverse problems of electrocardiography. EMBS Mag. **17**(5), 84–101 (1998)
14. MacLeod, R., Weinstein, D., de St. Germain, J.D., Brooks, D., Johnson, C., Parker, S.: SCIRun/BioPSE: integrated problem solving environment for bioelectric field problems and visualization. In: IEEE International Symposium Biomedical Imaging (ISBI), pp. 1–3. IEEE (2004)
15. Parker, S., Weinstein, D., Johnson, C.: The SCIRun computational steering software system. In: Arge, E., Bruaset, A., Langtangen, H. (eds.) Modern Software Tools in Scientific Computing, pp. 1–40. Birkhauser Press, Boston (1997). http://www.sci.utah.edu/publications/Par1997a/Parker_SCIRun1997.pdf
16. Rupp, L.C., et al.: Using uncertainSCI to quantify uncertainty in cardiac simulations. In: Computing in Cardiology, vol. 47 (2020)
17. Sapp, J.L., Dawoud, F., Clements, J.C., Horáček, B.M.: Inverse solution mapping of epicardial potentials: quantitative comparison with epicardial contact mapping. Circ. Arrhythm. Electrophysiol. **5**(5), 1001–1009 (2012). https://doi.org/10.1161/CIRCEP.111.970160. http://circep.ahajournals.org/content/5/5/1001
18. Swenson, D., Geneser, S., Stinstra, J., Kirby, R., MacLeod, R.: Cardiac position sensitivity study in the electrocardiographic forward problem using stochastic collocation and BEM. Annal. Biomed. Eng. **30**(12), 2900–2910 (2011)
19. Tate, J.D., Zemzemi, N., Good, W.W., van Dam, P., Brooks, D.H., MacLeod, R.S.: Effect of segmentation variation on ECG imaging. In: Computing in Cardiology, vol. 45 (2018). https://doi.org/10.22489/CinC.2018.374
20. Tate, J.D., Zemzemi, N., Good, W.W., van Dam, P., Brooks, D.H., MacLeod, R.S.: Shape analysis of segmentation variability. In: Computing in Cardiology, vol. 47 (2020)
21. Wiener, N.: The Homogeneous Chaos. Amer. J. Math. **60**(4), 897–936 (1938)
22. Xiu, D.: Numerical Methods for Stochastic Computations: A Spectral Method Approach. Princeton University Press (2010)
23. Xiu, D., Karniadakis, G.: The wiener-askey polynomial chaos for stochastic differential equations. SIAM J. Sci. Comput. **24**(2), 619–644 (2002). https://doi.org/10.1137/S1064827501387826

Spiral Waves Generation Using an Eikonal-Reaction Cardiac Electrophysiology Model

Narimane Gassa[1,2,3]($\boxtimes$), Nejib Zemzemi[1,2,3]($\boxtimes$), Cesare Corrado[4]($\boxtimes$), and Yves Coudière[1,2,3]($\boxtimes$)

[1] University of Bordeaux, IMB, Bordeaux, France
[2] National Institute of Mathematics and Informatics, Inria Bordeaux, Talence, France
[3] IHU-Lyric, Bordeaux, France
{narimane.gassa,nejib.zemzemi,yves.coudiere}@inria.fr
[4] School of Biomedical Engineering and Imaging Sciences, Kings College London, London, UK
cesare.corrado@kcl.ac.uk

Abstract. *Aim:* Computer models enabled the study of the fundamental mechanisms responsible for arrhythmias and have the potential of optimizing the clinical procedure for an individual patients pathology. The model complexity and the computational costs affecting computer models hamper their application on a routinely performed procedure. In this work, we aim to design a computer model suitable for clinical time scales. *Methods:* We adopt a (multi-front) eikonal model that adapts the conduction velocity to the underlying electrophysiology; we describe the diffusion current using a parametrised form, fitted to reproduce the monodomain profile. *Results:* We simulated spiral waves on a 3D tissue slab and bi-atrial anatomy. We compared the numerical results obtained with a monodomain formulation with those obtained with the new method. Both models provided the same pattern of the spiral waves. While the monodomain model presented slower propagation fronts, the eikonal model captured the correct value of the conduction velocity CV even using a coarse resolution. *Conclusion:* The eikonal model has the potential of enabling computer-guided procedures when adapts the conduction velocity to the underlying electrophysiology and characterises the diffusion current with a parametrised form.

Keywords: Cardiac electrophysiology · Eikonal model · Spiral wave · Atrial fibrillation

1 Introduction

Cardiac ablation is widely used to treat cardiac atrial arrhythmias, and sometimes also ventricular arhythmias. In these heart conditions, the heartbeat could be slow in the case of bradycardia and it could be fast in cases of tachycardia,

© Springer Nature Switzerland AG 2021
D. B. Ennis et al. (Eds.): FIMH 2021, LNCS 12738, pp. 523–530, 2021.
https://doi.org/10.1007/978-3-030-78710-3_50

flutter or fibrillation. In these latter conditions, cardiac ablation techniques are used in order to restore a normal heart rhythm. This technique allows isolating trigger areas from the rest of the heart. In order to accurately identify the regions to ablate, catheter are introduced inside the heart chambers and measurements of the electrical signals allow to localize the abnormal regions. This invasive method is performed in the operating room where the patient is under local anesthesia. It is long and sometimes inaccurate, mainly because of the heart movement. The recent alternative to this approach is the non invasive electrocardiography imaging (ECGI), this method allows to reconstruct electrical informations on the heart surface using a set of electrodes measuring the electrical potential on the body surface and geometrical information about the patient extracted from CT-scans [10]. In this method an unstable mathematical inverse problem is solved and the results are not very satisfactory in clinical application [3]. One of the approaches, that could have a high potential to solve the ECGI inverse problem is to use propagation models in the cardiac domain and parametrize these models in order to personalize the model by fitting it to the electrical measurements. During the last decades, advances have been made in electrocardiology. The bidomain and monodomain models [1] are widely used as they are considered to accurately describe the electrical propagation in the heart. In spite of being applied successfully, these phenomenological models based on reaction-diffusion equations are known to be computationally very expensive. Thus their use in clinical application is challenged and non-practical. The eikonal equation, on the other hand, is a one non-linear partial differential equation that arises in problems of wave propagation and can be very fast to compute. Despite its low computational complexity, the eikonal model is less reliable in arrhythmia prediction than the reference models, which can be a major limit for studying arrhythmias. In particular, in the literature many studies have been dedicated to the use of the eikonal model in cardiac electrophysiology [2,4,6]. In *Neic et al.* (2017) [6] a new reaction-eikonal model has been introduced, but, as the authors stated in the discussion, this model is not suitable for reentries. In all these studies no spiral waves have been generated with the different formulations of eikonal models. This is mainly due to the fact that the eikonal models provides the activation times in the depolarization phase but does not accurately describe the repolorization of the cardiac cells. In this work, we present a novel approach wherein we enhance the conduction velocity adapted eikonal model (EK-CV) presented in [2] with a time-varying parametric function that describes the diffusive current. We then fit this expression to the diffusive current we obtained by solving the monodomain model. This new formulation allows the EK-CV model to simulate the dynamics of the spiral waves. We compare the generated waves with those simulated using the monodomain model.

2   Methods

2.1   The Monodomain Mitchell-Schaeffer Model

The Mitchell-Schaeffer (MS) ionic model [5] describes the ionic currents that are flowing through the cell membrane. It is characterised by two state variables, the transmembrane voltage V_m and a gating variable h. Given a cardiac domain $\Omega \subset \mathbb{R}^d (d = 2, 3)$ the mondomain in Ω coupled with the MS model can be written as follows:

$$
\begin{cases}
A_m(C_m \partial_t V_m + \beta I_{\text{ion}}(V_m, h)) - \text{div}\left(\underline{\underline{D}}_m \underline{\nabla} V_m\right) = A_m I_{app}, & \text{in } \Omega \times \mathbb{R}^+, \\
\partial_t h + g(V_m, h) = 0, & \text{in } \Omega \times \mathbb{R}^+, \\
\underline{\underline{D}}_m \nabla V_m \cdot n = 0, & \text{on } \partial\Omega
\end{cases}
\tag{1}
$$

with an ionic model given by,

$$
I_{\text{ion}}(V_m, h) = -\frac{h V_m^2 (1 - V_m)}{\tau_{\text{in}}} + \frac{V_m}{\tau_{\text{out}}}
$$

$$
g(V_m, h) =
\begin{cases}
\dfrac{h - 1}{\tau_{\text{open}}}, & \text{if } V_m \leq V_{\text{gate}}, \\
\dfrac{h}{\tau_{\text{close}}}, & \text{if } V_m > v_{\text{gate}},
\end{cases}
$$

2.2   The EK-CV Model

The eikonal model in cardiac electrophysiology has been proposed in the early nineties [4] as an alternative to the reaction-diffusion bidomain to compute the activation time $T_{\text{act}}(x)$ representing the arrival of the depolarization wavefront at each point $x \in \Omega$. The activation time $T_{act}(x)$ solves the eikonal equation:

$$
\begin{cases}
F\sqrt{(\underline{\nabla} T_{act})^T \underline{\underline{D}} \, \underline{\nabla} T_{act}} = 1, & \text{on } \Omega, \\
T_{act} = 0, & \text{on } \Gamma,
\end{cases}
\tag{2}
$$

where $\underline{\underline{D}}$ is a dimensionless tensor that describes the tissue anisotropy, the scalar function F is the velocity of the front and Γ is the set of the first activated sites. The EK-CV model [2] makes the conduction velocity F dependent on the underlying electrical state as follows:

$$
F = \alpha \sqrt{\frac{2\sigma_m^l}{\tau_{in}}},
\tag{3}
$$

$$
\alpha(DI_n) = \frac{1}{4}(3\sqrt{h(DI_n) - h_{min}} - \sqrt{h(DI_n)}).
$$

Here DI_n stands for the diastolic interval at the n-th beat and σ_m is the diffusivity tensor, defined as follows:

$$
\sigma_m = \frac{\underline{\underline{D}}_m}{A_m C_m},
\tag{4}
$$

with σ_m^l is the longitudinal component parallel to the fibre direction $\vec{a}$ and σ_m^t is the transverse component:

$$\sigma_m = \sigma_m^t \underline{\underline{I}} + (\sigma_m^l - \sigma_m^t) \vec{a} \otimes \vec{a} = \sigma_m^l (\sigma_m^t/\sigma_m^l \underline{\underline{I}} + (1 - \sigma_m^t/\sigma_m^l) \vec{a} \otimes \vec{a}) = \sigma_m^l \underline{\underline{D}}. \quad (5)$$

Once discretised in time, within each time interval, we first evaluate F using (3); next, we solve Eq. (2) using the fast marching method [9] and finally, we solve the ode system of the MS model:

$$\begin{cases} \partial_t V_m + I_{\text{ion}}(V_m, h)) = I_{\text{stim}}, & \text{in } \Omega \times \mathbb{R}^+, \\ \partial_t h + g(V_m, h) = 0, & \text{in } \Omega \times \mathbb{R}^+, \end{cases} \quad (6)$$

where the function I_{ion} and g are the same functions as in the MS ionic model and the stimulation current I_{stim} is computed using the activation time information and will be detailed in the following section.

2.3   Parametrization of the Diffusion Current

The method proposed in [2] describes the diffusive current with a stimulus of constant intensity and a given duration. This simple description of the diffusive term limits the study of spiral waves. In this work, we characterise the diffusive current with a time-varying parametric function, and we constrain the characterising parameters by fitting the diffusive term of the monodomain model. We parameterise the profile of the diffusive current with the sum of two Gaussian functions:

$$I_{\text{stim}}(t) = A_1 \cdot \exp{-\frac{(t - \mu_1)^2}{2\sigma_1^2}} - A_2 \cdot \exp{-\frac{(t - \mu_2)^2}{2\sigma_2^2}} \quad (7)$$

where the parameters A_1, A_2 are the magnitudes, μ_1, μ_2 are the times representing the centers of each Gaussian bump and σ_1, σ_2 are the standard deviations of the time in the Gaussian functions. The profile of (7) is shown in Fig. 1B and is consistent with the profiles obtained from the monodomain model simulations (Fig. 1A). We post-processed the diffusive current from the monodomain model simulations as follows:

$$\text{div}\left(\underline{\underline{D}}_m \nabla V_m\right) = I_{\text{stim}} = C_m \partial_t V_m + I_{\text{ion}}. \quad (8)$$

and we centred around the activation time evaluated as the instant the action potential exceeds the threshold of 0.15 with a positive slope. Finally, we fit the function (7) to the mean value of all diffusion the currents shown in Fig. 1A using a Levenberg-Marquardt algorithm [7] (Fig. 1A). We obtained $A_1 = 0.126$, $A_2 = 0.143$, $\mu_1 = 1.81\,\text{ms}$, $\mu_2 = 3.83\,\text{ms}$, $\sigma_1 = 2.499$, $\sigma_2 = 2.188$ with relative error on the fitting of the mean value of the diffusion current is equal to 0.028. In this work, inspired by the findings of *Neic et al.* (2017) [6], the stimulation current that will be introduced to the MS EK-CV model is the approximation of the

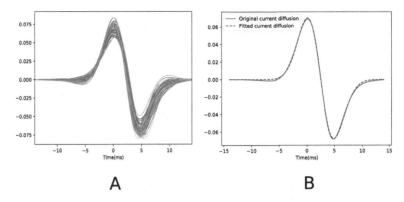

Fig. 1. Time course of the diffusion current. A: the diffusion currents computed with the monodoamin model at each node of the mesh. B: comparison between the mean value of the diffusion currents computed with the monodomain model and the fitted function I_{stim}.

diffusion current relative to a R-D monodomain model during the depolarization. I_{stim} will be added, in space and in time, as follows:

$$I_{stim}(t, x) = \begin{cases} I_{stim}(t), & \text{if } t \in [T_{act}(x) - t_{stim}, T_{act}(x) + t_{stim}] \\ 0, & else. \end{cases} \quad (9)$$

3   Numerical Results

We compare the simulated numerical results of spiral waves obtained using the monodoamin model and the EK-CV model with the parametrization of the diffusion current (EK-CV-DIFF). We consider two different geometries: a $(5 \times 5 \times 0.1\,\text{cm})$ 3D tissue slab and an anatomically detailed bi-atrial domain. On all the test cases, we adopt the Mitchell-Schaeffer ionic model parameters shown in Table 1. The simulations of propagation on the bi-atrial geometry during 400ms with the EK-CV-DIFF model required 21.89 s while the monodomain model required 1048.64 s on the same machine. We induce spiral waves in the simulations by applying S1-S2 protocol.

Table 1. Parameters of the MS ionic model.

τ_{in}	τ_{out}	τ_{open}	τ_{close}	v_{gate}
0.3 ms	6 ms	120 ms	80 ms	0.13

3.1   Tissue Slab

In this section, we show the simulation results obtained using the tissue slab geometry (dimensions $5 \times 5 \times 0.1$ cm). This was used to easily investigate the initiation of reentries and qualitatively compare the new approach to the actual state of the art. The number of vertices in the mesh is 20,490, the number of triangles is 15,256 and the number of tetrahedra is 99,420 with a characteristic mesh size $h = 1.4$ mm. The conductivity of the tissue is equal to 0.001 S/cm. The tissue is initially at rest, we apply a first stimulus on the left edge with a duration of t_{stim}. The wave front then propagates from left to right of the tissue slab and we apply a second stimulus in a rectangular region ($x < 3$ and $y < 2.5$) at time $S2 = 300$ ms. Figure 2 shows the simulated action potentials at time $t = 600$ ms, wtih the parametrised diffusive current (right) and with the constant expression adopted in [2](left). Both models generated a spiral wave with the same conduction velocity. The EK-CV model, however, produced a strong discontinuity on the action potential in the depolarized region (Fig. 2, left), not shown in the EK-CV-DIFF model (Fig. 2, right). In that follows, we adopt the EK-CV-DIFF model. Figure 3 shows the action potential at $t = (100, 400, 600, 800$ ms), simulated with the EK-CV-DIFF model (top row) and with the monodomain MS model (bottom row). Both models initiate a stable spiral wave; the monodomain model, however, presents an activation front that propagates slower,

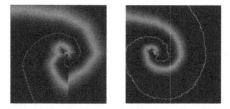

Fig. 2. Two snapshots of the transmembrane voltage V_m at $t = 600$ ms. On the left a constant stimulation current was applied and on the right we stimulate with the parameterised function of stimulus.

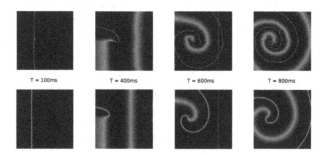

Fig. 3. Snapshots of the transmembrane voltage V_m for the EK-CV-DIFF model (top row) and the monodomain model (bottom row) at different time steps.

when compared with the EK-CV-DIFF model. The slow propagation is caused by the coarse resolution ($h = 1.4\,\mathrm{mm}$) here adopted, [8].

3.2 Atrial Geometry

In this section, we present the simulation results obtained using the anatomical atrial geometry discretised here with a triangular surface mesh with 141,314 nodes and 275,515 elements. The conductivity of the tissue is equal to $0.09\,\mathrm{S/cm}$. The tissue is initially at rest, we stimulate in the wall of the right atrium with an intensity I_{stim}. A second stimulus is then applied at the bottom of the atrium ($y > 21.1$ and $z < 17.6$) at time $S2 = 240\,\mathrm{ms}$. In Fig. 4, we show snapshots of the action potential at different times (t $= 10, 250, 390, 430\,\mathrm{ms}$). In the top row of the figure, we show the obtained simulations with the EK-CV-DIFF model. In the bottom row we present the simulation results obtained with the monodomain model. When comparing both simulations, we remark that two spiral waves occur in the same region for both models. We can also say that we have qualitatively the same pattern of reentry. On the other hand, we notice that the wave generated with the monodomain model is delayed with respect to the eikonal model: the pattern of the EK-CV-DIFF solution at time 390 ms corresponds to that for the monodomain solution at time 430 ms.

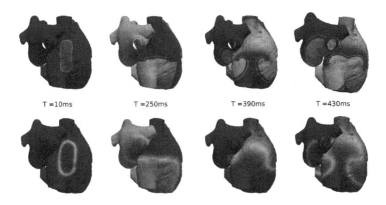

Fig. 4. Snapshots of the action potential V_m of the EK-CV-DIFF model (top row) and the monodomain model (bottom row) at different time steps.

4   Discussion and Conclusion

In this work, we used the conduction velocity adapted eikonal model [2] in order to simulate reentry spiral waves. The new model that we present here combines the CV-adapted eikonal model with a specific parametrized stimulation current obtained from the diffusion term of the monodomain model. The monodomain model is computed once off-line and the generated diffusion current has been approximated, parametrized and used to stimulate the MS ode system. The preliminary results showed that with this model we have been able to simulate spiral waves with the same pattern as the monodomain model. To the best of our knowledge, this is the first work able to generate spiral waves using the eikonal model. Differently from the monodomain model, [8], the EK-CV-DIFF formulation captures the correct value of the conduction velocity even using a coarse resolution. This eikonal model is easy to parametrize and thus could be a good candidate to conduct personalized simulations in order to simulate the electrical wave of a given patient. Our future work will be to parametrize this model in order to provide a new electrocardiography imaging tool especially in atrial fibrillation conditions.

Acknowledgment. This Project has received funding from the European Unions Horizon research and innovation programme under the Marie Skodowska-Curie grant agreement No. 860974 and by the French National Research Agency, grant references ANR-10-IAHU04- LIRYC and ANR-11-EQPX-0030.

References

1. Clements, J.C., Nenonen, J., Li, P., Horáček, B.M.: Activation dynamics in anisotropic cardiac tissue via decoupling. Ann. Biomed. Eng. **32**(7), 984–990 (2004)
2. Corrado, C., Zemzemi, N.: A conduction velocity adapted eikonal model for electrophysiology problems with re-excitability evaluation. Med. Image Anal. **43**, 186–197 (2018)
3. Duchateau, J., et al.: Performance and limitations of noninvasive cardiac activation mapping. Heart Rhythm **16**(3), 435–442 (2019)
4. Franzone, P.C., Guerri, L., Rovida, S.: Wavefront propagation in an activation model of the anisotropic cardiac tissue: asymptotic analysis and numerical simulations. J. Math. Biol. **28**(2), 121–176 (1990)
5. Mitchell, C.C., Schaeffer, D.G.: A two-current model for the dynamics of cardiac membrane. Bull. Math. Biol. **65**(5), 767–793 (2003)
6. Neic, A., et al.: Efficient computation of electrograms and ECGs in human whole heart simulations using a reaction-eikonal model. J. Comput. Phys. **346**, 191–211 (2017)
7. Newville, M., Stensitzki, T., Allen, D.B., Rawlik, M., Ingargiola, A., Nelson, A.: LMFIT: Non-linear least-square minimization and curve-fitting for python. Astrophysics Source Code Library, pp. ascl-1606 (2016)
8. Niederer, S.A., Kerfoot, E., Benson, A.P., Bernabeu, M.O., et al.: Verification of cardiac tissue electrophysiology simulators using an n-version benchmark. Philos. Trans. Roy. Soc. A Math. Phys. Eng. Sci. **369**(1954), 4331–4351 (2011)
9. Sethian, J.A.: A fast marching level set method for monotonically advancing fronts. Proc. Nat. Acad. Sci. **93**(4), 1591–1595 (1996)
10. Wang, Y., Cuculich, P.S., Zhang, J., Desouza, K.A., Vijayakumar, R., et al.: Noninvasive electroanatomic mapping of human ventricular arrhythmias with electrocardiographic imaging. Sci. Transl. Med. **3**(98), 98ra84–98ra84 (2011)

Simplified Electrophysiology Modeling Framework to Assess Ventricular Arrhythmia Risk in Infarcted Patients

Dolors Serra[1], Pau Romero[1], Miguel Lozano[1,2], Ignacio García-Fernández[1,2], Alejandro Liberos[1,3], Miguel Rodrigo[1,3], Antonio Berruezo[4], Alfonso Bueno-Orovio[5], and Rafael Sebastian[1,2(✉)]

[1] Computational Multiscale Simulation Laboratory (COMMLAB), Valencia 46100, Spain
[2] Department of Computer Science, Universitat de Valencia, Valencia, Spain
rafael.sebastian@uv.es
[3] Department of Electronics, Universitat de Valencia, Valencia, Spain
[4] Cardiology Department, Heart Institute, Teknon Medical Center, Barcelona, Spain
[5] Department of Computer Science, University of Oxford, Oxford, UK

Abstract. Patients that have suffered a myocardial infarction are at lifetime high risk for sudden cardiac death (SCD). Personalized 3D computational modeling and simulation can help to find non-invasively arrhythmogenic features of patients' infarcts, and to provide additional information for stratification and planning of radiofrequency ablation (RFA). Currently, multiscale biophysical models require high computational resources and long simulations times, which make them impractical for clinical environments. In this paper, we develop a phenomenological solver based on cellular automata to simulate cardiac electrophysiology, with results comparable to those of biophysical models. The solver can run simulations in the order of seconds and reproduce rotor dynamics, and ventricular tachycardia in infarcted patients, using a virtual pacing protocol. This model could be use to plan RFA intervention without the time constrains of complex models.

Keywords: Cellular automata · Cardiac electrophysiology simulation · Therapy planning · Arrhythmia

1 Introduction

Patients that have suffered a myocardial infarction are at lifetime high risk for sudden cardiac death (SCD) [1], which accounts for 50% of ischemic heart disease related deaths. Current methods for predicting SCD are mainly based on the left ventricular ejection fraction, but show limited clinical accuracy, not identifying many patients at high risk of death. Infarct-related ventricular tachycardias (VTs) are commonly associated with the so-called slow conducting channels (SCC), which are pathways composed of surviving myocytes across the

© Springer Nature Switzerland AG 2021
D. B. Ennis et al. (Eds.): FIMH 2021, LNCS 12738, pp. 531–539, 2021.
https://doi.org/10.1007/978-3-030-78710-3_51

infarct scar, responsible for VT initiation and maintenance. MRI-based substrate characterization has started to be used for SCCs delineation, and planning and guiding radio-frequency ablation (RFA) procedures in patients suffering infarct-related VTs [6,12]. However, an step forward would be to perform preventive substrate ablation (i.e. abolition of SCC) of patients at highest risk of developing lethal arrhythmias, based on the information of their MRI.

Advanced techniques based on personalized 3D computational modeling and simulation can help to find non-invasively arrhythmogenic features of patients' infarcts, and to provide additional information for stratification and planning of RFA [4,8,13]. Although such in-silico approaches have significant potential, it is still unclear what is the level of detail required by the model to be clinically useful and cast accurate predictions. Previous computational studies aiming to predict VT and SCD risk have mostly been based on a detailed modeling of human electrophysiology, i.e. multiscale biophysical models including specific cellular-level ionic modeling [2,9,13]. Within those detailed models, the most relevant differences are in the way the border zone of the infarct region is modeled: including fibrosis patches [9], or electrical remodeling [10]. Some of those models were able to predict the electrical circuits that sustained the tachycardia, and to induce the VT using a virtual pace-mapping protocol. However, the computational cost associated to these studies is very high (in the order of days using a high performance computing), and requires a very advanced expertise to build and run the simulations. Along these lines, alternative in-silico approaches, able to reduce model complexity but still providing accurate results, and with higher computational efficiency, are necessary to promote clinical uptake and impact [3,4,11]. Finally, fast computer simulations could be also used to create large databases of simulations with the aim of training artificial intelligence systems that help electrophsyiologist in therapy planning [5,7]. In this work, we propose the use of advanced cellular automata, encoding the dynamic properties of human myocardium in healthy and post-infarct conditions, in order to efficiently simulate the electrical activity in patients that have suffered a myocardial infarction. First, we calibrate the cellular automata and show its behavior in a simplified geometrical tissue model. Next, we use the calibrated automata in a patient specific model, including an infarct region, where we mimic the results of a detailed biophysical model.

2   Materials and Methods

We have developed a fast cardiac electrophysiology simulator based on cellular automata to assess arrhythmia risk, as described below. We compare the performance and results of simulations carried out with our cellular automata with respect to a detailed multi-scale biophysical model based on the monodomain formulation and the ten Tusscher ionic cellular model of human ventricular electrophysiology [14].

2.1   Electrophysiology Modeling with Cellular Automata

To model the electrophysiology and conduction of human cardiac tissue, we use a spatially extended, event-based, asynchronous cellular automata. Cells of the automaton, defined as hexahedral elements, are distributed within the region of space occupied by cardiac tissue. The neighborhood of each cell is formed by all adjacent cells including diagonals, making a total of 26 neighbors per node (extension to 3D of Moore's neighborhood). A cell of the automaton represents a portion of cardiac tissue that can be in two main states: active (depolarized) and inactive (repolarized). The behavior we want to model can be described as follows: when a cell that is inactive is excited, it activates; during the activation process it can depolarize the neighboring cells; it remains active during its action potential duration (APD), and does not accept further activation until this time expires. Once the duration of the APD is completed, the cell repolarizes and returns to the inactive state. The time between the deactivation and the eventual next activation is the diastolic interval (DI). When a cell activates, the excitation to the neighbors takes a finite, positive time that depends on the conduction velocity (CV) and the distance between the active and the inactive cells.

In order to incorporate the physiological response of myocytes into the model, we use restitution curves for APD (APDR) and CV (CVR), which depend on the DI after each activation. In particular, we make use of the APDR and CVR from the model of Ten Tusscher [14], differentiating endocardial, mid-myocrdial and epicardial cells. The curves have been obtained from the interpolation of the data published in the corresponding paper. Note that we can incorporate the data from any published model, so that the user can choose which values to use in their simulations.

We also include some additional terms, such as APD memory, which ensures smooth changes in APD in response to large cycle length gradients (e.g. S1–S2 protocols). Therefore, when the DI is updated between two consecutive activations, the resulting APD is obtained from from a weighted average between the previous APD and the one resulting from the APDR. The weight used in the model is 0.8 for the previous APD and 0.2 for the actual one, but can be tune by the user. In addition, we consider the electrotonic coupling effects, especially important in regions where pathological and healthy cells are in contact, by performing a weighted average of the cell APD and its directly-connected neighbors. Axisymmetric fiber orientation is included as well in the model by considering a ratio between longitudinal and transversal CVs along the fiber and across fiber directions.

2.2 Geometrical Models

To validate the cellular automata and the propagation on graphs, we built a 3D slab of tissue of $100 \times 100 \times resolution$ mm, composed of hexahedral elements at different resolutions (0.2, 0.4, 0.8 mm). The different meshes were used to analyze the errors in propagation due to discretization and mesh resolution. No fiber orientation was considered, leading to isotropic propagation.

Next, to evaluate the cellular automata model in a realistic human geometry, we have made use of the patient-specific anatomical model developed in [9], from a monomorphic VT patient due to a large chronic myocardial infarction. In brief, the model corresponds to a biventricular human heart anatomy segmented from a DE-MRI sequence, and meshed with hexahedral elements with an average length of 0.4 mm. It includes fiber orientation, based on the Streeter model, and segmentation of the infarct area (core and border zone), based on MRI gray intensity level (border zone gray level intensity between mean + 2 × SD). In this study, we only made use of the left ventricle (2 million nodes).

3 Results

We carried out electrophysiology simulations in tissue slabs and in a detailed left ventricular model, and compared the results of our proposed human-based cellular automata model with biophysically-detailed simulations using the monodomain model.

3.1 Simplified 3D Model

We performed a S1-S2 stimulation protocol to study the feasibility to simulate spiral wave activity in a healthy slab of tissue. The tissue considered human APDR and CVR, based on the ten Tusscher model [14], which allowed us to obtain stable reentry. Figure 1 shows the patterns simulated after applying five flat S1 stimuli on the Y axis with a basic cycle length (BCL) of 500 ms, followed by a single S2 stimulus coupled to the action potential tail with a BCL of 400 ms. The period of the spiral wave rotation stabilized at 250 ms, and showed some localized meandering. As it can be observed in the second row, where colors represent the APD, near the tip of the rotor APDs were shorter, due to the shorter DI. This gradient was observed through time, although the APDs shortened as the rotor stabilized. APD memory included in the cellular automaton prevents large changes in APD in short time periods, requiring more cycles to converge to a stable APD map. The same simulation protocol was carried out for the three mesh resolutions considered, resulting in identical results (not shown), demonstrating that the method is largely insensitive to the underlying mesh.

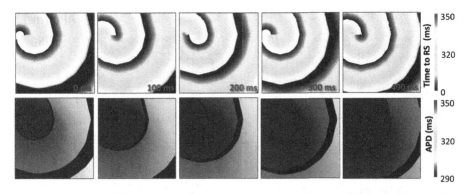

Fig. 1. Simulation of rotor activity in a slab of tissue of $100 \times 100 \times 0.2$ mm. Consecutive time instants observed every 100 ms, for total simulation time of 400 ms. First row shows color-coded reentrant activity by time since the activation (red tones), to the resting state (RS) (blue tones). Second row shows APD spatial maps, with shorter APDs observed near the rotor tip, which evolve over time as rotor stabilizes. (Color figure online)

3.2 Ventricular Model

In the anatomical model, an inducibility study was conducted using biophysical simulations (solving modomian equation) to replicate the patient's monomorphic VT circuit, using a virtual sequential pacing protocol triggered from multiple epicardial locations, as detailed in [9]. The post-infarcted ventricular model considered regions of healthy tissue, border zone, and (non-conductive) core zone. Human cellular electrophysiology was modeled in the healthy and border zone regions using the ten Tusscher model [14]. Following [10], the border zone included cellular electrical remodeling (I_{Na} reduced to 38% of its normal value, I_{CaL} to 31%, and I_{Kr} and I_{Ks} to 30% and 20%, respectively), together with a decrease of 75% in tissue CV. We did not model fibroblast content in the border zone region. In the cellular automata, the aforementioned post-infarction changes in human cardiac function were introduced in the model by means of specific APDR and CVR curves.

As in [9], large APD gradients were observed when the models were paced with a S1-S2 protocol. Figure 2 shows epicardial views of APD maps after a train of 6 stimuli at a BCL of 600 ms. The biophysical model showed an important spatial variability of APD, but the distribution and heterogeneity were very similar between models. This highlights that the proposed human cellular automata model is able to successfully capture, for each given DI, not only APD magnitude but dynamic APD memory effects, as well as the electrotonic properties of tissue coupling.

Using the cellular automata model, we applied an S1–S2 protocol to mimic the pace-mapping procedure usually conducted in the EP room. We paced the heart from the epicardial wall (see Fig. 3, first row), with a BCL of 600 ms for 10 cycles. This was followed by pacing with a BCL of 300 ms from the same

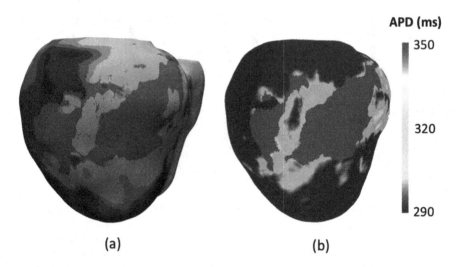

Fig. 2. Comparison of simulated APDs. APDs corresponding to simulations using (a) a biophysical ionic model, and (b) a cellular automata, after pacing at the epicardium six stimuli with a BCL of 600 ms.

location, which produced a conduction block at the entrance of a SCC mostly formed by border zone tissue. Figure 3 shows the simulation at specific time points after the S1 ans S2 stimuli, where colors represent the time remaining to resting state (to finish the current action potential) of each model node. Note that only the border zone region is displayed in the snapshots for visualization purposes, the healthy tissue and scar tissue are transparent, and the endocardial surface is displayed in gray color. The VT circuit stabilized after three cycles to a period of 350 ms (compared to 506 ms for the biophysical simulation), more close resembling clinical observations of 340 ms. This implies that CVs in healthy and BZ tissues were properly encapsulated by the proposed cellular automata model.

Regarding efficiency of the proposed solver, simulations on the 2.6 million element left ventricular model were performed on a laptop computer (2.9 GHz Intel i5 dual-core, with 8 GB 1867 MHz DDR3). Two beats (BCL of = 600 ms) required 76 s to be simulated, which is a large cost reduction (×311 faster) compared to 7 h, which corresponds to the use of a biophysical solver on a cluster computer with 64 cores at 2.3 GHz (4 x AMD Opteron 6376 processors). New approached based on GPU computing could also reduce the simulations times, however, they are very dependent on the hardware which makes them difficult to maintain or scale due to memory limitations.

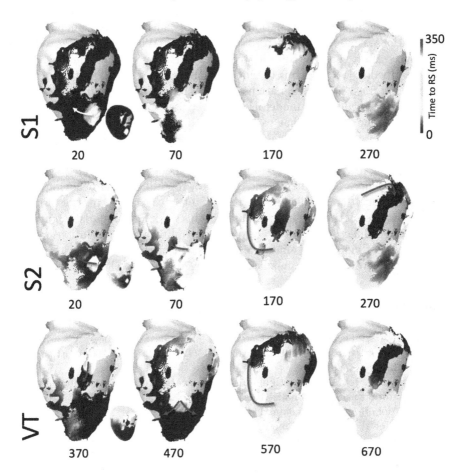

Fig. 3. Simulation time snapshots (ms) showing the border zone region in proposed cellular automata model. Episode of VT obtained after applying a S1–S2 protocol. First row corresponds to an activation using a BCL of 600 ms, second and third rows are the results after coupling an S2 stimulus of 300 ms. Colors correspond to the time to resting potential. Conduction block can be observed at 170 ms after S2. VT is subsequently sustained across the conduction channel. Arrows show the direction of the wavefront.

4   Discussion and Conclusions

We have developed a fast electrophysiology simulator based on cellular automata and graph-based propagation. The cellular automata incorporates cell dynamic behavior thanks to the consideration of APD and CV restitution properties of the ten Tusscher biophysical model, but can include the properties of any other desired model or experimental data. The electrical tissue propagation based on a graph and a priority event list makes the result independent from the mesh resolution, and uses a dynamic discretization time to calculate exact diffusion times. The inclusion of APD memory into the cells, and the electrotonic effects in the neighboring, make the simulations more stable, although they might require some additional time to converge to stationary states. With the software, we were able to successfully simulate rotor activity in a slab of tissue, with similar behavior as described in [14], which is the model we have based on our automata. The wavefront obtained in the simulations did not show a smooth curvature, which was due to the neighboring model used and the fact that the mesh was regular. In the future, the approach has to be improved by considering the direction of the wavefront propagation, plus a weighted contribution of neighbors activation times.

Our results in a ventricular computational model that included scar and border zone regions showed that the automata reproduced the behavior of the biophysical model, and could induce the VT observed in the patient. Finally, although high-resolution meshes were used in this work for comparison purposes, the proposed cellular automata model does not require of high resolutions meshes as in the case of biophysical simulations. This will further reduce its associated computational cost, bringing in-silico technologies closer to clinical practice.

References

1. Adabag, A.S., Therneau, T.M., Gersh, B.J., Weston, S.A., Roger, V.L.: Sudden death after myocardial infarction. JAMA **300**(17), 2022–2029 (2008). https://doi.org/10.1001/jama.2008.553
2. Arevalo, H.J., et al.: Arrhythmia risk stratification of patients after myocardial infarction using personalized heart models. Nat. Commun. **7**, 11437 (2016). https://doi.org/10.1038/ncomms11437
3. Chen, Z., et al.: Biophysical modeling predicts ventricular tachycardia inducibility and circuit morphology: a combined clinical validation and computer modeling approach. J. Cardiovasc. Electrophysiol. **27**(7), 851–860 (2016). https://doi.org/10.1111/jce.12991
4. Corral-Acero, J., et al.: The 'digital twin' to enable the vision of precision cardiology. Eur. Heart J. **41**(48), 4556–4564 (2020). https://doi.org/10.1093/eurheartj/ehaa159
5. Doste, R., et al.: In silico pace-mapping: prediction of left vs. right outflow tract origin in idiopathic ventricular arrhythmias with patient-specific electrophysiological simulations. Europace **22**(9), 1419–1430 (2020). https://doi.org/10.1093/europace/euaa102
6. Fernández-Armenta, J., et al.: Three-dimensional architecture of scar and conducting channels based on high resolution ce-CMR: insights for ventricular tachycardia ablation. Circ. Arrhythm. Electrophysiol. **6**(3), 528–537 (2013). https://doi.org/10.1161/CIRCEP.113.000264
7. Godoy, E.J., et al.: Atrial fibrosis hampers non-invasive localization of atrial ectopic foci from multi-electrode signals: a 3d simulation study. Front Physiol. **9**, 404 (2018). https://doi.org/10.3389/fphys.2018.00404
8. Lopez-Perez, A., Sebastian, R., Ferrero, J.M.: Three-dimensional cardiac computational modelling: methods, features and applications. Biomed Eng. Online **14**, 35 (2015). https://doi.org/10.1186/s12938-015-0033-5

9. Lopez-Perez, A., Sebastian, R., Izquierdo, M., Ruiz, R., Bishop, M., Ferrero, J.M.: Personalized cardiac computational models: from clinical data to simulation of infarct-related ventricular tachycardia. Front Physiol. **10**, 580 (2019). https://doi.org/10.3389/fphys.2019.00580

10. McDowell, K.S., Arevalo, H.J., Maleckar, M.M., Trayanova, N.A.: Susceptibility to arrhythmia in the infarcted heart depends on myofibroblast density. Biophys. J. **101**(6), 1307–1315 (2011). https://doi.org/10.1016/j.bpj.2011.08.009

11. Relan, J., et al.: Coupled personalization of cardiac electrophysiology models for prediction of ischaemic ventricular tachycardia. Interface Focus **1**(3), 396–407 (2011). https://doi.org/10.1098/rsfs.2010.0041

12. Soto-Iglesias, D., et al.: Cardiac magnetic resonance-guided ventricular tachycardia substrate ablation. JACC Clin. Electrophysiol. **6**(4), 436–447 (2020). https://doi.org/10.1016/j.jacep.2019.11.004

13. Trayanova, N.A., Pashakhanloo, F., Wu, K.C., Halperin, H.R.: Imaging-based simulations for predicting sudden death and guiding ventricular tachycardia ablation. Circ. Arrhythm Electrophysiol. **10**(7) (2017). https://doi.org/10.1161/CIRCEP.117.004743

14. ten Tusscher, K.H.W.J., Noble, D., Noble, P.J., Panfilov, A.V.: A model for human ventricular tissue. Am. J. Physiol. Heart Circ. Physiol. **286**(4), H1573–H1589 (2004). https://doi.org/10.1152/ajpheart.00794.2003

Sensitivity Analysis of a Smooth Muscle Cell Electrophysiological Model

Sanjay R. Kharche[1,2]([✉]) [iD], Galina Yu. Mironova[3], Daniel Goldman[2],
Christopher W. McIntyre[1], and Donald G. Welsh[3]

[1] Lawson Health Research Institute, London N6A 5W9, Canada
Sanjay.Kharche@lhsc.on.ca
[2] Department of Medical Biophysics, Western University, London N6A 5C1, Canada
[3] Robarts Research Institute, Western University, London N6A 5C1, Canada

Abstract. Cardiac smooth muscle cell mathematical models are increasingly being used in clinical decision making and drug testing. The cell models also have the potential to assist interpretation and extending of our multi-scale experimental findings. Components of the models interact with each other to regulate model behavior in a non-linear manner. To permit meaningful deployment of the models, it is therefore a necessity to understand the regulatory significance of model components' parameters on the model's behavior. In this study, the regulation of mean intra-cellular calcium and mean membrane potential (model behavior) by underlying model parameters (regulators) in a smooth muscle cell mathematical model was quantified using two sensitivity analysis methods. It was found that extracellular electrolytes and gating kinetics are prime model behavior regulators. A representative case relevant to widespread hypertension focusing on the L-type channel's parameters is presented. This sensitivity analysis will guide our future data driven modelling efforts.

Keywords: Smooth muscle cell · Mathematical model · Coronary vasculature · Sensitivity analysis

1 Introduction

Electrically active building blocks of blood vessels called smooth muscle cells (SMCs) effect vascular tone [1, 2] in arteries, including those in cerebral and coronary arteries. We think that debilitating conditions such as hypertension and renal failure alter L-type calcium ion channel (I_{CaL}) function and promote accumulation of intracellular calcium ($[Ca^{2+}]_i$) in SMCs [3] (see Fig. 1). In accordance with experimental protocols, SMC model behavior is characterized by $[Ca^{2+}]_i$ and membrane potential, V_m. An understanding of how I_{CaL} parameters regulate SMC mathematical model behavior will guide our future experimental-modelling investigations. Specifically, the parameter ranking obtained from sensitivity analysis will reduce the degrees of freedom in our optimization based model fitting to ion channel patch clamp data, thus accelerating model refinement.

© Springer Nature Switzerland AG 2021
D. B. Ennis et al. (Eds.): FIMH 2021, LNCS 12738, pp. 540–550, 2021.
https://doi.org/10.1007/978-3-030-78710-3_52

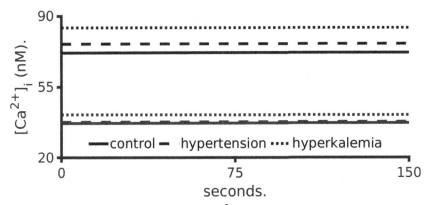

Fig. 1. Effect of cell size on SMC intracellular $[Ca^{2+}]_i$. Control (capacitance of 25 pF) is shown as black lines while small SMC (capacitance of 15 pF) is illustrated using blue lines. Hypertension (simulated as an increase of inactivation time constant) is represented by dashed lines while hyperkalemia (simulated using $[K^+]_o = 10$ mM) is represented by dotted lines. (Color figure online)

In multiple SMC modelling studies, the regulation of model behavior by either maximal transport rates and conductances [4, 5] or biochemical availability [6] has previously been demonstrated. However, other modelling parameters that are pathophysiologically relevant also play a vital role in model behavior (see Fig. 2). We have previously observed that the inherently coupled and non-linear nature of mathematical model components (e.g. coupled ion channel gating and intracellular $[Ca^{2+}]_i$ processes) necessitates the inclusion of all model parameters in meaningful sensitivity analysis [7]. The sensitivity analysis may provide limited insights if a subset of the model's control parameters were to be examined, and is essentially a local analysis in parameter space.

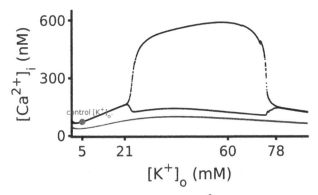

Fig. 2. Bifurcation diagram illustrating regulation of $[Ca^{2+}]_i$ dynamics by extracellular potassium, $[K^+]_o$, in control (black) and small (blue) SMCs. Under control conditions, low concentrations of $[K^+]_o$ (less than 21 mM), the SMC model generated a steady state $[Ca^{2+}]_i$. For larger values of $[K^+]_o$ (21 mM to 78 mM), $[Ca^{2+}]_i$ oscillations were observed. In the small SMC, the $[Ca^{2+}]_i$ oscillations' amplitude was found to be small. (Color figure online)

Furthermore, smooth muscle cells size, quantified using capacitance, is known to be smaller in the arterioles of multiple organs [1, 8, 9] as compared to the existing smooth muscle model. As a result of the different sizes, ion current densities that regular intracellular $[Ca^{2+}]_i$ dynamics in smooth muscle cells are expected to be tissue specific. Our *in silico* studies, including the presented preliminary sensitivity analysis of the smooth muscle cell model electrophysiology, take into account multiple modelling parameters in light of experimental findings where $[Ca^{2+}]_i$ increase has been attributed to augmented ion channel expression density [10], altered ion channel gating pathways independently of conductance alterations [11], as well as cell size. The functional relevance of parameters in cardiomyocyte and neuronal models has been widely studied. The sensitivity of one model output (e.g. membrane potential) to a parameter of interest (e.g. L-type channel conductance), termed as local analysis, has been explored using derivative [12] and individual parameter perturbation representing experimentally controlled conditions [13, 14]. Relevant to this work, regression based methods have been used to generate experimentally testable predictions in the study of calcium sparks [15]. However, perturbation of one parameter is unrepresentative of system wide alterations caused by diseases such as hypertension. To overcome the limitation, a widening literature has adopted regression based methods that elicit model information from outputs obtained using simultaneously perturbing several model regulators (parameters, also called inputs) [16, 17]. One such global method relies on partial rank correlation coefficients (PRCC) [18] that assumes monotonic input-output relationships driven by the inherent correlation driven analysis. To encompass non-monotonic relationships formed of potentially inter-dependent parameters [19], we used information theory to compute entropy (i.e. uncertainty of univariate data) as well as mutual information (MI, i.e. redundancy in multi-variate data) [7, 20, 21]. PRCC and MI present complementary understanding of the system's behaviour and may not be simplistically comparable measures of sensitivity. Other methods [22–24] will be deployed in future studies.

In this preliminary work, an established smooth muscle cell model [4] was used to rank the regulatory impact of all parameters in a SMC model [4] on its mean intracellular $[Ca^{2+}]_i$ and mean membrane potential, V_m. To do so, two complementary sensitivity analysis methods [7, 18] were adopted. As a case of interest, findings focused on parameters related to I_{CaL} and cell capacitance are presented. The complementary information obtained from PRCC and MI are presented.

2   Methods

The smooth muscle cell, SMC, model [4] adopted in this work is a system of 27 coupled ordinary differential equations. The model computer program was obtained from CellML repository [25] and programmed into our robust ordinary differential equations solver based on an advanced external library [26]. The model has 48 parameters that consist of all conductance values, maximum transport fluxes, steady state gating parameters, time constant scaling factors, extracellular electrolyte concentrations, and the capacitance. Parameters' descriptions and acronyms, as well as their control values relevant to this work are provided in Table 1. Model behavior was defined as mean values of $[Ca^{2+}]_i$ and membrane potential, V_m.

Table 1. Relevant model parameters (model inputs).

Parameter.	Description.	Control value.
Extracellular electrolytes and capacitance.		
$[Ca^{2+}]_o$	Calcium.	2 mM.
$[Na^+]_o$	Sodium.	140 mM.
$[K^+]_o$	Potassium.	5 mM.
$[Cl^-]_o$	Chlorine.	129 mM.
C_m	Cell capacitance.	25 pF.
Membrane current conductance values.		
P	L-type calcium channel conductance.	1.88×10^{-5} cm/s.
g_K	Delayed rectifier potassium channel conductance.	1.35 nS.
Intra-cellular calcium release/uptake.		
I_{SERCA}	Maximum $[Ca^{2+}]_i$ uptake by sarcoplasmic reticulum (SR).	6.68 pA.
K_{leak}	SR $[Ca^{2+}]_i$ leak parameter.	1.07×10^{-7} (unitless).
L-type calcium channel gating.		
τ_a	Time constant of activation.	seconds. (fast).
τ_i	Time constant of inactivation.	seconds (slow).
$V_{1/2,a}$	Half voltage of steady state activation.	0 mV.
$V_{1/2,i}$	Half voltage of steady state inactivation.	-42 mV.
k_a	Slope of steady state activation.	8.3 mV.
k_i	Slope of steady state inactivation.	9.1 mV.

To permit sensitivity analysis, a control model population of 10^5 instances was constructed. To generate the population, 48 modelling parameter were each randomly sampled simultaneously from Gaussian distributions using a non-repetitive Mersenne Twister random number generator [27]. The sampling was constrained using Latin Hypercube Sampling [28]. The mean values for each parameter's Gaussian distribution was obtained from the original model [4], while a coefficient of variation of 1.5 was adopted. The adopted coefficient of variation provided a large ($\pm 100\%$) range sampling for each parameter. A model instance in the population was defined by one sampling, and was used in the model to generate mean $[Ca^{2+}]_i$ and V_m. Each model instance required 9.5 s to generate a simulated temporal activity of 100 s using one Linux processor. The population of 10^5 models was generated using 128 processors within four hours using open source GNU tools. In addition to the control population, a second population where cell capacitance was reduced to 15 pF was also generated. The model parameters and model outputs were stored and further processed using two sensitivity analysis methods.

Sensitivity analysis which ranked parameters according to their impact on model behavior was performed using partial rank correlation coefficients (PRCC) [18] as well as mutual information [7].

To compute PRCC, first the normally distributed parameters (x_i) as well as the observed outputs (y_i) were rank transformed. Then, the linear effects of other additional variables are accounted for by expressing each as a linear regression of the inputs,

$$\hat{x}_j = a_o + \sum_{\substack{k=1 \\ k \neq j}}^{N} a_k x_k, \text{ and } \hat{y}_j = b_o + \sum_{\substack{k=1 \\ k \neq j}}^{N} b_k x_k. \tag{1}$$

Using residuals defined as $r_{x_j} = x_j - \hat{x}_j$ and $r_{y_j} = y_j - \hat{y}_j$, PRCC is defined as the correlation among these residuals normalized using their respective variances, i.e.

$$PRCC(x_i, y_j) = \frac{Cov(r_{x_i}, r_{y_j})}{Var(r_{x_i})Var(r_{y_j})}. \tag{2}$$

As evident in Eq. 1, PRCC assumes an underlying statistical model that is linear (regression), and assumption of monotonicity provides the strength of the linear relationship between a given pair of parameter and output [18, 29]. The range of PRCC indices is from -1 to $+1$ by its mathematical definition in Eq. 2.

The use of information theory for sensitivity analysis was motivated to overcome PRCCs assumptions of linearity, monotonicity, and parameter independence. To construct an information theoretic motivated mutual information index [30, 31], the probability mass functions of each of the inputs and outputs are constructed using the discreet data generated above. Entropy, H, of a random variable, x_i, is then defined as $H(x_i) = \sum_k -p_j \ln p_j$. Mutual information is defined using the individual probabilities of inputs and outputs, $p(x_i)$ and $p(y_i)$ respectively, along with the joint probabilities, $p(x_i, y_j)$, as

$$I(x_i; y_j) = \sum_k p(x_{i,k}, y_{j,k}) \ln\left(\frac{p(x_{i,k}, y_{j,k})}{p(x_{i,k})p(y_{j,k})}\right) = H(x_i) + H(y_j) - H(x_i, y_j) \tag{3}$$

which provides a measure of the amount of mutual dependence of a given parameter on a chosen model output and vice-versa. Equation 3 also shows the relationship of mutual information to individual statistical variable entropy (i.e. uncertainty). Paraphrased, mutual information is the reduction in uncertainty of the modelling parameter, x_i, given that model behaviour, y_j ($[Ca^{2+}]_i$ or V_m) have been observed experimentally. It can be appreciated from Eq. 3 that mutual information is symmetric, *i.e.* measurement of either the modelling parameter or model behaviour is sufficient to provide the information content of one in the other. Normalization of the mutual information (Eq. 3) using the sum of input and output entropies provides a normalized mutual information index (NMII), which permits comparison across parameters and populations. The NMII ranges between 0 (no information of parameter in the observed output) and $+$ 1 (model parameter can be absolutely deduced from the model behaviour observation) when mutual information is computed in bits.

Regression based methods such as PRCC account for the linear contributions of other variables in contrast to NMII. Further, NMII does not assume monotonicity but does take the model's non-linear behavior into account by means of variable entropy. These factors imply that PRCC and NMII provide distinct sensitivity information.

Whereas the sensitivity analysis considered all model parameters (48 in total), Table 1 provides the 15 parameters whose sensitivities to $[Ca^{2+}]_i$ and V_m are presented in this study.

3 Results

As illustrated in Fig. 3, the maximal PRCC values regarding intracellular calcium ($[Ca^{2+}]_i$) belong to extracellular calcium, cell capacitance, and I_{CaL} activation gating parameter, k_a. The I_{CaL} gating parameters, $V_{1/2,a}$ and $V_{1/2,i}$, have larger numerical values for the PRCC of $[Ca^{2+}]_i$, in comparison to the ion channel conductances, P and g_K. $[Ca^{2+}]_i$ transport rates, represented by I_{SERCA} and K_{leak}, have a low intracellular PRCC ranking.

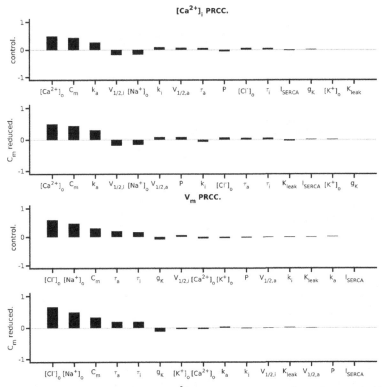

Fig. 3. PRCCs for intracellular calcium ($[Ca^{2+}]_i$, top two panels) and membrane potential (V_m, bottom two panels) each under control and reduced capacitance (C_m reduced) conditions. Symbols are described in Table 1.

The maximal membrane potential (V_m) PRCC values belong to extracellular chloride, sodium, as well as cell capacitance. The I_{CaL} gating parameters, τ_a and τ_i, are ranked higher than the I_{CaL} conductance by the PRCC sensitivity analysis. The $[Ca^{2+}]_i$ transport processes were seen to have minimal ranking. The PRCC ranking of the potassium conductance, g_K, is higher for membrane potential as compared to intra-cellular calcium. Reduction of capacitance altered the numerical values of PRCCs, but rankings were largely unaffected.

Normalized mutual information indices (NMIIs) (Fig. 4) show that model behavior (i.e. both $[Ca^{2+}]_i$ and V_m) depend maximally on extracellular sodium, potassium, I_{CaL} inactivation gating parameters (τ_i and $V_{1/2,i}$), and capacitance. NMIIs ranked all I_{CaL} gating parameters higher than ion channel conductances. Among intracellular calcium transport rates, the uptake process (I_{SERCA}) was consistently ranked higher than the leak process, regulated by parameter K_{leak}. Similar to PRCC, reduction of capacitance altered NMII numerically, but rankings were largely unaffected.

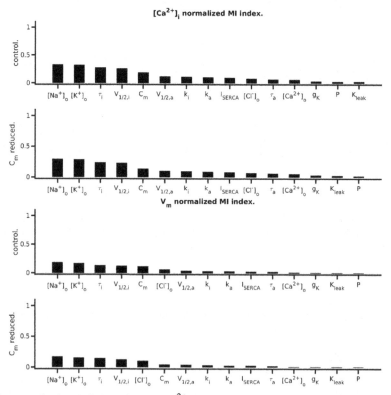

Fig. 4. NMII for intracellular calcium ($[Ca^{2+}]_i$, top two panels) and membrane potential (V_m, bottom two panels) under control and reduced capacitance (C_m reduced) conditions. Symbols are described in Table 1.

4   Conclusions and Discussion

Both sensitivity analysis measures, PRCC as well as NMII, suggest that the prime regulators of model $[Ca^{2+}]_i$ are electrolytes and cell capacitance rather than calcium induced calcium release processes within the Kapela model [11], which is in line with experimental protocols that use external biochemical and biophysical stimulation to study $[Ca^{2+}]_i$ dynamics [1, 32]. In addition, both analyses demonstrated that ion channel gating processes supersede conductances in their impacts on $[Ca^{2+}]_i$ and V_m, which is also aligned with experimental findings related to I_{CaL} alterations under hypertension [11]. The SMC model used in this study [4], similarly to several cardiac models [13, 14, 33], rely upon a membrane I_{CaL} calcium induced sarcoplasmic calcium release mechanism to represent $[Ca^{2+}]_i$ dynamics. The higher impact of I_{CaL} in comparison to intracellular calcium transporter processes was reflected by both, PRCC and NMII.

PRCC and NMII provide very different insights regarding model properties. The partial regression based coefficients (PRCC) provide a quantitative manner in which model parameters can be estimated in relation to experimental data using appropriate methods [34]. It can be appreciated that PRCC based parameter estimation is likely to be in the vicinity of the baseline model, after which new PRCC values are required. Additionally, PRCC is expected to provide better assessments if the statistical variables are independent of each other. As illustrated in Fig. 5, the outputs are tightly coupled to each other thus rendering PRCC as a local rather than global sensitivity method. On the other hand, mutual information provides a quantitative index providing co-dependence information. Paraphrased, the information index provides a measure for the number of likely distinguishable system states (values of $[Ca^{2+}]_i$ and V_m) for a given value of model parameters. A knowledge of experimentally, and eventually clinically, relevant outputs will provide information and permit fine tuning of the highly ranked parameters. Thus, mutual information informs the modelling regarding the most important subset of parameters that should be estimated, thus reducing computation effort considerably without compromising accuracy of the fitting.

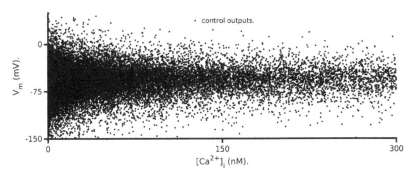

Fig. 5. Co-dependence of outputs. At small and physiological values of $[Ca^{2+}]_i$ (below 60 nM concentrations), a large number of V_m values are allowed by the SMC model system. At high values of $[Ca^{2+}]_i$ (over 150 nM concentrations) corresponding to hypertension, a relatively small number of V_m values were observed.

We note that both indices depend on the distributions used to sample parameters. This study assumed that all model parameters were derived from experimental data, and therefore follow a Gaussian distribution. However, parameters that are based on scant data should be sampled using the maximum entropy principle, and those that are estimated should be sampled from uniform distributions. In general, using uniform distributions will provide optimal sensitivity indices in both PRCC and NMII [7]. Furthermore, potential parameter co-dependence as well as high order statistics will be addressed in future studies. An important part of capturing physiological processes is the underlying modelling structure within which parameters operate and outputs emerge. An assessment of the validity of the model structure is relevant for future model development [35].

As such, disease conditions are known to affect model parameters, and the combination of the two complementary methods used in this preliminary work offer a reliable approach to guide our data driven SMC modelling. Other methods such as derivate based [22–24] and variance based [36, 37] global sensitivity analysis will also be considered in the future.

Acknowledgements. This work was supported by Canada Canarie Inc. (RS3-111, CWM, DG, SRK), Canadian Heart and Stroke funding (G-20-0028717, CWM) and (R4081A03, DG). The Welsh lab (DGW and GM) are funded by the Canadian Institute of Health Research. SRK thanks Compute Canada for HPC resources. All authors approved the manuscript.

References

1. Abd El-Rahman, R.R., et al.: Identification of L- and T-type Ca2+ channels in rat cerebral arteries: role in myogenic tone development. Am. J. Physiol. Heart Circ. Physiol. **304**, H58–H71 (2013)
2. Hansen, P.B., et al.: Functional importance of L- and P/Q-type voltage-gated calcium channels in human renal vasculature. Hypertension **58**, 464–470 (2011)
3. Navedo, M.F., et al.: Increased coupled gating of L-type Ca2+ channels during hypertension and Timothy syndrome. Circ. Res. **106**, 748–756 (2010)
4. Kapela, A., Bezerianos, A., Tsoukias, N.M.: A mathematical model of Ca2+ dynamics in rat mesenteric smooth muscle cell: agonist and NO stimulation. J. Theor. Biol. **253**, 238–260 (2008)
5. Morotti, S., Nieves-Cintron, M., Nystoriak, M.A., Navedo, M.F., Grandi, E.: Predominant contribution of L-type Cav1.2 channel stimulation to impaired intracellular calcium and cerebral artery vasoconstriction in diabetic hyperglycemia. Channels (Austin) **11**, 340–346 (2017)
6. Jacobsen, J.C., Aalkjaer, C., Nilsson, H., Matchkov, V.V., Freiberg, J., Holstein-Rathlou, N.H.: A model of smooth muscle cell synchronization in the arterial wall. Am. J. Physiol. Heart Circ. Physiol. **293**, H229–H237 (2007)
7. Kharche, S., Lüdtke, N., Panzeri, S., Zhang, H.: A global sensitivity index for biophysically detailed cardiac cell models: a computational approach. In: Ayache, N., Delingette, H., Sermesant, M. (eds.) FIMH 2009. LNCS, vol. 5528, pp. 366–375. Springer, Heidelberg (2009). https://doi.org/10.1007/978-3-642-01932-6_40
8. Bowles, D.K., Hu, Q., Laughlin, M.H., Sturek, M.: Heterogeneity of L-type calcium current density in coronary smooth muscle. Am. J. Physiol. **273**, H2083–H2089 (1997)
9. Ulyanova, A.V., Shirokov, R.E.: Voltage-dependent inward currents in smooth muscle cells of skeletal muscle arterioles. PLoS ONE **13**, e0194980 (2018)

10. Pratt, P.F., Bonnet, S., Ludwig, L.M., Bonnet, P., Rusch, N.J.: Upregulation of L-type Ca2+ channels in mesenteric and skeletal arteries of SHR. Hypertension **40**, 214–219 (2002)
11. Ghosh, D., et al.: Dynamic L-type CaV1.2 channel trafficking facilitates CaV1.2 clustering and cooperative gating. Biochimica et biophysica acta. Mol. Cell Res. **1865**, 1341–1355 (2018)
12. Nygren, A., et al.: Mathematical model of an adult human atrial cell: the role of K+ currents in repolarization. Circ. Res. **82**, 63–81 (1998)
13. Fabbri, A., Fantini, M., Wilders, R., Severi, S.: Computational analysis of the human sinus node action potential: model development and effects of mutations. J. Physiol. **595**, 2365–2396 (2017)
14. Kharche, S., Yu, J., Lei, M., Zhang, H.: A mathematical model of action potentials of mouse sinoatrial node cells with molecular bases. Am. J. Physiol. Heart Circ. Physiol. **301**, H945–H963 (2011)
15. Lee, Y.S., Liu, O.Z., Hwang, H.S., Knollmann, B.C., Sobie, E.A.: Parameter sensitivity analysis of stochastic models provides insights into cardiac calcium sparks. Biophys. J. **104**, 1142–1150 (2013)
16. Sobie, E.A.: Parameter sensitivity analysis in electrophysiological models using multivariable regression. Biophys. J. **96**, 1264–1274 (2009)
17. Britton, O.J., et al.: Experimentally calibrated population of models predicts and explains intersubject variability in cardiac cellular electrophysiology. Proc. Natl. Acad. Sci. U.S.A. **110**, E2098–E2105 (2013)
18. Marino, S., Hogue, I.B., Ray, C.J., Kirschner, D.E.: A methodology for performing global uncertainty and sensitivity analysis in systems biology. J. Theor. Biol. **254**, 178–196 (2008)
19. Tondel, K., et al.: Hierarchical cluster-based partial least squares regression (HC-PLSR) is an efficient tool for metamodelling of nonlinear dynamic models. BMC Syst. Biol. **5**, 90 (2011)
20. Ludtke, N., et al.: Information-theoretic sensitivity analysis: a general method for credit assignment in complex networks. J. Roy. Soc. Interface Roy. Soc. **5**, 223–235 (2008)
21. Tremblay, P., Deschamps, I., Baroni, M., Hasson, U.: Neural sensitivity to syllable frequency and mutual information in speech perception and production. Neuroimage **136**, 106–121 (2016)
22. Pasma, J.H., Boonstra, T.A., van Kordelaar, J., Spyropoulou, V.V., Schouten, A.C.: A sensitivity analysis of an inverted pendulum balance control model. Front. Comput. Neurosci. **11**, 99 (2017)
23. Zhang, H.X., Dempsey Jr., W.P., Goutsias, J.: Probabilistic sensitivity analysis of biochemical reaction systems. J. Chem. Phys. **131**, 094101 (2009)
24. MacGregor, R.J., Tajchman, G.: Theory of dynamic similarity in neuronal systems. J. Neurophysiol. **60**, 751–768 (1988)
25. Lloyd, C.M., Lawson, J.R., Hunter, P.J., Nielsen, P.F.: The CellML model repository. Bioinformatics **24**, 2122–2123 (2008)
26. Hindmarsh, A.C., et al.: SUNDIALS: suite of nonlinear and differential/algebraic equation solvers. ACM Trans. Math. Softw. **31**, 33 (2005)
27. Matsumoto, M., Nishimura, T.: Mersenne twister: a 623-dimensionally equidistributed uniform pseudo-random number generator. ACM Trans. Model. Comput. Simul. **8**, 3–30 (1998)
28. Malone, B.P., Minansy, B., Brungard, C.: Some methods to improve the utility of conditioned Latin hypercube sampling. PeerJ **7**, e6451 (2019)
29. Hamby, D.M.: A comparison of sensitivity analysis techniques. Health Phys. **68**, 195–204 (1995)
30. Shannon, A.G., Hogg, J.M., Ollerton, R.L., Luzio, S., Owens, D.R.: A mathematical model of insulin secretion. IMA J. Math. Appl. Med. Biol. **11**, 245–266 (1994)

31. Critchfield, G.C., Willard, K.E., Connelly, D.P.: Probabilistic sensitivity analysis methods for general decision models. Comput. Biomed. Res. Int. J. **19**, 254–265 (1986)
32. Rahman, A., Matchkov, V., Nilsson, H., Aalkjaer, C.: Effects of cGMP on coordination of vascular smooth muscle cells of rat mesenteric small arteries. J. Vascul. Res. **42**, 301–311 (2005)
33. Kurata, Y., Hisatome, I., Imanishi, S., Shibamoto, T.: Dynamical description of sinoatrial node pacemaking: improved mathematical model for primary pacemaker cell. Am. J. Physiol. Heart Circ. Physiol. **283**, H2074–H2101 (2002)
34. Krishna, N.A., Pennington, H.M., Coppola, C.D., Eisenberg, M.C., Schugart, R.C.: Connecting local and global sensitivities in a mathematical model for wound healing. Bull. Math. Biol. **77**, 2294–2324 (2015)
35. Babtie, A.C., Kirk, P., Stumpf, M.P.: Topological sensitivity analysis for systems biology. Proc. Natl. Acad. Sci. U.S.A. **111**, 18507–18512 (2014)
36. Saltelli, A., Ratto, M., Tarantola, S., Campolongo, F.: Sensitivity analysis for chemical models. Chem. Rev. **105**, 2811–2828 (2005)
37. Saltelli, A.: Sensitivity analysis for importance assessment. Risk Anal.: Off. Publ. Soc. Risk Anal. **22**, 579–590 (2002)

A Volume Source Method for Solving ECGI Inverse Problem

Mohamadou Malal Diallo[1,2,3(✉)], Yves Coudière[1,2,3], and Rémi Dubois[1,4,5]

[1] IHU-Liryc, Foundation Bordeaux Université, Pessac, France
mohamadou-malal.diallo@inria.fr
[2] CARMEN Research Team, Inria Bordeaux – Sud-Ouest, Talence, France
[3] Université Bordeaux, IMB, UMR 5251, CNRS, Talence, France
[4] Université Bordeaux, CRCTB de Bordeaux, U1045, Bordeaux, France
[5] INSERM, CRCTB, U1045, Bordeaux, France

Abstract. Electrocardiographic Imaging (ECGI) is a non-invasive procedure that allows to reconstruct the electrical activity of the heart from body surface potential map (BSPM). In this paper, we present a volume model to solve the electrocardiography inverse problem capable to take into account structural informations obtained by imaging techniques. Thedirect problem maps a volume current in the cardiac muscle (ventricles) to the body surface electrical measures. The model is based on coupling bidomain heart model with torso conduction. The corresponding inverse problem is solved with the Tikhonov regularization. Simulated database are used for the evaluation of this method and we compared them to standard method of fundamental solutions (MFS). The sensitivity to noise is also assessed. The correlation coefficients (CC) and the relative error (RE) of activation times were computed. Results show that the CC (respectively RE) median is respectively 0.75 for the volume model and 0.4 for the MFS (respectively 0.31 vs 0.35) on the epicardium. On the endocardium, the CC and the RE median are 0.65 and 0.33 for the volume method. In conclusion, the volume method performs better than the method of fundamental solutions (MFS) for any noise level, and reconstruct in addition endocardial information.

Keywords: Inverse problem · Scar · ECGI · Electrocardiography

1 Introduction

Electrocardiographic imaging (ECGI) is a non-invasive technique that is used to reconstruct the electrical activity of the heart from body surface electrical potential maps (BSPM), and the geometry of the heart and torso.

The most common approach to compute this reconstruction follows the electrostatic theory of perfect volume conductors. It assumes that the electrical field outside of the heart (or just the ventricles in our case) is perfectly defined by the outward current flux or potential distribution on its surface, considered as enclosing a finite volume. In such an approach, the distribution of charges in the enclosed

© Springer Nature Switzerland AG 2021
D. B. Ennis et al. (Eds.): FIMH 2021, LNCS 12738, pp. 551–560, 2021.
https://doi.org/10.1007/978-3-030-78710-3_53

volume is not relevant. Hence the Laplace equation is set on the volume between the epicardial and the body surfaces, and defines a direct mapping from epicardial potentials to body surface potentials. The inverse problem consists in finding the epicardial potentials from the body surface potentials, assuming the additional zero flux condition on the body surface. This is a famous Cauchy problem, notably ill-posed [1]. It is commonly solved as an optimization problem with a Tikhonov regularization. The method of fundamental solutions (MFS) with Tikhonov regularization is commonly used in this case [8]. In clinical care, structural images of the patient are often available, that may show details about the electrical properties inside the cardiac volume. A major question is therefore how to integrate this information in order to drive the inverse problem. The classical approach is limited because it cannot take into account scars inside the heart volume, and because it looks for the electrical field on the epicardial surface.

The homogenized bidomain equations describe the electrical behavior of the heart within the torso in a different manner, since they consider two superimposed electrical fields, the intracellular one u_i supposed to be defined in the myocardium only, with a no flux boundary condition, and the extracellular one u_e, that prolong as the extracardiac one outside the heart. The homogenized equations are strongly coupled, since both electrical fields are found to be defined by an anisotropic Laplace equation with a right hand side equal to $\xi \left(C_m \partial_t (u_i - u_e) + I_{\text{ion}} \right)$, that is the transmembrane current per unit of surface rescaled by the ratios of surface of membrane per unit volume of tissue. This right-hand side is a current per unit of volume in A.m^{-3}.

In this paper, we consider the torso as a heterogeneous volume conductor for the electrical potential u equal to the extracellular potential field in the myocardium, and to the extracardiac one elsewhere. This volume conductor includes the intracavitary blood, the heart and the remaining torso volume. The electrical potential u is controlled by the data of the current per unit of volume defined above. We recall that a similar model is used in other studies like [5,7], where the authors looked to reconstruct the transmembrane potential (TMP) source in the specific case of ischemia.

Our direct problem maps the current per unit of volume, called *volume current source*, in the heart to the BSPM. We can hence try to reconstruct the volume current source, and we are able to take into account different electrical conductivity values in each of the regions, and in particular in scars. We use the approach based on transfert matrix for solving the inverse problem, and it is computed by using a standard finite element method (FEM).

The problem is generally ill-posed and underdetermined, hence we apply the Tikhonov regularization technique. The regularization applies to the volume current source, while it applies on virtual charges spread around the torso volume for the MFS. We will refer to this method as the volume method (VM). We wanted to test the method on public datasets, and therefore used the ECGI consortium database EDGAR [6]. Among EDGAR datasets, only one contains enough data to test our method (KIT-20-PVC-Simulation-1906-10-30-EP-EndoEpi). The dataset is computed on realistic human-based anatomical model. The activation times (AT) recovered by the standard MFS and VM method were compared to the reference AT obtained from the model. The noise sensitivity of the model was also assessed.

2   Methods

2.1   Mathematical Model

The mathematical model derives from the standard bidomain equations, and has the form of a Laplace equation for the potential field $u(t, x)$ in the blood, heart, and torso domains, respectively denoted by Ω_B, Ω_H, and Ω_T (Fig. 1). It reads

$$- \nabla \cdot (\sigma(x) \nabla u(t, x)) = F(t, x) \quad \text{in } \Omega, \tag{1}$$

(where $\Omega = \Omega_B \cup \Omega_H \cup \Omega_T$) with the no flux boundary condition $-\sigma \nabla u \cdot n = 0$ on the torso surface $\partial \Omega$. In this model, the electrical conductivity is the function $\sigma(x)$, piecewise constant:

$$\sigma(x) = \begin{cases} \sigma_H & \text{if } x \in \Omega_H \backslash \Omega_{\text{scar}}, \\ \rho \sigma_H & \text{if } x \in \Omega_{\text{scar}}, \\ \sigma_B & \text{if } x \in \Omega_B, \\ \sigma_T & \text{if } x \in \Omega_T, \end{cases} \tag{2}$$

where the factor $\rho \leq 1$ may be used to decrease the conductivity in the scar volume $\Omega_{\text{scar}} \subset \Omega_H$ (possibly $\Omega_{\text{scar}} = \emptyset$).

The source term $F(t, x)$ accounts for the total ionic and diffusion currents, and has the form

$$F(t, x) = \begin{cases} f(t, x) & \text{if } x \in \Omega_H, \\ 0 & \text{if } x \in \Omega_B \cup \Omega_T. \end{cases} \tag{3}$$

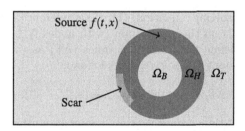

Fig. 1. Sketch of the model with a scar.

Thus, the geometry and the conductivity coefficients are parameters of our model. The equation derives from the bidomain one, and the potential field

$u(t, x)$ is interpreted as the extracellular in the heart Ω_H and the extracardiac potential in the rest of the domain. The body surface potential map is the trace of $u(t, x)$ on the boundary of Ω, denoted by $u_T := u_{|\partial\Omega}$. At each time instant, the forward problem maps the volume current source $f(t, x)$ to the BSPM $u_T(t, x)$. The inverse problem consists in finding an inverse mapping from BSPM data d to a volume current source f in the heart, which direct image best matches d.

In the present work, we put the factor $\rho = 1$ because there is no scar area in dataset.

2.2   Numerical Resolution

Equation (1) was discretized by the P1 Lagrange finite element method (FEM) using the FreeFem++ software [4]. The matrix of the direct problem, called the transfer matrix, is an $N_T \times N_H$ matrix denoted by M, such that $u_T = Mf$, where N_T is the number of electrode nodes on the body surface $\partial\Omega$, and N_H the number of mesh nodes in the heart domain Ω_H. The problem is generally ill-posed and underdetermined, and it need to be regularized for instance by a Tikhonov term. Given data $d \in \mathbb{R}^{N_T}$ on electrode torso nodes, we reconstructed the volume current source $f \in \mathbb{R}^{N_H}$ as the optimum

$$f = \arg\min_{g \in \mathbb{R}^{N_H}} \left\{ \|Mg - d\|^2_{l_2(\mathbb{R}^{N_T})} + \lambda^2 \|g\|^2_{l_2(\mathbb{R}^{N_H})} \right\}, \tag{4}$$

where λ is the regularization parameter. The solution of problem (4) is finally computed by solving the normal equation

$$\left(M^T M + \lambda^2 I\right) f = M^T d, \tag{5}$$

where I is the identity matrix. The CRESO method was used to chose the regularisation parameter λ and the median of all time step regularization parameter was chosen for global parameter.

Given a BSPM time sequence $d = (d_i^n)_{i \in \{0, \cdots, N_T\}}$ with $n \in \{0, \cdots, T\}$, we finally have reconstructed a volume current source time sequence $f = (f_j^n)_{j \in \{0, \cdots, N_H\}}$, and are able to reconstructed an extracellular time sequence $(u_j^n)_{j \in \{0, \cdots, N_H\}}$ by solving equation (1). The reference MFS gives acces a sequence of potentials on the epicardium (with nodes N_e) $(w_k^n)_{k \in \{0, \cdots, N_e\}}$. For each sequence, we define the activation times (AT) as the time of maximum negative derivation along time. It reads as follows:

$$AT_i = \arg\min_{n \in \{0, \cdots, T\}} (s_i^{n+1} - s_i^n),$$

where s_i^n is the signal time sequence at each point P_i on the surface (epicardium or endocardium) at time n. We used also the method proposed by Duchateau et al. [3] to smooth the AT maps. It consists to estimate the delays in activation for neighboring signal locations.

2.3   Simulated Data and Evaluation

Datasets were provided by the ECGI consortium database EDGAR, and were simulated in Karlsruhe Institute of Technology (KIT) [6]. The simulations pacing were performed on a voxel-based grid using a cellular automaton, then the BSPM of these simulated beats were interpolated on tetrahedal mesh and forward-calculated using a finite element method and a bidomain model. The extracellular potentials were extracted from the tetrahedra mesh at 163 electrode positions and at the nodes of the endo- and epicardial surface mesh. The heart model consisted of left and right ventricles and it was based, with thorax model, on real human anatomy MRI data. The data include 8 sets obtained from simulations of single pacing sites distributed as follows: center of septum (SEPTCENTER), left ventricle apex (LVAPEX), left ventricle lateral (LVLAT), left ventricle anterior (LVANT), left ventricle lateral epicardium (LVLATEPI), left ventricle lateral endocardium (LVLATENDO), right ventricle posterior (RVPOST) and right ventricle anterior (RVANT).

Reference activation times (AT) were computed directly from the value of simulated extracellular potential. Using the simulated BSPM, the volume current source were computed with the VM. Afterwards AT were calculated using the method proposed by Duchateau [3] directly for the volume current source. Using the same data, the extracellular epicardial potentials were computed with the classical MFS, and AT calculated.

To evaluate the methods, we compute the correlation coefficients (CC) and the relative error (RE) on the activation times (AT) defined as follows:

$$\text{CC} = \frac{\sum_{i=1}^{N}(\text{AT}_i^{ref} - \overline{\text{AT}}^{ref})(\text{AT}_i^c - \overline{\text{AT}}^c)}{\sqrt{\sum_{i=1}^{n}(\text{AT}_i^{ref} - \overline{\text{AT}}^{ref})^2 \sum_{i=1}^{N}(\text{AT}_i^c - \overline{\text{AT}}^c)^2}} \text{ and } \text{RE} = \sqrt{\frac{\sum_{i=1}^{N}(\text{AT}_i^{ref} - \text{AT}_i^c)^2}{\sum_{i=1}^{N}(\text{AT}_i^{ref})^2}},$$

where AT^{ref} and AT^c denote respectively the reference AT and the computed ones. The number N may be epicardical or endocardial nodes. The numbers $\overline{\text{AT}}^{ref}$ and $\overline{\text{AT}}^c$ are the mean values of AT^{ref} and AT^c respectively.

Correlation coefficients (CC) and Relative errors (RE) between the reconstructed AT and reference ones were computed. The AT obtained with both the VM and the MFS were compared to the reference, simulated, ones. In addtition we could also compare the AT obtained with the VM to reference ones on the endocardium, including septum. The robustness of the method was analyzed by adding different level of noise to the data. We added a gaussian noise to the data with different signal to noise ratio (SNR).

3   Results

In this section, we present the results obtained with the VM and compare them to classical MFS ones using these simulated datasets.

3.1   Data Without Noise

At first, we tested our VM described in Sect. 2 and compared them with MFS on the datasets EDGAR. We report in Table 1 the correlation coefficient (CC) and the relative error (RE) of the activation times (AT) for both methods on all datasets. In addition, a boxplot of these results are depicted in Fig. 3. The Fig. 2 shows an example of the behavior of the reconstructed volume current source.

We see in Table 1 that the VM gives generally a better result in term of correlation coefficient (CC) and relative error (RE) than the MFS in these cases. Not that both the MFS and the VM are inaccurate on the SEPTCENTER dataset. For the data RVPOST, the MFS gives a very bad CC and RE. The best CC and RE are obtained for the data LVLAT for the VM (on the epicardium and the endocaedium) while for the MFS it was for the data LVLATEPI. In Fig. 3, we observe that the VM is better than for the MFS (median CC: 0.75 for the VM and 0.4 for the MFS and median RE: 0.31 for the VM and 0.35 for the MFS) on the epicardium. In addition, the VM could reconstruct the activation times (AT) on the endocardium with acceptable CC and RE (median CC: 0.65 and median RE: 0.33). For the VM, the reconstruction on the epicardium is slightly better than on the endocardium (median CC: 0.75 vs 0.65 and median RE: 0.31 vs 0.33). For the datasets LVLAT and SEPTCENTER, we represent the AT maps (Figs. 4 and 6) and the scatter plot (Figs. 5 and 7). We can see that AT maps for the VM was smoother than for the MFS ones and we observe more clustering of points along horizontal lines for the MFS method. We remark also that late activation times were not well reconstructed for the dataset LVLAT (Fig. 5). For dataset SEPT, the scatter plot (Fig. 7) shows more dispersion that indicates the bad correlation between AT reference and reconstructed ones for both methods.

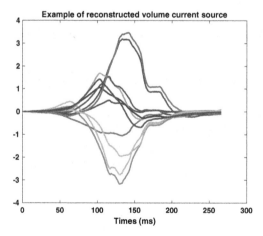

Fig. 2. Example of the behavior of the reconstructed volume current source over time.

Table 1. CC and RE on the AT for the MFS and the VM in all dataset (best result in bold font).

| | VM | | | | MFS | | | |
| | Epi | | Endo | | Epi | | Endo | |
	CC	RE	CC	RE	CC	RE	CC	RE
SEPTCENTER	0.26	0.32	0.36	0.40	0.17	0.40	–	–
LVLAT	**0.83**	**0.28**	**0.79**	**0.26**	0.69	0.31	–	–
LVAPEX	0.70	0.28	0.63	0.30	0.34	0.39	–	–
LVANT	0.74	0.29	0.66	0.36	0.47	0.33	–	–
RVPOST	0.75	0.37	0.37	0.54	−0.05	0.70	–	–
RVANT	0.67	0.37	0.70	0.38	0.10	0.38	–	–
LVLATEPI	0.78	0.35	0.73	0.28	**0.75**	**0.32**	–	–
LVLATENDO	0.79	0.31	0.66	0.29	0.70	0.30	–	–

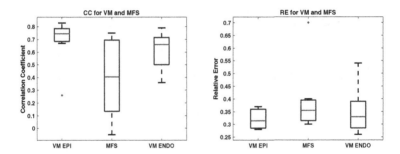

Fig. 3. Boxplot of CC and RE on the AT for the VM and the MFS

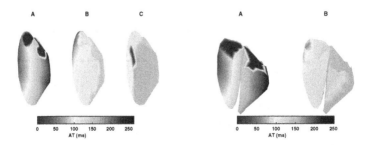

Fig. 4. AT maps for dataset LVLAT. A: reference. B: VM method. C: MFS method. Epicardium (left) and endocardium (right).

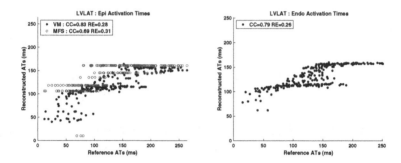

Fig. 5. Scatter plot of AT reference versus reconstructed ones for LVLAT dataset. Left: Comparison between VM and MFS on the epicardium. Right: VM on the endocardium.

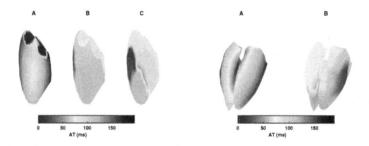

Fig. 6. AT maps for dataset SEPT. A: reference. B: VM method. C: MFS method. Epicardium (left) and endocardium (right).

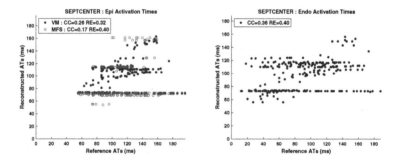

Fig. 7. Scatter plot of AT reference versus reconstructed ones for SEPT dataset. Left: Comparison between VM and MFS on the epicardium. Right: VM on the endocardium.

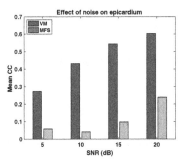 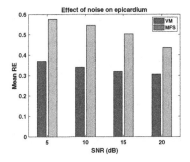

Fig. 8. Mean of CC and RE with respect to the SNR for VM and MFS on the epicardium

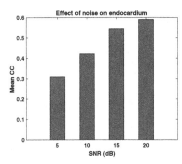 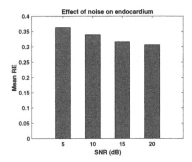

Fig. 9. Mean of CC and RE with respect to the SNR for VM on the endocardium

3.2 Data with Noise

Next, we analyzed the sensitivity of both models with respect to noise by adding Gaussian noise with SNR from 5 to 20 dB to the original BSPM data. Figure 8 shows the mean of the correlation coefficient (CC) and the relative error (RE) obtained using different SNR on the epicardium with both methods. Figure 9 presents the results obtained on the endocardium for the VM. In Fig. 8, we observe a degradation of the correlation coefficient (CC) and the relative error (RE) for both methods on the epicardium when the SNR decreases. The CC (respectively the RE) vary from 0.27 to 0.60 (respectively 0.31 to 0.37) for the VM and from 0.05 to 0.24 (respectively 0.44 to 0.58) for the MFS. However, the VM is more robust than for MFS. We think that the classical torso formulation of the ECGI problem is more ill-posed than the formulation proposed. On the endocardium, we observe in Fig. 9 a similar behaviour for the VM. The CC (respectively RE) vary from 0.31 to 0.59 (respectively 0.31 to 0.37).

4   Discussion and Conclusions

In this paper, we presented a volume method to solve inverse problem in electro-cardiography. The proposed volume method reconstructs a whole-heart volume current source, instead of the epicardial potential field, as in the MFS. The VM is quite different from the MFS since it directly searches a volume current source instead of a potential, and also because it accounts for the electrical conductivity in the heart, and its possible variation in a scar, but also in the intracavitary blood, while the MFS only accounts for the torso conductivity.

For the datasets presented in Sect. 3, the VM performed better than the MFS. In addition, the VM method could reconstruct acceptable endocardial activation (Fig. 3). Both methods degrade when the SNR decreases, but the VM always remains better than the MFS (Fig. 8).

All tests were completed with in silico data. It will be interesting to apply the VM to experimental or clinical data. It is also important to apply the VM method on more datasets with scars, as tested in a preliminary work [2]. At last, it would be interesting to study more regularization parameter chose for the VM method.

Acknowledgments. This study received financial support from the French Government as part of the "Investments for the Future" program managed by the National Research Agency (ANR), Grant reference ANR-10-IAHU-04, and also from European Research Council (ERC), Grant ECSTATIC (715093).

References

1. Belgacem, F.B.: Why is the cauchy problem severely ill-posed? Inverse Problems **23**, 823 (2007). https://doi.org/10.1088/0266-5611/23/2/020
2. Diallo, M.M., Potse, M., Dubois, R., Coudière, Y.: Solving the ECGI problem with known locations of scar regions. In: Computing in Cardiology 2020, Rimini, Italy (2020). https://doi.org/10.22489/CinC.2020.237
3. Duchateau, J., Potse, M., Dubois, R.: Spatially coherent activation maps for electro-cardiographic imaging. IEEE Trans. Biomed. Eng. **64**(5), 1149–1156 (2017). https://doi.org/10.1109/TBME.2016.2593003
4. Hecht, F.: New development in freefem++. J. Numer. Math. **20**(3–4), 251–266 (2012). https://doi.org/10.1515/jnum-2012-0013
5. Nielsen, B.F., Cai, X., Lysaker, M.: On the possibility for computing the transmembrane potential in the heart with a one shot method: an inverse problem. Math. Biosci. **210**(2), 523–553 (2007). https://doi.org/10.1016/j.mbs.2007.06.003
6. Schulze, W.H., et al.: A simulation dataset for ECG imaging of paced beats with models for transmural, endo-and epicardial and pericardial source imaging. In: ECG Imaging 2015, Germany (2015). https://doi.org/10.13140/RG.2.1.1946.8568
7. Wang, D., Kirby, R.M., MacLeod, R.S., Johnson, C.R.: Inverse electrocardiographic source localization of ischemia: an optimization framework and finite element solution. J. Comput. Phys. **250**, 403–424 (2013). https://doi.org/10.1016/j.jcp.2013.05.027
8. Wang, Y., Rudy, Y.: Application of the method of fundamental solutions to potential-based inverse electrocardiography. Ann. Biomed. Eng. **34**(8), 1272–1288 (2006). https://doi.org/10.1007/s10439-006-9131-7

Fast and Accurate Uncertainty Quantification for the ECG with Random Electrodes Location

Michael Multerer[ID] and Simone Pezzuto[✉][ID]

Center for Computational Medicine in Cardiology, Euler Institute,
Università della Svizzera italiana, Via la Santa 1, 6962 Lugano, Switzerland
{michael.multerer,simone.pezzuto}@usi.ch

Abstract. The standard electrocardiogram (ECG) is a point-wise evaluation of the body potential at certain given locations. These locations are subject to uncertainty and may vary from patient to patient or even for a single patient. In this work, we estimate the uncertainty in the ECG induced by uncertain electrode positions when the ECG is derived from the forward bidomain model. In order to avoid the high computational cost associated to the solution of the bidomain model in the entire torso, we propose a low-rank approach to solve the uncertainty quantification (UQ) problem. More precisely, we exploit the sparsity of the ECG and the lead field theory to translate it into a set of deterministic, time-independent problems, whose solution is eventually used to evaluate expectation and covariance of the ECG. We assess the approach with numerical experiments in a simple geometry.

Keywords: Random electrodes location · Uncertainty quantification · Lead field · Electrophysiology · Forward bidomain model

1 Introduction

The standard ECG is a routinely acquired recording of the torso electric potential [14]. It provides valuable information on the electric activity of the heart and, when combined with imaging data of the anatomy, it can be used for non-invasive personalization of sophisticated patient-specific models [10,17]. In these inverse ECG models, the ECG is rarely computed from the state-of-the-art bidomain model [6], otherwise the computational cost would be prohibitive. Commonly, the bidomain model is replaced by a "decoupled" version, called forward bidomain [19] or pseudo-bidomain [3,15] model, in which the transmembrane potential in the heart is computed independently from the extracellular potential in the torso. The resulting model still compares favourably to the coupled bidomain model and, more importantly, the ECG can be evaluated very efficiently and *exactly* by employing the lead field theory [16,18].

Obviously, when dealing with real data, as in patient-specific modeling, model parameters are subject to unavoidable uncertainty. This uncertainty should be

© Springer Nature Switzerland AG 2021
D. B. Ennis et al. (Eds.): FIMH 2021, LNCS 12738, pp. 561–572, 2021.
https://doi.org/10.1007/978-3-030-78710-3_54

accounted for in the forward and inverse ECG model [5]. Several sources of uncertainty may be considered, e.g., related to the segmentation process of the anatomy [7], the electric conductivities [1], or the fiber distribution [20]. Particularly relevant in the context of inverse ECG modeling is the uncertainty in the electrodes' locations, which has shown to yield sensible morphological changes in the precordial signals even with a displacement as low as 2 cm [13].

The present work focuses on the problem of estimating the expectation and the covariance of the surface ECG, if electrodes' locations are subject to uncertainty and the ECG is simulated with the forward bidomain model. In principle, given the torso potential, the statistical moments are readily available with little additional cost, as the solution of the UQ problem amounts to a simple integration over the torso domain. In spite of its simplicity, the computational cost of this approach grows linearly with the number of time steps and the number of evaluations of the forward model. Moreover, it relies on the full torso potential, despite the fact that the electrodes' locations may be very localized. We propose a computationally very efficient methodology to solve the UQ problem *without* the need of solving the full forward problem. Our method is still based on the lead field theory and it is an *exact* representation of the true ECG. Specifically, it exploits a low-rank approach to decouple the correlation problem into a small set of elliptic problems for different right hand sides [12]. Remarkably, the overall computational cost is drastically reduced and comparable to the solution of a few elliptic problems, *independently* of the number of time steps and forward evaluations.

This paper is organized as follows: in Sect. 2, we review the forward bidomain model for the ECG, the lead field approach and describe our method. In Sect. 3, we validate the approach on a simple geometry. We conclude in Sect. 4 with a brief discussion and outlook.

2   Methods

2.1   The Forward Bidomain Model

The electric potential $u_0(\mathbf{x}, t)$ in the torso $\Omega_T \subset \mathbb{R}^d$, and consequently the ECG, can be modelled from the transmembrane potential $V_m(\mathbf{x}, t)$ in the active myocardium $\Omega_H \subset \mathbb{R}^d$, with the time-dependent *forward bidomain model* [19], which reads as follows:

$$
\begin{cases}
-\nabla \cdot ((\mathbf{G}_i + \mathbf{G}_e)\nabla u_e(\mathbf{x}, t)) = \nabla \cdot (\mathbf{G}_i \nabla V_m(\mathbf{x}, t)), & \text{in } \Omega_H \times [0, \infty), \\
-\nabla \cdot (\mathbf{G}_0 \nabla u_0(\mathbf{x}, t)) = 0, & \text{in } \Omega_T \times [0, \infty), \\
-\mathbf{G}_0 \nabla u_0(\mathbf{x}, t) \cdot \mathbf{n} = 0, & \text{on } \Sigma \times [0, \infty), \\
u_e(\mathbf{x}, t) = u_0(\mathbf{x}, t), & \text{on } \Gamma \times [0, \infty), \\
-(\mathbf{G}_i + \mathbf{G}_e)\nabla u_e(\mathbf{x}, t) \cdot \mathbf{n} + \mathbf{G}_0 \nabla u_0(\mathbf{x}, t) \cdot \mathbf{n} = \mathbf{G}_i \nabla V_m(\mathbf{x}, t) \cdot \mathbf{n}, & \text{on } \Gamma \times [0, \infty).
\end{cases}
$$

$$(1)$$

Herein, $\Gamma = \bar{\Omega}_H \cap \bar{\Omega}_T$ is the heart-torso interface, $\Sigma = \partial\Omega_T \setminus \Gamma$ is the body surface, $u_e(\mathbf{x}, t)$ is the extra-cellular potential in the heart, $\mathbf{G}_i$ and $\mathbf{G}_e$ are respectively intra- and extra-cellular conductivity of the heart, $\mathbf{G}_0$ is the torso conductivity, and $\mathbf{n}$ is the outward normal for both Γ and Σ. For the sake of simplicity in the notation, we define

$$
\mathbf{G} := \begin{cases} \mathbf{G}_i + \mathbf{G}_e & \text{in } \Omega_H, \\ \mathbf{G}_0 & \text{in } \Omega_T, \end{cases} \qquad u(\mathbf{x}, t) := \begin{cases} u_e(\mathbf{x}, t) & \text{in } \Omega_H, \\ u_0(\mathbf{x}, t) & \text{in } \Omega_T, \end{cases}
$$

and assume, without loss of generality, that $u(\cdot, t) \in H^1(\Omega)$, where $\Omega = \Omega_H \cup \Omega_T$. In this case, the variational formulation for Eq. (1) can be written according to

For every $t \in \mathbb{R}$, find $u(\cdot, t) \in H^1(\Omega)$ such that

$$
\int_\Omega \mathbf{G} \nabla u(\mathbf{x}, t) \cdot \nabla v \, d\mathbf{x} = - \int_{\Omega_H} \mathbf{G}_i \nabla V_m(\mathbf{x}, t) \cdot \nabla v \, d\mathbf{x} \tag{2}
$$

for all $v \in H^1(\Omega)$. The well-posedness of the problem follows from standard application of the Riesz Theorem [9], given that Ω_H, Ω_T are Lipschitz domains and $V_m(\cdot, t) \in H^1(\Omega_H)$. We remark that the formulation in Eq. (2) is equivalent to Eq. (1) when the restriction of the solution $u|_{\Omega_i}$ belongs to $H^2(\Omega_i)$, $i \in \{H, T\}$, see e.g. [2, 4] for a more comprehensive treatment of interface problems.

The ECG is a set of so-called *leads*, typically 12 in the standard ECG. Each lead reads as follows:

$$
V(t, \boldsymbol{\xi}_1, \ldots, \boldsymbol{\xi}_L) = \sum_{\ell=1}^{L} a_\ell u(\boldsymbol{\xi}_\ell, t), \tag{3}
$$

where $\boldsymbol{\Xi} := \{\boldsymbol{\xi}_\ell\}_{\ell=1}^{L}$ is the set of electrodes and $\mathbf{a} = [a_1, \ldots, a_L]^\top$ is a zero-sum vector of weights defining the lead. For instance, a limb lead is the potential difference of 2 electrodes, whereas a precordial lead involves 4 electrodes (3 are used to build the Wilson Central Terminal, that is the reference potential). It is worth noting that Eq. (3) is valid only if $u(\cdot, t) \in C^0(\Sigma)$, which is not true for $u(\cdot, t) \in H^1(\Omega)$ and $d \geq 2$. For a rigorous discussion, see [6].

In this work, we are interested in computing statistics of $V(t, \boldsymbol{\Xi}(\omega))$ when the electrode positions $\boldsymbol{\Xi}(\omega) := \{\boldsymbol{\xi}_\ell(\omega)\}_{\ell=1}^{L}$ are not known exactly. Here, we denote by $\boldsymbol{\xi}_\ell(\omega)$ the random variable associated to the ℓ-th electrode and assume that the joint distribution is given by the density $\rho(\mathbf{X}) = \rho(\mathbf{x}_1, \ldots, \mathbf{x}_L)$ with respect to the surface measure $d\sigma_\mathbf{X} = d\sigma_{\mathbf{x}_1} \cdots d\sigma_{\mathbf{x}_L}$ on Σ^L. According to the definition in Eq. (3), the lead $V(t, \boldsymbol{\Xi})$ is a random field as well, with expectation and correlation respectively reading as follows:

$$
\mathbb{E}[V](t) = \int_{\Sigma^M} V(t, \mathbf{X}) \rho(\mathbf{X}) d\sigma_\mathbf{X}, \tag{4}
$$

$$
\text{Cor}[V](t, s) = \int_{\Sigma^M} V(t, \mathbf{X}) V(s, \mathbf{X}) \rho(\mathbf{X}) d\sigma_\mathbf{X}. \tag{5}
$$

In summary, the UQ problem for the random electrodes locations consists in solving the forward bidomain model Eq. (1) for $u(\mathbf{x}, t)$, given $V_m(\mathbf{x}, t)$, for every time t, and then computing the integrals in Eq. (4) and Eq. (5).

2.2   Lead Field Formulation of the UQ Problem

Clearly, in general it is not convenient to compute the ECG from Eq. (1), because the ECG is only a very sparse evaluation of $u(\mathbf{x}, t)$. Moreover, in a patient-specific or personalization context, the ECG needs to be simulated several times with different instances of $V_{\mathrm{m}}(\mathbf{x}, t)$, with no changes in the left hand side of Eq. (1). A better approach is based on Green's functions, also known as *lead fields* in the electrocardiographic literature [18]. In fact, it is possible to show that $V(t, \boldsymbol{\Xi})$ has the following representation [6]:

$$V(t, \boldsymbol{\Xi}) = \int_{\Omega_{\mathrm{H}}} \mathbf{G}_{\mathrm{i}}(\mathbf{x}) \nabla V_{\mathrm{m}}(\mathbf{x}, t) \cdot \nabla Z(\mathbf{x}, \boldsymbol{\Xi}) \, \mathrm{d}\mathbf{x}, \tag{6}$$

where $Z(\mathbf{x}, \boldsymbol{\Xi})$ is the weak solution of the elliptic problem:

$$\begin{cases} -\nabla \cdot \mathbf{G} \nabla Z(\mathbf{x}, \boldsymbol{\Xi}) = 0, & \text{in } \Omega, \\ -\mathbf{G} \nabla Z(\mathbf{x}, \boldsymbol{\Xi}) \cdot \mathbf{n} = \sum_{\ell=1}^{L} a_\ell \delta_{\boldsymbol{\xi}_\ell}, & \text{on } \Sigma, \end{cases} \tag{7}$$

where $\delta_{\boldsymbol{\xi}_\ell}$ is the $(d-1)$-dimensional Dirac delta centered at $\boldsymbol{\xi}_\ell(\omega)$. Therefore, given that all measurement locations are fixed, Eq. (7) is only solved once, at the cost of a single time step of Eq. (1), and then used to compute $V(t, \boldsymbol{\Xi})$ for any choice of $V_{\mathrm{m}}(\mathbf{x}, t)$.

Here, we exploit Eq. (6) to compute the the expectation and correlation of V, according to Eq. (4) and (5). Substituting Eq. (6) into Eq. (4), we obtain by the linearity of the expectation that

$$\mathbb{E}[V](t) = \int_{\Omega_{\mathrm{H}}} \mathbf{G}_{\mathrm{i}}(\mathbf{x}) \nabla V_{\mathrm{m}}(\mathbf{x}, t) \cdot \nabla \mathbb{E}[Z](\mathbf{x}) \, \mathrm{d}\mathbf{x}. \tag{8}$$

Again by linearity, the equation for the expected lead field $\mathbb{E}[Z]$ follows from Eq. (7) and reads as follows:

$$\begin{cases} -\nabla \cdot \mathbf{G} \nabla \mathbb{E}[Z](\mathbf{x}) = 0, & \text{in } \Omega, \\ -\mathbf{G} \nabla \mathbb{E}[Z](\mathbf{x}) \cdot \mathbf{n} = \sum_{\ell=1}^{M} a_\ell \rho_\ell(\mathbf{x}), & \text{on } \Sigma, \end{cases} \tag{9}$$

where ρ_ℓ is the marginal distribution of ρ with respect to $\boldsymbol{\xi}_\ell$, that is

$$\rho_\ell(\mathbf{x}_\ell) := \int_{\Sigma^{L-1}} \rho(\mathbf{X}) \, \mathrm{d}\sigma_{\mathbf{x}_1} \cdots \mathrm{d}\sigma_{\mathbf{x}_{\ell-1}} \mathrm{d}\sigma_{\mathbf{x}_{\ell+1}} \cdots \mathrm{d}\sigma_{\mathbf{x}_L}. \tag{10}$$

To show this, we observe that:

$$\mathbb{E}\left[\sum_{\ell=1}^{L} a_\ell \delta_{\boldsymbol{\xi}_\ell} \right] = \sum_{i=1}^{L} a_\ell \int_{\Sigma^L} \delta_{\mathbf{x}_\ell} \rho(\mathbf{X}) \, \mathrm{d}\sigma_{\mathbf{x}} = \sum_{\ell=1}^{L} a_\ell \rho_\ell.$$

Therefore, the cost of computing the average ECG is equivalent to that for solving for the point-wise ECG, i.e., one solution of the elliptic problem in Eq. (9).

We observe that both Eq. (7) and Eq. (9) are well-posed, since the right hand side has zero average over Σ in both cases. In particular, for every ω, $Z(\mathbf{x}, \boldsymbol{\Xi}(\omega))$ and $\mathbb{E}[Z](\mathbf{x})$ are only defined up to a constant.

The natural continuation of the above argument yields the correlation for the ECG according to

$$\operatorname{Cor}[V](t,s) = \int_{\Sigma^2} (\mathbf{G}_i \nabla \otimes \mathbf{G}_i \nabla) V_m(\mathbf{x}, t) V_m(\mathbf{x}', s) : (\nabla \otimes \nabla) \operatorname{Cor}[Z] \, \mathrm{d}\sigma_\mathbf{x} \mathrm{d}\sigma_{\mathbf{x}'},$$

where the tensor product is $[\mathbf{u} \otimes \mathbf{v}]_{ij} = u_i(\mathbf{x}) v_j(\mathbf{x}')$ and the inner product between tensors is $\mathbf{A} : \mathbf{B} = \sum_{ij} [A]_{ij} [B]_{ij}$. The problem for the correlation $\operatorname{Cor}[Z]$, obtained as above from Eq. (7), reads as follows:

$$\begin{cases} (\nabla \cdot \mathbf{G}\nabla \otimes \nabla \cdot \mathbf{G}\nabla) \operatorname{Cor}[Z] = 0, & \text{in } \Omega \times \Omega, \\ (\mathbf{n} \cdot \mathbf{G}\nabla \otimes \nabla \cdot \mathbf{G}\nabla) \operatorname{Cor}[Z] = 0, & \text{on } \Sigma \times \Omega, \\ (\nabla \cdot \mathbf{G}\nabla \otimes \mathbf{n} \cdot \mathbf{G}\nabla) \operatorname{Cor}[Z] = 0, & \text{on } \Omega \times \Sigma, \\ (\mathbf{n} \cdot \mathbf{G}\nabla \otimes \mathbf{n} \cdot \mathbf{G}\nabla) \operatorname{Cor}[Z] = R, & \text{on } \Sigma \times \Sigma, \end{cases} \tag{11}$$

where the correlation $R(\mathbf{x}, \mathbf{x}')$ of the Neumann data in Eq. (7) is

$$R(\mathbf{x}, \mathbf{x}') = \operatorname{Cor}\left[\sum_{\ell=1}^L a_\ell \delta_{\boldsymbol{\xi}_\ell}, \sum_{\ell'=1}^L a_{\ell'} \delta_{\boldsymbol{\xi}_{\ell'}}\right]$$
$$= \sum_{\ell=1}^L a_\ell^2 \rho_\ell(\mathbf{x}) \delta_\mathbf{x}(\mathbf{x}') + \sum_{\ell \neq \ell'}^L a_\ell a_{\ell'} \rho_{\ell,\ell'}(\mathbf{x}, \mathbf{x}'), \tag{12}$$

with $\rho_{\ell,\ell'}(\mathbf{x}, \mathbf{x}')$ being the marginal distribution of ρ with respect to $(\boldsymbol{\xi}_\ell, \boldsymbol{\xi}'_\ell)$ and defined as follows:

$$\rho_{\ell,\ell'}(\mathbf{x}_\ell, \mathbf{x}_{\ell'}) := \int_{\Sigma^{L-2}} \rho(\mathbf{X}) \, \mathrm{d}\sigma_{\mathbf{x}_1} \cdots \mathrm{d}\sigma_{\mathbf{x}_{\ell-1}} \mathrm{d}\sigma_{\mathbf{x}_{\ell+1}} \cdots \mathrm{d}\sigma_{\mathbf{x}_{\ell'-1}} \mathrm{d}\sigma_{\mathbf{x}_{\ell'+1}} \cdots \mathrm{d}\sigma_{\mathbf{x}_L}. \tag{13}$$

We observe that, when $\boldsymbol{\xi}_\ell$ and $\boldsymbol{\xi}'_\ell$ are independent, $\rho_{\ell,\ell'}(\mathbf{x}_\ell, \mathbf{x}_{\ell'})$ factorizes into the product of the marginals $\rho_\ell(\mathbf{x}_\ell)$ and $\rho_{\ell'}(\mathbf{x}_{\ell'})$.

As the computation of $\operatorname{Cor}[Z]$ requires the solution of a tensor product boundary value problem, it is computationally rather expensive. In what follows, we will exploit the particular structure of $R(\mathbf{x}, \mathbf{x}')$ to significantly reduce the computational cost and implementation effort. We remark that also Eq. (11) is well-posed because, by construction, $R(\mathbf{x}, \mathbf{x}')$ is such that $\langle R, 1 \otimes v \rangle_{\Sigma^2} = \langle R, v \otimes 1 \rangle_{\Sigma^2} = 0$ for all $v \in \mathrm{H}^{1/2}(\Sigma)$, where $\langle \cdot, \cdot \rangle_{\Sigma^2}$ is the duality pairing in $\mathrm{L}^2(\Sigma^2)$.

2.3   Numerical Discretization

The variational formulation of the averaged lead field problem Eq. (8) resembles Eq. (2) with a different right hand side. With $Y = \mathrm{H}^1(\Omega)$, the problem is:

Find $\mathbb{E}[Z] \in Y$ such that

$$\int_\Omega \mathbf{G}\nabla\mathbb{E}[Z] \cdot \nabla v \, \mathrm{d}\mathbf{x} = \int_\Sigma \sum_{\ell=1}^L a_\ell \rho_\ell v \, \mathrm{d}\mathbf{x}, \quad \text{for all } v \in Y.$$

The Galerkin approximation in the space $Y_h \subset Y$, with $Y_h = \text{span}\{\phi_k\}_{k=1}^{N_h}$, reads as follows:

$$\mathbf{K}\mathbf{z} = \mathbf{g}, \tag{14}$$

where $\mathbf{z}$ is the solution vector, that is $\mathbb{E}[Z] \approx Z_h = \sum_k [\mathbf{z}]_k \phi_k$ and

$$[\mathbf{K}]_{k\ell} = \int_\Omega \mathbf{G}\nabla\phi_\ell \cdot \nabla\phi_k \, d\mathbf{x}, \tag{15}$$

$$[\mathbf{g}]_k = \int_\Sigma \sum_{\ell=1}^M a_\ell \rho_\ell(\mathbf{x})\phi_k(\mathbf{x}) \, d\mathbf{x}. \tag{16}$$

For the correlation in Eq. (11), the variational formulation is as follows:

Find $\text{Cor}[Z] \in Y \otimes Y$ such that

$$\int_{\Omega^2} (\mathbf{G}\nabla \otimes \mathbf{G}\nabla)\,\text{Cor}[Z] : (\nabla \otimes \nabla)v \, d\mathbf{x}d\mathbf{x}' = \int_{\Sigma^2} Rv \, d\mathbf{x}d\mathbf{x}'$$

for all $v \in Y \otimes Y$. The corresponding Galerkin formulation on $Y_h \times Y_h$ is:

$$(\mathbf{K} \otimes \mathbf{K})\mathbf{Z} = \mathbf{R}, \tag{17}$$

where $\text{Cor}[Z] \approx \sum_{k,\ell}[\mathbf{Z}]_{k\ell}\phi_k \otimes \phi_\ell$ and

$$\begin{aligned}
[\mathbf{R}]_{pq} &= \int_{\Sigma^2} R\phi_p\phi_q d\mathbf{x}d\mathbf{x}' \\
&= \sum_{\ell=1}^L a_\ell^2 \int_\Sigma \rho_\ell\phi_p\phi_q d\mathbf{x} + \sum_{\ell\neq\ell'}^L a_\ell a_{\ell'} \int_{\Sigma^2} \rho_{\ell,\ell'}\phi_p\phi_q d\mathbf{x}d\mathbf{x}'.
\end{aligned} \tag{18}$$

As the number of degrees of freedom for the correlation problem is N_h^2, it may easily become computationally prohibitive. However, assuming that the marginal densities ρ_ℓ, $\ell = 1,\ldots,L$ are strongly localized, the right hand side in (17) may be represented by a low-rank approximation according to

$$\mathbf{R} \approx \sum_{k=1}^K \boldsymbol{r}_k \otimes \boldsymbol{r}_k, \quad \boldsymbol{r}_k \in \mathbb{R}^{N_h},$$

with $K \ll N_h$. In this case, we also expect a low-rank solution, that is

$$\mathbf{Z} \approx \sum_{k=1}^K \boldsymbol{\zeta}_k \otimes \boldsymbol{\zeta}_k, \quad \boldsymbol{\zeta}_k \in \mathbb{R}^{N_h},$$

with $K \ll N_h$. Then, due to the tensor product structure of (17), there simply holds

$$\mathbf{K}\boldsymbol{\zeta}_k = \boldsymbol{r}_k, \qquad k = 1,\ldots,K. \tag{19}$$

In practice, we compute the low-rank approximation by a diagonally pivoted, truncated Cholesky decomposition, see [11].

Finally, for the computation of statistics of the ECG, we insert the computed Galerkin approximations into Eq. (4) and Eq. (5) and obtain

$$\mathbb{E}[V](t) \approx \mathbf{V}(t) \cdot \mathbf{z}, \tag{20}$$

$$\mathrm{Cor}[V](t,s) \approx \sum_{k,m=1}^{K} \big(\mathbf{V}(t) \cdot \boldsymbol{\zeta}_k\big)\big(\mathbf{V}(s) \cdot \boldsymbol{\zeta}_m\big), \tag{21}$$

where

$$[\mathbf{V}(t)]_j = \int_{\Omega_{\mathrm{H}}} \mathbf{G}_{\mathrm{i}} \nabla V_{\mathrm{m}}(t) \cdot \nabla \phi_j \, \mathrm{d}\mathbf{x}.$$

The computational cost for the proposed approach is dominated by the solution of $K+1$ systems (one for the expectation and K for the correlation) of the form of Eq. (14). It is therefore independent on the number of time steps N_t or the choice of V_{m}, oppositely to the solution of the forward bidomain model in Eq. (1), which requires N_t solutions for each choice of V_{m}.

We summarize below the proposed procedure to evaluate expectation and correlation of a single lead of the ECG, defined with $L \geq 2$ coefficients $\mathbf{a}$ as in Eq. (6) and with random electrodes locations $\{\boldsymbol{\xi}_\ell\}_{\ell=1}^{L}$ with density $\rho(\mathbf{X})$. We assume as above that $V_{\mathrm{m}}(\mathbf{x},t)$ is given and computed elsewhere.

1. Compute $\rho_\ell(\mathbf{x})$ with Eq. (10) and assemble $\mathbf{g}$ with Eq. (16);
2. Assemble $\mathbf{K}$ with Eq. (15) and solve Eq. (14) to find $\mathbf{z}$;
3. Compute $\mathbb{E}[V](t)$ from $\mathbf{z}$ and V_{m} with Eq. (20);
4. Compute $\rho_{\ell,\ell'}(\mathbf{x},\mathbf{x}')$ with Eq. (13) and assemble $\mathbf{R}$ with Eq. (18);
5. Compute the low-rank Cholesky decomposition $\{\mathbf{r}_k\}_{k=1}^{K}$ of R;
6. For each $k = 1,\ldots,K$, solve Eq. (19) for $\boldsymbol{\zeta}_k$;
7. Compute $\mathrm{Cor}[V](t,s)$ from $\{\boldsymbol{\zeta}_k\}_{k=1}^{K}$ and V_{m} with Eq. (21).

3   Numerical Assessment

We tested the proposed approach on a idealized heart-torso geometry in 2-D, as depicted in Fig. 1. The anatomy consists of an ellipsoidal torso with major semi-axis of $T_y = 15\,\mathrm{cm}$, vertically oriented, and minor axis of $T_x = 10\,\mathrm{cm}$. The heart was an annulus centered at $\mathbf{x}_{\mathrm{h}} = (-4\,\mathrm{cm}, 2\,\mathrm{cm})$ with respect to the center of the torso, and with inner (endocardium) and outer (epicardium) radius respectively equal to $2\,\mathrm{cm}$ and $3\,\mathrm{cm}$. The domain was split into 3 distinct regions, namely blood pool, myocardium and torso (see Fig. 1).

For this test, we considered an ECG with 2 leads obtained from 4 random electrodes $\boldsymbol{\xi}_\ell(\omega)$, $\ell = \{\mathrm{VL, VR, VF, V1}\}$, see Fig. 1. The leads, II and V1, were

$$V_{\mathrm{II}}(t,\boldsymbol{\Xi}) = u(\boldsymbol{\xi}_{\mathrm{VF}},t) - u(\boldsymbol{\xi}_{\mathrm{VL}},t),$$

$$V_{\mathrm{V1}}(t,\boldsymbol{\Xi}) = u(\boldsymbol{\xi}_{\mathrm{V1}},t) - \frac{1}{3}\Big(u(\boldsymbol{\xi}_{\mathrm{VL}},t) + u(\boldsymbol{\xi}_{\mathrm{VR}},t) + u(\boldsymbol{\xi}_{\mathrm{VF}},t)\Big),$$

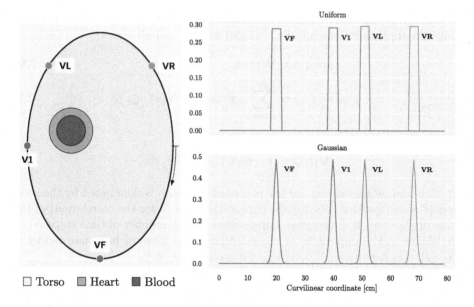

☐ Torso   ▨ Heart   ■ Blood

Fig. 1. Geometrical configuration for the numerical test. On the left, the domain is represented with electrodes locations on the boundary and tissue properties. On the right, the probability density function for both uniform and Gaussian-like cases is reported. The x-axis is the curvilinear coordinate, and colors of the curves refer to the electrodes on the left. (Color figure online)

respectively corresponding to $\mathbf{a}_{II} = (-1, 0, 1, 0)$ and $\mathbf{a}_{V1} = (-\frac{1}{3}, -\frac{1}{3}, -\frac{1}{3}, 1)$. The average position $\mathbf{x}_\ell$ of the electrodes was conveniently defined using the formula $\mathbf{x}_\ell = [T_x \cos(\theta_\ell), T_y \sin(\theta_\ell)]^\top$ with $\theta_{VL} = \frac{3}{4}\pi$, $\theta_{VR} = \frac{1}{4}\pi$, $\theta_{VF} = \frac{3}{2}\pi$ and $\theta_{V1} = \pi$.

In all tests, we evaluated the deterministic ECG, computed from Eq. (6), the average ECG when electrodes were randomly located, and the variance from the formula $\text{Var}[V](t) = \text{Cor}[V](t, t) - \big(\mathbb{E}[V](t)\big)^2$.

The random electrode locations, independent from each other, were either uniformly distributed or with a Gaussian-like distribution, both defined on the outer boundary Σ of the torso (the "chest"), see Fig. 1. In the case of the uniform distribution, the marginal density ρ_ℓ for each electrode was the characteristic function of the set $\Sigma_\ell = \{\mathbf{x} \in \Sigma : d(\mathbf{x}_\ell, \mathbf{x}) \leq r_\ell\}$, that is the r_ℓ-neighborhood of $\mathbf{x}_\ell$ with respect to the geodesic distance d on the curve Σ. We also considered a Gaussian-like distribution computed, after a normalization, by solving the heat equation on the boundary curve Σ with diffusion defined along the arc-length, initial datum $\delta(\mathbf{x} - \mathbf{x}_\ell)$ on Σ, and solved for a total time T_ℓ, see Fig. 1. We selected $r_\ell = 1.5\,\text{cm}$ and $T_\ell = \sqrt{3}/3 r_\ell$, so that both distributions have the same variance.

For convenience, we report the full expression $R(\mathbf{x}, \mathbf{x}')$ for lead II, obtained by assuming that $\boldsymbol{\xi}_{VF}$ and $\boldsymbol{\xi}_{VL}$ were independent:

$$R(\mathbf{x}, \mathbf{x}') = \big(\rho_{VF}(\mathbf{x}) + \rho_{VL}(\mathbf{x})\big)\delta_{\mathbf{x}}(\mathbf{x}') - \rho_{VF}(\mathbf{x})\rho_{VL}(\mathbf{x}') - \rho_{VL}(\mathbf{x})\rho_{VF}(\mathbf{x}').$$

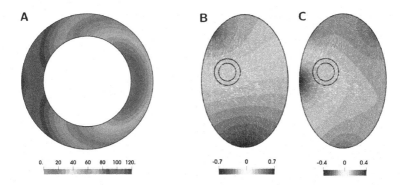

Fig. 2. On the left (panel A), the activation map computed with the eikonal solution, in ms. On right, lead fields (in mV) for the average problem are reported for lead II (panel B) and lead V1 (panel C).

In particular, the assembly of the tensor $\mathbf{R}$ simplifies as well, with no need of evaluating a double integral. In fact,

$$[\mathbf{R}]_{pq} = \int_{\Sigma} (\rho_{\mathrm{VF}} + \rho_{\mathrm{VL}})(\phi_p \phi_q - \phi_p - \phi_q)\mathrm{d}\sigma_{\mathbf{x}}.$$

The electric conductivities were uniform and isotropic in the torso and in the blood pool, and respectively set to $2\,\mathrm{mS\ cm}^{-2}$ and $6\,\mathrm{mS\ cm}^{-2}$. The myocardium was assumed transversely isotropic, with fibers $\mathbf{f}$ circularly oriented and of unit length. Specifically:

$$\mathbf{G}_{\mathrm{i}} = \sigma_{\mathrm{i,t}}\mathbf{I} + (\sigma_{\mathrm{i,l}} - \sigma_{\mathrm{i,t}})\mathbf{f} \otimes \mathbf{f},$$
$$\mathbf{G}_{\mathrm{e}} = \sigma_{\mathrm{e,t}}\mathbf{I} + (\sigma_{\mathrm{e,l}} - \sigma_{\mathrm{e,t}})\mathbf{f} \otimes \mathbf{f},$$

and values set to $\sigma_{\mathrm{i,l}} = 3\,\mathrm{mS\ cm}^{-2}$, $\sigma_{\mathrm{i,t}} = 0.3\,\mathrm{mS\ cm}^{-2}$, $\sigma_{\mathrm{e,l}} = 3\,\mathrm{mS\ cm}^{-2}$ and $\sigma_{\mathrm{e,t}} = 1.2\,\mathrm{mS\ cm}^{-2}$.

For sake of simplicity, the transmembrane potential $V_{\mathrm{m}}(\mathbf{x}, t)$ was modelled by shifting a fixed action potential template at given activation times $\tau(\mathbf{x})$, according to the formula

$$V_{\mathrm{m}}(\mathbf{x}, t) = U(t - \tau(\mathbf{x})), \qquad U(s) = V_{\mathrm{rest}} + \frac{V_{\mathrm{dep}} - V_{\mathrm{rest}}}{2}\left(1 + \tanh\left(\frac{s}{\varepsilon}\right)\right).$$

The activation map $\tau\colon \Omega_{\mathrm{H}} \to \mathbb{R}$ was simulated with the eikonal model

$$\begin{cases} \sqrt{\mathbf{D}(\mathbf{x})\nabla\tau \cdot \nabla\tau} = 1, & \mathbf{x} \in \Omega_{\mathrm{H}} \setminus \{\mathbf{x}_{\mathrm{s}}\}, \\ \tau(\mathbf{x}_{\mathrm{s}}) = 0. \end{cases}$$

The conductivity tensor was set proportional to the monodomain conductivity and such to yield a conduction velocity along the fibers of $65\,\mathrm{cm\ s}^{-1}$, that is,

$$\mathbf{D} = \alpha \cdot \mathbf{G}_{\mathrm{m}},$$

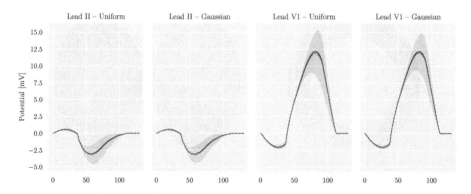

Fig. 3. ECG results for all tests. In the plots, the dashed black curve is the deterministic ECG, the solid blue curve is the average ECG, and the shaded blue area corresponds to the 95% confidence interval, that is $\mathbb{E}[V](t) \pm 1.96\sqrt{\text{Var}[V](t)}$. (Color figure online)

where $\mathbf{G}_m = \mathbf{G}_i(\mathbf{G}_i + \mathbf{G}_e)^{-1}\mathbf{G}_e$ and $\alpha \approx 2.82 \times 10^{-3}\,\text{cm}^4\text{ms}^2\text{mS}^{-1}$. The other parameters were as follows: $V_{\text{rest}} = -85\,\text{mV}$, $V_{\text{dep}} = 30\,\text{mV}$, $\varepsilon = 0.4\,\text{ms}$, and $\mathbf{x}_s = (-2\,\text{cm}, 2\,\text{cm})$.

The computational domain was approximated by a triangular mesh $\mathcal{T}_h$ with median edge size of 0.04 cm in Ω_H and 0.5 cm in the rest of the domain, thus totalling 27820 nodes and 55476 cells. All quantities were represented by linear finite elements on $\mathcal{T}_h$. The eikonal equation was solved with an anisotropic version of the heat distance method [8], with $\Delta t = 4\,\text{ms}$. The implementation in FENICS is publicly available[1] and complemented with additional tests and comparison to the monodomain and bidomain models.

The activation map and the average lead fields for lead II and lead V1, as computed from Eq. (8), are reported in Fig. 2. In the correlation problem, the low-rank representation counted 17 (resp. 33) modes for lead II with uniform (resp. Gaussian) distribution of electrodes, and 33 (resp. 59) modes for lead V1. A lower number of modes for the uniform distribution was expected, as its support was compact and highly localized. The resulting ECGs are reported in Fig. 3. In both leads, the deterministic and average ECGs were very close, with an absolute error between 0.083 mV (lead II) and 0.28 mV (lead V1). The uncertainty was significantly higher in the late part of the QRS-complex. Maximum standard deviation was as high as 2.07 mV in lead V1 and 0.83 mV in lead II. In lead V1, the morphological variations were limited but the maximum amplitude changed significantly. In lead II, morphological differences were present in the second half of the ECG. No appreciable differences in ECGs were noted when comparing uniform and Gaussian-like distributions.

Finally, we compared the proposed method against the solution of the forward bidomain model, see Fig. 4. Differences between ECGs computed from our approach were essentially matching those derived from the forward bidomain

---

[1] See https://github.com/pezzus/fimh2021.

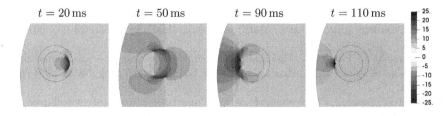

Fig. 4. Excerpt of extracellular potential (in mV) of the forward bidomain simulation.

model, with an absolute error less than 0.01mV in all cases and quantities of interests. The total cost of the forward simulation was from 2 to 8-fold higher than the lead field approach.

4   Discussion and Conclusions

In this work, we have solved the problem of quantifying the uncertainty in the ECG when uncertainty in the electrode positions is taken into account. Our method recasts the problem into a fully deterministic setting by using the lead field theory and a low-rank approximation for the correlation.

The computational advantage is significant over the standard forward simulation of the bidomain model. In fact, the number of lead fields to be computed in the proposed approach, for both the expectation and the correlation, does not depend on neither the transmembrane potential nor time, oppositely to the bidomain model. The method is therefore suitable to compute ECGs for long simulations, e.g., arrhythmic events, and it is even more advantageous in the context of inverse ECG approaches. Finally, the lead fields are smoother than the extracellular potential, especially within the heart, where potential gradients are strong along the activation front. A much coarser resolution may be employed for computing the lead fields, with no significant loss in accuracy [18].

While formulated for the chest electrodes, the presented theory also applies with minimal changes to assess the uncertainty of intracardiac electrogram recordings, widely employed in clinical electrophysiological studies. As a matter of fact, the formulation is flexible enough to address other relevant problems, such as quantifying the uncertainty in the ECG due to, e.g., uncertain transmembrane potential or torso-heart segmentation, hence leading to more robust simulation results.

References

1. Aboulaich, R., Fikal, N., El. Guarmah, E., Zemzemi, N.: Stochastic finite element method for torso conductivity uncertainties quantification in electrocardiography inverse problem. Math. Model. Nat. Phenom. **11**(2), 1–19 (2016)

2. Ammari, H., Chen, D., Zou, J.: Well-posedness of an electric interface model and its finite element approximation. Math. Models Methods Appl. Sci. **26**(03), 601–625 (2016)
3. Bishop, M.J., Plank, G.: Bidomain ECG simulations using an augmented monodomain model for the cardiac source. IEEE Trans. Biomed. Eng. **58**(8), 2297–2307 (2011)
4. Chen, Z., Zou, J.: Finite element methods and their convergence for elliptic and parabolic interface problems. Numer. Math. **79**(2), 175–202 (1998)
5. Clayton, R.H., et al.: An audit of uncertainty in multi-scale cardiac electrophysiology models. Philos. Trans. R. Soc. Lond. A **378**(2173), 20190335 (2020)
6. Colli Franzone, P., Pavarino, L.F., Scacchi, S.: Mathematical Cardiac Electrophysiology. M, vol. 13. Springer, Cham (2014). https://doi.org/10.1007/978-3-319-04801-7
7. Corrado, C., et al.: Quantifying atrial anatomy uncertainty from clinical data and its impact on electro-physiology simulation predictions. Med. Image Anal. **61**, 101626 (2020)
8. Crane, K., Weischedel, C., Wardetzky, M.: Geodesics in heat: a new approach to computing distance based on heat flow. ACM Trans. Graph. **32**(5), 1–11 (2013)
9. Evans, L.C.: Partial Differential Equations. 2nd edn., American Mathematical Society, Providence, Rhode Island (2010)
10. Giffard-Roisin, S., et al.: Transfer learning from simulations on a reference anatomy for ECGI in personalized cardiac resynchronization therapy. IEEE. Trans. Biomed. Eng. **66**(2), 343–353 (2019)
11. Harbrecht, H., Peters, M., Schneider, R.: On the low-rank approximation by the pivoted Cholesky decomposition. Appl. Numer. Math. **62**, 28–440 (2012)
12. Harbrecht, H., Li, J.: First order second moment analysis for stochastic interface problems based on low-rank approximation. ESAIM Math. Model. Numer. Anal. **47**(5), 1533–1552 (2013)
13. Kania, M., et al.: The effect of precordial lead displacement on ECG morphology. Med. Biol. Eng. Comput. **52**(2), 109–119 (2014)
14. Malmivuo, J., Plonsey, R.: Bioelectromagnetism-Principles and Applications of Bioelectric and Biomagnetic Fields. Oxford University Press, New York (1995)
15. Neic, A., et al.: Efficient computation of electrograms and ECGs in human whole heart simulations using a reaction-Eikonal model. J. Comput. Phys. **346**, 191–211 (2017)
16. Pezzuto, S., Kal'avský, P., Potse, M., Prinzen, F.W., Auricchio, A., Krause, R.: Evaluation of a rapid anisotropic model for ECG simulation. Front. Physiol. **8**, 265 (2017)
17. Pezzuto, S., et al.: Reconstruction of three-dimensional biventricular activation based on the 12-lead electrocardiogram via patient-specific modelling. EP Europace **23**(4), 640–647 (2021)
18. Potse, M.: Scalable and accurate ECG simulation for reaction-diffusion models of the human heart. Front. Phys. **9**, 370 (2018)
19. Potse, M., Dubé, B., Richer, J., Vinet, A., Gulrajani, R.M.: A comparison of monodomain and bidomain reaction-diffusion models for action potential propagation in the human heart. IEEE Trans. Biomed. Eng. **53**(12), 2425–2435 (2006)
20. Quaglino, A., Pezzuto, S., Koutsourelakis, P.S., Auricchio, A., Krause, R.: Fast uncertainty quantification of activation sequences in patient-specific cardiac electrophysiology meeting clinical time constraints. Int. J. Numer. Method. Biomed. Eng. **34**(7), e2985 (2018)

Cardiovascular Flow: Measures and Models

Quantitative Hemodynamics in Aortic Dissection: Comparing *in Vitro* MRI with FSI Simulation in a Compliant Model

Judith Zimmermann[1,2]([envelope])[iD], Kathrin Bäumler[1][iD], Michael Loecher[1,3],
Tyler E. Cork[1][iD], Fikunwa O. Kolawole[5], Kyle Gifford[1], Alison L. Marsden[4],
Dominik Fleischmann[1], and Daniel B. Ennis[1,3]

[1] Department of Radiology, Stanford University, Stanford, USA
judith.zimmermann@tum.de
[2] Department of Computer Science, Technical University of Munich, Munich,
Germany
[3] Division of Radiology, VA Palo Alto Health Care System, Palo Alto, USA
[4] Department of Pediatrics, Stanford University, Stanford, USA
[5] Department of Mechanical Engineering, Stanford University, Stanford, USA

Abstract. The analysis of quantitative hemodynamics and luminal
pressure may add valuable information to aid treatment strategies and
prognosis for aortic dissections. This work directly compared *in vitro*
4D-flow magnetic resonance imaging (MRI), catheter-based pressure
measurements, and computational fluid dynamics that integrated fluid-
structure interaction (CFD FSI). Experimental data was acquired with a
compliant 3D-printed model of a type-B aortic dissection (TBAD) that
was embedded into a flow circuit with tunable boundary conditions. *In
vitro* flow and relative pressure information were used to tune the CFD
FSI Windkessel boundary conditions. Results showed overall agreement
of complex flow patterns, true to false lumen flow splits, and pressure
distribution. This work demonstrates feasibility of a tunable experimental setup that integrates a patient-specific compliant model and provides
a test bed for exploring critical imaging and modeling parameters that
ultimately may improve the prognosis for patients with aortic dissections.

Keywords: Aortic dissection · CFD FSI · 4D-flow MRI

1 Introduction

An aortic dissection is a life-threatening vascular disorder in which a focal tear
develops within the inner aortic wall layer. This leads to subsequent formation
of a secondary channel ('false lumen', FL) that is separated from the primary
channel ('true lumen', TL) by a dissection flap. [13] Patients with type-B aortic
dissection (TBAD, i.e. without involvement of the ascending aorta) often receive
pharmacologic treatment and frequent monitoring is used in an attempt to predict late adverse events. Prognosis of late adverse events is largely informed by

© Springer Nature Switzerland AG 2021
D. B. Ennis et al. (Eds.): FIMH 2021, LNCS 12738, pp. 575–586, 2021.
https://doi.org/10.1007/978-3-030-78710-3_55

morphologic imaging features, but conflicting results have been reported among several predictors [17].

To improve prognosis several hemodynamic quantities are of potential interest and may confer added sensitivity of individual risk. Recent studies have suggested high FL outflow [15] as strong predictor for late adverse events, and FL ejection fraction [5] as indicator for false lumen growth rate.

To retrieve these hemodynamic markers, computational fluid dynamics (CFD) frameworks provide simulated patient-specific flow fields at high spatio-temporal resolution [11]; and those that integrate fluid-structure interaction (FSI) at deformable walls are expected to amplify the realism of patient-specific modeling even further. If simulations were able to reliably replicate hemodynamics, it would further enable non-invasive prediction of risk related to pathological changes (e.g. tear size).

While CFD FSI approaches show great potential, a direct validation with measured data in highly controlled, but realistic environments is missing. Previous comparisons between simulations and *in vivo* 4D-flow MRI [1,6,14] are challenged by: the assumption of a rigid aortic wall and dissection flap; a lack of information on accurate patient-specific hemodynamic conditions; and/or an unknown patient-specific aortic wall and dissection flap compliance.

Herein, we compare qualitative and quantitative TBAD hemodynamics based on: (1) simulations that use a recently proposed FSI framework [1], and (2) *in vitro* MRI including catheter-based pressure mapping. We utilized a patient-specific, compliant TBAD model embedded into a highly-controlled flow circuit. Uniaxial tensile testing of the compliant material, image-based flow splits and catheter-based pressure recordings informed simulation tuning.

2 Methods

2.1 Patient-Specific Aortic Dissection Model

A 3D computed tomography angiogram (CTA) of a patient (31 y/o, female) with TBAD was selected from our institution's database. A proximal intimal 'entry' tear was present distal to the left subclavian artery and an 'exit' tear was located proximal to the celiac trunk. Each tear measured 2.3 cm^2 in area size. The CTA study was approved by the institutional review board and written consent was obtained prior to imaging.

The lumen of the thoracic aorta was segmented using the active contour algorithm with manual refinements (itk-SNAP v3.4, Fig. 1a). Two tetrahedral meshes were generated (Fig. 1b): the 'fluid domain' representing the full aortic lumen; and the 'wall domain' (as extruded fluid domain) representing the outer aortic wall and dissection flap that separates TL and FL with uniform thickness ($h = 2$ mm). The wall domain mesh was further refined with (i) cylindrical caps that facilitated tubing connections, and (ii) visual landmarks to define image analysis planes. Meshing and refinements were done using SimVascular (release 2020-04) [19] and Meshmixer (v3.5, Autodesk). Further details on model generation are given in [1].

The wall model was 3D-printed using a novel photopolymer technique (Poly-Jet J735, Stratasys Inc.), as shown in Fig. 1d. The print material underwent uniaxial tensile testing as described in [21] and proved to be comparable to *in vivo* aortic wall compliance (tangent Young's modulus $E_{y,t} = 1.3$ MPa).

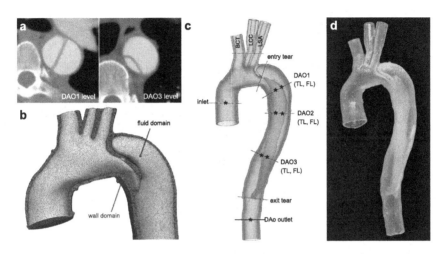

Fig. 1. (a) CTA images with lumen segmentation. (b) Tetrahedral meshes of fluid (gray) and wall domains with dissection flap (blue). (c) Cross-sectional landmarks and pressure mapping points (*). 'Entry' and 'exit' tear cover sections with combined TL and FL flow. Landmarks DAO1, DAO2, and DAO3 consist of a TL and FL cross-section. (d) Photograph of finished 3D-printed model. (Color figure online)

2.2   MRI Experiments

Imaging was performed on a 3 T MRI machine (Skyra, Siemens). An MRI-compatible flow circuit that includes a programmable pump (CardioFlow 5000 MR, Shelley) was engineered to provide controllable flow and pressure conditions similar with target values within the physiological range (Fig. 2a). Details were recently published in [21]. Glycerol-water (ratio = 40%/60%) with contrast (ferumoxytol) was used as a blood-mimicking fluid; and a typical aortic flow waveform (Fig. 2b) was applied (heart rate = 60/min, stroke volume = 74.1 mL/s, total flow = 4.45 L/min).

The circuit was tuned on the scanner table prior to image acquisition, targeting a flow split of 70%/30% (DAo outlet vs. arch branches), and luminal systolic pressure (at model inlet) of 120 mmHg. The pulse pressure was controlled via capacitance elements—designed as sealed air compression chambers—at the DAo outlet ($C1$) and at the merged arch branches ($C2$). A pressure transducer (SPR-350S, Millar) was inserted through ports at the model inlet and DAo outlet, and luminal pressures were recorded at eight points (Fig. 1c). Ultrasonic flow and pressure signals were fed into PowerLab (ADInstruments) for analysis.

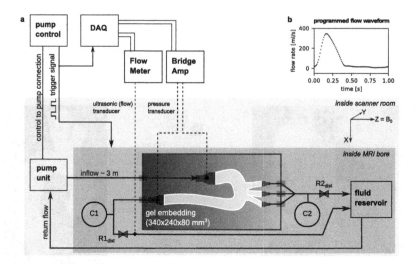

Fig. 2. (a) Schematic of the flow circuit setup. Pressure transducers were inserted through ports at the model inlet and DAo outlet. (b) Flow rate waveform that was programmed into the pump. $C1, C2$: 'capacitance' air-compression chambers, $R1, R2$: 'resistance' flow clamps. (Color figure online)

2D-Cine and 2D-PC MRI. Two-dimensional (2D) acquisitions at landmarks (Fig. 1c) included: (1) 2D cine gradient echo (2D-cine) with pixel size $= 0.9 \times 0.9$ mm^2, slice thickness $= 6$ mm, $T_E/T_R = 3/4.75$ ms, flip angle $= 7°$, avg. $= 2$, retro. gating (40 frames); and (2) 2D phase-contrast (2D-PC) with pixel size $= 1.1 \times 1.1$ mm^2, slice thickness $= 6$ mm, $T_E/T_R = 3/5.25$ ms, flip angle $= 25°$, avg. $= 2$, $V_{enc} = 90 - 120$ cm s^1, retro. gating (40 frames).

4D-Flow MRI. A four-point encoded Cartesian 4D-flow sequence was acquired as follows: FoV $= 340 \times 236 \times 84$ mm^3, matrix $= 220 \times 156 \times 56$, voxel size $= 1.5 \times 1.5 \times 1.5$ mm^3, $T_E/T_R = 2.7/5.6$ ms, flip angle $= 15°$, parallel imaging (GRAPPA, R $= 2$), $V_{enc} = 120$ cm s^{-1}, lines/seg. $= 2$, retro. gating (20 frames).

Image Analysis. Lumen contours were automatically tracked through time based on 2D-cine data via image-based deformable registration, which provided values of cross-sectional area and served as the boundary for net flow calculation. 2D-PC images were corrected for phase offsets (via planar 2nd order fitting) and then processed to retrieve the inlet flow and net flow splits across outlets.

4D-flow MRI data was corrected for (i) Maxwell terms, (ii) gradient non-linearity distortion [10], and (iii) phase offsets (via 3rd order fitting). Five landmarks along the dissected region were used for analysis (Fig. 1c). 4D-flow MRI offset correction, flow calculations, and streamline visualization were done using MEVISFlow (v11.2, Fraunhofer Institute) and ParaView (v5.7); quantitative results were exported as numeric files for comparison with simulation results.

2.3   CFD FSI Simulations

Governing Equations. The governing equations for fluid flow and structural mechanics were solved in the fluid and wall domain, respectively. In the fluid domain, the working fluid was considered incompressible and Newtonian ($\varrho_f = 1100$ kg m^{-3}, $\mu_f = 0.00392$ Pa s). Momentum and mass balance were described by the Navier-Stokes Equations in arbitrary Lagrangian Eulerian formulation to account for motion. The structural material was modeled with a Neo-Hookean model for homogeneous, isotropic hyperelastic materials ($E_y = 1.3$ MPa, $\varrho_s = 1450$ kg/m^3). Both domains were coupled at the interface via kinetic and dynamic interface conditions. A detailed mathematical description can be found in [1].

CFD FSI Boundary Conditions. The 2D-PC derived flow waveform was prescribed at the model inlet as a Dirichlet boundary condition, assuming a parabolic velocity profile. Three-element Windkessel boundary conditions were applied at fluid outlets and coupled to the 3D domain with the coupled multidomain method [7]. The catheter-based pressure values at the inlet of the model used as simulation tuning targets were: 119 mmHg, 42 mmHg, and 77 mmHg for the systolic (P_s), diastolic (P_d) and pulse pressure ($\triangle P$), respectively. Additionally, the 2D-PC derived flow splits informed the Windkessel parameter tuning, and were measured as 78.4%, 12.3%, 3.0%, and 5.2% for the DAo outlet, BCT, LCC, and LSA, respectively. The tuning of the Windkessel parameters (a distal and proximal resistance and capacitance at each of the model outlets) was then carried out in an iterative and manual process, by which a total resistance R_T and total capacitance C_T are distributed across all model outlets according to the measured flowsplits and a pre-prescribed ratio of distal to proximal resistance ($k_d = 0.9$). Details of the tuning process can be found in [1].

Wall domain outlets were fixed in space via a homogeneous Dirichlet condition for the displacement and a homogeneous Neumann boundary condition was prescribed at the outer wall of the vessel domain. This is in contrast to patient-specific simulations where a non-homogeneous Robin type boundary condition can be prescribed to account for external tissue support of the vessel. Likewise, the outer wall of the vessel domain was assumed to not be under prestress, contrary to a typical *in vivo* environment.

Numerical Formulation. The numerical simulations were performed with the SVFSI finite element solver, as implemented in SimVascular [19]. SVFSI features linear elements for velocity and pressure and is based on the "Residual Based Variational Multiscale" method. The fluid and wall domain were solved in a monolithic approach and backflow stabilization was applied at the fluid outlets. To avoid mesh degeneration, a nodal mesh smoothing was performed after each time step. Details of the numerical formulation are given in [1]. For details about the numerical discretization we refer to [2,3,8,18].

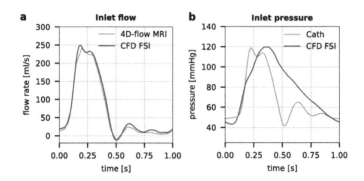

Fig. 3. CFD FSI (blue) tuning conditions, showing (a) flow rate and (b) pressure waveform at TBAD model inlet, in comparison to 4D-flow and catheter measurements (green). While the inlet flow rate waveforms match well, CFD FSI shows a slower diastolic pressure decay without oscillations. (Color figure online)

Discretization and Simulation Setup. Tetrahedral meshes of fluid and wall domain were sampled with a spatial resolution of $\triangle h = 1.3$ mm (1.6×10^6 tetrahedral elements) which was found to be a sufficiently fine resolution [1]. The temporal resolution was set to 4×10^3 timesteps per cardiac cycle ($\triangle t = 0.25$ ms). The simulation achieved cycle-to-cycle periodicity within 5 iterative runs. Compute time was 12 h per cycle on a high performance computing cluster.

CFD FSI Analysis. Time-resolved parameters were extracted from the last simulation cycle: (i) flow rate, (ii) area change, and (iii) pressure. We extracted data from every 50th simulated time step, which totaled 80 incremental results with an effective temporal resolution $\triangle t = 12.5$ ms. Quantitative metrics were analyzed at cross-sectional landmarks (Fig. 1c) using ParaView (v5.7) and exported as numeric files for direct 4D-Flow MRI comparison.

3   Results

Boundary Conditions. Inlet flow (Fig. 3a) for CFD FSI was directly prescribed based on 2D-PC results and agreed well with 4D-flow MRI. CFD-FSI flow splits across model outlets 78.7%, 12.7%, 3.2%, and 5.5% for DAo outlet, BCT, LCC, and LSA, respectively) aligned well with 2D-PC splits (78.4%, 12.1%, 3.0%, and 5.2%). After tuning, CFD FSI pressure (Fig. 3b) matched catheter measurements within the pre-defined 10% error margin (119.6 mmHg, 43.2 mmHg, and 76.4 mmHg for simulated P_s, P_d and $\triangle P$, corresponding to a relative error of $\leq 4\%$). While catheter-based measurements showed oscillations and a fast pressure drop at end-systole ($t = 0.4$ s), CFD FSI pressure decayed slower and without oscillation. As a results, mean pressure differed by 15.8% (78 mmHg for CFD FSI compared to 68 mmHg for catheter-based measurements).

Flow Patterns and Velocities. Qualitative flow visualizations (Fig. 4) showed well-matched flow patterns between CFD FSI and 4D-flow MRI. Particularly, streamlines depicted helical flow in FL aneurysm during systole and distal FL during diastole, as well as increased velocities through the proximal FL entry tear and along the distal TL. Overall, velocities were higher in CFD FSI, but the intra-model spatial distribution of velocities matched well.

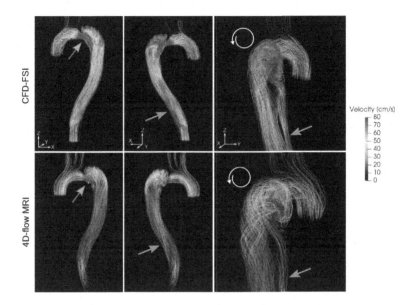

Fig. 4. Streamline visualization at systole ($t = 0.2$ s) for CFD FSI (top) and 4D-flow MRI (bottom) data. CFD FSI shows higher velocities, but intra-modality flow patterns and velocity distribution is consistent. Increased velocities through entry tear (blue arrows) and true lumen (green arrows). A helical flow pattern is visible in the false lumen aneurysm (white arrows). (Color figure online)

Pressure, Area, and Flow. Systolic TL pressure exceeded FL pressure (Fig. 5a) for both simulation and catheter measurements. At peak systole, the TL-FL presure difference was greater for CFD FSI data at landmarks DAO1 and DAO2, but matched well at DAO3. During diastole, the TL-FL difference was close to zero for CFD FSI, but was 1 to 2 mmHg for the catheter measurements. Cross-sectional area (Fig. 6, dashed lines) expanded most in FL cross-sections with up to 11% based on CFD FSI and up to 5% based on 2D-cine MRI.

Net flow volumes (Fig. 6) revealed a FL to TL flow split of 78%/22% for CFD FSI and 73%/27% for 4D-flow MRI measurements. Flow waveform shapes (Fig. 6, solid lines) aligned well, particularly regarding the peak flow timepoint, systolic upslope ($t = 0.1$ s), and oscillatory lobes in diastole. CFD FSI flow rates were higher in systole and lower in diastole when compared to 4D-flow values.

Pressure-area loops showed a steeper slope for *in vitro* data (Fig. 5b). FL peak flow preceded peaks of pressure and area change. This temporal delay was longer for CFD FSI, which was consistent for all DAO landmarks (Fig. 5c).

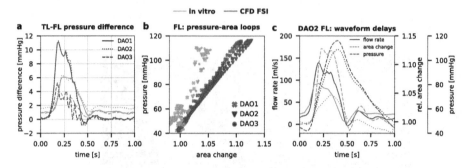

Fig. 5. (a) The TL-FL pressure difference was higher in proximal and lower in distal region. (b) FL pressure-area loops. (c) Flow rate peaks preceding both pressure and area peaks, with greater delay times for CFD FSI. (Color figure online)

4   Discussion

This study leveraged compliant 3D-printing as well as a highly-controlled MRI-compatible flow circuit setup to directly compare CFD FSI and MRI results with regards to flow and pressure dynamics in a patient-specific TBAD model. The aorta's secondary lumen and proximal FL aneurysm presented complex flow patterns with a large velocity range. These characteristics were well captured by both modalities and streamline visualizations were in very good agreement.

Our approach links measured luminal pressure with CFD FSI boundary conditions, which presents a major advantage over previous comparisons with *in vivo* data that usually lack invasive pressure measurements. During simulation tuning, pressure targets (P_s, P_d) were met, but pressure waveform shapes differed—i.e. faster and oscillatory pressure decay in catheter measurements versus slower and steady decline in CFD FSI. We note that a slower and steady pressure decline in diastole is desirable and would resemble *in vivo* pressure shapes of the arterial system [12].

To further investigate this mis-match, additional exploratory pressure data were recorded on the benchtop. Moreover, additional CFD FSI simulations with varying configurations of boundary conditions were computed. We identified three aspects to better match the measured and simulated pressure conditions. First, increasing the ratio between the distal and proximal resistance—described by parameter k_d in the three-element Windkessel model—is the key factor to improve the pressure shape towards a slower diastolic decline with its minimum at end-diastole (Fig. 7a). In practice, we increased k_d by decreasing proximal

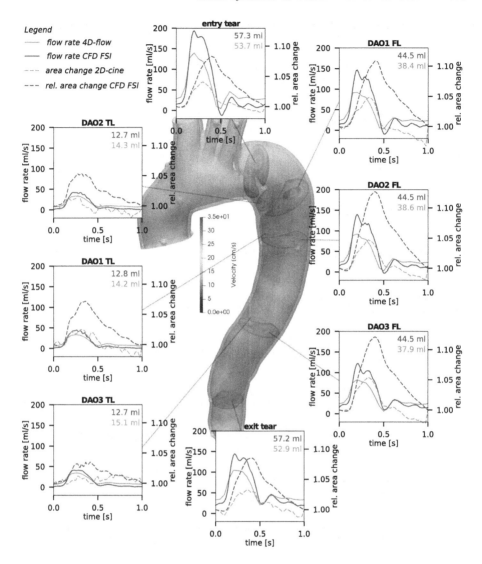

Fig. 6. Flow rate and area change (w.r.t area of first frame) at eight landmarks. Net flow values for CFD FSI (blue) and 4D-flow MRI (green) are given. CFD-FSI showed increased FL flow (78%) compared to 4D-flow (73%); and increased maximum area expansion (11% for CFD FSI vs. 5% for 4D-flow). (Color figure online)

resistance, and thus also decreased total system resistance which led to reduced $\triangle P$. Second, exploratory benchtop experiments also suggested that the characteristic pressure oscillations originate from wave reflections at multiple branching points. Other previous works that deployed flow circuits in model-based studies reported similar pressure waveform shapes [4,9,16,20]. Further engineering efforts should be made to minimize wave reflections at non-smooth boundaries

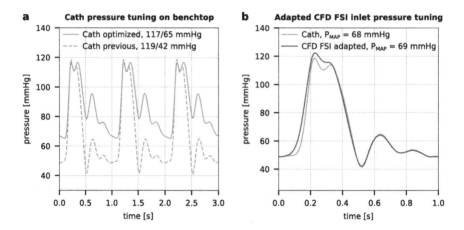

Fig. 7. (a) Exploratory benchtop data showing improvements for the setup's pressure condition. By drastically reducing proximal resistance, pulse pressure is reduced from 78 to 52 mmHg, and—if wave reflections were to be neglected—diastolic pressure declines steadily towards its minimum at end-diastole. (b) CFD FSI (blue) vs catheter-based (green) pressure at the inlet face in an adapted simulation with rigid walls and inhomogeneous pressure boundary condition at the BCT outlet. By tuning the simulation using the full pressure waveform, inlet pressure conditions closely resemble measured values, including wave reflections. (Color figure online)

as much as possible. Third, additional exploratory simulation runs suggested that the measured pressure conditions can be closely approximated by directly prescribing pressure data as inhomogeneous Neumann boundary condition at one of the outlets. To do so, we prescribed the catheter-based pressure data that was measured at the model inlet as pressure boundary condition at the brachiocephalic trunk outlet. Resulting well-matched pressure waveforms are shown in Fig. 7b. We seek to adapt both our experimental and simulation setup in future studies regarding these three aspects.

Overall, our presented results showed similar tendencies of flow and pressure parameters in the dissected region between MRI, catheter measurements, and FSI simulation. TL-FL pressure differences were comparable such that they were almost consistently positive, and that the most distal landmark (DAO3) showed a smaller difference compared to the two proximal points (DAO1, DAO2). Interestingly, CFD FSI TL-FL pressure difference briefly dropped to negative ($t = 0.4$ s) and then to zero ($t > 0.65$ s), while catheter measurements showed preserved positive TL-FL differences at all locations and times. Moreover, with only 5% difference between modalities, results suggest a well-matched TL-FL flow split.

Multiple results indicate that the performed tensile testing underestimated $E_{y,t}$: a steeper slope of the pressure-area loop for *in vitro* data, shorter flow-pressure-area waveform delays, and consistently lower outer wall expansion. To address this mis-match, future CFD FSI experiments should iteratively increase the value for $E_{y,t}$ until *in vitro* wall deformations are sufficiently replicated.

Moving toward clinical deployment of simulation-based treatment decision support, future work should also investigate uncertainties of pressure and flow conditions and their impact on hemodynamic quantities. In particular, if pressure data are unavailable, it should be investigated how approximations of pressure boundary conditions (e.g. two-point systolic to diastolic cuff pressure) propagate errors into hemodynamic quantities. The presented highly-controlled *in vitro* setup is well suited to investigate these effects.

In conclusion, this work presents valuable information on hemodynamic similarities and differences as retrieved from CFD FSI, *in vitro* MRI, and catheter-based pressure measurements in a patient-specific aortic dissection model.

Acknowledgements. We thank the Stanford Research Computing Center for computational resources (Sherlock HPC cluster), Dr. Anja Hennemuth for making available software tools, and Nicole Schiavone for technical advice. Funding was received from DAAD scholarship program (to J.Z.) and NIH R01 HL131823 (to D.B.E).

References

1. Bäumler, K., et al.: Fluid-structure interaction simulations of patient-specific aortic dissection. Biomech. Model. Mechanobiol. **19**, 1607–1628 (2020)
2. Bazilevs, Y., Calo, V.M., Cottrell, J.A., Hughes, T.J., Reali, A., Scovazzi, G.: Variational multiscale residual-based turbulence modeling for large eddy simulation of incompressible flows. Comput. Meth. Appl. Mech. Eng. **197**(1–4), 173–201 (2007)
3. Bazilevs, Y., Calo, V.M., Hughes, T.J., Zhang, Y.: Isogeometric fluid-structure interaction: theory, algorithms, and computations. Comput. Mech. **43**(1), 3–37 (2008)
4. Birjiniuk, J., et al.: Pulsatile flow leads to intimal flap motion and flow reversal in an in vitro model of type B aortic dissection. Cardiovasc. Eng. Techn. **8**(3), 378–389 (2017)
5. Burris, N.S., et al.: False lumen ejection fraction predicts growth in type B aortic dissection: preliminary results. Eur. J. Cardio-thoracic Surg. **57**(5), 896–903 (2020)
6. Dillon-Murphy, D., Noorani, A., Nordsletten, D., Figueroa, C.A.: Multi-modality image-based computational analysis of haemodynamics in aortic dissection. Biomech. Model. Mechanobiol. **15**(4), 857–876 (2015). https://doi.org/10.1007/s10237-015-0729-2
7. Esmaily-Moghadam, M., Bazilevs, Y., Marsden, A.L.: A new preconditioning technique for implicitly coupled multidomain simulations with applications to hemodynamics. Comput. Mech. **52**(5), 1141–1152 (2013). https://doi.org/10.1007/s00466-013-0868-1
8. Esmaily-Moghadam, M., Bazilevs, Y., Marsden, A.L.: A bi-partitioned iterative algorithm for solving linear systems arising from incompressible flow problems. Comput. Meth. Appl. Mech. Eng. **286**, 40–62 (2015)
9. Gallarello, A., et al.: Patient-specific aortic phantom with tunable compliance. J. Eng. Science Med. Diagn. Therapy **2**(4), 041005 (2019)
10. Markl, M., et al.: Generalized reconstruction of phase contrast MRI: analysis and correction of the effect of gradient field distortions. Magn. Reson. Med. **50**(4), 791–801 (2003)

11. Marsden, A.L.: Simulation based planning of surgical interventions in pediatric cardiology. Phys. Fluids **25**, 101303 (2013)
12. Mills, C.J., et al.: Pressure-flow relationships and vascular impedance in man. Cardiovasc. Res. **4**(4), 405–417 (1970)
13. Nienaber, C.A., et al.: Aortic dissection. Nat. Rev. Dis. Primers. **2**, 16053 (2016)
14. Pirola, S., et al.: 4D flow MRI-based computational analysis of blood flow in patient-specific aortic dissection. IEEE Trans. Biomed. Eng. **66**(12), 3411–3419 (2019)
15. Sailer, A.M., et al.: Computed tomography imaging features in acute uncompli-cated stanford type-B aortic dissection predict late adverse events. Circ. Cardio-vasc. Imaging **10**(4), e005709 (2017)
16. Schiavone, N.K., Elkins, C.J., McElhinney, D.B., Eaton, J.K., Marsden, A.L.: In vitro assessment of right ventricular outflow tract anatomy and valve orientation effects on bioprosthetic pulmonary valve hemodynamics. Cardiovasc. Eng. Technol. **12**(2), 215–231 (2021). https://doi.org/10.1007/s13239-020-00507-6
17. Spinelli, D., et al.: Current evidence in predictors of aortic growth and events in acute type B aortic dissection. J. Vasc. Surg. **68**(6), 1925–1935 (2018)
18. Tezduyar, T.E., Mittal, S., Ray, S.E., Shih, R.: Incompressible flow computations with stabilized bilinear and linear equal-order-interpolation velocity-pressure ele-ments. Comput. Meth. Appl. Mech. Eng. **95**(2), 221–242 (1992)
19. Updegrove, A., Wilson, N.M., Merkow, J., Lan, H., Marsden, A.L., Shadden, S.C.: SimVascular: an open source pipeline for cardiovascular simulation. Ann. Biomed. Eng. **45**(3), 525–541 (2016). https://doi.org/10.1007/s10439-016-1762-8
20. Urbina, J., et al.: Realistic aortic phantom to study hemodynamics using MRI and cardiac catheterization in normal and aortic coarctation conditions. J. Magn. Reson. Imaging **44**(3), 683–697 (2016)
21. Zimmermann, J., et al.: On the impact of vessel wall stiffness on quantitative flow dynamics in a synthetic model of the thoracic aorta. Sci. Rep. **11**, 6703 (2021)

3-D Intraventricular Vector Flow Mapping Using Triplane Doppler Echo

Florian Vixège[1], Alain Berod[2,3], Franck Nicoud[3] (iD), Pierre-Yves Courand[1,4] (iD),
Didier Vray[1] (iD), and Damien Garcia[1](✉) (iD)

[1] CREATIS (Centre de Recherche en Acquisition et Traitement de l'Image pour la Santé),
UMR5220, U1294 Lyon, France
damien.garcia@inserm.fr
[2] Sim&Cure, Montpellier, France
[3] IMAG, University of Montpellier, Montpellier, France
[4] Department of Cardiology, HCL, Lyon, France
https://www.biomecardio.com

Abstract. We generalized and improved our clinical technique of two-dimensional intraventricular vector flow mapping (2D-*i*VFM) for a full-volume three-component analysis of the intraventricular blood flow (3D-*i*VFM). While 2D-*i*VFM uses three-chamber color Doppler images, 3D-*i*VFM is based on the clinical mode of triplane color Doppler echocardiography. As in the previous two-dimensional version, 3D-*i*VFM relies on mass conservation and free-slip endocardial boundary conditions. For sake of robustness, the optimization problem was written as a constrained least-squares problem. We tested and validated 3D-*i*VFM *in silico* through a patient-specific heart-flow CFD (computational fluid dynamics) model, as well as *in vivo* in one healthy volunteer. The intraventricular vortex that forms during left ventricular filling was deciphered. After further validation, 3D-*i*VFM could offer clinically compatible 3-D echocardiographic insights into left intraventricular hemodynamics.

Keywords: Ultrasound imaging · Triplane doppler echo · 3-D vector flow imaging · Full-volume intracardiac blood flow

1 Introduction

1.1 The Intraventricular Diastolic Vortex

During the filling of the left ventricle (diastole), the valvular mitral inlet constrains the intraventricular blood flow to form a vortex. Although the geometry and orientation of the mitral valve contribute significantly to vortex formation and dynamics, the shape and relaxing deformations of the left ventricular cavity are the major contributors to vortex properties [1]. When heart filling is impaired (diastolic dysfunction), a change in blood flow occurs, with a significant impact on the intraventricular vortex [2]. Recent studies highlighted that quantification of the intraventricular blood flow

© Springer Nature Switzerland AG 2021
D. B. Ennis et al. (Eds.): FIMH 2021, LNCS 12738, pp. 587–594, 2021.
https://doi.org/10.1007/978-3-030-78710-3_56

could improve the diagnosis of diastolic dysfunction, which often remains uncertain as standard echocardiographic indices may lead to divergent conclusions.

The clinical accessibility of echocardiography and its ability to provide non-invasive, real-time information make ultra-sound the preferred technique for analyzing the intra-cardiac blood flow at the bedside [3]. There are two main classes of methods for the analysis of blood flow by ultra-sound imaging: (1) techniques based on speckle tracking [4], such as echo-PIV (particle image velocimetry), and (2) approaches based on color Doppler, such as *i*VFM (intraventricular Vector Flow Mapping) [5]. Since speckle-tracking methods generally (but not necessarily [6]) require the injection of micro-bubbles [7], techniques derived from color-Doppler are preferable in clinical routine.

1.2 Full-Volume Three-Component by Conventional Color Doppler

Doppler echocardiography provides a scalar velocity field: it returns the velocity pro-jections along the ultrasound beams. Numerical methods were recently introduced to recover the three-dimensionality of intracardiac blood flow from color Doppler veloci-ties, such as multi-angle [8] or model-based [9] methods. Multi-angle techniques can be limited in a clinical context, as they require a series of acquisitions with different acous-tic windows. Model-based techniques rely on the optimization of an objective function based on hemodynamics (e.g., mass conservation). These 3-D studies were based on volume color Doppler, whose acquisitions are constrained by a compromise between spatial and temporal resolutions.

In the following section, we describe how we generalized our most recent version [10] of two-dimensional intraventricular vector flow mapping (2D-*i*VFM) in 3-D. The current 2D-*i*VFM technique is limited to the three-chamber apical long-axis view; it does not allow recovering the three-dimensional structure of intraventricular blood flow. With the objective of getting a full-volume three-component field:

(1) We used (and simulated) triplane Doppler echocardiographic images of a GE (Vivid e95) scanner, which gives access to three color Doppler planes (Fig. 2a) at a good temporal resolution.

(2) We modified the 2D-*i*VFM equations [10] for 3D-*i*VFM and made the numerical problem more robust by using a least-squares method with physics-based equality constraints.

(3) We determined the smoothing parameter automatically through the *L*-curve method [11] to make the regularization unsupervised and optimize operator independence.

2   Method – from Scalar Doppler to 3-D Vector Doppler

2.1   The Constrained Minimization Problem

The objective of 3D-*i*VFM is to reconstruct the actual 3-D intraventricular velocity field from triplane color Doppler. Triplane color Doppler returns three apical long-axis planes (color Doppler + B-mode) separated by an angle of 60° (Fig. 2a). We chose this triplane mode (rather than 3-D color Doppler) since (1) the ultrasound data before scan-conversion are available through EchoPAC, and (2) the temporal resolution is higher. The Doppler (velocities) and B-mode (8-bit ultrasound image) data returned by the EchoPAC soft-ware are given in a spherical $\{r,\ \theta,\ \varphi\}$ coordinate system. Doppler velocity measured by echocardiography, $u_D(r,\ \theta,\ \varphi)$, is related to the radial velocity component, $v_r(r,\ \theta,\ \varphi)$, with a negative sign and additive noise η:

$$v_r(r,\theta,\varphi) = -u_D(r,\theta,\varphi) + \eta(r,\theta,\varphi). \tag{1}$$

The negative sign takes into account the color Doppler convention, which stipulates positivity when blood is directed towards the ultrasound array. The goal of 3D-iVFM is to determine the angular and azimuthal components in this spherical coordinate system, $v_\theta(r,\ \theta,\ \varphi)$ and $v_\varphi(r,\ \theta,\ \varphi)$, by using physics-based and smoothness constraints. The algorithm for 3D-iVFM is based on a generalized and improved version of the 2D-iVFM algorithm introduced by Assi $et\ al.$ [10]. We rewrote the problem as a constrained least-squares problem, which was solved by the Lagrange multiplier method. We used hemodynamic properties based on fluid dynamics to constrain the problem, such as mass conservation for an incompressible fluid (divergence-free flow), and free-slip boundary conditions on the endocardial wall. The minimization problem was mathematically written as

$$\vec{v}_{VFM} = \operatorname*{argmin}_{\vec{v}} J(\vec{v}), \quad \text{with} \quad J(\vec{v}) = \int_\Omega (v_r + u_D)^2 \, d\Omega + \alpha \mathcal{L}(\vec{v}) \tag{2}$$

subject to:

$$\begin{cases} \vec{v}_{VFM} \cdot \vec{n}|_{wall} - \vec{v}_{wall} \cdot \vec{n}|_{wall} = 0 \quad \text{on } \partial\Omega, \\ \operatorname{div}(\vec{v}_{VFM}) = 0 \quad \text{on } \Omega. \end{cases} \tag{3}$$

The velocity vectors $\vec{v}_{VFM}$ are the velocities recovered by color-Doppler Vector Flow Mapping, i.e. they are the solutions of the constrained minimization problem. The coordinates in the spherical system are $\{v_r(r,\theta,\varphi), v_\theta(r,\theta,\varphi), v_\varphi(r,\theta,\varphi)\}$. The second term in the cost function $J(\vec{v})$ in (2), namely $\mathcal{L}(\vec{v})$, stands for a spatial smoothing function that uses second-order partial derivatives, with cross-terms, with respect to the radial and angular coordinates (r, θ). The parameter α is the smoothing parameter (a scalar). Ω represents the intracavitary (intraventricular) region of interest, and $\partial\Omega$ is its endocardial (wall) boundary. $\vec{n}|_{wall}$ is the vector normal to the wall, and $\vec{v}_{wall}$ is the endocardial wall velocity. In the echocardiographic laboratory, the myocardial velocities can be derived, for example, by standard speckle tracking [4]. The first constraint equality in (3) describes the free-slip (no penetration) boundary conditions. Note that no-slip conditions are not recommended because the spatial resolution of color Doppler is not fine enough to capture the boundary layer. The second constraint equality in (3) represents the mass conservation for an incompressible fluid (divergence-free flow).

We applied the Lagrange multiplier method [12] and a finite-difference discretization to convert this constrained least-squares problem into a matrix form. To minimize operator dependence, we selected the smoothing parameter (α) automatically through the L-curve method [11]. A too-large parameter induces too much smoothing, a too-low parameter generates a noisy vector field. The best bias-variance compromise must be determined objectively.

2.2 Periodic Interpolation Along the Azimuthal Direction

Recalling that 3D-iVFM uses three long-axis planes (six azimuthal half-planes), we used a periodic trigonometric interpolation along the azimuthal direction to express the velocity vectors $\vec{v} = \{v_r, v_\theta, v_\varphi\}$ in (2). We thus wrote the three velocity components as follows:

$$v_k = a0_k(r, \theta) + a1_k(r, \theta)\cos(\varphi)$$
$$+a2_k(r, \theta)\cos(2\varphi) + a3_k(r, \theta)\cos(3\varphi)$$
$$+a4_k(r, \theta)\sin(\varphi) + a5_k(r, \theta)\sin(2\varphi), \tag{4}$$

with $k \in \{r, \theta, \varphi\}$.

It follows that the minimization problem (2) yields the coefficients aN_k, with $N = 0 \dots 5$. These series of six coefficients correspond to the six half-planes given by triplane echocardiography. This interpolated expression ensures well-posedness of the problem and smoothness of each velocity component in the azimuthal direction. It also makes the numerical algorithm easily manageable in terms of memory and computation time (less than half a minute with a personal computer in Matlab language).

2.3 In Silico and in Vivo Analyses

We tested the 3D-iVFM algorithm (1) *in silico*, in a patient-specific 3-D dynamic model of intraventricular flow based on CFD (computational fluid dynamics) [13] (Fig. 1a), and (2) *in vivo*, in one healthy volunteer.

In Silico. The CFD model was already used to validate the previous 2D-iVFM version [10]. We used a patient-specific CFD cardiac model introduced by Chnafa *et al.* [13, 14]. In this CFD model, the heart chambers and wall dynamics were retrieved from 4D images acquired by computed tomography. An arbitrary Lagrangian-Eulerian (ALE) framework was adopted to handle the large amplitude motion of the cardiac tissues (endocardium and valves). This CFD model is detailed in several publications [13–15].

For the present study, triplane Doppler echo images were simulated by adding noise in the radial velocity components. Full-volume three-component velocity fields were estimated by 3D-iVFM (Eqs. 2–4), from these triplane Doppler velocities, and compared with the actual CFD velocity fields. Absolute errors and errors relative to the maximum speed were calculated (ground-truth CFD vs. 3D-iVFM).

In Vivo. As a clinical proof-of-concept, the left ventricle of one healthy volunteer was scanned by a cardiologist with a clinical ultrasound scanner (Vivid e95 GE), using the triplane color-Doppler mode (Fig. 2a). The Doppler and B-mode data before scan-conversion were extracted through the EchoPAC software (Fig. 2b). Full-volume three-component velocity fields were generated with 3D-iVFM (Eqs. 2–4).

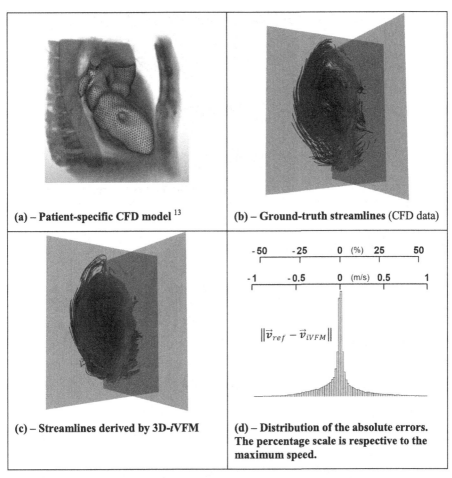

Fig. 1. *In silico results*: the streamlines (here, during filling) recovered after 3D-*i*VFM (c) were similar to the actual streamlines (b). The distribution of the absolute errors showed a narrow peak around zero (d).

3   Discussion

We successfully generalized 2D-*i*VFM for triplane Doppler echocardiography and introduced 3D-*i*VFM. Our *in silico* and *in vivo* one-case studies tend to show that 3D-*i*VFM could recover the 3-D structure of the intraventricular flow that forms during diastole. Although the 3-D intraventricular flow could not be recovered at small scales because of the smoothing regularizer present in the algorithm, our *in silico* results seem to show that a macroscopic 3D-flow analysis would be possible. For example, the global intraventricular circulation might reflect cardiac filling function. This remains to be demonstrated on a sufficiently large cohort of patients once the 3D-iVFM technique is finalized. The algorithm that we developed is fast and unsupervised, which makes it compatible with a

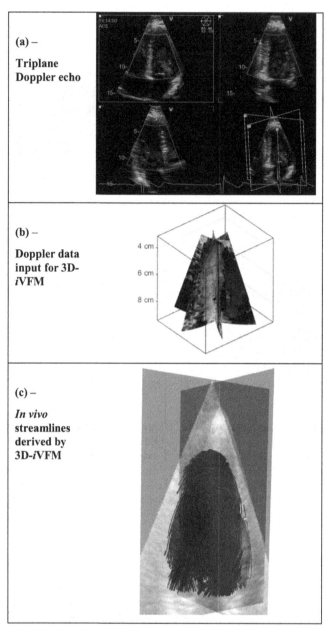

Fig. 2. *In vivo observation*: our innovative 3D-*i*VFM clinical method uses triplane Doppler echocardiography (a). The streamlines recovered by 3D-*i*VFM from the triplane Doppler velocities (b) depicted the large vortex that formed during filling (c).

standard examination in the echocardiographic laboratory. It depends on a single regularization parameter, which was selected automatically. This numerical aspect is important to minimize operator dependence. Further *in vivo* investigation is planned to compare 3D-*i*VFM against CMR (cardiac magnetic resonance) velocimetry in patients, as in [16].

This study complements our *in vivo* analyses of intra-ventricular flow by color Doppler, whether by conventional [10, 17], or high-frame-rate echocardiography [18, 19]. In the present study, we opted for standard ultrasonography to facilitate a possible future clinical study. For this reason, we used a triplane mode. This mode has two specificities: (1) the temporal resolution is higher than that of 3-D Doppler, and (2) the acquired data can be extracted through EchoPAC for post-processing by *i*VFM (as opposed to 3-D data) [10, 20]. Although this needs to be quantified, it is likely that access to volume data, rather than triplane data, could reduce the bias of the reconstructed velocities. Whether this difference in bias is significant from a physiological point of view needs to be analyzed.

4   Conclusion

Intracardiac velocimetry by 3D-*i*VFM could be an effective tool for quantifying the intraventricular vortex and for assessing blood flow during heart filling. Fast and clinically compatible, 3D-*i*VFM could offer new echocardiographic insights into left ventricular hemodynamics

Acknowledgments. This work was supported by the LABEX CeLyA (ANR-10-LABX-0060) of Université de Lyon, within the program «Investissements d'Avenir» (ANR-16IDEX-0005) operated by the French National Research Agency (ANR).

References

1. Charonko, J.J., Kumar, R., Stewart, K., Little, W.C., Vlachos, P.P.: Vortices formed on the mitral valve tips aid normal left ventricular filling. Ann. Biomed. Eng. **41**(5), 1049–1061 (2013). https://doi.org/10.1007/s10439-013-0755-0
2. Bermejo, J., Martínez-Legazpi, P., del Álamo, J.C.: The clinical assessment of intraventricular flows. Annu. Rev. Fluid Mech. **47**(1), 315–342 (2015). https://doi.org/10.1146/annurev-fluid-010814-014728
3. Sengupta, P.P., Pedrizzetti, G., Kilner, P.J., et al.: Emerging trends in CV flow visualization. JACC: Cardiovascul. Imaging **5**(3), 305–316 (2012). https://doi.org/10.1016/j.jcmg.2012.01.003
4. Garcia, D., Saloux, E., Lantelme, P.: Introduction to speckle tracking in cardiac ultrasound imaging. In: Handbook of Speckle Filtering and Tracking in Cardiovascular Ultrasound Imaging and Video. Institution of Engineering and Technology (2017)
5. Garcia, D., del Álamo, J.C., Tanné, D., et al.: Two-dimensional intraventricular flow mapping by digital processing conventional color-Doppler echocardiography images. IEEE Trans. Med. Imaging **29**(10), 1701–1713 (2010). https://doi.org/10.1109/TMI.2010.2049656
6. Nyrnes, S.A., Fadnes, S., Wigen, M.S., Mertens, L., Lovstakken, L.: Blood speckle-tracking based on high–frame rate ultrasound imaging in pediatric cardiology. J. Am. Soc. Echocardiogr. (2020). https://doi.org/10.1016/j.echo.2019.11.003

7. Hong, G.-R., Pedrizzetti, G., Tonti, G., et al.: Characterization and quantification of vortex flow in the human left ventricle by contrast echocardiography using vector particle image velocimetry. JACC: Cardiovascul. Imaging **1**(6), 705–717 (2008). https://doi.org/10.1016/j.jcmg.2008.06.008

8. Gomez, A., Pushparajah, K., Simpson, J.M., Giese, D., Schaeffter, T., Penney, G.: A sensitivity analysis on 3D velocity reconstruction from multiple registered echo Doppler views. Med. Image Anal. **17**(6), 616–631 (2013). https://doi.org/10.1016/j.media.2013.04.002

9. Grønli, T., Wigen, M., Segers, P., Lovstakken, L.: A fast 4D B-spline framework for model-based reconstruction and regularization in vector flow imaging. In: 2018 IEEE International Ultrasonics Symposium (IUS), pp. 1–9 (2018). https://doi.org/10.1109/ultsym.2018.8579767

10. Assi, K.C., Gay, E., Chnafa, C., et al.: Intraventricular vector flow mapping-a Doppler-based regularized problem with automatic model selection. Phys. Med. Biol. **62**(17), 7131–7147 (2017). https://doi.org/10.1088/1361-6560/aa7fe7

11. Hansen, P.C.: The L-curve and its use in the numerical treatment of inverse problems. In: Johnston, p. (ed.) Computational Inverse Problems in Electrocardiology. Advances in Computational Bioengineering, pp. 119–142. WIT Press (2000)

12. Kalman, D.: Leveling with Lagrange: an alternate view of constrained optimization. Math. Mag. **82**(3), 186–196 (2009)

13. Chnafa, C., Mendez, S., Nicoud, F.: Image-based large-eddy simulation in a realistic left heart. Comput. Fluids **94**, 173–187 (2014). https://doi.org/10.1016/j.compfluid.2014.01.030

14. Chnafa, C., Mendez, S., Nicoud, F.: Image-based simulations show important flow fluctuations in a normal left ventricle: what could be the implications? Ann. Biomed. Eng. **44**(11), 3346–3358 (2016). https://doi.org/10.1007/s10439-016-1614-6

15. Chnafa, C., Mendez, S., Moreno, R., Nicoud, F.: Using image-based CFD to investigate the intracardiac turbulence. In: Quarteroni, A. (ed.) Modeling the Heart and the Circulatory System. M, vol. 14, pp. 97–117. Springer, Cham (2015). https://doi.org/10.1007/978-3-319-05230-4_4

16. Faurie, J., Baudet, M., Assi, K.C., et al.: Intracardiac vortex dynamics by high-frame-rate Doppler vortography – in vivo comparison with vector flow mapping and 4-D flow MRI. IEEE Trans. Ultrason. Ferroelectr. Freq. Control **64**(2), 424–432 (2017)

17. Mehregan, F., Tournoux, F., Muth, S., et al.: Doppler vortography: a color Doppler approach to quantification of intraventricular blood flow vortices. Ultrasound Med. Biol. **40**(1), 210–221 (2014). https://doi.org/10.1016/j.ultrasmedbio.2013.09.013

18. Faurie, J., Baudet, M., Poree, J., Cloutier, G., Tournoux, F., Garcia, D.: Coupling myocardium and vortex dynamics in diverging-wave echocardiography. IEEE Trans. Ultrason. Ferroelectr. Freq. Control **66**(3), 425–432 (2019). https://doi.org/10.1109/TUFFC.2018.2842427

19. Garcia, D., Kadem, L., Savéry, D., Pibarot, P., Durand, L.-G.: Analytical modeling of the instantaneous maximal transvalvular pressure gradient in aortic stenosis. J. Biomech. **39**(16), 3036–3044 (2006). https://doi.org/10.1016/j.jbiomech.2005.10.013

20. Muth, S., Dort, S., Sebag, I.A., Blais, M.-J., Garcia, D.: Unsupervised dealiasing and denoising of color-Doppler data. Med. Image Anal. **15**(4), 577–588 (2011). https://doi.org/10.1016/j.media.2011.03.003

The Role of Extra-Coronary Vascular Conditions that Affect Coronary Fractional Flow Reserve Estimation

Jermiah J. Joseph[1,2] (iD), Ting-Yim Lee[2], Daniel Goldman[2],
Christopher W. McIntyre[1,2], and Sanjay R. Kharche[1,2(✉)] (iD)

[1] Lawson Health Research Institute, London, Canada
Sanjay.Kharche@lhsc.on.ca
[2] Department of Medical Biophysics, Western University, London, Canada

Abstract. The treatment of coronary stenosis relies on invasive high risk surgical assessment to generate the fractional flow reserve, a ratio of distal to proximal pressures in respect of the stenosis. Non-invasive methods are therefore desirable. Non-invasive imaging-computational methodologies call for robust and calibrated mathematical descriptions of the coronary vasculature that can be personalized. In addition, it is important to understand extra-coronary co-morbidities that may affect fractional flow estimates. In this preliminary theoretical work, a 0D human coronary vasculature model was implemented, and used to demonstrate the distinct roles of focal and extended stenosis (intra-coronary), as well as microvascular disease and atrial fibrillation (extra-coronary) on fractional flow reserve estimation. It was found that the right coronary artery is maximally affected by diffuse stenosis and microvascular disease. The model predicts that the presence, rather than severity, of both microvascular disease and atrial fibrillation affect coronary flow deleteriously. The model provides a computationally inexpensive instrument for future *in silico* coronary blood flow investigations as well as clinical-imaging decision making. The framework provided is extensible as well as can be personalized. Furthermore, it provides a starting point and crucial boundary conditions for future 3D computational hemodynamics flow estimation.

Keywords: Coronary vasculature · Lumped parameter model · Fractional flow reserve · Computational cardiology

1 Introduction

Coronary artery disease is a major clinical concern, and its diagnosis remains invasive and high risk. The current use of angiographic imaging permits localization of coronary stenosis but provides is suboptimal to generate percutaneous coronary intervention recommendations. Fractional flow reserve (FFR) is an increasingly used composite that quantifies the severity of stenosis [1]. Clinically, FFR is based upon pressures obtained in the vicinity of a stenosis using invasive catheters. A FFR value of nil (0) indicates complete stenosis and very close to unity (1) indicating an absence of stenosis. Current

© Springer Nature Switzerland AG 2021
D. B. Ennis et al. (Eds.): FIMH 2021, LNCS 12738, pp. 595–604, 2021.
https://doi.org/10.1007/978-3-030-78710-3_57

clinical practice recommends angiographic interventions such as revascularization upon finding a FFR of less than 0.8 [2, 3]. However, invasive assessment of FFR is resource intensive and poses surgical complications risk. Surgical and pharmacological sensitivity is increasingly relevant where adverse events often occur in critically ill patients such as those with renal failure [4, 5].

In this study, a theoretical modelling methodology was developed. It was used to demonstrate the distinct effects of focal and diffuse stenosis (structural factors affecting FFR), as well as extra-coronary co-morbidities of microvascular and cardiac dysfunction (functional factors affecting FFR). The approach is based on tangible 0D models of human hemodynamics that permit uncovering relevant cause-effect relationships [6, 7]. Circulation models represented by ordinary differential equations (0D models) have been widely used in the literature. A previously developed model [8] was used to elicit uncover the causes of hemodynamic instability due to hypertension [6] and atrial fibrillation [9]. A further detailed and calibrated biophysical model [10, 11] has been widely used. ODE models of organs [12], feedback mechanisms such as baroreflex [11], and surgical instruments such as dialyzer machines [13, 14] have also been used to improve our understanding of processes underlying clinical practice. Others have used composite models to assess the efficacy of surgeries such as dialysis [15].

Whereas the presented model is theoretical in nature, its design will provide expectation and boundary conditions in future spatially extended modeling. Further, since the validation of the 3D models will provide a better understanding of pathophysiological processes, doing so for the presented model was considered as a first step. Since this model has high manipulability and extensibility, future studies will use it for personalization driven coronary vasculature health status assessment.

Recent advances permit CT imaging driven development of 3D coronary geometries to perform hemodynamic simulations, thus offering an inter-disciplinary approach to non-invasively assess coronary artery disease [16]. The lumped parameter approach of modelling cardiovascular systems using the three-element Windkessel representation of vessels has been widely studied to investigate their utility in simulating hemodynamic physiology [17]. By using structured trees to represent small artery bifurcations, outflow boundary conditions can be estimated using 0D models [18, 19]. Coupling 3D finite element models with 0D and 1D models of the systemic circulation have been used to investigate their interactions and effects on coronary blood flow simulations [20].

Computationally efficient 0D modelling for study of coronary blood flow dynamics remains limited [21]. In this work, we further developed an existing model coronary vasculature [21] and used it to demonstrate important factors that regulate FFR. The model development's specific purpose for this study was to qualitatively demonstrate that FFR potentially depends on other co-morbidities such as micro-vascular disease and atrial fibrillation in light of two prevalent conditions consisting of focal and diffuse coronary constriction. For this purpose, a 0D modelling approach was found to be suitable as our goal was to understand coronary flow in the presence of pathological conditions. It can be appreciated that an in-depth validation, although highly desirable, was not essential in this theoretical study.

2 Methods

A recent model of the coronary circulation [21] was adapted. The lumped coronary system was further developed by incorporating a detailed four chamber heart description (Fig. 1, A) [11]. The connectivity, diameters (D_n) and lengths (l_n) for each vessel were obtained from the literature [21] (Fig. 1, B), and used to assign resistance (R_n), compliance (C_n), and inductance (L_n) values. The microvasculature terminal impedances (Z_i) were estimated using a structured tree model by Olufsen [22] as,

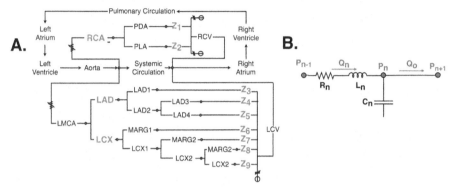

Fig. 1. Electrical analogue of the coronary vasculature model. A: schematic representation and connectivity coronary vessels. The vessels and impedances in red were used in this study. B: an electrical analog of a blood vessel. (Color figure online)

$$Z_i = \frac{8\mu\lambda(2\gamma^3)^{-(N+1)} - 1}{(2\gamma^3)^{-1} - 1}.$$ (1)

where $\gamma = 2^{-\frac{1}{\epsilon}}$ and ϵ represents the daughter vessel radius taper exponent, λ is the ratio of microvascular length to its diameter, and r_0 is the root vessel radius of the structured tree. N represents the number of generations for each structured tree where the radius is greater than 0.001 cm.

Simulations were designed to explore the effects of stenosis severity with varying (a) stenosis length; (b) microvascular disease (increase of terminal impedance); and (c) atrial fibrillation. Atrial fibrillation (AF) was simulated by reducing both atrial systolic elastances to their diastolic values, mimicking the absence of atrial systole. Further, the irregularity in the heartbeat that occurs during AF was simulated by randomly sampling the heart rate from a Gaussian distribution with mean around the control heart rate ($\mu = 65$ bpm) and a suitably large coefficient of variation ($\sigma = 15.6$) [9].

Using the parameters given in Table 1, and impedances calculated using Eq. 1, pressure at each node of the model (Fig. 1) was computed as

$$\frac{dP_n}{dt} = \frac{Q_n - Q_0}{C_n}.$$ (2)

Table 1. Model parameter values. See Fig. 1 for vessel connectivity.

vessel	R (mmHg s ml^{-1})	C (ml mmHg^{-1} x 10^{-3})	L (mmHg ml^{-1})
LMCA	0.2299	2.9	0.00228
LAD	0.4662	1.6	0.0298
LAD1	0.5729	1.6	0.0342
LAD2	1.7077	3.4	0.0916
LAD3	3.7484	1.3	0.1115
LAD4	3.2930	0.4	0.0716
LCX	0.3929	1.2	0.0241
LCX1	0.4730	0.7	0.0231
LCX2	1.0264	0.7	0.0380
LCX3	3.2342	1.1	0.0944
MARG1	1.7351	1.2	0.0655
MARG2	2.9195	0.8	0.0787
MARG3	3.0683	1	0.0896
RCA	1.8302	6.3	0.1171
PLA	2.4412	1.1	0.0799
PDA	1.2571	1.8	0.0596

and the flow through each vessel (resistance) was calculated as

$$\frac{dQ_n}{dt} = \frac{P_{n-1} - P_n - R_n Q_n}{L_n}. \tag{3}$$

Further, the flow through each of the impedances was calculated as

$$Q_{z,n} = \frac{P_{n-1} - P_n}{Z_n}. \tag{4}$$

Stenosis in three large vessels, namely the left anterior descending artery (LAD), the left circumflex artery (LCX), and the right coronary artery (RCA), was investigated. Simulations were performed by imposing focal or diffuse stenosis in a given large vessel. In addition, downstream microvascular disease and atrial fibrillation were imposed. FFR was estimated as the ratio of the time averaged distal pressure, P_d, to the time averaged proximal pressure, P_a, with respect to the stenosis:

$$FFR = \frac{P_d}{P_a}. \tag{5}$$

The model is a system of 36 coupled stiff ordinary differential equations. Pressures and flows were computed as state variables according to Eqs. 2–4. The four chambered heart's elastances generated time varying aortic pressure that provided inlet boundary conditions at the coronary sinus. Constant flow terminal boundary conditions were applied. The system was solved using our robust implicit solver [6]. The method used in the solver is based on implicit backward difference formulae that provides $O(dt^6)$ accuracy. A maximum user time step of 0.005 s gave stable solutions which remained unaffected when the maximum time step was halved. Simulations were performed on local and national clusters. Each instance of the model is a serial run that took 15 s. To construct results in the presented work, a large number of model instances for predefined values of physiologically relevant parameters were executed. The trivially parallel simulations were performed using GNU Utilities [23]. The simulation outputs were post-processed, which included figure construction, using a combination of UNIX and MATLAB scripts.

3 Results

Each simulation generated biological activity of 500 s, and the last 2 s of activity were used to generate results.

In the first experiment, FFR estimates in the three major coronary arteries (LAD, LCX, and RCA) were obtained. To do so, a control simulation was first performed and the distal and proximal pressures in each of the vessels were used to confirm model derived FFR values to be closed to 1. Subsequently, each of LAD, LCX, and RCA were either stenosed (by 90%), the downstream microvascular impedances increased (by 50%), or atrial fibrillation imposed (by 75% reduction of atrial elastances). The results are illustrated in Fig. 2. In the control case (Fig. 2, **first row**), the FFR values were virtually unity (i.e. more than 0.8) for all three coronary vessels. When there was a full-length stenosis (Fig. 2, **second row**; $\alpha = 90\%$), the FFR values reduced to 0.56 for the LAD, 0.52 for the LCX, and 0.50 for the RCA. When microvascular disease was simulated (Fig. 2, **third row**), the FFR values were 1 for all three vessels. With atrial fibrillation (Fig. 2, **third row**; AF = 75%) and intermediate stenosis, the FFR values reduce to 0.9, 0.9, and 0.65 for the LAD, LCX, and RCA, respectively.

The dependence of FFR on simultaneous presence of reduced vessel diameters (focal stenosis) and diffuse stenosis extended through the length of a vessel (length stenosis) were quantified (Fig. 3). To simulate focal stenosis, the blood vessel was divided into two and its biophysical parameters (Table 1) were revised using

$$
\begin{aligned}
R_s &= R_o \alpha^{-2} \\
C_s &= C_o \alpha^{3/2} \\
L_s &= L_o \alpha^{-1}
\end{aligned}
\tag{6}
$$

where the stenosis severity, α, is given by

$$
\alpha = \frac{A_s}{A_o}.
\tag{7}
$$

To simulate diffuse stenosis extended through a certain length percentage $x_s (0 \leq x_s \leq 1)$ of a vessel, the revised parameters were calculated as

$$
\begin{aligned}
R &= R_s x_s + R_o (1 - x_s) \\
L &= L_s x_s + L_o (1 - x_s)
\end{aligned}
$$

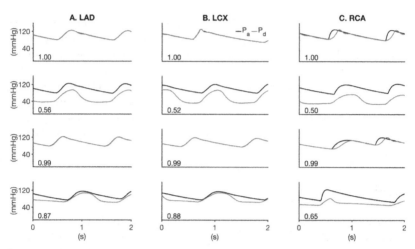

Fig. 2. FFR estimates in the LAD (left column), LCX (middle column), and RCA (right column) under multiple disease conditions. FFR values are provided in each panel. In all panels black lines represent proximal while red lines represent distal pressure. Top row shows non-stenosed model behavior. Second row shows the result of focal stenosis ($\alpha = 90\%$). The third row shows the result of downstream microvascular disease and no focal stenosis ($\epsilon = 2.33$). Bottom row shows the effect of atrial fibrillation (75% AF) with intermediate stenosis ($\alpha = 50\%$). (Color figure online)

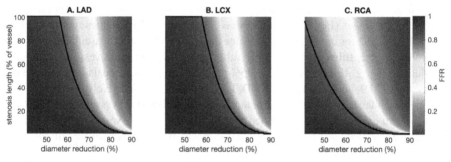

Fig. 3. Dependence of FFR on stenosis length and vessel diameter. The color coding shows FFR as a function of diameter reduction (focal stenosis, horizontal axis) and stenosis length (diffuse stenosis, vertical axis). The black line represents 0.8 FFR isoline.

$$C = C_s x_s + C_o(1 - x_s) \qquad (8)$$

and used in Eqs. 2–4. Progressive focal stenosis in the RCA (Fig. 3, C) caused a larger reduction of FFR as compared to focal stenosis in the LAD (Fig. 3, A) as well as in the LCX (Fig. 3, B). Further, diffuse stenosis in the absence of focal stenosis (vertical axis in Fig. 3) did not reduce FFR. Clinically significant reduction of FFR due to diffuse stenosis was observed only in the RCA. Simultaneous presence of focal and diffuse stenosis caused the most severe reduction of FFR in the RCA, followed by in the LAD and LCX.

The FFR values of simultaneously reducing diameters and imposing varying levels of respective downstream microvascular disease are shown in Fig. 4. Microvascular disease was simulated by varying the daughter vessel radius taper exponent $\in$ from 2.76 (0% microvascular disease) which represents the baseline case to 2.33 (100% microvascular disease) which represents turbulent flow [22]. At diameter reductions above 0.8, microvascular exacerbates the effect of the stenosis on FFR values. However, an almost unique value of diameter reduction for each, LAD, LCX, and RCA, was observed to characterize a clinically significant FFR transition to below 0.8 in the presence of an arbitrary severity microvascular disease. While the diameter reduction was 0.7 for LAD and LCX, it was seen to be a much lower 0.55 in case of the RCA.

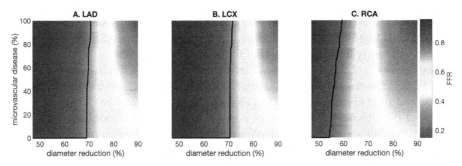

Fig. 4. Dependence of FFR on microvascular disease and vessel diameter. While FFR are color coded as shown in the color bar, the black line represents FFR = 0.8 threshold. Panels A, B, and C show FFR for LAD, LCX, and RCA, respectively.

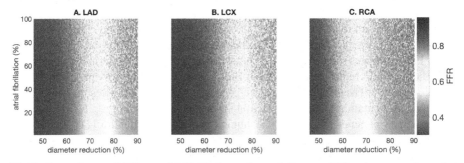

Fig. 5. Dependence of FFR on atrial fibrillation and vessel diameter. FFR values are color coded as shown in the color bar. Panels A, B, and C show FFR data for LAD, LCX, and RCA, respectively.

The FFR values of simultaneously varying full-length stenosis severity in each vessel and varying levels of AF are shown in Fig. 5. Atrial fibrillation (AF) was simulated by reducing both atrial systolic elastances from their baseline case (0% AF) to their diastolic values (100% AF), mimicking the absence of atrial systole. Further, the irregularity in heartbeat that occurs during AF was simulated by sampling the heart rate from a gaussian distribution ($\mu = 65\,bpm$, $\sigma = 15.6$) [9]. Regardless of the degree of AF severity, a diameter reduction of approximately 0.7 in the LAD and LCX resulted in FFR reducing to under the threshold value. In contrast, a diameter reduction of 0.6 in the RCA was seen to reduce FFR below threshold values. As is evident in all panels of Fig. 5, FFR estimation also depends on the erratic heart rates.

4 Conclusions and Discussion

Focal and diffuse coronary stenosis were both observed to modulate FFR (Fig. 2). However, our simulations indicate that FFR estimation must consider other conditions such as AF and microvascular disease, both of which are routinely diagnosed among patients using non-invasive techniques. Furthermore, it appears that blood flow to the right ventricle is more severely affected due to the extra-coronary and RCA stenosis conditions (Figs. 3, 4 and 5). Novel imaging protocols that account for cardiac chamber to chamber diastole will fortify further refinement of the diagnostic instrument.

As seen in Fig. 2, focal as well as diffuse stenosis reduces FFR relative to the control case. However, it can also be seen that extra-coronary conditions such as microvascular disease and atrial fibrillation also affect FFR estimates. It is therefore clear that consideration of the effects of co-morbidities is essential in FFR estimation. The result also indicates that our approach is suitable for ranking the severity of co-morbidities. Specially, Fig. 2, indicates that microvascular disease alone does affect FFR estimation (see definition of FFR). Furthermore, the left and right heart's coronary are affected differentially. Whereas imaging studies are optimized to provide information regarding left coronaries, the model suggests that the right coronaries should also be considered. Our model suggests that stenosis may not be an exclusive focal or diffuse phenomenon. As Fig. 3 shows, consideration of a combination of the two natures of stenosis is essential, especially in our future 3D modelling (see Fig. 6). In such future studies, the 0D models in this detailed investigation will further be useful as boundary conditions to 3D model computational fluid dynamics [24]. In addition to detailed geometry, Fig. 4 indicates that *a priori* knowledge of microvascular health status will permit 3D models to provide better FFR estimates. Finally, our preliminary results indicate that atrial fibrillation, as an isolated co-morbidity, may not affect FFR significantly (Fig. 5) which may be the model's attribute. The presented model suggests that in detailed 3D investigations a more accurate representation of atrial fibrillation is warranted.

A lumped parameter model of the human coronary vasculature [21] was further developed in this study. The model is capable of personalization based on clinical measurements of aortic pressure waves, imaging based vascular geometry (lengths, radii, and morphometry), as well as cardiac wall motion kinematics [25]. As such, the model permits imaging-clinical data assessment as a computationally efficient instrument, prior to detailed 3D computational fluid dynamics simulations. This theoretical study illuminated the relative relevance of focal and diffuse stenosis. It also suggests that knowledge of co-morbidities will improve our clinical diagnostics. Furthermore, it pre-informed our upcoming 3D investigation regarding the clinical data that will permit both validation as well as prediction.

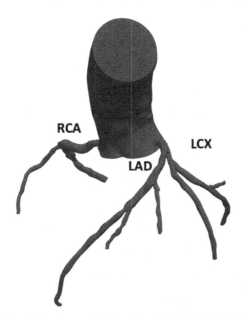

Fig. 6. Representative imaging derived solid model.

Acknowledgements. This work was supported by Canada Canarie Inc. (RS-111), Canada Heart and Stroke Foundation grant (G-20-0028717), Canada NSERC operational grant (R4081A03), and NSERC graduate scholarship. We thank Compute Canada for high performance computing resources. We thank Dr. Kapiraj Chandrabalan for editing support.

References

1. Fearon, W.F., Tonino, P.A., De Bruyne, B., Siebert, U., Pijls, N.H.: Rationale and design of the fractional flow reserve versus angiography for multivessel evaluation (FAME) study. Am. Heart J. **154**, 632–636 (2007)
2. Pijls, N.H., et al.: Percutaneous coronary intervention of functionally nonsignificant stenosis: 5-year follow-up of the DEFER study. J. Am. Coll. Cardiol. **49**, 2105–2111 (2007)
3. Tonino, P.A., et al.: Fractional flow reserve versus angiography for guiding percutaneous coronary intervention. N. Engl. J. Med. **360**, 213–224 (2009)
4. Odudu, A., Francis, S.T., McIntyre, C.W.: MRI for the assessment of organ perfusion in patients with chronic kidney disease. Curr. Opin. Nephrol. Hypertens. **21**, 647–654 (2012)
5. McIntyre, C.W., et al.: Hemodialysis-induced cardiac dysfunction is associated with an acute reduction in global and segmental myocardial blood flow. Clin. J. Am. Soc. Nephrol.: CJASN **3**, 19–26 (2008)
6. Altamirano-Diaz, L., Kassay, A.D., Serajelahi, B., McIntyre, C.W., Filler, G., Kharche, S.R.: Arterial hypertension and unusual ascending aortic dilatation in a neonate with acute kidney injury: mechanistic computer modeling. Front. Physiol. **10**, 1391 (2019)
7. Kharche, S.R., So, A., Salerno, F., Lee, T.Y., Ellis, C., Goldman, D., McIntyre, C.W.: Computational assessment of blood flow heterogeneity in peritoneal dialysis patients' cardiac ventricles. Front. Physiol. **9**, 511 (2018)

8. Shi, Y., Korakianitis, T., Bowles, C.: Numerical simulation of cardiovascular dynamics with different types of VAD assistance. J. Biomech. **40**, 2919–2933 (2007)
9. Anselmino, M., Scarsoglio, S., Saglietto, A., Gaita, F., Ridolfi, L.: A computational study on the relation between resting heart rate and atrial fibrillation hemodynamics under exercise. PLoS ONE **12**, (2017)
10. Heldt, T., Mukkamala, R., Moody, G.B., Mark, R.G.: CVSim: an open-source cardiovascular simulator for teaching and research. Open Pacing Electrophysiol. Ther. J. **3**, 45–54 (2010)
11. Heldt, T., Shim, E.B., Kamm, R.D., Mark, R.G.: Computational modeling of cardiovascular response to orthostatic stress. J. Appl. Physiol. **1985**(92), 1239–1254 (2002)
12. Debbaut, C., Monbaliu, D.R., Segers, P.: Validation and calibration of an electrical analog model of human liver perfusion based on hypothermic machine perfusion experiments. Int. J. Artif. Organs **37**, 486–498 (2014)
13. Coli, L., et al.: Evaluation of intradialytic solute and fluid kinetics. Setting up a predictive mathematical model. Blood Purif. **18**, 37–49 (2000)
14. Ursino, M., Coli, L., Brighenti, C., Chiari, L., de Pascalis, A., Avanzolini, G.: Prediction of solute kinetics, acid-base status, and blood volume changes during profiled hemodialysis. Ann. Biomed. Eng. **28**, 204–216 (2000)
15. Lim, K.M., Choi, S.W., Min, B.G., Shim, E.B.: Numerical simulation of the effect of sodium profile on cardiovascular response to hemodialysis. Yonsei Med. J. **49**, 581–591 (2008)
16. Sankaran, S., Esmaily Moghadam, M., Kahn, A.M., Tseng, E.E., Guccione, J.M., Marsden, A.L.: Patient-specific multiscale modeling of blood flow for coronary artery bypass graft surgery. Ann. Biomed. Eng. **40**, 2228–2242 (2012)
17. Olufsen, M.S., Nadim, A.: On deriving lumped models for blood flow and pressure in the systemic arteries. Math. Biosci. Eng.: MBE **1**, 61–80 (2004)
18. Olufsen, M.S., Peskin, C.S., Kim, W.Y., Pedersen, E.M., Nadim, A., Larsen, J.: Numerical simulation and experimental validation of blood flow in arteries with structured-tree outflow conditions. Ann. Biomed. Eng. **28**, 1281–1299 (2000)
19. Olufsen, M.S.: A one-dimensional fluid dynamic model of the systemic arteries. Stud. Health Technol. Inform. **71**, 79–97 (2000)
20. Kim, H.J., Vignon-Clementel, I.E., Coogan, J.S., Figueroa, C.A., Jansen, K.E., Taylor, C.A.: Patient-specific modeling of blood flow and pressure in human coronary arteries. Ann. Biomed. Eng. **38**, 3195–3209 (2010)
21. Duanmu, Z., Yin, M., Fan, X., Yang, X., Luo, X.: A patient-specific lumped-parameter model of coronary circulation. Sci. Rep. **8**, 874 (2018)
22. Olufsen, M.S.: Structured tree outflow condition for blood flow in larger systemic arteries. Am. J. Physiol.-Heart C **276**, H257–H268 (1999)
23. Tange, O.: GNU parallel - the command-line power tool. USENIX Mag. **36**(1), 42–47 (2011)
24. Marsden, A., Moghadam, M.E.: Multiscale modeling of cardiovascular flows for clinical decision support. Appl. Mech. Rev. **67**(3), 030804, 11 p. (2015)
25. Crowley, L.E., McIntyre, C.W.: Remote ischaemic conditioning—therapeutic opportunities in renal medicine. Nat. Rev. Nephrol. **9**, 739–746 (2013)

In-Silico Analysis of the Influence of Pulmonary Vein Configuration on Left Atrial Haemodynamics and Thrombus Formation in a Large Cohort

Jordi Mill[1]([envelope]), Josquin Harrison[2], Benoit Legghe[3], Andy L. Olivares[1], Xabier Morales[1], Jerome Noailly[1], Xavier Iriart[3], Hubert Cochet[3], Maxime Sermesant[2], and Oscar Camara[1]

[1] Physense, BCN Medtech, Department of Information and Communication Technologies, Universitat Pompeu Fabra, Barcelona, Spain
jordi.mill@upf.edu
[2] Inria, Université Côte d'Azur, Epione Team, Sophia Antipolis, France
[3] Hôpital de Haut-Lévêque, Bordeaux, France

Abstract. Atrial fibrillation (AF) is considered the most common human arrhythmia. Around 99% of thrombi in non-valvular AF are formed in the left atrial appendage (LAA). Studies suggest that abnormal LAA haemodynamics and the subsequently stagnated flow are the factors triggering clot formation. However, the relation between LAA morphology, the blood pattern and the triggering is not fully understood. Moreover, the impact of structures such as the pulmonary veins (PVs) on LA haemodynamics has not been thoroughly studied due to the difficulties of acquiring appropriate data. On the other hand, in-silico studies and flow simulations allow a thorough analysis of haemodynamics, analysing the 4D nature of blood flow patterns under different boundary conditions. However, the reduced number of cases reported on the literature of these studies has been a limitation. The main goal of this work was to study the influence of PVs on left atrium (LA) and LAA haemodynamics. Computational fluid dynamics simulations were run on 52 patients, the largest cohort so far in the literature, where different parameters were individually studied: pulmonary veins orientation and configuration; LAA and LA volumes and its ratio; and flow velocities. Our computational analysis showed how the right pulmonary vein height and angulation have a great influence on LA haemodynamics. Additionally, we found that LAA with great bending with its tip pointing towards the mitral valve could contribute to favour flow stagnation.

Keywords: Pulmonary veins · Computational fluid dynamics · Left atrium haemodynamics and thrombus formation

© Springer Nature Switzerland AG 2021
D. B. Ennis et al. (Eds.): FIMH 2021, LNCS 12738, pp. 605–616, 2021.
https://doi.org/10.1007/978-3-030-78710-3_58

1 Introduction

Atrial fibrillation (AF) is considered the most common of human arrhythmias. Approximately 2% of people younger than age 65 have AF, rising to about 9% of people aged 65 years or more [1]. AF is currently seen as a marker of an increased risk of stroke since it favours thrombus formation inside the left atrium (LA). Around 99% of thrombi in non-valvular AF are formed in the left atrial appendage (LAA) [2]. LAA shapes are complex and have a high degree of anatomical variability among the population. Thus, researchers have sought to classify LAA morphologies and relate them to the risk of thrombus formation [3]. However, no classification has achieved a consensus due to their subjective interpretation and contradictions in their relationship with thrombus formation. Blood flow hemodynamics is another relevant factor for thrombogenesis, following Virchow's triad principles [4]; low velocities and stagnated flow have been associated with the triggering of the inflammatory process and, therefore, the risk of thrombus generation [5].

The pulmonary veins' (PVs) configuration and orientation play a key role in radiofrequency ablation therapy (e.g., PVs isolation) since they are a preferential origin of ectopic foci in atrial fibrillation. On the other hand, there are not robust studies on the relation between the PVs configuration and the hemodynamics in the LA, including the LAA and, as a consequence, the thrombus formation process [6]. Some large-scale studies have classified PVs configurations into different anatomical categories [7] but they have never been related with haemodynamics and thrombus formation. Moreover, there is high anatomical variability among the population, where most of humans have 4 PVs but there are reported cases of 3, 5, 6 or even 7 PVs present. Even more, the orientation of how the PVs are inserted into the LA can differ substantially from patient to patient.

In daily clinical practice, LA haemodynamics is mainly studied using echocardiographic images, usually simplified to a single blood flow velocity value at one point in space and time (e.g. center of LAA ostium at end-diastole) [8]. Advanced imaging techniques such as 4D flow Magnetic Resonance Images (MRI), allowing a more complete blood flow analysis, are emerging but they still provide limited information in the left atria [8]. Therefore, the question if the PVs configuration changes LA hemodynamics remains open. Recently [9], researchers have investigated the dependence of blood flow entering to the LAA and related thrombogenic indices with different PVs configurations, but only in synthetic cases. At this juncture, patient-specific models based on computational fluid dynamics can provide a better haemodynamic characterization of the LA and LAA, deriving in-silico indices of the blood flow at each point of the geometry over time. In the last decade there have been several attempts to develop simulation frameworks for the blood flow analysis of the human LA and LAA [10], but only applied to a very limited number of patient-specific cases (<10) and independently of morphological parameters [11,12]. In this study, we built patient-specific computational models of 52 patients with atrial fibrillation and used computational fluid dynamics (CFD) simulations to study their hemodynamics in relation with PVs configurations, constituting the largest cohort of LA-based fluid simulations in the literature.

2   Methods

2.1   Clinical Data

The clinical data used in this work were provided by Hospital Haut-Lévêque (Bordeaux, France), including AF patients that underwent a left atrial occlusion (LAAO) intervention and with available pre-procedural high-quality Computed Tomography (CT) scans. 52 patients were selected for this study. During the data processing pipeline, clinical decisions and patient outcomes were hidden to the researchers setting up the in-silico simulations. Cardiac CT studies were performed on a 64-slice dual source CT system (Siemens Definition, Siemens Medical Systems, Forchheim, Germany). Images were acquired using a biphasic injection protocol: 1 mL/kg of Iomeprol 350 mg/mL (Bracco, Milan, Italy) at the rate of 5 mL/s, followed by a 1 mL/kg flush of saline at the same rate. 31 CT images were acquired in systole (15 with stroke history) and 21 in diastole (10 with stroke history). The study was approved by the Institutional Ethics Committee, and all patients provided informed consent.

Regarding PVs configuration the subjects were grouped as follows: 2 cases with 3 PVs; 23 with 4 PVs; 19 with 5 PVs; 6 with 6 PVs; and 2 with 7 PVs. For LAA volume, the distribution of the patients stands as follows: 4 cases between 1–5 mL; 14 between 10–15 mL; 18 between 10–15 mL; 12 between 15–20 mL; and 4 cases with more than 20 mL. Regarding thrombus formation, 27 cases were in the control group while 25 had a history of stroke or a thrombus was found in the LAA.

2.2   In-Silico Computational Model

The LA geometries were segmented from the CT images by a different member who did not participate in the subsequently modelling process in order to maintain the study blind to the modellers. The final volumetric meshes were between $8-9 \times 10^5$ elements, depending on the volume of the LA after performing a mesh convergence study up to 1M elements.

A velocity curve at the mitral valve (MV) was obtained from Doppler ultrasound from a patient of the whole database to impose as outlet boundary condition. The same pressure wave from an AF patient was used as inlet boundary conditions at the pulmonary veins. The movement of the mitral valve ring plane was defined according to literature [13] and was diffused through the whole LA with a dynamic mesh approach based on the spring based method implemented on the CFD solver in Ansys Fluent 19 R32 (ANSYS Inc, USA). The edges between any two nodes were idealised as a network of springs and the Hooke's Law was applied in each node. Physiologically speaking, the method tries to mimic the longitudinal movement passively produced in the LA by the contraction of the left ventricle (LV). The lack of radial movement tries to mimic the lack of active contraction since the patient suffers from AF. All boundary conditions were synchronised with the patient's electrocardiogram. Post-processing and visualisation of simulation results were performed using ParaView 5.4.13[1]. The blood was treated as Newtonian fluid, with a density of 1060 Kg/m^3 and a viscosity of 0.0035 Pa/s. Three heart beats were calculated with a 0.01 s time step size.

2.3 Haemodynamic Descriptors

To assess the origin of the flow entering the LAA, 50 seeds were placed at the LAA, from which streamlines were computed. A streamline can be thought of as the path a mass-less particle takes flowing through a velocity field (i.e., vector field) at a given instant in time. In order to have a complete understanding of the blood flow patterns under study, the streamlines were computed in different time frames over the heartbeat in order to gather more representative samples. Specifically, the chosen time frames in the cardiac cycle were the following: 1) at the beginning of the ventricular systole; 2) just before MV opens; 3) when the maximum velocity is reached within the E wave (early diastole, after MV opens); 4) middle time point between the E and A wave; and 5) when the maximum velocity of the A wave is reached (end diastole, just before MV closes to start the cycle again). This process was repeated for three beats. The streamlines were also used to localise the position where the main collision between the PVs flows was produced (e.g., from the right and left PVs), that is the moment that the flow coming from each PVs crosses with each other (see Fig. 2). Furthermore, blood flow patterns within the LA were studied placing additional 50 streamline seeds in each PVs.

The flow rate was measured from the fluid simulations at the entrance of the LAA, with a robust criterion considering the large anatomical variability. We selected a 2D plane below the first lobe before the LAA bending (see Fig. 1), considering flow entering the LAA as positive (and negative with outgoing flow). Two beats were analysed since the first was used to reach convergence and a more stable flow solution.

Subsequently, the obtained flow rate curve was integrated over time to compute the final volume crossing the selected 2D plane. A zero value in the integration would mean that all flow entering the LAA leaves at the end of the beats. On the other hand, a large positive value would indicate that a lot of flow goes inside the LAA without leaving, i.e., potentially signalling flow stagnation. In order to consider the large LAA volume variation in the population, the obtained values were estimated as a percentage of the LAA volume for each case, i.e., obtaining the amount of flow volume staying in the LAA with respect to its volume.

2.4 Morphological Descriptors

To extract morphological features from the LA, the LAA and each pulmonary vein were labelled; their barycenter was computed together with the center of intersection of each label, producing a two-point representation of each veins

---

[1] https://www.paraview.org/.

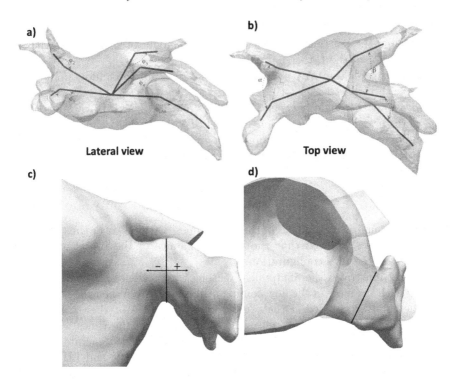

Fig. 1. a) Lateral view of the left atrium with the diffrerent ϕ angles computed; b) Angles α, β, γ and the distances L_α and L_β;c) 2D plane (black line) for the flow rate computation, being placed below the first lobe of the left atrial appendage (LAA) bending. Flow was considered positive if entering the LAA and negative, otherwise.

(which we will call a branch). Additionally, for the LAA we added one more point: the LAA was cut at the barycenter with the normal plane to the LAA centerline; , followed by the computation of the barycenter of the outermost half. Finally, we added the barycenter of the body of LA (the part that was left unlabelled). Overall, we obtained a *skeleton* view of the LA as shown in black in Fig. 1). To labelise every mesh, we first applied a diffeomorphic registration (using the deformetrica software[14]) from a chosen template shape to the rest of the population. Thanks to this we could transfer the labels so the skeleton representation can be extracted. T From the points we choose to compute the following morphological features:

- The angle α is the angle between the right inferior pulmonary vein (RIPV), to and the right superior pulmonary vein (RSPV);
- Angle β, equivalent to angle alpha but for the left superior pulmonary vein (LSPV) and left inferior pulmonary vein (LIPV);
- L_α is the length between RSPV ostium and RIPVs ostium;
- L_β is the length between LSPV ostium and LIPVs ostium;
- α/β and L_α/L_β being their respective ratios;

- Angle ϕ is the angle at the PVs intersection between the centre of the LA cavity and the PV centre, characterising the amount of fold between the vein and the rest of the LA;
- ϕ_{LAA} measures the LAA bending angle described at the middle point in the LAA with respect to its ostium and its tip;
- γ measures the angle between the LSPVs and the LAA. To compute it, we perform a rigid transformation of the LSPVs branch to the LAA to have common ostium point and compute the angle at this ostium with respect to the translated point and the middle point of the LAA.

3 Results

A thorough qualitative analysis of the streamlines' patterns, over the three beats and for all cases, showed that the haemodynamics of the LA changed a lot depending on the PVs configuration.

Figure 2 illustrates the anatomical reasons why the left side of the LA contributed with the most of the flow coming inside the LAA. When the flow coming from the left PVs collided at the superior part of the LA, it then went directly to the LAA ostium (interface with LA), which most of the times was located just under the LSPV. Actually, we realised that the angle generated between the PVs of the same side (α, the angle between the LSPV and LIPV and β, the angle between the RSPV and RIPV in Fig. 2 in the case of 4PVs) had a considerable effect on blood flow patterns. Additionally, it was observed that the orientation of the right side PVs, their inclination with respect to the LA, varied substantially among the studied population, whereas it was quite stable for the left side PVs.

Scenario 1 shown in Fig. 2 was the most common configuration of PVs and blood flow patterns. In LA with 4 pulmonary veins, the collision between PVs flows usually took place at centre of the LA. If the angle between the right PVs (α) increased, the right side flow could go laterally through the sides of the LA (Scenarios 2 and 3 in Fig. 2), reaching the LAA. In our cohort, the angle between left PVs (β) did not vary as much as α. Additionally, the inclination (ϕ in Figs. 1 and 2) of the right PVs with respect to the LA and other PVs determined whether the blood went vertically towards the MV or towards the centre of the LA (dashed lines in Fig. 2). Thus, the right side PVs configuration was the main determinant factor for the whole LA haemodynamics.

Cases with 5 PVs shared some common morphological traits: 1) the right inferior pulmonary vein (RIPV) usually was more inclined (ϕ in Fig. 2, and Scenario 5); and 2) the presence of the right central pulmonary vein (RCPV) shifted the RSPV be shifted towards a more transverse position. The haemodynamics in the LA in these cases were mainly influenced by the α and ϕ angles (Scenarios 5 and 6). If the RSPV were shifted to the right and pushed to a more

Table 1. Distribution of cases in the control and thrombotic groups based on the number of pulmonary veins (PVs), together with volumes of the left atrial appendage (LAA) and estimations of retained flow for chicken wing (CW) and non-CW LAA morphologies. Values in brackets correspond to the standard deviation. Significative differences are highlighted in bold.

	Control	Thrombotic
3 PVs (# cases/LAA volume)	2/7.68 mL	0
4 PVs (# cases/LAA volume)	11/12.33 mL	12/16.5 mL
5 PVs (# cases/LAA volume)	9/10.45 mL	10/12.33 mL
6 PVs (# cases/LAA volume)	4/9.92 mL	2/21.45 mL
7 PVs (# cases/LAA volume)	1/11.27 mL	1/14.95 mL
LAA volume (std)	11.10 mL (4.20)	**15** mL (10.5)
LAA/LA ratio (std)	6.55 (2.7)	8.11 (2.83)
Flow retained in LAA, all shapes (std)	11.57% (9.25)	9.68% (11.54)
Flow retained in LAA, CW shapes (std)	15% (11.11)	2.12% (2)
Flow retained in LAA, non-CW shapes (std)	6.69% (1.87)	**8.81%** (5.87)

transverse position with respect to the main LA body, the flow crossed the LA, preventing the remaining right PVs flows to collide with the left ones (Scenario 5 in Fig. 2). On the other hand, if the RCPV was located in a higher position than the RSPV, RCPV was the one colliding with the left PV ones, thus RSPV flow going under RCPV one (Scenario 6). If the right PVs were very inclined, the collision point was shifted towards the right side; in the case that α was also large, then the collision was clearly shifted to the left side; the flow from the PVs could then reach the walls from the right side (Scenario 7 in Fig. 2). Blood flow patterns observed in cases with 6 and 7 PVs were qualitatively similar to 5 PVs (Scenarios 5 and 6 in Fig. 2).

The computed values of the angles α, β, ϕ and the longitudes L_α and L_β are reported in Table 2 for control and thrombotic groups. In general, we did not found clear relationships between those angles and the formation of thrombi. The only remarkable difference was observed in the angles γ, α being larger in patients who suffered a thrombotic event and ϕ_1 and ϕ_{LAA} being smaller smaller in patients who suffered a thrombotic event. Similarly ϕ_{LAA} and γ were also different between the control and the stroke group in patients who had 4 PVs and 6 PVs. Patients with 5 PVs who suffered stroke had a smaller ϕ_4 in comparison with the control group (Table 3).

There were not direct correlations between the risk of thrombotic event and the number of pulmonary veins (Table 1). Stroke/thrombus patients with six pulmonary veins had a lower ratio of flow inside the LAA than non-stroke cases, but differences between the LA volumes in both groups were very large and the number of samples too small (4 vs 2 cases) to draw any conclusion. The LA volume and LAA/LA ratio were higher in patients who suffered from a thrombotic event. Studying the LAA shape, we could find them well distributed between thrombus vs non-thrombus cases, except for the chicken wing (CW) morphologies: four chicken wings were found in the stroke group (interestingly, the four of them having more than 4 pulmonary veins, which is the most common configuration), whereas 9 where in the control group. Nevertheless, we did not found flow or anatomical differences in thrombus vs non-thrombus cases

Table 2. Morphological descriptors (e.g. angles) from the skeleton representation of the left atria, including its appendage (LAA), for the control and thrombogenic cases. Values are reported as the average and standard deviation (in brakets). Most significant differences are highlighted in bold. The angle α is the angle between the right inferior pulmonary vein (RIPV), to and the right superior pulmonary vein (RSPV); angle β, equivalent to angle alpha but for the left superior pulmonary vein (LSPV) and left inferior pulmonary vein (LIPV); α/β; L_α/L_β being their respective ratios; angle ϕ is the angle at the PVs intersection between the centre of the LA cavity and the PV centre, characterising the amount of fold between the vein and the rest of the LA; ϕ_{LAA} measures the LAA bending angle described at the middle point in the LAA with respect to its ostium and its tip; γ measures the angle between the LSPVs and the LAA.

	Systole		Diastole	
	Control	Thrombotic	Control	Thrombotic
α	50.0 (13.1)	46.5 (11.9)	**42.5 (11.6)**	**52.5 (13)**
β	82.2 (15.7)	80.3 (10.4)	87.5 (16.7)	76.6 (13.9)
α/β	0.6 (0.2)	0.6 (0.2)	**0.5 (0.2)**	**0.7 (0.2)**
L_α	22.0 (4.8)	21.1 (6.4)	**23.8 (4.6)**	**20.2 (2.8)**
L_β	19.5 (4.5)	17.9 (6.0)	19.6 (4.2)	20.2 (3.4)
L_α/L_β	1.15 (0.2)	1.2 (0.2)	**1.2 (0.1)**	**1.0 (0.1)**
ϕ_1	4.7 (1.8)	5.3 (3.1)	**6.1 (2.4)**	**4.6 (2.9)**
ϕ_2	4.8 (2.0)	4.1 (2.8)	3.4 (2.0)	3.1 (1.5)
ϕ_3	15.0 (5.5)	13.3 (5.3)	13.3 (4.6)	13.4 (5.7)
ϕ_4	18.7 (5.8)	19.1 (5.3)	19.1 (3.7)	17.5 (5.1)
ϕ_{LAA}	12.8 (6.4)	13.4 (3.8)	**16.1 (5.7)**	**10.0 (3.9)**
γ	138.0 (15.1)	142.5 (18.4)	**141.6 (10.0)**	**152.4 (12.3)**

Table 3. Morphological descriptors for control and thrombotic groups, separated by the number of pulmonary veins (PVs). Values are reported as average and standard deviation (in brackets). Significant values are highlighted in bold. Morphological descriptors for control and thrombotic groups, separated by the number of pulmonary veins (PVs). Values are reported as average and standard deviation (in brackets). Significant values are highlighted in bold. α is the angle between the right inferior pulmonary vein (RIPV), to and the right superior pulmonary vein (RSPV); angle β, equivalent to angle alpha but for the left superior pulmonary vein (LSPV) and left inferior pulmonary vein (LIPV); α/β; L_α/L_β being their respective ratios; angle ϕ is the angle at the PVs intersection between the centre of the LA cavity and the PV centre, characterising the amount of fold between the vein and the rest of the LA; ϕ_{LAA} measures the LAA bending angle described at the middle point in the LAA with respect to its ostium and its tip; γ measures the angle between the LSPVs and the LAA.

	4 PVs		5 PVs		6 PVs	
	Control	Thrombotic	Control	Thrombotic	Control	Thrombotic
α	47.1 (16.8)	50.3 (15.1)	48.5 (9.9)	47.1 (13.3)	50.7 (6.0)	50.8 (8.4)
β	85.1 (13.3)	78.8 (11.5)	83.2 (7.7)	78.9 (13.9)	81.2 (21.7)	83.0 (5.7)
α/β	0.57 (0.24)	0.63 (0.16)	0.58 (0.11)	0.63 (0.30)	0.66 (0.20)	0.61 (0.05)
L_α	**24.3 (4.2)**	**20.9 (4.3)**	22.5 (5.5)	21.0 (3.0)	**18.6 (1.8)**	**27.0 (0.2)**
L_β	20.8 (4.8)	19.7 (3.9)	19.3 (3.1)	18.5 (3.9)	**18.4 (2.5)**	**23.4 (2.9)**
L_α/L_β	**1.19 (0.19)**	**1.06 (0.16)**	1.16 (0.16)	1.17 (0.27)	1.03 (0.21)	1.16 (0.15)
ϕ_1	4.9 (2.4)	4.4 (2.7)	6.3 (1.5)	5.9 (2.7)	**4.1 (2.7)**	**1.5 (0.6)**
ϕ_2	3.9 (2.4)	3.5 (1.2)	4.3 (2.2)	3.4 (2.0)	**5.2 (1.5)**	**2.3 (0.2)**
ϕ_3	14.0 (3.7)	12.1 (5.7)	14.4 (5.0)	13.5 (4.7)	16.4 (8.9)	15.2 (4.4)
ϕ_4	18.3 (4.8)	18.0 (5.3)	**21.4 (5.7)**	**16.7 (3.0)**	17.9 (3.8)	22.1 (1.6)
ϕ_{LAA}	**16.7 (5.8)**	**11.9 (3.9)**	10.1 (1.8)	10.6 (2.7)	**12.2 (4.9)**	**9.9 (0.6)**
γ	**129.8 (17.8)**	**143.4 (15.2)**	123.7 (25.4)	120.9 (30.5)	**133.5 (12.2)**	**139.3 (8.8)**

with CW morphologies. As for the relation of blood flow in the LAA with the stroke/thrombus event, fluid simulations did not show differences between control and thrombogenic groups. Nevertheless, when analysing CW vs non-CW separately, the non-CW patients with stroke had a poorer washing than controls (see Table 1).

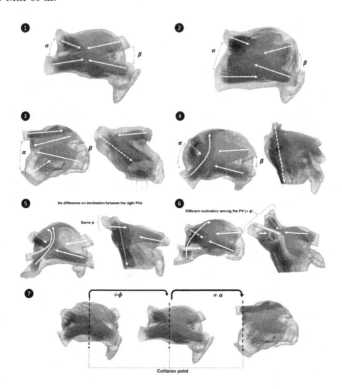

Fig. 2. Different scenarios of pulmonary veins (PVs) configurations and angles. The angle α is the angle between the right inferior pulmonary vein (RIPV), to and the right superior pulmonary vein (RSPV); angle β, equivalent to angle alpha but for the left superior pulmonary vein (LSPV) and left inferior pulmonary vein (LIPV); angle ϕ is the angle at the PVs intersection between the centre of the LA cavity and the PV centre, characterising the amount of fold between the vein and the rest of the LA. Each colour represents flow coming from the same PVs. The solid white line defines the direction of the flow coming from the PVs. The white dashed line defines the flow from the PV with the higher inclination (high ϕ). Scenarios 1–4 had 4 PVs, while Scenario 5 and 6 had 5 and 6 PVs, respectively. In Scenario 7 the black dashed line represents where point of the flow collision. The snapshots were taken at end diastole when the maximum velocity of the A wave is reached just before MV closing.

4   Discussion and Conclusions

The presented study analyses how the PVs configuration influences LA haemodynamics, being the first time these factors are related to thrombus risk due to potential LAA flow stagnation. In order to do so, we have created the largest database of LA fluid simulations in the literature so far; until now, less than 10 different real LA anatomies were processed in each study, a number substantially lower than the 52 cases employed in this work. In general, LA haemodynamics were mostly influenced by morphological characteristics of the right PVs. Their angles and inclinations determined the point of collision of the multiple PV flows in the LA. These were key to understand how blood patterns evolved and traversed the whole LA main cavity, in particular how they were reaching the LAA (key for thrombus formation risk estimation). According to the obtained results, the number of the PVs is not directly related to the probability to suffer a thrombotic event. In agreement with the literature, significant differences were found between LAA volume and LAA/LA ratio between control and thrombotic cases. Also, CW morphologies were less likely to generate thrombus. Nevertheless, in the few CW cases with stroke, the number of PVs was larger than four, which is the most frequent configuration. Despite not finding a strong relationship between thrombus formation and the number of PVs or LAA flow washing in the whole population, the independent analysis of the non-chicken wing cases showed a better LAA flow washing in the control vs thrombotic groups. Finally, some PV angle values were significantly different in controls vs thrombotic cases (γ, α,ϕ_1, ϕ_{LAA} and ϕ_4) although the relation was not strong. These results complement the qualitative results obtained with the streamlines where RSPVs (ϕ_1 and α) had a great influence on LA haemodynamics. The angle ϕ_{LAA} was also mentioned by [15] as possible new risk factor of thrombus formation in LAA.

One limitation in our study was the absence of any Doppler data acquired during an AF event, therefore, the A wave was present in the boundary conditions of the fluid simulations. However, an AF scenario was replicated by imposing a pressure curve at the mitral valve as well as omitting the LA radial (i.e., active) contraction. In the future, we will explore the use of pathlines for a better assessment of LA flow patterns.

Funding. This work was supported by the Agency for Management of University and Research Grants of the Generalitat de Catalunya under the Grants for the Contracting of New Research Staff Programme - FI (2020 FI_B 00608) and the Spanish Ministry of Economy and Competitiveness under the Programme for the Formation of Doctors (PRE2018-084062), the Maria de Maeztu Units of Excellence Programme (MDM-2015-0502) and the Retos Investigación project (RTI2018-101193-B-I00). Additionally, this work was supported by the H2020 EU SimCardioTest project (Digital transformation in Health and Care SC1-DTH-06-2020; grant agreement No. 101016496) and the European project PARIS (ID35).

References

1. Rahman, F., Kwan, G.F., Benjamin, E.J.: Global epidemiology of atrial fibrillation. Nat. Rev. Cardiol. **11**(11), 639–654 (2014)
2. Cresti, A., et al.: Prevalence of extra-appendage thrombosis in non-valvular atrial fibrillation and atrial flutter in patients undergoing cardioversion: a large transoesophageal echo study. EuroIntervention : Journal of EuroPCR in collaboration with the Working Group on Interventional Cardiology of the European Society of Cardiology **15**(3), e225–e230 (2019)

3. Di Biase, L., et al.: Does the left atrial appendage morphology correlate with the risk of stroke in patients with atrial fibrillation? Results from a multicenter study. J. Am. Coll. Cardiol. **60**(6), 531–538 (2012)

4. Watson, T., Shantsila, E., Lip, G.Y.H.: Mechanisms of thrombogenesis in atrial fibrillation: Virchow's triad revisited. Lancet **373**(9658), 155–166 (2009)

5. Naser, A., et al.: Left atrial blood stasis and Von Willebrand factor-ADAMTS13 homeostasis in atrial fibrillation. Arterioscler. Thromb. Vasc. Biol. **31**(11), 2760–2766 (2011)

6. Cronin, P., et al.: Normative analysis of pulmonary vein drainage patterns on multidetector CT With Measurements of pulmonary vein ostial diameter and distance to first bifurcation. Acad. Radiol. **14**(2), 178–188 (2007)

7. Marom, E.M., Herndon, J.E., Kim, Y.H., McAdams, H.P.: Variations in pulmonary venous drainage to the left atrium: implications for radiofrequency ablation. Radiology **230**(3), 824–829 (2004)

8. Beigel, R., Wunderlich, N.C., Ho, S.Y., Arsanjani, R., Siegel, R.J.: The left atrial appendage: Anatomy, function, and noninvasive evaluation. JACC: Cardiovasc. Imaging **7**(12), 1251–1265 (2014)

9. Markl, M., et al.: Assessment of left atrial and left atrial appendage flow and stasis in atrial fibrillation. J. Cardiovasc. Magn. Reson. **17**(1), M3 (2015)

10. Guadalupe, G.I., et al.: Sensitivity analysis of geometrical parameters to study haemodynamics and thrombus formation in the left atrial appendage. Int. J. Numer. Meth. Biomed. Eng. Acepted(ja), 122–140 (2018)

11. Fumagalli, I., et al.: An image-based computational hemodynamics study of the Systolic Anterior Motion of the mitral valve. Comput. Biol. Med. **123**, 103922 (2020)

12. Otani, T., et al.: A computational framework for personalized blood flow analysis in the human left atrium. Ann. Biomed. Eng. **44**(11), 3284–3294 (2016). https://doi.org/10.1007/s10439-016-1590-x

13. Veronesi, F., et al.: Quantification of mitral apparatus dynamics in functional and ischemic mitral regurgitation using real-time 3-dimensional echocardiography. J. Am. Soc. Echocardiograph. **21**(4), 347–354 (2008)

14. Bône, Alexandre, Louis, Maxime, Martin, Benoît, Durrleman, Stanley: Deformetrica 4: an open-source software for statistical shape analysis. In: Reuter, Martin, Wachinger, Christian, Lombaert, Hervé, Paniagua, Beatriz, Lüthi, Marcel, Egger, Bernhard (eds.) ShapeMI 2018. LNCS, vol. 11167, pp. 3–13. Springer, Cham (2018). https://doi.org/10.1007/978-3-030-04747-4_1

15. Yaghi, S., et al.: The left atrial appendage morphology is associated with embolic stroke subtypes using a simple classification system: a proof of concept study. J. Cardiovasc. Comput. Tomograph. **14**(1), 27–33 (2020)

Atrial Microstructure, Modeling, and Thrombosis Prediction

Shape Analysis and Computational Fluid Simulations to Assess Feline Left Atrial Function and Thrombogenesis

Andy L. Olivares[1], Maria Isabel Pons[1], Jordi Mill[1], Jose Novo Matos[2],
Patricia Garcia-Canadilla[3], Inma Cerrada[2], Anna Guy[4],
J. Ciaran Hutchinson[5], Ian C. Simcock[4,6], Owen J. Arthurs[4,6],
Andrew C. Cook[7], Virginia Luis Fuentes[2], and Oscar Camara[1(✉)]

[1] PhySense, Department of Information and Communication Technologies,
Universitat Pompeu Fabra, Barcelona, Spain
oscar.camara@upf.edu
[2] Royal Veterinary College, London, UK
[3] Institut d'Investigacions Biomèdiques August Pi i Sunyer, Barcelona, Spain
[4] Department of Radiology, Great Ormond Street Hospital for Children,
NHS Foundation Trust, London, UK
[5] National Institute for Health Research Great Ormond Street Hospital Biomedical
Research Centre, London, UK
[6] Department of Histopathology, Great Ormond Street Hospital for Children,
NHS Foundation Trust, London, UK
[7] Institute of Cardiovascular Science, University College London, London, UK

Abstract. In humans, there is a well-established relationship between atrial fibrillation (AF), blood flow abnormalities and thrombus formation, even if there is no clear consensus on the role of left atrial appendage (LAA) morphologies. Cats can also suffer heart diseases, often leading to an enlargement of the left atrium that promotes stagnant blood flow, activating the clotting process and promoting feline aortic thromboembolism. The majority of pathological feline hearts have echocardiographic evidence of abnormal left ventricular filling, usually assessed with 2D and Doppler echocardiography and standard imaging tools. Actually, veterinary professionals have limited access to advanced computational techniques that would enable a better understanding of feline heart pathologies with improved morphological and haemodynamic descriptors. In this work, we applied state-of-the-art image processing and computational fluid simulations based on micro-computed tomography images acquired in 24 cases, including normal cats and cats with varying severity of cardiomyopathy. The main goal of the study was to identify differences in the LA/LAA morphologies and blood flow patterns in the analysed cohorts with respect to thrombus formation and cardiac pathology. The obtained results show significant differences between normal and pathological feline hearts, as well as in thrombus vs non-thrombus cases and asymptomatic vs symptomatic cases, while it was not possible to discern in congestive heart failure with thrombus and from non-thrombus cases. Additionally, in-silico fluid simulations demonstrated lower LAA blood flow velocities and higher thrombotic risk in the thrombus cases.

© Springer Nature Switzerland AG 2021
D. B. Ennis et al. (Eds.): FIMH 2021, LNCS 12738, pp. 619–628, 2021.
https://doi.org/10.1007/978-3-030-78710-3_59

Keywords: Feline hearts · Thrombus formation · Congestive heart failure · Left atrial appendage · Computational fluid dynamics

1   Introduction

Cardiac diseases affect most species in the animal kingdom. Atrial fibrillation (AF) is the most frequent arrhythmia in humans, which can lead to irregular contraction of the left atria (LA) and subsequent blood stagnation triggering the formation of thrombi, usually in the left atrial appendage (LAA). In other species such as cats, AF rarely occurs due to the small size of their hearts. However, other cardiac diseases such as hypertrophic cardiomyopathy (HCM) is common in felines, with a prevalence of approximately 10–15% [1]. The natural consequence of HCM over time is the enlargement of the left atria, followed by blood flow abnormalities [3] that can potentially lead to blood stasis and thrombus formation. Therefore, LA size has a prognostic importance in cats with cardiomyopathy, being identified as an important risk factor for feline arterial thromboembolism (FATE), which affects around 48% [3] of felines with HCM. Veterinary professionals often use echocardiographic images to assess cardiac function by measuring basic morphological descriptors with 2D standard imaging tools. For instance, it has been established that the maximum diameter of the LA (LAD) is around 16 mm for normal heart cats, while constituting severe risk for LAD larger than 25 mm [4]. Left atrial volumes (LAV) have also been reported in some studies [2,5], showing differences between healthy and congestive heart failure (CHF) cats (average minimum/maximum LAV in mL): 0.53/1.77 for healthy hearts; and 2.72/3.87 for congestive heart failure cats.

However, to better understand the complex relationship of feline heart morphology and haemodynamics with thrombus formation, more sophisticated descriptors are required. There are several studies in humans using advanced computational techniques to better describe LA and LAA morphologies in relation with the risk of thrombus formation, either with qualitative categories (e.g., chicken-wing et al.) or with more quantitative metrics characterising the ostium (i.e., interface between the LA and the LAA) and LAA shape (e.g., centreline [6]). Additionally, computational fluid simulations are emerging as a powerful tool to investigate the 4D nature of blood flow patterns in the LA, furnishing in-silico indices of thrombogenic risk [7,8]. Unfortunately, these state-of-the-art techniques have never been applied for the assessment of feline cardiac function. The main goal of this study was to derive advanced 3D shape and haemodynamic descriptors of the feline LA from the analysis and computational modelling of micro-computed tomography (micro-CT) data, available in 24 cats, to identify differences between normal (8 cats) and pathological (16 cats, including congestive heart failure or asymptomatic HCM, and with/without thrombus) cases. The pipeline of the developed methodology is illustrated in Fig. 1.

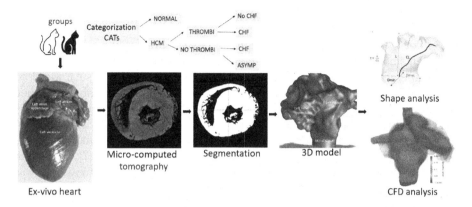

Fig. 1. Computational pipeline to derive morphological and haemodynamic descriptors of the feline hearts. Ex-vivo isolated feline hearts were evaluated with micro-computed tomography (micro-CT) images, from which the left atria were segmented prior to create 3D surface meshes. Then, shape and Computational Fluid Dynamics (CFD) analyses were performed to derive morphological and haemodynamic descriptors. HCM: hypertrophic cardiomyopathy; CHF: congestive heart failure; ASYMP: asymptomatic.

2   Materials and Methods

2.1   Sample Description

The present prospective observational post-mortem study was approved by the Clinical Research Ethical Review Board of the Royal Veterinary College (RVC) (URN 2016-1638-3). Ex-vivo isolated feline heart specimens were evaluated by micro-CT at the Great Ormond Street Hospital (GOSH), followed by standard gross and histopathology examinations performed by a board-certified veterinary pathologist at the RVC as the gold standard. Twenty-four hearts were finally processed, including 8 normal (N) hearts and 16 presenting different cardiac diseases: 5 with congestive heart failure (CHF); 5 asymptomatic HCM; and 6 with thrombus[1]).

Micro-computed tomography imaging and staining protocols were performed as previously described for feline hearts [9]. Micro-CT scans were carried out using a Nikon XTH225 ST and a Med-X micro-CT scanner (Nikon Metrology, Tring, UK) based at GOSH.

Reconstructions were carried out using modified Feldkamp filtered back projection algorithms with proprietary software (CTPro3D; Nikon Metrology) and post-processed using VGStudio MAX (Volume Graphics GmbH, Heidelberg, Germany). Whole heart micro-CT scans were reconstructed into an isotropic volume. Isotropic voxel sizes ranged between 19–24 µm, furnishing approximately 900 2D slices per case.

---

[1] Note that two cases had both congestive heart failure and thrombus.

2.2   3D Surface Reconstruction of the Left Atrium

An in-house script consisting on automatic Otsu's thresholding techniques, 3D Gaussian smoothing (7 pixels per dimension) and noise reduction with morphological operations implemented in Matlab (R2018b Academic license[2]), was used to segment the whole heart in the micro-CT scans. The LA chamber was then manually extracted from the whole heart segmentation. The resulting LA binary masks (see Fig. 1) were introduced into the marching cubes algorithm to build 3D surface meshes. The computational pipeline to create the volumetric meshes required for the fluid simulations was based on Aguado et al. [7]. A Taubin smoothing filter (with scale factors $\lambda = 0.6$, $\mu = -0.53$ and 10 iterations) was applied to the surface meshes using MeshLab[3]. Afterwards, the pulmonary veins (PVs) and the mitral valve (MV) were synthetically placed using MeshMixer[4] to set up the boundary conditions. Finally, the Gmsh[5] software was used to create the volumetric mesh needed for the fluid solver.

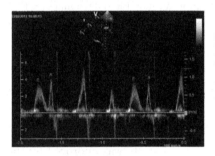
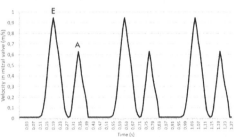

Fig. 2. Doppler echocardiographic recording from a feline heart (left) and transmitral flow velocity applied as outlet boundary condition in the fluid simulations of this study (right)

2.3   Morphological and Haemodynamic Descriptors

The surface meshes of all feline hearts were analysed to derive several shape descriptors aiming at characterising the morphology of the LA and LAA, as can be seen in Fig. 1. First, volumes of the LA and LAA as well as their ratio (i.e., LAA/LA volume ratio) were estimated. Then, the ostium was manually selected to compute its area (AO) and their maximum/minimum diameters (Dmax, Dmin, respectively in Fig. 1), which have been related to thrombus formation in humans [10]). Moreover, the LAA centreline (CL in Fig. 1) was defined as the line going from the ostium centre to the furthest geodesic point in the

---

[2] https://mathworks.com/products/matlab.html.
[3] https://www.meshlab.net.
[4] https://www.meshmixer.com.
[5] https://www.gmsh.info.

LAA, as in [6]. Finally, the centreline length and tortuosity (τ in Fig. 1, dividing the length by the height of the LAA, L) were estimated. Several morphological descriptors were normalised by the LA and LAA volumes, as well as for the cat heart weight, to compensate for different heart sizes.

Due to computational reasons, in-silico haemodynamics descriptors were only evaluated in a subset of the studied dataset, with a total of 10 computational fluid dynamics (CFD) simulations, two for each one of the following categories: normal, asymptomatic HCM with/without thrombus and CHF with/without thrombus. The CFD analysis was performed using the finite volume element method (FVM) in Ansys Fluent 2019 solver (Ansys Inc., USA)[6]. Blood was modelled as an incompressible Newtonian fluid (density of $\rho = 1060\,\text{kg/m}^3$; dynamic viscosity of $\mu = 0.0035\,\text{Pa·s}$), using the Navier-Stokes and continuity equations. A generic boundary condition for a cat with 140 bpm was simulated over three cardiac cycles. A velocity wave was applied in the mitral valve (MV), as shown in Fig. 2, and a pressure wave was imposed in the pulmonary veins (PVs) [2]. In addition, the motion of the mitral annular plane excursion in cats with HCM estimated in Spalla et al. [3] was imposed with a dynamic mesh approach, using the spring-based method available in Fluent, being guided by a MV ring function from Mill. et al. [8]. Several haemodynamics descriptors were obtained from CFD simulations, including velocity curves in the LAA ostium and in-silico indices to estimate the risk of thrombus formation. In particular, we computed the endothelial activation potential (ECAP) as the ratio between the oscillatory shear index (OSI) and the time average wall shear (TAWSS), since it highlights regions with low velocities and high complexity of blood flow, thus prone to thrombus generation [7].

Morphological descriptors were analysed with statistical student t-test, Mann-Whitney - Wilcoxon and χ^2 tests when appropriate, with a level of significance of 0.05. The main objective of the statistical study was to identify shape differences between the following cohorts (number of cases): normal (8) vs cardiomyopathy (16); thrombus (6) vs non-thrombus (10); asymptomatic HCM (5) vs symptomatic (CHF-Thrombus) (11); and CHF with thrombus (2) vs CHF non-thrombus (5).

3   Results and Discussion

Table 1 shows the clinical and morphological descriptors comparing normal and cardiomyopathy feline hearts. There were statistically significant differences between normal and pathological cases in the majority of volumetric indices (both in the LA and the LAA) and in the ostium area, with larger values for cardiomyopathy cases, as expected. The LAA centrelines were larger and less tortuous for pathological cats, reaching statistical difference for the length of the centreline.

Significant differences between thrombus and non-thrombus cases were found for the LAA and LA volumes, ostium area, centreline length and tortuosity, as

---

[6] http://www.ansys.com.

Table 1. Descriptors of normal (n = 8) vs cardiomyopathy (n = 16) cats. HW: Heart weight (g); LAv/LAAv (mL): Left Atrial and Appendage volumes; AO (mm^2): Area of Ostium; CL (mm), τ: LAA centreline length and tortuosity, respectively; Age (years). Results are presented as mean ± standard deviation. *Data was not available for all cases.

Descriptors	Normal	Cardiomyopathy	p-value
HW	15.15 ± 2.67	22.41 ± 6.62	**0.001**
LAAv	0.23 ± 0.11	1.07 ± 0.85	**p < 0.001**
LAv	1.13 ± 0.50	3.78 ± 1.99	**p < 0.001**
LAAv/LAv (%)	21.78 ± 8.58	27.59 ± 14.79	0.70
AO	37.60 ± 14.48	117.60 ± 55.08	**p < 0.001**
AO/LAv	34.54 ± 14.38	33.75 ± 10.11	0.89
AO/LAAv	165.24 ± 55.43	134.54 ± 45.52	0.20
AO/Heart weight	2.57 ± 1.16	5.60 ± 3.03	**0.002**
CL (mm)	8.09 ± 1.77	14.46 ± 5.60	**0.01**
τ	0.84 ± 0.07	0.73 ± 0.24	0.31
Gender*	Female (3) (37.50%)	Female (5) (33.33%)	
	Male (5) (62.50%)	Male (10) (66.67%)	1.00
Neutered*	Yes (2) (33.33%)	Yes (12) (80%)	
	No (4) (66.67%)	No (3) (20%)	0.13
Age*	6.34 ± 4.83	7.93 ± 3.75	0.46

Table 2. Descriptors of thrombus (n = 6) vs non-thrombus (n = 10) cats. HW: Heart weight (g); LAv/LAAv (mL): Left Atrial and Appendage volumes; AO (mm^2): Area of Ostium; CL (mm), τ: LAA centreline length and tortuosity, respectively; Age (years). Results are presented as mean ± standard deviation. *Data was not available for all cases.

Descriptors	Thrombus	Non-thrombus	p-value
HW	24.07 ± 8.50	21.42 ± 5.48	0.52
LAAv	1.76 ± 1.07	0.67 ± 0.30	**0.02**
LAv	5.23 ± 2.08	2.92 ± 1.39	**0.04**
LAAv/LAv (%)	35.02 ± 23.15	23.14 ± 2.79	0.49
AO	157.92 ± 65.66	93.41 ± 30.45	**0.04**
AO/LAv	31.56 ± 11.75	35.06 ± 9.42	0.55
AO/LAAv	105.50 ± 38.32	151.97 ± 41.69	**0.04**
AO/Heart weight	7.06 ± 3.31	4.72 ± 2.62	0.18
CL (mm)	19.42 ± 4.31	11.49 ± 3.97	**0.004**
τ	0.60 ± 0.26	0.81 ± 0.20	**0.02**
Gender*	Female (3) (50%)	Female (2) (22.22%)	
	Male (3) (50%)	Male (7) (77.78%)	0.58
Neutered*	Yes (4) (66.67%)	Yes (8) (88.89%)	
	No (2) (33.33%)	No (1) (11.11%)	0.53
Age*	8.33 ± 3.08	7.67 ± 4.30	0.73

Table 3. Descriptors of asymptomatic (n = 5) vs symptomatic (CHF-Thrombus) (n = 11) cats. HW: Heart weight (g); LAv/LAAv (mL): Left Atrial and Appendage volumes; AO (mm^2): Area of Ostium; CL (mm), τ: LAA centreline length and tortuosity, respectively; Age (years). Results are presented as mean ± standard deviation. *Data was not available for all cases.

Descriptors	Asymptomatic	Symptomatic	p-value
HW	20.16 ± 6.47	23.44 ± 6.73	0.38
LAAv	0.57 ± 0.31	1.31 ± 0.93	**0.05**
LAv	2.31 ± 1.34	4.45 ± 1.90	**0.03**
LAAv/LAv (%)	24.61 ± 3.16	28.95 ± 17.82	0.66
AO	92.66 ± 40.67	128.94 ± 58.62	0.15
AO/LAv	42.34 ± 6.76	29.84 ± 9.02	**0.01**
AO/LAAv	175.38 ± 44.94	115.98 ± 32.98	**0.04**
AO/Heart weight	5.19 ± 3.71	5.78 ± 2.85	0.51
CL (mm)	9.41 ± 2.03	16.76 ± 5.18	**0.001**
τ	0.89 ± 0.06	0.65 ± 0.26	**0.01**
Gender*	Female (1) (25%)	Female (4) (36.36%)	
	Male (3) (75%)	Male (7) (63.64%)	1.00
Neutered*	Yes (3) (75%)	Yes (9) (81.82 %)	
	No (1) (25%)	No (2) (18.18%)	1.00
Age*	8.75 ± 6.60	7.64 ± 2.50	0.76

well as for the ostium area normalised by the LAA volume (see Table 2). In general, we observed larger LA and LAA volumes, as well as ostium areas in thrombus vs non-thrombus cases, in agreement with values reported in human studies [10]. The possible hypothesis is that large ostia imply lower velocities in the LAA and then potential blood stasis. However, the ostium areas, when normalised by the LA and LAA volumes, were higher for non-thrombus cases, which has also been reported [11]. In this scenario, we are roughly considering the amount of flow getting into the LAA, accounting for the LA volume. A potential alternative hypothesis for thrombus generation would be that smaller ostia, compared to LA and LAA volumes, would prevent a proper washing of the blood flow out of the LAA, also promoting stasis. Finally, larger LAA centreline length and smaller tortuosity values were also found from cats with thrombus, in agreement with this second hypothesis.

Table 3 summarises clinical and morphological descriptors comparing asymptomatic HCM (A-HCM) and symptomatic HCM cases. The LA volume values in our study were comparable with those reported by Duler et al. [5]. LA and LAA volumes and ostium areas were larger for symptomatic cases, also having larger and less tortuous LAA centrelines. As mentioned above, when normalising the ostium area with the LA and LAA volumes (the former at the level of statistical significance, p = 0.05), symptomatic cases presented smaller values,

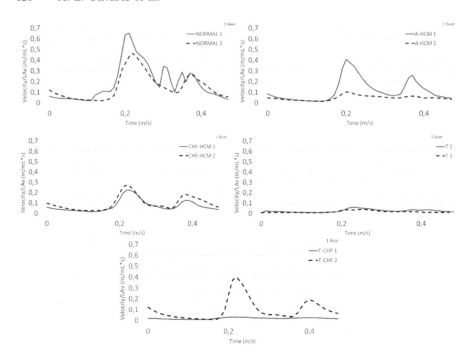

Fig. 3. Blood velocity profiles at the left atrial appendage (LAA) ostium (normalised by the LA volume) from computational fluid simulations for two cases representing each feline heart category. CHF: congestive heart failure; A-HCM: asymptomatic hypertrophic cardiomyopathy; T: thrombus.

following the washing hypothesis for blood stasis. We did not observe significant differences when comparing the clinical and morphological descriptors of CHF cases with and without thrombus, likely due to the small number of samples (two and five, respectively).

Figure 3 shows the fluid simulation results in the ten studied cases. It can be observed how blood flow velocity profiles at the LAA ostium, normalised by the LA volume, showed the expected behaviour: higher and lower values for normal and thrombus cases, respectively, with the non-thrombus CHF somewhere in the middle. On the other hand, the asymptomatic HCM cases had different velocity magnitudes in the two analysed cases, being difficult to draw any conclusion.

The ECAP distribution is also illustrated in Fig. 4, showing larger areas with higher values of ECAP (i.e., complex flows and low velocities; red-green areas) in cats with thrombus.

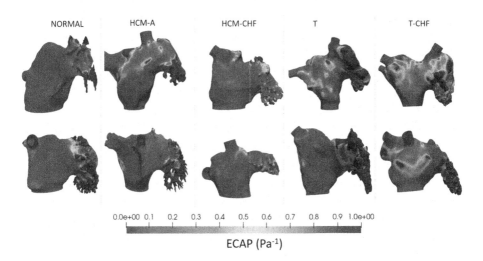

Fig. 4. Endothelial Cell Activation Potential (ECAP) maps from the in-silico fluid simulations, for one case of each feline heart category. Red-green areas represent areas with high ECAP values, (lower blood flow velocities and more complex flow), thus more prone to thrombus formation. CHF: congestive heart failure; A-HCM: asymptomatic hypertrophic cardiomyopathy; T: thrombus. (Color figure online)

4   Conclusions

The present research study is the first attempt to applying advanced computational tools on high-resolution images of normal and cardiomyopathy feline hearts to better understand the complex relation between left atrial morphology, blood flow patterns and thrombus formation. Morphological differences were found in several descriptors between normal and pathological hearts, as well as in thrombus vs non-thrombus cases, even if further investigations are required to reconcile different hypothesis for the role of LAA ostium in blood stasis. Nevertheless, the trends observed in the studied feline hearts (e.g., larger LA volumes in cardiomyopathies vs normals or CHF vs asymptomatic, larger ostium area for thrombus vs non-thrombus) are similar to those found in humans. In-silico indices from fluid simulations confirmed differences in blood flow patterns in these cohorts, with lower velocities in thrombus vs non-thrombus cases. On the other hand, the asymptomatic HCM and CHF hearts were not possible to clearly differentiate from the hemodynamics descriptors. Future work will focus on analysing a larger population of cases, including data from other species such as dogs; these studies could contribute to improve the health of our loved companions as well as to create new knowledge that can potentially be translated to human data.

References

1. Kiatsilapanan, A., et al.: Assessment of left atrial function in feline hypertrophic cardiomyopathy by using two- dimensional speckle tracking echocardiography. BMC Vet. Res. **3**, 1–10 (2020). https://doi.org/10.1186/s12917-020-02557-3

2. Schober, K.E., et al.: Echocardiographic evaluation of left ventricular diastolic function in cats: hemodynamic determinants and pattern recognition. J. Vet. Cardiol. **17**, 102–133 (2015)

3. Spalla, I., et al.: Prognostic value of mitral annular systolic plane excursion and tricuspid annular plane systolic excursion in cats with hyper- trophic cardiomyopathy. J. Vet. Cardiol. **20**, 154–164 (2018)

4. Johns, S.M., et al.: Left atrial function in cats with left-sided cardiac disease and pleural effusion or pulmonary edema. J. Vet. Intern. Med. **26**, 1134–1139 (2012)

5. Duler, L., et al.: Left atrial size and volume in cats with primary cardiomyopathy with and without congestive heart failure. J. Vet. Cardiol. **24**, 36–47 (2019)

6. Genua, I., et al.: Centreline-based shape descriptors of the left atrial appendage in relation with thrombus formation. In: Pop, M., et al. (eds.) STACOM 2018. LNCS, vol. 11395, pp. 200–208. Springer, Cham (2019). https://doi.org/10.1007/978-3-030-12029-0_22

7. Aguado, A.M., et al.: In silico optimization of left atrial appendage occluder implantation using interactive and modeling tools. Front. Physiol. **10**, 237 (2019)

8. Mill, J., et al.: Impact of flow dynamics on device-related thrombosis after left atrial appendage occlusion. Can. J. Cardiol. **36**, 968-e13 (2020)

9. Simcock, I.C., et al.: Investigation of optimal sample preparation conditions with potassium triiodide and optimal imaging settings for microfocus computed tomography of excised cat hearts. Am. J. Vet. Res. **81**, 326–333 (2020)

10. Lee, Y., et al.: Comparison of morphologic features and flow velocity of the left atrial appendage among patients with atrial fibrillation alone, transient ischemic attack, and cardioembolic stroke. Am. J. Cardiol. **119**, 1596–1604 (2017)

11. Khurram, I.M., et al.: Relationship between left atrial appendage morphology and stroke in patients with atrial and fibrillation. Heart Rhythm **10**(12), 1843–1849 (2013)

Using the Universal Atrial Coordinate System for MRI and Electroanatomic Data Registration in Patient-Specific Left Atrial Model Construction and Simulation

Marianne Beach, Iain Sim, Arihant Mehta, Irum Kotadia, Daniel O'Hare,
John Whitaker, Jose Alonso Solis-Lemus, Orod Razeghi, Amedeo Chiribiri,
Mark O'Neill, Steven Williams, Steven A. Niederer, and Caroline H. Roney$^{(\boxtimes)}$

Department of Biomedical Engineering, King's College London, London, UK
`caroline.roney@kcl.ac.uk`

Abstract. Current biophysical atrial models for investigating atrial fibrillation (AF) mechanisms and treatment approaches use imaging data to define patient-specific anatomy. Electrophysiology of the models can be calibrated using invasive electrical data collected using electroanatomic mapping (EAM) systems. However, these EAM data are typically only available after the catheter ablation procedure has begun, which makes it challenging to use personalised biophysical simulations for informing procedures. In this study, we first aimed to derive a mapping between LGE-MRI intensity and EAM conduction velocity (CV) for calibrating patient-specific left atrial electrophysiology models. Second, we investigated the functional effects of this calibration on simulated arrhythmia properties. To achieve this, we used the Universal Atrial Coordinate (UAC) system to register LGE-MRI and EAM meshes for ten patients. We then post-processed these data to investigate the relationship between LGE-MRI intensities and EAM CV. Mean atrial CV decreased from 0.81 ± 0.31 m/s to 0.58 ± 0.18 m/s as LGE-MRI image intensity ratio (IIR) increased from IIR < 0.9 to $1.6 \le$ IIR. The relationship between IIR and CV was used to calibrate conductivity for a cohort of 50 patient-specific models constructed from LGE-MRI data. This calibration increased the mean number of phase singularities during simulated arrhythmia from 2.67 ± 0.94 to 5.15 ± 2.60.

Keywords: Atrial fibrillation · Patient-specific modelling · Atrial imaging · Registration · Conduction velocity

1 Introduction

Current biophysical atrial models for investigating atrial fibrillation (AF) mechanisms and treatment approaches typically use imaging data to define patient-specific anatomy [1]. These models may also include representations of fibrotic remodelling according to fibrosis distributions estimated from late-gadolinium enhancement magnetic resonance imaging (LGE-MRI) data [2]. Recent studies have demonstrated the importance of electrophysiology, including the conduction velocity (CV) of electrical propagation, on AF

© Springer Nature Switzerland AG 2021
D. B. Ennis et al. (Eds.): FIMH 2021, LNCS 12738, pp. 629–638, 2021.
https://doi.org/10.1007/978-3-030-78710-3_60

mechanisms [3]. The electrophysiological properties of models can be calibrated using invasive electrical data from electroanatomic mapping (EAM) systems to construct personalised models [4]. However, these EAM data are typically only available after the catheter ablation procedure has begun, which makes it challenging to use personalised biophysical simulations for informing procedures.

Recent studies have identified a spatial correlation between the distribution of LGE-MRI intensities and CV. For example, Fukumoto et at. demonstrated reduced CV in regions of increased gadolinium uptake that may indicate regions of fibrosis [5]. In addition, Ali et al. showed that the degree of association between LGE-MRI intensity and CV depends on the measurement scale used for this analysis [6]. Caixal et al. demonstrated that areas with higher LGE exhibit lower voltage and slower CV in sinus rhythm [7]. While these studies show that spatial correlations between LGE-MRI and electrophysiological properties (including CV) can be identified, these relationships have not been used for calibrating cohorts of patient-specific models. Building personalised models by calibrating CV to LGE-MRI intensities means that patient-specific models can be constructed pre-procedure, using non-invasive data, and utilized to predict patient-specific treatment response. Furthermore, the functional implications of these relationships have not been fully investigated. For example, the effects of fibrosis derived changes in CV on arrhythmia mechanisms, including the number of stability of re-entries, are unknown [8].

In this study, first we aimed to derive a mapping between LGE-MRI intensity and EAM CV. Second, we aimed to investigate the functional effects of calibrating patient-specific models using this mapping on simulated arrhythmia properties. We previously developed a technique for registering atrial datasets: the Universal Atrial Coordinate (UAC) system [9]. We will use the UAC system to register LGE-MRI to EAM data, and to investigate the relationship between LGE-MRI intensities and EAM bipolar peak-to-peak voltage and CV. We will then use the measured relationship to calibrate conductivity across a cohort of patient-specific left atrial models constructed from LGE-MRI data. Finally, we will investigate the effects of conductivity personalisation on predicted arrhythmia dynamics.

2   Methodology

2.1   Data Modalities, Mesh Processing and Universal Atrial Coordinates

Personalised anatomical models were constructed from LGE-MRI data for ten patients undergoing first-time ablation at St Thomas' Hospital, UK. Seven of these patients had paroxysmal AF and 3 patients had persistent AF. MRI data included contrast-enhanced magnetic resonance angiogram (CE-MRA) scans and LGE-MRI data. The CE-MRA data was used to define the left atrial endocardial wall and the LGE-MRI data was used to estimate the patient-specific distribution of atrial fibrosis. Imaging data were obtained using previously described methods [10]. Ethical approval was given by both the local institutional review board (17/LO/0150; IRAS ID: 217417) and local research and ethics committee (15/LO/1803). Segmentation of CE-MRA data was performed semi-automatically [11] using CemrgApp software [12].

The process of constructing personalised anatomical models involves several steps, which are shown in Fig. 1. Initial meshes (see Fig. 1A), generated from either CE-MRA data segmentations or EAM data, were processed using Paraview. Artefacts were removed and then the data were refined using MeshLab. Filters were applied using Meshlab, including Poisson surface reconstruction, marching cubes and quadric collapse edge decimation (see Fig. 1B). The atrial meshes were then clipped at the pulmonary vein (PV) openings and the mitral valve (see Fig. 1C). Then each of the following structures were labelled using Paraview (see Fig. 1D): the right superior and inferior and left superior and inferior PVs, and the left atrial appendage (LAA).

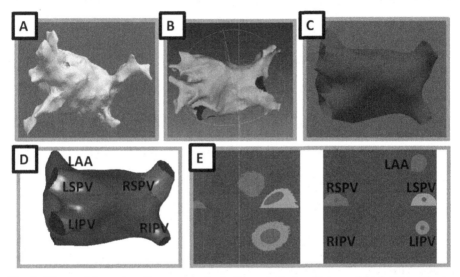

Fig. 1. The model construction process, including calculating Universal Atrial Coordinates from EAM/LGE-MRI surface meshes. (A) Input data (LGE-MRI or EAM). (B) Anatomical meshes refined using MeshLab filters. (C) Mitral valve opening and vessel clipping using Paraview. (D) Labelling of right superior and inferior and left superior and inferior pulmonary veins and left atrial appendage. (E) Conversion from 3D surface to 2D format, ready for further processing. Abbreviations: LSPV: left superior pulmonary vein, LIPV: left inferior pulmonary vein, RSPV: right superior pulmonary vein, RIPV: right inferior pulmonary vein, LAA: left atrial appendage

To calculate UAC, three landmarks were selected as follows: two on the roof at the junction of the left atrial body with the left superior and right superior PVs, and one on the septal wall at the fossa ovalis location. These landmarks were used to calculate geodesic paths used as boundary conditions for a series of Laplace solves. These boundary conditions fix the locations of the junctions of the LA body with each of four PVs and LAA to standard locations in the unit square (see Fig. 1E), using two coordinates: an anterior-posterior coordinate and a lateral-septal coordinate [9, 13]. The contours of the PV openings were also fixed to standard circles in UAC.

2.2 Calculating Conduction Velocity

EAM data exported from the Carto EAM system were read into Matlab using the OpenEP analysis platform (available open source at: openep.io [14]) and post-processed to estimate atrial CV. Local activation times (LATs) assigned by the Carto system were post-processed, together with the electrogram recording locations; see Fig. 2A for an example. For each electrogram recording location, the CV was estimated using the cosine-fit method [15]. Specifically, recording locations together with LATs within a 1 cm radius of the target electrogram recording location were fit to a planar wave and CV was estimated [16]. Locations with fewer than 5 points used for CV estimation were excluded. This calculation is demonstrated in Fig. 2B.

Fig. 2. Calculating electroanatomic mapping conduction velocity from local activation time maps. (A) Example local activation time (LAT) map and (B) example conduction velocity map.

2.3 Registering Imaging and Electrical Data

UAC were calculated as described in Sect. 2.1 for both LGE-MRI meshes and EAM meshes. Each electrode location, together with its CV and bipolar peak-to-peak voltage value, was mapped to UAC. These UAC locations were then mapped to the LGE-MRI mesh using the LGE-MRI UAC mapping. This enabled registration of the EAM data to the LGE-MRI mesh. For each EAM data point that was registered to the LGE-MRI mesh, the closest LGE-MRI node was located for analysis, and data were considered as paired LGE-MRI intensities with EAM measurements (bipolar peak-to-peak voltage and CV). This registration is demonstrated in Fig. 3.

2.4 Constructing an Average Atlas

To investigate the distribution of LGE-MRI intensities, CV and bipolar voltage across patients, we constructed average maps. All scalar fields were mapped to UAC and the average was calculated at UAC nodes corresponding to the first LGE-MRI mesh.

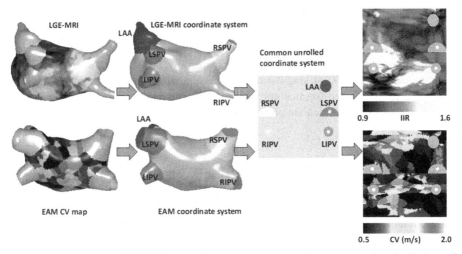

Fig. 3. Registration of LGE-MRI and electroanatomic mapping meshes using the Universal Atrial Coordinate system. Abbreviations: LSPV: left superior pulmonary vein, LIPV: left inferior pulmonary vein, RSPV: right superior pulmonary vein, RIPV: right inferior pulmonary vein, LAA: left atrial appendage, IIR: image intensity ratio, EAM: electroanatomic mapping, CV: conduction velocity.

2.5   Binning Electrical Data by LGE-MRI Intensity

To investigate the relationship between LGE-MRI intensity values, CV and bipolar peak-to-peak voltage, we first post-processed the LGE-MRI intensity field to calculate the image intensity ratio (IIR) by dividing the intensity field by the blood pool mean [17]. Each EAM electrode location was assigned to a bin depending on the LGE-MRI intensity value of its registered location on the MRI mesh. LGE-MRI intensity bins were as follows: IIR < 0.9 ('healthy'), 0.9 < IIR < 1.1 ('mild fibrotic remodelling'), 1.1 < IIR < 1.4 ('mild-moderate'), 1.4 < IIR < 1.6 ('moderate-severe'), 1.6 < IIR ('severe'). For each bin, we considered CV and bipolar voltage values for each individual mesh, as well as for all meshes together.

2.6   Personalising Electrophysiology in Patient-Specific Models

We used a previously constructed cohort of 50 patient-specific bilayer left-atrial anatomies [2] with endocardial and epicardial fibre fields mapped to each anatomy from a DT-MRI dataset using UAC [13]. We calibrated each patient-specific model based on their LGE intensity map. Specifically, we assigned longitudinal conductivity values to LGE-MRI intensities within each of the bins depending on the average CV for that bin. The mapping used between CV and longitudinal conductivity was estimated for the Courtemanche et al. AF atrial cell model [18] by pacing in a 1D cable. Longitudinal conductivity was assigned following this mapping, and transverse conductivity was assigned to be the longitudinal conductivity divided by four, to give a homogeneous anisotropy ratio, following previous studies [19, 20]. In addition, ionic changes were included in regions with IIR > 1.22 to include the effects of transforming growth factor-β1: maximal ionic conductances were rescaled as 50% g_{K1}, 60% g_{Na} and 50% g_{CaL} [1, 21].

Simulations were performed using the Cardiac Arrhythmic Research Package software (available at opencarp.org [22]). For each anatomy, arrhythmia was induced using a technique that seeds phase singularities, as in our previous study [2]. Arrhythmia simulations were performed for anatomies with or without calibration. These simulations were post-processed to calculate phase singulrity maps [21].

3   Results

3.1   Anatomical, Structural and Electrical Maps – Distributions of LGE-MRI Intensities, CV and Peak-to-Peak Bipolar Voltage

We analysed 7 paroxysmal patients and 3 persistent patients; see Table 1 for characteristics. Figure 4 A shows LGE-MRI distributions for each patient.

Table 1.   Patient characteristics.

Age (years)	63.1 ± 9.8
Female	5
BMI	25.8 ± 2.7
Mean CHADS2 score	1.8 (0–4)
Mean LVEF (%)	62.8 ± 5.6
Hypertension	4

An area of increased intensity on the posterior wall close to the left inferior pulmonary vein was observed for all of the patients. Consequently, this area was also seen in the average LGE-MRI intensity map in Fig. 4B. Calculating the average map allows identification of any trends across patients. Benito et al. also observed a preferential distribution of increased LGE-MRI intensity close to the LIPV across a cohort of 113 patients [23]. The average CV distribution is shown in Fig. 4C and the average peak-to-peak bipolar voltage distribution in Fig. 4D. The mean CV calculated as a mean value across the left atrium for each case varied in the range: 0.69–1.00 m/s.

3.2   Binning Electrical Data by LGE-MRI Intensity

Table 2 gives the mean and standard deviation of the CV values and bipolar peak-to-peak voltage values for each of the LGE-MRI intensity bins. A decrease in mean peak-to-peak bipolar voltage was seen with increasing IIR, with a lower mean peak-to-peak bipolar voltage in the $1.4 \leq IIR < 1.6$ and $1.6 \leq IIR$ bins. Mean CV decreased when fibrosis IIR increased from $IIR < 0.9$ to $0.9 \leq IIR < 1.1$, and then further decreased for $IIR > 1.6$. Longitudinal conductivity values used for calibrating simulations are given for each IIR bin in the final column.

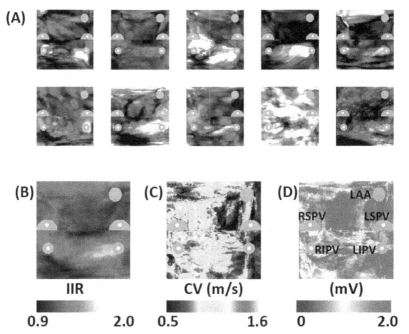

Fig. 4. Average distributions of LGE-MRI image intensity ratio, conduction velocity and peak-to-peak bipolar voltage. (A) Image intensity ratio (IIR) distributions displayed in UAC across the ten cases. (B) Average IIR map. (C) Average conduction velocity (CV) map. (D) Average peak-to-peak bipolar voltage map.

Table 2. LGE bins and mean conduction velocity and bipolar peak-to-peak voltage for each bin Values are given as mean and standard deviation.

LGE IIR	Bipolar voltage (mV)	CV (m/s)	Conductivity (S/m)
IIR < 0.9	1.29 ± 1.11	0.81 ± 0.31	0.4
0.9 ≤ IIR < 1.1	1.42 ± 1.14	0.74 ± 0.32	0.31
1.1 ≤ IIR < 1.4	1.13 ± 1.07	0.74 ± 0.32	0.31
1.4 ≤ IIR < 1.6	0.72 ± 0.71	0.71 ± 0.30	0.28
1.6 ≤ IIR	0.50 ± 0.35	0.58 ± 0.18	0.19

3.3 Patient-Specific Arrhythmia Simulations

Simulations without calibration had a longitudinal conductivity of 0.4S/m, and a transverse conductivity of 0.1S/m, equivalent to an average CV of 0.81 m/s. Calibrated simulations used these conductivity values for regions of IIR < 0.9, while for IIR > 0.9, longitudinal conductivities followed Table 2. Arrhythmia simulations were post-processed to compare properties such as the number of phase singularities. The mean number of phase singularities increased following calibration to LGE-MRI intensity values from

2.67 ± 0.94 for models without calibration to 5.15 ± 2.60 for models with calibration. This was significant with $p < 0.001$, paired t-test. The correlation between phase singularity maps calculated for models with and without calibration was low: 0.11 ± 0.14 (range: -0.06–0.76). This shows that patient-specific calibration has a large effect on predicted arrhythmia properties.

4 Discussion and Conclusions

Our methodology may be used for constructing personalised atrial models with patient-specific anatomy and conductivity from imaging and electrical data. We demonstrated that the measured relationship between LGE-MRI image intensity ratio and electroanatomic mapping CV can be used for calibrating atrial model CV in the case that only LGE-MRI data are available. This means that patient-specific models can be constructed using non-invasive data and utilized to predict patient-specific response to a range of treatment approaches before the patient undergoes a catheter ablation procedure.

The key limitation of our study is that it is a proof-of-concept study applied to ten patients only. Applying this methodology across a larger cohort of paroxysmal and of persistent patients is the subject of ongoing investigation, together with validation of the mapping. Image intensity ratio values may depend on the MRI scanner used. The choice of values used for binning IIR could be improved by comparing with expert visual assessment, or by choosing values to best separate populations of patients or types of tissue (e.g. healthy, pre-ablation fibrosis and post-ablation scar). Future work will determine the effects on the results of EAM recording resolution, MRI resolution and the choice of mesh used for calculating CV. In addition, recent inverse eikonal models that infer tissue anisotropy from EAM data [24] could be used for improved model calibration.

Acknowledgement. CR is funded by an MRC Skills Development Fellowship (MR/S015086/1). SN acknowledges support from the EPSRC (EP/M012492/1, NS/A000049/1, and EP/P01268X/1), the British Heart Foundation (PG/15/91/31812, PG/13/37/30280), and Kings Health Partners London National Institute for Health Research (NIHR) Biomedical Research Centre. SW acknowledges a British Heart Foundation Fellowship (FS 20/26/34952). This work was supported by the Wellcome/EPSRC Centre for Medical Engineering (WT 203148/Z/16/Z).

References

1. Boyle, P.M., et al.: Computationally guided personalized targeted ablation of persistent atrial fibrillation. Nat. Biomed. Eng. **3**, 870–879 (2019). https://doi.org/10.1038/s41551-019-0437-9
2. Roney, C.H., et al.: In silico comparison of left atrial ablation techniques that target the anatomical, structural, and electrical substrates of atrial fibrillation. Front. Physiol. **11** (2020). https://doi.org/10.3389/fphys.2020.572874
3. Lalani, G.G., Schricker, A., Gibson, M., Rostamian, A., Krummen, D.E., Narayan, S.M.: Atrial conduction slows immediately before the onset of human atrial fibrillation: a bi-atrial contact mapping study of transitions to atrial fibrillation. J. Am. Coll. Cardiol. **59**, 595–606 (2012). https://doi.org/10.1016/j.jacc.2011.10.879

4. Corrado, C., Williams, S., Karim, R., Plank, G., O'Neill, M., Niederer, S.: A work flow to build and validate patient specific left atrium electrophysiology models from catheter measurements. Med. Image Anal. **47**, 153–163 (2018). https://doi.org/10.1016/j.media.2018.04.005

5. Fukumoto, K., et al.: Association of left atrial local conduction velocity with late gadolinium enhancement on cardiac magnetic resonance in patients with atrial fibrillation. Circ. Arrhythmia Electrophysiol. **9**, 1–7 (2016). https://doi.org/10.1161/CIRCEP.115.002897

6. Ali, R.L., et al.: Left atrial enhancement correlates with myocardial conduction velocity in patients with persistent atrial fibrillation. Front. Physiol. (2020). https://doi.org/10.3389/fphys.2020.570203

7. Caixal, G., et al.: Accuracy of left atrial fibrosis detection with cardiac magnetic resonance: correlation of late gadolinium enhancement with endocardial voltage and conduction velocity. EP Eur. **23**, 380–388 (2021). https://doi.org/10.1093/europace/euaa313

8. Haissaguerre, M., et al.: Intermittent drivers anchoring to structural heterogeneities as a major pathophysiological mechanism of human persistent atrial fibrillation. J. Physiol. **594**, 2387–2398 (2016). https://doi.org/10.1113/JP270617

9. Roney, C.H., et al.: Universal atrial coordinates applied to visualisation, registration and construction of patient specific meshes. Med. Image Anal. **55**, 65–75 (2019). https://doi.org/10.1016/j.media.2019.04.004

10. Sim, I., et al.: Reproducibility of Atrial Fibrosis Assessment Using CMR Imaging and an Open Source Platform. JACC Cardiovasc. Imaging **12**, 65–75 (2019). https://doi.org/10.1016/j.jcmg.2019.03.027

11. Sim, I., et al.: Reproducibility of atrial fibrosis assessment using CMR imaging and an open source platform. JACC Cardiovasc. Imaging **12**, 2076–2077 (2019). https://doi.org/10.1016/j.jcmg.2019.03.027

12. Razeghi, O., et al.: CemrgApp: an interactive medical imaging application with image processing, computer vision, and machine learning toolkits for cardiovascular research. SoftwareX **12**, (2020). https://doi.org/10.1016/j.softx.2020.100570

13. Roney, C.H., et al.: Constructing a human atrial fibre atlas. Ann. Biomed. Eng. (2020). https://doi.org/10.1007/s10439-020-02525-w

14. Williams, S.E., et al.: OpenEP: a cross-platform electroanatomic mapping data format and analysis platform for electrophysiology research. Front. Physiol. **88**, 105–121 (2021)

15. Roney, C.H., et al.: An automated algorithm for determining conduction velocity, wavefront direction and origin of focal cardiac arrhythmias using a multipolar catheter. In: 2014 36th Annual International Conference of the IEEE Engineering in Medicine and Biology Society, pp 1583–1586. IEEE (2014)

16. Roney, C.H., et al.: A technique for measuring anisotropy in atrial conduction to estimate conduction velocity and atrial fibre direction. Comput. Biol. Med. **104**, 278–290 (2019). https://doi.org/10.1016/j.compbiomed.2018.10.019

17. Khurram, I.M., et al.: Magnetic resonance image intensity ratio, a normalized measure to enable interpatient comparability of left atrial fibrosis. Hear Rhythm **11**, 85–92 (2014). https://doi.org/10.1016/j.hrthm.2013.10.007

18. Courtemanche, M., Ramirez, R.J., Nattel, S.: Ionic targets for drug therapy and atrial fibrillation-induced electrical remodeling: insights from a mathematical model. Cardiovasc. Res. **42**, 477–489 (1999)

19. Bayer, J.D., Roney, C.H., Pashaei, A., Jaïs, P., Vigmond, E.J.: Novel radiofrequency ablation strategies for terminating atrial fibrillation in the left atrium: a simulation study. Front. Physiol. **7**, 108 (2016). https://doi.org/10.3389/fphys.2016.00108

20. Roney, C.H., et al.: A technique for measuring anisotropy in atrial conduction to estimate conduction velocity and atrial fibre direction. Comput. Biol. Med. **104**, 278–290 (2019). https://doi.org/10.1016/j.compbiomed.2018.10.019

21. Roney, C.H., et al.: Modelling methodology of atrial fibrosis affects rotor dynamics and electrograms. Europace (2016). https://doi.org/10.1093/europace/euw365
22. Plank, G., et al.: The openCARP Simulation Environment for Cardiac Electrophysiology. bioRxiv, 1–22 (2021). https://doi.org/10.1101/2021.03.01.433036
23. Benito, E.M., et al.: Preferential regional distribution of atrial fibrosis in posterior wall around left inferior pulmonary vein as identified by late gadolinium enhancement cardiac magnetic resonance in patients with atrial fibrillation. Europace **20**, 1959–1965 (2018). https://doi.org/10.1093/europace/euy095
24. Grandits, T., Pezzuto, S., Lubrecht, Jolijn M., Pock, T., Plank, G., Krause, R.: PIEMAP: Personalized Inverse Eikonal Model from Cardiac Electro-Anatomical Maps. In: Puyol Anton, E., Pop, M., Sermesant, M., Campello, V., Lalande, A., Lekadir, K., Suinesiaputra, A., Camara, O., Young, A. (eds.) STACOM 2020. LNCS, vol. 12592, pp. 76–86. Springer, Cham (2021). https://doi.org/10.1007/978-3-030-68107-4_8

Geometric Deep Learning
for the Assessment of Thrombosis Risk
in the Left Atrial Appendage

Xabier Morales[1(✉)], Jordi Mill[1], Guillem Simeon[1], Kristine A. Juhl[2],
Ole De Backer[3], Rasmus R. Paulsen[2], and Oscar Camara[1]

[1] Physense, BCN Medtech, Department of Information
and Communications Technologies, Universitat Pompeu Fabra, Barcelona, Spain
xabier.morales@upf.edu
[2] DTU Compute, Technical University of Denmark, Kongens Lyngby, Denmark
[3] Heart Center, Rigshospitalet, Copenhagen, Denmark

Abstract. The assessment of left atrial appendage (LAA) thrombogenesis has experienced major advances with the adoption of patient-specific computational fluid dynamics (CFD) simulations. Nonetheless, due to the vast computational resources and long execution times required by fluid dynamics solvers, there is an ever-growing body of work aiming to develop surrogate models of fluid flow simulations based on neural networks. The present study builds on this foundation by developing a deep learning (DL) framework capable of predicting the endothelial cell activation potential (ECAP), linked to the risk of thrombosis, solely from the patient-specific LAA geometry. To this end, we leveraged recent advancements in Geometric DL, which seamlessly extend the unparalleled potential of convolutional neural networks (CNN), to non-Euclidean data such as meshes. The model was trained with a dataset combining 202 synthetic and 54 real LAA, predicting the ECAP distributions instantaneously, with an average mean absolute error of 0.563. Moreover, the resulting framework manages to predict the anatomical features related to higher ECAP values even when trained exclusively on synthetic cases.

Keywords: Geometric deep learning · Left atrial appendage ·
Thrombus formation · Computational fluid dynamics

1 Introduction

Atrial fibrillation (AF) is the most common clinically significant arrhythmia, which can lead to irregular contraction and wall rigidity of the left atrium (LA). This often results in atrial blood stagnation promoting the formation of thrombi within the LA, thereby, increasing the risk of cerebrovascular accidents [14]. In fact, non-valvular AF is responsible for 15 to 20% of all cardioembolic ischemic strokes, 99% of which originate in the left atrial appendage (LAA) [3]. As a result, there have been several attempts at characterizing LA haemodynamics

© Springer Nature Switzerland AG 2021
D. B. Ennis et al. (Eds.): FIMH 2021, LNCS 12738, pp. 639–649, 2021.
https://doi.org/10.1007/978-3-030-78710-3_61

either through transesophageal echocardiography (TEE) or computational fluid dynamics (CFD). Yet, ultrasound imaging is quite ill-suited to characterize complex three-dimensional haemodynamics, while the latter suffers from tediously long computing times and demands huge computational resources [9].

In this regard, deep learning (DL) has made its way into fluid flow modelling, resulting in highly accurate surrogate models that can be evaluated with significantly less computational resources [7]. That being said, many of the most widespread DL models are not well adapted to non-Euclidean domains, such as graphs and meshes, in which medical data is often best represented [5]. As a response, a set of methods have emerged under the umbrella term Geometric DL, that have succeeded in generalizing models such as convolutional neural networks (CNN) to non-Euclidean data [2].

Hence, seeking to improve upon prior studies [11], we leveraged Geometric DL to develop a CFD surrogate capable of learning the complex relationship between the heterogeneous LAA geometry and the endothelial cell activation potential, parameter linked to an increased thrombosis risk. More specifically, we employed a spline-based spatial convolution operator, which enables extracting features from the underlying anatomy without the need for mesh correspondence [5], i.e., they extend properties that have made classical CNNs so successful (local connectivity, weight sharing and shift invariance), without the need to convert LAA meshes to Euclidean representation. We show that our model not only is accurate, but also generalizes well from synthetic to real patient data.

2   Methods

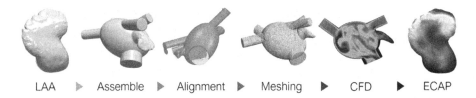

LAA   ▶   Assemble   ▶   Alignment   ▶   Meshing   ▶   CFD   ▶   ECAP

Fig. 1. Pipeline to generate ground truth ECAP maps. LAA: left atrial appendage, CFD: computational fluid dynamics, ECAP: endothelial cell activation potential.

The pipeline of the study involved, at first, the generation of the ground truth data through in-silico CFD simulations of the entire LA, requiring prior assembly of the geometries as shown in Fig. 1. Subsequently, the meshes derived from the simulations were converted to graph format, suitable for the training of the geometric neural network. Finally, the model was trained seeking to learn the complex non-linear relationship between the LAA anatomy and the ECAP maps.

2.1   Data

The employed dataset consisted of 256 LAA, combining 202 synthetic and 54 real patient geometries. The synthetic geometries and their corresponding simulations were borrowed from a previous study [11]. More specifically, the synthetic dataset stems from a statistical shape model (SSM) based on 103 patient LAA surfaces [13]. All cases were reconstructed from computed tomography (CT) images provided by the Department of Radiology of Rigshospitalet, Copenhagen.

For the time being, we have just considered the geometry of the LAA as incorporating the highly heterogeneous LA anatomy would qualitatively increase the inter-subject variability of the hemodynamic parameters. Thus, prior to the simulations, all LAA were assembled to an oval approximation of the LA [6] to ensure that ECAP variability solely depended on individual anatomical differences of the appendage. Finally, since the employed framework does not require any sort of mesh correspondence, all the synthetic data were remeshed to ensure that the network was only able to learn from geometric features.

2.2   In-Silico Thrombosis Risk Index - ECAP

The endothelial cell activation potential (ECAP), defined by Di Achille et al. [4], was the parameter chosen to evaluate the risk of thrombosis in the LAA. Since the pathophysiology of thromboembolism in AF is based upon the formation of mural thrombi, the calculation of ECAP is based upon haemodynamics in the proximity of the vessel wall, more precisely, as the ratio between the oscillatory shear index (OSI) and the time averaged wall shear stress (TAWSS).

$$ECAP = \frac{OSI}{TAWSS} \qquad (1)$$

High ECAP values result from low TAWSS and high OSI values, indicating the presence of low velocities and high flow complexity, which is associated with endothelial susceptibility and risk of thrombus formation. The ground truth ECAP distributions were obtained through CFD simulations performed on Ansys Fluent 19.2[1] and automated through the MATLAB R2018b Academic license[2]. Simulation setup was performed accordingly to the preceding study [11].

2.3   Geometric Deep Learning Framework

The model was constructed by leveraging PyTorch Geometric (PyG)[3], a Geometric DL extension of PyTorch[4]. PyG offers a broad set of convolution and pooling operations that extend the capabilities of traditional CNN to irregularly structured data such as graphs and manifolds. With this in mind, the

---

[1] https://www.ansys.com/products/fluids/ansys-fluent.
[2] https://es.mathworks.com/products/matlab.html.
[3] https://github.com/rusty1s/pytorch_geometric.
[4] https://pytorch.org/.

mesh dataset resulting from the simulations had to be converted into individual graphs. Together with PyVista[5], we converted each mesh to a graph represented by $\mathcal{G} = (\mathcal{V}, \mathcal{E})$, with $\mathcal{V} = 1, ..., N$ being the set of nodes, and $\mathcal{E}$ corresponds to the set of edges of the triangular faces. For each vertex the curvature and surface normal vectors were computed, totaling 4 input feature channels.

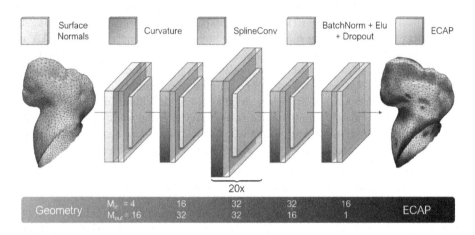

Fig. 2. General overview of the network architecture. The input vertex features consisted on the vertex-wise curvature and normal vectors. The spatial relations between the nodes were stored as edge attributes through cartesian pseudo-coordinates. We employed 24 consecutive SplineConv layers [5] with a kernel size of $k = 5$.

Among all the available graph CNN layers, we opted for SplineCNN [5], since being a spatial method, it offers several advantages when dealing with meshes. In particular, it avoids the need of establishing mesh correspondence. Additionally, defining the spatial relations between vertex features becomes trivial by employing pseudo-coordinates. In our use case, pseudo-coordinates were obtained by computing the relative distance in Cartesian coordinates between the vertices of each edge. During the training process, these edge attributes define the way in which the input features will be aggregated in the neighborhood of a given node. Lastly, we also tested the residual and dense layers for graph neural networks developed by Li et al. [8], aiming to reduce vanishing gradients in deep layers.

2.4 Experimental Setup and Hyperparameter Tuning

The schematic representation of the model architecture is shown in Fig. 2. A thorough grid search was carried out to fine tune the model by iteratively swapping several hyperparameters, sequentially increasing model depth from 5 to 25 layers and including dense blocks of different sizes with a fixed random seed. The highest accuracy was obtained when employing 20 consecutive SplineConv

---

[5] https://docs.pyvista.org/.

hidden layers with 32 feature channels per layer. Besides, the ideal amount of transition layers from input-output to the hidden layers was also tested. The inclusion of one transition layer in both ends, as observed in Fig. 2, yielded the best performance. Moreover, several configurations of residual connections were evaluated, with the best results attained using dense blocks of depth = 4. Various pooling and U-Net like models were tested, aiming to improve multi-scale feature extraction, but so far to no avail.

In regards to the parameters of the SplineConv layer, a B-spline basis of degree 1 and a kernel size of $k = 5$ were chosen, following suggestions by the authors [5]. Concerning general hyperparameters, the exponential linear unit (ELU) provided the best results among all activation functions, always coupled with batch normalization and a dropout of 0.1. In addition, the training loop was carried out through 300 epochs with a batch size of 16 and a learning rate of 0.001. Adam was employed as an optimizer with a weight decay of 0.05 when training in synthetic only. Finally, the L_1 loss was chosen for regression.

Given the limited availability of comparable models in similar tasks, the performance of the model was benchmarked against one of our earlier studies [11]. As opposed to the novel graph-based network, this study relied on conventional fully connected layers (FCN) and therefore it required thorough preprocessing of the input meshes. Two separate experiments were completed. In the first, a 10-fold cross-validation was performed with the whole dataset, meaning that the model was given both synthetic and patient data during training and testing. In the latter, we trained the model solely on synthetic data and tested the accuracy of the model on the real cases to test its generalization capabilities.

3   Results

The accuracy results with the final model are given in Table 1 in terms of the mean absolute error (MAE), which indicates that the geometric DL network significantly outperforms the conventional fully connected network in both tasks. Furthermore, a small batch of 5 testing geometries from the first experiment is shown in Fig. 3. Cases in row 1–3 are derived from the SSM model while the

Table 1. Prediction accuracy results in terms of mean absolute error (MAE) for different training setups. FCN: Fully connected network.

Training	Geometric	FCN
Synthetic+Real	0.563	0.608
Only Synthetic	0.621	0.808

Ground truth   Geo prediction   FCN prediction

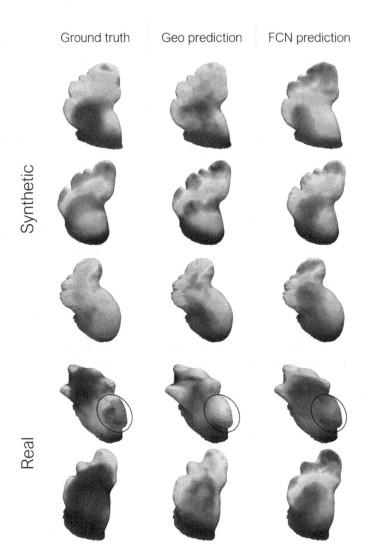

Fig. 3. From left to right: in-silico index (endothelial activation cell potential, ECAP) ground-truth from fluid simulations; prediction obtained with the geometrical deep learning model (Geo); and prediction from the fully connected network (FCN) [11]. The ECAP values are colored from low values (0, blue) to higher than 6 (red areas), the latter indicating a higher risk of thrombus formation. The FCN model struggles to identify the highlighted lobe in the circle. (Color figure online)

remaining two represent real patient cases. Additional test subjects are provided in Appendix A.1 and A.2. As only the areas of high ECAP values are said to be related to increased risk of thrombosis, a binary classification was performed with a positive condition of ECAP > 4, being the 90th percentile of the distribution.

Once again, the Geometric DL model outperformed its counterpart with a true positive rate of 73.1% against 67.5% in the FCN model.

4 Discussion

Careful inspection of the results in Table 1 indicate not only that the geometric DL model outperforms the conventional network but also that it has a higher generalisation potential. While the accuracy of the graph-based network decreases by just 9% when training solely in the synthetic data, the accuracy in the latter falls by almost 30%. The drop in accuracy was to be expected as the real geometries present far higher heterogeneity than its synthetic counterparts. In this sense, the inclusion of a weight decay turned out be crucial in avoiding over-fitting the model to the synthetic cases.

Our hypothesis, although difficult to ascertain due to the "black box" nature of neural networks, is that the graph CNN model is probably better able to exploit the anatomical features in the vicinity of each node and, consequently, is capable of predicting higher ECAP values in areas with fluctuating curvature and normal vectors, which reflect the lobes and cavities of the LAA where blood tends to stagnate. Therefore, even though the network has only been provided with synthetic geometries during the training process, when tested on real cases it is able to recognise anatomical features such as bulges and gaps more proficiently, ultimately leading to improved accuracy. This is best exemplified in case 4 shown in Fig. 3, as the FCN network completely fails to recognise the bulge (encircled in the figure), being in a region where the synthetic population rarely shows high ECAP values, while the graph-based network shows moderate success.

In spite of the results, this study has several limitations that must be addressed before it can be of any use in a clinical setting. First, the choice of ECAP as an index of thrombosis risk may be debatable, as its validity in the LAA has not yet been demonstrated in any clinical study. Nonetheless, although the ECAP index was originally developed in carotid and abdominal aorta fluid models [4], the underlying mechanisms of thrombus formation are analogous to those in the LAA, which typically involve some degree of blood stagnation or re-circulation at low velocities that the ECAP should be able to reflect. In fact, it has already seen some use in clinical studies exploring device-related thrombus formation in LAA occlusion surgeries [1,10].

Secondly, the hemodynamic variability arising from the heterogeneous anatomy of the LA has been completely neglected for the sake of simplicity. Nonetheless, since the chosen deep learning framework does not involve mesh correspondence it should be fairly trivial to include the complete LA anatomy. Moreover, the network should be capable of learning the ECAP fluctuations caused by factors such as the interaction of pulmonary vein orientation [6].

Lastly, at the moment, the model is completely agnostic to flow dynamics and boundary conditions that play a key role in the process of thrombogenesis. To address this challenge, we intend on capitalising on the rapid advances in the field of physics-informed neural networks, with examples such as the study by

Pfaff et al. [12], enabling the full exploration 4D flow MRI and CFD data that may pave the way towards the prediction of the velocity vector field in the LA.

5 Conclusion

In the present study we have successfully leveraged recent advances in graph neural networks to instantaneously predict the ECAP mapping in the LAA, solely from its anatomical mesh, effectively skipping the need to run CFD simulations. Furthermore, we have significantly improved the results from our previous model with a framework that no longer requires mesh correspondence. These results could lay the foundation for real-time monitoring of LAA thrombosis risk in the future and open exciting avenues for future research in cardiological mesh data.

Funding. This work was supported by the Agency for Management of University and Research Grants of the Generalitat de Catalunya under the Grants for the Contracting of New Research Staff Programme - FI (2020-FI-B-00690) and the Spanish Ministry of Economy and Competitiveness under the Programme for the Formation of Doctors (PRE2018-084062), the Maria de Maeztu Units of Excellence Programme (MDM-2015-0502) and the Retos Investigación project (RTI2018-101193-B-I00). Additionally, this work was supported by the H2020 EU SimCardioTest project (Digital transformation in Health and Care SC1-DTH-06-2020; grant agreement No. 101016496).

Conflict of Interest Statement. The authors declare that the research was conducted in the absence of any commercial or financial relationships that could be construed as a potential conflict of interest.

Appendix

A.1 Real LAA results

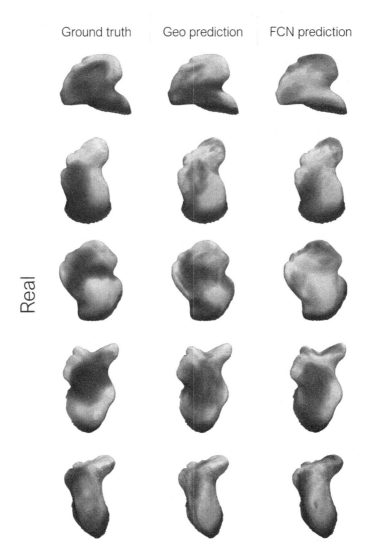

Fig. 4. Predictions over real LAA, belonging to the same cross-validation experiment as Fig. 3. From left to right: in-silico index (endothelial activation cell potential, ECAP) ground-truth from fluid simulations; prediction obtained with the geometrical deep learning model (Geo); and prediction from the fully connected network (FCN) [11]. The ECAP values are colored from low values (0, blue) to higher than 6 (red areas), the latter indicating a higher risk of thrombus formation. (Color figure online)

A.2 Synthetic LAA results

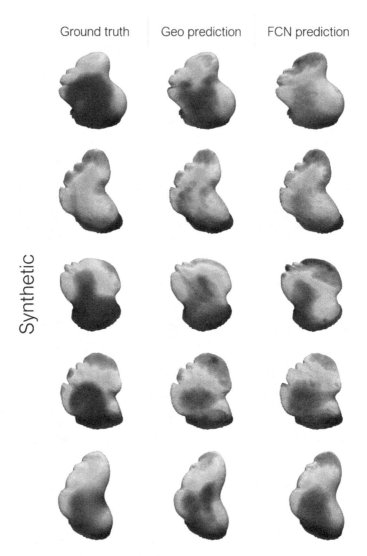

Fig. 5. Predictions over synthetic LAA, belonging to the same cross-validation experiment as Fig. 3. From left to right: in-silico index (endothelial activation cell potential, ECAP) ground-truth from fluid simulations; prediction obtained with the geometrical deep learning model (Geo); and prediction from the fully connected network (FCN) [11]. The ECAP values are colored from low values (0, blue) to higher than 6 (red areas), the latter indicating a higher risk of thrombus formation. (Color figure online)

References

1. Aguado, A.M., et al.: In silico optimization of left atrial appendage occluder implantation using interactive and modeling tools. Front. Phys. **10**, 237 (2019)
2. Bronstein, M.M., Bruna, J., LeCun, Y., Szlam, A., Vandergheynst, P.: Geometric deep learning: Going beyond Euclidean data. IEEE Sig. Process. Mag. **34**(4), 18–42 (2017)
3. Cresti, A., et al.: Prevalence of extra-appendage thrombosis in non-valvular atrial fibrillation and atrial flutter in patients undergoing cardioversion: a large transoesophageal echo study. EuroIntervention **15**(3), e225–e230 (2019)
4. Di Achille, P., Tellides, G., Figueroa, C.A., Humphrey, J.D.: A haemodynamic predictor of intraluminal thrombus formation in abdominal aortic aneurysms. Proc. Royal Soc. A Math. Phys. Eng. Sci. **470**(2172), 20140163 (2014)
5. Fey, M., Lenssen, J.E., Weichert, F., Müller, H.: Splinecnn: fast geometric deep learning with continuous b-spline kernels. In: Proceedings of the IEEE Conference on Computer Vision and Pattern Recognition (CVPR) (2018)
6. García-Isla, G., et al.: Sensitivity analysis of geometrical parameters to study haemodynamics and thrombus formation in the left atrial appendage. Int. J. Numer. Methods Biomed. Eng. **34**(8), e3100 (2018)
7. Hennigh, O.: Lat-Net: Compressing Lattice Boltzmann Flow Simulations using Deep Neural Networks (2017)
8. Li, G., Xiong, C., Thabet, A., Ghanem, B.: Deepergcn: All you need to train deeper gcns (2020)
9. Liang, L., Liu, M., Martin, C., Sun, W.: A deep learning approach to estimate stress distribution: a fast and accurate surrogate of finite-element analysis. J. Royal Soc. Interface **15**(138), 20170844 (2018)
10. Mill, J., et al.: Impact of flow dynamics on device-related thrombosis after left atrial appendage occlusion. Can. J. Cardiol. **36**(6), 968.e13–968.e14 (2020)
11. Morales, X., et al.: Deep learning surrogate of computational fluid dynamics for thrombus formation risk in the left atrial appendage. In: Pop, M., et al. (eds.) STACOM 2019. LNCS, vol. 12009, pp. 157–166. Springer, Cham (2020). https://doi.org/10.1007/978-3-030-39074-7_17
12. Pfaff, T., Fortunato, M., Sanchez-Gonzalez, A., Battaglia, P.W.: Learning mesh-based simulation with graph networks (2021)
13. Slipsager, J.M., et al.: Statistical shape clustering of left atrial appendages. In: Pop, M., et al. (eds.) STACOM 2018. LNCS, vol. 11395, pp. 32–39. Springer, Cham (2019). https://doi.org/10.1007/978-3-030-12029-0_4
14. Watson, T., Shantsila, E., Lip, G.Y.: Mechanisms of thrombogenesis in atrial fibrillation: Virchow's triad revisited. Lancet **373**(9658), 155–166 (2009)

Learning Atrial Fiber Orientations and Conductivity Tensors from Intracardiac Maps Using Physics-Informed Neural Networks

Thomas Grandits[1,2]($\boxtimes$), Simone Pezzuto[3], Francisco Sahli Costabal[4,5,6],
Paris Perdikaris[7], Thomas Pock[1,2], Gernot Plank[2,8], and Rolf Krause[3]

[1] Institute of Computer Graphics and Vision, TU Graz, Graz, Austria
{thomas.grandits,pock}@icg.tugraz.at
[2] BioTechMed-Graz, Graz, Austria
[3] Center for Computational Medicine in Cardiology, Euler Institute, Università della Svizzera italiana, Lugano, Switzerland
{simone.pezzuto,rolf.krause}@usi.ch
[4] Department of Mechanical and Metallurgical Engineering, School of Engineering, Pontificia Universidad Católica de Chile, Santiago, Chile
fsc@ing.puc.cl
[5] Institute for Biological and Medical Engineering, Schools of Engineering, Medicine and Biological Sciences, Pontificia Universidad Católica de Chile, Santiago, Chile
[6] Millennium Nucleus for Cardiovascular Magnetic Resonance, Santiago, Chile
[7] Department of Mechanical Engineering and Applied Mechanics, University of Pennsylvania, Philadelphia, Pennsylvania, USA
pgp@seas.upenn.edu
[8] Gottfried Schatz Research Center - Division of Biophysics, Medical University of Graz, Graz, Austria
gernot.plank@medunigraz.at

Abstract. Electroanatomical maps are a key tool in the diagnosis and treatment of atrial fibrillation. Current approaches focus on the activation times recorded. However, more information can be extracted from the available data. The fibers in cardiac tissue conduct the electrical wave faster, and their direction could be inferred from activation times. In this work, we employ a recently developed approach, called physics informed neural networks, to learn the fiber orientations from electroanatomical maps, taking into account the physics of the electrical wave propagation. In particular, we train the neural network to weakly satisfy the anisotropic eikonal equation and to predict the measured activation times. We use a local basis for the anisotropic conductivity tensor, which encodes the fiber orientation. The methodology is tested both in a synthetic example and for patient data. Our approach shows good agreement in both cases, with an RMSE of 2.2 ms on the in-silico data and outperforming a state of the art method on the patient data. The results show a first step towards learning the fiber orientations from electroanatomical maps with physics-informed neural networks.

© Springer Nature Switzerland AG 2021
D. B. Ennis et al. (Eds.): FIMH 2021, LNCS 12738, pp. 650–658, 2021.
https://doi.org/10.1007/978-3-030-78710-3_62

1   Introduction

The fiber structure of the cardiac tissue has a prominent role on its function. From an electrophysiological viewpoint, electrical conduction along the fiber direction is generally higher than the cross-fiber direction [4]. The overall propagation of the action potential is therefore anisotropic, with preferential pathways in the activation. In the atria, fiber orientations and the resulting conductivity are however known only with large uncertainty, rendering the construction of accurate computational models more difficult [12].

Electroanatomical mapping, a keystone diagnostic tool in cardiac electrophysiology studies, can provide high-density maps of the local electric activation. Since conduction properties of the tissue and the activation times are physiologically correlated, electroanatomical maps (EAMs) can be potentially the basis for a parameter identification procedure of the former [10]. Subsequently, local fiber orientations may be extrapolated from the fastest conductivity direction.

The problem of identifying the conductivity of the tissue from point-wise recordings of the activation time has been already addressed by others. Local approaches estimate the front velocity at every point of the domain only by using the local geometrical and electrical information, that is the location and the activation time of neighboring points [17]. Local fiber direction can be estimated from multiple pacing sequences by comparing the velocity in the propagation directions [13]. Similarly, the EAM may be interpolated into a smooth activation map, e.g., with linear or radial basis functions [6]. Local methods are general and do not account for the physics: there is no guarantee that activation and conduction will satisfy a given model. Moreover, they may fail at those locations where activation is not differentiable, such as collision lines and breakthroughs. Model-based approaches are not novel either [2,7]. In these methods, the mismatch between observed and simulated (by the model) activation is minimized by optimizing the local conduction velocity of the tissue.

In the presented work, we solve the problem of learning the anisotropic structure of the conductivity tensor from electroanatomical maps by imposing to the identification problem a physiological constraint, encoded in the anisotropic eikonal equation. For this purpose, we extend our recent work into this problem [8,14] by combining Physics-informed Neural Networks (PINNs) and the anisotropic eikonal equation. In the presented approach, a set of neural networks representing the conductivity tensor and the activation times are fitted to the data by weakly imposing the eikonal model through a penalization term. PINNs are particularly suitable for recovering complex functions from sparse and scarce data, as in the present case [11]. Moreover, the weak imposition of the model does not require its explicit solution. In contrast to prior learning methods, aiming at learning Finite Element Method (FEM) solutions [16], PINNs learn a single (albeit complex) continuous function to estimate the solution on the whole domain. Therefore, PINNs can model functions on domains irrespective of a FE discretization, i.e. mesh. The function can in this setting also be evaluated outside the domain, where it is not required to weakly satisfy the model partial differential equation (PDE).

We can also omit boundary conditions, which for the eikonal model are the sites of early activation, usually unknown. The proposed method, fully implemented in TensorFlow[1], performed well with synthetic data, being able to recover ground-truth fiber orientations in limited regions, just with a single activation sequence. Additionally, we applied the method to EAMs from a patient who underwent a mapping procedure prior ablation, with promising results. Both models were compared to our previous classical optimization approach [8] and performed comparably in all cases.

This manuscript is organized as follows: In Sect. 2 we formulate the problem of estimating conductivities as a physics-informed neural network, and we introduce its numerical solution. In Sect. 3, we show two numerical experiments to test the accuracy of the method. We end this manuscript with a discussion and future directions in Sect. 4.

2   Methods

2.1   Anisotropic Eikonal Model for Cardiac Activation

The cardiac tissue is electrically active, in the sense that a sufficiently strong stimulus can trigger the propagation of a travelling wave, called action potential. The time of first arrival of the action potential, usually set as the point of crossing of a threshold potential, is called activation time. Herein, we denote as $\phi \colon \Omega \to \mathbb{R}$ the activation map in a embedding domain $\Omega \subset \mathbb{R}^3$ (e.g., a box containing the atria), that is $\phi(\mathbf{x})$ is the activation time at $\mathbf{x} \in \Omega$. Neglecting curvature effects of the wave on the propagation speed [5], the activation map may be well described by the anisotropic eikonal equation, which reads as follows:

$$\sqrt{\mathbf{D}(\mathbf{x})\nabla\phi(\mathbf{x}) \cdot \nabla\phi(\mathbf{x})} = 1, \quad \mathbf{x} \in \Omega, \tag{1}$$

where $\mathbf{D} \in \mathcal{C}(\bar{\Omega}; \mathcal{P}(3))$ is a continuous tensor field from $\bar{\Omega}$ to $\mathcal{P}(3)$, the set of 3×3 symmetric positive-definite real matrices. The eikonal equation is not explicitly solved, hence there is no need to enforce boundary conditions. We rather consider the model residual:

$$R_{\mathrm{m}}[\phi](\mathbf{x}) := \sqrt{\max\{\mathbf{D}(\mathbf{x})\nabla\phi(\mathbf{x}) \cdot \nabla\phi(\mathbf{x}), \varepsilon\}} - 1, \tag{2}$$

for a sufficiently small $\varepsilon > 0$ to avoid infeasible gradients, as a metric of pointwise model discrepancy for a given pair of activation ϕ and conductivity tensor $\mathbf{D}$.

2.2   Representation of the Conductivity Tensor

The conductivity tensor $\mathbf{D}$ shall be defined through a parameter vector in a way that ensures its symmetry, positive-definiteness, and zero velocity orthogonal to

---

[1] https://www.tensorflow.org/.

the atrial surface $\mathcal{S} \subset \Omega$ in all cases. Analogous to our previous method in [8], we therefore consider the parameter vector $\mathbf{d}(\mathbf{x}) = [d_1(\mathbf{x}), d_2(\mathbf{x}), d_3(\mathbf{x})]^\top$ and define $\mathbf{D}$ through $\mathbf{d}$ as follows:

$$\mathbf{D}(\mathbf{x}) := e^{\mathbf{P}(\mathbf{x})\mathbf{D}_2(\mathbf{d}(\mathbf{x}))\mathbf{P}(\mathbf{x})^\top}, \qquad \mathbf{D}_2(\mathbf{d}(\mathbf{x})) := \begin{bmatrix} d_1(\mathbf{x}) & d_2(\mathbf{x}) \\ d_2(\mathbf{x}) & d_3(\mathbf{x}) \end{bmatrix} \qquad (3)$$

where $\mathbf{P}(\mathbf{x}) \in \mathbb{R}^{3 \times 2}$ is a matrix whose columns contain two orthonormal vectors in the tangent plane at $\mathbf{x} \in \mathcal{S}$, computed using the vector heat method [15]. The smooth tangent bases generated by the vector heat method for both the in-silico model, as well as the in-vivo measured EAM can be seen in Fig. 1.

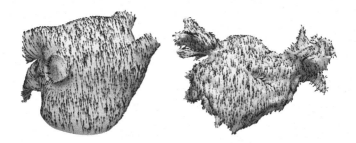

Fig. 1. Smooth generated manifold bases using the vector heat method for both considered models. These bases are used to provide a smooth 2D map across the manifold and are a useful foundation for computing TV in 2D.

Please note that we consider the matrix exponential in Eq. (3) rather than component-wise exponentiation. This choice corresponds to the Log-Euclidean metric of the space $\mathcal{P}(3)$ [1].

The fiber direction is defined as the direction of fastest propagation, that is the eigenvector associated to the largest eigenvalue of $\mathbf{D}$. By the definition of matrix exponential, such direction is also the maximum eigenvector of $\mathbf{P}\mathbf{D}_2\mathbf{P}^\top$.

2.3   Physics-Informed Neural Network

In the considered experiments, we are given a set of points, each composed by a location $\mathbf{x}_i$ and a recorded activation time $\hat{\phi}_i : \Gamma_\mathcal{S} \to \mathbb{R}$ for $\Gamma_\mathcal{S} \subset \mathcal{S}$ representing the EAM locations and timings of the recordings. The objective is therefore to identify a conductivity tensor field $\mathbf{D}$ such that the corresponding activation map ϕ, as resulting from Eq. (1), will closely reproduce the observed data. The tensor $\mathbf{D}$ can then be reconstructed by means of using $\mathbf{d}$.

For this purpose, we approximate both the activation map $\phi(\mathbf{x})$ and the conductivity vector $\mathbf{d}(\mathbf{x})$ with a feed-forward neural network $\mathrm{NN}_{n,m,\boldsymbol{\theta}} \colon \mathbb{R}^n \to \mathbb{R}^m$ with n inputs to m outputs and characterized by a vector $\boldsymbol{\theta}$ containing weights and biases, as was initially promoted in [11]. The used architecture of the networks is shown in Fig. 2. Specifically, we have $\phi(\mathbf{x}) \approx \phi_{\mathrm{NN}}(\mathbf{x}, \boldsymbol{\theta}_\phi) = \mathrm{NN}_{3,1,\boldsymbol{\theta}_\phi}(\mathbf{x})$ and $\mathbf{d}(\mathbf{x}) \approx \mathbf{d}_{\mathrm{NN}}(\mathbf{x}, \boldsymbol{\theta}_\mathbf{d}) = d_{\max} \cdot \tanh\!\big(\mathrm{NN}_{3,3,\boldsymbol{\theta}_\mathbf{d}}(\mathbf{x})\big).$ where tanh is meant component-wise, and $d_{\max}$ is an upper limit for the components of $\mathbf{d}_{NN}$, meant to avoid over- and underflows in the numerical calculations. This construction of ϕ and $\mathbf{d}$ enables us to use standard machine learning methods and frameworks to efficiently calculate the gradients $\nabla\phi$ and $\nabla\mathbf{d}$, used in the chosen PDE model (2) and inverse regularization. The usage of these gradients in the optimization necessitates at least second order smooth activation functions in the neurons, achieved by the use of tanh functions.

Similar to the original PINN algorithms in [11], we define a loss function to train our model as the sum of a data fidelity, a PDE model fidelity term and two regularization terms:

$$
\mathcal{L}(\boldsymbol{\theta}_\phi, \boldsymbol{\theta}_\mathbf{d}) := \int_{\Gamma_\mathcal{S}} \left(\phi_{NN}(\mathbf{x}) - \hat{\phi}(\mathbf{x})\right)^2 \mathrm{d}\mathbf{x} + \alpha_\mathrm{m} \int_\mathcal{S} \left(R_\mathrm{m}[\phi_{NN}](\mathbf{x})\right)^2 \mathrm{d}\mathbf{x}
$$
$$
+ \alpha_\theta \left(\|\boldsymbol{\theta}_\phi\|^2 + \|\boldsymbol{\theta}_\mathbf{d}\|^2\right) + \alpha_\mathbf{d} \int_\mathcal{S} H_\delta\left(\nabla\mathbf{d}_{NN}(\mathbf{x})\right) \mathrm{d}\mathbf{x},
$$
(4)

for the three weighting parameters $\alpha_\mathrm{m}, \alpha_\mathbf{d}, \alpha_\theta$. Regularization is both applied to the weights of the networks as well as on the inverse parameter estimation. The latter regularization term is an Huber-type, approximated Total Variation regularization for the conductivity vector parameters $\mathbf{d}$. Specifically,

$$
H_\delta(\mathbf{x}) = \begin{cases} \frac{1}{2\delta}\|\mathbf{x}\|^2, & \text{if } \|\mathbf{x}\| \leq \delta, \\ \|\mathbf{x}\| - \frac{1}{2}\delta, & \text{otherwise} \end{cases}
$$
(5)

with $\delta = 5 \cdot 10^{-2}$ for our experiments. Note that by construction, $\mathbf{D}$ has zero velocity in directions normal to the manifold (see Sect. 2.2) and thus allows us to neglect the additional normal penalization used in [14].

2.4 Numerical Implementation

The domain $\mathcal{S}$ is discretized using a triangular mesh, usually obtained directly from the mapping system, along with point-wise evaluations of the activation times, that is $\Gamma_\mathcal{S} = \{\mathbf{x}_1, \ldots, \mathbf{x}_N\}$. In all experiments, the integrals were approximated using a point-wise evaluation for both domains: On the vertices for the approximation of $\mathcal{S}$ and on the discrete measurements for $\Gamma_\mathcal{S}$.

For the optimization, we experimentally selected the hyper-parameters as $\alpha_\mathrm{m} = 10^4$ for the model atria and $\alpha_\mathrm{m} = 10^3$ for the EAM. The other two hyperparameters are the same for both experiments: $\alpha_\theta = 10^{-4}$, $\alpha_\mathbf{d} = 10^{-3}$. The two neural networks for ϕ and $\mathbf{d}$ had 7 and 5 hidden fully connected layers respectively. All hidden layers consisted of 20 neurons for ϕ_{NN} and 5 neurons for $\mathbf{d}_{NN}$, with the weights being initialized using Xavier initialization. This choice of neural network architecture was inspired by the work in [14]. We opted for adding a regression layer to ϕ_{NN}, since this allows us to model arbitrary ranges of ϕ. Optimization is performed by first using the ADAM [9] optimizer for 10^4 epochs with a learning rate of 10^{-3}, followed by a L-BFGS optimization [3] until convergence to a local minimum is achieved. Each experiment took no longer than 1.5 h on a desktop machine with an Intel Core i7-5820K CPU with 6 cores of each 3.30 GHz, 32 GB of working memory and a NVidia RTX 2080 GPU.

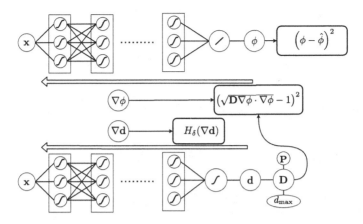

Fig. 2. Structural view of the proposed PINN architecture, containing the two NNs ϕ_{NN} and $\mathbf{d}_{NN}$. Nodes containing a curve indicate a tanh activation function. The final layer of ϕ_{NN} is a linear layer. $\mathbf{D}$ is computed using (3). $\nabla\phi$ and $\nabla\mathbf{d}$ can be obtained by means of backpropagation (reverse arrows). The bold rectangular boxes show the three major loss terms from (4): Data fidelity, eikonal (PDE) and TV loss (top to bottom).

3   Numerical Experiments

Herein, we consider two experiments: a synthetic example, with ground-truth on a realistic anatomy of the left atrium, and an example with patient-specific geometry and data. The first example is optimized and tested against different levels of i.i.d. normal noise: $\tilde{\phi}(\mathbf{x}) = \hat{\phi}(\mathbf{x}) + \mathcal{N}(0, \sigma_{\mathcal{N}})$. We measure the performance of the synthetic model in terms of the root-mean-square error (RMSE) over the whole surface here denoted as $\text{RMSE}_{\mathcal{S}}$. Errors directly on the measurement points, employed in the optimization, are used to compute RMSE_O. In the patient specific example, we randomly split $\Gamma_{\mathcal{S}}$ into Γ_O, used for optimization/training, and Γ_T, for testing.

The results of our method on the in-silico model atria, and a comparison to PIEMAP [8], are presented in Table 1. Both methods are comparable in terms of RMSE, with both methods achieving less than 5 ms of RMSE for all levels of noise. Additionally, we tested the presented PINN on an EAM, achieving a RMSE of the activation times on the test set of $\text{RMSE}_T \approx 5.59\,\text{ms}$. The RMSE on the measurements used in the optimization was only slightly lower at $\text{RMSE}_O \approx 4.82\,\text{ms}$, indicating that α_m was chosen in a proper range to avoid overfitting to the data. In the patient-specific test (not shown in the table), our method was able to outperform PIEMAP, which reported $\text{RMSE}_T \approx 6.89\text{ms}$ and $\text{RMSE}_O \approx 1.18\text{ms}$ on test and optimization/training set respectively, showing a slight overfit to the data used in the optimization.

Figure 3 shows the qualitative results of using this method on the two chosen models (the model atria in the noise-less case). We can nicely fit the activation encountered at the surface and create an eikonal-like activation. The initiation sites are automatically deduced by the PINN algorithm with only soft eikonal

Table 1. Evaluation of the presented PINN approach compared to PIEMAP [8] for different levels of noise (given in standard deviation and signal-to-noise ratio) on the in-silico model atria. The result of the noiseless scenario (∞ dB) is visualized in Fig. 3 on the left side.

		RMSE$_S$/RMSE$_O$ PINN	RMSE$_S$/RMSE$_O$ PIEMAP
σ_N/PSNR	0 ms/∞ dB	2.20/1.38	1.04/0.83
	0.1 ms/64.1 dB	4.28/2.08	1.02/0.83
	1 ms/43.9 dB	3.32/1.39	1.09/0.83
	5 ms/29.9 dB	3.76/1.85	1.90/0.84

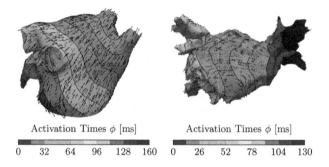

Activation Times ϕ [ms]   Activation Times ϕ [ms]

0   32   64   96   128   160   0   26   52   78   104   130

Fig. 3. Results of the PINN method on an in-silico (left) and an in-vivo (right) EAM model with the overlayed measurements as points and fibers as arrows. The underlying contour lines and colors on the mesh itself represents the activation of the PINN, sampled at each vertex.

and data constraints. The smooth basis generated with the vector heat method, together with the TV regularization give us a smooth, fiber field.

4   Discussion and Outlook

In this work, we present a novel approach to the recovery of fiber orientations from activation time measurements. We train a neural network that aims to approximate the data while also satisfy the eikonal equation. In this way, we can identify the conductivity tensor that best explains the data. Our approach has some advantages compared to methods that exactly solve the forward problem of the eikonal equation. First, we only weakly enforce the solution of the PDE, which may be convenient when the model is not exactly satisfied. Here the eikonal equation is an approximation to the much complex process of cardiac electrophysiology. Second, we do not need to impose boundary conditions, or define the number of early activation sites, which are unknown beforehand. Third, our approach scales wells both with the number of measurements and the size of the mesh. We take advantage of mini-batch optimization strategies to utilize all the data while keeping the computational cost manageable. Finally, the implementation is compact and easily portable, taking advantage of the automatic differentiation capabilities of modern machine learning languages.

Our work also has some limitations and disadvantages. First, the PINN optimization in (4) is highly non-linear and exposes many local minima. A direct use of a simple convex optimization algorithm is therefore often not advised as usually only very poor local minima are found. ADAM can overcome some of these minima, but the final solution is very dependent on the initialization. This could be mitigated by training multiple neural networks at a negligible cost, which at the same time can be used to quantify uncertainty [14]. Future research could mitigate some of these problems by using different loss functionals, or alternative optimization algorithms. Second, our method showed worse performance in terms of error when compared to PIEMAP in a synthetic example. However, this trend was reversed when we used activation times coming from a patient. This indicates that further testing is necessary, with a larger dataset, to determine the accuracy of this method. Finally, our approach introduces multiple hyper-parameters that need to be tuned. In the future, we plan to use a more systematic approach to determine these parameters.

Overall, our work represent a first step to learning the fiber orientations from activation times in cardiac electrophysiology with physics-informed neural networks.

Acknowledgements. This work was financially supported by the Theo Rossi di Montelera Foundation, the Metis Foundation Sergio Mantegazza, the Fidinam Foundation, the Horten Foundation and the CSCS–Swiss National Supercomputing Centre production grant s1074. We'd also like to additionally acknowledge funding from the BioTechMed ILearnHeart project, the EU grant MedalCare 18HLT07, the grant FONDECYT-Postdoctorado 3190355, as well as the Swiss National Science Foundation for their support under grant 197041 "Multilevel and Domain Decomposition Methods for Machine Learning".

References

1. Arsigny, V., Fillard, P., Pennec, X., Ayache, N.: Geometric means in a novel vector space structure on symmetric positive-definite matrices. SIAM J. Matrix Anal. Appl. **29**(1), 328–347 (2007)
2. Barone, A., Carlino, M.G., Gizzi, A., Perotto, S., Veneziani, A.: Efficient estimation of cardiac conductivities: a proper generalized decomposition approach. J. Comp. Phys. **423**, 109810 (2020)
3. Byrd, R.H., Lu, P., Nocedal, J., Zhu, C.: A limited memory algorithm for bound constrained optimization. SIAM J. Sci. Comput. **16**(5), 1190–1208 (1995)
4. Clerc, L.: Directional differences of impulse spread in trabecular muscle from mammalian heart. J. Physiol. **255**(2), 335–346 (1976)
5. Colli Franzone, P., Guerri, L., Rovida, S.: Wavefront propagation in an activation model of the anisotropic cardiac tissue: asymptotic analysis and numerical simulations. J. Math. Biol. **28**(2), 121–176 (1990)
6. Coveney, S., et al.: Gaussian process manifold interpolation for probabilistic atrial activation maps and uncertain conduction velocity. Philos. Trans. R. Soc. A **378**(2173), 20190345 (2020)

7. Grandits, T., et al.: An inverse Eikonal method for identifying ventricular activation sequences from epicardial activation maps. J. Comp. Phys. **419**, 109700 (2020)

8. Grandits, T., Pezzuto, S., Lubrecht, J.M., Pock, T., Plank, G., Krause, R.: PIEMAP: personalized inverse Eikonal model from cardiac electro-anatomical maps. In: Puyol Anton, E., et al. (eds.) STACOM 2020. LNCS, vol. 12592, pp. 76–86. Springer, Cham (2021). https://doi.org/10.1007/978-3-030-68107-4_8

9. Kingma, D.P., Ba, J.: Adam: A method for stochastic optimization. arXiv preprint arXiv:1412.6980 (2014)

10. Maagh, P., Christoph, A., Dopp, H., Mueller, M.S., Plehn, G., Meissner, A.: High-density mapping in ventricular Tachycardia Ablation: a Pentaray® study. Cardiol. Res. **8**(6), 293–303 (2017)

11. Raissi, M., Perdikaris, P., Karniadakis, G.E.: Physics-informed neural networks: A deep learning framework for solving forward and inverse problems involving nonlinear partial differential equations. J. Comp. Phys. **378**, 686–707 (2019)

12. Roney, C.H., et al.: Constructing a human atrial fibre atlas. Ann. Biomed. Eng. **49**(1), 233–250 (2020). https://doi.org/10.1007/s10439-020-02525-w

13. Roney, C.H., et al.: A technique for measuring anisotropy in atrial conduction to estimate conduction velocity and atrial fibre direction. Comput. Biol. Med. **104**, 278–290 (2019)

14. Sahli Costabal, F., Yang, Y., Perdikaris, P., Hurtado, D.E., Kuhl, E.: Physics-informed neural networks for cardiac activation mapping. Front. Phys. **8**, 42 (2020)

15. Sharp, N., Soliman, Y., Crane, K.: The vector heat method. ACM Trans. Graph. **38**(3), 1–19 (2019)

16. Takeuchi, J., Kosugi, Y.: Neural network representation of finite element method. Neural Netw. **7**(2), 389–395 (1994)

17. Verma, B., Oesterlein, T., Loewe, A., Luik, A., Schmitt, C., Dössel, O.: Regional conduction velocity calculation from clinical multichannel electrograms in human atria. Comput. Biol. Med. **92**, 188–196 (2018)

The Effect of Ventricular Myofibre Orientation on Atrial Dynamics

Marina Strocchi[1]([✉]), Christoph M. Augustin[2], Matthias A. F. Gsell[2],
Elias Karabelas[7], Aurel Neic[3], Karli Gillette[2], Caroline H. Roney[1],
Orod Razeghi[1], Jonathan M. Behar[1,6], Christopher A. Rinaldi[1,6],
Edward J. Vigmond[4,5], Martin J. Bishop[1], Gernot Plank[2],
and Steven A. Niederer[1]

[1] King's College London, London, UK
marina.strocchi@kcl.ac.uk
[2] Medical University of Graz, Graz, Austria
[3] NumeriCor GmbH, Graz, Austria
[4] IHU Liryc, Electrophysiology and Heart Modeling Institute, Bordeaux, France
[5] University of Bordeaux, Bordeaux, France
[6] Guy's and St Thomas' NHS Foundation Trust, London, UK
[7] University of Graz, Graz, Austria

Abstract. Cardiac output is dependent on the tight coupling between atrial and ventricular function. The study of such interaction mechanisms is hindered by their complexity, and therefore requires a systematic approach. We have developed a four-chamber closed-loop cardiac electromechanics model which, through the coupling of the chambers with a closed-loop cardiovascular system model and the effect of the pericardium, is able to capture atrioventricular interaction. Our model simulates electrical activation and contraction of the atria and the ventricles coupled with a closed-loop model based on the CircAdapt framework. We include the effect of the pericardium on the heart using normal springs, scaling the local spring stiffness based on image-derived motion. The coupled model was used to study the impact of ventricular myofibre orientation on atrial dynamics by varying ventricular fibre orientation from $-40°/+40°$ to $-70°/+70°$. We found that steeper fibres increase atrioventricular valve plane motion from 1.0 mm to 14.0 mm, leading to a lower minimum left atrial (LA) pressure (-0.4 mmHg vs -1.1 mmHg) and greater venous return (LA maximum volume: 168 mL vs 182 mL), and that fibres angles $-50°/+50°$ were consistent with a physiological atrial contraction and filling pattern. Our framework is capable of capturing complex interaction dynamics between the atria, the ventricles and the circulatory system accounting for the effect of the pericardium. Such simulation platform represents a useful tool to study both systolic and filling phases of all cardiac chambers, and how these get altered in diseased states and in response to treatment.

Keywords: Four-chamber model · Electromechanics · Pericardium · Atrioventricular interaction

© Springer Nature Switzerland AG 2021
D. B. Ennis et al. (Eds.): FIMH 2021, LNCS 12738, pp. 659–670, 2021.
https://doi.org/10.1007/978-3-030-78710-3_63

1   Introduction

The left atrium (LA) affects left ventricular (LV) dynamics, especially during ventricular filling. During ventricular systole, when the mitral valve is closed, the LA acts as a blood reservoir, and fills with blood due to venous return. When LV contraction ends and the LV pressure decays below the LA pressure, the mitral valve opens and blood passively flows from the LA to the LV. As the LV starts filling, the LA volume and pressure decrease. The LA then starts contracting, pushing more of the blood into the LV before the mitral valve closes. The multiple LA functions (blood reservoir, conduit and active pump) are present in the LA pressure-volume (PV) relationship, which typically has a figure-of-eight shape, with the left (or *v-loop*) and the right (or *a-loop*) loops representing passive and active atrial function, respectively (Fig. 1). Atrial function both impacts and is driven by ventricular function. Therefore, understanding how these two systems interact requires an integrative approach.

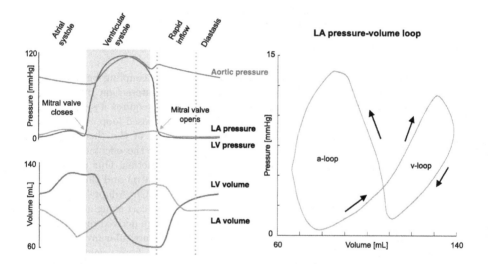

Fig. 1. Left heart pressure and volume. Left: LV, LA and aortic pressure and LV and LA volumes plotted over time. Right: LA pressure-volume loop.

Computational models have been used in the past to provide insight into physiology and pathophysiology of the heart. Recently, cardiac electromechanics models have been moving from ventricular to four-chamber models [7, 12–14, 21, 22], which offer a wider spectrum of simulated dynamics. Previous four-chamber heart studies showed the importance of the pericardium to reproduce physiological atrioventricular (AV) plane displacement and therefore to capture interaction mechanisms between the atria and the ventricles [9, 18, 22]. Additional interaction mechanisms between the four chambers are also mediated by the circulatory system, which connects the atria to the ventricles through blood flow across the systemic and the pulmonary circulations. Therefore, a four-chamber heart model needs to account for both the pericardium and the coupling with the circulatory system represented as a closed loop to simulate complex interaction mechanisms within the heart.

In this paper, we present our framework for four-chamber electromechanics coupled with a zero-dimensional closed-loop model. The effect of the pericardium on the heart is represented with normal springs with locally varying stiffness, derived from motion data. This four-chamber model is used to study the effect of ventricular myofibre orientation on atrial pressure and volume dynamics by running simulations with different ventricular endocardial and epicardial fibre angles.

2   Methods

All electromechanics simulations were run with the cardiac arrhythmia research package (CARP) [4,23], using a four-chamber geometry from a previous study, generated from a female heart failure patient recruited as part of a cardiac resynchronisation therapy (CRT) clinical trial [22]. ECG-gated whole heart CT images (in-plane resolution: 0.36 mm, slice thickness: 0.5 mm) were acquired over 10 frames throughout a cardiac cycle. The mesh was generated using the end-diastolic CT frame. The mesh is inclusive of ventricular and atrial myofibre orientation [22]. Ventricular fibre orientation was assigned with a Laplace-Dirichlet rule-based method by Bayer et al. [5]. The base of the ventricles was defined as the intersection between the mitral and the tricuspid valve planes and the ventricles, while the apex was selected as the LV epicardial point furthest from the LA. Additionally, the endocardial surface of the left and the right ventricles and the epicardium of the ventricles were extracted from the mesh to define different Laplace solutions that were then used to compute the myofibre orientation. We defined endocardial/epicardial fibre and sheet angles of $-50°/+50°$ and $-65°/+25°$ [5], respectively. Epicardial and endocardial atrial fibres were mapped from a human *ex-vivo* DTMRI dataset using the universal atrial coordinates [20]. The fibre orientation was then linearly interpolated between the epicardium and the endocardium.

To study the effect of ventricular fibres on atrial dynamics, the baseline simulation was repeated with the following epicardial and endocardial ventricular fibre angles, while atrial fibres were left unaltered: $-40°/+40°$, $-60°/+60°$ and $-70°/+70°$.

In the next sections, we briefly describe our electromechanical simulation framework. In particular, we focus on the spring boundary conditions to represent the effect of the pericardium on the atria. CT motion data from a previously published cohort [21] were used to define a scaling map for spring stiffness, that was then applied to one single patient.

2.1 Electrophysiology Simulation

The electrical activation of the heart was simulated with a reaction-eikonal model [16], which solves for activation time distribution $t_a(\mathbf{x})$ as a function of node location $\mathbf{x} \in \Omega$. Electrical activation of the atria was initiated next to the superior vena cava, to simulate sinoatrial node activation. The RV was stimulated based on the location of the CRT lead. The patient's LV lead was not considered in the simulation, as we wanted to simulate only RV pacing baseline conditions. Under these conditions, we had clinical invasive LV pressure measurements and the volume transient from the CT scans, which were used to retrieve the stress free configuration (see Sect. 2.2). The atria and the ventricles were separated by a 1 mm thick passive layer to prevent an unphysiological spread of the activation from atria to ventricles. This also allowed controlling the AV delay in the simulation. The RA and the RV were stimulated with a basic cycle length of 850 ms (corresponding to a heart rhythm of 70 bpm) for six beats to reach a steady state, and the AV delay was set to 120 ms.

Atrial and ventricular tissue were simulated as transversely isotropic conduction media. The atria were assigned with a conduction velocity (CV) of 1.7 m/s and 0.68 m/s in the fibre and transverse directions, respectively, to achieve full atrial activation within 100 ms. The ventricles were assigned with a fibre and transverse CV of 0.6 m/s and 0.24 m/s [22]. To simulate fast endocardial activation, we defined a one-element thick layer at the LV endocardium with two-times faster CV compared to the ventricular myocardium.

2.2 Mechanics Simulation

The atria and ventricles were simulated as hyperelastic, transversely isotropic and nearly incompressible materials with the Guccione law [10]. Parameters were set according to literature values estimated for healthy subjects [15]. All the other tissues (wall of the aorta, pulmonary artery, veins and valve planes) were modelled with a neo-Hookean model, with parameters set based on previous studies [21, 22]. Incompressibility of all tissues was enforced with a penalty method.

Active tension generation in the atria and ventricles was simulated with a phenomenological active tension model, where active tension rise is triggered by local activation time $t_a(\mathbf{x})$ [17]:

$$S_a(\mathbf{x}, t) = T_{\mathrm{peak}}\phi(\lambda)\tanh^2\left(\frac{t_s}{\tau_r}\right)\tanh^2\left(\frac{t_{\mathrm{dur}} - t_s}{\tau_d}\right), \qquad 0 < t_s < t_{\mathrm{dur}} \,,$$

$$\phi(\lambda) = \tan(\mathrm{ld}(\lambda - \lambda_0)) \,, \qquad t_s = t - t_a(\mathbf{x}) - t_{\mathrm{emd}} \,.$$

where T_{peak} is the peak in active tension and t is time. Timing parameters t_{emd}, τ_r, τ_d and t_{dur} represent electromechanical delay, rising time, decay time and twitch duration, respectively. Parameters λ_0 and ld represent stretch below which no active tension is generated and the degree of length dependence, respectively. T_{peak} was set to 60 kPa and 120 kPa for the atria and ventricles, respectively, based on lower peak in active tension in the atria compared to the ventricles [14]. Timing parameters for the ventricles were set as t_{emd}=20 ms and t_{dur}=450 ms. These parameters were adapted for the atria based on [2]: t_{emd}=10 ms and t_{dur}=300 ms. τ_r, τ_d, ld and λ_0 were set to 100 ms, 50 ms, 6.0 and 0.7, respectively, for all four chambers.

Boundary Conditions. We simulated the effect of the pericardium on the whole heart using normal springs boundary conditions [22]. Spring stiffness was scaled on the ventricles according to a map previously derived from motion data to apply maximum and minimum constraint at the base and at the apex, respectively [21]. We defined a similar scaling map for the LA and the RA based on motion data available from a cohort of twenty-three four-chamber heart meshes previously published [21]. The LA and RA of each patient were divided in eighteen regions (Fig. 2, centre) each by applying the universal ventricular coordinates algorithm [6] and treating each atrium as an upside down LV. Although this approach is not an atrial-based reference system as opposed to other methods in the literature, this method allowed us to define a coordinate system throughout the cohort similar to the one defined on the ventricles. The displacement was computed from the CT images similarly to [22], and the frame with maximum epicardial displacement normal to the surface was selected out of the ten or twenty available CT frames for each patient. The displacement was normalised between 0 and 1 over the whole epicardial surface of the LA, and the average computed for each atrial region for each patient. We took the average over the whole cohort (Fig. 2, left), which showed that the roof of the atria moves the least while regions close to the AV plane move the most due to ventricular shortening. Based on this, a scaling map for spring stiffness was defined to apply maximum constraint at the roof of the atria and no constraint at the AV plane (Fig. 2, right). A maximum spring stiffness of 10 kPa/mm was scaled according to the ventricular, LA and RA penalty maps to simulate local constraints the pericardium exerts on the outer surface of the patient's heart.

The atrial and ventricular mechanical contraction were coupled with a closed-loop model for the circulatory system based on CircAdapt, as this provides physiological preload and afterload of all cardiac chambers [3,24]. Adaptation rules were not included. Parameters for the mitral valve and the aortic valve orifices area were measured from the end-diastolic CT images and set to 920.0 mm^2 and 345.0 mm^2, respectively. Since the leaflets of the tricuspid and the pulmonary valves were not visible on the CT, the tricuspid and the pulmonary valve orifices areas were set equal to the mitral and the aortic valve orifices areas. The systemic and the pulmonary veins orifice cross-sectional area were set to 300 mm^2, similarly to the aortic valve orifice area. Systemic and pulmonary artery wall cross-sectional areas were computed assuming a wall thickness of 2 mm, resulting in wall cross-section of 144.0 mm^2. Systemic and pulmonary vein walls were assumed to have 1 mm thickness, resulting in a wall cross-sectional area of 64.0 mm^2. Systemic and pulmonary veins reference pressure and the systemic and the pulmonary resistance scale factors were adjusted to obtain physiological

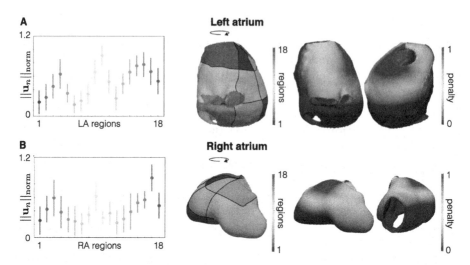

Fig. 2. Boundary conditions on the atria. Left: image-derived mean average displacement over the cohort on the eighteen atrial regions (centre) shown as mean±standard deviation. Right: penalty map for the atria derived from the motion data.

ranges for pressure in the ventricles and in the atria. LV free wall, RV free wall, septum, LA and RA wall volumes were computed from the tetrahedral mesh and set to 90.0 mL, 37.0 mL, 33.0 mL, 26.0 mL and 23.0 mL. All the other parameters were set to the default values [24].

The stress-free geometry was estimated using the image-derived motion and the patient's clinical PV relationship. First, the image-derived displacement was interpolated over time to obtain a hundred frames over one cardiac cycle, using as input the ten acquired CT frames. Then, the clinically measured PV loop was used to find diastasis, the state of the heart where pressure loads are minimum, making it the best image-based estimate of the stress-free configuration of the heart. The simulation was initialised at diastasis. The geometry was not preloaded, assuming zero pressure in all chambers at the initial time. The pressure in the systemic and the pulmonary veins was initialised at their reference pressure of 2.7 mmHg and 7.5 mmHg, respectively. The positive pressure drop across the atrial inlets triggers an inflow in the LA and the RA which partially inflates the atria. Then, atrial contraction starts and the ventricles are further filled with blood by atrial contraction until ventricular end-diastole is reached. The ventricles then contract, causing venous return in the atria and the cycle starts again.

3   Results

3.1   Baseline Simulation

Figure 3 shows results for the baseline simulation, run with $-50°/+50°$ ventricular fibres. Our model simulated LV end-diastolic pressure (EDP) systolic peak pressure (PP), maximum pressure derivative (dP/dt_{max}) and ejection fraction (EF) of 8.5 mmHg, 99.7 mmHg, 935 mmHg/s and 34%, respectively, which fall within ranges in values measured in heart failure patients [11]. Simulated RV EDP, PP, dP/dt_{max} and EF were 5.8 mmHg, 34.1 mmHg and 314 mmHg/s and 37%, respectively, which are also consistent with measurements in heart failure patients [11]. The simulated motion (Fig. 3, middle) replicates AV plane displacement (Fig. 3, bottom), with upwards motion during atrial contraction (2nd frame). Once atrial contraction ends, ventricular systole starts, the ventricles shorten and the base moves downwards (4th and 5th frame). The ventricles then relax and the heart returns to diastasis. It is worth noting that our model results in unrealistic inward motion of the RV free wall, especially during relaxation (see Sect. 4.1 for model limitations).

3.2   The Effect of Ventricular Fibre Angle on the Atria

The effect of ventricular fibres on atrial dynamics was studied by running simulations with four different fibre orientations, ranging from $-40°/+40°$ to $-70°/+70°$. All the other parameters and boundary conditions, including spring stiffness scaling, were left unchanged. Steep ventricular fibres (orange and green curves in Fig. 4) cause larger downward motion of the base (4th and 9th columns, Table 1), leading to a larger increase in atrial volume (5th and 10th columns, Table 1) and smaller atrial pressure during atrial relaxation (6th and 11th columns, Table 1), which in turn increases venous return due to a bigger pressure drop across the vein-atria inlets.

Figure 4 shows corresponding simulated LA and RA PV loops. As described in the introduction, reservoir, conduit and pumping functions of the atria result in a figure-of-eight shape in the atrial PV loops. However, the changes in atrial pressure and volume induced by different ventricular fibres cause the atrial PV loops to lose the eight shape characteristic of the atria. $-40°/+40°$ ventricular fibre angles (red curves, Fig. 4) also cause unphysiological atrial PV loops due to a lower increase of atrial volume during ventricular systole, as more circumferential fibres reduce atrial stretch due to ventricular shortening. For this patient geometry, $-50°/+50°$ fibres were the only fibre angles out of the values we considered in this study that led to realistic LA and RA PV loops, reproducing a figure-of-eight shape that was not observed with more shallow or steeper fibre orientation.

3.3   Computing Time

The four-chamber heart model for this study had 417863 nodes and 1988945 linear tetrahedra. All simulations were run on TOM2, a high performance com-

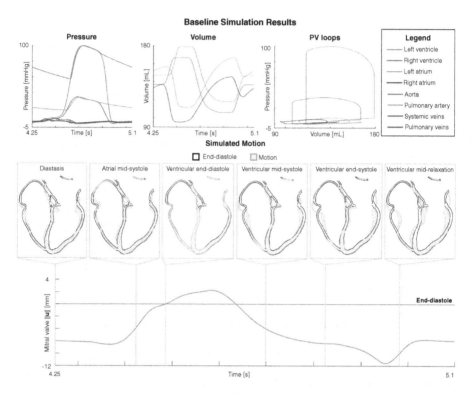

Fig. 3. Baseline simulation results. Top: pressure and volume results. Middle: simulated motion. Bottom: magnitude of displacement vector **u** of the mitral valve plane projected normal to the surface. Positive and negative displacements represent upwards and downwards motion from end-diastole (black line), respectively. All results are shown for the last simulated beat.

puting facility at King's College London, with a total number of 42 nodes with 64 AMD EPYC 7551 CPUs and 256 Gb memory each. The total time on 512 cores for one six beat simulation was 2.6 h.

4 Discussion

In this study, we presented a framework for cardiac electromechanics to simulate contraction of the atria and the ventricles, the interaction with the circulatory system and the effect of the pericardium. The effect of ventricular myofiber orientation on atrial function was studied by varying epicardial and endocardial fibre angles from $-40°/+40°$ to $-70°/+70°$. Steeper ventricular myofiber orientation increases atrial filling due to increased AV plane displacement, while shallow fibres lead to small and unphysiological AV plane downward motion.

Ventricular myofiber orientation can change in pathological conditions. Von Deuster et al. measured LV transmural fibre angle variations in controls and

Table 1. The effect of ventricular fibres on the atria. Abbreviations: ejection fraction (EF), maximum and minimum pressure (p_{max}, p_{min}), LV systolic AV plane displacement (AVPD) and maximum volume (V_{max}) during atrial relaxation.

	Chambers Pressure and Volume									
	Left heart					Right heart				
	Ventricle			Atrium		Ventricle			Atrium	
	EF %	p_{max} mmHg	AVPD mm	V_{max} mmHg	p_{min} mm	EF %	p_{max} mmHg	AVPD mm	V_{max} mL	p_{min} mmHg
-40°/+40°	32	91.1	1.0	157	-0.4	32	31.5	4.5	143	0.7
-50°/+50°	34	99.7	6.7	168	-0.7	37	34.2	9.1	145	0.1
-60°/+60°	35	103.9	11.2	176	-0.9	39	36.2	12.0	150	-0.3
-70°/+70°	33	103.2	14.0	182	-1.1	40	37.0	13.9	152	-0.7

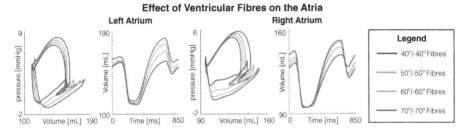

Fig. 4. The effect of ventricular fibres on the atria. Simulated LA and RA PV loops and volume transients are shown for different ventricular fibre orientations. (Color figure online)

in dilated cardiomyopathy (DCM) patients [8]. At diastole, DCM patients had greater transmural fibre angle changes compared to controls. However, during systole fibre angles got steeper compared to diastole in controls, while systolic fibre angles did not change in DCM patients. *Ex-vivo* measurements on porcine hearts showed that RV dilation following pulmonary insufficiency leads to shallower fibres in the RV [1]. Pathological ventricular remodelling can induce changes in ventricular myofiber arrangement, and can therefore worsen ventricular function. Our results suggest that such alterations can induce further changes in atrial function.

4.1 Limitations

The main limitation of this study is that the effect of model parameters other than ventricular fibre angles on atrial dynamics was not considered, despite the fact that changes in other factors such as AV timing might also affect atrial filling dynamics. The model was validated using metrics in the literature, although PV relationship for the LV were available for the patient. However, a more in-depth parameter study or fitting the model to patient-specific data would have required a computationally intense sensitivity analysis, that was not the within the scope of this study.

Only symmetric ventricular fibre angles (e.g. endo/epi $+\alpha/-\alpha$) were considered, while asymmetric ventricular fibre angles are also reported in the literature. Furthermore, the rule-based method from Bayer et al. [5] used in this study to define ventricular myofiber orientation was shown not to be ideal for the RV [19]. This together with fast RV relaxation might have partly caused unrealistic motion of the RV free wall in our model (Fig. 3).

We used a phenomenological active tension model for both the atria and the ventricles, with atrial parameters for electromechanical delay, twitch duration and peak in active tension based on a previous study. The phenomenological active tension model we used does not account for the effect of calcium transient dynamics, despite intracellular calcium being the main determinant of myocyte shortening. To represent differences in active tension development at the cellular level, a more complex active tension model accounting for cross-bridges kinetics and electromechanical coupling through the calcium transient would have been needed.

5   Conclusions

The coupled four-chamber heart framework we presented is able to account for complex interaction mechanisms between the atria and the ventricles. The pericardium boundary conditions allow the model to capture mechanical interaction of the chambers, while the coupling with a closed-loop circulatory system provides physiological preload and afterload for all four chambers. We used this framework to study the consequences of altered ventricular fibres on atrial dynamics. Although significant challenges remain in model parametrisation and personalisation for such complex models, our model represents a useful tool to study bilateral interaction mechanisms between the cardiac chambers and the circulatory system.

Acknowledgement. This study received support from the UK Engineering and Physical Sciences Research Council (EP/M012492/1, NS/A000049/1, EP/L015226/1, and EP/P01268X/1), the Wellcome EPSRC Centre for Medical Engineering (NS/A000049/1 and WT 203148/Z/16/Z), the British Heart Foundation (PG/15/91/31812 and PG/13/37/30280), the National Institute of Health (NIH R01-HL152256), the European Research Council (ERC PREDICT-HF 864055) and King's Health Partners London National Institute for Health Research (NIHR) Biomedical Research Centre. Miss Marina Strocchi was supported by an unrestricted Abbott educational grant through the Centre for Doctoral Training in Medical Imaging at King's College London. Dr Neic is employed by NumeriCor GmbH, Graz, Austria. All other authors have reported that they have no conflicts relevant to the contents of this paper to disclose.

References

1. Agger, P., et al.: Changes in overall ventricular myocardial architecture in the setting of a porcine animal model of right ventricular dilation. J. Cardiovas. Magn. Reson. **19**(1), 1–16 (2017). https://doi.org/10.1186/s12968-017-0404-0

2. Augustin, C.M., et al.: The impact of wall thickness and curvature on wall stress in patient-specific electromechanical models of the left atrium. Biomech. Model. Mechanobiology **19**(3), 1015–1034 (2019). https://doi.org/10.1007/s10237-019-01268-5

3. Augustin, C.M., Gsell, M.A.F., Karabelas, E., et al.: Validation of a 3D–0D closed-loop model of the heart and circulation - modeling the experimental assessment of diastolic and systolic ventricular properties. arXiv e-prints p. arXiv:2009.08802 (2020)

4. Augustin, C.M., Neic, A., Liebmann, M., et al.: Anatomically accurate high resolution modeling of human whole heart electromechanics: a strongly scalable algebraic multigrid solver method for nonlinear deformation. J. Comput. Phys. **305**, 622–646 (2016)

5. Bayer, J.D., Blake, R.C., Plank, G., et al.: A novel rule-based algorithm for assigning myocardial fiber orientation to computational heart models. Ann. Biomed. Eng. **40**(10), 2243–2254 (2012)

6. Bayer, J., Prassl, A.J., Pashaei, A., et al.: Universal ventricular coordinates: A generic framework for describing position within the heart and transferring data. Medical image analysis **45**, 83–93 (2018)

7. Bucelli, M., Salvador, M., Quarteroni, A., et al.: Multipatch isogeometric analysis for electrophysiology: simulation in a human heart. Comput. Methods Appl. Mech. Eng. **376**, 113666 (2021)

8. von Deuster, C., et al.: Studying dynamic myofiber aggregate reorientation in dilated cardiomyopathy using in vivo magnetic resonance diffusion tensor imaging. Circ. Cardiovasc. Imaging **9**(10), e005018 (2016)

9. Fritz, T., Wieners, C., Seemann, G., Steen, H., Dössel, O.: Simulation of the contraction of the ventricles in a human heart model including atria and pericardium. Biomech. Model. Mechanobiology **13**(3), 627–641 (2013). https://doi.org/10.1007/s10237-013-0523-y

10. Guccione, J.M., McCulloch, A.D., Waldman, L.K., et al.: Passive material properties of intact ventricular myocardium determined from a cylindrical model. J. Biomech. Eng. **113**(1), 42–55 (1991)

11. Hyde, E.R., Behar, J.M., Crozier, A., et al.: Improvement of right ventricular hemodynamics with left ventricular endocardial pacing during cardiac resynchronization therapy. Pacing Clin. Electrophysiol. **39**(6), 531–541 (2016)

12. Jafari, A., Pszczolkowski, E., Krishnamurthy, A.: A framework for biomechanics simulations using four-chamber cardiac models. J. Biomech. **91**, 92–101 (2019)

13. Kariya, T., et al.: Personalized perioperative multi-scale, multi-physics heart simulation of double outlet right ventricle. Ann. of Biomed. Eng. **48**(6), 1740–1750 (2020). https://doi.org/10.1007/s10439-020-02488-y

14. Land, S., Niederer, S.A.: Influence of atrial contraction dynamics on cardiac function. Int. J. Numer. Method Biomed. Eng. **34**(3), e2931 (2018)

15. Nasopoulou, A., Nordsletten, D.A., Niederer, S.A., Lamata, P.: Feasibility of the estimation of myocardial stiffness with reduced 2D deformation data. In: Pop, M., Wright, G.A. (eds.) FIMH 2017. LNCS, vol. 10263, pp. 357–368. Springer, Cham (2017). https://doi.org/10.1007/978-3-319-59448-4_34

16. Neic, A., Campos, F.O., Prassl, A.J., et al.: Efficient computation of electrograms and ECGs in human whole heart simulations using a reaction-eikonal model. J. Comput. Phys. **346**, 191–211 (2017)

17. Niederer, S.A., et al.: Length-dependent tension in the failing heart and the efficacy of cardiac resynchronization therapy. Cardiovasc. Res. **89**(2), 336–343 (2010)

18. Pfaller, M.R., et al.: The importance of the pericardium for cardiac biomechanics: from physiology to computational modeling. Biomech. Model. Mechanobiology **18**(2), 503–529 (2018). https://doi.org/10.1007/s10237-018-1098-4

19. Piersanti, R., et al.: Modeling cardiac muscle fibers in ventricular and atrial electrophysiology simulations. Comput. Methods Appl. Mech. Eng. **373**, 113468 (2021)

20. Roney, C.H., et al.: Variability in pulmonary vein electrophysiology and fibrosis determines arrhythmia susceptibility and dynamics. PLoS Comput. Biol. **14**(5), e1006166–e1006166 (2018)

21. Strocchi, M., Augustin, C.M., Gsell, M.A., et al.: A publicly available virtual cohort of four-chamber heart meshes for cardiac electro-mechanics simulations. PLoS One **15**(6), e0235145 (2020)

22. Strocchi, M., Gsell, M.A., Augustin, C.M., et al.: Simulating ventricular systolic motion in a four-chamber heart model with spatially varying Robin boundary conditions to model the effect of the pericardium. J. Biomech. **101**, 109645 (2020)

23. Vigmond, E.J., Aguel, F., Trayanova, N.A.: Computational techniques for solving the bidomain equations in three dimensions. IEEE Trans. Biomed. Eng. **49**(11), 1260–1269 (2002)

24. Walmsley, J., Arts, T., Derval, N., et al.: Fast simulation of mechanical heterogeneity in the electrically asynchronous heart using the multipatch module. PLoS Comput. Biol. **11**(7), e1004284 (2015)

Intra-cardiac Signatures of Atrial Arrhythmias Identified by Machine Learning and Traditional Features

Miguel Rodrigo[1,2](✉) 🔘, Benjamin Pagano[1], Sumiran Takur[1], Alejandro Liberos[2] 🔘, Rafael Sebastián[2] 🔘, and Sanjiv M. Narayan[1] 🔘

[1] Cardiovascular Department, Stanford Universtiy, Stanford, CA 94305, USA
mrodrigo@stanford.edu
[2] Universitat de València, 46100 Burjassot, VA, Spain

Abstract. Intracardiac devices separate atrial arrhythmias (AA) from sinus rhythm (SR) using electrogram (EGM) features such as rate, that are imperfect. We hypothesized that machine learning could improve this classification.

In 71 persistent AF patients (50 male, 65 ± 11 years) we recorded unipolar and bipolar intracardiac EGMs for 1 min prior to ablation, providing 50,190 unipolar and 44,490 bipolar non-overlapping 4 s segments. We developed custom deep learning models to detect SR or AA, with 10-fold cross-validation, compared to classical analyses of cycle length (CL), Dominant Frequency (DF) and autocorrelation.

Classical analyses of single features were modestly effective with AUC ranging from 0.91 (DF) to 0.70 for other rate metrics. Performance increased by combining features linearly (AUC 0.991/0.987 for unipolar/bipolar), by Bagged Trees (0.995/0.991) or K-Nearest Neighbors (0.985/0.991). Convolutional deep learning of raw EGMs with no feature engineering provided improved AUC of 0.998/0.995 to separate AA from SR.

Deep learning of raw EGMs outperforms classic rule-based classifiers of SR or AA. This could improve device diagnosis, and the logic developed by deep learning could shed novel insights into EGM analyses beyond current classification based on EGM features and rules.

Keywords: Atrial arrhythmias · Machine learning · Signal features

1 Introduction

Identifying atrial arrhythmias from atrial electrograms is critical to diagnosis and choice of anticoagulation or anti-arrhythmic therapy, yet is suboptimal [1]. Despite the plethora of increasingly available atrial signal data from wearables, implanted recorders, pacemakers and defibrillators, devices often still identify atrial fibrillation (AF) or flutter (AFL) by irregular ventricular rates or rapid atrial rates (atrial high-rate events, AHRE). This has suboptimal accuracy [2] in classifying sinus tachycardia or premature atrial contractions as AF, and may lead to diagnostic and therapeutic errors [3, 4]. We hypothesized that machine learning (ML) could identify atrial electrogram signatures to identify

© Springer Nature Switzerland AG 2021
D. B. Ennis et al. (Eds.): FIMH 2021, LNCS 12738, pp. 671–678, 2021.
https://doi.org/10.1007/978-3-030-78710-3_64

atrial arrhythmias, improving accuracy over traditional approaches. ML is a provocative, rapidly developing branch of computer science, which can reveal hidden data structures in complex data [5, 6]. ML has been applied to ECG recordings to diagnose arrhythmias [7–9], yet has rarely been applied to intracardiac AF data.

We set out to develop a deep ML approach to classify intracardiac recordings into either Atrial Arrhythmias (AA) or Sinus Rhythm (SR), and comprehensively assess whether this approach provides higher accuracy than traditional statistical analyses. One limitation of ML is that it is often considered a 'black box' that cannot be probed to explain its decisions [5, 6], which reduces confidence in its clinical use. We thus further hypothesized that the classification of atrial arrhythmias by ML could provide insights into how electrogram signatures, such as waveform shape, differ between rhythms. A secondary goal was to prove how deep ML separates atrial signals between rhythms.

2   Methods

We recruited 86 patients from the COMPARE registry (NCT02997254) presenting with atrial arrhythmias who had a successful ablation to sinus rhythm. Cases were performed after written informed consent under protocols approved by the Human Research Protection Program at each center. Intracardiac electrograms were reviewed by a panel of 3 experts and classified as Sinus Rhythm (SR), Atrial Fibrillation (AF) or as organized atrial tachycardia or flutter. We included N=71 patients in whom atrial arrhythmias and sinus rhythm were both recorded, for a total of 142 episodes. Patients with no Atrial Arrhythmia or Sinus Rhythm recordings were discarded (N=15). Atrial arrhythmias comprised atrial flutter (N=28) or fibrillation (N=43).

2.1   Electrogram Collection and Export

An electrophysiology study was performed in each patient after discontinuing antiarrhythmic medications for 5 half-lives. Catheters were introduced transvenously to the right atrium, coronary sinus and transseptally to the left atrium. A 64-pole basket catheter (FIRMap, Abbott Electrophysiology, Menlo park, CA) was advanced to first map AF in the right, and following the left atria. We exported unipolar electrograms from our electrophysiological recorder (Prucka, GE Marquette, Milwaukee, WI), filtered at 0.05-500 Hz for ML and statistical classification. We analyzed unipolar electrograms in durations of 4000 ms, which were resampled to 400 samples per second to reduce signal dimensionality. Bipolar EGM signals were calculated as the difference of neighboring electrodes across the same basket spline.

Balanced datasets of 50,190 unipolar and 44,490 bipolar EGM signals (4-seconds length) were segmented into 10-fold cross-validation sets comprising 80% of patients (N=57) for training and 20% (N=14) for validation.

2.2   Classical Machine Learning Methods

Raw electrograms have a high dimensionality (voltage at 1600 timepoints) that reduces the power of traditional statistical comparisons. We thus used standard approaches for dimensionality reduction using a variety of featurization approaches. This included 49 intuitive features such as electrogram rate, regularity and morphology, amplitude and derivative statistics, number of local maxima, autocorrelation metrics and dominant frequency measures (Fig. 1A). Moreover, we also used other 750 signal analysis features extracted from a standard signal analysis library (*tsfresh*, Python). Electrograms were classified as sinus rhythm or AA using these features, applied first individually then in combination as trainable classifiers. Three feature-based classifiers were applied:

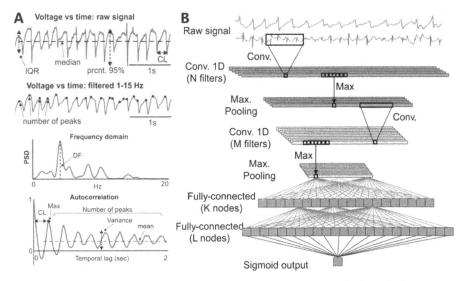

Fig. 1. A. Featurization of Electrograms by Intuitive Features. **B.** CNN design.

- Linear combination. This was constructed by combining all parameters into one linear function ($y = a_0 + a_1 f_1 + a_2 f_2 + \ldots + a_N f_N$), a_i representing the linear parameters to be trained and f_i the features used. The linear model was trained using a logistic binomial regression using function *fitglm* from Matlab® (Mathworks, Natick, MA).
- Bagged Trees (Forest). In this approach, the averaged output of N decision trees is provided, based on binary thresholding of individual parameters. An ensemble of 200 decision trees were trained using function *TreeBagger* from Matlab®.
- K-Nearest Neighbor (KNN). The predicted output is calculated through the K nearest neighbors in the domain of size N, where N is the number of features. Neighbors are points/features combinations used for training. K=20 neighbors were considered in our analysis using function *fitcknn* from Matlab®.

2.3 Deep Learning

We developed a CNN in Python using *Keras* library able to perform the classification using the EGM raw signal (1600 timepoints). The CNN architecture comprises of 2 one-dimensional convolutional layer and 2 fully-connected dense layers, an input layer, an output layer, as well as multiple max-pooling layers. The 2 convolutional layers and 2 dense layers included 256 filters each and the convolutional lengths were 32 and 8 for the convolutional layers. The dropout values for the 2 convolutional and 2 dense layers were 0.3, 0.2, 0.1 and 0.0, respectively. Previous experiments explored the best CNN configuration between using 1–3 convolutional and dense layers and 16-1024 filters.

The architecture was applied to 10 cross-validation cohorts. This design is similar to CNN models we have developed for prior arrhythmia applications [10, 11]. Figure 1B shows each layer and its characteristics.

3   Results

3.1   Classification of Arrhythmias by Single Features

We used intuitive and clinically validated features of the activation rate, that is faster for AF and AFL than sinus rhythm. These metrics (AUC, Sens. Spec.) were evaluated across the whole database and not using the cross-validation scheme. Rate was assessed by Dominant Frequency analysis, cycle length derived from autocorrelation (CL, = 60000/rate in milliseconds) and DF (in Hz). Figure 2 shows that rate cutpoints only modestly separated atrial arrhythmias from sinus rhythm. Dominant Frequency provided AUC of 0.91 and a sensitivity of 0.93 and specificity of 0.86 ($p < 0.0001$ between SR and AA groups). Notably, CL measures were substantially less effective. The number of peaks in unit time was significantly higher for arrhythmia than sinus rhythm, but provided an AUC of 0.70. Autocorrelation-derived CL was significantly longer for sinus rhythm than arrhythmias ($p < 0.001$), but with an AUC of only 0.60.

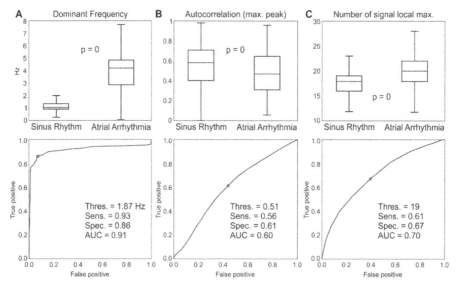

Fig. 2. Performance of rate cutpoints. A. Dominant Frequency. **B.** Maximal value of autocorrelation. **C.** Number of local maxima.

3.2 Statistical Classification Using Machine Learning

We progressed to more sophisticated featurization. Key features that emerged from training of sinus rhythm or atrial arrhythmias were ranked by their individual AUC. The most critical features were the number of signal peaks after 30–60 Hz filtering, and metrics of signal amplitude (25 percentile, 50 and 75 percentile), supporting the importance of atrial beat rate and signal amplitudes in separating AAs from SR. We incorporated features in the classifiers sequentially, in order, from highest AUC (c-statistic) to those with lower AUCs for predicting atrial arrhythmias, until validation classification plateaued. The optimal number of features were used for classification which, for unipolar electrograms, were 32, 34 and 12 features for linear, Bagged Trees and K-Nearest Neighbor classifiers, respectively. For bipolar electrograms, 63, 28 and 23 features were selected for linear, Bagged Trees and K-Nearest Neighbor classifiers, respectively.

Figure 3 compares the performance of 3 feature-based classifiers of electrograms to separate atrial arrhythmia from sinus rhythm, for the average of the 10-cross validation sets. For unipolar electrograms (Fig. 3A), the Bagged-Trees classifier (34 features) provided an accuracy of 0.963 ± 0.0267, a sensitivity of 0.967 ± 0.0247, a specificity of 0.968 ± 0.019 and an AUC of 0.995 ± 0.006. Other statistical classifiers were less accurate. For Bipolar EGM classification (Fig. 3B), there was a similar performance of feature-based classifiers, with Bagged-Trees (28 features) providing the best outcomes with an accuracy of 0.955 ± 0.0246, a sensitivity of 0.959 ± 0.0197, a specificity of 0.957 ± 0.0291 and an of AUC 0.991 ± 0.009. Figures 3C, D summarize the AUCs of feature-based classifiers (Linear, Bagged Trees and KNN) by pooling all validation classifications from the different cross-validation set in a single curve.

3.3 Deep Machine Learning Classification of Raw Electrograms

Figure 3 shows that the CNN classifier improved classification accuracy, averaged across cross-validation sets, compared to classical ML-based classifiers. For unipolar electrograms, Fig. 3A shows accuracy of $0.987 \pm 0.01\%$, a sensitivity of 0.988 ± 0.009, a specificity of 0.986 ± 0.011 and an AUC of 0.998 ± 0.002, higher than feature-based classifiers. For bipolar electrograms (Fig. 3B), CNN classification showed similar performance (0.972 ± 0.025 accuracy, 0.975 ± 0.021 sensitivity, 0.971 ± 0.027 specificity and 0.995 ± 0.007 AUC), again better than all three ML-based classifiers. Figure 3C, D shows ROC curves across the 10-cross validation cohorts for both unipolar and bipolar signals. Deep learning was able to use raw electrograms to classify atrial arrhythmias from sinus rhythm without requiring user defined features.

3.4 Probing CNN to Identify Critical Signatures of Atrial Arrhythmias

We used simulated electrograms in which we varied a controlled feature as a novel approach to probe CNN to explain their classification. Electrogram sequences were constructed from actual EGMs from the validation cohorts, varying EGM rate (from 200 ms to 1200 ms, Fig. 4A).

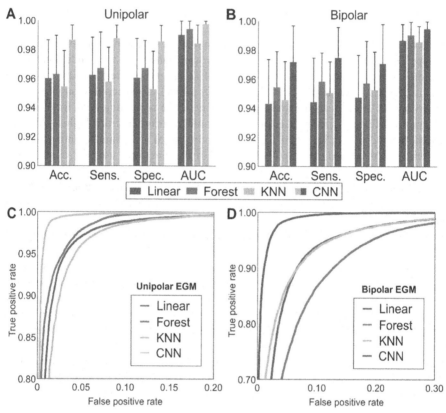

Fig. 3. Classifying SR versus AA with Classical Machine Learning. Accuracy, sensitivity, specificity and Area Under the Curve (AUC) across cross-validation sets for Unipolar (**A**) and Bipolar (**B**) EGM classifiers. Receiver operating characteristic curves for Unipolar (**C**) and Bipolar EGM classifiers (**D**).

Bagged trees and CNN architectures already trained from previous sections were used to classify these synthetic signals (Fig. 4B, C). Applying the best statistical classifier (Bagged Trees) to reconstructed simulations of identically shaped electrograms, shortening the CL was more likely to be classified as atrial arrhythmias. In signals #1 and #3, the signal at CL=260 ms was classified as AA. However, signals at CL=1100 ms were classified as either SR (#2) or AA (#4). Conversely, CNNs better classified these signals, with >98% of signals classified as sinus rhythm until CL fell <400 ms (>150 beats/min), with all classified as AA for CL < 400 ms as SR, showing the accurate EGM classification based on rate by CNN.

4 Discussion

Classification of intracardiac electrograms greatly depends on the proposed classification architecture. The extraction of classic features of the signal for the differentiation

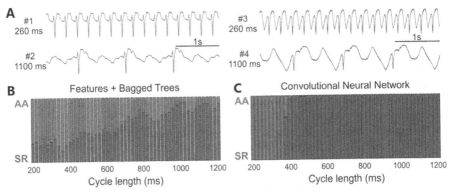

Fig. 4. Classification of Synthetic EGM Signals. Examples of synthetic EGM signals (**A**) and their classification by Bagged Trees (**B**) and CNN (**C**) based on their rate.

between sinus rhythm and atrial arrhythmias, such as cycle length or dominant frequency, showed reasonable predictive value with AUC values ranging from 0.70 to 0.90. Using comprehensive feature extraction, statistical classifiers improved AUC. However, CNN on raw signals with no manual feature selection provided the optimal AUC of 0.998 and 0.995 for unipolar and bipolar EGM, respectively.

While feature engineering has been classically used to diagnose atrial arrhythmias, it greatly depends on algorithms for parameter extraction. In this article we tested a total of 799 features in temporal and frequency domains, as well as classical indexes of EGMs. While this approach was effective, it is likely that this may differ for each classification problem, such as sinus rhythm versus atrial arrhythmias or AF versus atrial flutter.

The use of non-feature-based deep learning classifiers, such as convolutional networks, provided better results than any feature-based classifier. The availability of a greater number of free parameters to be trained in whose values the prediction is made, together with the algorithms that allow the training of these networks, enable these architectures to provide better classification. The existence of a sufficiently large database allows these trained architectures to cover a larger range of characteristics of the two groups of arrhythmias.

Finally, we investigated how these deep learning architectures operated, by introducing synthetic signals that varied in only one parameter at a time. In this way, we observed that deep learning architectures classified more accurately, using more than simply rate, compared to traditional feature-based classifiers.

4.1 Conclusions

Deep learning of raw EGMs improves the classification outcomes of classic rule-based classifiers in the sinus rhythms vs. atrial arrhythmia classification. These deep learning architectures could improve device diagnosis using raw signal classification, as well as other signal classification problems without the need to perform an expert feature identification. The development and understating of the deep learning classification architectures could shed novel insights into EGM analyses beyond current classification based on EGM features and rules.

References

1. Hindricks, G., Pokushalov, E., Urban, L., et al.: Performance of a new leadless implantable cardiac monitor in detecting and quantifying atrial fibrillation: results of the XPECT trial. Circ. Arrhythm Electrophysiol. **3**(2), 141–147 (2010)
2. Kaufman, E.S., Israel, C.W., Nair, G.M., et al.: Positive predictive value of device-detected atrial high-rate episodes at different rates and durations: an analysis from ASSERT. Heart Rhythm **9**(8), 1241–1246 (2012)
3. Bertaglia, E., Blank, B., Blomstrom-Lundqvist, C., et al.: Atrial high-rate episodes: prevalence, stroke risk, implications for management, and clinical gaps in evidence. Europace **21**(10), 1459–1467 (2019)
4. Tomson, T.T., Passman, R.: Management of device-detected atrial high-rate episodes. Card. Electrophysiol. Clin. **7**(3), 515–525 (2015)
5. Krittanawong, C., Johnson, K.W., Rosenson, R.S., et al.: Deep learning for cardiovascular medicine: a practical primer. Eur. Heart J. **40**(25), 2058–2073 (2019)
6. Topol, E.J.: High-performance medicine: the convergence of human and artificial intelligence. Nat. Med. **25**(1), 4456 (2019)
7. Bumgarner, J.M., Lambert, C.T., Hussein, A.A., et al.: Smartwatch algorithm for automated detection of atrial fibrillation. J. Am. Coll. Cardiol. **71**(21), 2381–2388 (2018)
8. Tison, G.H., Sanchez, J.M., Ballinger, B., et al.: Passive detection of atrial fibrillation using a commercially available smartwatch. JAMA Cardiol. **3**(5), 409–416 (2018)
9. Rajpurkar, P.H.A., Haghpanahi, M., Bourn, C., Ng, A.: Cardiologist-Level Arrhythmia Detection with Convolutional Neural Networks (2017)
10. Alhusseini, M.I., Abuzaid, F., Rogers, A.J., et al.: Machine learning to classify intracardiac electrical patterns during atrial fibrillation: machine learning of atrial fibrillation. Circ. Arrhythm Electrophysiol. **13**(8), (2020)
11. Rogers, A.J., Selvalingam. A., Alhusseini. M.I., et al: Machine learned cellular phenotypes predict outcome in ischemic cardiomyopathy. Circ. Res. (2020)

Computational Modelling of the Role of Atrial Fibrillation on Cerebral Blood Perfusion

Timothy J. Hunter[1,2], Jermiah J. Joseph[1,2], Udunna Anazodo[1,2],
Sanjay R. Kharche[1,2(✉)], Christopher W. McIntyre[1,2], and Daniel Goldman[1]

[1] Department of Medical Biophysics, University of Western Ontario, London,
ON N6A 3K7, Canada
Sanjay.Kharche@lhsc.on.ca, dgoldma2@uwo.ca
[2] Lawson Health Research Centre, London, ON N6A 5W9, Canada

Abstract. Atrial fibrillation is a prevalent cardiac arrhythmia, and may reduce cerebral blood perfusion augmenting the risk of dementia. It is thought that geometric variations in the cerebral arterial structure called the Circle of Willis play an important role influencing cerebral perfusion. The objective of this work is to use computational modelling to investigate the role of variations in cerebral vascular structure on cerebral blood flow dynamics during atrial fibrillation.

A computational blood flow model was developed by coupling whole-body and detailed cerebral circulation models, modified to represent the most common variations of the Circle of Willis. Cerebral blood flow dynamics were simulated in common Circle of Willis variants, with imposed atrial fibrillation conditions. Perfusion and its heterogeneity were quantified using segment-wise hypoperfusion events and mean perfusion at terminals.

It was found that cerebral perfusion and the rate of hypoperfusion events strongly depend on Circle of Willis geometry as well as atrial fibrillation induced stochastic heart rates. The missing ACA1 variant had a 25% decrease in hypoperfusion events compared to normal, while the missing PCA1 and PCoA variant had a 550% increase. A similar trend was observed in flow heterogeneity. The hypoperfusion events were specific to particular arteries within each variant. Our results, based on biophysical principles, suggest that cerebral vascular geometry plays an important role influencing cerebral hemodynamics during atrial fibrillation. Additionally, our findings suggest potential clinical assessment sites. Further work will be conducted using spatially resolved 1D modelling.

Keywords: 0D model · Atrial fibrillation · Cerebral blood flow

1 Introduction

Atrial fibrillation (AF) is known to reduce cerebral perfusion [1]. Ongoing imaging research strongly suggests that a disrupted cerebral blood flow promotes debilitating early dementia [2]. The effects of AF on cerebral perfusion may be modulated by cerebral vascular geometry, and specifically by common congenital Circle of Willis (CoW) variants [3]. The function of a complete CoW is to ensure consistent distribution of blood

© Springer Nature Switzerland AG 2021
D. B. Ennis et al. (Eds.): FIMH 2021, LNCS 12738, pp. 679–686, 2021.
https://doi.org/10.1007/978-3-030-78710-3_65

flow to all regions of the brain. In cases with missing segments in the CoW, regions of the brain may be more susceptible to harmful altered hemodynamics.

Multi-scale hemodynamic modelling has been used to study cerebral circulation and gain insights into patient-specific hemodynamics [4]. While 3D modelling provides realistic and accurate patient-specific assessment, current methods remain computationally resource intensive. In contrast, lumped parameter (0D) models are known to provide clinically relevant information using significantly less time and computational resources [5, 6]. We have previously used 0D models to gain insights into the causes of pediatric hypertension [7] and investigate therapeutic hypothermia [8]. 0D models have also been used to understand the interplay between AF and cerebral hemodynamics [9]. However, the role of cerebral vascular structural variants, i.e. CoW variants, in AF-cerebral perfusion relationship remains underexplored. As the CoW is known to play an important role in the distribution of blood flow to the brain, common variants should be considered while studying the interplay between AF and cerebral hemodynamics. In this study, a composite 0D model of human circulation with detailed cerebral vasculature was developed to discover the effects of AF on cerebral perfusion in cases with common CoW variants.

2   Methods

An established 0D lumped parameter whole body circulation model [10, 11] was coupled to a baroreflex control model [12] in a recent study, and was further developed to incorporate detailed cerebral vasculature [5, 6] descriptions (Fig. 1).

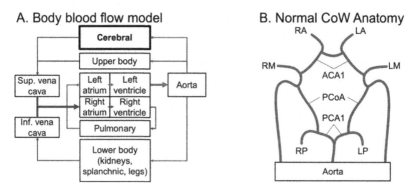

Fig. 1. Components of the hemodynamic model. A: Caricature of the whole-body blood flow model. The cerebral model (black box, top) is expanded in the right (B) panel. B: A representative cerebral arterial architecture consisting of Circle of Willis. RA: Right anterior artery; LA: left anterior artery; RM: right middle artery; LM: left middle artery; RP: right posterior artery; LP: left posterior artery; ACA1: pre-communicating anterior cerebral artery; PCoA: posterior communicating artery; and PCA1: pre-communicating posterior cerebral artery

The model simulates blood flow using the Windkessel approach, with resistances (blood viscosity, blood vessel length and diameters) and capacitances (blood vessel elasticity). The heart was modelled as a four-chamber system, with each chamber providing a pumping function based on respective time-varying elastances [9, 10]. Blood flow from the body to the cerebral model was allowed by connecting the inlets of the cerebral circulation, the basilar and internal carotid arteries, to the aortic pressure node as shown in Fig. 1, B. Similarly, the cerebral outlet vein was connected to the low-pressure superior vena cava via a pressure boundary condition. (Fig. 1, A). The baroreflex control mechanism was modelled as a system of first-order low-pass filters that used aortic pressure to modulate all systemic resistances, capacitances, and maximum ventricular elastances [11].

This work considers the three common variants of the Circle of Willis found in the cerebral vasculature. The first variant, termed normal, is represented in Fig. 1, B. The second variant has ACA1 artery missing. The third has PCoA and contralateral PCA1 (PCoA + P1) missing [12]. CoW variants were modelled by blocking flow in relevant arterial segments in the normal model (Fig. 1, B).

Simulations were run under AF and control conditions. Control was defined as normal sinus rhythm (NSR) with stochastic RR intervals sampled from a gaussian distribution [13]. AF was modelled by assigning stochastic RR intervals sampled from an exponentially modified gaussian distribution around a mean heart rate [13–15], modifying ventricular elastances (contractility) [14, 15], and assigning nil atrial contractility [9, 14, 15]. Each simulation was performed at three different intrinsic heart rates (50, 70 and 90 bpm) in accordance with clinical practice [16]. The probability distribution functions underlying the RR intervals and the representative RR interval time series are illustrated in Fig. 2.

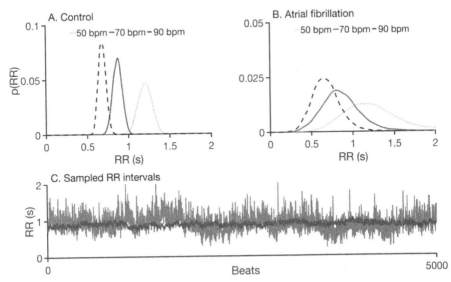

Fig. 2. Stochastic RR interval assignment. Top row: probability distribution functions for sampled RR intervals in NSR (A), and AF (B) at shown heart rates. C: Sampled RR intervals at 70 bpm over the span of 5000 beats under NSR (blue line) and AF (red line) conditions. (Color figure online)

Each simulation was performed under normal sinus rhythm (NSR) and AF conditions. In each simulation, the number of hypoperfusion events were recorded to represent cerebral perfusion deficit. Derived measurements were the number of hypoperfusion events in each vascular bed over the 5000 beats of the simulation. A hypoperfusion event in any vascular bed was defined as a heartbeat in which the mean blood flow through the vascular bed falls below the 5th percentile of blood flow in the corresponding NSR experiment.

The model used in this study has 83 coupled stiff ordinary differential equations (ODEs). An in-house ODE solver [17] was deployed to generate stable and accurate numerical solutions. The maximum integration timestep in the adaptive and implicit solver was 0.001 s, which was found to provide the same solution when timestep was halved. The solutions were obtained using a relative tolerance of 10^{-6}, with an accuracy of $O(dt^6)$. Each instance of the simulation could be processed by available computing resources running Red Hat Linux within 2 min.

3 Results

It is observed that large variations in blood pressure are propagated through the large arterial circulation and have a high impact on small vessels downstream. This effect is demonstrated in Fig. 3 where a drop in blood pressure due to a long RR interval is associated with two consecutive hypoperfusion events in the left middle (LM) artery.

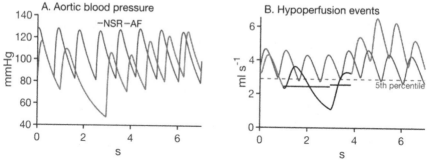

Fig. 3. Hemodynamic outputs of a simulation of AF (red) and NSR (blue) at 70 bpm in the normal CoW. A: Aortic blood pressures. B: Blood flow through the left middle distal artery with hypoperfusion events shown in black. (Color figure online)

The heart rate and vascular geometry dependence of hypoperfusion events is illustrated in Fig. 4. Three heart rates representing bradycardia (50 bpm), baseline (70 bpm), and tachycardia (90 bpm) were used. The following percent changes in hypoperfusion event frequency are given relative to the respective baseline value. Under bradycardia, 33% fewer cerebral hypoperfusion events were observed (Fig. 4, A). In contrast, under tachycardia, 362% more hypoperfusion events were observed (Fig. 4, A). In the missing ACA1 variant, bradycardia reduced hypoperfusion events by 47%, but tachycardia increased the same by 260% (Fig. 4, B). In case PCoA as well as PCA1 are missing, bradycardia reduced hypoperfusion events by 82%, but tachycardia increased the same by 192% (Fig. 4, C). Additionally, at the baseline heart rate, it was found that the missing ACA1 variant had 25% less hypoperfusion events than the normal variant, while the missing PCoA and PCA1 variant had an increase of 550%.

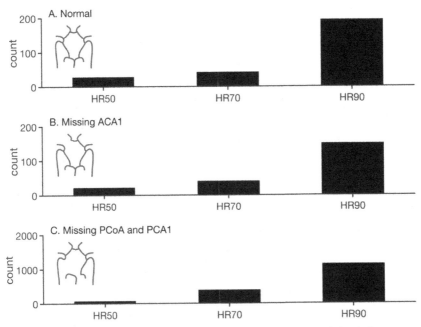

Fig. 4. Absolute frequencies of hypoperfusion events in the distal cerebral circulations at varying heart rates under AF conditions. Histogram stacks are coded according to the region they occur in: LM, RM, LA, RA, LP, RP from bottom to top. A: normal CoW. B: ACA1 variant. C: PCoA and PCA1 variant.

Figure 5 illustrates alterations in cerebral blood flow heterogeneity between the three variants. Under AF conditions (Fig. 5, A), terminals of the left middle, left anterior, and left posterior arteries experience a balanced outflow, indicating virtually uniform cerebral perfusion. A similar behavior was observed when the variant with ACA1 artery was missing (Fig. 5, B). However, when PCoA along with PCA1 were missing, the blood flow to terminals was highly heterogeneous (Fig. 5, C). Additionally in the case of the missing PCoA and PCA1 variant, blood flow to the left anterior region oscillates out of phase with other regions. Moreover, the transient effect of the abnormal heartbeat at 2 s is notably different between the left anterior and other regions, with flow increasing during the heartbeat but decreasing for a short time (2 s) immediately after, in contrast to the other regions.

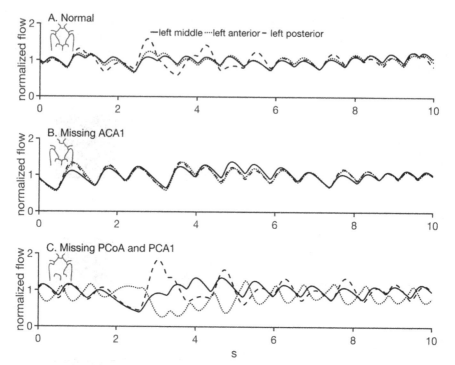

Fig. 5. Perfusion to various parts of the brain, represented by outflow at three distinct vessel terminals. In all panels solid line represents flow at the terminal of left middle artery, dotted lines represent flow at the terminal of left anterior artery, and dashed line represents flow at the terminal of left posterior artery. A: normal CoW. B: ACA1 variant. C: PCoA and PCA1 variant

4   Conclusions and Discussion

Variations from regular blood pressure in large arteries due to AF were shown to be associated with large changes in blood flow in the distal circulation of the brain. These changes lead to occurrences of critical hypoperfusion events in the brain, which over time, may lead to damage of the brain tissue. It is clear from the present investigation of different CoW variants that individual vascular structure plays a role in the impact of AF on the cerebral circulation. In the case of a missing ACA1 segment, the brain was found to be less prone to hypoperfusion events than normal, while a missing PCoA and contralateral PCA1 led to approximately 5 times the frequency of events. It was also shown that certain structures could lead to increased heterogeneity in cerebral blood flow, with increased blood flow in some regions, and decreased in others. The results demonstrate the need for increased study of cerebrovascular disease with respect to structure. The 0D model has been informative in that it has revealed these phenomena in the general case, where current in vivo methods could not.

While current treatment methods for AF, namely heart rate control and atrial ablation, are assessed based on treatment mortality, there is growing evidence that other factors, such as impact on cognitive function, should be considered [1]. As research continues in this field the results of the present study suggest that cerebrovascular structure should be considered in treatment planning to ensure better clinical outcomes.

The present work represents general cases of congenitally missing arteries in order to demonstrate the potential implications of varied vascular geometries in AF. Further investigation is needed to accurately determine the implications for populations affected by AF, with widely varying CoW structures. Previously used techniques for representing populations using 0D models will be employed to elucidate the impacts of varied cerebrovascular structures [8].

In a clinical setting it is critical for computational models to be applicable on a patient-specific basis. Methods for the incorporation of imaging data into 0D blood flow models are currently in development and will be used to further assess the impact of variant vascular structures using patient specific data [5, 18]. Such methods will also be effective in the clinic, opening up the possibility of patient-specific assessments for persistent AF patients. Further investigation will be done using spatially resolved 1D modelling to investigate the impacts of these phenomena on the blood vessels as well as the surrounding tissue in greater detail.

Acknowledgements. This work was supported by Canada Canarie Inc. (RS-111), Canada Heart and Stroke Foundation grant (G-20-0028717), Canada NSERC operational grant (R4081A03), and NSERC graduate scholarship. We thank Compute Canada for high performance computing resources. We thank Dr. Kapiraj Chandrabalan for editing support.

References

1. Gardarsdottir, M., et al.: Atrial fibrillation is associated with decreased total cerebral blood flow and brain perfusion. Europace **20**, 1252–1258 (2018)
2. Anazodo, U.C., Shoemaker, J.K., Suskin, N., St Lawrence, K.S.: An investigation of changes in regional gray matter volume in cardiovascular disease patients, pre and post cardiovascular rehabilitation. NeuroImage Clin. **3**, 388–395 (2013)
3. Steinman, D.A., Poepping, T.L., Tambasco, M., Rankin, R.N., Holdsworth, D.W.: Flow patterns at the stenosed carotid bifurcation: effect of concentric versus eccentric stenosis. Ann. Biomed. Eng. **28**, 415–423 (2000)
4. Antiga, L., Piccinelli, M., Botti, L., Ene-Iordache, B., Remuzzi, A., Steinman, D.A.: An image-based modeling framework for patient-specific computational hemodynamics. Med. Biol. Eng. Comput. **46**, 1097–1112 (2008)
5. Ursino, M., Giannessi, M.: A model of cerebrovascular reactivity including the circle of willis and cortical anastomoses. Ann. Biomed. Eng. **38**, 955–974 (2010)
6. Altamirano-Diaz, L., Kassay, A.D., Serajelahi, B., McIntyre, C.W., Filler, G., Kharche, S.R.: Arterial hypertension and unusual ascending aortic dilatation in a neonate with acute kidney injury: mechanistic computer modeling. Front. Physiol. **10**, 1391 (2019)
7. Anselmino, M., Scarsoglio, S., Saglietto, A., Gaita, F., Ridolfi, L.: Transient cerebral hypoperfusion and hypertensive events during atrial fibrillation: a plausible mechanism for cognitive impairment. Sci. Rep. **6**, 28635 (2016)
8. Joseph, J.J., Hunter, T.J., Sun, C., Goldman, D., McIntyre, C.W., Kharche, S.R.: Using a human circulation mathematical model to simulate the effects of hemodialysis and therapeutic hypothermia. Front. Physiol. (2021, Submitted)

9. Heldt, T., Shim, E.B., Kamm, R.D., Mark, R.G.: Computational modeling of cardiovascular response to orthostatic stress. J. Appl. Physiol. **1985**(92), 1239–1254 (2002)
10. Heldt, T., Mukkamala, R., Moody, G.B., Mark, R.G.: CVSim: an open-source cardiovascular simulator for teaching and research. Open Pacing Electrophysiol. Ther. J. **3**, 45–54 (2010)
11. Lin, J., Ngwompo, R.F., Tilley, D.G.: Development of a cardiopulmonary mathematical model incorporating a baro chemoreceptor reflex control system. Proc. Inst. Mech. Eng. Part H J. Eng. Med. **226**, 787–803 (2012)
12. Alastruey, J., Parker, K.H., Peiro, J., Byrd, S.M., Sherwin, S.J.: Modelling the circle of Willis to assess the effects of anatomical variations and occlusions on cerebral flows. J. Biomech. **40**, 1794–1805 (2007)
13. Hennig, T., Maass, P., Hayano, J., Heinrichs, S.: Exponential distribution of long heart beat intervals during atrial fibrillation and their relevance for white noise behaviour in power spectrum. J. Biol. Phys. **32**, 383–392 (2006)
14. Scarsoglio, S., Guala, A., Camporeale, C., Ridolfi, L.: Impact of atrial fibrillation on the cardiovascular system through a lumped-parameter approach. Med. Biol. Eng. Comput. **52**, 905–920 (2014)
15. Anselmino, M., Scarsoglio, S., Saglietto, A., Gaita, F., Ridolfi, L.: A computational study on the relation between resting heart rate and atrial fibrillation hemodynamics under exercise. PLoS ONE **12**, e0169967 (2017)
16. Pianelli, M., et al.: Delaying cardioversion following 4-week anticoagulation in case of persistent atrial fibrillation after a transcatheter ablation procedure to reduce silent cerebral thromboembolism: a single-center pilot study. J. Cardiovasc. Med. **12**, 785–789 (2011)
17. Hindmarsh, A.C., et al.: SUNDIALS: suite of nonlinear and differential/algebraic equation solvers. ACM Trans. Math. Softw. **31**, 363–396 (2005)
18. Joseph, J.J., Lee, T., Goldman, D., Mcintyre, C.W., Kharche, S.R.: The role of extra-coronary vascular conditions that affect coronary fractional flow reserve estimation. Lect. Notes Comput. Sci. (2021, Submitted)

Author Index

Printed in the United States
by Baker & Taylor Publisher Services